Proceedings of the International Conference on
Electrical Engineering and Automation (EEA2016)

ELECTRICAL
ENGINEERING
AND AUTOMATION

Proceedings of the International Conference on
Electrical Engineering and Automation (EEA2016)

ELECTRICAL ENGINEERING AND AUTOMATION

Hong Kong, China   24 – 26 June 2016

editor

Xiao-xing Zhang

Wuhan University, China

World Scientific

W JERSEY · LONDON · SINGAPORE · BEIJING · SHANGHAI · HONG KONG · TAIPEI · CHENNAI · TOKYO

Published by

World Scientific Publishing Co. Pte. Ltd.

5 Toh Tuck Link, Singapore 596224

USA office: 27 Warren Street, Suite 401-402, Hackensack, NJ 07601

UK office: 57 Shelton Street, Covent Garden, London WC2H 9HE

British Library Cataloguing-in-Publication Data
A catalogue record for this book is available from the British Library.

ELECTRICAL ENGINEERING AND AUTOMATION
Proceedings of the International Conference on Electrical Engineering and Automation (EEA2016)

ISBN 978-981-3220-35-5

Preface

2016 International Conference on Electrical Engineering and Automation (EEA2016) was held in Hong Kong, China during June 24th-26th, 2016. The objective of EEA2016 is to provide a platform to leading academic scientists, researchers, scholars and students around the world to get together to share their results and compare notes on their research discovery in the development of Electrical Engineering and Automation. EEA2016 features a wide spectrum of topics covering Electronics Engineering and Electrical Engineering; Materials and Mechanical Engineering; Control and Optimization; Modeling and Simulation; Testing and Imaging; as well as Robotics and Automation.

The conference program includes two keynote speakers, and a total of 445 submissions from various parts of the world. However, after peer-review, only 128 articles will be included in the proceedings.

In here, on behalf on the organizing committee, we would like to express our sincere gratitude to all the members of Technical Program Committee for their support and effort to plan to ensure the smooth running of this conference. We would also like to thank all the authors, speakers and participants for their great contributions to the success of EEA2016. Finally, from the bottom of our hearts, we like to express our great appreciation to all the volunteers and staff who help to run the operation of the conference like a clock work; without them, this conference would be impossible.

We shall be looking forward to seeing you again in our next conference.

General Chair
Xiao-xing Zhang
EEA2016 Organizing Committee

2016 International Conference on Electrical Engineering and Automation [EEA2016]

General Chair
Professor, Xiao-xing Zhang, Wuhan University, China

Editor
Professor, Xiao-xing Zhang, Wuhan University, China

COMMITTEE OF EEA2016

General Chair
Professor, Xiao-xing Zhang, Wuhan University, China

Editor
Professor, Xiao-xing Zhang, Wuhan University, China

Technical Program Committee
Professor, Ming-Chun Tang, Chongqing University, China
Lecturer, Jun Ye, Sichuan University of Science & Engineering, China
Associate Professor, Bin Liu Dalian University of Technology, China
Lecturer, Liang Hong, Xi`an Polytechnic University, China
Lecturer, Yun-feng Wen, Chongqing University, China
Professor, Lei Zhang, Chongqing University, China
Professor, Jian-hua Wu, Nanchang University, China
Associate Professor, Zhan-ying Li, Dalian Polytechnic University, China
Associate Professor, Yun-liang Zhang, Institute of Scientific and
Technical Information of China, China
Professor, Feng Gao, Chongqing University, China
Professor, Yu-dong Zhang, Nanjing Normal University, China
Professor, He-suan Hu, Xidian University, China
Associate Professor, Bo Huang, Nanjing University of Science and
Technology, China
Professor, Shu-kai Duan, Southwest University, China
Lecturer, Bo Zhou, Beijing Institute of Technology, China
Dr. Bin-gang Xu, The Hong Kong Polytechnic University, Hong Kong
Professor, Fu-Yun Zhao, Wuhan University, China
Professor, Bo Cheng, Beijing University of Posts and
Telecommunications, China

Dr. Yuan-yuan Wang, Xinjiang University, China
Dr. M. Pecht, University of Maryland, USA
Dr. Ai Fong, City University of Hong Kong, China
Dr. P. Selvam, Indian Institute of Technology Madras, Indian
Dr. Xin-suo Luo, Guangxi Normal University, China

Keynote Speakers of EEA2016

Keynote Speakers of 2016 International Conference on Electrical Engineering and Automation

Keynote Speaker I: Prof. Jian-hua Zhang
Affiliation: Department of Automation, East China University of Science and Technology
Editor in Chief of international journal Recent Patents in Signal Processing (Bentham Science)

Keynote Speech: Pattern Recognition of Instantaneous Mental Workload: A Comprehensive Empirical Comparison of Different Feature Reduction and Classification Methods

Abstract- Accurate recognition and assessment of the temporal variation in mental workload (MWL) of human operator is crucial to prevent human errors or accidents due to operator cognitive overload and inattention in a large class of real-world safety-critical human-machine systems.

In this talk, we will develop a Mental Workload (MWL) recognition system based on electrophysiological data to assess temporal variation in the level of MWL in an objective and noninvasive manner. Furthermore, considering the generally poor generalization ability of subject-specific MWL classifiers, we will propose a cross-subject MWL recognition framework based on the strategy of transfer learning, which is capable of transferring classification rules (or knowledge) from the source domain (i.e., 5 experimental participants) to the target one (i.e., another participant).

The EEG features will be derived by using fuzzy mutual-information-based wavelet-packet transform approach. In order to find the optimal MWL classification scheme, we will make a comprehensive comparison of different combinations of 4 classification and 9 electroencephalographic (EEG) feature reduction techniques in terms of their respective MWL classification accuracies. Extensive comparative results revealed that the combination of two feature reduction algorithms, viz. Kernel Spectral Regression (KSR) and Transferable Discriminative Dimensionality Reduction (TDDR), and two classifiers, namely k-Nearest Neighbor (KNN) and Support Vector Machine (SVM), can effectively improve the accuracy of MWL recognition.

Brief Introduction:
Prof. Jian-Hua Zhang received Dr.-Ing. degree in Electrical Engineering from Ruhr-University Bochum, Germany in 2005. From 2005 to 2006 he was a Postdoctoral Research Associate at University of Sheffield, UK. Since 2007 he has been a Full Professor at East China University of Science and Technology. He held a Guest Professorship at TU Berlin, Germany in 2011, 2012, 2014, and 2015. His research interests are in computational intelligence, machine learning, pattern recognition, biomedical signal processing, modeling and control of biomedical systems, human-machine systems, and brain-machine interactions. He has published over 100 peer-reviewed journal and conference papers. He is a member of IFAC TC on Human Machine Systems, Large Scale Complex Systems, Biological and Medical Systems, and on Transportation Systems. He served as IPC Vice-Chair of IFAC

LSS2013, Technical Associate Editor of IFAC World Congress 2014, and IPC Co-Chair of IFAC HMS2016. He won a full scholarship from German Academic Exchange Service in 2002. He received Governmental Scholarship and Senior Research Fellowship from China Scholarship Council in 2001 and 2012, respectively. He was selected into Shanghai Municipal Pujiang Talents Program in 2007. He received a Senior Research Fellowship from Max-Planck Institute for Dynamics of Complex Technical Systems in 2011.

Keynote Speaker II: Prof. Ting Yang
Affiliation: School of Electrical Engineering and Automation, Tianjin University, Tianjin, China

Keynote Speech: Cyber Physical System in Smart Grid–Communication Technology

Abstract- CPS, Cyber Physical Systems, one of key techniques in smart grid, makes the electric power system become more intelligent and easier to be controlled. And then the system can provide better quality of electric power service for users, great efficiently utilization, green energy, and decrease carbon emission. In CPS, all of electrical information should be exchange among the dispatching center, digital equipment and sensors/actuators. Effective communication technology is necessary for different application and scenarios. This presentation will from the following 5 points to discuss the communication technology of CPS applying in smart grid: main communication application in smart grid, communication requirements from electrical power system, communication layers in CPS, communication technologies, and communication standards.

With the detailed analysis from communication technologies to standards, a clear picture of CPS is proposed, which will be of great interest to potential readers.

Brief Introduction:
Dr. Ting Yang is currently a Chair professor of Theory and Advanced Technology of Electrical Engineering, at the School of Electrical Engineering and Automation, Tianjin University, China. He was the cooperative research staff of Imperial College London (2008), University of Sydney, Australia (2011, 2015). Prof. Yang is

the winner of the "New Century Excellent Talents in University Award" from Chinese Ministry of Education. He is the leader of tens of research grant projects, including the International S&T Cooperation Program of China, the National High-Tech Research and Development Program of China (863 Program), the National Natural Science Foundation of China, and so on. Prof. Yang is the chairman of two workshops of IEEE International Conference, and the editor in Chief of one of Special Issues of the international journal of DSN. He is the author/co-author of four books, more than one hundred publications in internationally refereed journals and conferences. Prof. Yang is a senior member of the Chinese Institute of Electronic, the fellow of Circuit and System committee, the fellow of Theory and Advanced Technology of Electrical Engineering, and the member of International Society for Industry and Applied Mathematics.

His research fields include Smart Grid, advanced metering infrastructure, and information and communication technologies in electric power system.

2016 International Conference on Electrical Engineering and Automation [EEA2016]

Contents

Chapter 2. Materials and Mechanical Engineering   159

Chapter 3. Electronics Engineering and Electrical Engineering   189

xxv

Chapter 4. Modeling and Simulation   531

Chapter 5. Testing and Imaging   711

Chapter 6.   Robotics   **783**

Chapter 1

Control and Automation

Fault Diagnosis of the Stator Inter-Turn Short Circuit for a High-Speed Train Motor

Chun-ping Tang[†], Fu-yang Chen and Lin-qiang Jin
*College of Automation Engineering, Nanjing University of Aeronautics
and Astronautics, China Department of Electrical and Computer Engineering
E-mail: [†]Tangchunping@nuaa.edu.cn
chenfuyang@nuaa.edu.cn*

In order to build the stator winding fault model of the asynchronous motor in the high-speed train system, anew state-space representation of the dynamic equations by using the Park's vector method is proposed. Compared with the flux linkage variable, the stator current as state variables is chosen to determine the stator current and diagnose the stator fault timely, which is important in the high-speed train. The simulation results prove the effectiveness of the approach.

Keywords: Asynchronous motor; Turn-to-turn short circuit; Park's vector; High-speed train.

1. Introduction

Due to the development of China's economic and social needs, rail transit industry develop rapidly in our country. In the modern transportation system, the high-speed train becomes the most important means of transportation with its large capacity, low energy consumption, light pollution, high speed, high security, and many other advantages. So, the fault diagnosis of a high-speed train is very meaningful.

The traction motor is one of the most important parts of the traction system, which provides driving force for high-speed trains. However, because of the change of traction motor's load, bad working environment and so on, traction motor often breaks down. So the study of fault diagnosis for AC traction motor is of great significance in engineering.

Before detecting the stator short circuit fault, we should obtain the model of the induction motor of a high-speed train first. Most studies use the machine-specific model[1] which is complex and indirect. As for the stator inter-turn fault, there are many methods to detect. We can get that the stator winding fault is diagnosed by using Fractional Fourier domain[2-4], current signature analysis, high frequency signal injection, d-q current trajectory and

Park's vector approach. We can know that the Park's vector method is useful in many motor faults[5].

This paper first analyze the Park's vector diagram under the normal operation of asynchronous motor. Then, establish stator inter-turn short circuit fault model, simulate and apply the Park's vector to analyze the signal of the stator current. According to the Park diagram, we can recognize and diagnose the fault severity and calculate the accuracy of fault diagnosis.

2. Park's vector method

The Park's vector is composed of two parts($i\alpha$ $i\beta$), and the two components are a function of the motor stator currents(ia ib ic). The formula is as follow:

$$i\alpha = \left(\frac{\sqrt{2}}{\sqrt{3}}\right)ia - \left(\frac{1}{\sqrt{6}}\right)ib - \left(\frac{1}{\sqrt{6}}\right)ic \tag{1}$$

$$i\beta = \left(\frac{1}{\sqrt{2}}\right)ib - \left(\frac{1}{\sqrt{2}}\right)ic \tag{2}$$

3. The transformation of high-speed train motor model

In the high-speed train system, the traction asynchronous motor system is a strong interference, coupling, high order and multi-variable nonlinear system, of which transient process is very complicated.

Assumes that three phase windings of the stator are with star connection, and failure occurs in A phase, as shown in figure 1.

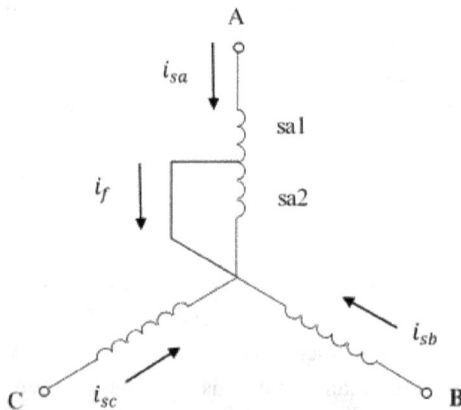

Figure 1. A phase inter-turn short circuit of the stator.

In figure 1, A phase winding of the stator is divided into two parts: the normal winding sa1 and the short winding sa2. Set $\mu = l_{sa1}/(l_{sa1}+l_{sa2})$ as short-circuit coefficient, among them, l_{sa1} denoting normal winding length and l_{sa2} denoting short winding length.

In order to simplify the model and highlight the fault feature, it is necessary to make the following assumptions:

(1) Assumes that the magnetic potential is distributed by the sine along air gap circle , and the three-phase winding of the motor is symmetrical;

(2) Set the stator-rotor mutual inductance and self-induction is linear, don't consider magnetic circuit saturation effect;

(3) Ignore the core loss;

(4) Ignore the electrical resistance varying with temperature and frequency.

With the Park's vector transformation, we can get the asynchronous motor voltage equation as follows:

The voltage equation:

$$\left.\begin{aligned}
u_{s\alpha} &= R_s i_{s\alpha} + P\psi_{s\alpha} - \frac{2}{3}\mu R_s i_f \\
u_{s\beta} &= R_s i_{s\beta} + P\psi_{s\beta} \\
0 &= R_r i_{r\alpha} + P\psi_{r\alpha} + w_r \psi_{r\beta} \\
0 &= R_r i_{r\beta} + P\psi_{r\beta} - w_r \psi_{r\alpha}
\end{aligned}\right\} \tag{3}$$

Flux equation:

$$\begin{bmatrix} \psi_{s\alpha} \\ \psi_{s\beta} \\ \psi_{r\alpha} \\ \psi_{r\beta} \end{bmatrix} = \begin{bmatrix} L_s & 0 & L_m & 0 \\ 0 & L_s & 0 & L_m \\ L_m & 0 & L_r & 0 \\ 0 & L_m & 0 & L_r \end{bmatrix} \begin{bmatrix} i_{s\alpha} \\ i_{s\beta} \\ i_{r\alpha} \\ i_{r\beta} \end{bmatrix} - \frac{2}{3}\mu \begin{bmatrix} L_s \\ 0 \\ L_m \\ 0 \end{bmatrix} i_f \tag{4}$$

For the short turns, voltage and flux linkage equations can be represented as:

$$u_{sa2} = R_f i_f = \mu R_s(i_{sa} - i_f) + P\psi_{sa2} \tag{5}$$

$$\psi_{sa2} = \mu L_{ls}(i_{s\alpha} - i_f) + \mu L_m\left(i_{s\alpha} + i_{r\alpha} - \frac{2}{3}\mu i_f\right) \tag{6}$$

The electromagnetic torque equation:

$$T_e = \frac{3}{2}n_p L_m(i_{s\beta} i_{r\alpha} - i_{s\alpha} i_{r\beta}) - n_p \mu L_{mi} f i_{r\beta} \tag{7}$$

According to equation (4–7), we can get the state equation that bases on the current state variables.

5

$$A = \begin{bmatrix} K_{10}K_{11} & -w_r K_1 L_m L_s K_{10} & -R_r K_1 L_s K_{10} & -w_r L_r L_s K_1 K_{10} & K_{12}K_{10} \\ -K_{13}w_r L_m^2 & -K_{13}R_s L_r & -K_{13}w_r L_r L_m & K_{13}R_r L_m & \dfrac{2K_{13}w_r \mu L_m^2}{3} \\ R_s L_m K_{13} & -w_r L_m L_s K_{13} & -R_r L_s K_{13} & -w_r L_r L_s K_{13} & \dfrac{-2K_{13}\mu L_s L_m}{3} \\ w_r L_m L_s K_{13} & R_s L_m K_{13} & w_r L_r L_s K_{13} & -R_r L_s K_{13} & \dfrac{-2w_r K_{13}\mu L_m L_s}{3} \\ K_{14}K_4 & -K_{14}K_5 & -K_{14}K_6 & -K_{14}K_7 & -K_{14}K_8 \end{bmatrix} \quad (8)$$

$$B = \begin{bmatrix} u_{s\alpha} & u_{s\beta} & 0 & 0 & 0 \end{bmatrix}^T \quad (9)$$

$$C = \begin{bmatrix} -\dfrac{K_1 L_m + 1}{K_{10}} & 0 & 0 & 0 & 0 \\ 0 & L_r K_{13} & 0 & 0 & 0 \\ -K_{13}L_m & 0 & 0 & 0 & 0 \\ 0 & -L_m K_{13} & 0 & 0 & 0 \\ -K_{14}K_9 & 0 & 0 & 0 & 0 \end{bmatrix} \quad (10)$$

$$Pi = \begin{bmatrix} P_{is\alpha} & P_{is\beta} & P_{ir\alpha} & P_{ir\beta} & P_{if} \end{bmatrix}^T \quad (11)$$

$$i = \begin{bmatrix} i_{s\alpha} & i_{s\beta} & i_{r\alpha} & i_{r\beta} & i_f \end{bmatrix}^T \quad (12)$$

$$Pi = Ai + CB \quad (13)$$

In this part, we improve the derivation process of the state space equation. We simplify the derivation by omitting the middle of the derivation process, which select the magnetic flux as state variables. We can directly choose the stator current as stator variables, and obtain the motor fault state equations quickly by using S function, which saves time.

$i_{s\alpha}$ $i_{s\beta}$ $i_{r\alpha}$ $i_{r\beta}$ i_f are respectively the stator phase current alpha, beta phase current of the stator and rotor of alpha beta phase current phase current, rotor, stator short-circuit current;

$\psi_{s\alpha}$ $\psi_{s\beta}$ $\psi_{r\alpha}$ $\psi_{r\beta}$ T_e are respectively to alpha of the stator flux and rotor flux, the stator beta phase alpha beta phase flux linkage, the rotor flux and electromagnetic torque;

$\psi_{s\alpha}$ $\psi_{s\beta}$ $\psi_{r\alpha}$ $\psi_{r\beta}$ T_e are respectively self inductance, mutual inductance and rotor stator self inductance and rotor pole logarithm, moment of inertia;

$u_{\alpha s}$ $u_{\beta s}$ $u_{\alpha r}$ $u_{\beta r}$ are respectively alpha of the stator voltage respectively, beta phase voltage of the stator and rotor alpha beta phase voltage phase voltage and rotor;

w_r T_l P are respectively the rotor speed, load torque, respectively differential operator;

$k_z(z=1,2,\ldots\ldots,13)$ are coefficients about factors of R_s R_r L_s L_r w_r L_m.

4. Test results

The simulation parameters settings: supply voltage $U_N = 220V$, voltage frequency $f = 50Hz$, the stator resistance $R_s = 0.144\Omega$, the stator leakage inductance $L_{s1} = 0.001417H$, the leakage inductance of the rotor $L_{r1} = 0.001294H$, the rotor resistance $R_r = 0.146\Omega$, stator-rotor mutual inductance $L_{m1} = 0.3279H$, moment of inertia $J = 10kg \cdot m^2$, pole-pairs $n_p = 2$, coefficient of short circuit μ.

Under the no-load operation, The stator inter-turn short circuit fault simulation results is as follows:

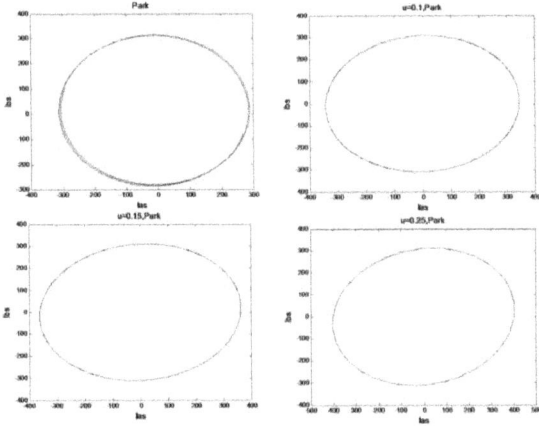

Figure 2. $\mu = 0 - 0.25$ Park's vector diagram.

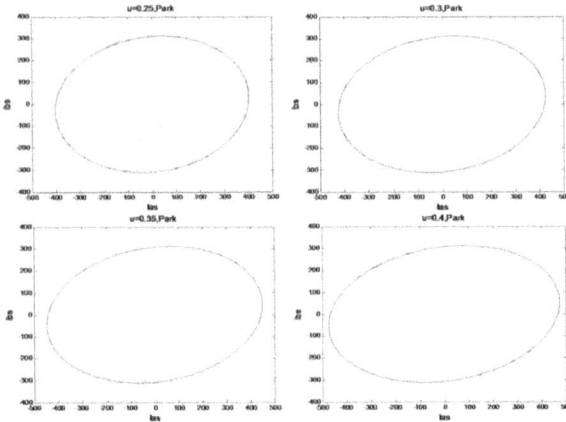

Figure 3. $\mu = 0.25 - 0.4$ Park's vector diagram.

According to figure 2 and figure 3, we can know that: When the motor runs in normal situation, analysis the graphics of the stator current, and the Park vector of stator current are round, which validate the motor model;

With gradually increasing the coefficient of short circuit, the short circuit fault is more and more serious, and the Park's vector graphics gradually become ellipse. When μ value is between 0 and 0.25, the angle between the long axis of the ellipse and the horizontal axis is bigger and bigger; When μ value is between 0.25 and 0.4, Ellipse is more and more flat;

Changing μ between 0 and 1 and simulating when μ change by 0.05, there will be obvious features in Park's vector diagram. So we can know that the Park's vector approach can detect the stator inter-turn short circuit fault.

5. Conclusion

In this paper, we improved the stator winding fault model by using the stator current as state variables, which derive process was simple and understandable compared with linkage variables. And we used a Park's vector method to detect the stator inter-turn fault, which can diagnosis the fault quickly and correctly when the induction motor stator winding fault occurred.

Acknowledgment

The project was funded by the National Natural Science Foundation of China (61490703).

References

1. Tallam R M, Habetler T G, Harley R G. Transient model for induction machines with stator winding turn faults [J]. Industry Applications IEEE Transactions on, 2002, 38(3):632-637.
2. Shashidhara S M, Raju D P S. Stator Winding Fault Diagnosis Of Three-Phase Induction Motor By Park's Vector Approach [J]. International Journal of Advanced Research in Electrical Electronics & Instrumentation Engineering, 2013.
3. Abitha, M. W., V. Rajini. "Park's vector approach for online fault diagnosis of induction motor." Information Communication and Embedded Systems (ICICES), 2013 International Conference on IEEE, 2013:1123-1129.

4. Nejjari, H., M. E. H. Benbouzid. "Monitoring and diagnosis of induction motors electrical faults using a current Park's vector pattern learning approach." IEEE Transactions on Industry Applications36.3 (2000):730-735.

5. Guo Q, Li X, Yu H, et al. Broken Rotor Bars Fault Detection in Induction Motors Using Park's Vector Modulus and FWNN Approach [J]. Lr No N Omr N, 2008, 5264:809-821.

The Design of High Frequency Induction Heating Power Supply Based on DSP and FPGA Dual Core Processors

Li-na Liu

BaoTou Light Industry Vocational Technical College, BaoTou, China
E-mail: Liulina_bt@126.com

DSP (XC2267) is the main control chip which does the increment PI algorithms and realizes the voltage and current double closed-loop control. Three-phase thyristor rectifier method based on the phase self-adjusting of FPGA is proposed. A method for swept-frequency start circuit is designed to implement the transition between the high-power and low-power. The invert system uses the compound control of improved PWM and frequency phase lock to improve the control accuracy and efficiency. Lastly, by using bench, the circuit is tested and the feasibility and effectiveness of the whole system are verified.

Keywords: Induction heating power system; DSP; FPGA; PLL; Frequent tracking.

1. Introduction

The method which uses the APLL to realize the frequent tracking exists many problems, such as the circuit is complicated, the system upgrade is hard to carry out, and so on. With the development of FPGA and DSP, it is possible to realize the digital intermediate frequency heating power and to improve the power regulation speed and resonance frequency tracking. The ability to deal with data and algorithm of 16-bit single chip microcomputer makes the whole control system more simplify and control the algorithm of the mediate frequency heating power system efficiently.

2. Principle and Structure

2.1. Principle of induction heating power supply

According to the principle of electromagnetic induction, the induction heating power supply heats the work pieces up quickly in the alternating magnetic field [1–4]. Through the work piece, the alternating magnetic field lines form the circuit and generate the induced current in the cross section. The induced current makes the part of work pieces generate heat instantaneously.

2.2. Block diagram of the induction heating power system

The induction heating power system consists of rectifier, inverter, filter, drive circuit, signal acquisition circuit, protective circuit, control circuit, PLL and so on. The three- phase full-bridge controlled rectifier circuit converts the three phase current to pulsating direct current which is to be a constant DC power supply through the large inductance filter. The constant direct current is inverted to a constant power of single-phase medium frequency voltage through the single-phase bridge inverter using SCR. The drive circuit mainly produces the drive signal to control the diode of Rectifier Bridge and SCR of Inverter Bridge. The intermediate frequency voltage signal which is gathered by the sensor is sent to the controller to do the improved PID and to track the frequency. The overall block diagram is shown in Fig. 1.

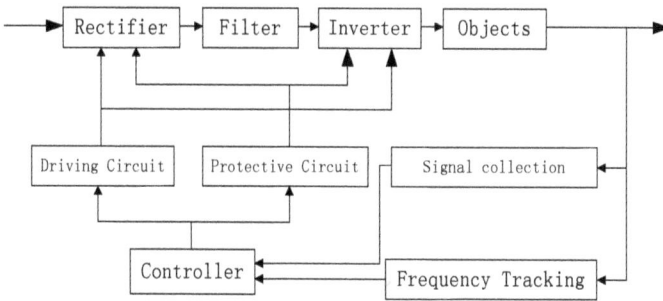

Fig. 1. The Overall Block Diagram.

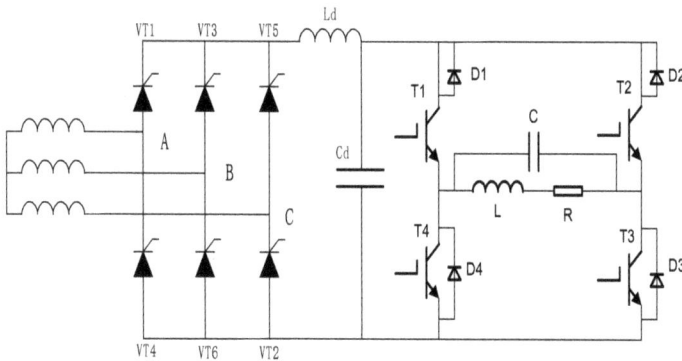

Fig. 2. Parallel-resonant Inverter.

11

2.3. Solution and hardware construction

The parallel-resonant inverter structure is shown in Fig. 2. The large inductance filter is used to suppress the surge current and to reduce the impact of the power grid by circuit. The main controller of this system is XC2267 which does the algorithm. The FPGA is used to generate the signal of all the driver circuit and to design PLL, which is shown in Fig. 3.

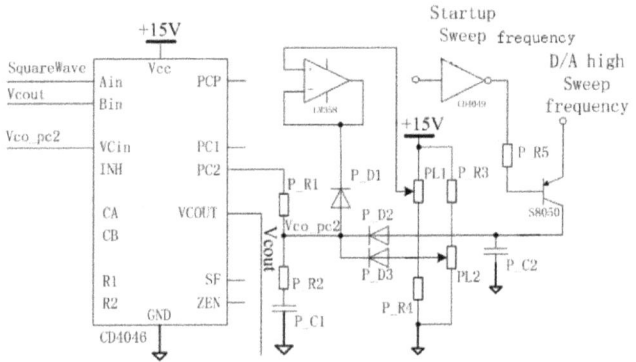

Fig. 3. Overall System Structure.

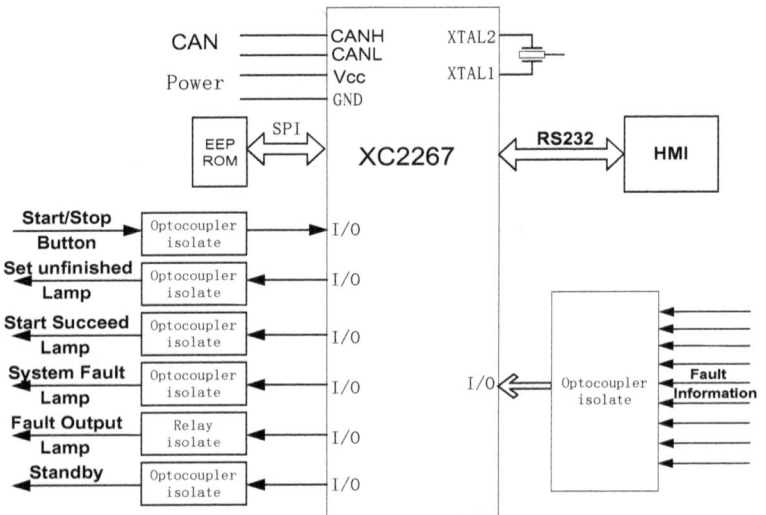

Fig. 4.   Frequency tracking and control system.

3. Frequency Tracking and Control

There are two phase discriminator in CD4046, the phase discriminator 1 is exclusive-OR gate, where there are 90° phase shift between Ain and Bin, the phase discriminator 2 is edge triggered flip flop, where there is 0° phase shift between Ain and Bin. In this schedule, phase discriminator 2 is used. Frequency tracking and control system is shown in Fig. 4.

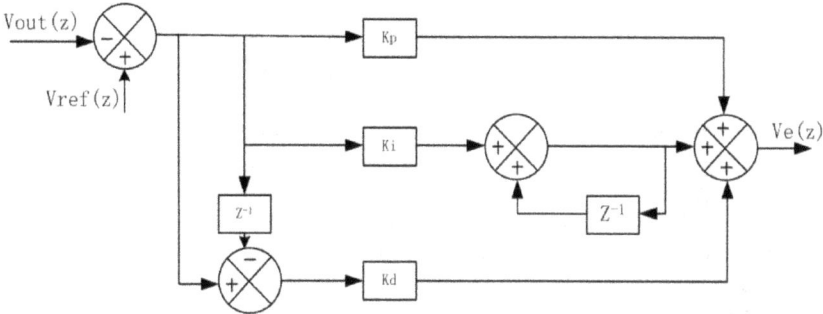

Fig. 5. Block diagram of a Digital implementation of PID Controller.

4. PID Control

In order to overcome the aforementioned limitations of conventional PID controller certain modifications of this kind of controller are proposed . The use of two different PID controllers is proposed resulting in two different bandwidth values, a lower bandwidth during steady state and a higher bandwidth which takes effect during transients. It might cause unnecessary oscillations due to a large sudden change in bandwidth. A nonlinear implementation of PID controller that behaves differently during load transient condition is proposed. Its implementation is not straight forward and requires calculation of many parameters. Moreover, the digital implementation requires more resources for the calculation of exponential terms. In order to make a smooth PID transition from steady state to transient state, an adaptive PID controller method is presented in this paper. This paper proposes a control strategy that results in reduction in the overshoot/undershoot and the settling time for the power converter output voltage during dynamic transients. The proposed control method does not require sensing any additional parameters that are not already available in any conventional PID controller and is relatively simple to realize.

The strategy of integral separation PID control is adopted. The basic idea of improved PID is as follow: When the deviation of controlled variable and set point is large, integral action is cancelled , on the other side, integral action is used to eliminate static error and improve the control precision. Follow the steps outlined below as Fig. 5.

5. Conclusion

All the test is based on 160 KW / 8 KHZ SCR medium frequency induction power supply. Controller board is shown in Fig. 4. The result of PLL test is as shown in Fig. 7 – Fig. 11. The XOR gate is used as PD. When the system is stable, the output signal is 90 degrees ahead of input signal. The blue line is output signal and the yellow line is input signal. The middle line is the basic time of 90 degrees and the work frequency is 20kHz. The controller board is shown in Fig. 6.

The result of the short-circuit test is as shown in Fig. 7. It shows that when there is a short circuit load, the current increases rapidly and appears overshoot. It turned out that improved PI can adjust the current effectively and protect the main circuit.

Fig. 6.   Controller Board.

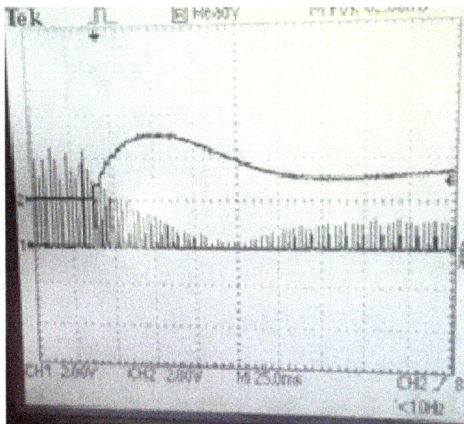

Fig. 7. Current Loop Response Curve.

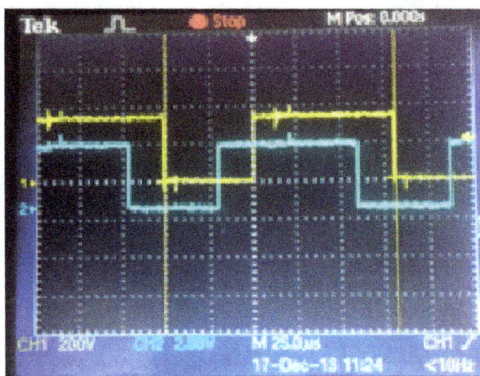

Fig. 8. Phase shift between Ain and Bin.

Fig. 9. Overlap time between upper and lower IGBT.

15

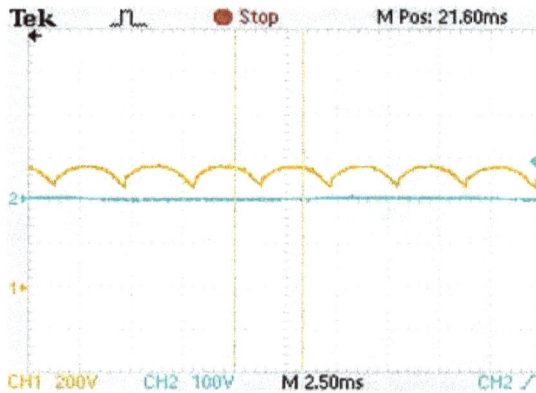

Fig. 10.   DC Bus Voltage Curve.

The waveform of Rectifier Bridge with pure impedance load is as shown in Fig. 9. The experiment steps are as follows: Remove the load of inverter, and connect a resistance of 500Ohm with DC bus in series. As there is no feedback, the phase-shifting angle of Rectifier Bridge is zero.

The waveform of current and voltage of load is as shown in Fig. 11. After the load oscillation and given the intermediate frequency voltage, the waveform of current is square wave and the waveform of voltage is sine wave.

The experiment indicated that: the hardware designed in this thesis meets the design requirements; the digital phase-locked loop increases the range of phase, and improves the power factor; frequency tracking is realized; improved PID also achieves great control effect [5].

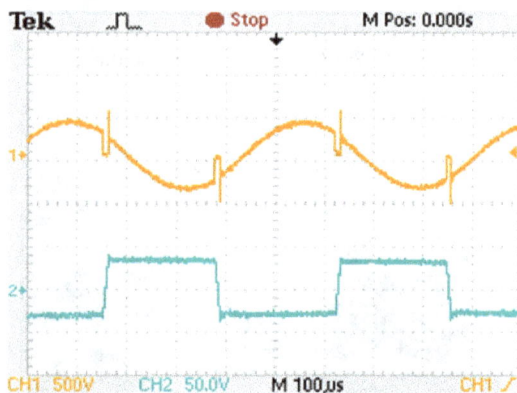

Fig. 11. AC side Current and Voltage Curve.

References

1. Su Yanping, Cui Jianfeng: Design of digital phase shift trigger of induction heating power system. (International Conference on Electronic, 2011).
2. Li Heming, Li Yabin: FPGA-Based All Digital Phase-Locked Loop Controlled Induction Heating Power Supply Operating at Optimized ZVS Mode. IEEE 1-4244-0549-1/06(2006).
3. PiaoXingzhe: The Research on 10kHz/150kW Mid-Frequency Heating Power Induction Heater. Shenyang University of Technology(2004).
4. Yang Xiaojing: The Key Technology Research of the Digital Intermediate Frequency Induction Heating Power Supply.Xi'an University of Architecture and Technology(2012).
5. D.H. Wolaver: Phase-Locked Loop Circuit Design. Prentice Hall. Englewwood Cliffs. NJ(1991).

The Interferometric Optical Fiber Perimeter Security System

Xu Wang

Beijing Institute of Radio Metrology and Measurement
Beijing, China
E-mail: alexeo@sina.com

This paper studies a fiber optic perimeter security system based on dual Mach Zehnder interferometer. The structure and principles of the system are analyzed. With the capture card, photoelectric detectors and other hardware circuit designed, this paper points out the key factors that cause the error of the system, and puts forward some measures to improve it. The experimental results show that the wavelet denoising and improved cross-correlation algorithm have significant effect to improve the positioning accuracy.

Keywords: dual Mach Zehnder interferometer, Perimeter security, Wavelet denoising, Improved cross-correlation algorithm, Positioning.

1. Introduction

Since the advent of optical fiber in 1960s, it has been applied to the transmission of images, detection technology, and other fields. With the application of optical communication, optical fiber technology has been developed rapidly[1]. Gradually, people come to realize that many of the properties of optical fiber can be used to detect various physical quantities. It can realize the stereo surveillance and continuous monitoring, quickly and accurately locate the displacement disturbance. So the optical fiber technology has obvious advantages in the security monitoring [2-4]. With the development of laser technology, the laser with kHz linewidth has come out. The coherence of laser source is becoming stronger and stronger, which makes it possible to develop a long range fiber interferometer [5-6]. Interferometric sensing positioning technology makes two interferometers sharing one sensing optical fiber through special optical structure. It uses the optical delay line to make optical path difference between the two interferometers. With the data from the two interferometers, the size and location of the data is obtained, which realizes the long distance distributed measurement. As the dual M-Z interferometric fiber optic sensor possess the advantages of distant position, fast response, no demand on absolute phase, simple follow-up treatment and wide application in

18

engineering, this paper selects the dual M-Z interferometric fiber optic sensor solution as a research program [7-8].

2. System Structure

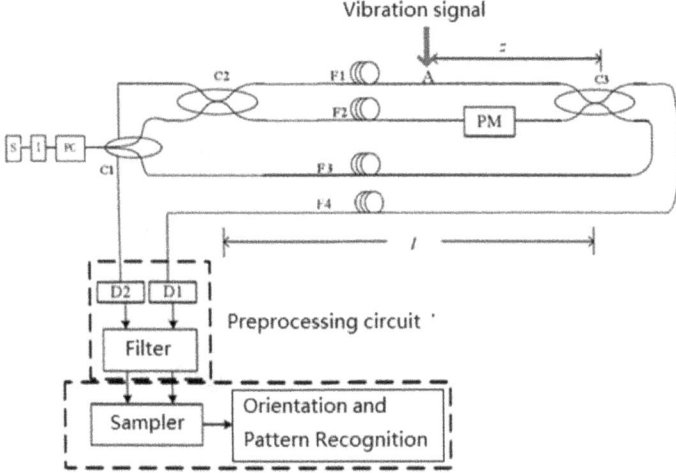

Fig. 1. A schematic of distributed sensing system based on dual M-Z interferometer

The distributed fiber optic sensing system that this paper intends to adopt is shown in figure 1. In the optical path system, S stands for narrow linewidth laser source; I for optical isolator; PC for mechanical fiber polarization controller; C1, C2 and C3 for 2*2 coupler; PM for PZT optical fiber phase modulator; D1 and D2 for photoelectric detector; F1 and F2 for sensing fiber; and finally F3 and F4 for conducting fiber. Among them, the laser source S, the coupler C1, C2, the optical fiber F1, F2, F4, the coupler C3 and detector D1 constitute the clockwise MZ interferometer. While the laser source S, the coupler C1, the optical fiber F3, the coupler C3, the optical fiber F1, F2, the coupler C2 and D2 constitute the counter clockwise MZ interferometer. The optical wave emitted by laser source S is divided into two channels, which respectively pass through the interferometer formed by the coupler C2, C3 and the sensing fiber F1, F2, along the clockwise and counterclockwise direction. When the disturbance signal is used in the sensing optical fiber F1 and F2, the modulation signal of the M-Z interferometer is converted into a light intensity signal, and then it is converted into the electrical signal by D1 and D2.Meanwhile, the phase modulated signal caused by the disturbance contains the position dependent delay information. Assuming the signal of the first road detector is as follow [9],

$$S_1(t) = s_1(t) + \omega_1(t) \tag{1}$$

19

After Δt, the signal of the second detector is as follow:

$$S_2(t + \Delta t) = s_2(t + \Delta t) + \omega_2(t + \Delta t). \tag{2}$$

Among them, $s_1(t)$ and $s_2(t)$ are the interference signal, while $\omega_1(t), \omega_2(t)$ are the noise signal. Since $s_1(t)$ and $s_2(t)$ come from the same source, the two are highly correlated. Hence based on the relevant algorithm, we can solve the correlation function on Δt of the two signals:

$$R_{S_1 S_2}(\Delta t) = \lim_{T \to \infty} \frac{1}{T} \int_0^T S_1(t) S_2(t + \Delta t) dt. \tag{3}$$

Considering that $s_1(t)$, $s_2(t)$ are not related to detect noise $\omega_1(t), \omega_2(t)$, it can easily get that:

$$R_{s_1 \omega_2}(\Delta t) = R_{s_2 \omega_1}(\Delta t) = R_{\omega_1 \omega_2}(\Delta t) = 0. \tag{4}$$

Therefore, the correlation function of two signals can be expressed as following:

$$R_{S_1 S_2}(\Delta t) = R_{s_1 s_2}(\Delta t) = \lim_{T \to \infty} \frac{1}{T} \int_0^T s_1(t) s_2(t + \Delta t) dt. \tag{5}$$

This shows that the correlation function of the detection signal from the two detectors is equivalent to the real signal. This means that when the correlation function $R_{S_1 S_2}(\Delta t)$ obtains its maximum value, the corresponding Δt represents the actual time delay of the two detectors. That is [10-11]:

$$\Delta t = aug \max\{R_{S_1 S_2}(\Delta t)\} = aug \max\left\{\lim_{T \to \infty} \frac{1}{T} \int_0^T s_1(t) s_2(t + \Delta t) dt\right\}. \tag{6}$$

The positioning of the disturbance can be realized through the positioning algorithm through the actual time delay of the two detectors.

3. Experimental Analysis

The whole localization experiment system includes three key subsystems indicated in figure 2. The first one is the distributed optical fiber interferometer system, which can output two channels of delay signals related to each other. The second one is photoelectric conversion and signal preprocessing system that converts the interference signal into electrical signal. The third is the high-speed signal acquisition system that digitalizes the analog electrical signal and the positioning software processing system based on filtering technology and cross correlation method.

Distributed fiber optic interferometer system	→	Fiber converter and signal processing	→	Data Acquisition and Processing Software

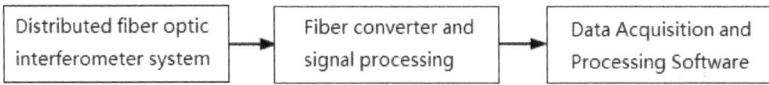

Fig. 2. Three key subsystems of dual M-Z interferometer's distributed sensing system

Firstly, we design the capture card, photoelectric detectors and other hardware circuit that the system needs. As the system positioning accuracy is related to the sampling frequency of the output signal of the interferometer, therefore, the higher the frequency is, the better the positioning accuracy is. This paper intends to adopt the 10MHz high-speed capture card designed by the team, and take samples of the two channels signals. The theoretical accuracy of the positioning can reach ±10m.Considering measurement error and signal noise interference, the pre-positioning accuracy is expected to be ±50m.When choosing the photoelectric conversion device, the following factors should be taken into consideration: the consistence between the working wavelength and the wavelength range of the light source, the working bandwidth of the detector, the receiving sensitivity, and whether in-built an amplifier circuit in the detector. In the positioning process, in order to improve the positioning accuracy, the working bandwidth of the photoelectric detector should be more than 2 times higher than the maximum frequency of the vibration signal. Normally, the vibration signal is concentrated between 2kHz and 20kHzs. Here this paper uses the detector of 1M bandwidth designed by the team.

Through the analysis of the results, we find that the noise of the optical path and circuit is coupled with the signal of the photoelectric detector. Therefore, the signal needs to be de-noised before the cross-correlation algorithm. As for the mixed noise signal, experiments show that the performance of noise and signal are different in the wavelet transform. The average amplitude of the noise is inversely proportional to the scale factor. The wavelet transform of the noise in different scales is highly uncorrelated, while the wavelet transform of the signal has a strong correlation. So we use wavelet de-noising method for signal processing. First we select the wavelet and wavelet decomposition levels to get the wavelet decomposition signal. Then we select the threshold for the high-frequency coefficients, choosing one threshold for each layer from layer 1 to layer N and processing the high frequency coefficients with soft threshold. At last, the wavelet reconstruction signal is obtained based on the low frequency coefficients of the N layer and the modified high frequency coefficients from layer 1 to layer N. At the same time, when the vibration power and the duration of the signal exceed the setting threshold, it is considered as an intrusion event. We make the improvement to the correlation function-when near the maximum power point, the similarity of the signal is best; the vibration frequency is also higher, and the accuracy of the correlation algorithm is higher. Therefore, in practical engineering applications, the vibration signal around the maximum power point should be selected to calculate the correlation function. At the same time, the signal around the maximum power point is segmented for several cross-correlation function calculations. Then the multiple positioning distance is statistically sorted; with the large error of the positioning value eliminated and the rest value processed averagely, the positioning accuracy is improved.

The polarized light in the fiber will produce double refraction effect, resulting in the decline of the polarization state of light transmission, which reduces the visibility of interference fringes output in interference structure. This reduces the positioning accuracy, or even makes it unable to locate. In order to solve the polarization fading, we add a mechanical polarization controller after the laser source and the isolator. By artificially sustained vibration, we adjust the position of the polarization controller's three wave plate; in the meantime, observe the signal of the detector in the oscilloscope. When the two signals come to its maximum amplitude, stop to adjust the plate. Through observation, it is found that polarization state of the system can be maintained about 50 minutes after adjusting the polarization state of the system. In the following experiment, this paper intends to use the General Photonics electronic control module PCD-M02 to realize the closed-loop control.

Now fiber warning model of 20 km has been completed. Figure 3 shows the software interface of the warning system without disturbance.

Fig. 3. the software interface of host computer without disturbance

4. Conclusion

This paper studies an optical fiber security system based on dual M-Z interferometer, and analyzes the structure and principle of the system. In this paper, we design the capture card, electro photonic detector and other hardware circuit that the system needs, point out the key factor causing positioning error and put forward the improvement measure.we carry out the experimental study on optical fiber security system with wavelet de-noising and improved mutual correlation algorithm. The results show that wavelet de-noising and improved cross-correlation algorithm can be applied to fiber perimeter security system, which could significantly improve the positioning accuracy. At last, in the experimental testing, the positioning error is within 50metre, and the stable experimental results have been obtained.

References

1. Zhang Chunxi, Liang Sheng, Feng Xiujuan. Rayleigh scattering and stimulated Brillouin distributed spectral characteristics of fiber optic sensors and its disturbing influence [J] Spectroscopy and Spectral Analysis, 2011, 31(7):1862-1867.
2. Li Qin, Wang Hongbo, Li Lijing. Fiber Distributed-based Michelson interferometer disturbance sensor [J]. Infrared and Laser Engineering, 2015,(1):205-209.
3. Zhou Yan, JinShijiu, Zhang Yun. Distributed optical fiber pipeline leakage detection and location technology [J]. Petroeum,2006,27(2):121-124.
4. Xu Weijun, Hou Jianguo, Li Duanyou. Application of distributed optical fiber temperature measurement system in Jinghong Hydropower Station dam concrete temperature monitoring [J]. Hydroelectric Engineering, 2007,26(1):97-101.DOI:10.3969/j.issn.1003-1243.2007.01.019.
5. Wu Yong hong, Shao Changjiang, ZhouWei. Intelligent fully distributed fiber optic strain sensing accuracy [J]. Tongji University (Natural Science Edition), 2010, 38(4):500-503,526.
6. Hang Lijun, He Cunfu, Wu Bin. Detection technology and new distributed optical fiber pipeline leakage location method [J]. SPIE, 2008,28(1): 123-127.
7. Cui Wenhua, Cheng Zhibin. Research on distributed optical fiber temperature monitoring and alarm systems [J]. Infrared and Laser Engineering, 2002,31(2):175-178.
8. Shi Yanxin, Zhang Qing, Meng Xianwei. Distributed Optical Fiber Sensing Technology in Landslide Monitoring [J]. Jilin University (Earth Science Edition),2008,38(5):820-824.
9. D.Tu, S. Xie, Z. Jiang, M. Zhang. Ultra long distance distributed fiber-optic system for intrusion detection, 2012, pp.85611W-85611W-85616.

10. T. Zhu, Q. He, X. Xiao, X. Bao, Modulated pulses based distributed vibration sensing with high frequency response and spatial resolution, Optics Express, 21 (2013) 2953-2963.
11. Zhou Yan, Zhu Gejingjing, Feng Hao. Detection of early warning technology research and distributed optical fiber pipeline leakage [J]. Journal of Scientific Instrument, 2008, 29(8):1588-1592.

Design and Simulation of Decoupling Control for Recycled-Water System

Meng-han Ao[†], Wei-chen Gu and Bo Zhang

Guiyang Industrial Technology Institute,
Guiyang, Guanshanhu area,
South Changling Road Number 31,
National Digital Content Industrial Park, China
E-mail: [†]13651818246@163.com

The recycled-water system is a complicated multi-input and multi-output process. Stable pressure of the pipe and the level of sump in recycled-water system are greatly needed in a concentration process, but there is strong coupling between them. Distributed PID and decoupling control algorithm are applied respectively, and simulation research is carried out based on the model. The results showed that the application of feedback compensation decoupling scheme achieved good results. It solved recycled-water into two independent systems.

Keywords: Recycled-water system, Coupling, Feedback compensation decoupling.

1. Introduction

During the mineral select processing, each link has to be involved in water, the reused water treated and recycled system called recycled-water system [1]. In order to ensure normal production, stable the quality of production and improve productivity, the primary task is to supply enough and stable pressure water. But recycled-water system is a complicated multi-input and multi-output process, there has interrelated and coupling phenomena between its control volume and the amount charged. In the recycled-water system, there is a strong coupling between water pressure of pipe and water level of pool. It is necessary to introduce decoupling device to make the multi-variable system become a single-variable system.

In industrial process control, the most common method used in decoupling are diagonal matrix decoupling synthesis method, matrix synthesis method and feed-forward compensation synthesis method. Feed-forward compensation method is the introduction of an integrated feed-forward control mind into the decoupling system, its network desire is very simple, it is easy to implement in engineering [2–4]. Based on the characters of recycled-water system, feed-forward compensation method is selected to decouple the system.

2. The Foundation of Recycled-Water System and the Analysis of Coupling

In order to design the control system, the established of a mineral processing mathematical recycled-water system model becomes necessary. In this paper, the closed-loop oscillation amplitude test model is used. Figure 1 is the factors of the simple closed-loop system, D(s) is regulator, G(s) is the generalized object, including the regulators and measuring instruments.

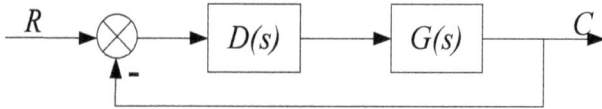

Figure 1. Simple closed-loop control system.

Based on the picture, the related function of the system equation (1) is known.

$$W(s) = \frac{D(s)G(s)}{1 + D(s)G(s)} \tag{1}$$

Its characteristic equation is shown as equation (2):

$$1 + D(s)G(s) = 0 \tag{2}$$

In order to simplify the experiment, making D(s)=Kp. When the regulators and objects keep closing, making the state variable on equal amplitude by adjust Kp, then record Kp and operation cycle time Ts. The mathematical model for the object can be deduced basis of the data.

Figure 2 is the mathematical model of the recycled-water system, which is founded by experimental method. The water level of the high pool is controlled by the frequency of pump in pumping station. The water pressure of the pipe is adjusted by the frequency of recycled pump. There is coupling phenomena between these two system. During the experiment, the cycle of the high pool regulator and ultrasonic level meter keeps closed; the cycle of pipe pressure keeps opening. When δi is in amplitude oscillation state, recording data is in Table 1.

Make the cycle of high pool keep opening, the pressure of cycled pipe keeps closed. When δp is in amplitude oscillation state, recording data in Table 2.

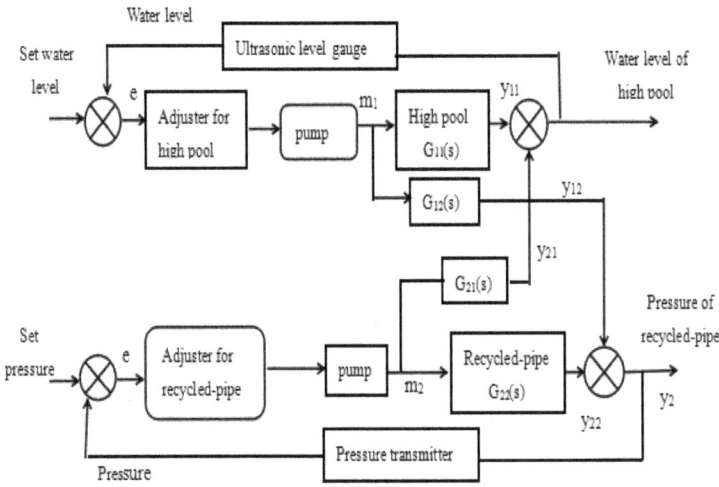

Figure 2. Coupling model of recycled-water system.

Table 1. Measured parameters when the cycle of the high pool keeps closed.

Measure items	Value
δ_i (system in amplitude oscillation state)	δ_i =48%
K_0 (the static gain of $G_{11}(s)$)	K_0=0.72
Ts (the cycle time of operation)	Ts=7.6 min
A (the amplitude ratio of y1 and m_1)	A=0.6
Φ (the delayed phase angle between y_2 and m_1)	Φ=360°×1.8÷7.6
τ (the initial insensitivity time y_2 respect to m_1)	τ=0.8min=48s

Table 2. Measured parameters when the cycle of the cycled pipe pressure keeps closed.

Measure items	Value
δ_p (system in amplitude oscillation state)	δ_p=35%
K_0 (the static gain of $G_{22}(s)$)	K_0=0.6
Ts (the cycle time of operation)	Ts=4.69min
A (the amplitude ratio of y_1 and m_2)	A=0.3
Φ (the delayed phase angle between y_1 and m_2)	Φ=360°×3÷4.69
τ (the initial insensitivity time y_1 respect to m_2)	τ=1.1min=66s

3. The Establish of Recycled-Water System

Based on experiments, the object model formula can be expressed as (3) [7]:

$$G(s) = \frac{K_0}{(T_s + 1)^n} \tag{3}$$

Equation (4) is simplified by function(3):

$$G(s) = \frac{K_0 e^{-\tau s}}{T_s + 1} \tag{4}$$

On the foundation of G11(s), G22(s), the simplified schematic model of oscillation amplitude and the measured data, the function of recycled-water system can be expressed as (5) and (6):

$$G_{11}(s) = \frac{0.72 e^{-51.6s}}{102s + 1} \tag{5}$$

$$G_{22}(s) = \frac{0.6 e^{-148s}}{222s + 1} \tag{6}$$

4. The Establish of Decoupling Model

$G_{12}(s)$ and $G_{21}(s)$ were fitted by using the simplest model, respectively. Such as function (7) and (8):

$$G_{12}(s) = \frac{K_{12} e^{-\tau s}}{T_1 s + 1} \tag{7}$$

$$G_{21}(s) = \frac{K_{21} e^{-\tau s}}{T_2 s + 1} \tag{8}$$

While the system is in a state of equal amplitude, the amplitude ratio and the delayed phase angle measured on the basis of the character of object input and output.

According to the measured parameters, the amplitude ratio and the relationship of phase angle, the simplified model of decoupling system is obtained, which can be expressed as (9) and (10):

$$G_{12}(s) = \frac{2.61 e^{-48s}}{190s + 1} \tag{9}$$

$$G_{21}(s) = \frac{0.7e^{-66s}}{213s+1} \tag{10}$$

5. The PID Control of Feed-Forward Compensation Synthesis Method

PID control is the abbreviation of proportional integral differential control. In the development procedure of automatic production process control, PID control is the oldest and the most vital form of control. It has a series of advantages, such as convenient to use, strong adaptability and robustness [5–7]. PID is the most widely used control algorithms; about 95 percent industry uses it. In this article, PID control algorithms are used in recycled-water system. Described by Figure 3 is the scene of system of PID control. yij is the object lap lace of column vector, ri is the object input lap lace transform column vector, G(s) is the transfer function matrix of controlled object, Gii(s) is the transfer function of the i-th channel, Gij(s) is the transfer function of the coupling channel, mc1 is the pressure of recycled-pipe, mc2 is the water level of high pool.

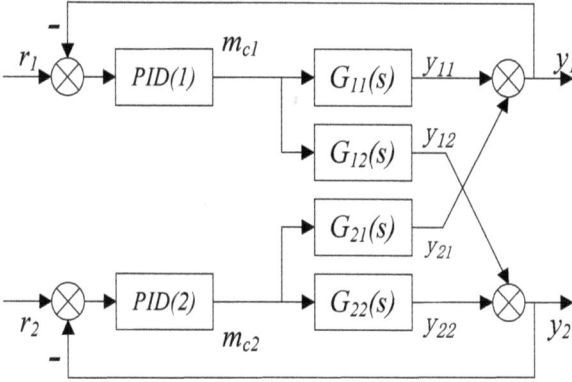

Figure 3. System of PID control.

Based on the picture, the transfer function can be expressed below:

$$G_{11}(s) = \frac{0.72e^{-51.6s}}{102s+1} \tag{11}$$

$$G_{22}(s) = \frac{0.6e^{-148s}}{222s+1} \tag{12}$$

29

$$G_{12}(s) = \frac{2.616e^{-48s}}{190s+1} \tag{13}$$

$$G_{21}(s) = \frac{0.7e^{-66s}}{213s+1} \tag{14}$$

There are strong coupling phenomena in the system. It is necessary to introduce the decoupling system. Figure 4 is the system of feed-forward decoupling.

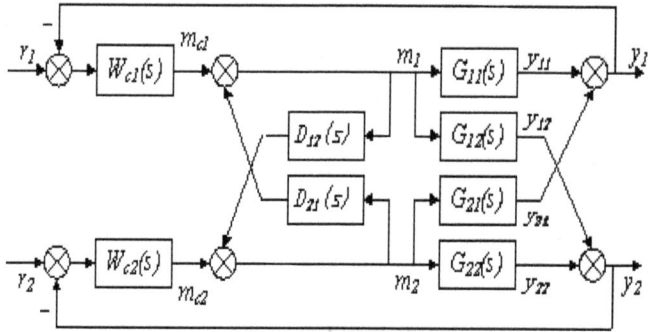

Figure 4. 2×2 PID control system of feed-forward decoupling.

Feed-forward compensation decoupling algorithm is by using the invariance principle to eliminate the coupling effects. The formula can be written like this:

$$y_{11} + y_{21} = 0 \quad (m_2 \neq 0) \tag{15}$$

$$y_{12} + y_{22} = 0 \quad (m_1 \neq 0) \tag{16}$$

The compensation condition is that:

$$G_{21}(s) + D_{21}(s)G_{11}(s) = 0 \tag{17}$$

$$G_{12}(s) + D_{12}(s)G_{22}(s) = 0 \tag{18}$$

So, the model formula can be expressed like this:

$$D_{12}(s) = -\frac{G_{12}(s)}{G_{22}(s)} \tag{19}$$

$$D_{21}(s) = -\frac{G_{21}(s)}{G_{11}(s)} \qquad (20)$$

6. The Results of Control

The application of feed-forward decoupling control system can eliminate the coupling that existed in recycled-water system. Making the system become two separate into two different noninterference circuits. The dynamic decoupling simulation model can be obtained:

$$D_{12}(s) = -\frac{G_{12}(s)}{G_{22}(s)} = -4.36\frac{222s+1}{190s+1} \qquad (21)$$

$$D_{21}(s) = -\frac{G_{21}(s)}{G_{11}(s)} = -0.97\frac{102s+1}{213s+1} \qquad (22)$$

The Matlab simulation control system is used in this paper. Figure 5 is the result without feed-forward compensation device. From the simulation result can be seen, when there is no decoupling device, the pressure of recycled-pipe is not stable, while the water-level of high pool response slowly. Based on the analysis of the structure of the system has found that the system exists serious coupling. Take measures to eliminate the coupling become necessary. The best way is to join decoupling device Figure 6 is the simulation result of the recycled-water system which added into decoupling equipment. When the system is introduced in the dynamic decoupling device, the water level and pressure of the pipe are stable to a certain value after a short adjustment. By the simulation results can be obtained that dynamic feed-forward decoupling is suitable for using in recycled-water system to eliminate the strong coupling.

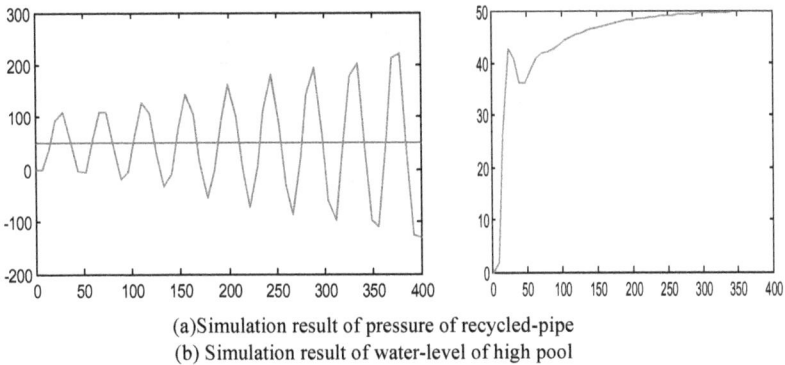

(a)Simulation result of pressure of recycled-pipe
(b) Simulation result of water-level of high pool

Figure 5. Simulation results without dynamics decoupling device.

31

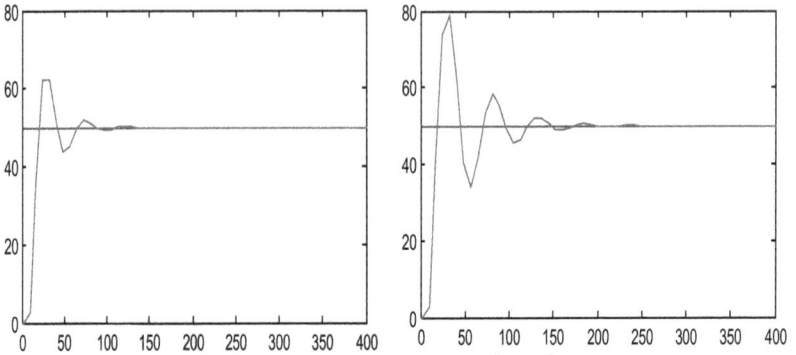

(a) Simulation result of pressure of recycled-pipe
(b) Simulation result of water-level of high pool
Figure 6. Simulation results which added into dynamics decoupling device.

7. Hardware Conditions of Recycled-Water System

In the light of the characters of recycled-water system, the PLC-SCADA system was used in mineral processing control system. The system includes on-site control, process monitoring and production management, as shown in Figure 7. The on-site control uses programming and distributing control system with high reliability. The process monitoring relied on computers and servers in operation station. These computers can backup and redundant each other to ensure data integrity and accurate. The servers can post the real-time parameters and pictures to the corporate network. Production management can seize the whole state of equipment operation and production through the computer. At the field control layer, Siemens series S7-300 PLC(CPU315-2DP) was selected as the main controller.

8. The Software Conditions

Siemens STEP7V5.2 control software program, which has a wealth system of organization blocks OBs, SFBs function blocks and system functions SFCs, makes the program structure is more simple. The transfer function of the block FB80(LEAD-LAG) can be written as formula (23):

$$W(s) = k \frac{T_1 s + 1}{T_2 s + 1}$$

$$(23)$$

Its Differential Equations:

$$u(k) = \frac{T_2}{T_2 + T_s} u(k-1) + K \frac{T_1 + T_2}{T_2 + T_s} f(k) - K \frac{T_1}{T_2 + T_s} f(k-1) \qquad (24)$$

u(k) is the control amount of k-time, f(k) is the enter the amount of interference of k-time, Ts is the sampling period.

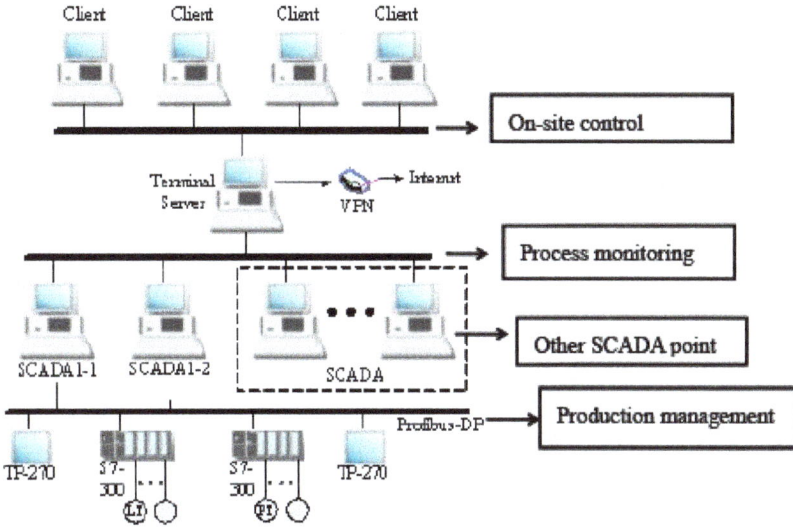

Figure 7. The control system of mineral processing.

9. Conclusion

In this paper, feed-forward compensation decoupling control method is used in recycled-water system to control water pressure and level. Simulation results show that this decoupling control can reach a good effect. The high pool water level and water pressure are stable at a certain value.

References

1. Long Yuanyuan. Design and simulation of decoupling control on multivariable coupling system [D]. Guangxi, Guangxi University, 2010.
2. Li Qiheng. Crushing and grinding [M]. Metallurgical Industry Press, 2004.
3. Pontus Nordfeldt. Decoupler and PID controller design of TITO systems [J]. Journal of Process Control, 2006,16:923–936.
4. Francesco Ticozzi. Dynamical decoupling in quantum control: A system theoretic approach [J]. Systems & Control Letters, 2006, 55:578–584.

5. X Zheng. An evaluation of different models of water recovery in flotation [J]. Minerals Engineering, 2006, 19:871–882.

6. R Toscano. Robust synthesis of a PID controller by uncertain multimodel approach [J]. Information Sciences, 2007,177:1441–1451.

7. Saeed Tavakoli. Tuning of decentralised PI (PID) controllers for TITO processes [J]. Control Engineering Practice, 2006,14:1069-1080.

Two Modified Newton-Type Iterative Methods for Solving Nonlinear Equations

Liang Fang[†], Xin Xue, Zhong-yong Hu and Rui Chen

*College of Mathematics and Statistics, Taishan University,
Tai'an, 271000, China*
[†]*E-mail: fangliang3@163.com*

With the rapid development of information science and engineering technology, nonlinear problems become an important research in the field of numerical analysis. In this paper, iterative methods for solving nonlinear equations are researched. Two modified Newton-type algorithms for solving nonlinear equations are proposed and analyzed, whose order of convergence are six and seven respectively. Both of the methods are free from second derivatives. The efficiency index of the presented methods are 1.431 and 1.476, respectively, which are all better than that of the classical Newton's method 1.414. Some numerical experiments demonstrate the performance of the presented algorithms.

Keywords: Iterative method; Nonlinear equation; Order of convergence; Newton's method; Efficiency index.

1. Introduction

Nonlinear problems are an important research field in numerical analysis. They have much applications in many fields such as engineering technology, applied information technology, computer, physics. The solution of nonlinear equations is one of the most investigated topics in numerical analysis, applied mathematics, and the problem of solving nonlinear equations by numerical methods has gained more and more importance than before, since many practical problems in the applied information technology, as well as in engineering technology, can build a suitable mathematical model, and then be transformed into nonlinear equations to solve.

In this paper, we consider iterative methods to find a simple root α of a nonlinear equation

$$f(x) = 0, \tag{1}$$

where

$$f : D \subseteq R \rightarrow R$$

for an open interval D is a scalar function and it is sufficiently differentiable in a neighborhood of α.

It is well known that classical Newton's method (NM) is a basic and important method for solving nonlinear equation [1] by the following iterative scheme

$$x_{n+1} = x_n - \frac{f(x_n)}{f'(x_n)} \tag{2}$$

which is quadratically convergent in the neighborhood of α.

In recent years, much attention has been given to develop iterative methods for solving nonlinear equations, and many literature has been produced [2-6].

Motivated and inspired by the activities in this direction, in this paper, we mainly study the iterative methods for solving nonlinear equations. We propose and analyze two modified Newton-type algorithms for solving nonlinear equations, whose order of convergence are six and seven respectively. Both of the methods are free from second derivatives. The efficiency index of the presented method s are 1.431 and 1.476, respectively, which are all better than that of the classical Newton's method 1.414. Some numerical experiments demonstrate the performance of the presented algorithms, present a sixth-order convergent iterative method. The proposed method is free from second derivatives. Several numerical results are given to illustrate the efficiency of the algorithm.

2. The Algorithms and Their Convergence Analysis

We consider the following iterative schemes.

Algorithm 1. For given x_0, we consider the following iteration scheme

$$y_n = x_n - \frac{f(x_n)}{f'(x_n)}, \tag{3}$$

$$z_n = y_n - \frac{3f'(x_n) - f'(y_n)}{f'(x_n) + f'(y_n)} \frac{f(y_n)}{f'(x_n)}, \tag{4}$$

$$x_{n+1} = z_n - \frac{3f'(x_n) + f'(y_n)}{-f'(x_n) + 5f'(y_n)} \frac{f(z_n)}{f'(x_n)}. \tag{5}$$

Theorem 1. Assume that the function $f : D \subseteq R \to R$ has a single root $\alpha \in D$, where D is an open interval. If $f(x)$ has first, second and third derivatives in the interval D, then Algorithm 1 is sixth-order convergent in a neighborhood of α and it satisfies error equation

36

$$e_{n+1} = 3c_2{}^5 e_n^6 + O(e_n^7),$$

where

$$e_n = x_n - \alpha, \ c_k = \frac{f^{(k)}(\alpha)}{k! f'(\alpha)}, \ k = 1, 2, \cdots.$$

Proof. Let α be the simple root of $f(x)$,

$$e_n = x_n - \alpha, c_k = \frac{f^{(k)}(\alpha)}{k! f'(\alpha)}, \ k = 1, 2, \cdots.$$

Consider the iteration function $F(x)$ defined by

$$F(x) = z(x) - \frac{3f'(x) + f'(y(x))}{-f'(x) + 5f'(y(x))} \frac{f(z(x))}{f'(x)} \tag{6}$$

where

$$y(x) = x - \frac{f(x)}{f'(x)},$$

$$z(x) = y(x) - \frac{3f'(x) - f'(y(x))}{f'(x) + f'(y(x))} \frac{f(y(x))}{f'(x)}.$$

By some computations using Maple we can obtain

$$F(\alpha) = \alpha, \ F^{(i)}(\alpha) = 0, i = 1, 2, 3, 4, 5, \quad F^{(6)}(\alpha) = \frac{135}{2} \cdot \frac{f^{(2)}(\alpha)^5}{f'(\alpha)^5} \tag{7}$$

Furthermore, from the Taylor expansion of $F(x_n)$ around α, we get

$$x_{n+1} = F(x_n) = F(\alpha) + F'(\alpha)(x_n - \alpha) + \frac{F^{(2)}(\alpha)}{2!}(x_n - \alpha)^2 + \frac{F^{(3)}(\alpha)}{3!}(x_n - \alpha)^3$$

$$+ \frac{F^{(4)}(\alpha)}{4!}(x_n - \alpha)^4 + \frac{F^{(5)}(\alpha)}{5!}(x_n - \alpha)^5 + \frac{F^{(6)}(\alpha)}{6!}(x_n - \alpha)^6$$

$$+ O((x_n - \alpha)^7). \tag{8}$$

Substituting (7) into (8) yields

$$x_{n+1} = \alpha + e_{n+1} = \alpha + 3c_2^5 e_n^6 + O(e_n^7).$$

Therefore, we have

$$e_{n+1} = 3c_2^5 e_n^6 + O(e_n^7)$$

which shows that Algorithm 1 is sixth-order convergent.

Similarly, we can obtain the following algorithms and results.

Algorithm 2. For given x_0, we consider the following iteration scheme

$$y_n = x_n - \frac{f(x_n)}{f'(x_n)}, \tag{9}$$

$$z_n = y_n - \frac{3f'(x_n) + f'(y_n)}{-f'(x_n) + 5f'(y_n)} \frac{f(y_n)}{f'(x_n)}, \tag{10}$$

$$x_{n+1} = z_n - \frac{3f'(x_n) - f'(y_n)}{f'(x_n) + f'(y_n)} \frac{f(z_n)}{f'(x_n)}. \tag{11}$$

Theorem 2. Assume that the function $f : D \subseteq R \to R$ has a single root $\alpha \in D$, where D is an open interval. If $f(x)$ has first, second and third derivatives in the interval D, then Algorithm 1 is seventh-order convergent in a neighborhood of α and it satisfies error equation

$$e_{n+1} = (3c_2^6 - 4c_2^4 c_3) e_n^7 + O(e_n^8),$$

where

$$e_n = x_n - \alpha, \ c_k = \frac{f^{(k)}(\alpha)}{k! f'(\alpha)}, \ k = 1, 2, \cdots.$$

Proof. The proof of the theorem is similar with that of Theorem 1. We omit it here for brevity.

To obtain an assessment of the efficiency of the proposed methods, we shall make use of efficiency index, according to which the efficiency of an iterative method is given by $p^{1/\omega}$, where p is the order of the method and ω is the number of function evaluations per iteration required by the method. It is not hard to see that the efficiency index of the Algorithm 1 is 1.431, and the

38

efficiency index of the Algorithm 2 is 1.476 which are all better than that of the classical Newton's method 1.414.

3. Numerical Results

Now, we employ Algorithm 1 and Algorithm 2 to solve some nonlinear equations and compare them with NM and the iterative method (PPM for short) Potra and Pták presented in [3]

$$x_{n+1} = x_n - \frac{f(x_n) + f(y_n)}{f'(x_n)} \tag{12}$$

which is cubically convergent with efficiency index 1.442, where

$$y_n = x_n - \frac{f(x_n)}{f'(x_n)}.$$

Displayed in Table 1 are the number of iterations (ITs) required such that

$$|f(x_n)| < 1.E - 15.$$

Table 1. Comparison of Algorithm 1, Algorithm 2, PPM and NM

Functions	x_0	NM	PPM	Algorithm1	Algorithm 2
f_1	1	5	3	3	2
	1.41	4	3	3	3
f_2	7.5	12	9	6	5
	-0.5	4	4	3	3
f_3	1.6	5	4	3	3
	2.1	6	5	4	3
f_4	2.4	5	4	4	4
	1.7	4	3	3	3

In Table 1, we use the following functions.

$$f_1(x) = x^3 + 4x^2 - 10, \quad \alpha = 1.36523001341410.$$

$$f_2(x) = (x+2)e^x - 1, \quad \alpha = -0.44285440096708.$$

$$f_3(x) = \sin^2(x) - x^2 + 1, \, \alpha = 1.40449164885154.$$

$$f_4(x) = 2\sin x - x, \, \alpha = 1.89549426703398.$$

The computational results in Table 1 show that Algorithm 1 and Algorithm 2 require less ITs than NM. Therefore, they are of practical interest and can compete with NM.

4. Conclusion

In order to obtain efficient iterative methods for the nonlinear equations which come from the practical problems in engineering technology and applied information technology field, in this paper, we present and analyze two modified Newton-type iterative method for nonlinear equations. Both of the methods are free from second derivatives. The efficiency index of the presented method s are 1.431 and 1.476, respectively, which are all better than that of the classical Newton's method 1.414. Some numerical experiments demonstrate the performance of the presented algorithms. The algorithm is free from second derivatives. Computational results demonstrate that it is more efficient and performs better than the classical NM.

References

1. J.F. Traub: Iterative Methods for Solution of Equations, Prentice-Hall, Englewood Clis, NJ(1964).
2. F.A. Potra, Potra-Pták: Nondiscrete induction and iterative processes, Research Notes in Mathematics, Vol. 103, Pitman, Boston, 1984.
3. A. N. Muhammad, I. N. Khalida: Modified iterative methods with cubic convergence for solving nonlinear equations, Applied Mathematics and Computation 184 (2007) 322-325.
4. C. Chun: Some fourth-order iterative methods for solvingnonlinear equations, Applied Mathematics and Computation 195 (2008) 454-459.
5. L. Fang, G. He: Some modifications of Newton's method with higher-order convergence for solving nonlinear equations, J. Comput. Appl. Math. 228 (2009), p. 296-303.
6. X. Feng, Y. He: High order iterative methods without derivatives for solving nonlinear equations, Appl. Math. Comput., 186(2007) 1617-1623.

Markov Process on Optimal Maintenance Strategy of Urban Rail Transit Equipment

Liu Cao[†], Dan-yu Zhang and Rui Hu

*Electrical and Information College, Jinan University,
Zhuhai, Guangdong Province, China
[†]E-mail: 156905347@qq.com*

Wei-hua Li

*School of Electric Power, South China University of Technology
Guangzhou, Guangdong Province, China
Electrical and Information College, Jinan University
Zhuhai, Guangdong Province, China*

The condition of urban rail transit equipment is an important factor affecting the health of its security operation, so an equipment maintenance strategy should be reasonably determined. This paper established Markov-Switching model to simulate the aging process of equipment as well as the recovery level of equipment maintenance, calculating the state transition probability matrix corresponding with the special strategy. The optimum strategy based on Markov decision-making stands on the point of economic effect and the reliability of equipment. Meanwhile, this paper used policy iteration method to obtain comprehensive expected maintenance cost. Further study claimed different loss cost of fault condition has different impacts on the optimal maintenance strategy, which proves this method is reasonable. It provides a theoretical basis for rail maintenance personnel on-site maintenance and decision making.

Keywords: Optimal Strategy; Markov-Switching Model; Maintenance Economic Effect; Reliability Of Equipment; Markov Processes; Policy Iteration Method.

1. Introduction

With the increasing complexity and precise of urban rail equipment, Plan Maintenance remains a high probability to cause work insufficient, inadequate or excess under the guidance of Wear and tear theory and Bath-rub curve theory [1]. With the increase of using years, the aging and damage of equipment affect the normal operation of rail transit. How to guarantee the reliability of rail transit equipment while the maintenance strategy is optimized, which improves the economic effect of equipment maintenance and management. It is an important research topic in the field of rail maintenance management.

2. Markov Decision Process

Markov decision [2–3] is a dynamic optimization process. It applies to sequential decision-making in dynamic random system with no aftereffect structure. In every moment, decision-makers select the appropriate measures for the current state according to the available information. When system state changes randomly with a period, decision-makers collect new information and make new decisions.

3. Markov-Switching Model of Equipment

3.1. State transition model without maintenance

Changes in the equipment condition is a state probability transition process that has no Markov Property. Equipment condition transition without maintenance is shown in Figure 1:

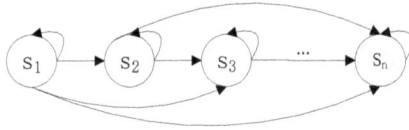

Fig. 1. Equipment condition transition without maintenance

The information of condition transition probabilities can be obtained by the actual rail transit equipment condition monitoring [4]. In these feedback data, assume there are n_i equipment in s_i, the number of a transition from s_i to s_j is n_{ij}. So the state transition probabilities is n_{ij}/n_i, namely p_{ij}.

Assume all p_{ij} make up matrix P_{ij}, P_{ij} is a Markov state transition matrix. According to maintenance experience, the condition of equipment usually becomes worse (or unchanged) after a natural period. $P_{ij} = \{p_{ij}\}, i, j \in \{1, 2, \cdots n\}$ is a $n \times n$ dimension upper triangular matrix.

3.2. State transition model with Condition Based Maintenance

According to Condition Based Maintenance, changes in equipment condition is also a Markov state transition process. Equipment condition transition diagram with maintenance is shown in Figure 2:

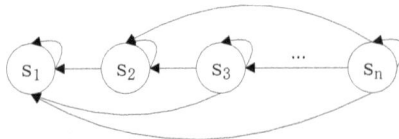

Fig. 2. Equipment condition transition with Condition Based Maintenance

In Figure 2, statistical data of equipment recovery condition transition probabilities can also be obtained through monitoring the actual power equipment. With Condition Based Maintenance, the probability that s_i transfers into s_j is q_{ij}. All q_{ij} make up matrix Q_{ij}. This paper assumes that worse transition due to technical or other reasons has been ignored. The equipment only transfers into a better (or unchanged) condition.

Thus, $Q_{ij} = \{q_{ij}\}, i, j \in \{1, 2, \cdots n\}$ is a $n \times n$ dimension lower triangular one.

4. Optimal Maintenance Markov Decision Strategies

4.1. Determine equipment condition levels

According to IEEE standard guidelines [5], they are normal condition, attentive condition, abnormal condition, serious condition and fault condition. These conditions progressively deteriorate into worse levels, which can be represented by s_i ($i = 1, 2, 3, 4, 5$).The greater i indicates, the worse degree of the condition will be.

4.2. Determine the set of measures

Assume the operating condition set of equipment is $S\{s_1, s_2, \cdots, s_n\}$. R means a set of maintenance measures. Suppose that there are m measures in all and $R\{R_1, R_2, \cdots, R_m\}$ means the whole maintenance measures. Among of them, R_1 means lowest level maintenance and R_m means the highest level maintenance, which costs the most. d_i means a maintenance measure for s_i. $D\{d_1, d_2, \cdots d_n\}$ means a maintenance strategy for whole conditions. d_i can be any measure of $R\{R_1, R_2, \cdots, R_m\}$ and only influenced by the condition of equipment.

4.3. Determine the remuneration function

While the strategy(D) is made, $C = c(s_i, D)(s_i \in S, D \in R)$ indicates that the system in a condition (s_i) will obtain an expected remuneration when it adopts D at current time. The remuneration function is not related with the history of system, no matter what initial condition of system transfers into s_i, it eventually obtains the same remuneration as long as D is in the same.

4.4. Determine equipment condition transition probability matrix

Assume the completion of maintenance is in an instant (t_M is greatly close to 0), in an actual period, the condition of equipment experiences two transitions. The first state transition probability matrixes is Q_{ij}. The second state transition probability matrixes is P_{ij}.In addition, suppose that equipment which experiences instantaneous maintenance has the same state transition probabilities as the case

43

of the equipment without maintenance after a natural period. These two transitions can be regarded as an equivalent period. So the equivalent state transition probability matrix is:

$$P_{ij}^{(D^{spe})} = Q_{ij} \bullet P_{ij} = \begin{bmatrix} p_{11}^{d_1^{spe}} & p_{12}^{d_1^{spe}} & \cdots & p_{1n}^{d_1^{spe}} \\ p_{21}^{d_2^{spe}} & p_{22}^{d_2^{spe}} & \cdots & p_{2n}^{d_2^{spe}} \\ \vdots & \vdots & \ddots & \vdots \\ p_{n1}^{d_n^{spe}} & p_{n2}^{d_n^{spe}} & \cdots & p_{nn}^{d_n^{spe}} \end{bmatrix} \tag{1}$$

Where $p_{ij}^{d^{spe}}$ means when the condition is s_i, d_i^{spe} is adopted to make s_i transfer into s_j and all d_i^{spe} make up D^{spe} which is a specific measure group, namely a strategy.

4.5. Determine the objective function

The objective function is to adopt a maintenance strategy so that the sum of expected cost for each condition achieves a minimum. It can be availably quantified by the cost of maintenance. Combined with remuneration function, the maintenance strategy not only reflects the current cost from corresponding measures, but also estimates the impact of its next strategy according to the next predicting equipment condition. Then the target function of total expected maintenance cost is:

$$y(D) = \min \sum_{i=1}^{n} W_i^{d_i} = \sum_{i=1}^{n} (c(s_i, D) + \beta \sum_{j=1}^{n} P_{ij}^{d_i} W_j^{d_i})$$

$$s_i, s_j \in S; D \in R \tag{2}$$

Where D is a maintenance strategy; $y(D)$ is a total expected cost of maintenance when it adopts the D. $W_i^{d_i}$ is the expected cost when the initial condition is i and its corresponding measure is d_i. $c(s_i, D)$ is remuneration function, which indicates the equipment maintenance cost for the current s_i and the costs is known according to the historical records. The probability that s_i transfers into s_j is $P_{ij}^{d_i}$. β is a discount factor, which indicates a current value of future cost. The comprehensive equipment maintenance cost is limited by time effect. So the range of β belongs to $[0,1]$.

4.6. Determine the optimal iteration strategy

The objective function indicates that the strategy is to select a group of measures so that the total expected cost of maintenance achieves a minimum. If there are n measures can be applied to each condition (suppose there are m conditions) in the calculation process, the total number of strategies is n^m. However, large number of conditions and measures leads too many calculations, which is not as convenient to obtain the consequence. Thus, this paper adopts policy iteration method.

5. Example Analysis

This paper lists a group data of locomotive transformers to determine the optimal strategy. According to the data from paper[6-7], this paper combines with the historical maintenance records and probability estimation method to obtain five conditions, they are s_1, s_2, s_3, s_4, s_5. The maintenance measures are R_1, R_2 and R_3. R_1 means no maintenance, R_2 means local maintenance and R_2 means complete maintenance. The state transition probability matrixes of them are as follows in Table1:

Table 1. The state transition probabilities

Condition	s_1	s_2	s_3	s_4	s_5
s_1	0.8/1/1	0.1/0/0	0.075/0/0	0.02/0/0	0.005/0/0
s_2	0/0.75/0.9	0.65/0.25/0.1	0.2/0/0	0.1/0/0	0.05/0/0
s_3	0/0.6/0.85	0/0.3/0.1	0.5/0.1/0.05	0.4/0/0	0.1/0/0
s_4	0/0.5/0.7	0/0.3/0.2	0/0.1/0.075	0.4/0.1/0.025	0.6/0/0
s_5	0/0.45/0.65	0/0.25/0.175	0/0.15/0.1	0/0.1/0.05	1/0.05/0.025

The equivalent state transition probability matrixes of local maintenance is:

$$P_{ij}^{(1)} = \begin{bmatrix} 0.8 & 0.1 & 0.075 & 0.02 & 0.005 \\ 0.6 & 0.2375 & 0.1063 & 0.04 & 0.0163 \\ 0.48 & 0.255 & 0.155 & 0.082 & 0.028 \\ 0.4 & 0.245 & 0.1475 & 0.12 & 0.0875 \\ 0.36 & 0.2075 & 0.1588 & 0.134 & 0.1397 \end{bmatrix} \tag{3}$$

45

The equivalent state transition probability matrixes of complete maintenance is:

$$P_{ij}^{(2)} = \begin{bmatrix} 0.8 & 0.1 & 0.075 & 0.02 & 0.005 \\ 0.72 & 0.155 & 0.0875 & 0.028 & 0.0095 \\ 0.68 & 0.15 & 0.1088 & 0.047 & 0.0143 \\ 0.56 & 0.2 & 0.13 & 0.074 & 0.036 \\ 0.52 & 0.1788 & 0.1337 & 0.0905 & 0.077 \end{bmatrix} \tag{4}$$

Equipment maintenance cost should include parts replacement cost, labor cost and downtime cost. The maintenance cost of a single equipment under different maintenance measures is shown in Table 2:

Table 2. The integrated maintenance cost of different condition

Measures/Conditions	s_1	s_2	s_3	s_4	s_5
R_1	0	0	0	0	-
R_2	-	4.5	6	9	-
R_3	-	8.5	14	21	30

Note: "-" means maintenance measures are not taken in this condition.
The unit of cost is Ten thousand RMB.

According to Table 2, if decision makers consider strategies based on economic effect. Assumed that the starting item of policy iteration method is $D\{R_1, R_1, R_1, R_1, R_3\}$ and the value of β is 0.85. The final results of the iteration is $[R_1, R_1, R_2, R_2, R_3]$. The total cost of the strategy is 110.28 ten thousand RMB. Obviously, this strategy has few consideration on equipment reliability.

However, if decision makers only consider equipment reliability. In order to minimize the probability of equipment fault condition, no matter what condition of equipment transfers into fault condition(suppose that normal condition takes no maintenance, the fault condition takes highest level of maintenance), then the objective function is:

$$P_{\min}^{sum} = \sum_{i=2}^{4} P_{s_i} \cdot P_{i5} \tag{5}$$

Where p_{s_i} means the probability of the current equipment of which the condition is s_i, P_{i5} means a probability that the equipment experiences a natural period from non-fault condition to a fault condition under some maintenance measure. So it is easy to obtain the strategy: $[R_1, R_3, R_3, R_3, R_3]$, which takes high maintenance cost. Thus, it is also unreasonable to adopt this strategy.

Considering economic effect and reliability of rail transit equipment, the data in Table 1 does not include extra loss cost due to fault condition (such as scrapped equipment). Then, it is assumed to regard equipment reliability as a constraint. No matter what non-fault condition is, when it transfers into fault condition with a period, the adding loss cost assumed to be 500,000 RMB (including maintenance cost and impairment charges etc.). So the current comprehensive maintenance cost for s_5 is 800,000 RMB. In addition, there are also extra loss cost for s_2, s_3 and s_4. The cost are respectively assumed to be 10,000 RMB, 20,000 RMB and 40,000 RMB.

With other assumptions remaining unchanged, the new maintenance decision iterative result is: $[R_1, R_2, R_2, R_2, R_3]$. And the expected maintenance cost of each condition is: $[12.68, 19.39, 22.93, 30.81, 100.08]$ (10,000 RMB).

According to the importance and value of equipment, if the loss cost of fault condition increases, the changes of maintenance measures in a strategy is shown in Figure 3:

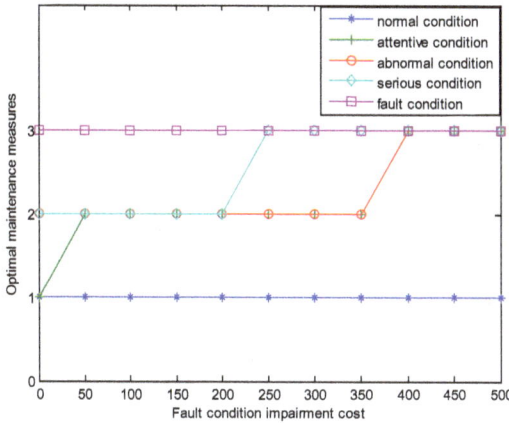

Fig. 3. Optimal maintenance strategy for the different loss cost of fault condition

From Figure 3, with the loss cost of equipment fault condition increasing, the level of maintenance measure for each condition to take becomes higher. When the fault condition loss cost reaches a certain value, all the conditions (except normal condition) need complete maintenance, which is corresponding with the strategy based on the highest reliability of the equipment.

6. Conclusion

This paper uses Markov decision to determine the optimal maintenance strategy for urban rail transit equipment, which considers both equipment reliability and economic effect of maintenance. The establishment of Markov state transition model is used to describe the actual condition of equipment. The expected maintenance cost include the current cost of measures corresponding with all the conditions and the loss cost of the impact on next equipment conditions based on the current strategy. Further analysis of how optimal maintenance strategy changes is based on the different loss cost of fault condition, which proves the rationality of this method. The strategy fully considers regularity of equipment operation and property of equipment, providing a more reasonable maintenance strategy in an effective way.

Acknowledgment

This work was funded by National Training Program of Innovation and Entrepreneurship for Undergraduates of Jinan University.

References

1. Wang Shuang Li. Research on Theory and Methods of Maintenance Decision Making for Equipment in Power Plant [D]. Tian Jin: Tian Jin University, 2007.
2. Guo Chao Zhong, Study of remaining life prediction based on Markov model[D]. Harbin: Harbin Institute of Technology, 2009.
3. Haussman W. Sequential decision problems a model to exploit existing forecasters [J]. Management Science, 1969,16(2):93-111.
4. Anders G J, Endrenyi J, Ford G L, et al. A probabilistic model for evaluating the remaining life of evaluating the remaining life of electrical insulation in rotating machines [J]. IEEE Trans on Energy Conversion, 1990, 5(4): 761-767.
5. IEEE Standard C57.104TM-2008. IEEE guide for the interpretation of gases generated in oil-immersed transformers [S]. 2008.
6. Guo Ji-Wei, Liu Gang, Tang Guo-Qing, Wang Ying, Markov Decision Process on Electrical Equipment Maintenance Optimization [J]. Proceedings of the CSU-EPSA, 2004.
7. Jiang Xue Lei, Maintenance optimization for power equipment based on Markov process [D]. Ji Nan: Shang Dong University, 2013.

Study of the Characters of Cold Air Activities in Sichuan

Yi-yu Gong[†] and Song-hai Fan

State Grid Sichuan Electric Power Research Institute
Chengdu, Sichuan, China
†E-mail: 290118787@qq.com

Based on the temperature data of 34 years in Sichuan Province from 1980 to 2013, the frequency of cold air activity interannual variability and regional space-time variation characteristics of cold air activities were analyzed. After the partition of Sichuan Province, the overall space-time distribution of cold air was focused. Based on the review, it is recognized that cold air activity average frequency in Sichuan decreases from west to east, which means that the cold air activity decreases from west to east. Because of the impacts of the circulation feature and the landscape terrain in winter months, it is colder in western Sichuan than central and eastern regions. With increasing time, the amount of cold air produces a fluctuating trend and the overall trend has weakened with occasional mutations.

Keywords: Sichuan partition; cold air activities; frequency; temporal variation.

1. Introduction

Cold air activity is a very common weather process in winter months occurring in East Asia which influences the people's production and life a lot. The development and progression of cold air which, along with the activity of rain, snow and other weather process causes the temperature suddenly dropped, resulting in an extremely important impact on people's life and production.

Since the 20th century, the temperature has increased significantly for the global warming. In the foreign country, Lau, N.C and K.M. Lau pointed that the cold wave of cold air activities has two types of disturbances to break out, and summarized the development of the law of its occurrence. [1] In China, Shi and Guo found that Siberian High and the East Asian winter monsoon is the main area system of coldair activities affecting China. [2-4] Chou etc. found that, since 1970, the strong cold air process (including strong cold air and cold wave) tends to decrease. [5-13]

Aimed at the special geographical situation of Sichuan and its circulation background, the daily temperature datum by intensity, frequency, regional area, etc. collected by 52 ground stations from 1980 to 2013 in Sichuan Province are

49

used to research and summarize Sichuan cold air's regional activities and temporal differences in the frequency variation.

2. Materials and Methods

China Meteorological Data sharing service network provides Jan. 1, 1980 - Dec. 31, 2013, Sichuan Province 52 stations daily minimum temperature map. Also, according to the degree of strength of the cold air, it is divided into four levels of criteria: weak cold, moderate cold, cold air and cold wave. Criteria for the classification of cold air is showed in Table 1. Process intensity is decided by the daily minimum temperature negative temperature change size within 48 hours and time t decides the temperature process. The cold air activities generally occur in the winter months, so in the winter months, the definition of the cold weather process activities is that the changing temperature is less than 0 within 48 hours.

Table 1: Cold strength standard levels

Level	48h variable temperature $\triangle$ t ($°C$)	cold strength
1	$\triangle$t <6 weak	weak cold air
2	$6 = <\triangle$ t <8	moderate cold air
3	$8 <= \triangle$ t <10	strong cold air
4	$\triangle$ t> = 10	cold wave

3. Sichuan Characteristic Frequency of Cold Air Activity

3.1 *Zoning*

In order to study the temporal change of cold air activities in entire region of Sichuan, the Province will be divided into the following three parts in eastern region, the central region, and the western region to facilitate statistical induction.

Table 2: Eastern, central and western stations and number

	Station	Number		Station	Number	Station	Number
	Guangyuan	57206		Ruoergai	56079	Kangding	56374
	Wanyuan	57237		Xiaojin	56178	Emei	56385
Eastern	Langzhong	57306	Central	Wenjiang	56187	Leshan	56386
	Bazhong	57313		Songpan	56182	Jiulong	56462
	Daxian	57328		Dujiangyan	56188	Yuexi	56475
	Suining	57405		Pingwu	56193	Yibin	56492
	Gaoping District	57411		Mianyang	56196	Neijiang	57504
				Ya'an	56287	Luzhou	57602
						Xuyong	57608

Table 2: Eastern, central and western stations and number (Cont.)

	Station	Number	Station	Number	Station	Number
	Shiqu	56338	Hongyuan	56173	Zhaojue	56479
	Dege	56144	Batang	56247	Leibo	56485
	Ganzi	56146	Xinlong	56151	Yanyuan	56565
Western	Seda	56152	Litang	56257	Xichang	56571
	Daofu	56167	Daocheng	56357	Panzhihua	56666
	Aba	56171	Derong	56441	Huili	56671
	Ma'erkang	56172	Muli	56459		

3.2 Three regions of cold air activities

Cold air activities mainly occurred in the winter months and summer months rarely, therefore, this study focuses on the activity in cold winter months.

Analysis of historical data from 1980 to 2013 in 34 years, cold winter activities throughout the eastern region of the average annual frequency is 78.21/a, the western region the average annual frequency is 89.1/a, while the central region, compared with 88.3/a. In addition, the number of cold air activity trends shown in Figure 1:

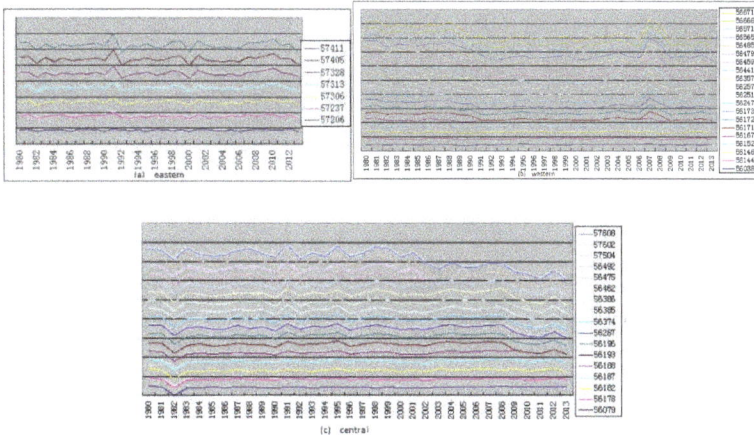

Fig. 1. 34-year change in the number of cold air activity trends

It could be seen in Figure 1 that from 1980 to 2013 in Sichuan the cold winter activities substantially stabilized, with occasional fluctuations in the central and east. In 1991, the number of cold air activity in central and eastern Sichuan reached a peak, and then stabilized. To 2000 the number of cold air activities in eastern fell valley, and the western and central relatively stable. While the western region in 2007 ushered in 34 years of cold air activity peak times, temperatures lower than in previous years. Moreover, since 2001,

51

although the central region of cold air activity occasional fluctuated, the overall continued to show a downward trend.

Time change in frequency of cold air activity average frequency of fluctuations varies a lot. As could be seen in the Figure 2, the eastern region of the frequency of cold air activity mainly changed between 80/a and 100/a; in the western region in the frequency of cold air activity changed between 85/ a and 93/a, while in the middle was 85/a to 93/a change. Variation in western and central was broadly consistent. The annual average frequency of cold air activities in eastern could be up to 100 / a (1990) and down to 83 / a (1992,2000). In 1990, western and central frequency of the average annual minimum was 85 / a, followed by the year 2000 86 / a, describing that a few years before and this two years with respect to a temperature rise of warmer; the highest value was in the year 2004, followed by 2007, were 93 / a, describing that cold air activities activated relatively frequently in the last two years and winter air temperature declined compared with adjacent years in western Sichuan. Over the last few years thereafter Midwest winter months compared with the year 2000 showed a relatively low temperature.

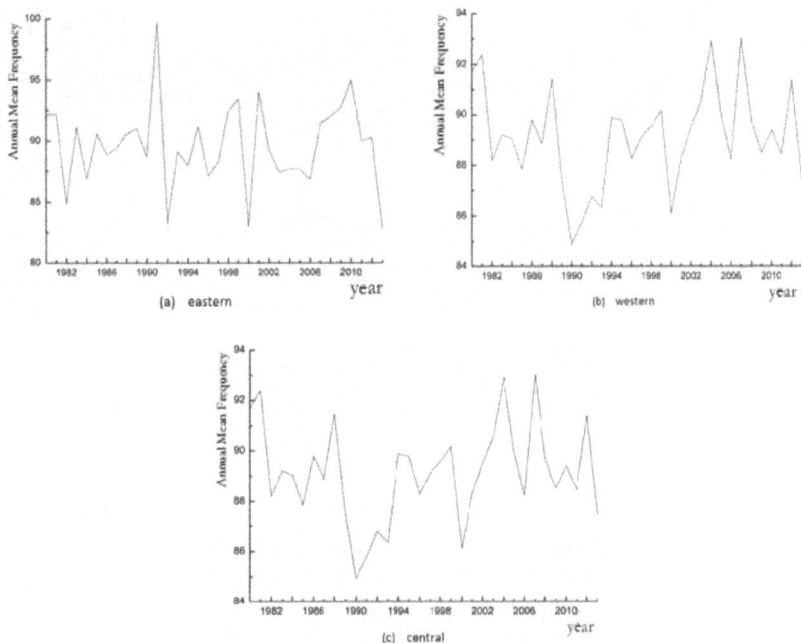

Fig. 2. Time changes of the average frequency of cold air activity times

4. Conclusion

According to the 34-years temperature data from 1980 to 2013 in Sichuan Province, the Sichuan Province was divided into eastern, central and west parts, exploring the regional differences of interannual variability of the cold air activity's frequency in Sichuan Province in 34 years, as well as the change of time and space activity characteristics, and the conclusions are as followed:

The average annual frequency of cold air activity from west to east gradually reduced in Sichuan, indicating that cold air activity gradually decreased from west to east; and because of the influence of circulation and topography terrain in winter months, temperatures in western Sichuan Province was lower than the central and eastern. Over time, the number of occurrences of cold air fluctuations showed signs of abating overall, with occasional mutations.

References

1. Lau, NC and KM Lau, The structure and energetics of mid latitude disturbances accompanying cold air outbreaks over East Asia, Mon, Wea, Rev. 112, 1309-1327, 1984.
2. XieLian, Shi Xiaohui, Xiangde trends interannual nearly 40 years of East Asian winter monsoon [J] Atmospheric Sciences, 2007, 31 (4): 747-756.
3. Yun Guo its East Asian Winter Monsoon unusual temperature relationship with China [J] Atmospheric Sciences, 1994, 5 (2): 218-225.
4. Gong DY, Wang Shaowu Siberian Studylong-term changes high and global warming may affect the [J] Geographical Science, 1999, 54 (2): 125-132.
5. Qiu Yong Yan, Chou Yongkang, Li Xiaodong features cold air activities and Eurasia snow cover China and its relationship [J] Journal of Applied Meteorology, 1992, 3 (2): 235-241.
6. Evolution of the Siberian high by former Chen Jun, Xie, Lu Ying winter monsoon [J] Atmospheric Sciences, 1992, 16 (6): 677-685.
7. Ding dissemination and the role of a planetary scale cold cold air sinks in East Asia [J] Applied Meteorology News, 1991, 2 (2): 124-132.
8. Chen Jun, Sun Shuqing East Asian winter monsoon and global change contrast abnormal atmospheric circulation anomalies I. winter monsoon strength [J]. Journal of Atmospheric Sciences, 1999, 23 (1): 101-111.
9. Ding Yihui. Build-up, air mass transformation and propagation of Siberian high high and its relation to cold surge in East Asia [J] Meteor Atoms Phys, 1990, 44: 281-292.
10. Chen Guangming, Cold high statistical research Zhangpei Zhong cold wave affecting China [J]. Journal of Meteorology, 1999, 57 (4): 493-501.

11. Jun Chen, Zhang Peizhong Asia and the Western Pacific extratropical cyclone climate portfolio [M] Beijing: China Meteorological Press, 1992.

12. Ding Yihui, Statistical study of East Asian winter monsoon [J], Tropical Meteorology, 1990, 6 (2): 119-128.

13. Li Feng, Jiao Meiyan, Ding Yihui, etc., Changes in the Arctic region for nearly 30 years and the impact on the circulation of cold air and strong China [J] Plateau Meteorology, 2006, 25 (2): 209-219.

Study on the Controllability in a Steady State for the Quasilinear Wave Equations with Weak Decay

Shu-xian Deng

College of Science, Henan University of Engineering,
Zhengzhou, 451191, China
E-mail: dshuxian@163.com

Xin-xin Ge

College of Management, Henan University of Engineering,
Zhengzhou, 451191, China
E-mail: gexinxin219@163.com

Zhi-jun Wang

College of Science, Zhengzhou Normal university,
Zhengzhou, 450044, China
E-mail: wzjzzsf@163.com

A new method of investigating the so-called quasilinear strongly-damped wave equations is proposed in this paper, concerned with controllability in a steady state for the Quasilinear Wave Equations with weak decay. A so-called energy perturbation method to establish weak controllability of solutions in terms of energy norm for a class of nonlinear functions is presented. This method establishes the existence and uniqueness of energy solutions and the existence of finite-dimensional global and exponential attractors for the solution semigroup associated with that equation and their additional regularity. Controllability in a steady state with the help of differential inequalities by estimating the relationship between energy inequalities and attenuating property of weak solutions is demonstrated. A small positive number is determined and derived differential inequalities by using a perturbation of energy.

Keywords: Steady; Controllability; Weak; State.

1. Introduction

We consider the control problem of weak decay to the following so-called quasi-linear wave equation in a smooth bounded domain $\Omega \subset \mathbb{R}^3$

$$u_{tt} - \lambda \Delta u_t + u \cdot \nabla u = f(x,t) \tag{1.1}$$

$$\text{div } u = 0 \tag{1.2}$$

$$u(x,0) = u_0, \quad u_t(x,0) = u_1. \tag{1.3}$$

Where u_0, u_1, f are given functions, Δu is a Laplacian with respect to the variable $x \in \Omega$, $u = u(t, x)$ is an unknown function, $\lambda > 0$ is a fixed positive number, f are given external forces, and satisfying the following conditions:

$$f \in L_2, \; u \in L_q\left(\mathbb{R}^+, L_p\right) \; and \; \frac{2}{p} + \frac{n}{q} \le 1. \tag{1.4}$$

For some positive $a, p \in [1/2, 2), C > 0,$ and $q > 0$, we have a weak solution, which fulfills additionally

$$d + a \mid x \mid^p \le f'(x,t) \le d(1 + |x|^q), \forall s \in \mathbb{R}^3 \tag{1.5}$$

and
$$u_{tt} - div \lambda \nabla u^2 - \Delta u_t = f(x, t). \tag{1.6}$$

We have several techniques to prove the existence of weak decay solutions with respect to the phase space; and have additional nice properties with energy inequality for almost all times or solutions with weak decay properties for $t \to \infty$, this has been studied recently by several people, e.g. YM. Qin, Ebihara, Xin Liu [1-6] etc.

As a model of quasi-linear wave equation, for $N = 1/2$, 1 and $f = 0$, 1 equation(1.1)-(1.3) admits a global weak decay solution as large initial data, which was proved by Y.M. Qin, Xin Liu, X.G. Yang, Lan Huang etc [1-2]. T.G Wang, Ming Zhang, M.J Wang simplified the above arguments and give the proof of control with exponential decay. From the perspective and background of physics, this represents an implementation of axial movement of the viscoelastic material, this cause the form of above equations, in the one-dimensional case, their model of longitudinal vibration of a uniform rod with nonlinear stress function f. In two, three-dimensional case, they describe the viscoelastic solid of anti-plane shear action. While $n = 1$ and $f = 0$, ST Li prove existence of weak periodic decay strong solution on the periodicity condition, X.K Su and J.L Zhang [3-6] proved the controllability of a smooth solution in the method of Cauchy problems in the case of smooth and small data.

Ulteriorly, while $n > 1/2$ and $f \ne 0$, M.J. Wang and X.G. Yang gave the proof of global controllability with smooth solution in the case of small initial data. Make use of combining L^p -theorem of Soblev space and semi group theorem of operators, Nakao [3-4] and A.F and H.B [5-6] devised certain decay rate of energy of global solutions with large data under a specific condition which is certainly satisfied if the mean curvature of the boundary $\partial\Omega$ is non-positive. For $n > 1$ and $f = 1$, nonlinear elliptic equation with periodicity conditions was studied [10],

$$d'(t) + k_1 \|w_1(t) - w_2(t)\|_p^v \, d(t) \le k_2 d(t)^{p+q-1} + \kappa d(t)^\mu \|w_1(t)\|_2^{v+1} \tag{1.7}$$

$$w(x,\, t + \omega) = w(x,\, t), \tag{1.8}$$

In the case of $\|w(t)\| \le k_0 (1+t)^{-\frac{\lambda}{2}}$, and $\|w(t)\|_p \le \tau(1+t)^{(\frac{\tau}{2} + \frac{n}{2}(\frac{1}{2} - \frac{1}{p}))}$.

Here, Ω is a bounded domain in $\mathbb{R}^n$ with a smooth, $\partial\Omega$ is said to be C^2 class boundary, which satisfies the following uniform hyperbolic assumption:

For some constants $\rho_0, M > 0$, $\tau \in H^4[0, +\infty)$ satisfies:

$$x'(t) + \tau_0 (1+t)^{(\ell+\frac{d}{2})} x(t)^{(2+\frac{\gamma}{p})} \le \sigma y(t)^{\frac{v}{2}} + \tau_1 (1+t)^\delta y(t)^\gamma \tag{1.9}$$

$$\tau(v^2) + 2\sigma_0(v^2)v + \tau''(v^2) \le M < \infty \tag{1.10}$$

Thus, we can use a so-called energy perturbation method to establish weak controllability of solutions in terms of energy norm for a class of nonlinear functions.

This method allows us to establish the existence and uniqueness of energy solutions. We also establish the existence of finite-dimensional global and exponential attractors for the solution semi group associated with that equation and their additional regularity. We will show the controllability in a steady state with the help of differential inequalities by estimating the relationship between energy inequalities and attenuating property of weak solutions. And furthermore, we will determine a small positive number and derive differential inequalities by using a perturbation of energy, and conclude some following results immediately.

2. Main Results

First, we focus on the control of weak decay stability, in order to describe the maneuverability; we define global weak solution with decay for (1.1)-(1.3):

If the initial data satisfies u_0, $u_1 \in H^2 \cap H_0^4$, and the function ω is said to be a weak solution of Problem (1.1)-(1.3), if it satisfies the following conditions:

(a) $\quad E(t) \equiv \dfrac{1}{2}\left\{\left\|\omega_t(t)\right\|^2 + \displaystyle\int_\Omega\int_0^{|\nabla \omega|^2}\tau(\xi)\,d\xi dy + \int_0^t(f,\omega(t))\,dsdy\right\}$ (2.1)

(b) $\quad (\omega_t,\mu)\big|_0^t + \displaystyle\int_0^t\left\{(\omega_t,\mu_t)-\big(\mathrm{div}\,(\tau\nabla\omega),\mu\big)-(\Delta\omega_t,\mu)-(f,\mu)\right\}=0$ (2.2)

It is well known that the existence of such a weak solution with decay for all times is assured. Once this is known, one can identify this solution with the global weak one and continue this process to get that

$$t^2\left\|\partial_t v\right\|_H^2 \leq \lambda t^2\left\|\partial_t v\right\|_{L^\infty}^{1/2}\left\|\partial_t v\right\|_{L^2}^{3/2} \leq \delta t^3\left\|\partial_t v\right\|_{L^2}^3 + \rho\left\|\partial_t\omega\right\|_{L^\infty}^2 \qquad (2.3)$$

Obviously, we will construct the controllability of the local weak solution with decay for the semi-group generated by weak energy solutions:

Theorem1 *Under the above hypothesis and suppose that $\omega(x,t)$ is a sufficiently regular weak solution of problem (1.1)-(1.3), the following estimates hold*

(a) $\quad \left\|\partial_t^2\omega(\xi)\right\|_H^2 \leq \delta t^3\left\|\partial_t v\right\|_{L^2}^3 + C^3\rho^2\delta^{-2}\left\|\partial_t\omega\right\|_{L^\infty}^2 +\lambda v(t)e^{\varepsilon(t-\alpha)}$ (2.4)

(b) Let the assumptions (1.2) and (1.3) be satisfied with $\theta = 2$, and $f(t)\leq \kappa e^{-2\rho t}$. Then, the local weak solution with attenuation $\omega(t)$ of problem (1.1)-(1.3) which satisfies the additional regularity, and exists constants $M > 0$, $V > 0$ such that

$$E(t)\leq Me^{-3vt} \qquad (2.5)$$

Theorem 2 *Let ω be a weak solution with decay of (1.1)-(1.3). Denote by $\omega_0(t)$ $= e^{-t\Delta}$ the solution of the wave equation and suppose $\left\|\omega_0(t)\right\|_2 \leq \lambda_0(1 +t)^{-\tau/2}$, $\omega_0 \in H^2(\Omega)\cap H_0^4(\Omega)$, $\omega_1 \in H_0^4(\Omega)$, and $f'(x, t) \in W_0^2$ then problem(1.1)-(1.3) hold a unique local weak solution $\omega(t)$ with the following estimate*

$$y'(t)+\lambda_0(1+t)^{\frac{\kappa}{p}}y(t)^{(1+\frac{\varepsilon}{2}\kappa)} \leq \lambda_1 y(t)^{\frac{\gamma}{2}\mu} + \lambda_2(1+t)^{-q}\nabla x\omega(t) \quad (2.6)$$

Remark *Obviously, from Theorem 1 and Theorem 2, it is easy to establish the control of stability for polynomial decay to (2.3)-(2.5). In comparison to (2.2)-(2.3), we give the strong stability estimates. Add the limit $t \to \infty$ to the dissipative estimate (2.4) for the approximations $\omega(t)$, and together with Sobolev embedding theorem, we can immediately conclude that the limit weak solution $\omega(t)$ also satisfies:*

$$\left\| \nabla \omega \right\|_2 + \left\| \partial_t u \right\|_{L^2}^2 \le \left\| \partial_t u_1 \right\|_{L^\infty}^2 + \int_0^t \left\| \nabla h(x) \right\|_{L^2} dx + \left(\partial_t v, v \right) \qquad (2.7)$$

3. Proofs of Main Results

Proof of Theorem 1 Without loss of generality, we consider the initial boundary value problem for the following nonlinear wave equation:

$$u_{tt} - \mathrm{div} \{ (a \mid \nabla u \mid 2r + 1) \nabla u \} - \Delta u_t + \Delta_x u = g(x,\ t) \qquad (3.1)$$

$$\left(g(u), \partial_t v \right) + \Delta \omega_n + m_0 \mid \nabla \omega_n \mid^q \le \varepsilon \mid v \mid^{2q} + C_\varepsilon \left\| \partial_t v \right\|_{H^2}^2 + m_0 \mid \nabla v_n \mid \quad (3.2)$$

Analogously, for some positive $\varepsilon, \upsilon, \rho$, together with the Hölder inequality and the interpolation

$$g'(t) + Cg(t)^{1 + \frac{\upsilon}{p}} \le C(1+t)^{-\rho} (1+t)^{\frac{\upsilon d}{2n}} g(t)^\upsilon \le \varepsilon \left\| \partial_t v \right\|_{L^2}^2 C_\varepsilon \left\| \partial_t v \right\|_{H^2}^{p+1} \qquad (3.3)$$

Similar as above, we derive the differential inequality as long as the local weak solution $u(t)$ exists, inserting the above estimates, and using the energy estimate for estimating the energy norms, one can get

$$\left\| \partial_t u \right\|_{L^2}^{np - \gamma} + \frac{\gamma}{2} \left\| \nabla_x u \right\|_{L^2}^{\frac{1}{\upsilon} + \frac{d}{2}} - a \left\| \nabla u \right\|_{L^2}^{2r} \le -C_2 E(t) + C_3 \left\| g \right\|^2 - \frac{C_4}{2} \Phi + \frac{\varepsilon^2}{4} M \left\| u \right\|^2 \quad (3.4)$$

On the other hand, by Young's inequality, choosing $\varepsilon > 0$ small enough and $E(t)$ is bounded, we finally deduce the following

$$G_0'(t) + C_3 G(t) \le C_4 M_1 (1 + t)^{-\rho_1} \qquad (3.5)$$

$$g'(t) + C_4 (1+t)^{\frac{\lambda \rho}{2}} g(t)^{\frac{1+\lambda}{p}} < C_5 (1+t)^{-\frac{1}{1+\gamma}} g(t)^\beta \qquad (3.6)$$

Using now estimate (3.2)-(3.4) together with the interpolation inequality, we infer from (3.5)-(3.6) that

$$g'(t) + C_0 g(t)^{p+2-\upsilon} + \left\| \xi_u(y) \right\|_{L^2}^{p+2} \le C_6 e^{K(L-s)} \left\| \xi_u(x) \right\|_{L^2}^{p+2-n} + C_7 g(t)^{\frac{n}{p(p+2)}} \quad (3.7)$$

Now fix L, S. Then, taking the smoothing property together with the obviously bounded $E(t)$

$$E(t) + C_7 \Phi(0) + G_0(t) \le 2\Phi(t) \le 2g(t) + 2KG(0)(1+t)^{-K} + C_8 M_3 e^{-\lambda t} \qquad (3.8)$$

Multiplying the both sides of (3.7) by e^{Ct} and integrate from 0 to t, we derive

$$C_6 \Phi(s) + \int_{B(x_0,r)} \tau^2 g(u_k) p(x) |\nabla u_k|^q \, dx \le \int_{B(x_0,r)} \delta^2 f(u_k) p(x) \Delta u_k \, dx \qquad (3.9)$$

We finally obtain $\quad E(u(t)) \le 2G(u(t)) \le 2C_\varepsilon e^{-C_\varepsilon t} \le C_\sigma E\left(e^{-\lambda_\sigma t}\right) \qquad (3.10)$

Hence, the theorem is completed. $\qquad\qquad\qquad\qquad\qquad\qquad$ □

Proof of Theorem 2 in fact, for $n = 3$ $\omega \in L_1 \cap L_p$, and $\left\| v(t) - v_0(t) \right\| \le \sigma(1+t)^{\frac{\gamma}{3}\left(1 - \frac{1}{q}\right)}$, note that the initial data is dense, hence $\left\| v_0(t) \right\|$ attenuates exponentially fast, Indeed, using $\omega |\omega|^{\frac{2p}{n(p-2)}}$ and $\partial_t \omega$ multiply the equation (1.7) and integrate by parts over $x \in \Omega$. One gets

$$\delta'(t) + \lambda_1 (1+t)^{\kappa} \frac{d}{2} \delta(t)^{\frac{1}{q}} \le \varepsilon(1+t)^{-(p-r)} y(t)^{\lambda + \frac{\kappa}{p}} + \lambda_2 \delta(t)^{\frac{n}{2}\left(\frac{1}{r} - \frac{1}{2}\right)} \qquad (3.11)$$

Where ε is a small positive number which will be fixed, then we arrive at

$$\frac{1}{2}(p - n)\delta'(t) + \int_\Omega |\nabla \omega^2|^p \, dx \le \sigma_\varepsilon \int_\Omega \left(|v_0|^4 + |\nabla \omega|^{p-2}\right) dx + \frac{1}{2} \upsilon(G(v), v) \qquad (3.12)$$

And notice that

$$E(u(t)) = \frac{1}{2}\alpha \left\{ \left\| \nabla \omega_1(t) \right\|^q + \int_\Omega \int_0^{|\nabla u_1|^2} \sigma(\xi) \omega(t) \, d\xi \, dx \right\} \qquad (3.13)$$

$$E(\xi_u(t)) = \frac{p-2}{p} \left\| \partial_t \omega \right\|_{L^2}^p + -\left(G(\omega), \frac{1}{q-1} \right) - \frac{pn+4}{p+4-n} \left\| \nabla \omega \right\|_{L^2}^2 \qquad (3.14)$$

Let λ, α be small enough, one gets

$$\frac{d}{dt} E(\xi_u) + \rho \left\| \nabla_x u \right\|_{H_0^2}^p - \alpha \left\| \partial_t u \right\|_{H_0^2}^2 = \mu\left(\varphi'(\nabla_x u), \nabla_x u \right) + \beta(g(u), \omega) \qquad (3.15)$$

And $\quad \int |u|^{\frac{pn}{n-2}} dx \le \|\omega\|_{H_0^2}^{\frac{p}{2}-1} + \sigma \left\| \nabla_x u \right\|_{L^p}^2 \le \frac{C_\varepsilon}{\beta} \zeta(\omega) + \lambda_\varepsilon \|\omega\|_{L^p}^\gamma + \delta E \|\xi_u(t)\| \qquad (3.16)$

Where the constant $C_\varepsilon, \lambda_\varepsilon$ depend only on the ε.

Hence by the standard Galerkin method, interpolation inequality, Young's and Sobolev's inequality, this section can be estimated by

$$\alpha \|u\|_{L^q}^{n+1} \left(\varphi'(\xi_2) + \|u\|_{L^2}^{\frac{1}{2}-\frac{1}{p}} \right) \le \alpha \|u_0\|_{L^{2p}}^{p+2} \|v_0\|_{L^p}^{p-1} - \varphi'(\xi_1) + C_\varepsilon \|u\|_{L^p}^{\lambda p} \quad (3.17)$$

By the properties of heat kernel, we deduce that

$$\partial_t \left[\frac{pn-2n+2p\alpha}{p+4-n} (\partial_t v, v) \right] + \tau_0 \left(|\nabla_x \omega_1| + |\nabla_x \omega_2| \right)^{\frac{pn}{n-2}} \le C_1 \varphi(t)^{\frac{1}{p}} + C_2 (1+t)^{-\mu\tau} |v|^{p+2} \quad (3.18)$$

$$\rho \|\partial_t u\|_{L^{2p}}^{p+\mu} + C \|u\|_{L^2}^p \le \frac{1}{2} \kappa \varphi'(\nabla v_1) \le C(1+t)^{-p\upsilon} + C_\varepsilon (1+t)^{-\frac{n}{2}\left(\frac{1}{q}-\gamma\right)} \quad (3.19)$$

Noting the uniqueness of energy solution and the Lipschitz continuity in a weak space, there holds (2.6). ☐

Acknowledgment

This work was funded by the PHD Foundation of Henan University of Engineering (No. D2010012).

References

1. Y. Qin. Global existence of a classical solution to a nonlinear wave equation [J], Acta Math. Sci.17 (2003), 121-128.
2. Yu-ming Qin, Xin Liu, Shu-xian Deng. Decay Rate of Quasilinear Wave Equation with Viscosity. Acta Math App Sinica, 17 (2010), 147-152.
3. Nakao, M. Energy decay for the quasilinear wave equation with viscosity. Math. Z., 219: 289–299 (2005).
4. Nakao, M. On strong solutions of some quasilinear wave equtions with viscosity. Advances in Mathematical Sci. Appl., 6: 267–278 (1996).
5. A. Friedman, J. Necas, Systems of nonlinear wave equations with nonlinear viscosity, Pacific J. Math. 135 (2008) 29–55.
6. H. Beirao da Veiga, Existence and asymptotic behaviour for strong solutions of the N-S equations. Indiana Univ. Math. J., 36(1987), 149-166.

Research of Dynamics of Structure of Rapid Charging Robot with 2-DOF

Bo-wen Ni[†], Cun-yun Pan, Hai-jun Xu and Wen-hao Wang

*College of Mechanics Engineering and Automation, National Univ. of
Defense Technology, Changsha, 410073, China*
[†]*E-mail: 892922379@qq.com*

The paper proposes to build physical model of locating mechanism that is made up of pointing mechanism and multi-parallel bars mechanism. It researches on its kinematic and dynamic characteristics and establishes its kinematic model and dynamic model. The characteristics of the whole system are determined through the research of its mathematical model and simulates the model on software platform and compares with the results obtained from the mathematical model. It also verifies the validity of the relevant mathematical model to achieve academic solutions such as kinematic and dynamic regular pattern and change law of the driving force. The research conclusion can be used for the engineering application of the mechanism.

Keywords: Locating Mechanism; multi-parallel bars Mechanism; pointing Mechanism; kinematic; Dynamic; software simulation.

1. Introduction

Since the rapid development of modern technology, Satellite provides our society with all kinds of information services such as communication, navigation, resources detection, disaster early warning and so on. Aerospace technology has made great contributions to the economic development. During this progress improve also the pointing technology rapidly.

Direction-pitching locating mechanism is the most commonly used in the pointing platform. Many enterprises and graduate schools have done a lot of research on it. And this mechanism is often applied in many kinds of platforms, such as satellite, telescope and so on.

Since the performance index of the pointing platform rises continually. The direction-pitching locating mechanic, which is widely seen in the traditional platform cannot meet the requirements. We need to develop new kind of pointing platform technology. The new platform should have higher pointing accuracy class, better load ability, quicker response, and light in weight.

Usually we divide the platform into several different kinds according to the number of rotational axis involved, such as two-axis, three-axis, four-axis etc.

Two-axis platform is relatively common in our daily life, and it is the research object in this article. It usually has two forms; the direction-pitching structure and the X-Y axis structure. The direction-pitching structure has better load performance and is widely used in heavy-load environment. But it doesn't have the ability to track object continually. Compare to the direction-pitching structure is the X-Y structure usually lighter in weight and can track object continually but only suitable for light-load condition.

In this article we design a new mechanism with 2-DOF, which is used for rapid charging robot. This mechanism has the advantage of continual track ability and has relatively satisfying load ability. Through the analysis of this platform, to establish the kinematic model of the new platform, then make dynamic analysis of it through reversed kinematic model and Lagrange Equation. Combined with specific index and the terminal track planed is the change of the driving force solved.

2. Structural analysis of locating mechanism

Figure 1 shows the schematic view of the locating mechanism that we make research of. Indeed, the relationship between the two linear motor is orthogonal. The rounded rack drives the spherical gear to rotate maximal 45 degrees to any direction. The spherical gear links to the frame with the X-Y structure. Then, the multi-parallel bars structure pointing to any direction, driving the flange plate and terminal load to locate at XOY plate [1–4].

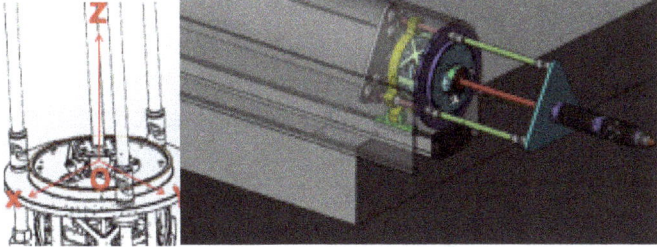

Figure 1. The schematic view of the locating mechanism.

From the figure we can figure out the working principle of this mechanism. The two motors make linear motion and the angle between them is orthogonal. Then there exists a kinematic pair between rounded rack and spherical gear. The rounded rack is fixedly connected to the upper linear motion and the spherical gear is linked to the frame outside. The transfer bar, which points to the specific direction is also connected to the spherical gear. The rational angle θ means the angle between z-axis and the pointing direction of the transfer bar. The directional angle α is the angle between the x-axis and the projection of the transfer bar on the XOY Plate.

According to the thesis of degrees of freedom for space mechanism, we can calculate the degree of freedom of the whole system.

The most complex kinematic pair in the system is the mechanism of rounded rack and the spherical gear [8 -10]. The number of degree of freedom this pair is 4. They are the two linear degrees of freedom of rounded rack (along x-axis and y-axis) and the two rotational degrees of freedom of spherical gear (rotate with x-axis and y-axis)

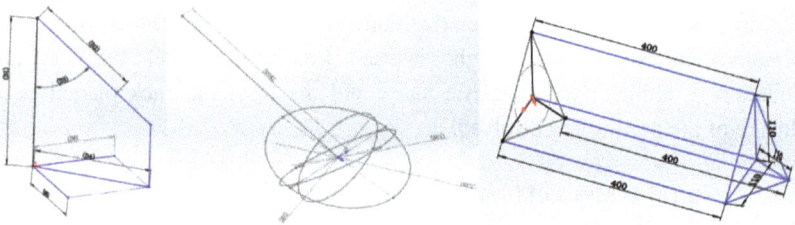

Figure 2. The schematic view of working thesis sub mechanism.

The calculation of the degrees of freedom of the locating mechanism:

$$f = 6(n-p) + \sum_{i=1}^{n} ip_i + L$$

n: the number of moving component
p: the number of kinematic pair
L: the number of independent moving loop

When we add parallel bar on the pointing mechanic one by one and calculate the number of degree of freedom in each different conditions.

We get the following results as the final results:

$$f = 6(7-10) + 8 + 2 + 4 + 4 + 2 = 2$$

The fourth parallel bar is redundant constraint. It does not affect the degrees of freedom. It can be improve the stress condition of the mechanism.

3. The kinematic research

The kinematic model of the mechanism has two main parts: the forward kinematic model and the reversed kinematic model. The forward kinematic model equals that we figure out the terminal movement locus with the already known driving forces.

3.1. The forward kinematic mode

$$
\begin{cases}
\sqrt{x^2 + y^2} = s = r\theta \\
\tan(\alpha) = \dfrac{y}{x} \\
P = \begin{bmatrix} l\sin\theta\cos\alpha \\ l\sin\theta\sin\alpha \\ l\cos\theta \end{bmatrix}
\end{cases}
$$

x: the moving distance of the linear motor along the x-axis
y: the moving distance of the linear motor along the y-axis
s: the moving distance of the rounded rack
r: the radius of the spherical gear
θ: the rotational angle
α: the directional angle
P: the space position of the terminal point
l: the length of the transfer bar
The space location of the terminal point:

$$
\begin{cases}
x_s = l\sin(\dfrac{\sqrt{x^2+y^2}}{r})\cos(\arctan(\dfrac{y}{x})) = f_1(l,x,y) \\[2mm]
y_s = l\sin(\dfrac{\sqrt{x^2+y^2}}{r})\sin(\arctan(\dfrac{y}{x})) = f_2(l,x,y) \\[2mm]
z_s = l\cos(\dfrac{\sqrt{x^2+y^2}}{r}) + H = f_3(l,x,y)
\end{cases}
$$

The space velocity of the terminal point:

$$
\begin{cases}
\dot{x}_s = f_1(l,x,y,\dot{x},\dot{y}) \\
\dot{y}_s = f_2(l,x,y,\dot{x},\dot{y}) \\
\dot{z}_s = f_3(l,x,y,\dot{x},\dot{y})
\end{cases}
$$

The space acceleration of the terminal point has the similar conclusion.

3.2. The reversed kinematic model

Use the known terminated movement locus to solve the unknown change rule of the driving force. When we get the moving condition of the terminated point, we can conclude the detailed condition of the driving force. When the space location of the terminated point determines, the direction angle and the rotational angle of the transfer bar are also determined. Then we can use them to deduce the moving condition of the driving motor.

The phase can be seen in the following thesis:

$$\begin{cases} x = r\theta \cos\alpha \\ y = r\theta \sin\alpha \end{cases}$$

Use the differential we can obtain the velocity and the acceleration of the driving motor:

$$\begin{cases} \dot{x} = f_1(r,\alpha,\dot{\alpha},\theta,\dot{\theta}) \\ \dot{y} = f_2(r,\alpha,\dot{\alpha},\theta,\dot{\theta}) \\ \ddot{x} = f_3(r,\alpha,\dot{\alpha},\ddot{\alpha},\theta,\dot{\theta},\ddot{\theta}) \\ \ddot{y} = f_4(r,\alpha,\dot{\alpha},\ddot{\alpha},\theta,\dot{\theta},\ddot{\theta}) \end{cases}$$

4. The dynamic model

Here we assume that the mechanism is in ideal condition. All of the components are rigid and the kinematic pairs are in good lubrication. And we ignore the damp here. The whole system has 2 degrees of freedom. We use the Lagrange equation to solve the dynamic analysis of the system [5 – 7].

The progress show in the following context:

$$\frac{d}{dt}(\frac{\partial E_k}{\partial \dot{q}_j}) - \frac{\partial E_k}{\partial q_j} + \frac{\partial E_p}{\partial q_j} = F_j$$

$$E_k = E_a + E_b + E_c + E_d + E_f + E_g$$

$$= \frac{1}{2}m_a \dot{x}^2 + \frac{1}{2}m_b(\dot{x}^2 + \dot{y}^2) + \frac{1}{2}J_c w_c^2 + \frac{1}{2}(I_{dx}w_{dx}^2 + I_{dy}w_{dy}^2 + I_{dz}w_{dz}^2)$$

$$+ 3 \times \frac{1}{2}I_f w_{sum}^2 + \frac{1}{2}m_g(w_{sum}l)^2$$

E_k: kinetic energy of the system
E_p: potential energy of the system
q_j: generalized coordinates
F_j: generalized Force
j=1, 2: show the 2 degrees of freedom
w_{sum}: the rotational velocity of component f

66

$$E_p = m_a g h_a + m_b g h_b + m_c g h_c \cos(\sqrt{x^2 + y^2}\,/r) +$$

$$m_d g h_d \cos(\sqrt{x^2 + y^2}\,/r) + \frac{3}{2} m_f g l \cos(\sqrt{x^2 + y^2}\,/r) + m_g g l \cos(\sqrt{x^2 + y^2}\,/r)$$

J_c: the rotational inertia of component c

I_{di}: the rotational inertia around axis i of component d

m_j: the mass of component j(j=a, b, c...g)

I_f: the rotational inertia of component f

$$\begin{cases} x = \dfrac{\dfrac{1}{2} F_1 t^2 + u_1 t + u_2}{m_a + m_b + J_c / r^2 + 2 I_{dy} / r^2 + 3 I_f / r^2 + m_g l^2 / r^2} \\[4mm] y = \dfrac{\dfrac{1}{2} F_2 t^2 + v_1 t + v_2}{m_b + 2 I_{dx} / r^2 + 3 I_f / r^2 + m_g l^2 / r^2} \end{cases}$$

u1, u2, u3, u4: structural index related to the whole system

5. Analysis on the result of the simulation

Here we use the reversed Dynamic Analysis to run the simulation experiment of the whole system that we make research of.

5.1. The movement locus of the terminal point

Below is the detailed structural index of the system

Table 1. Detailed structural index of the mechanism.

m_a/kg	m_b/kg	r/mm	J_c/kg.mm^{-2}	I_{dx}/kg.mm^{-2}	I_{dy}/kg.mm^{-2}
1.2314	0.9477	25.5	191.5112	78.1982	82.9216
I_f/kg.mm^{-2}	m_f/kg	m_g/kg	l/mm	h_d/mm	
1900	0.105	0.6225	400	33.6	

The movement locus of the terminal point shows below:

$$\begin{cases} \alpha = 2\pi \cdot \sin[(\pi/4) \cdot t] \\ \theta = 0.85 \cdot \sin[(\pi/2) \cdot t] \end{cases}$$

According to the calculated mathematical results from the Dynamic Model, the change rule of the driving force shows in the below Figure 3. F1 and F2 show respectively the driving force of the lower driving motor and the upper driving motor.

5.2. Dynamic simulation experiment

According to the calculated mathematical results from the Dynamic Model, if only the space location, velocity and acceleration of the terminal point determine, we can also figure out the driving force of the two linear driving forces. In the ADAMS software platform, we use the system model to run the simulation experiment. In Figures 4 and 5, we can see the comparison of the theoretical force and the simulation calculated force of the upper and lower driving motor.

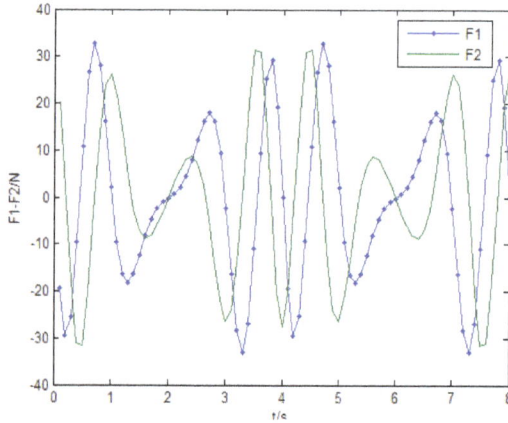

Figure 3. The theoretical F1 and F2.

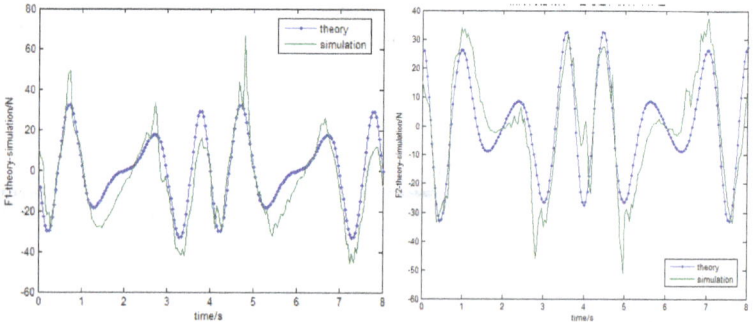

Figure 4. the comparison of the theoretical force and the simulation calculated F1 and F2.

5.3. *The analysis on the results*

The figures show respectively the comparison of the theoretical force and the simulation calculated F1 (lower driving motor) and F2 (upper driving motor). From the above figures, we can find that the curve of theoretical force can fit the simulation calculated result satisfyingly. Prove the validity of the dynamic Model that we build. But there also exists obvious deviation at the peak of the theoretical force curve. On the whole, the condition of F2 is worse than F1.

In this case, we consider kinematic and potential energy of all components in the mechanic. But the interval, friction and vibration are not involved in this model. We can plus these factors into the system model in the research in the future.

6. Conclusion

1) Introduce a new kind of locating mechanic and set up the relevant kinematic model and dynamic model.

2) Based on the reversed kinematic Model, build dynamic mode of the locating system through the Lagrange equation. And calculate the change rule of theoretical force.

3) Consider the structural index, solve the reversed Dynamic Model. Using the known terminal position's space Location, velocity and acceleration to solve the driving force of linear motor.

4) The accuracy of upper linear motor has a bigger influence on the whole system, so it need better control.

References

1. Yang S C, Chen C K, Li K Y. A Geometric model of a spherical gear with a double degree of freedom [J]. Journal of Materials Processing Technology, 2002, 123(2):219-224.
2. Li Ting, Pan Cunyun. On grinding manufacture technique and tooth contact and stress and analysis of ring-involute spherical gears [J]. Mechanism and Machine Theory, 2009, 3: 1-19.
3. Chao L C, Tsay C B. Contact characteristics of spherical gears [J]. Mechanism and Machine Theory, 2008, 43: 1317-1331.
4. Yang S C. A rack-cutter surface used to generate a spherical gear with discrete ring-involve teeth [J]. The International Journal of Advanced Manufacturing Technology, 2005, (27): 14-20.
5. Zhang Li-Jie. Kinematics and dynamics of a novel bionic sheelled-legged-fused mechanism [D]. Changsha: National University of Defense Technology, 2008. (in Chinese)

6. Yang Yi-Yong, Jin De-Wen. Dynamics of mechanic systems [M]. Beijing: Tsinghua University Press, 2009. (in Chinese)

7. Chen Li-Ping, Zhang Yun-Qing, Ren Wei-Qun, etc. The tutorial of mechanic systems using ADAMS [M]. Beijing: Tsinghua University Press, 2005. (in Chinese)

8. Li Ting, Pan Cunyun, Li Qiang, etc. Analysis of assembly error affecting on directing precision of spherical gear attitude adjustment mechanism [J]. Acta Armamentarii, 2009, 30 (7): 962-966. (in Chinese)

9. Pan Cunyun, Wen Xisen. Profile formula derivation of the in-volute spherical gears [J]. Journal of National University of Defense Technology, 2004, 26(4): 93-98. (in Chinese)

10. Pan Cunyun, Wen Xisen. Research on transmission principle and kinematic analysis for in-volute spherical gear [J]. Journal of Mechanical Engineering, 2005, 41(5): 1-9. (in Chinese)

The Risk of Cascading Failure Base on Queuing Theory Model M[k]/M/1

Yu-qi Yang[†] and Lei Zhu
Institute of Communications Engineering, PLA
University of Science and Technology
Nanjing, Jiangsu Province, China
[†]E-mail: wpyangyq@163.com

Modern system has become increasingly complex to design and build. Once a system has been built, it is also difficult to detect if it is a stabilizing when some error happened. Predict the cascading failure and decrease the probability of it is a central issue in the research of complex network. Predicting the cascading failure before it happens is still a popular issue. The paper has made a model to describe the cascading failure base on a classical queuing theory model M[k]/M/1 to compare with the model Coupled Map Lattices (CML). The proposed model not only can describe the process and the scale of the failure as CML, but it can also give a prediction for the failure. It serves as a useful guide before building a complex system. It can also give a standard for the stability of the proposed network.

Keywords: Cascading failure; Complex network; Queue theory M[k]/M/1.

1. Introduction

Since the scale-free network being introduced in last century, the network science has attracted a lot of attention. The research of cascading failure is a important area of complex network [1]. And it has been proved very useful in various fields especially in power grid and communication [1–5]. Most of those studies base on the assumption that the networks are isolate. And it is difficult to build a standard to assess the scale of the failure. There are some kinds of model try to describe the failure. The first one is the capacity-load (C-L) model [6]. This model has been built by Motter give a intuitive description for the failure base on the principle. The second one is the mixture dynamic model base on the node and line [7]. It has been great used in the power grid. They use the average of the efficient network to explain the scale of the damage of the network [8]. The third one is CML. The CML model is a great success to describe the cascading failure in small-world and scale-free network [9–11]. In order to research the traffic network, the studiers build the new CML model with the time delay.

There are only two states of each node, normal and down, in the model of C-L and CML. They can give a great characterization for the cascading failure. But they cannot give alarm before the failure happen especially for the important nodes. The mixture dynamic model is a nice tool to measure the scale of the failure. However, it cannot describe the change of each node.

This paper makes a model base on the queue theory model $M^{[K]}/M/1$. $M^{[K]}/M/1$ is a classical model to describe a queue process with one serve and the number of the customer arrive is a random value. The dynamic process the network has have been expressed by the speed of the customer arrive. In the end, the model uses the probability of each node down at the time series as the standard. It will give a great help for the alarm the cascading failure.

2. Model for node

As we know the cascading failure of a complex network is connected with the load and capacity of each node. When the value of load is over the capacity we define the node is down. When one node is down, the load of the node will be shared by the network. It will increase the load of another node and increase the risk that another node is down. It is the description of the cascading failure. The model we make for the node is try to describe the statement change with time goes by. It also can explain that the change of the load when the cascading failure is broken up.

2.1. *Physic topology and relationship topology*

In our world, we define network as two sets. One for nodes is defined as *Node*, the other for line is defined as *Line*. If two members of *Node*, which is defined as n_1 and n_2, has some relationship, then we define $<n_1,n_2>$ is the member of *Line*. In communication network, the relationship can be defined as the two node have can contact directly with some physic line or channel. At that time, we research the cascading failure of computer network, the physic topology is the real topology network whose node is the computer or road and the line is physic line or channel.

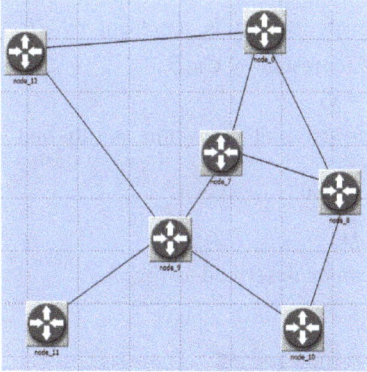

Fig. 1. Physic topology for a network   Fig. 2. Relationship topology for network Fig. 1

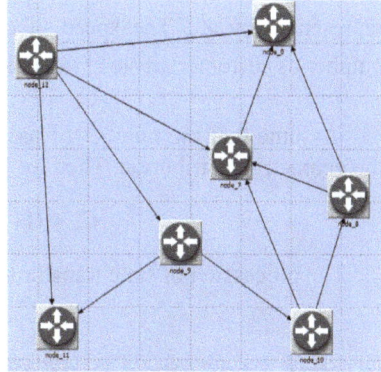

In order to research the cascading failure of a computer network we define the relationship topology as follow. If two nodes n_1 and n_2 from set *Node* have relationship like n_1 down will make n_2 down, then we define $<n_1,n_2>$ is a member from *Line*. We should understand that the relationship topology is directed graph, the physic topology is indirect graph.

2.2. Model base on Queue theory

The process of information propagating at a network can be divide into a number of sub-process. First the information will be divided and transform into a number of data transmission. And each node will build cache to defer the data transmissions and build a queue to manage them. As we can see from the Fig. 1, at time of t, one node can get a lot data transmission from its neighbor which need to propagate data. At time of t, the number of data transmission is a random variable. Define the number of data transmission is the load of the node, and the size of cache is the capacity. We can describe the node statement by $M^{[k]}/M/1$ model.

$M^{[K]}/M/1$ is a classical model in queue theory. It describes the process that has one serve and the number of arriving customer is a random variable. The customer arrive is satisfied the Poisson Process. For the research, the probability of the load is over the capacity should be concentrated on, represented as $P\{Lo(t)>Ca\}$, which the $Lo(t)$ is represent the load for a node at time of t and Ca is represent the capacity. We use the probability to represent the state of the node.

$$S_i(t) = P\{Lo(t) > Ca\} \qquad (1)$$

$S_i(t)$ is represent the state of node i at time of t. There are some important parameters in classical $M^{[K]}/M/1$ model. The speed of customers arriving is

73

represented by λ. The speed of customers leaving is represent by μ. The number of customer arrive at same time is represent by $Cu(t)$.

$$C_k = P(Cu(t) = k) \tag{2}$$

Assume that the number of passenger arrive at same time is satisfied with the geometric distribution. Then we get that:

$$C_k = (1 - \frac{1}{De})\frac{1}{De}^{k-1} \tag{3}$$

The procession of state transforming can be described as Fig. 3.

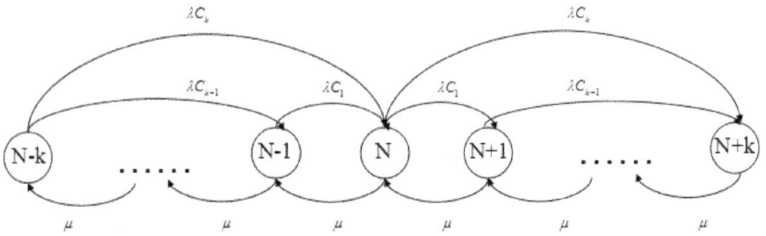

Fig. 3. State transforming process

As we know, the information propagation is a dynamic process. It means that the parameter will be different with time. But the relationship of them do not change. Now the alphabet $\lambda(t)$ represent the speed of customer arrive at time of t. Notice that the speed of customer leaving is independent with time t. According the relationship described by the Fig. 3 and the principle that the speed of flowing into is equal the speed of flowing out we have the recurrence formula as follows.

$$\lambda(t)p(t)_n + \mu p(t)_n = \mu p(t)_{n+1} + \sum_{j=1}^{k} p(t)_{n-j} C_j \lambda(t)$$
$$\lambda p(t)_0 = \mu p(t)_1 \tag{4}$$

$p(t)_n$ represent that the probability that the size of the queue is n at time of t. As we describe before, the size of the queue is the load of the node. The probability $P\{Lo(t) > Ca\}$ we used to describe the state of the node can be calculated.

$$S_i(t) = P\{Lo(t) > Ca\} = \sum_{j=Ca}^{\infty} p(t)_j \tag{5}$$

In the real network, the state will be a limited value. So the superior limit of function(5) cannot be infinity. We choice N represent the superior limit for function(5)

$$S_i(t) = P\{Lo(t) > Ca\} = \sum_{j=Ca}^{N} p(t)_j \tag{6}$$

74

2.3. The speed of arriving for queue model

The parameter $\lambda(t)$ is a variable with time. In order to describe the state transforming clearly, the process can be explained by the discrete dynamic system. The parameter $\lambda(t)$ represents the speed of customer arriving. The $Lo(t)$ represents the load of a node at time of t. They should be satisfied the follow relationship.

$$\lambda(t) = \frac{Lo(t) - Lo(t - \varepsilon)}{t - (t - \varepsilon)}, \quad 0 \le \lambda(t) \le \max(Ca_i) = Ca \ (i = 1 \ldots N) \tag{7}$$

In order to make the function groups (4) have stable solution, it must have relationship between the λ and μ as follows:

$$\frac{\lambda(t)}{\mu} < 1 \tag{8}$$

Assume that $\mu = 1$, we get the $\lambda(t)$ by unitary it as follows:

$$\lambda(t) = \frac{Lo(t) - Lo(t - \varepsilon)}{\varepsilon Ca} \tag{9}$$

The load state at time of t have connection with the neighbor state. And it also be influenced by itself. After assuming $\varepsilon = 1$ the function about $\lambda(t)$ can be built as follows.

$$\lambda(t) = \frac{Lo(t) - Lo(t - 1)}{Ca} = f(t) + \sum_{j=1}^{|N|} a_{ij} (\frac{Lo(t-1)_j}{Ca}) P\{Lo(t-1)_j > Ca_j\} \tag{10}$$

$$Lo(0)_i = r_i$$

$a_{i,j}$ is the element from the adjacency matrix A which describe the relationship of physic topology. r_i represent the value which we begin to record the load of the i-th node.

3. Model for network

There are a lot of model, such as CML, have provide a great exhibition of the whole network run. And as Fig. 4 shows us that in different complex network the cascading failure will have a different stationary point. If our model can predict the probability, it will help a lot.

In the model CML. They give the state function for each node as follows [10]:

$$X(t+1)_i = |(1-\varepsilon)g(X(t)_i) + \varepsilon \sum_{j=1, j \neq i}^{N} a_{i,j}(X_j(t))/De_i| + R \tag{11}$$

The $X(t)_i$ is the state value of the i-th node at time of t. It belongs to the domain [0,1] if the node is running well. The g(x) is the logistic map $g(x)=4x(1-x)$. R is a turbulence.

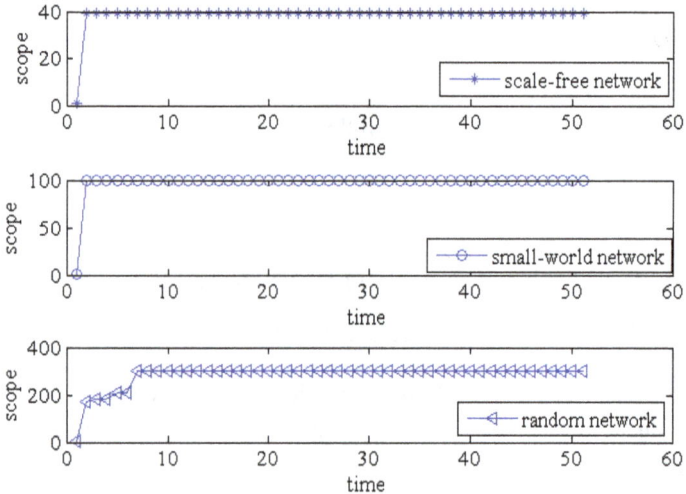

Fig. 4. The scope of cascading failure described by CML with $\varepsilon = 0.5, R = 20$

Now we assume that the network is made up by the node that satisfied with the model we made in section 2. If the capacity of each node is big enough, the node in the network can be thought as a classic $M^{[k]}/M/1$ model. The parameter $\lambda(t)$ only depend on the f(t) which in the function (10). Then we can get the moment generating function(MGF) as follows:

$$P(Z) = (1-\rho_1)\{1 + \sum_{n=1}^{\infty}[\frac{1}{De} + \rho_1(1-\frac{1}{De})]^{n-1}\rho_1(1-\frac{1}{De})z^n\}$$

$$\rho_1 = \frac{f(t)}{1-\frac{1}{De}} \tag{12}$$

After transforming the MGF P(Z) into Taylor series, we can get the value of p_n for one point in the network.

$$p_n = \frac{P^{(n)}(0)}{n!} = \rho_1(1-\frac{1}{De})(1-\rho_1)[\frac{1}{De} + \rho_1(1-\frac{1}{De})]^{n-1} \tag{13}$$

The probability described by function (9) is give as a probability that the load of each node over the capacity without a turmoil about the load. If there are some turmoil, the $\lambda(t)$ will be different.

$$\lambda(t) = f(t) + \sum_{j=1}^{|N|} a_{ij}(Lo_j(t-1) - Ca_j) P\{Lo_j(t-1) > Ca_j\} \tag{14}$$

The new MGF can be built as follows:

$$P(Z) = (1-\rho_2)\{1 + \sum_{n=1}^{\infty}[\frac{1}{De} + \rho_2(1-\frac{1}{De})]^{n-1}\rho_2(1-\frac{1}{De})z^n\}$$
$$\rho_1 = \frac{\lambda(t)}{1-\frac{1}{De}} \tag{15}$$

The probability of p_n can be shown as follows:

$$p_n = \frac{P^{(n)}(0)}{n!} = \rho_2(1-\frac{1}{De})(1-\rho_2)[\frac{1}{De} + \rho_2(1-\frac{1}{De})]^{n-1} \tag{16}$$

The function (13) and (16) have shown the possibility of p_n have great relation with the degree De and adjacency matrix. The load of each time for each node have the relationship as follows:

$$Lo(t) = \frac{\partial P(Z,t)}{\partial Z}, Z = 1$$
$$Lo(t) = (1-\rho_2)\{1 + \sum_{n=1}^{\infty}[\frac{1}{De} + \rho_2(1-\frac{1}{De})]^{n-1}\rho_2(1-\frac{1}{De})^n\} \tag{17}$$
$$\rho_2 = \frac{\lambda(t)}{1-\frac{1}{De}}$$

Define the function to describe the scope of cascading failure as follows:

$$Sco(t) = \sum_{i \in Node} S_i(t) \tag{18}$$

4. Simulation

Now we build different kinds of network and compare with the CML mode to check the probability mode the paper provide is satisfied with the actual problem. The scale-free network, random network and small world network is be made according to book [12].

From function (5)(16) It is easy found that the probability of node have relationship with the capacity and the speed of customer arrive λ. The sensitivity analysis for the two variables is showed as follow tables.

Table 1. The probability of Cascading failure at different capacity.

Probability of different capacity	Scare-free network	Small world network	Random network
Capacity=100	0.0965	0.0910	0.0981
Capacity=150	0.0965	0.0909	0.0980
Capacity=200	0.0965	0.0909	0.0980
Capacity=250	0.0965	0.0909	0.0980
Capacity=300	0.0965	0.0909	0.0980

It is clear that the more capacity each node has the less probability the cascading failure happen.

Table 2. The probability of Cascading failure at different capacity.

Probability of different capacity	Scare-free network	Small world network	Random network
$\lambda = 0.1$	0.9997	0.9997	0.9997
$\lambda = 0.6$	0.9995	0.9995	0.9996
$\lambda = 1.1$	0.9995	0.9995	0.9996
$\lambda = 1.6$	0.9995	0.9995	0.9996
$\lambda = 2.1$	0.9995	0.9995	0.9996

It shows that the relationship between the parameter λ and the probability.

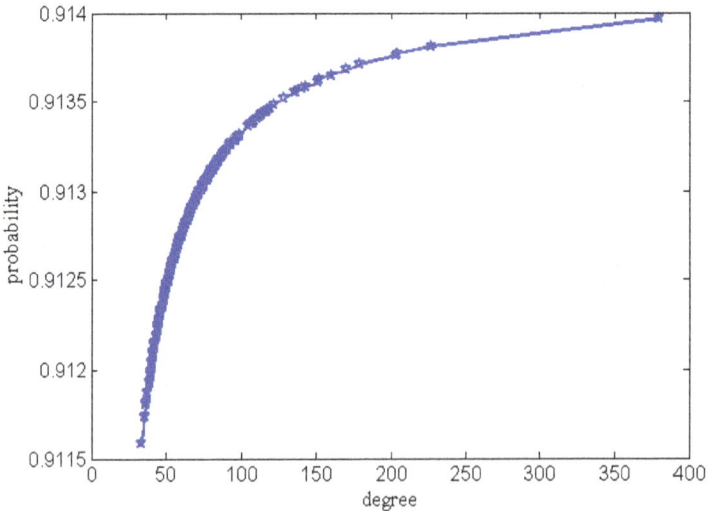

Fig. 5. Distribution of probability in a network

Fig. 5 shows the function (17) describes the relationship between the probability and degree with the capacity equal 250 and the speed λ equal 0.9.

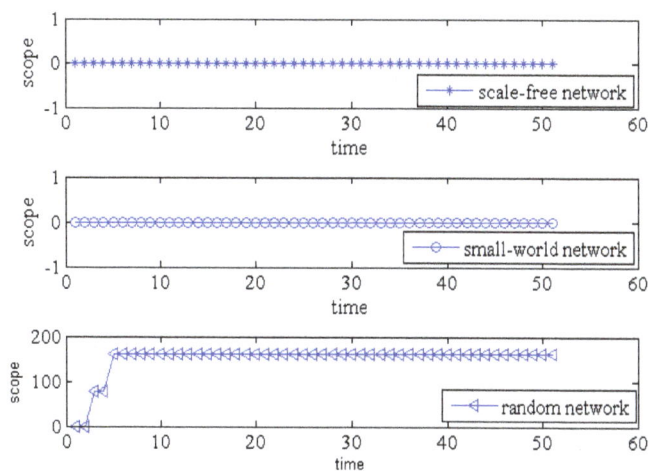

Fig. 6. The scope of cascading failure described by CML with $\varepsilon = 0.5, R = 0$

Fig. 6 shows the CML model with the $\varepsilon = 0.5, R = 0$ at different network.

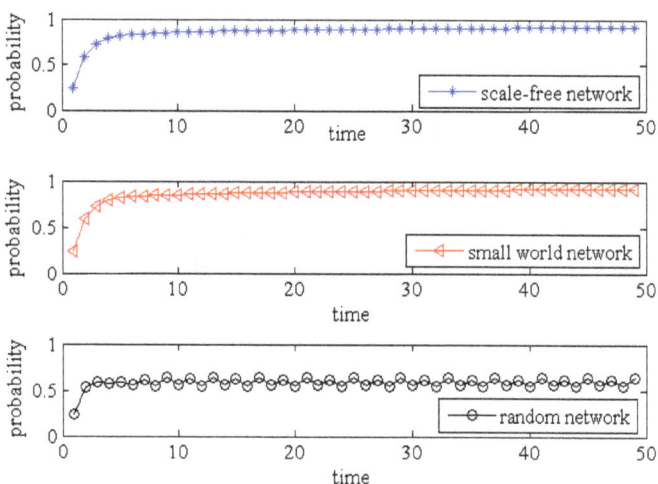

Fig. 7. The probability of one node down in different network with passing time

Fig. 7 shows the probability of the node cascading failure in the scale-free network.

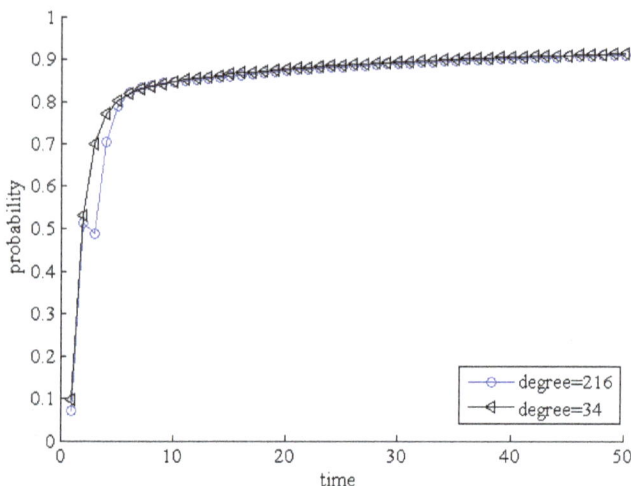

Fig. 8. The probability of different node down at same network

Fig. 8 shows the probability of the node cascading failure in the scale-free network.

5. Conclusion

In summary, we have made a model to help give a standard for cascading. The model has its own advantage. Firstly, it gives a description for the procession of the cascading failure. It is found that the mode can do this job will by comparing with the model CML. Secondly, the mode find that the probability of cascading is close to the capacity, customer arrive speed and the degree of the node. If the dynamic equation can be found before, we can use the model to research each node independently. And it can calculate the probability of the node break down with time goes by. It is important to study the cascading failure and predict the failure before.

References

1. Z. Zhao, P. Zhang, and H. Yang, "Cascading failure in interconnected network with dynamical redistribution of load," *Physica A*, vol. 433, pp. 204–210, 2015.
2. V. Cupac, J. T. Lizier, and M. Prokopenko, "Compare dynamic of cascading failures between network-centric and power flow model," *Electrical Power and Energy Systems*, vol. 49, pp. 369–379, 2013.

3. J. C. Zunshui Cheng, "Cascade of failures in interdependent network coupled by different type network," *Physic A*, vol. 430, pp. 193–200, 2015.
4. Y. Koc, M. Warnier, R. E. Kooij, and F. M. T. Brazier, "An entropy-based metric to quantify the robustness of power grids against cascading failures," *Safety Science*, vol. 59, pp. 126– 134, 2013.
5. Z. Chen, W. Du, X. Cao, and X. Zhou, "Cascading failure of interdependent network with different coupling preference under targeted attack," *Chaos Solitons & Fractals*, vol. 80, pp. 7–12, 2015.
6. Motter. A. E, Nishikawa. T, and L. YC, "Cascade-based attacks on complex network," *Phys. Rev. E*, vol. 66, 2002.
7. Crucitti. P, Latora. V, and M. M, "Model for cascading failures in complex networks," *Phys. Rev. E*, vol. 69, 2004.
8. X. Peng, H. Yao, J. Du, Z. Wang, and C. Ding, "Invulnerability of scale-free network against critical node failures based on a renewed cascading failure model," *Physica A*, vol. 421, 2015.
9. X. Fang, Q. Yang, and W. Yan, "Modeling and analysis of cascading failure in directed complex network," *Safety Science*, vol. 65, 2014.
10. Gade. P. M and Hu. C. K, "Synchronous chaos in coupled map lattices with small-world interactions," *Phys. Rev. E*, vol. 62, 2000.
11. Wang. X. F and Xu. J, "Cascading failures in coupled map lattices," *Phys. Rev. E*, vol. 70, 2004.
12. R. C. S. H, *Complex networks structure, robustness and function.* Cambridge University Press, 2010.

Design and Realization of the Automatic Test System for the Directivity of Loudspeaker

Zhi-kai Zhang
Zhejiang Province Institute of Metrology
Hangzhou, China 310023
zzkhello520@163.com

Jian-min Qiu
Zhejiang Province Institute of Metrology
Hangzhou, China 310023

In order to meet the performance test demand for the directivity of loudspeaker, this paper compared and analyzed the directivity of the different loudspeakers. The automatic test system for the directivity of loudspeaker had been designed. The structure and main functions were introduced in detail. The test method for the loudspeaker directivity in the free acoustic field was also researched. The directivities of the different loudspeakers were tested and the directivity patterns were drawn. The experimental result shows that this system is reliable and easy to operate. To the greatest extent, it reduced the error in the test process because of the reflection, better practical application value to the directivity of loudspeakers and performance test to another acoustic measuring instrument.

Keywords: Loudspeaker; Directivity; Automatic test system; Free acoustic field; Performance test.

1.   Introduction

As an important part of the acoustic field, the loudspeaker is widely used in the acoustic measurement, and the number of its main performance index is about ten [1-3]. Frequency and directivity characteristic are two important properties of the loudspeaker. Directivity is not only used to describe loudspeaker's ability to radiate the sound waves to space, it's also used to characterize the distribution characteristic of sound pressure in all directions which radiated by loudspeaker [4]. In the national standard of loudspeaker GB/T 12060.5-2011*Sound system equipment-Part 5: Methods of measurement for main characteristics of loudspeakers* [5], manufacturers should be suggested to give out the directivity of loudspeakers, which brings a lot of difficulties to the selection of loudspeakers for professional acoustic environment. For example, during the calibration and acceptance of different sizes of anechoic room, semi-anechoic room, the directivity of acoustic source for testing should be verified first. Currently, there is no single loudspeaker can meet the requirements of full-band testing, one of the reasons is that the directivity of loudspeakers is not ideal, another is the lack of stable directivity test system for loudspeakers.

Under this background, the automatic test system for the directivity of loudspeaker had been designed in this paper; the test method of the loudspeaker directivity in the free acoustic field was researched. The experimental result shows that this system is reliable and easy to operation, better practical application value to the directivity of loudspeakers.

2. System structure

The automatic test system for directivity of loudspeaker has two parts, the software system and the hardware system, the block diagram of system is shown in Fig. 1.

The hardware is mainly made up of turntable, loudspeaker system and signal processing system. Firstly, install the standard baffle with fixed loudspeakers on the turntable. Secondly, realize the real-time control of the turntable by VC++ program, and the high precision stepping motor drives the turntable to rotate in the anechoic room. Then, control the Pulse analysis system to generate signal, the signal will be transmitted to the loudspeaker through the power amplifier, and the loudspeaker will send out the acoustic signal. Finally, the Pulse analysis system transfers the data measured by microphone to PC software, and realizes the test of loudspeaker's directivity.

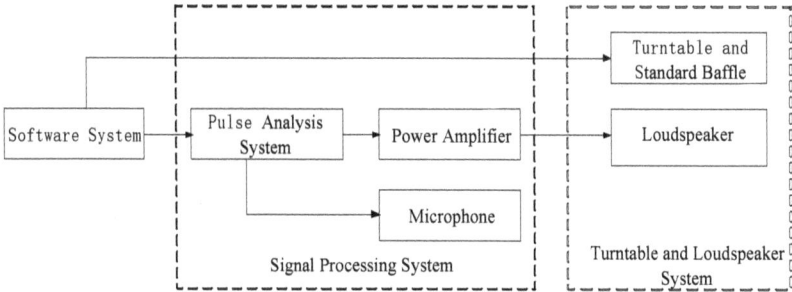

Fig. 1   Block diagram of the automatic test system.

The software can realize the function of real-time control, real-time signal collection, signal processing, and data processing with the support of hardware, what's more, the system can monitor the test of loudspeakers and the running state of system online, draw the directivity test pattern automatically.

83

2.1. Turntable and loudspeaker system

According to the requirements of the test environment, turntable and loudspeaker system is placed in anechoic room to reduce the interference caused by environment noise. Turntable and loudspeaker system consists of MCU, stepping motor, turntable, standard baffle, and loudspeaker. The block diagram of the system is shown in Fig. 2.

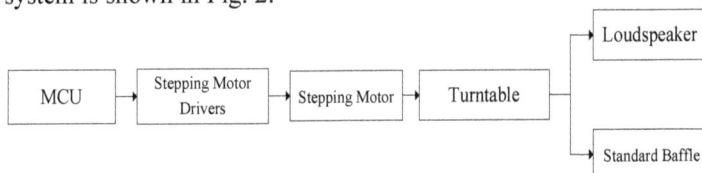

Fig. 2 Block diagram of the turn table and loudspeaker.

When measuring the directivity of loudspeaker, the directivity characteristics will change if the baffle is moved, so the test must be measured under the same conditions to judge different types of loudspeaker. Therefore, design and produce a set of standard baffle, as shown in Fig. 3. In this system, the stepper motor driver receives the pulse signal and direction signal sent by the microcontroller [7,8], enable the turntable rotate clockwise and counterclockwise by changing the frequency of the sent pulses to change the steering and speed of the stepper motor. The step accuracy of motor is up to 1°, which meet the testing requirements of loudspeaker.

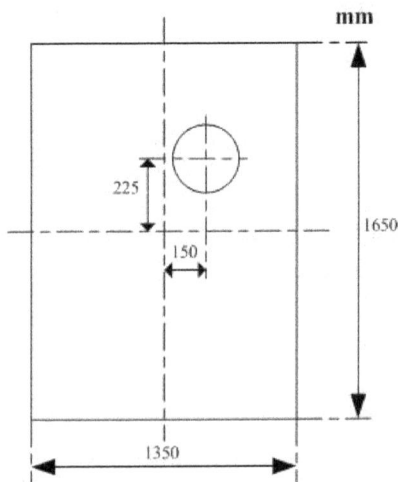

(a) Standard baffle   (b) Standard baffle with inclined plane

Fig. 3 Size figure of the standard baffle.

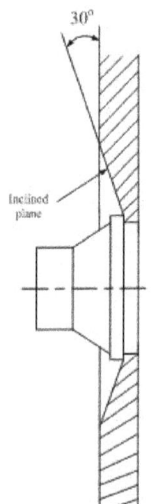

84

2.2. Signal processing system

Signal processing system consists of Pulse analysis system[9], power amplifier and microphone unit, etc. As the Fig. 1 shows that, the output terminal produce standard acoustic signal, then the signal is transmitted to the loudspeaker through the power amplifier, and the acoustic signal produced by the loudspeaker will be input to the Pulse system. Finally, the Pulse analysis system will transfer the processed signals to the software system, and the real-time directivity pattern can be drawn with these signals, which makes the directivity test more efficient, convenient and intuitive.

2.3. Software System

As one of the most important part, the quality of the software system affects the function of the whole automatic test system for the directivity of loudspeaker. The software is used to control the stepping motor's rotational period, rotational speed, step precision, data collection and drawing of loudspeaker's directivity pattern. It is mainly made up of Pulse analysis system and stepping motor control system, the flowchart of software control system is shown in Fig. 4.

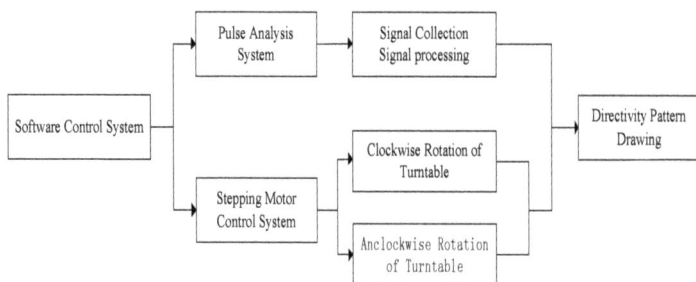

Fig. 4 The flowchart of software control system.

To control the input and output of the standard acoustic source produced by Pulse analysis system, the interface module of Pulse system has been embedded in the software system, and system can read the signal collected by microphone, which ensures the rapidity and accuracy of the test.

Stepping motor control software can be used to control the rotational speed and stepping precision of the stepping motor, control the rotation direction of turntable, which ensures the stability of rotation of turntable.

3. Test results and analysis system

According to the above analysis, the loudspeaker directivity of different sizes and frequency-band have been tested and corresponding directivity pattern has been plotted by this automatic test system, including loudspeakers of 4-inch, 6-inch, 8-inch and other specifications.

Notice:

- The center of microphone and the center of loudspeaker installed on the standard coaxial baffle plate should be coaxial;
- The rotational shaft center of standard baffle and the center of loudspeaker should be located in the same vertical plane;
- The distance of the center of microphone and the center of loudspeaker center should be 1.0 m;
- The muffler pat should be added to microphone support bar in order to eliminate the impact brought about by reflection as far as possible.

During the test, the Pulse analysis system generate fixed frequency and the stepper motor rotate at 5°stepping, directivity pattern is automatically drawn on 360°polar coordinates. The test takes single test frequency at points: 500Hz, 1000 Hz, 2000Hz, 4000 Hz and 8000 Hz. The 4-inch loudspeaker directivity result is shown in Fig. 5, and so as the 6-inch and 8-inch loudspeaker.

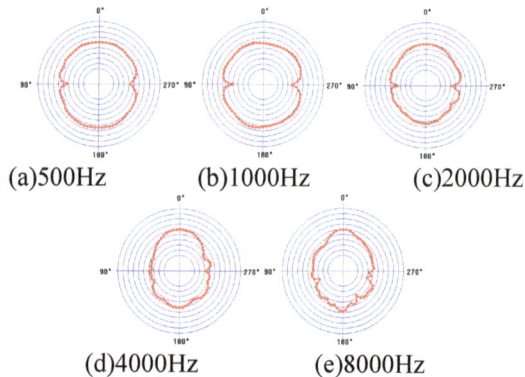

(a)500Hz   (b)1000Hz   (c)2000Hz

(d)4000Hz   (e)8000Hz

Fig. 5 The 5 frequency points directivity test pattern.

Summary has been made through the analysis of three different types of loudspeaker directivity test:

- The symmetry of loudspeaker directivity diagram is obvious;
- The directivity of loudspeaker is more obvious with increasing frequency;
- The treble loudspeaker is more sensitive to directivity compared to bass loudspeaker;
- The loudspeaker directivity test of different types verifies the stability of the system from graphical and data.

4. Conclusion

The automatic test system for directivity of loudspeaker designed by this paper can test the directivity for different types of loudspeaker in free acoustic field, and save data and draw directivity pattern automatically. This system is reasonable, stable and easy to operate, and furthest reducing the possible influence of errors due to the reflection during the test, and has good practical value for loudspeaker performance testing. What's more, this system can be applied to the microphone directivity, the sound level meter directivity, and the performance test systems of other sound source, and can offer certain practical value in the performance testing for acoustic instrument.

Acknowledgment

This project was funded by Zhejiang Bureau of Quality and Technical Supervision, project 20150217. Authors are also greatful to Professor Hu Fei of Shanghai Jiaotong University for her advice on software development.

References

1. Y. Z. Wang, G.W. Wu and S.G. Zhang, *Loudspeaker System* (National Defense Industry Press, Beijing, 2009).
2. Y. Z. Wang, *Practical Manual Loudspeaker Technology* (National Defense Industry Press, Beijing, 2003).
3. J. Y. Yu, *Loudspeaker Design and Production* (Guangdong Science and Technology Press, Guangzhou, 2006).
4. K.A. Chen, X.Y. Zeng and H. Y. Li, *Acoustic Measure* (Guangdong Science and Technology Press, Guangzhou, 2006).
5. General Administration of Quality Supervision, Inspection and Quarantine of the PRC, GB/T 12060.5-2011 *Sound System Equipment-Part 5: Main Method for Measuring Performance of Loudspeaker* (China Standard Press, Beijing, 2012).
6. J. Han, B. Jiang, *The New Parameters and Measurement of Loudspeaker Directivity Characteristic* Electro-acoustic Technology, 33(2009)).
7. X.M. Wang, *Single Chip Microcomputer Control of Motor* (Beijing University of Aeronautics and Astronautics Press, 2007).
8. G. Ma, *Principles and Applications of Single Chip Microcomputer* (China Machine Press, 2008).
9. L.Y ao, J.M. Qiu and J. Chen (eds.), *The Application of Pulse Systems in the Measurement Field of Acoustics and Vibration* (The National Academic Conference of Acoustics Design and Noise Vibration Control Engineering as supporting equipment, 2010).

The Baseline-Length Comparing Method for Testing of Dual Frequency Wide-Lane Observation for Ambiguity Solution in Dynamic Positioning

Jie Wang[†] and Guan-nan Xu

*Department of Navigation Engineering, Naval University of Engineering,
Wuhan, 430033, China
†Email: Wangjie7312@163.com*

The traditional GNSS relative positioning precision test is unitary, so it is applied in limited scope. On the basis of analyzing linear combination of dual-frequency wide lane observation method of ambiguity solution and the traditional method of relative positioning precision test, the paper discusses the method based on the comparison of the real-time relative baseline-length between two vehicles. Precision test methods are proposed respectively in the conditions of static positioning and dynamic positioning. In addition, the characteristics of each method is compared. For validity, some experimental data is provided and analyzed, in which the dual frequency wide-lane observation for ambiguity solution is used in dual dynamic positioning.

Keywords: Dynamic positioning, Position precision test, Relative baseline, Ambiguity.

1. Introduction

In the dynamic relative positioning based GNSS, the method of ambiguity solution is the key to precious positioning. Comparing with the traditional LAMBDA method, the linear combination of dual-frequency wide lane observation of ambiguity solution is improved in the solution rate. The positioning accuracy test is the judgment for the performance of methods used in positioning. The real-time relative baseline of two receiver station changing cause the receiver signal loss of carrier phase tracking, so the traditional test method for real-time positioning accuracy, most are unitary, such as test is carried in the condition of static baseline length is known before, and the test method of comparing real-time positioning results with results processed afterwards, but these methods cannot completely reflect the positioning model effect of respective algorithm [1-2].

2. The Model of the Linear Combination of Dual-Frequency Wide Lane Observation Method

2.1. Dual-Frequency Wide Lane Combination Equation

If the two of the satellites as i and j can be observed both at the base station and the rover station at same time, and for a particular epoch time and two stations about B_1 carrier, wide lane combination observation equation is:

$$\lambda_W \varphi_{12,W}^{ij} = \rho_{12,W}^{ij} + \lambda_W N_{12,W}^{ij} + \varepsilon_{12,W}^{ij} \tag{1}$$

In this equation, λ is for the carrier wavelength, φ as the carrier phase observations, ρ as the distance between the star and receiver, N as the ambiguity, ε as the carrier phase observation noise, W as Wide lane combination.

Wide lane carrier ambiguity least squares equation is followed:

$$\begin{bmatrix} A^{T}C^{-1}A & A^{T}C^{-1}B \\ B^{T}C^{-1}A & B^{T}C^{-1}B \end{bmatrix} \begin{bmatrix} X \\ N \end{bmatrix} = \begin{bmatrix} A^{T}C^{-1}L \\ B^{T}C^{-1}L \end{bmatrix} \tag{2}$$

In this equation, X as the baseline vector, $N_{\alpha\beta}$ as the combination of double difference ambiguity vector, A, B as coefficient matrix respectively, and C^{-1} as weight matrix. According to the coefficient of combination, have established under type:

$$\varphi_{\alpha\beta} = \alpha\varphi_{L1} + \beta\varphi_{L2} \tag{3}$$

$N_{\alpha\beta}$ can be determined according to the least squares method, because α, β as only one set of coefficient. The mathematical relationship is unable to correctly solve their ambiguity, so if any search for a new set of coefficients as η, γ, as the type of combination, in accordance with the following combination equation of solving ambiguity and correctly.

$$\begin{bmatrix} N_{\alpha\beta} \\ N_{\eta\gamma} \end{bmatrix} = \begin{bmatrix} \alpha & \beta \\ \eta & \gamma \end{bmatrix} \begin{bmatrix} N_1 \\ N_2 \end{bmatrix} \tag{4}$$

2.1 OVT inspection

For n consecutive epochs, the dual-difference data of wide lane combination is solved to get a set of ambiguity integer solutions as $\overline{N}_i^*$, $i = 1, 2, ..., m$.

If the following conditions are validated in Consistency test of the ambiguity vector, we think the ambiguity is solved correctly.

$$\overline{N}_1^* = \overline{N}_2^* = ... = \overline{N}_n^*$$

3. The Traditional Static Test

Static test method is one of the most common methods in positioning accuracy test, that is choosing two test points for long time static observation, and the relative baseline between two points can be solved as the standard data.

In static test method, its significant advantages is that test results can more accurately reflect the positioning accuracy, but defect is that in dynamic testing system, the positioning accuracy cannot be reflected fully. Reasons has the following two points: first, the static test method is usually adopts the static measurement method, in this kind of ideal under the condition of test result, it is difficult to reflect the actual move in the applications of relative positioning precision testing; Secondly static testing requires constant known accurate coordinates of points or relative to the baseline standards, and in some complex area measurement is difficult to achieve.

4. Dynamic Detection Method

4.1. A fixed base length relative detection

Such as base station and rover receiver antenna fixed on a motor carrier for real-time dynamic positioning, Data is collected and for post-processing, not using the known information of the baseline length, and difference between the real time calculating length and the known baseline length reflects the actual dynamic precision and performance of application method.

4.2. The instantaneous dynamic detection

When tested for accuracy, at a specific point in course ,such as turning point, on the set of relative position precisely known more testing point, base station and rover station at the same time their movement to the above on a set of testing point, record the number of the big dipper for calculating the moment, later for comparing the calculating results under the epoch, which is compared with the pure static and dynamic test, the method can reflect the motion state of the receiver, deficiency is only able to detect the motion state of for some time, to continuously reflect the system positioning precision.

5. Experiment and Data Analysis

5.1. *The static test*

To test the static test result, static relative positioning tests now. Due to relative positioning in the actual application of baseline will not very long, now choose two known relative coordinates accurately check points, one of the ultra short baseline, measured the length of 15.254 m, in addition to a baseline length is slightly longer, to 1550.682 m. Testing process, the synchronous observation of the baseline of each time is about 1 hour, the sampling time for 1 second. Real-time decoding software calculated according to the data collected by independent each epoch GPS and BDS baseline results, statistics of the corresponding internal and external precision of solution. Table 1 for the static test results of two baselines, among them, and direction deviation is centered on base station northeast days' rectangular coordinates.

Table 1 The results of each GPS and BDS tests.

The baseline length mtype		L1		L2	
		Within	Outside	Within	Outside
GPS	N	0.004	0.005	0.015	0.040
	E	0.005	0.005	0.008	0.012
	U	0.021	0.025	0.038	0.049
	Plane	0.009	0.008	0.014	0.042
BDS	N	0.010	0.011	0.010	0.035
	E	0.008	0.015	0.008	0.015
	U	0.030	0.039	0.032	0.045
	Plane	0.012	0.018	0.018	0.045

Can be seen from the table above, BDS real-time baseline calculating results in three direction and plane direction, and precision are better than 4 cm, among them, the first calculation of GPS baseline, the result is slightly better than the BDS, the second baseline, BDS and precision of calculating result of GPS, and whether the BDS system or GPS system, the calculating results in the direction of the plane precision is higher than the height direction.

5.2. *Dynamic test*

Due to the dynamic test 1 and 3 two methods are compared with the results of standard, the static tests has been analyzed, thus the following main do test for the reduced.

Fig. 1 Photo of the experiment equipment.

Test before the start of the antenna of base station and rover separately respectively fixed in a threaded joint on both ends of the tube, to place it in the respective test car, push the car movement about 10 min in the measurement area, data sampling rate is 1 s, synchronization acquisition and save the two data. It is calculated respectively using real-time decoding software BDS and GPS baseline of each epoch time as a result, after the event respectively statistics BDS and internal and external of GPS real-time calculating precision. After data processing, accurate fixed base length measured value of 0.910 m. Precision testing, 1 m long baseline for upper limit, are listed in Table 2.

Table 2. The test results of the fixed based length results.

Positioning system	GPS	BDS
The total number of observation epoch	560	560
Transfinite number/bias > 0.1m	2	10
Outer precision RMS/mm	3.5	9.2
Qualified percentage epoch /%	99.64	98.21

By above test results of the BDS calculating the baseline outside precision is within 1 cm long, compared with the GPS, BDS long baseline data calculating precision and data effectively lower success rate.

6. Conclusion

Based on the traditional based on the analysis of the relative positioning accuracy detection method, this paper studies the move under the condition of the positioning accuracy of the detection method and application. Finally using the experimental data proves the effectiveness of the methods, points out the respective application conditions, etc. Through the study of this article, in the actual location accuracy tests can choose according to the above methods advantages and disadvantages, in order to achieve the optimal detection result.

References

1. Ge Maorong, Xie Baotong. Real–time Relative Positioning for Two Moving GPS Receivers [J]. Journal of Geotechnical study & Surveying, 1998, 4:57-59 (GeMaoRong Xie Baotong. Dynamic of dynamic GPS real-time differential Positioning [J]. Journal of engineering survey, 1998, 4:57-59)
2. Liu Lilong. The Research on The Precise KINRTK found and Its Applications [D]. Wuhan: Wuhan University, 2005 (li-long Liu. Dynamic Research on dynamic GPS Precise positioning Theory and application [D]. Wuhan: Wuhan University, 2005).
3. Tang Weiming. Research on Techniques of Large Area and Long Range GNSS Network RTK and Developing the Network RTK Software [D]. Wuhan: Wu han University, 2006 (wei-ming Tang. A wide Range of long-distance GNSS Network RTK technology Research and the Software implementation [D]. Wuhan University, 2006).
4. Huang Xiaorui, Tian Wei. Establish resplendence ardently GPS receiver autonomous integrity monitoring algorithm study [J]. Journal of telemetry remote control, 2003, 24 (1) : 1-3
5. Yan-rui Geng, Wei-wei Zhang, Cui Zhong. Based on SINS assisted GPS integrity monitoring method [J]. Journal of Beijing university of aeronautics and astronautics, 2001, 27 (2): 164-166
6. GeMaoRong, static Jun. GPS relative navigation application in spacecraft rendezvous and docking [J]. Bulletin of surveying and mapping, 1998, (5): 6-7.
7. Guang-jun Liu, Ceng Jibin. GPS dynamic relative navigation for spacecraft rendezvous and docking studies and OTF decoding method [J]. Journal of aircraft measurement and control, 2000, 12 (2): 86-93.
8. El - Mowafy a. Performance Analysis of the RTK Techique in an Urban Environment [J]. Journal of Australian Surveyor, 2000, (1): 47-54
9. Feng Y, Wang j. GPS RTK Performance Characteristics and Analysis [J]. Journal of Global Positioning Systems, 2008, 7 (1): 1-8
10. Gao Y, Wojciechowski a. High Precision Kinematic Positioning Using Single Dual Frequency GPS Receiver [J]. Journal of Remote Sensing and Spatial Information Sciences, 2004:1-5

Imbalanced Data Classification via a Cost-Sensitive Majority Weighted Minority Oversampling Approach

Xi-qing Cui and Jian-hua Zhang[†]
School of Information Science and Engineering, East China University
of Science and Technology, Shanghai, 200237, China
†E-mail: zhangjh@ecust.edu.cn

Ru-bin Wang
School of Sciences, East China University of Science and Technology
Shanghai, 200237, China

To address the imbalance classification tasks, a method for Cost-sensitive Majority Weighted Minority Oversampling (CS-MWMOS) technique was proposed. The between-class and inner-class imbalance problem of the sample distribution are synthetically considered. Imbalance ratio was introduced to alleviate between-class imbalance issue. The inner-class imbalance problem is also reduced by allocating different weights for minority sample. Extensive experiments on 20 UCI imbalance datasets showed that the proposed method can effectively address the class imbalance problem in terms of the geometric mean and average accuracy assessment metrics.

Keywords: Imbalanced Multi-Classification; Cost Sensitive; Imbalance Ratio; Ensemble Framework.

1. Introduction

In traditional classification tasks, the class imbalance problem[1] appears in the training data when the samples of one class are seriously outnumbered by the samples from other classes.

A wide number of methods have been proposed to address class imbalance problem among which are data level approaches and algorithmic level ones[2]. Data level approach is generally based on resampling methods in which the training samples are modified as to generate a more balanced data distribution [3], such as random oversampling (ROS) and under sampling (RUS) technique, while algorithmic level approach is implemented by assigning unequal weights for different classes, such as cost sensitive learning(CSL)[4]. Heuristically, both sampling and training processes can be combined to achieve better performance. Such as DyROS [5], weighting method, such as MWMOTE[6].

This paper is organized as follows. In Section 2, the proposed method is described in details. Latter, detailed experimental results of the proposed method and the empirical study methods are compared and analyzed in Section 3 and Section 4. Finally, the results and conclusion can be found in Section 5.

2. The Proposed Framework

2.1. *Introduction to CS-MWMOS*

In class imbalance issues, the between-class imbalance and the inner-class imbalance problem of the dataset need to be considered. The implementation steps of the method is mainly described below.

1) For each minority sample $x_i \in S_{min}$, compute the nearest neighbor set $NN(x_i)$ which consists of the nearest k_1 neighbors of x_i according to euclidean distance.

2) Construct the filtered minority set S_{minf} by removing those minority examples which have no minority example in their neighborhood.

3) For each $x_i \in S_{minf}$, compute the nearest majority set $N_{maj}(x_i)$ which consists of the nearest k_2 majority examples from x_i by euclidean distance.

4) Find the borderline majority set S_{bmaj} as the union of all $N_{maj}(x_i)$ s ,i.e.,
$$S_{bmaj} = \bigcup_{x_i \in S_{minf}} N_{maj}(x_i).$$

5) For each majority sample $y_i \in S_{bmaj}$, compute the nearest minority set $N_{min}(y_i)$ which consists of the nearest k_3 minority samples from y_i according to euclidean distance.

6) Find the informative minority set S_{imin} as the union of all $N_{min}(y_i)$ s, i.e.,
$$S_{imin} = \bigcup_{y_i \in S_{bmaj}} N_{min}(y_i).$$

7) For each $y_i \in S_{bmaj}$ and each $x_i \in S_{imin}$, compute the informative weight $I_w(y_i, x_i)$.

8) For each $x_i \in S_{imin}$, compute the selection weight $S_w(x_i)$ as
$$S_w(x_i) = \sum_{y_i \in S_{bmaj}} I_w(y_i, x_i).$$

9) Convert each $S_w(x_i)$ into selection probability $S_p(x_i)$ according to
$$S_p(x_i) = S_w(x_i) / \sum_{z_i \in S_{imin}} S_w(z_i).$$

10) Assign different classification cost for each class according to the imbalance ratio and compute the number of new synthetic samples N_{min}.

11) Find the clusters of S_{min}. Let M clusters are formed which are $L_1, L_2, ..., L_M$.

12) Initialize the set , $S_{omin} = S_{min}$.

95

13) Do for $j = 1, ..., N_{min}$

    a) Select an example x from S_{imin} according to probability distribution $\{S_p(x_i)\}$. Let x is a member of the cluster L_k, $1 \leq k \leq M$.

    b) Randomly select another example y from samples of the cluster L_k.

    c) Generate one synthetic data x_{new} according to $x_{new} = x + \alpha \times (y - x)$, where α is a random number in the range [0,1].

    d) Add x_{new} to $S_{omin} : S_{omin} = S_{omin} \cup \{x_{new}\}$.

14) End Loop.

Output the oversampled minority set S_{omin}

2.2. Cost-sensitive algorithm

Cost sensitive strategy is used to assign different costs for each class based on the class imbalance ratio. Specifically, the imbalance ratio r is computed. $Cost[i, c]$ implies that the sample which belongs to class i is misclassified as class c. $Cost[i, i]$ is generally set to 0. Thus $Cost[maj, min] < Cost[min, maj]$ $= random(r)$. $Cost[i]$ is defined as the misclassification cost of class i. Finally, the sample number of new minority set, N_{newmin}, can be computed by Eq. (1).

$$N_{newmin} = \frac{Cost[min]}{Cost[maj]} * N_{maj}$$

(1)

2.3. Ensembling framework

A multi-classification framework is designed by using CS-MWMOS for every two-class samples extracted from multi-class dataset, as is shown in Fig. 1.

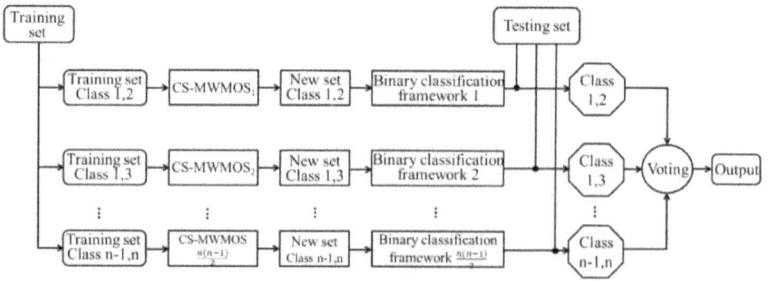

Fig. 1. Multi-classification framework for imbalanced data.

2.4. Performance metrics

Geometric mean is used as the assessment metric which is shown in Eq. (2).

$$Gmean = (\prod_{i=1}^{c} \frac{tr_i}{n_i})^{\frac{1}{c}}$$ (2)

Where c is the class number of the dataset, n_i is the number of samples in class i and tr_i is the number of correctly classified samples in class i. Furthermore, average accuracy of each class is used as is described in Eq. (3).

$$Acc_i = \frac{tr_i}{n_i}$$ (3)

3. Binary Classification Results and Analysis

The validation process is considered to verify the superiority of the proposed method compared with MWMOTE based on 14 datasets extracted from [6]. 20 times 10-fold CV (cross validation) is used. Fig. 2 has shown an obvious improvement of Gmean on 14 dataset. For example, the gmean result is improved by nearly 58% compared with that of MWMOTE on *Abalone*.

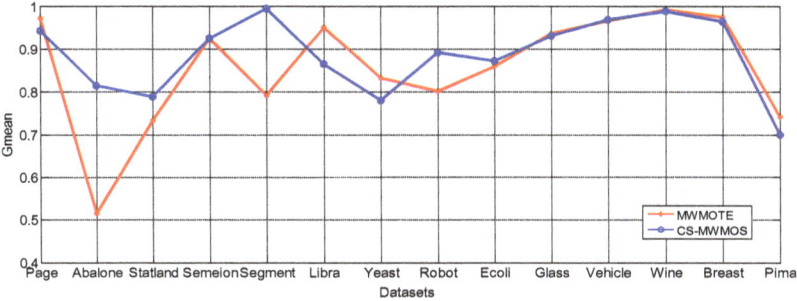

Fig. 2. G-mean results for binary classification tasks with different methods.

4. Multi-label Classification Results and Analysis

4.1. UCI datasets

Furthermore, 20 UCI multi-class datasets were selected to validate the performance of our proposed method which were described in Table 1.

97

Table 1. Description of UCI Multi-Class Datasets.

No.	Name	Size	# Atts	#Class	Sample size of each class
1	Allrep	2800	28	4	2713/35/29/23
2	Autompg	398	7	3	249/70/79
3	Balance-scale	625	8	3	49/288/288
4	Car	1728	6	4	1210/384/69/65
5	Contraceptive	1473	9	3	629/333/511
6	Dermatology	360	34	6	111/60/71/48/50/20
7	DNA	2000	180	3	464/485/1051
8	Ecoli	327	6	5	143/77/52/35/20
9	Flag	194	27	6	31/17/35/52/39/20
10	Glass	192	9	4	70/76/17/29
11	Hypothyroid	3770	28	3	95/194/3481
12	Landsat	2000	36	6	461/224/397/211/237/470
13	Machine	203	7	5	21/135/29/11/7
14	New-thyroid	215	5	3	150/35/30
15	Nursery	12958	8	4	4266/4320/328/4044
16	Page-blocks	5473	10	5	4913/329/28/88/115
17	Primary-tumor	380	17	10	14/14/16/20/24/24/28/29/39/84
18	Satellite	4435	36	6	1072/479/961/415/470/1038
19	Wine	178	13	3	59/71/48
20	Yeast	1484	8	10	463/429/244/163/51/44/35/30/20/5

4.2. Classification Results and Discussion

10 times 5-fold CV was used. Several other methods are used to compared with our proposed method. Basic method is the sole BP method, CSOS and CSUS are cost sensitive random oversampling and under sampling technique, respectively. Statistical histogram is shown in Fig. 3. Classification performance of the Basic method performs the worst. Our proposed method performs better on 16 datasets when compared with CSOS. Meanwhile, CS-MWMOS performs better than CSUS and DyROS methods on almost all the datasets in terms of Gmean.

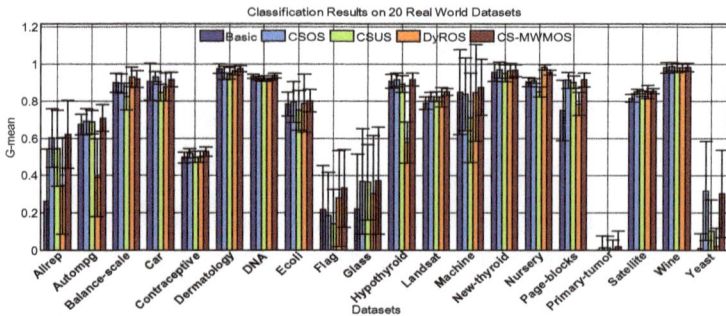

Fig. 3. G-mean results for multi-classification tasks with different methods.

The imbalance classification problem is different from normal classification tasks. To analyze the average class accuracy of the imbalanced datasets, the average class accuracy of each dataset is computed, shown in Fig. 4. Take *Allrep* and for example, the horizontal axis represents for each class with class number in the bracket. The classification accuracy for class 2, 3,4 are 0.7086, 0.5 and 0.6043, respectively. There is an obviously improvement compared with Basic method which are 0.5, 0.2310 and 0.4435. For *Satellite*, the classification accuracy has been improved on class 2,4,5. Fig. 4 has shown the superiority and genearalization performance of the CS-MWMOS method.

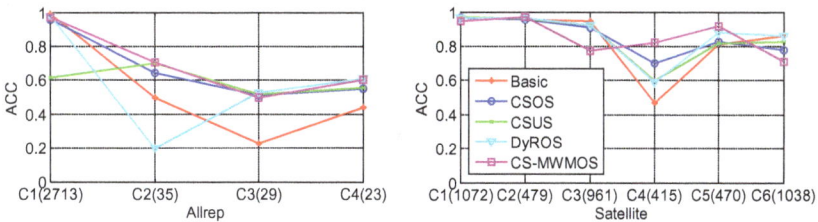

Fig. 4. Average Accuracy of each class on *Allrep* and *Satellite*.

5. Conclusion

In this paper, an effective method(CS-MWMOS) was proposed to address multi-class imbalance problem. Both between-class imbalance and inner-class imbalance problem were considered. The proposed method was used as the basic sampling method to design an effective multi-classification framework. Extensive experiment has shown its superiority and generalization performance compared with other imbalance classification methods.

Acknowledgment

This work was funded by the National Natural Science Foundation of China under Grant No. 61075070 and Key Grant No. 11232005.

References

1. Sun Y, Wong A K C, Kamel M S. Classification of imbalanced data: A review [J]. International Journal of Pattern Recognition and Artificial Intelligence, 2009, 23(04):687-719.
2. He H, Garcia E A. Learning from imbalanced data [J]. Knowledge and Data Engineering, IEEE Transactions on, 2009,21 (9):1263-1284.

3. Cateni S, Colla V, Vannucci M.A method for resampling imbalanced datasets in binary classification tasks for real-world problems [J]. Neurocomputing, 2014, 135:32-41.
4. Zhou Z H, Liu X Y. Training cost-sensitive neural networks with methods addressing the class imbalance problem [J]. Knowledge and Data Engineering, IEEE Transactions on, 2006, 18(1): 63-77.
5. Lin M, Tang K, Yao X. Dynamic sampling approach to training neural networks for multiclass imbalance classification [J]. Neural Networks and Learning Systems, IEEE Transactions on, 2013, 24(4): 647-66.
6. Barua S, Islam M M, Yao X, et al. MWMOTE--majority weighted minority oversampling technique for imbalanced data set learning [J]. Knowledge and Data Engineering, IEEE Transactions on, 2014, 26(2): 405-425.

Application of Control and Automation in the 500 W@4.5 K Helium Refrigerator for ADS Project

Zhi-wei Zhou†, Qi-yong Zhang, Xiao-fei Lu, Gen-hai Xia and Liang-bing Hu

Institute of Plasma Physics, Chinese Academy of Sciences
Hefei, Anhui 230031, China
†E-mail: zzw@ipp.ac.cn

The 500W@4.5K helium refrigerator was designed to provide forced-flow cooling of 4.5 K supercritical helium for the superconducting coils testing or as a helium liquefier with the capacity of 150 L/hr liquefaction rate. Its control system was designed and developed based on the Programmable Logic Controller (PLC) to achieve automatic process supervisory and control in each defined operational mode. Via the analysis of process flow and control requirements, all operational modes with their logic control were designed. Through the dynamic simulation, the dynamic behaviors of cryogenic process were revealed and the process parameters control loops were designed and simulated for optimizing the cool-down process. The cryogenic instrumentation and control as well as the cryogenic commissioning results will be presented in this paper. It shows that the application of control and automation in the home-made helium refrigerator of hundred-watt class is successful.

Keywords: Cryogenic Instrumentation and Control; PLC; Helium Refrigerator.

1. Introduction

With the development of big science projects in China, the large-scale helium refrigerator has been widely applied in the research fields such as fusion and high energy physics. The large-scale helium refrigerator is usually equipped with a set of control system to achieve automatic process control, which would not only improve its operation stability and reduce the risk of manual operation but also promote the operation efficiency. At present, many of the helium refrigerators used in the big science devices were ordered from LINDE or AIR LIQUIDE Corporation, such as the 9 kW@4.5 K helium refrigeration system for the KSTAR in Korea and 1.35 kW@4.5 K helium refrigerator for the SST-1 in India both from AIR LIQUIDE [1], while two sets of 500 W@4.5 K helium refrigerators for the BEPC II in China from LINDE [2], and so on. They are all equipped with the control system and have a high degree of automation.

The Institute of Plasma Physics, Chinese Academy of Sciences (CAS) undertook a sub-project of ADS (Accelerator Driven Subcritical) system, which was supported by the Special Fund for strategic pilot technology CAS. A 500

W@4.5 K helium refrigerator and its PLC control system for ADS project was designed and constructed for the forced flow cooling, which could be used for the superconducting coils testing or as a helium liquefier. It can be counted as the first home-made 4.5K helium refrigerator of hundred-watt class in China.

2. Cryogenic Process Control Flow

The helium refrigerator was designed according to the refrigeration mode, which can produce the 500 W@4.5 K refrigeration power at the design working condition, while in the liquefaction mode it can produce liquid helium (LHe) with the 150 L/hr liquefaction rate. The simplified cryogenic process flow schematic is shown as Figure 1. The refrigerator adopted the standard Claude refrigeration cycle, which was designed to have five cooling levels. They are liquid nitrogen (LN2) pre-cooling, two levels of expand cooling with turbines T1 and T2, and the last two levels of throttle. It reserves a stream of 80K helium gas to meet the requirement of superconducting magnets pre-cooling. Depending on the return gas temperature, it provides three gas return paths of different temperature levels, which are 80K, 20K and 4.5K respectively.

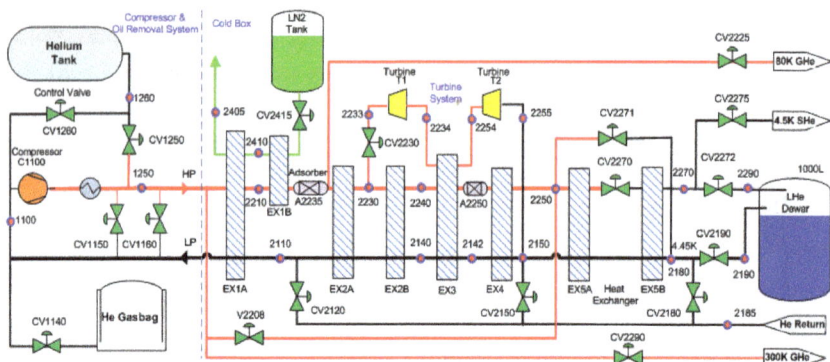

Fig. 1. The simplified process flow schematic of 500 W@4.5 K helium refrigerator.

3. PLC Control System for Helium Refrigerator

The helium refrigerator has the properties of typical industrial process system, so its control system adopts the mature control structure used for industrial process system. However, it also has some unique characteristics of cryogenic process. Its control system should have a high reliability, compatibility and expandability.

3.1. Control network structure

The refrigerator was developed on the SIMATIC S7-300 PLC control system from Siemens Corporation. It can be divided into four subsystems, compressor system, oil-removal system, cold box and turbine system. The distributed input/output (I/O) control network structure based on the Field-bus as the Figure 2 shows was adopted. It has a three-layer structure, which are respectively field device layer, control layer and supervisory layer.

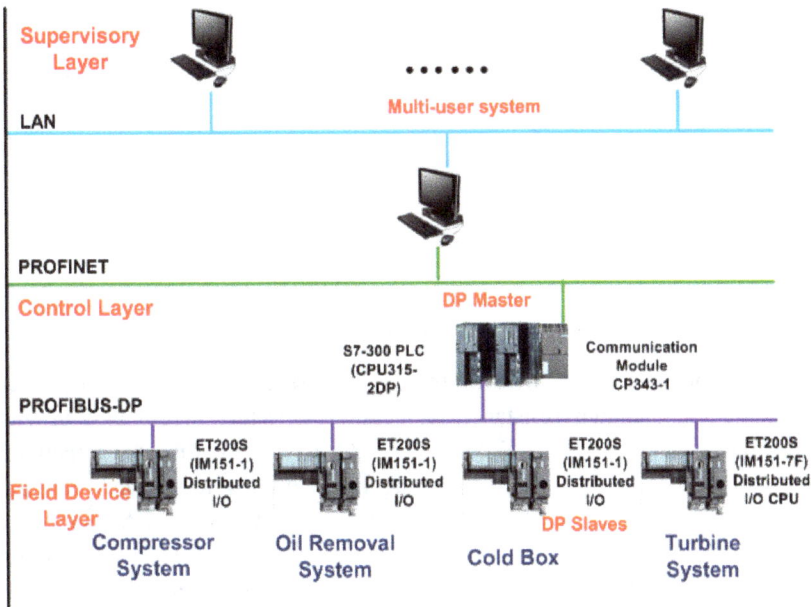

Fig. 2. The distributed I/O control network structure of the 500W@4.5K refrigerator.

The four subsystems as the DP slaves adopt the distributed interface module and I/O modules of ET200S series, which have their own control cabinets in filed and communicate with the main controller through DP protocol. The controller chose the CPU 315-2DP which is suitable for processing the programs of medium or large capacity and has a PROFIBUS DP master/slave interface. The CPU as a DP master as well as the Ethernet communication module and some I/O modules were installed in the main control cabinet, which are used to execute the control programs for the refrigerator. The supervisory layer was built with a multi-user system, which communicates with the PLC through PROFINET.

103

3.2. Cryogenic instrumentation

In the field layer of control system, there are many instrument for measuring pressure, temperature, mass flow rate, liquid level, rotation speed and purity, and many actuators such as cryogenic valves, switches and heaters. The I/O numbers which have been used in each control cabinet are listed in Table 1. There are still spare I/O modules for future extension.

Table 1. Input and output numbers of the 500W@4.5K helium refrigerator.

Area / I/O	Main Control Cabinet	Compressor peripheral & Oil removal system	Cold box	Turbine System	Total
AI	26	13	50	13	102
AO	6	5	20	12	43
DI	11	3	12	21	47
DO	13	5	8	7	33
Total	56	26	90	53	225

There are 45 temperature sensors, of two types, are used in the refrigerator, including the platinum resistance temperature detectors Pt100 used in high temperature range from 73 K to 573 K, the Cryogenic Linear Temperature Sensor (CLTS) used in low temperature range from 4 K to 296.9 K. Each temperature sensor as a 4-wire resistor was equipped with a transducer to convert the resistance to the standard 4-20 mA linear signal. The nominal resistance of CLTS is 290.0 ohms ±0.5 %. With proper instrumentation, a resolution of 0.01 K can be easily achieved.

The 42 pressure gauges are used on cryogenic pipes and vessels, which are displayed with absolute pressure through the pressure transducer. The precision is maximum ±0.5 %. The venture flow meters are adopted to measure the mass flow rate of the helium cryogenic process, which are calculated by some parameters including helium density, differential pressure and geometric dimensioning in a OPC (OLE for Process Control) client program. The 32 pneumatic control valves with intelligent positioners are driven by the analog output signals of 4-20 mA from PLC.

4. Control Software for Helium Refrigerator

4.1. *Control programs function module*

The control software development for the helium refrigerator include the STEP 7 control programs and WinCC supervisory control interfaces. Figure 3 illustrates the control software function module design for the helium refrigerator. The Structured Control Language (S7-SCL), Structured Text Language (S7-STL) and Continuous Function Chart (S7-CFC) were adopted to develop the control programs. Making use of SCL and STL generates the standard CFC predefined block, which could be directly called in the CFC programming. This programming method is suitable for the large-scale industrial system and system integration like the helium refrigerator.

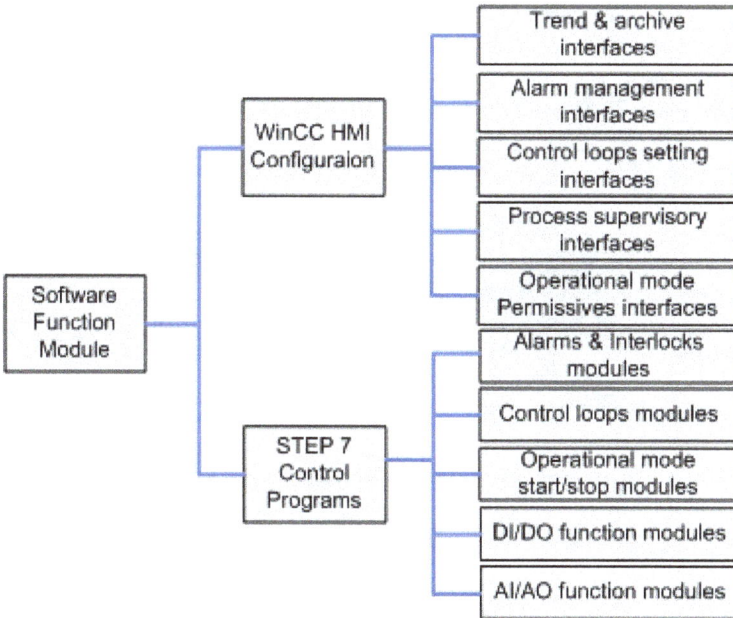

Fig. 3. The control software function module design for the helium refrigerator.

The STEP 7 control programs realize the function of system input/output, process parameters control loops in each operational mode, alarm and interlocks for the whole refrigerator. The WinCC human machine interface (HMI) configuration realizes the function of process supervisory and control, alarm management, trend and historical data archiving and analyzing.

105

4.2. Operational modes

The helium refrigerator has the control requirement that it can be put into operation in automatic mode or manual mode, which can be shifted to each other without any interruption. The operational modes have been defined and designed for the automatic process control. In each operational mode, a set of start permissive, running conditions, shutdowns, alarms and interlocks and process parameters control loops are activated and executed.

The operational modes were defined into main modes and sub modes. The main modes represent the main cryogenic control flows, while the sub modes represent the process unit or device's operation. Figure 4 illustrates operation sequence of the operational modes and their logic shift relationship. Each operational mode has its permissive conditions to start and stop or shutdown, which is supervised by the automatic control programs, also need to be activated by the operator.

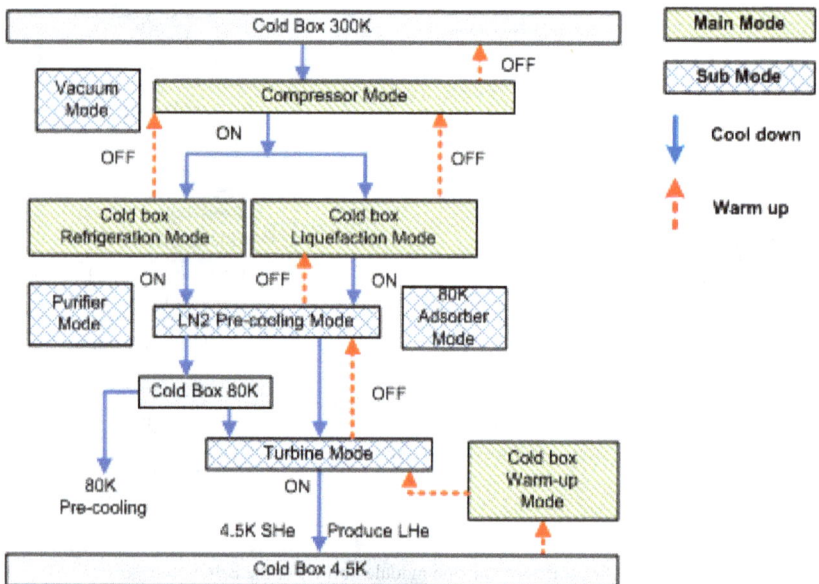

Fig. 4. Operation sequence of the operational modes and their logic shift relationship.

4.3. Process parameters control loops

The dynamic modeling for the refrigerator has been developed based on Aspen HYSYS [3] to reveal the cryogenic dynamic behaviors. The process parameters control loops were designed and simulated for optimizing the cool-down operation. The main control loops in Liquefaction mode are listed in Table 2.

Table 2. Main control loops in Liquefaction mode.

Control loops	Control requirement	Control actuators	Control method
LP pressure PC1100	PI1100=1.05barA	CV1150; CV1160	PI1100>SP, close CV1150; PI1100<SP, open CV1150; CV1160fine control.
HP pressure PC1250	PI1250_SP=10.6-13.6barA	CV1250; CV1260	PI1250>SP, open CV1250; PI1250<SP, open CV1260; PID split range control.
LN2 pre-cooling TC2405_TC2410	TI2405_SP=268K; TI2410_SP=300~80K	CV2415	TI2410>SP, open CV2415; TI2405>SP, open CV2415; Select min PID output
T2 outlet temperature TC2255	TI2255_SP=11K(LN2 mode ON)	CV2271; CV2272	TI2255 high, open CV2271; TI2255<SP, open CV2272;
JT pressure PC2270	PI2270_SP=5barA	CV2270	PI2270<SP, open CV2270; PI2270>SP, close CV2270
Dewar return pressure PC2190	PI2190_SP=1.25barA	CV2190	PI2190<SP, close CV2190; PI2190>SP, open CV2190

Each control loop is usually set in the automatic control mode, which can be activated automatically in the operational mode when it reaches its start/stop conditions, but can be converted to the manual mode by the operator in any time. Most of the control loops coupled with other loop so make up the multi-variables control loops. Aimed at the cryogenic behavior of the long delay, big inertial and non-linear, the PID control algorithm with dead band is used in most of the control loops. And the controller output is gradually with a different ramp for the control valve to avoid the process shock and valve abrasion.

5. Control and Automation Commissioning

Up till now, the helium refrigerator and its PLC control system have performed the 4.5 K cryogenic commissioning twice in Liquefaction mode. It was cooled down successfully and gained a liquefaction rate of 165L/h. The communication and control of the PLC control network as well as part of the automatic control loops and interlocks have already been tested. Some problems found in the test have already been solved. It showed that the design scheme of automatic control is feasible and the HMI is friendly to the operator. However, some control parameters still need to be retuned for optimized control based on the actual experimental operation. Figure 5 is the Turbine supervisory control interface in the real cryogenic test.

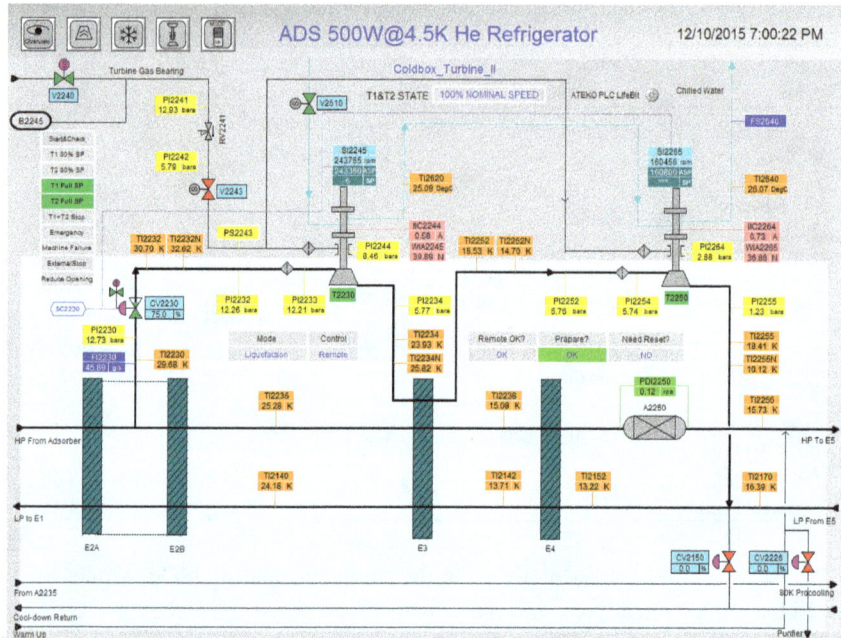

Fig. 5. The Turbine supervisory control interface in the cryogenic commissioning .

6. Conclusion

In this paper, the instrumentation and control for the 500W@4.5K refrigerator has been presented in detail. Adopting the modular control idea with hierarchy, the refrigerator was design to be operated automatically either in Refrigeration mode or Liquefaction mode. The distributed PLC control system based on the DP Field-bus was used to develop the automatic control strategies. Its modular programming is helpful to the extension and transplantation of the control programs. A friendly supervisory control interface has been developed and tested, which lays the foundation for the automatic operation of the whole refrigerator. The test result shows that the application of control and automation in the home-made helium refrigerator of hundred-watt class is successful. This year, the refrigerator will be tested and optimized for the full automation degree.

Acknowledgments

This project was funded by the National Natural Science Foundation of China (Grant No. 11505237) and the Application & Development Project of the Institute of Plasma Physics, Chinese Academy of Sciences (Y35ETY130G) are gratefully acknowledged.

108

References

1. P. Dauguet, G.M. Gistau-Baguer, M. Bonneton, J.C. Boissin, E. Fauve, J.M. Bernhardt, J. Beauvisage and F. Andrieu, Cryogenics for fusion, *Advances in Cryogenic Engineering: Transactions of the Cryogenic Engineering Conference*, (Tennessee, USA, 2008).
2. G. Li, K. Wang, J. Zhao, K. Yue, M. Dai, Y. Huang and B. Jiang, The cryogenic control system of BEPC II, *Chinese Physics C* **32**, 4(2008).
3. X. Lu, Z. Zhou, M. Zhuang and L. Qiu, Process modeling and control simulation for a 500W@4.5K helium refrigerator, *Journal of Harbin Institute of Technology*, (2015).

Multi-Station Network Layout Optimization of Workspace Measuring and Positioning System

Zhi Xiong[†], Chong Yue and Jun Tu

School of Mechanical Engineering, Hubei University of Technology
Wuhan, Hubei, China
[†]*E-mail: xiongzhi0611@163.com*

Xue Bin

School of Marine Science and Technology, Tianjin University
Tianjin, China
E-mail: xuebin@tju.edu.cn

The most prominent advantage of workspace Measuring and Positioning System (wMPS) is to solve the confliction between measurement range and measurement accuracy in large-scale measurement field. Due to the layout of system network has a great impact on measurement range, accuracy and cost, system layout optimization method using intelligent algorithms was researched in this paper. Firstly, layout optimization model was established on the basis of the above three objectives. Then the simulated annealing algorithm and particle swarm algorithm were combined to get the global optimal deployment, which was able to realize complete coverage of the measurement range and meet the measurement accuracy requirements at the lowest possible cost. Finally, the effectiveness of the optimization method was demonstrated by experiment, of which the result shows that using simulated annealing-particle swarm algorithm, the overall positioning error of ten-station system has been improved 0.04mm averagely. This method is also suitable for the layout optimization of angle intersection measuring system and not limited by the number of stations.

Keywords: wMPS; Layout Optimization; Simulated annealing - particle swarm algorithm.

1. Introduction

Workspace measuring and positioning system(wMPS) is a kind of large-scale system, which dependents on the multi-station synergy to achieve the coordinate measuring, so the station layout optimization is a common but important problem[1]. wMPS layout optimization method is divided into two categories, one kind is the typical layouts, which Robert Schmitthe[2] analyzed in robot localization and tracking, the results show that the standard type of measurement is the best. In 2013, the author[3] studied the influence of the typical layout on the positioning error of the wMPS network layout, the results show that the overall measurement accuracy of O_4 layout is the highest. The other kind is using intelligent algorithms to get global optimal deployment. Authors[4–5] used genetic algorithm to study the influence of the different intersection of the light plane of the wMPS emission station on the measurement stability, established the mathematical model to perform layout evaluation, the simulation results show that the measuring stability of four station system is better than three stations and two station system.

The above research on the layout of wMPS network aims to a small number of stations or specific layouts, which has no general expansion when faced with more stations. This paper designed a multi-station deployment method based on simulated annealing — particle swarm algorithm, which took positioning accuracy, coverage and cost as objective functions. This method is suitable for more station deployment. Finally, The effectiveness of the optimization method was demonstrated by experiment.

2. Multi-objective Optimization Model of wMPS

The purpose of optimizing station layout is to reach maximum coverage area, meet the measurement accuracy requirement and cost less. The relationship between these three goals and the station layouts was studied in the following.

2.1. wMPS positioning accuracy

This paper adopts the positioning error model in the authors' previous work [6–7], station distribution geometry for any point P_k on the space geometry of GDOP (Geometric Dilution of Precision) is expressed as:

$$GDOP_{P_k} = \sqrt{tr(D_{P_k})} \tag{1}$$

Due to the positioning accuracy and coverage, cost belongs to different dimension, so it is necessary to carry out the normalization as:

$$O_1 = \begin{cases} 1 - \dfrac{GDOP_{P_k}}{PDOP_{\lim}} & \max(GDOP_{P_k}) \le PDOP_{\lim} \\ 0 & others \end{cases} \tag{2}$$

Where $PDOP_{\lim}$ is measurement accuracy for the user.

2.2. Coverage Area Analysis

The coverage problem of the station is defined as the ratio of the number of the measured points to the total number of the measured points:

$$O_2 = \frac{sum(k == 1)}{Total} \qquad (sumk == 1) \le Total \tag{3}$$

2.3. Cost Analysis

This paper only considers station after completing the investment cost of measurement task, regardless of operating costs during the measurement process. After all stations are deployed, costs under the layout model can be expressed as:

$$O_3 = 1 - \frac{N_{act}}{N_{max}} \qquad N_{act} \leq N_{max} \qquad (4)$$

2.4. Definition of Multi-objective Optimization Function

Based on the analysis of the above objective function, giving a certain weight to each objective, the multi- objective optimization of wMPS is transformed into a single objective optimization problem, and the optimization function is constructed as:

$$\max f(x) = K_1 O_1 + K_2 O_2 + K_3 O_3 \qquad (5)$$

Where K_i (i=1, 2... n) is weight, and $\sum_{i=1}^{n} K_i = 1$.

3. Layout Optimization Algorithm Design

Particle swarm algorithm is a stochastic optimization method based on groups, each possible solution is expressed as a population of particles, each particle has the speed and position of their own, and there is a fitness determined by the objective function. All particles fly at a certain speed in a given search space, followed by the current optimal value to find the global optimal [8]. So the particle swarm algorithm is faster in the early convergence rate, but it is easy to be affected by the parameter selection and the nature of the objective function, which results in slow convergence speed and easy to fall into local optimum.

In order to avoid the particle swarm algorithm falling into local optimum, this article combines simulated annealing algorithm with particle swarm algorithm [9], adopts the simulated annealing in the particle update process and uses Eq.(6) and Eq.(7) to update the particle velocity and position. Metropolis acceptance criteria [10] was used. If the difference between two fitness function value exp(−Δf/T)>r and then accept the new solution, continue "create new solutions - Judgment - accept or discard process". At the same time, use the Eq. (8) to set the initial temperature and annealing temperature, adjustment annealing temperature adaptively, so that the algorithm can jump out from the local optimal value and converge to the global optimal solution.

112

$$v_{id}(k+1) = \chi[v_{id}(k) + c_1 r_{1d}(k)(p_{id}(k) - x_{id}(k)) + c_2 r_{2d}(k)(p_{gd}(k) - x_{id}(k))] \quad (6)$$

$$x_{id}(k+1) = x_{id}(k) + v_{id}(k+1) \quad (7)$$

$$\begin{cases} T_k = -fitness(P_g)/\log(0.2) \\ T_{k+1} = C \cdot T_k \end{cases} \quad (8)$$

Where $v_{id}(k+1)$ is the velocity of particles in k+1 time, $v_{id}(k)$ is the velocity of particles in k time, $p_{id}(k)$ is the best location for the individual in k time, $p_{gd}(k)$ for the current global optimal location. c_1 and c_2 are learning factors, r_1 and r_2 are random number between 0-1. $\chi = \dfrac{2}{|2 - C - \sqrt{C^2 - 4C}|}$ is compression factor, and $C = c_1 + c_2$, $C > 4$.

Simulated annealing- particle swarm algorithm process which is applied to wMPS station deployment was shown in the Fig. 1.

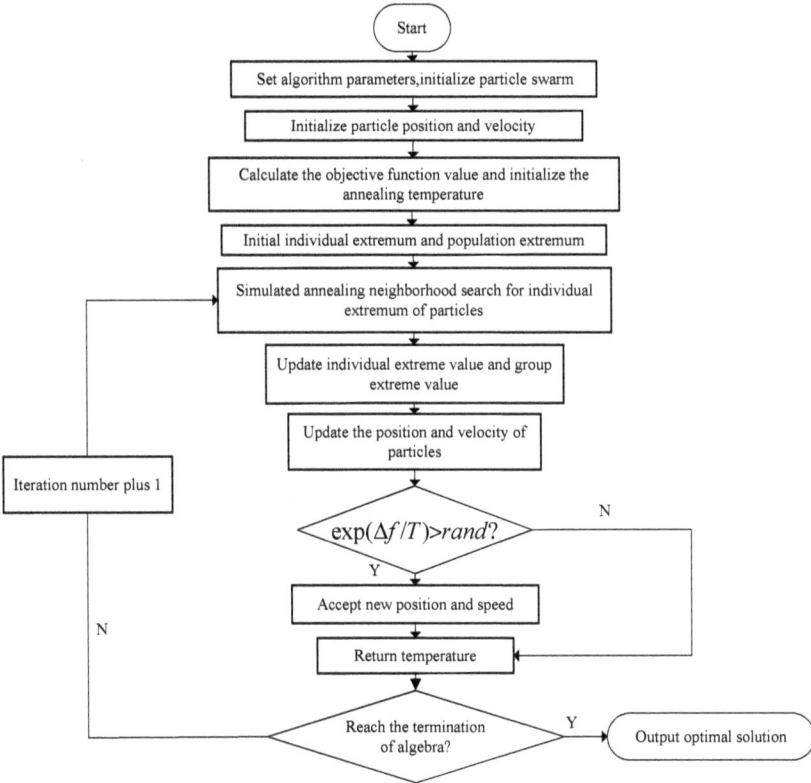

Fig. 1. Simulated annealing- particle swarm algorithm process of wMPS network layout.

4. Experiment

To validate the algorithm proposed in this paper, experiment of ten-station layout optimization was designed on the existing platform in the laboratory. Equipment was shown in Fig. 2, including the transmitting station, receivers and calibration accessories [11–12].

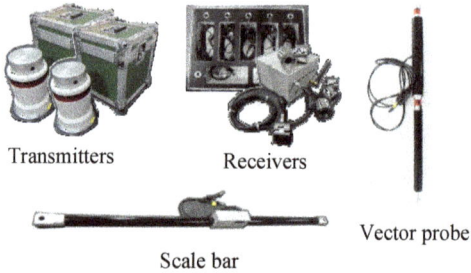

Fig. 2. Experiment equipment.

Before optimization, the number of stations was determined to meet the requirement of full coverage of measured range, and then the deployment area of each station was divided and described quantitatively. At last simulated annealing - particle swarm algorithm was applied to look for optimized layout to meet the measurement accuracy requirement. Experiment layout was shown as Fig. 3.

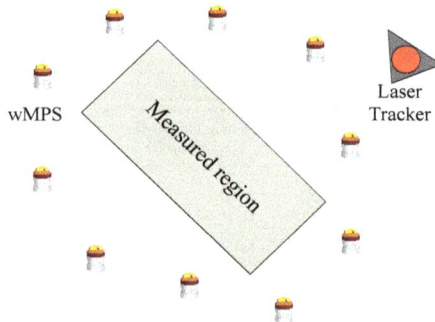

Fig. 3. Experiment layout.

20 test points were randomly selected. The measurement results before and after optimization were compared with laser tracker measurement results, which were shown in Fig. 4.

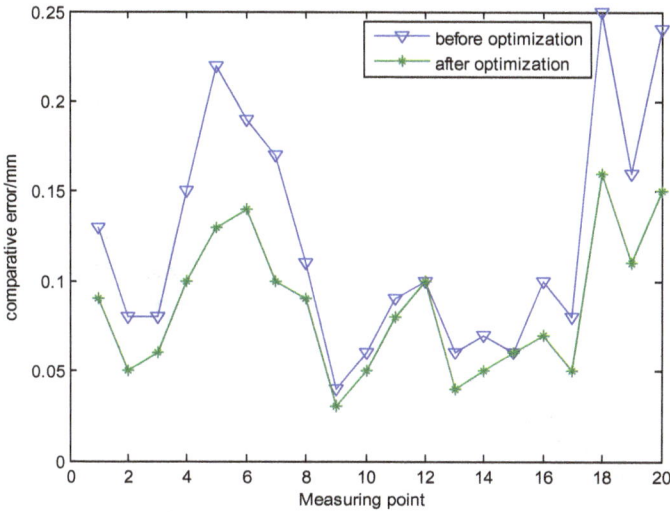

Fig. 4. Experiment layout.

From Fig. 4 it can be seen that using simulated annealing-particle swarm algorithm proposed in the paper, the overall positioning error of ten station has been greatly improved. Through comparison with laser tracker, measurement error of wMPS after optimization is reduced 0.09mm maximally and 0.04mm averagely.

5. Conclusion

A new method to solve the optimal deployment of wMPS station layout based on simulated annealing-particle swarm algorithm is proposed. This method takes positioning accuracy, coverage and cost into consideration as the objective function of optimization. Simulated annealing-particle swarm algorithm preserved the global search ability of simulated annealing algorithm and local search ability of particle swarm algorithm. The experiment results show that the proposed method is effective to optimize station deployment of wMPS.

Acknowledgment

This work was funded by the National Natural Science Foundation of China (51305130, 61505140) and Hubei University of Technology Doctoral start up funded projects (BSQD12123).

References

1. Ye Shenghua, Zhu Jigui, Zhang Zili, et al., Status and development of large-scale coordinate measurement research [J]. ACTA Metrological Sinica, 2008, 29(4A):1-6.
2. Schmitt R, Nisch S, Schönberg A, et al., Performance Evaluation of iGPS for Industrial Applications [J], Proceedings pf International Conference on Indoor Positioning and Indoor Navigation (IPIN), Zürich, Switzerland, 2010.
3. Xiong Zhi, Zhu Jigui, Xue Bin, et al., Typical deployments of workspace measurement and positioning system [J]. Optics and Precision Engineering, 2013,09:2354-2363.
4. Xue Bin, Zheng Yingya, Xiong Zhi, Zhu Jigui, Simulation for network deployment of indoor workspace measurement and positioning system [J]. Journal of Mechanical Engineering, 2015,08:1-8.
5. Zheng Yingya, Zhu Jigui, Xue Bin, Lin Jiarui, Network deployment optimization of indoor workspace measurement and positioning system [J]. Opto-Electronic Engineering, 2015, 05:20-26.
6. Xiong Z, Zhu J G, Zhao Z Y., Workspace measuring and positioning system based on rotating laser planes [J]. Mechanika, 2012,18(1):94-98.
7. Xiong Zhi, Zhu Jigui, Geng Lei, et al., Verification of Angle Measuring Uncertainty for Workspace Measuring and Positioning System [J]. Chinese Journal of Sensors and Actuators, 2012, 25(2): 229-235.
8. Mo Simin, Zeng Jianchao, Particle swarm optimization based on self-organization topology [J]. Journal of System Simulaition, 2013,03:445-450.
9. Meng Xiangtao, Wang Wei, Dynamic compensation of FOG navigation system based on particle swarm optimization and simulated annealing algorithm [J]. Infrared and Laser Engineering, 2014, 05:1555-1560.
10. Shao Wei, Zuo Yijun, Simulated annealing for higher dimensional projection depth [J]. Computational Statistics & Data Analysis, 2012, 56(12):4026-4036.
11. Xiong Z., Zhu J. G., Ren Y. J., et al., Analysis and Design Of the Best Layout Based on the Network Measurement of WMPS [J], Proceedings of the 10th International Symposium on Measurement and Quality Control, Osaka, Japan, 2010.

12. Xue B, Zhu J, Zhao Z, et al., Validation and mathematical model of workspace Measuring and Positioning System as an integrated metrology system for improving industrial robot positioning [J], Proceedings of the Institution of Mechanical Engineers Part B: Journal of Engineering Manufacture, 2014, 228(3): 422-440.

Fuzzy Task Assignment Model for Web Services Supplier

Jian Sun[1,2,†], Xiu-yan Peng[1], Ying Xu[1,3] and Na-ji Ma[1,2]

[1]College of Automation, Harbin Engineering University,
Harbin, 150001, China
[2]Center of Information and Network,
Heilongjiang University of Science and Technology
Harbin, 150001, China
[3]College of Science,
Heilongjiang University of Science and Technology,
Harbin, 150001, China
†E-mail: sj_1666@163.com

In view of collaborative development environment web services supplier in ability, cost, time, supplier relations and component relevance information under uncertainty problems. The fuzzy multi-objective task assignment model of web services are built in a collaborative development environment. Using α cut sets and extension principle to simplify the fuzzy multi-objective assignment mode, we get the solution of simplified assignment model via the Genetic and simulated-annealing algorithm. Finally, the simulation results verify the feasibility of the proposed method, which can ensure the suppliers' tasking in successive software project in collaborative development environment.

Keywords: Collaborative Development; Fuzzy Task Assignment; Extension Principle.

1.   Introduction

In the process of collaborative software development based on web services, software designers and management personnel will consider the collaborative development of web services providers. By giving full play to the providers' capability and advantage of design and development, they will ensure the services quality, reduce the overall cost of development and even shorten the development cycle [1, 2]. In spite of those advantages, there are some problems such as how to assign tasks for the suppliers in the collaborative development and how to select supplier scientifically with uncertain information. Therefore, it is necessary to put forward a scientific and reasonable method and establish a fuzzy task assignment model for web services under the collaborative development environment.

In view of the above problems, The fuzzy multi-objective task assignment model of web services supplier is proposed in collaborative development environment, which meet the suppliers' need of high reliability, low cost and short time in the process of software project development and provide the task optimization assignment for software development enterprises, which can guarantee the collaborative development environment software suppliers involved in the successful completion of the project. In the process of implementation in the model, we simplify the complex fuzzy multi-objective problem by the α sets and the extension principles and set up single objective optimization model decomposed with genetic and simulated-annealing algorithm [6].

2.   Model Construction

Assuming that the software maker has a large software projects demanding high reliability, low cost, the shortest time to market. The marker will decompose the project into several web services and choose some web services suppliers to participate in the project development, in order to make full use of their technology and resources. Assuming that the project decomposition has been completed, the number of components is m, the number of the web services supplier involved in the software development is n. There is a time sequence in the web services development process, so time process development is w.

i :the i web services development tasks, $i = 1..m$; j :the j the web services suppliers, $j = 1..n$; x_{ij} :0-1 variables, the i web services is assigned to the j supplier; p_{ij} :the ability of j supplier development of i web services, as a fuzzy number range is [0,1]; t_{oij} :in the o development sequence, the j web services suppliers development the i WEB services of the time, as a fuzzy number range is [0,1]; t_i : the longest development time of the i web services; q_{oi} : in the o development sequence, the start time of the i web services, $o = 1..w$; c_{ij} :the costs of j supplier development of i web services, as a fuzzy number range is [0,1]; e_{ik} :degree of information dependence between the i web services and the k web services, as a fuzzy number range is [0,1]; $d_{j,f(k)}$: the coordination degree between the j supplier and bear the k web services supplier $f(k)$,as a fuzzy number range is [0,1].

During the web services supplier collaborative software development process, it is essential to determine the collaborative working time of each WEB services supplier. Each WEB services completion time depends on the three parts: web services started collaborative time, each supplier to the WEB services and WEB services between development time. By using the methods mentioned in the literature [3], there are dependencies among the web services, so the collaborative time supplier j required in the development of the web services i can be expressed as:

$$T_{com}(j) = \sum_{f(i)=j} \sum_{k=1}^{m} \frac{e_{ik}}{d_{j,f(k)}}. \tag{1}$$

e_{ik} indicates the degree of information dependence between the i web services and the k web services, but the web services information dependence is difficult to represent specific quantitative data, so there are expressed by fuzzy numbers. $d_{j,f(k)}$ represents the coordination degree between the i supplier and the j supplier. When $f(k) = j$, the coordination degree of up to 1. Each supplier to the web services development time is expressed as $t_{oij} \times t_i$. According to the web services of development time and each supplier to the web services, the j supplier to develop collaborative development time of all web services can be expressed as:

$$T_{total}(j) = \sum_{f(i)=j} (\sum_{o=1}^{w} (q_{oi} + t_{oij} \times t_i) + \sum_{k=1}^{m} (\frac{e_{ik}}{d_{j,f(k)}})). \tag{2}$$

Because of the existence of information dependence between different web services in the process of web services development, there is a procedure problem when the web services assign tasks for suppliers. The overall time of software project development needs the completion of each web services, so it inevitably depends on the total time of suppliers who take the most development time in concurrent development process [4, 5]. Therefore, software project development time shortest is seeking a minimal model, expressed as:

$$\min \max(T_{total}(j))(j = 1..n). \tag{3}$$

Fuzzy task assignment problem of web services supplier in collaborative development environment can be described as a fuzzy multi-objective assignment problem as follows:

$$G_1 = \max(\sum_{i=1}^{n} \sum_{j=1}^{m} p_{ij} x_{ij}). \tag{4}$$

120

$$G_2 = \min(\sum_{i=1}^{n}\sum_{j=1}^{m} c_{ij}x_{ij}). \tag{5}$$

$$G_3 = \min \max_{j=1,..,n}(\sum_{f(i)=j} (\sum_{o=1}^{w}(q_{oi}+t_{oij}\times t_i)+\sum_{k=1}^{m}(\frac{e_{ik}}{d_{j,f(k)}})x_{ij})$$

$$s.t. \sum_{i=1}^{n} x_{ij} = 1, j = 1,..,m;\ q_{oi}+t_{oij}\times t_i \le q_{o,i+1}, o = 1,..,w; i = 1,..,m; j = 1,..,n;$$

$$x_{ij} = 0 \text{ or } 1, i = 1,..,m; j = 1,..,n. \tag{6}$$

3. Algorithm Design

According α cut sets and extension principle, we have decomposed the fuzzy multi-objective task assignment model into a linear goal programming model.
MODEL FMOAP_1:

$$Z_\alpha^L = \max(\beta\times\mu_p(x)+\delta\times\mu_c(x)+\gamma\times\mu_t(x))$$
$$s.t.$$

$$\mu_p(x) = \frac{\sum_{i=1}^{m}\sum_{j=1}^{n}(p_{ij})_\alpha^L x_{ij} - G_p^{\min}}{G_p^{\max}-G_p^{\min}}, \mu_c(x) = \frac{G_c^{\max}-\sum_{i=1}^{m}\sum_{j=1}^{n}(c_{ij})_\alpha^U x_{ij}}{G_c^{\max}-G_c^{\min}}$$

$$\mu_t(x) = \frac{G_t^{\max}-\left\{\max_{j=1,..,n}(\sum_{f(i)=j}^{m}(\sum_{o=1}^{w}(q_{oi}+(t_{oij})_\alpha^U\times t_i)+\sum_{k=1}^{m}(\frac{(e_{ik})_\alpha^U}{(d_{j,f(k)})_\alpha^L}))x_{ij})\right\}}{G_t^{\max}-G_t^{\min}}$$

$$\sum_{j=1}^{n} x_{ij} = 1, j = 1,..,m;\ q_{oi}+(t_{oij})_\alpha^U\times t_i \le q_{o,i+1}, o = 1,..,w; i = 1,..,m; j = 1,..,n;$$

$$x_{ij} = 0 \text{ or } 1, i = 1,..,m; j = 1,..,n;\quad \beta+\delta+\lambda = 1. \tag{7a}$$

MODEL FMOAP_2:

$$Z_\alpha^U = \max(\beta \times \mu_p(x) + \delta \times \mu_c(x) + \gamma \times \mu_t(x))$$

$s.t.$

$$\mu_p(x) = \frac{\sum\limits_{i=1}^{m}\sum\limits_{j=1}^{n}(p_{ij})_\alpha^U x_{ij} - G_p^{\min}}{G_p^{\max} - G_p^{\min}}, \mu_c(x) = \frac{G_c^{\max} - \sum\limits_{i=1}^{m}\sum\limits_{j=1}^{n}(c_{ij})_\alpha^L x_{ij}}{G_c^{\max} - G_c^{\min}}$$

$$\mu_t(x) = \frac{G_t^{\max} - \left\{ \max\limits_{j=1,..,n}(\sum\limits_{f(i)=j}^{m} (\sum\limits_{o=1}^{w}(q_{oi} + (t_{oij})_\alpha^L \times t_i) + \sum\limits_{k=1}^{m}(\frac{(e_{ik})_\alpha^L}{(d_{j,f(k)})_\alpha^U}))x_{ij}) \right\}}{G_t^{\max} - G_t^{\min}}$$

$$\sum_{j=1}^{n} x_{ij} = 1, j = 1,..,m; \ q_{oi} + (t_{oij})_\alpha^L \times t_i \le q_{o,i+1}, o = 1,.., w; i = 1,..,m; j = 1,..,n;$$

$$x_{ij} = 0 \ \text{or} \ 1, i = 1,..,m; j = 1,..,n; \quad \beta + \delta + \lambda = 1. \tag{7b}$$

The assignments of web services supplier in the collaborative development are interactive, therefore it is necessary to select suitable heuristic algorithm. By using the genetic and simulated-annealing algorithm to solve the task assignment model, the basic process of solving algorithm are as follows:

(1) Describe the algorithm parameters

m: the number of web services; n: the number of supplier; T:initial temperature; W:mutation rate; Gen: the number of generation cycle.

(2) Algorithm description

① Code

Using binary code (0 represent the supplier is not assigned, 1 represent the supplier is assigned), the length of the string is the number of web services.

② The initial population

Randomly generating the required number of population, the length of each population for the web services number is m. The population of nodes are represented by binary, each individual in the population represents whether the supplier should be assigned to complete the corresponding web services.

③ Select operation

Generating the offspring group from the parent group, then randomly select individual of i and j both from the parent and offspring group, the i and j is competitive into the next generation of probability for:

$$\exp(\frac{f(i) - f(j)}{T}).$$

④ Crossover operation

The random part structure of two parent individual is replaced and reorganized and then generates new individual by using the multi-point crossover operator.

⑤ Mutation operation

The random number 0-1 and the comparison between the ways of mutation, if the random number is less than W, the selected parent population by random mutation to generate new population.

4. Conclusion

Because there is no certain relevant information of component suppliers' ability, cost, time, supplier relations and web services in collaborative development environment, we put forward the fuzzy multi-objective task assignment model of web services in this paper. In this way, we not only realize the demand of high reliability, low cost and short time of web services in the process of software development projects, but also provide a task optimization assignment for software Development Company. In the process of solving the model optimal solution, we simplify the complex fuzzy multi-objective problem into a single objective optimization problem using α set and extension of the principle. Finally, we get the simplified single objective optimization model with the Genetic and simulated-annealing algorithm.

Acknowledgements

The authors wish to acknowledge the funding support from The Education Department of Heilongjiang province science and technology research project no. 12543064.

References

1. Chi-Jen Lin, Ue-Pyng Wen, Pei-yi Lin, M.S.: Advanced sensitivity analysis of the fuzzy assignment problem. Applied Soft Computing. 11(8), 5341-5349(2011).
2. Feyzan Arikan, M.S.: A fuzzy solution approach for multi objective supplier selection. Expert Systems with Application. 40(6), 947-952(2013).
3. Zhang Wan-jun, Liu Wei, Zhang Zi-jian, M.S.: Task assignment for suppliers participation in collaborative product development. Computer Integrated Manufacturing Systems. 15(6), 1231-1236(2009).
4. Chen L.H, M.S.: A fuzzy model for exploiting quality function deployment. Mathematical and Computer Modelling, 38(5-6), 559-570(2003).

5. Kao C, M.S.: Fuzzy efficiency measures in data envelopment analysis. Fuzzy Sets and Systems. 113(3), 427-438(2000).

6. Zadeh L A, M.S.: Fuzzy sets as a basis for a theory of possibility. Fuzzy Sets and Systems. 113(3), 427-438(2000).

Duty Cycle Estimation Algorithm for
Pulsed Interference Based on Binary Integration

Bing Wu[†] and Lin Pang
Zhengzhou Institute of Air Defense Forces,
Zhengzhou, 450052, China
[†]E-mail: 13633833568@139.com

Shu-min Huo
Information Engineering University
Zhengzhou, 450002, China
E-mail: huoshumin123@126.com

In order to estimate the duty cycle of pulsed interference in Global Navigation Satellite System (GNSS), a simple but effective algorithm is proposed based on the consecutive mean excision and binary integration techniques. The proposed algorithm has less computational and logic complexity compared to the localization algorithm based on double-thresholding with adjacent cluster combining (LAD-ACC). Simulation results show that the proposed algorithm can compute the duty cycle of pulsed interference as accurately as the LAD-ACC algorithm does.

Keywords: Pulsed Interference; Duty Cycle Estimation; Binary Integration; Consecutive Mean Excision; Global Navigation Satellite System (GNSS).

1. Introduction

It is definitely helpful to locate the interference source quickly by monitoring the interference, estimating the parameters of power, carrier frequency and bandwidth for interference in Global Navigation Satellite System (GNSS) bands, including transport control systems, measurement and control systems and, applications systems. Then we can take the right measures to deal with system problems and helps to eliminate interference guiding equipment, to provide the necessary safeguards [1] for the successful construction of satellite navigation systems and normal operation. Pulsed interference is a popular interference pattern in the navigation system bands, resulting in a potential serious impact on navigation system [2]. Therefore, the pulsed interference parameters should be monitored and suppressed by taking appropriate techniques to ensure the normal operation of the system. The time domain blanking [3] and transform domain filtering [4] are two types of popular pulsed interference mitigation techniques and, the duty cycle parameter of pulsed interference is an important reference factors for which type of algorithm selected. For pulsed interference with larger duty cycles, the transform domain filtering is a better choice, whereas the time blanking technique should be adopted under low duty cycle pulsed interference scenarios [5].

The estimation of duty cycle for pulsed interference can be calculated by the ratio between the number of samples over the detection threshold and the total number of input samples, including the backward consecutive mean excision (BCME) algorithm, the forward consecutive mean excision algorithm (FCME), the localization algorithm based on double-thresholding (LAD) and the LAD with adjacent cluster Combining (LAD-ACC) algorithm [6-10]. The BCME-based algorithm is suitable for low-duty cycle pulsed interference scenarios. For high duty cycle pulsed interference scenarios, the FCME-based algorithm is more effective, but needs to hold the sample sorting process with a larger calculation amount compared to the former. The detection threshold is usually calculated based on the probability of false alarm, then the duty cycle of the estimation results is larger than the real value.

In order to reduce the false alarm probability FCME algorithm,[7-8] proposed LAD and LAD-ACC algorithm based on the FCME algorithm. LAD algorithm uses high and low thresholds with FCME algorithm to calculate the number of samples over the thresholds, respectively. Then the samples exceeded the low threshold of several adjacent are classified as a set. If the largest sample exceeds the high threshold, the set is assumed to be disrupted by pulsed interference. Finally, the ratio between the numbers of all disrupted sets samples with the total number of samples can be used for the estimate ion for pulsed interference duty cycle. Due to the fact that the LAD-ACC algorithm is based on LAD algorithm and adds some processes of "collection" between the adjacent sets, so that some of the samples below the threshold of interference can be detected correctly. But the LAD and LAD-ACC algorithm involves double threshold determination and collection of samples of their various operations such as calculation and logic complexity has improved to some extent [10].

In this paper, a novel duty cycle estimation algorithm of the pulsed interference based on binary integration which is popular in radar signal detection [11] and CME algorithm is proposed. It uses the BCME or FCME preliminary results algorithm pulse interference detection, and then use the accumulated binary module for further filtering to give the final duty cycle estimation results.

2. Proposed Algorithm

Let the input data vector is represented as $x = [x_1, x_2, ..., x_N]^T$, where N is the total number of samples. S1 is the subscript set for "disrupted" samples and S2 is the subscript set for "interference-free" samples detected by the BCME or FCME algorithm. Obviously, the following equation holds

$$S1 \cup S2 = \{1, 2, ..., N\} \tag{1}$$

Then, we introduce yi which can be calculated by the following expression

$$y_i = 1, i \in S1 \text{ else } y_i = 0, i \in S2 \tag{2}$$

Then the vector of $[y1, y2, ..., y_N]^T$ can be used for filtering and comparison operation. K is assumed the filtering widow size with $K \ll N$. Then, the N-K filtering results in the middle can be used to estimate the duty cycle, $z = [z_{K/2 +}$ $1, ..., z_{N-K/2}]^T$. Finally, the total numbers of elements with values of 1 in z are used to estimate the duty cycle pulsed interference. The proposed algorithm's description is shown in Fig. 1.

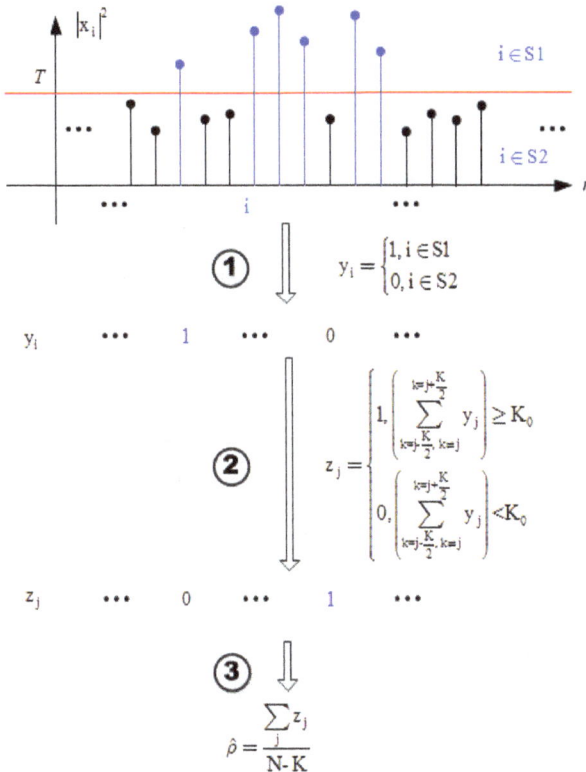

Fig. 1. Description of algorithm processing steps.

127

Compared to BCME (or FCME) algorithm, the proposed algorithm adds a small amount of computation, including the (K-1)*(N-K) adding operation and (N-K) comparison operations, which can usually be ignorable for the popular processors.

3. Simulation Results

Consider the following two typical scenarios with different interference duty cycle, denoted as $\rho 0$, of 10% and 70%, respectively. The estimated result is defined as ρ and $e_\rho=|(\rho-\rho_0)/\rho_0|$ is characterized bias with respect to the ideal values. Each point of the simulated results is obtained averaging 100 times Monte Carlo simulation runs. The "Enhanced-BCME" and "Enhanced-FCME" indicate the proposed algorithm in Fig. 2 and Fig. 3. The simulation parameters and corresponding values are given in TABLE 1.

Table 1. Simulation Parameters and Values.

Parameters	Values	Parameters	Values
Sampling rate	4.092MHz	Thresholds of LAD	6.901, 2.995
Noise level	2W	N	4092
C/N0	40dBHz	Period of Pulse	1e-3s
INR	[-10,30]dB	K	8
Threshold of BCME/FCME	4.6	K_0	3

First, it can be seen from the figure "BCME algorithm" and "FCME algorithm" "Scene I" (low duty cycle) performance is almost the same, and because of the presence of a single threshold detection probability of false alarm and fixed, when the interfere-to-noise ratio (INR) is sufficiently large (e.g., greater than 20dB), the duty cycle of the interference scenario is lower than INR, e_ρ is more greater.

Second, the "BCME algorithm" and proposed "BCME- Enhanced Algorithm" in "Scene II" were unable to get an accurate estimate of the duty cycle, which is determined by the "BCME algorithm" backward iteration characteristics of the decision. While the others based on the FCME algorithm with the increase in INR, can get a more accurate estimated results.

Third, the proposed "BCME-Enhanced" algorithm and "FCME-Enhanced" algorithm have similar performances in the "Scenario I" compared with the LAD algorithm and LAD-ACC algorithm. At the same time, "FCME-Enhanced" algorithm in "Scenario II" has the similar performance with respect to the LAD algorithm and LAD-ACC algorithm, and reaches to the values as low as 10^{-2}.

128

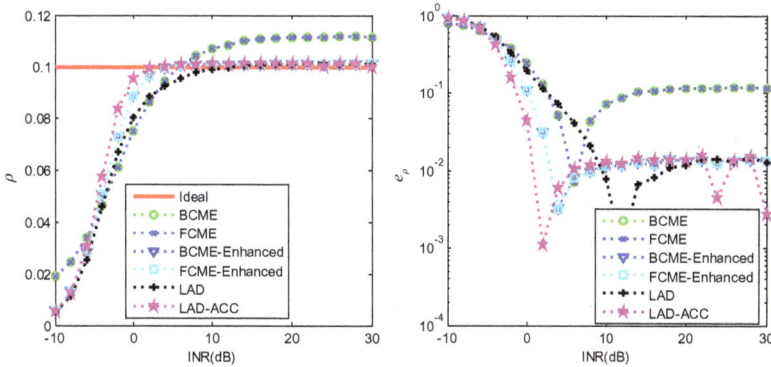

Fig. 2. Comparison for $\rho_0=10\%$ (Scenario I).

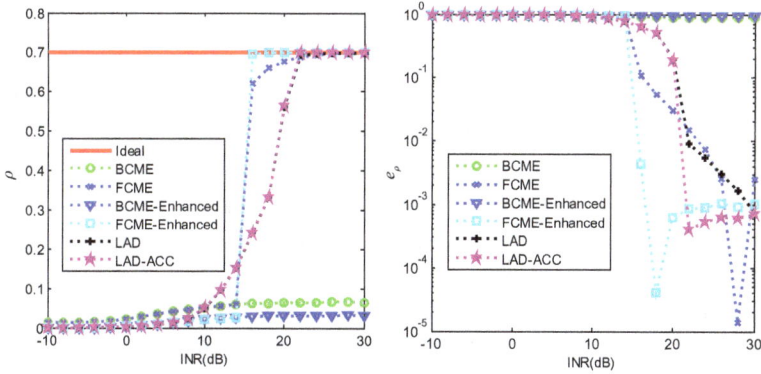

Fig. 3. Comparison for $\rho_0=10\%$ (Scenario II).

4. Conclusion

In this paper, a novel and low-complexity duty cycle estimation algorithm is proposed for pulsed interference. It can be used for pulsed interference monitoring and mitigation technique selection in GNSS.

References

1. Qiwei Han, Xianghua Zeng, Zhengrong Li, Feixue Wang. Recent development and prospect of interference monitoring for GNSS bands. *Aerospace Electronic Warfare*, 6: 17−19,29 (2009) .

2. E.D. Kaplan, C. J. Hegarty. *Understanding GPS: Principles and Applications*, 2nd Ed. (Artech House, 2005)

3. Grabowski J., Hegarty C., Characterization of L5 receiver performance using digital pulse blanking in *Proceeding of The Institute of Navigation GPS Meeting* (2002).

4. Musumeci L, Dovis F, A Comparison of Transformed-Domain Techniques for Pulsed Interference Removal on GNSS Signals in *Int. Conf. Localization and GNSS* (2012).

5. Anyaegbu, E., Brodin, G., Cooper, J., Aguado, E., Boussakta, S., An integrated pulsed interference mitigation for GNSS receivers, *J. Navig.*, 61 (2): 239–255 (2008).

6. P. Henttu and S. Aromaa. Consecutive mean excision algorithm in *IEEE 7th International Symposium on Spread Spectrum Techniques and Application*, Prague, Czech Republic (2002).

7. J. Vartiainen, J. Lehtomaki, S. Aromaa and H. Saarnisaari. Localization of multiple narrowband signals based on the FCME algorithm in *Proc. Nordic Radio Symp.*(2004).

8. J. Variainen, H. Sarvanko, J. Lehtomaki, M. Juntti, and M. LatvaAho. Spectrum sensing with LAD-based Methods in *Proc. IEEE Int. Symp. Personal, Indoor, and Mobile Radio Communications*(2007).

9. Janne Lehtomaki, Johanna Vartiainen, Markku Juntti and Harri Saarnisaari. Analysis of the LAD Methods, *IEEE Signal Processing Letters*, 15:237-240(2008).

10. Johanna Vartiainen, Janne Lehtomäki, Harri Saarnisaari and Markku Juntti. Limits of detection for the consecutive mean excision algorithms in *Proc. Int. Conf. Cognitive Radio Oriented Wirel. Network Comm.*, Cannes, France (2010).

11. Xiangwei Meng, Performance of Rank Quantization (RQ)Nonparametric Detector in Weibull Background. *Acta Electronic Sinica*, 37(9): 2030-2034(2009).

Rejection of General Periodic Disturbances for a Class of Uncertain Linear Systems

Peng-nian Chen

College of Mechatronics Engineering, China Jiliang University,
Hangzhou, Zhejiang 310018, China
E-mail: pnchen@cjlu.edu.cn

This paper deals with the problem of adaptive control of a class of linear systems. The linear system contains unknown parameters and matched general periodic disturbance. The period of the disturbance is assumed to be known. A novel adaptive control method is presented, which consists of differential adaptive control laws and difference adaptive control laws. The adaptive control method can guarantee that all signals in the closed loop system are bounded and the state vector converges to zero.

Keywords: Rejection of Disturbances, Periodic Disturbances, Uncertain Linear Systems.

1. Introduction

Over the past decade, the problem of rejection of general periodic disturbances has been attracted great attention (see, for instance, [1–6] and the references therein). The asymptotic rejection of general periodic disturbances for output feedback nonlinear systems [1-3], in which the ware patterns of the periodic disturbances to be rejected are assumed to be known. In [4,5], adaptive tracking problems are dealt with for a class of nonlinear systems with general periodic disturbances, in which the ware patterns of the periodic disturbances are not known. However, the control method in [4,5] can only guarantee that the tracking errors converge to zero in the sense of $L_2[0, T]$, where T is the period of the periodic disturbance in the system. The adaptive rejection of unmatched general periodic disturbances for a class of output feedback nonlinear systems is considered, in which the ware patterns of the periodic disturbances are not known and the control method can guarantee that all signals in closed loop system are bounded and the tracking error converges to zero. [6]

In this paper, we deal with the problem of rejection of periodic disturbances for a class of uncertain linear systems. A novel adaptive control method is presented, which can guarantee that all signals in the closed loop system are bounded and the state vector converges to zero. Compared with the existing results, the main feature of this study is that the system considered here contains unknown parameters, while the systems in [1-3, 6] do not contain unknown parameters. In the case that the system contains unknown parameters, the control methods [1-3, 6] are not applicable.

2. Problem Formulation

Consider the uncertain linear system

$$\dot{x} = Ax + b\left(c^{\mathrm{T}}x + u + w(t)\right),$$ (1)

where $x = (x_1, x_2, \cdots, x_n)^T$ is the state vector, u is the input; $c = (c_1, c_2, \cdots, c_n)^T$ is an unknown parameter vector, $w(t)$ is a periodic disturbance; A and b have the following forms:

$$A = \begin{pmatrix} 0 & 1 & 0 & \cdots & 0 \\ 0 & 0 & 1 & \cdots & 0 \\ \vdots & \vdots & \vdots & \ddots & \vdots \\ 0 & 0 & 0 & \cdots & 1 \\ 0 & 0 & 0 & \cdots & 0 \end{pmatrix}, \quad b = \begin{pmatrix} 0 \\ \vdots \\ 0 \\ 1 \end{pmatrix}.$$ (2)

We make the following assumptions for system (1).

Assumption 1: The state vector $x(t)$ is available.

Assumption 2: The $w(t)$ is a continuously differentiable function with known period T.

Assumption 3: There exists a known constant $m > 0$ such that $\|c\| < m$.

The objective of control is design an adaptive repetitive learning control for system (1) such that in the closed loop system, all signals are bounded and

$$\lim_{t \to +\infty} x(t) = 0.$$ (3)

3. Design of Controller

Let

$$s = x_n + a_1 x_{n-1} + a_2 x_{n-2} + a_{n-1} x_1,$$ (4)

where $a_i > 0$, $j = 1,, 2, \cdots, n-1$, are chosen such that

$$P(\lambda) = \lambda^{n-1} + a_1 \lambda^{n-2} + \cdots + a_{n-2} \lambda + a_{n-1}$$

is a Hurwitz polynomial.

By a straightforward calculation, using (1), we have

$$\dot{s} = c^T x + u + \sum_{j=1}^{l} w_j(t) + a_1 x_n + a_2 x_{n-1} + \cdots + a_{n-1} x_2. \quad (5)$$

Let

$$u = -\alpha x_n + v, \quad (6)$$

where $\alpha > 0$ is a constant and v is a new input. Since $\|c\| < m$ and a_1 is known, we can choose α such that

$$\alpha - c_n - a_1 > 1. \quad (7)$$

Then system (5) can be written as

$$\dot{s} = -\beta s + v + \sum_{i=1}^{n-1} d_i x_i + w(t), \quad (8)$$

where $\beta = \alpha - c_n - a_1$, $d_1 = c_1 - \beta a_{n-1}$, and

$$d_i = c_i + a_{n-i=1} - \beta a_{n-i}, \quad j = 1, 2, \cdots, n-1.$$

where β, d_i, $i = 1, 2, \cdots, n-1$, are all unknown.

Let

$$V_0 = s^2. \quad (9)$$

Differentiating V_0 results in

$$\dot{V}_0 = 2\dot{s}s = -2\beta V_0 + 2s \left(v + \sum_{i=1}^{n-1} d_i x_i + w(t) \right). \quad (10)$$

In view of (10), we present the following adaptive control law

$$v = -\sum_{i=1}^{n-1} \hat{d}_i(t) x_i - \hat{w}(t), \quad (11)$$

where $\hat{d}_j(t)$ is an estimate of d_i, $i = 1, 2, \cdots, n-1$, and $w(t)$ is an estimate of $\hat{w}(t)$. The estimate $\hat{d}_i(t)$ is defined by

$$\dot{\hat{d}}_i = \gamma_i x_i s, \; i = 1, 2, \cdots, n-1, \tag{12}$$

where $\gamma_i > 0$ is a constant. The estimate $\hat{w}(t)$ is defined by

$$\hat{w}(t) = 0, t \in [-T, 0],$$
$$\hat{w}(t) = \hat{w}(t - T) + \varepsilon \gamma(t) s, t > 0, \tag{13}$$

where $\varepsilon > 0$ is a constant, and $\gamma(t)$ is a smooth function with the property that $0 \le \gamma(t) \le 1, t \ge 0$, and

$$\gamma(t) = \begin{cases} 1, & t > (2/3)T, \\ 0, & 0 \le t < (1/3)T. \end{cases} \tag{14}$$

For the construction of $\gamma(t)$, see, for instance, [6]. The controller for system (1) consists of (6), (11), (12) and (13). The closed loop system is consists of (1), (6), (11), (12) and (13).

It can be proven that the solutions $x(t), \hat{d}_i(t)(i = 1, 2, \cdots, n-1)$ and $\hat{w}(t)$ of the closed loop system exist on $[0, \infty)$.

4. Stability Analysis

The following theorem is the main result of the paper.

Theorem 4.1. Consider *the closed loop system consisting of (1), (6), (11), (12) and (13). In the closed loop system,*

$$\lim_{t \to +\infty} x(t) = 0. \tag{15}$$

Proof. Let

$$V = V_0 + \sum_{i=1}^{n-1} \gamma_i^{-1} \tilde{d}_i^2(t) + \varepsilon^{-1} \int_{t-T}^{t} \tilde{w}^2(\tau) d\tau, \tag{16}$$

where $\tilde{d}_i(t) = d_i - \hat{d}_i$, $\tilde{w}(t) = w(t) - \hat{w}(t)$. By a straightforward calculation, we have

$$\dot{V} = \dot{V}_0 - 2 \sum_{i=1}^{n-1} \gamma_i^{-1} \tilde{d}_i \dot{\hat{d}}_i(t) + \varepsilon^{-1} (\tilde{w}^2(t) - \tilde{w}^2(t - T)). \tag{17}$$

134

By (10) *and* (11), we have

$$\dot{V}_0 = -2\beta V_0 + 2s\left(\sum_{i=1}^{n-1}\tilde{d}_i x_i + \tilde{w}(t)\right).\tag{18}$$

Substituting (18) and (12) into (17) yields

$$\dot{V} = -2\beta V_0 + 2s\tilde{w}(t) + \varepsilon^{-1}(w^2(t) - \tilde{w}^2(t-T)).\tag{19}$$

By (13), using $w(t) = w(t-T)$, we obtain

$$\tilde{w}^2(t) - \tilde{w}^2(t-T) = 2\tilde{w}(t)(-\hat{w}(t) + \hat{w}(t-T))$$
$$-|\hat{w}(t) - \hat{w}(t\_T)|^2.\tag{20}$$

Since, by (14), $\gamma(t) = 1$, for $t \geq (2/3)T$, it follows from (14) that, for $t \geq (2/3)T$,

$$\hat{w}(t) - \hat{w}(t-T) = \varepsilon s.\tag{21}$$

The equations (20) and (21) imply that

$$\tilde{w}^2(t) - \tilde{w}^2(t-T) = -2\varepsilon\tilde{w}(t)s - |\varepsilon s|^2.\tag{22}$$

Substituting (22) into (19) results in

$$\dot{V} \leq -2\beta V_0.\tag{23}$$

This implies that $s^2(t)$, $\hat{d}_i(t)$ ($i = 1,2,\cdots,n$) and $\int_{t-T}^{t}\tilde{w}(\tau)d\tau$ are bounded on $[0,\infty)$ and $s^2(t)$ is integrable on $[0,\infty)$, that is,

$$\int_0^{+\infty} s^2(t)dt < +\infty.\tag{24}$$

Furthermore, we can prove that $\int_{t-T}^{t}\dot{s}^2(\tau)d\tau$ is bounded on $[0,\infty)$, which implies that $s(t)$ is uniformly continuous. Thus by (25), we have

$$\lim_{t\to+\infty} s(t) = 0.$$

This implies that (15) holds, since $P(\lambda)$ is a Hurwitz polynomial. □

Remark: Theorem 4.1 only shows that the state vector of the closed loop system converges to zero. In fact, we can also prove that all signals in the closed loop system are bounded, that is, $x(t), \hat{d}_i(t)(i = 1,2,\cdots,n-1)$ and $\hat{w}(t)$ are all bounded on $[0,\infty)$. Due to the limitation of space, the proof is omitted.

5. Conclusion

In this paper, we studied the problem of adaptive control of linear systems, with unknown parameters and matched general periodic disturbance. An adaptive control model, capable of coping with differential control constraints are put forwarded. The stability analysis has shown that all signals in the closed-loop systems are bounded, with the state vector converges to zero.

References

1. Z. Ding, Asymptotic rejection of general periodic disturbances in output-feedback nonlinear systems, *IEEE Trans. Autom. Control*, **51**, 303(2006).
2. Z. Ding, Asymptotic rejection of asymmetric periodic disturbances in output-feedback nonlinear systems, *Automatica*, **43**, 555(2007).
3. Z. Ding, Asymptotic rejection of unmatched general periodic disturbances in a class of nonlinear systems, IET Control Theory Appl. **2**, 269(2008).
4. J. Xu, A new periodic adaptive control approach with known periodicity, IEEE Trans. Autom. Control, **49**, 579(2004).
5. J. Xu and J. Xu, Observer based learning control for a class of nonlinear systems with time-varying parametric uncertainties, IEEE Trans. Autom. Control, **49**,273(2004).
6. P. Chen, M. Sun, Q. Yan and X. Fang, Adaptive asymptotic rejection of unmatched general periodic disturbances in output-feedback nonlinear systems, IEEE Trans. Autom. Control, **57**, 1056(2012).

A Fault Detection Technology for De-Energized Distribution Lines

Ke Zhu and Yu-tao Song†
Key Laboratory of Power System Intelligent Dispatch and Control
of Ministry of Education, Shandong University
Jinan, Shandong, China
†*Email: syt798@126.com*

Energizing a de-energized distribution feeder safely has received particular attention by power companies. Hence, a fault detection technology for de-energized distribution lines is proposed in this paper to improve reliability of the closing. An inverter is connected to the low voltage side of any transformer of the distribution feeders by the thyristor bridge. When the de-energized line needs to be restored, the electrician can control the thyristor to apply instantaneous high-voltage to the line. Then, the current responses are analyzed to detect whether a fault exists. The effectiveness of the proposed technology has been verified with theoretical calculation and simulation.

Keywords: Fault Detection; Distribution Lines; Thyristor.

1. Introduction

After a distribution line loses power for some time because of the events, such as storms, fault, scheduled outages, or repair, the fault may still exist or humans may contact with the line without knowing it. Energizing or closing in such a situation will result in the accident. Thus, energizing a de-energized distribution line safely has received particular attention by power companies.

Considering the operation blindness of energizing or traditional automatic reclosing technology, scholars have proposed a technology with the capability to identify the existence of any fault. However, the research of the proposed technology only focuses on detecting faults in an energized system. The judgement of fault types are based on the electrical characteristics reflected by the releasing process of the residual electromagnetic energy. Nevertheless, this technology owns no applicability to the distribution lines which lose power [1, 3]. Based on the power electronics perturbation technique, a detection technique for de-energized distribution lines is proposed in this paper, considering the limitations of existing technology and the condition transformer winding Y_n/Y_n connection is widely used in North America. Theory, simulation and experiments all verify the effectiveness of this method.

2. Scheme

2.1. Operating Principle

An inverter is connected to the low voltage side of any transformer of the distribution feeders by the thyristor bridge. When the feeder operates in normal condition, the inverter is in floating charge. After a line loses power for a period due to some events and the de-energized line needs to be restored, the electrician can control to turn on the thyristor at several degrees before the voltage crosses zero, applying instantaneous high-voltage to de-energized lines and generating a transient current flowing into lines as well. Then the current responses are analyzed to detect whether a fault exists. A typical embodiment is shown in Fig. 1. Initially, the conduction angle of thyristor representing the strength of voltage should be small, so that human contacting with the lines can sense it and avoid electric shock. Then, the strength of voltage should increase gradually until it can puncture the short-circuit hitch if there is a high impedance fault.

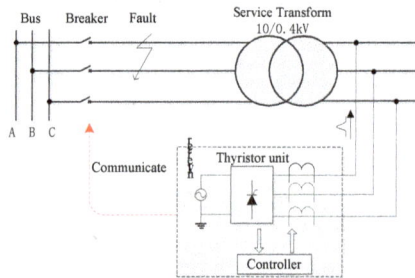

Fig. 1 A typical embodiment.

The detection scheme for different types of faults depends on gating control logic of the thyristor unit. Generally, Two control modes for phase-to-ground faults and phase-to-phase faults detection are as shown in Fig. 2.

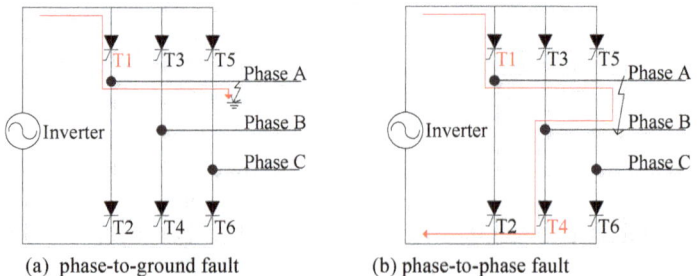

(a) phase-to-ground fault   (b) phase-to-phase fault

Fig. 2 Gating control logic in two fault detection modes.

138

As shown in Fig. 2(a), in phase-to-ground fault mode, when an upper thyristor (T1) is fired and others are off, a voltage pulse used for detection is injected to phase A, creating a corresponding fault current if there is a phase-to-ground fault in phase A. In another mode as shown in Fig. 2(b), if there is a phase A-to phase B fault, a fault current path is provided from thyristor T1 and T4. In conclusion, the thyristor control strategy is illustrated in table 1.

Table 1 Thyristor control logic.

Step No.	Control Logic	Detected Phase
Step I	T1, T3, T4 On, T2, T4, T6 Off.	Phase-to-ground fault
Step II	T1, T4 On, Others Off.	Phase A to Phase B fault
Step III	T1, T6 On, Others Off.	Phase A to Phase C fault
Step IV	T3, T6 On, Others Off.	Phase B to Phase C fault

2.2. Establishment of Theory Criterion

2.2.1. Phase-to-ground fault

Simplified equivalent circuit of the embodiment is shown in Fig. 3. Z_{load} is the equivalent impedance of the load and $Z_{load}=R_{load}+j\omega L_{load}$. R_f is the resistance representing the phase-to-phase fault. The open and close of the switch respectively correspond to the healthy phase and fault phase.

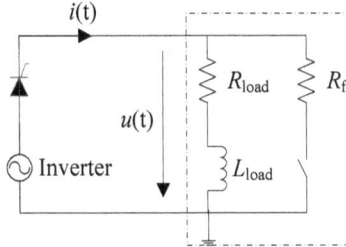

Fig. 3 Simplified circuit.

Suppose that the equivalent impedance of the system $Z=R+L$

$$u(t) = \sqrt{2}E\sin(\omega t+\delta)$$
$$= L\frac{di(t)}{dt} + Ri(t)$$
$$\omega t \in [0, 2\pi - 2\delta] \quad (1)$$

139

The current can be expressed as

$$i(t) = \frac{\sqrt{2}E}{\sqrt{R^2 + (\omega L)^2}}[\sin(\omega t + \delta - \tan^{-1}\frac{\omega L}{R})$$

$$-e^{-Rt/L}\sin(\delta - \tan^{-1}\frac{\omega L}{R})] \tag{2}$$

So, the current magnitude can calculated as

$$i(t) = \frac{\sqrt{2}E}{|Z|}[\sin(\omega t + \delta_1) - e^{-Rt/L}\sin\delta_1] \tag{3}$$

where $\delta_1 = \delta - \tan^{-1}\frac{\omega L}{R}$

The impedance of healthy phase is

$$Z = R_{\text{load}} + L_{\text{load}} \tag{4}$$

The impedance of fault phase is shown in (5), which is smaller than that of healthy phase.

$$Z' = (R_{\text{load}} + L_{\text{load}}) / / R_f < Z \tag{5}$$

Thus, the peak current flow in the fault phase is higher according to (3) and (5).

2.2.2. Phase-to-phase fault

Simplified equivalent circuit of the embodiment is shown in Fig. 4.

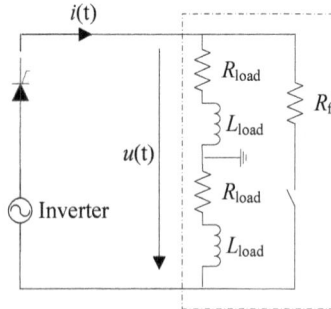

Fig. 4 Simplified circuit.

The expression of injected current magnitude for phase-to-phase fault is shown as expression (3) as well.

140

The impedance of healthy interphase is

$$Z = 2(R_{\text{load}} + L_{\text{load}}) \tag{6}$$

The impedance of faulty interphase is shown in (7), which is smaller than that of healthy phase.

$$Z' = 2(R_{\text{load}} + L_{\text{load}}) \,/\, R_f < Z \tag{7}$$

Thus, the peak current flow in the fault interphase is higher according to (3) and (7).

In a conclusion, combining the phase-to-ground fault with the phase-to-phase fault detection strategy, the overall decision strategy is summarized in Fig. 5.

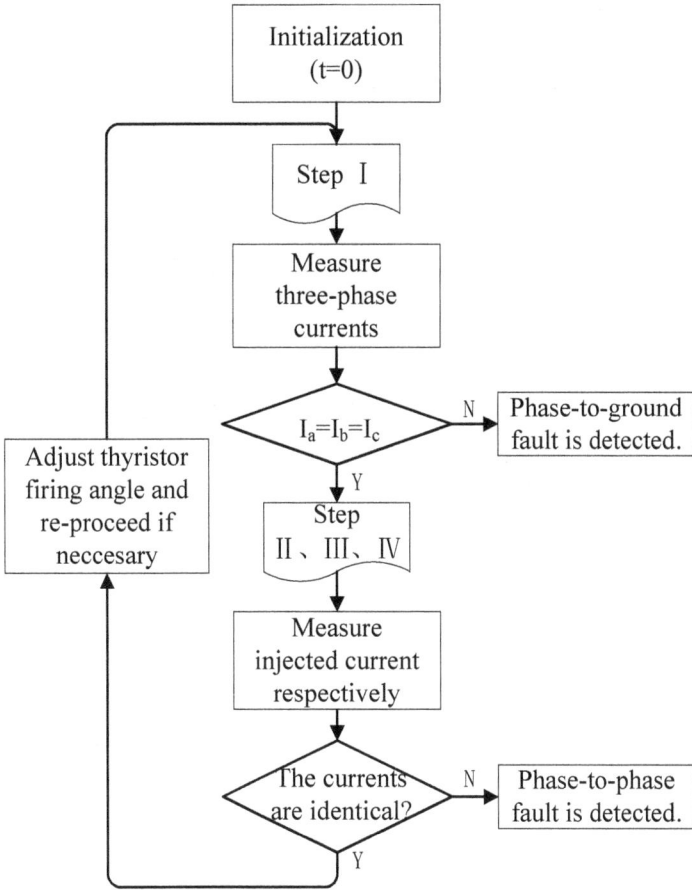

Fig. 5 Decision logic flow.

3. Simulation

The simulation model is shown in Fig. 1. The parameter is shown in Table 2.

Table 2 Simulation parameters.

Voltage Level	10kV/380V
Load Capacity	4MVA
Service transformer	10kV/380V, 1.25MVA, P_k=10.3kW, U_k%=4.5%
10kV Circuit	10km, r_{line}=0.45Ω/km, l_{line}=1.17×10^{-3}H/km
Load	Before power cut: 3MVA, p.f.=0.9。 30% RL loads, 60% motors, 10% power electronic loads; After power cut: Motors and power electronic loads take 90% of its own capacity off the grid, respectively.

If a fault exists between phase A and ground, the current of each phase in Step I will be different. As shown in Fig. 6, the current of phase A is obviously larger.

If a fault exists between phase A and phase B, the injected current in Step II, Step III and Step IV will change. As shown in Fig. 7, the current differs significantly.

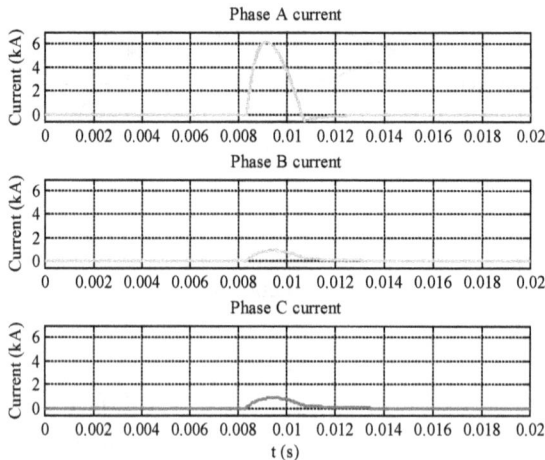

Fig. 6 The current of each phase in step I .

142

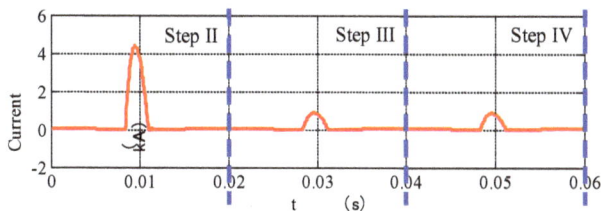

Fig. 7 The injected current in step II, Step III and Step IV.

4. Conclusion

This paper proposes a fault detection technology for de-energized distribution lines, based on the current responded by applying instantaneous high-voltage to the low voltage side of de-energized lines. The proposed method has been verified by theoretical calculation, computer simulations.

References

1. Liang Z, et al. Research review of adaptive reclosure in transmission lines [J]. Power System Protection and Control, 2013, 41(6): 140-147.
2. Liu Y, Zhang Z, Yin X, et al. A Combination Criterion on Adaptive Reclosure for Double Transmission Lines [J]. Automation of Electric Power Systems, 2011 (2): 56-61.
3. Shao W, Song G, Suonan J, et al. Identification of Permanent Faults for Three-phase Adaptive Reclosure of the Transmission Lines With Shunt Reactors [J]. Proceedings of the CSEE, 2010 (4): 91-98.

Personal Decision-Making Plan Under Hazardous Gas Leakage

Nan Zhang, Bo-ni Su and Hong Huang[†]

Institute of Public Safety Research, Department of Engineering Physics,
Tsinghua University,
Beijing, 100084, China
[†]E-mail: hhong@tsinghua.edu.cn

Knowledge on the characteristics of regional evacuation in hazardous gas leakage in metropolises plays a critical role. A comprehensive risk analysis model for personal decision-making is established in order to guide residents to make appropriate personal emergency response plan in hazardous gas leakage. The model was developed considering eight influencing factors, type and flow rate of hazardous gas, location of leakage source, wind speed and direction, information acquirement time, leakage duration, state of window (open/closed), and personal inhalation. Using Beijing as a case study, the risk of all grids and people based on standard condition were calculated and also obtained the three dimensional special risk distribution. Through the microcosmic personal evacuation simulation in different condition, detailed data were obtained to analyze personal decision-making. The results provided useful references for efficient regional evacuation planning in hazardous gas leakage scenarios in metropolises.

Keywords: Risk Based Emergency Response Plan; Hazardous Gas Leakage; Decision-Making; Regional Evacuation.

1. Introduction

In recent years, we have seen more hazardous gas leakage due to rapid economic development, an increase in the number of industries, complex traffic conditions, old gas pipes lacking of maintenance, and poor design etc. In China, 1,400 gas leakage accidents occurred during 2006 to 2011[1]. These hazardous gas leakage accidents usually affect a great number of people and a large sphere of air and land. Therefore, research on hazardous chemical leakage in urban areas is important to public health.

The accidental severity, death tolls, and property losses, which were caused by hazardous gas leakage, are related with several influencing factors including individual exposure duration[2], governmental pre-warning time[3], population density[4], speed of response of disaster carrier, evacuation plan, and some other factors.

Along with the development of computers and software technology, current models of gas diffusion are also highly developed. The Gaussian Dispersion

Model is usually used to generally simulate gas diffusion. Moreover, short simulation time is very important in emergency response. In this paper, we used the Gaussian Diffusion Model to simulate the trajectory of hazardous gas diffusion.

A vast majority of emergency response research focuses on evacuation based on disasters such as earthquake, hurricanes, high-rise fires, flood and some other disasters. However, middle-scale regional evacuation under the condition of hazardous gas leakage is still lacked[5].

In this paper, according to the time of residents to require the disaster's information, hazardous gas leakage duration, and atmospheric environment, a risk analysis model is developed and a dynamic real-time personal emergency response plan is provided. This risk analysis model and evacuation plan can be helpful in reducing unnecessary losses derived from hazardous gas leakage and achieving an effective decision-making.

2. Methodology

Hazardous gas leakage can be monitored by governments or relevant agencies. Utilizing different social media and interpersonal communication, emergency information can be quickly disseminated. Residents will obtain an optimized personal emergency plan by government based on dynamic risk for hazardous gas. The main goal is to reduce the total risk of hazardous gas leakage for every resident.

2.1. Gaussian Diffusion Model

Gaussian Diffusion Model is typically used for gas leakage calculation. The classical Gaussian Diffusion Model, which is shown in Eq. 1, is a steady-state model based on continuous gas release.

$$C_{(x,y,z,t)} = \frac{Q}{(2\pi)^{\frac{3}{2}} \cdot \sigma_x \cdot \sigma_y \cdot \sigma_z} e^{\frac{-(x-ut)^2}{2\sigma_x^2}} \cdot e^{\frac{-y^2}{2\sigma_y^2}} \cdot \left[e^{\frac{-(z+H)^2}{2\sigma_z^2}} + e^{\frac{-(z-H)^2}{2\sigma_z^2}} \right] \quad (1)$$

Based on our research, Q is mass of released hazardous gas per second (unit: kg/s); H is the height of leakage source (unit: m); x, y, z is coordinate of measuring point (unit: m); u is the average wind speed (unit: m/s), and t is the time since start of hazardous gas leakage (s); σ_x, σ_y, and σ_z are dispersion coefficients of x, y and z direction, respectively. According to the Klug method[6], when we hypothesized atmospheric stability is C, three dispersion coefficients can be calculated by Eq. 2 and 3.

$$\sigma_y = \sigma_x = 0.23x^{0.855} \tag{2}$$

$$\sigma_z = 0.076x^{0.879} \tag{3}$$

2.2. Pedestrian evacuation in hazardous gas leakage

Different evacuation plans are needed according to different disasters. In the hazardous gas leakage, residents should evacuate along with the direction which is perpendicular with direction of wind.

Considering the lethal concentration of a hazardous gas, evacuees have to avoid the grid has lethal concentration (Concentration of toxic gas is lethal). Combining with the dijkstra method, we obtained the optimal evacuation route for every evacuee. However, some people need not evacuate from their home because the concentration of hazardous gas in home is low when all windows are closed. Therefore, the government should provide emergency a response plan noting whether residents should stay at home based on leakage duration, emergency information acquirement time, wind speed, and some relevant factors.

During evacuation, the grids whose risk are less than 6.25 kg/m³ are set to safe ends. Residents who need to evacuate have to reach to these grids avoiding lethal-risk route as soon as possible.

3. Parameters setting

(1) Hazardous gas

We used carbon monoxide (CO) as an example. According to Chinese standard of CO[7], personal health will be influenced when concentration of CO is more than 5 ppm ($\approx 6.25 \times 10$-6 kg/m3). A lethal concentration of CO is 6400 ppm ($\approx 8 \times 10$-3 kg/m3).

(2) Location of leakage source

In order to influence the study area as much as possible, the leakage source is set at the left of the middle part (relative coordinate x=0 m, y=575 m, z=0 m.).

(3) Speed and direction of wind

According to historical data of wind in Beijing, the average wind speed is 2.3 m/s. In addition, westerly wind is hypothesized because in this way, study area can be influenced with greatest extent.

(4) Information acquirement time

Information acquirement time can be obtained via information dissemination and monitoring time. An efficient monitoring and information dissemination system can assure a short information acquirement time.

(5) Leakage duration

The duration of a gas leakage directly determines its severity. Governments and relevant agencies should monitor the gas leakage quickly and control it in a short time to protect residents around the leakage source from toxic.

(6) Concentration rate of hazardous gas between outdoor and indoor

When residents get information about hazardous gas leakage, they can choose to close the door and window and stay at home to protect themselves hazardous gas exposure. Through experimental analysis, ventilation coefficient of a typical Chinese room, which considers doors and windows, are closed 0.1-0.2 h^{-1}[8]. For the simulation, we used an average value of 0.15 h^{-1}.

(7) Personal inhalation

Personal inhalation influences the intake of hazardous gas. Data shows that the inhalation rate of a Chinese adult is 5.5 L/min at and 32.9 L/min during heavy physical activities[9]. We hypothesized that evacuation is a heavy physical activity and staying at home is considered the state at rest.

Table 1. Parameters setting for standard value

Parameters	Value
Type of hazardous gas	CO (carbon monoxide)
Leak rate	10 kg/s
Location of leakage source	relative coordinate : (0 m, 575 m, 0 m)
Average wind speed	2.3 m/s (normal distribution)
Wind direction	Westerly
Information acquirement time	8 minutes
Duration time	40 minutes
Concentration rate when window is closed	0.15 h^{-1}
Personal inhalation in normal time	5.5 L/min
Personal inhalation in heavy physical activities	32.9 L/min

According to parameters introduction above, standard values of each parameter are set in Table 1.

4. Results

We defined four kinds of risks to analyze personal emergency response plans during an gas leakage in order to guide the government's emergency response plan and assist the public in taking communication decisions during the emergency . Grids risk (R_g) is the concentration of hazardous gas (unit: kg/m^3; lethal risk: 8.0×10^{-3} kg/m^3); Personal risk (R_p) is the total personal inhalation of hazardous gas per second (unit: kg/s; lethal risk: 7.3×10^{-7} kg/s); Cumulative grid risk (R_{gc}) is the total number of weight of hazardous gas in each grid since the start of leakage (unit: kg); Cumulative personal risk (R_{pc}) is the total personal inhalation of hazardous gas (kg).

4.1. *Spatio-temporal dynamic risk distribution*

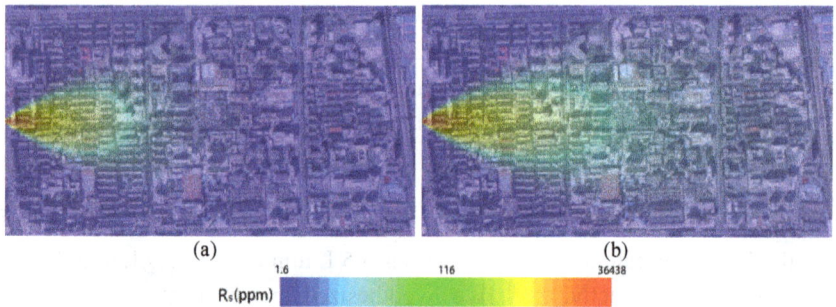

(a)   (b)

1.6   116   36438

R_s(ppm)

Fig. 1. Spatio-temporal dynamic risk distribution under different time:
(a) 300s; (b) 600s;

According to normal weather in Beijing, average wind speed is 2.3 m/s. Fig. 1 shows the spatial-temporal dynamic risk distribution of wind speed at different times. At 300 seconds, 3336 (14.7%) grids had risk ($R_g > 6.25 \times 10^{-6}$ kg/m^3) and 12 grids had lethal risk ($R_g > 8 \times 10^{-3}$ kg/m^3). As time goes on, 6808 (29.9%), 9376 (41.2%), 11242 (49.4%) grids had risk at 600, 900 and 1200 seconds, respectively. However, after 600 seconds, the total number of grids which had lethal risk keeps the same (14 grids with lethal risk), therefore, when a hazardous gas leaks, an area of 140 m^2 near leak source is extremely dangerous.

4.2. *Regional evacuation based on risk of hazardous gas leakage*

Fig. 2 shows the risk of hazardous gas leakage on during regional evacuation. A red star represents a leakage source and green stars are safe ends for evacuees. Blueish areas represents the roads with grid risk and gray roads are safe. A, B, C, and D are four residents. In this case, we hypothesized that all residents got the information 20-minutes post gas leak and government controlled the gas diffusion at 40 minutes. In this condition, we recommend that the government should make different emergency plans for residents A, B, C, and D.

148

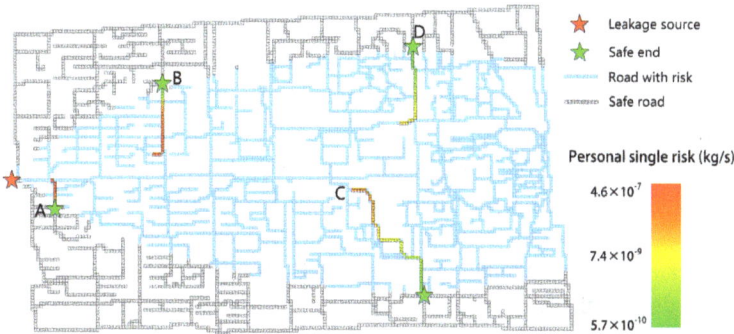

Fig. 2. Risk of hazardous gas leakage on residents evacuation.

In this case, we hypothesized that residents got emergency information at 5 minutes and wind speed is 1 m/s. A slow wind speed for gas diffusion which will lead to higher concentration of hazardous gas. Fig. 3 shows the personal risk in regional evacuation. Resident A cannot evacuate because the peak value of personal risk on route is more than lethal risk ($7.33 \times 10\text{-7}$ kg/s). Because of very short distance, indoor and outdoor concentration of hazardous gas is the same when resident received the information. Therefore, resident A is in danger because of its disadvantageous location. Because the personal cumulative risk between evacuation plan and optimal-stay plan is almost the same, and risk is not very high on evacuee route, resident B can decide either to evacuate or to stay at home.

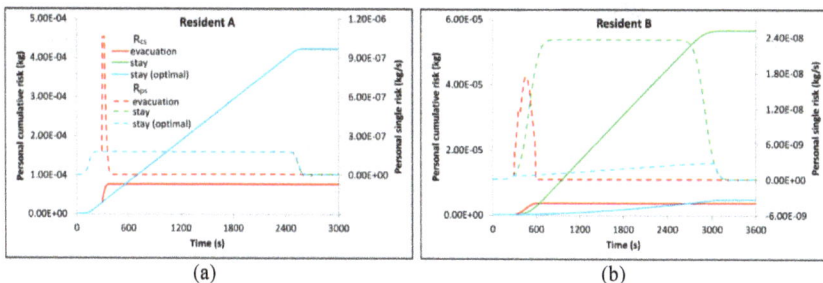

Fig. 3. Personal risk in regional evacuation under different locations (stay optimal): residents who stay at home will open the window when outdoor concentration of hazardous gas is higher than indoor concentration, otherwise, close the window): (a) Resident A; (b) Resident B.

5. Conclusion

The personal risk analysis model considering optimal information dissemination and regional evacuation was developed in this study, combining 8 influencing factors to analyze personal risk in hazardous gas leakage.

Optimal information dissemination can reduce information spreading and acquirement time. Through information dissemination analysis, we found that residents who stay near to the leakage source had better stay at home because of high concentration of hazardous leakage on their evacuation route. Instead of evacuation, staying at home and adopting optimal stay plan is very efficient if residents can receive the information before the hazardous gas totally dispersed. For people who lived far from leakage source, evacuation is usually a good choice because they have longer time to avoid high-concentration hazardous gas.

According to the obtained results, government can make efficient response plan in hazardous gas leakage in metropolises.

Acknowledgment

This work was funded by the National Natural Science Foundation of China (Grant No. 71473146, 71173128) and the Ministry of Science and Technology of the People's Republic of China under Grant No. 2015BAK12B01.

References

1. Y. Li, H. Ping, Z.H. Ma and L.G. Pan, Statistical analysis of sudden chemical leakage accidents reported in China between 2006 and 2011, *Environ. Sci. Pollut. Res.* **21**, 5547 (2014).
2. J.L. Adgate, B.D. Goldstein and L.M. McKenzie, Potential public health hazards, exposure and health effects from unconventional natural gas development, *Environ. Sci. Technol.* **48**, 8307 (2014).
3. D. Dong, H. Wang and P. Jia, Mine gas concentration pre-warning based monitoring data relational analysis, *Adv. Mater. Res.* **634-638**, 3655 (2013).
4. G. Salihoglu, Industrial hazardous waste management in Turkey: Current state of the field and primary challenges, *J. Hazard. Mater.* **177**, 42 (2010).
5. N. Zhang, H. Huang, B. Su and H. Zhang, Population evacuation analysis: considering dynamic population vulnerability distribution and disaster information dissemination, *Nat. Hazards,* **69**, 1629 (2013).
6. W. Klug, A method for determining diffusion conditions from synoptic observations, Staub-Reinhalt. Luft, 29 (1969) 14-20.
7. Specification for design of combustible gas and toxic gas detection and alarm for petrochemical industry, GB 50493-2009, Zhong Guo Ji Hua Chu Ban She, Chinese Edition 2009, Beijing, China.
8. Y. Li and X. Li, Field testing of natural ventilation of residential buildings in Beijing in summer, *Heat. Vent. Air Cond.* **43**, 46 (2013).
9. P. Liu, B.B. Wang, X.G. Zhao, X.L. Duan, Y.T. Chen and L.M. Wang, Research on inhalation rate of Chinese adults, *J. Environ. Health,* **31**, 953 (2014).

Simulation Model for Fuel Spreading on Complex Topography Based on Hydrodynamics

Bo-ni Su, Hong Huang† and Nan Zhang
Institute of Public Safety Research, Department of Engineering Physics,
Tsinghua University,
Beijing, 100084, China
†E-mail: hhong@tsinghua.edu.cn

Accidents of fuel leakage occur frequently during transport and storage. Numerical simulation of fuel spreading is very helpful for risk analysis and accidents prevention. However, most of current studies on fuel spreading paid attention to water surface or flat ground. Therefore, it is almost impossible to simulate fuel spreading accurately under real accident scenarios due to the lack of simulation model on complex topography. In this paper, a simulation model for fuel spreading on complex topography was established based on hydrodynamics. Many details were improved to make the model suitable for simulation of fuel spreading. The effectiveness of model is verified using experimental data. The model is applied to an actual oil depot for oil leakage simulation, then the spatial-temporal distributions of oil depth were obtained. The simulation model plays a critical role in risk analysis of overland fuel leakage.

Keywords: Fuel Spreading; Complex Topography; Shallow-Water Equations; Numerical Simulation.

1. Introduction

Accidents of fuel leakage, which waste oil resources, pollute environment, and may cause serious fire, occur frequently during transport and storage. Numerical simulation of fuel spread is helpful for risk analysis and accident prevention.

Currently, most of studies on fuel spreading paid attention to spreading on water surface. In contrast, studies on fuel spreading on ground are still preliminary. Bradley[1] studied the steady state after fuel spread based on the equilibrium of surface tension and pressure, and found that contact angle plays an important role in fuel film thickness. Grimaz et al.[2] studied fuel film spreading on ideal (smooth and impervious) solid surface and obtained the relationship between spreading area and spreading time. Simmons[3] studied fuel spreading on incline flat surfaces using experiments. The results showed that even 1 degree of incline slope will influence spreading significantly, indicating that topography is an important factor in fuel spreading. Currently, it is almost impossible to simulate fuel spread accurately under real accident scenarios due to the lack of simulation model on complex topography.

In this paper, a simulation model for fuel spreading on complex topography is established based on hydrodynamics. The model is verified using experimental data. The model is applied to an actual oil depot for oil leakage simulation.

2. Methods

Shallow-Water Equations (SWEs) are widely used for flood simulation in hydraulics. Diffusive wave approximation can be applied when variation of flow is not violent. Most of fuel spreading cases satisfy this condition. There are many diffusive wave approximation models for flood simulation, such as LISFLOOD-FP[4]. Theoretically, these models have the potential to be applied to fuel spreading simulation. However, as these models are designed for flood simulation, many details must be improved before simulation for fuel spreading.

Firstly, the calculation formula of flow velocity should be modified. Most of hydraulic models use Manning Formula to calculate flow velocity (Eq. 1):

$$u = \frac{k \cdot d^{2/3} \cdot S^{1/2}}{n}. \tag{1}$$

Where u is flow velocity (m/s); k is unit conversion factor (= 1 $m^{1/3}s^{-1}$); d is water depth (m); S is water surface slope (dimensionless); n is the Manning coefficient (dimensionless).

However, this formula is not suitable for fuel spreading. Fuels are often with high viscosity (e.g. crude oil). Depth and velocity of fuel spreading are also very small. These lead to low Reynolds numbers, and spreading flows are likely to be laminar flows. On the other hand, Manning Formula is an empirical formula for fully turbulent flows, and is not applicable for laminar flows. In order to solve this problem, Manning Formula can be replaced with Darcy-Weisbach Equation (Eq. 2):

$$u = \sqrt{\frac{8 \cdot g \cdot d \cdot S}{f}} \tag{2}$$

Where g is gravitational acceleration (= 9.8 m/s^2); f is the Darcy coefficient (dimensionless).

The calculation formulas of the Darcy coefficient under different Reynolds number conditions can be obtained according to Yen's study[5]. In this paper, calculation method is simplified under the hypothesis that transitional Reynolds number (Re) is a constant of 500.

The Reynolds number here is defined as (Eq. 3):

$$\mathrm{Re} = \frac{u \cdot d}{v} \tag{3}$$

Where v is fluid's kinematic viscosity (m^2/s).

When the flow is a laminar flow (Re<500), Darcy f satisfies (Eq. 4):

$$f = 24 / \mathrm{Re} \tag{4}$$

When the flow is a turbulence flow (Re>500), Darcy f satisfies (Eq. 5):

$$\frac{1}{\sqrt{f}} = -K_1 \cdot \lg\left(\frac{k_s}{K_2 \cdot d} + \frac{K_3}{4 \cdot \mathrm{Re} \cdot \sqrt{f}} \right) \tag{5}$$

Where k_s is surface roughness (m); K_1, K_2, K_3 are constants (dimensionless). In this study, these constants are selected according to Graf's research[6], $K_1 = 2.0$, $K_2 = 12.9$, and $K_3 = 2.77$.

When the flow is a laminar flow, substituting Eq. (3) and Eq. (4) into Eq. (2):

$$u = \sqrt{\frac{8 \cdot g \cdot d \cdot S \cdot \mathrm{Re}}{24}} = \sqrt{\frac{g \cdot d^2 \cdot S \cdot u}{3}} \tag{6}$$

Thus u can be solved:

$$u = \frac{g \cdot d^2 \cdot S}{3 \cdot v} \tag{7}$$

When the flow is a turbulence flow, according to Eq. (2), Eq. (3), and Eq. (5):

$$\frac{u}{\sqrt{8 \cdot g \cdot d \cdot S}} = -K_1 \cdot \lg\left(\frac{k_s}{K_2 \cdot d} + \frac{K_3 \cdot v}{4 \cdot u \cdot d} \frac{u}{\sqrt{8 \cdot g \cdot d \cdot S}} \right) \tag{8}$$

Thus u can be solved:

$$u = -K_1 \cdot \sqrt{8 \cdot g \cdot d \cdot S} \cdot \lg\left(\frac{k_s}{K_2 \cdot d} + \frac{K_3 \cdot v}{4 \cdot d \cdot \sqrt{8 \cdot g \cdot d \cdot S}} \right) \tag{9}$$

In this study, when calculating flow velocity, laminar flow velocity u_L and turbulence flow velocity u_T are calculated according to Eq. (7) and Eq. (9) respectively. If $\dfrac{u_L \cdot d}{v} < 500$, the flow is indeed a laminar flow, flow velocity

$u = u_L$. If $\dfrac{u_T \cdot d}{v} > 500$, the flow is indeed a turbulence flow, flow velocity

$u = u_T$. Otherwise the flow is a transition flow, according to the hypothesis that

transitional Reynolds number is a constant of 500, $u = \dfrac{500 \cdot v}{d}$.

Except for flow velocity calculation method, another modification that is needed is the special treatment of spreading boundary. In fuel spreading simulation, it is important to obtain the spreading boundary, which is not a necessity in flood simulation. In this study, Hussein's treatment method[7] is adopted. A limitation is added, that fuel can only flow out of a grid cell if its depth is greater than a threshold h_{min}.

$$h_{min} = h_d + \sqrt{\dfrac{2\sigma(1-\cos\varphi)}{\rho g}} \qquad (10)$$

Where h_d is ponding depth (m); σ is surface tension of fuel (N/m); φ is contact angle between fuel and solid surface; ρ is fuel density (kg/m³). For simplicity, it is suggested that $h_d \approx k_s$ in this paper.

In this study, a simulation model for fuel spreading on complex topography is established based on LISFLOOD-FP model[4] with modifications mentioned above.

3. Results and Discussions

3.1. Model Verification

Lister[8] studied the spreading of viscous fluids on an inclined plane through experiments. A group of experimental data (Experiment 1) is used for simulation model verification. In this case, silicon oil (with kinematic viscosity $v = 1.13 \times 10^{-3}$ m²/s) leaks at constant flowrate (q= 1.48×10^{-6} m³/s) on an inclined glass plane with slope of 2.5 degrees. Simulation is carried out according to experimental parameters. Study area has the size of 1.2 m × 0.6 m, and size of grid cells is selected as 5 mm × 5 mm. The simulation results and measured data at different time are compared in Figure 1.

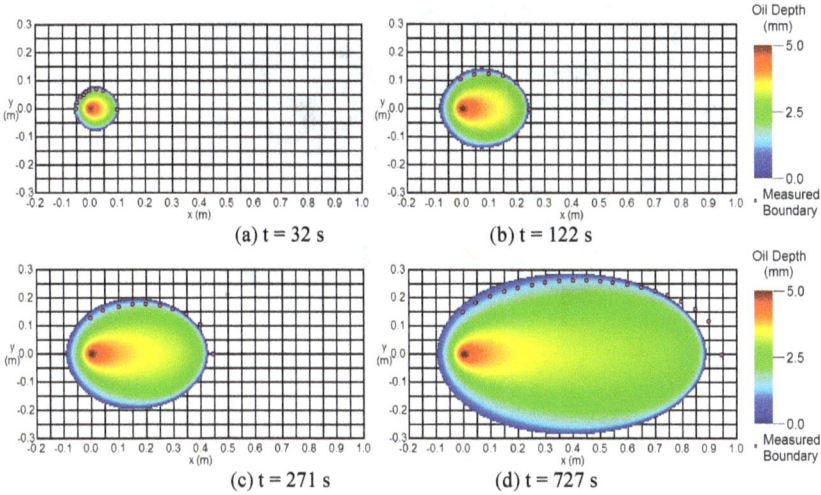

(a) t = 32 s (b) t = 122 s

(c) t = 271 s (d) t = 727 s

Fig. 1. Comparison of Simulation Results and Measured Data.

It can be seen that simulation results and experimental measured data are in good agreement. However, as time goes by, spreading along down slope direction is slightly faster than simulation results, and spreading perpendicular to slope direction is slightly slower than simulation results. These may be caused by experimental factors such as errors of plane slope.

The simulation model is verified through these simulation and comparison.

3.2. Application in actual oil depot

An oil depot is selected as the study area. Figure 2(a) shows the region of study area. The Geographic Information System (GIS) data of study area are provided by the oil depot, which are shown in Figure 2(b). The study area is stretching about 540 meters from north to south and about 600 meters from east to west. There are 23 oil tanks in this oil depot. All tanks are filled with crude oil. Currently, the oil depot uses pools to prevent disastrous oil leakage (as shown in Figure 3a). An alternative plan is using walls to prevent leakage (as shown in Figure 3b). Building oil tanks directly on flat ground is forbidden (as shown in Figure 3c). In this study, all these three plans are tested. An artificial accident scenario is simulated. In this scenario, oil tanks #9, #10, #11, and #12 leak at the same time due to external damage (if only one tank leaks, pools or walls can easily stop the spreading). Parameters are selected according to handbooks and experiences, $k_s =$ 0.0015 m, $v = 1.8 \times 10^{-4}$ m^2/s, $\varphi = 30°$, $\sigma = 3 \times 10^{-2}$ N/m, $\rho = 950$ kg/m^3. Size of simulation grid cell is chosen as 2.5 m $\times$ 2.5 m. The simulation results under different plans at different time are shown in Figure 4.

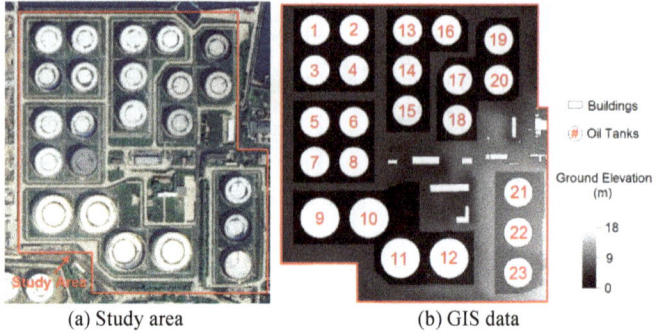

(a) Study area   (b) GIS data

Fig. 2. Study Area.

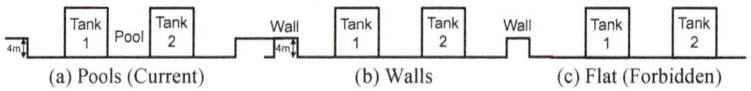

(a) Pools (Current)   (b) Walls   (c) Flat (Forbidden)

Fig. 3. Schematic Diagram of Three Plans.

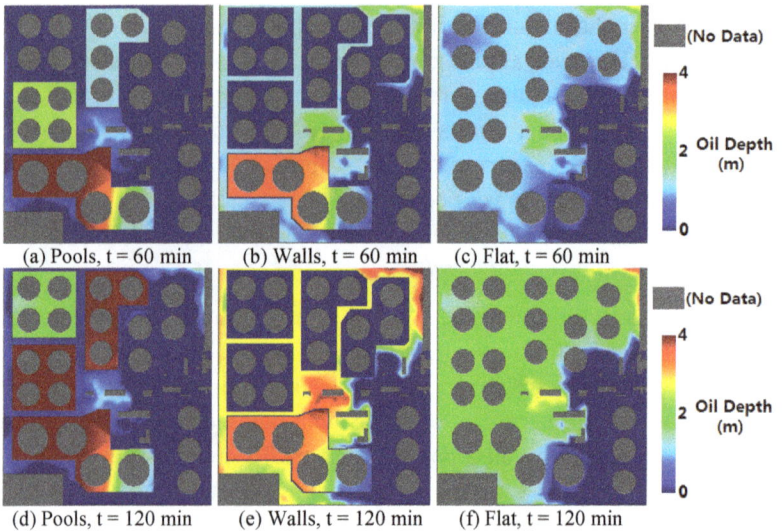

(a) Pools, t = 60 min   (b) Walls, t = 60 min   (c) Flat, t = 60 min

(d) Pools, t = 120 min   (e) Walls, t = 120 min   (f) Flat, t = 120 min

Fig. 4. Simulation Results in Actual Oil Depot.

In "Pools" case, oil gradually filled the pool, then overflowed from the pool and started to spread to low-lying areas. However, situation in "Pools" case is much better than that in "Flat" case (the spreading area in "Pools" case is much smaller than that in "Flat" case). Indicating that, pools can mitigate disastrous oil leakage. The spreading area in "Walls" case is small, and other tanks are not affected by fuel spreading, meaning that using walls rather than pools may be more effective to mitigate oil leakage. However, walls are not as reliable as pools, and are vulnerable in some disasters such as earthquakes. Oil depot should consider such conditions to make a proper plan.

4. Conclusion

In this study, a simulation model for fuel spreading on complex topography is established based on hydrodynamics. Many details were improved to make the model suitable for fuel spread simulation. The model is verified using existing experimental data. The model is applied to an actual oil depot for oil leakage simulation, and the spatial-temporal distributions of oil depth were obtained. The simulation model is helpful for risk analysis of overland fuel leakage.

Acknowledgment

This work was funded by the Ministry of Science and Technology of the People's Republic of China (Grant No. 2015BAK12B01), the National Natural Science Foundation of China (Grant No. 71473146), State Key Laboratory of Safety and Control for Chemicals, and China Clean Development Mechanism Foundation (Grant No. 2013049).

References

1. D. Bradley. *Model for Pool Fires following Chemical Agent Spills* (Science Applications International Corporation, 2002).
2. S. Grimaz, S. Allen, J. Stewart, et al. Predictive evaluation of surface spreading extent for the case of accidental spillage of oil on the ground. *Chemical Engineering*, **11**, 389(2007).
3. C. S. Simmons, J. M. Keller, and J. L. Hylden. *Spills on Flat Inclined Pavements* (United States. Department of Energy, 2004).
4. P. D. Bates, and A. P. J. De Roo. 2000. A simple raster-based model for flood inundation simulation. *Journal of Hydrology* **236**, 54(2000).
5. B. C. Yen. Open Channel Flow Resistance. *Journal of Hydraulic Engineering*, **128**, 20(2002).
6. W. H. Graf. *Hydraulics of Sediment Transport* (McGraw-Hill, New York, 1971).
7. M. Hussein, M. Jin, and J. W. Weaver. Development and verification of a screening model for surface spreading of petroleum. *Journal of Contaminant Hydrology*, **57**, 281(2002).
8. J. R. Lister. Viscous flows down an inclined plane from point and line sources. *Journal of Fluid Mechanics*, **242**, 631(1992).

Chapter 2

Materials and Mechanical Engineering

Multi-Disciplinary Robust Optimal Design of Mechanical Structure Based on Interval Model

Jing Chen, Yong-cheng An, Wen-bo Wang and Hao Yan
*College of Mechanical and Control Engineering, Guilin University
of Technology, Guilin, Guangxi, China
E-mail: jingc812@163.com*

Considering the design variables and the uncertainty of design parameters, a robust optimization design method of two stage mechanical structure model based on multi-disciplinary interval is proposed. The first stage simplifies multi-disciplinary uncertainty optimization three nested loop optimization architectures, to determine two nested loop optimization architectures. The second stage is to use the nearest value method to continuous variables discretization and obtain the optimized dimensions of the discrete structure with tolerance. The engineering example shows that the proposed method has strong feasibility and validity with practical value.

Keywords: Interval model; Mechanical Structure; Tolerance; Robust Optimization.

1. Introduction

In the design of complex mechanical structure, it is inevitable to involve the coupling problem of multi subjects. Multidisciplinary design optimization (MDO) is one of the methods to solve these problems. Its purpose is to get the overall optimal solution through making full use of the interaction between the various disciplines, taking into account the interplay of various factors, the effective application of the optimal design strategy [1]. At the same time, robust optimal design is commonly used in engineering problems [2].

At present in the multi-disciplinary robust design optimization of mechanical structure, there is a complex and huge computation architecture optimization difficult. At the same time, because there is no consideration of tolerance, the engineering practicability of the optimization results is poor. Considering the design variables and the uncertainty of design parameters, a robust optimization design method of two stage mechanical structure model based on multi-disciplinary interval is proposed. In the first stage, the continuous variables are optimized. In the second stage discrete variables are optimized.

2. Introduction The Mechanical Structure Mathematical Model of Multi-disciplinary Robust Optimization Design Based on Interval

2.1. *The optimization model*

Considering structure size tolerance, using the upper deviation and lower deviation expression structure size tolerance, expressed in the interval with the structure of the tolerance size, based on interval model containing multi-disciplinary optimization model of uncertain parameters are as follows:

$$\min f_i\left(Z, x_i, y_i, p_i\right)$$

$$s.t.\ g_i\left(Z, x_i, y_i, p_i\right) \le V_i^I; H_i(Z, y_i, y_{ji}\left(Z, x_j, y_j, p_j\right)) = 0; V_i^I = \left[V_i^L, V_i^R\right]; p_i \in \Gamma = \left[p_i^L, p_i^R\right]; \quad (1)$$

$$\left[x_i + \Delta x_i^L, x_i + \Delta x_i^U\right] \subseteq \left[x_{\min}, x_{\max}\right]; \left[Z + \Delta Z^L, Z + \Delta Z^U\right] \subseteq \left[Z_{\min}, Z_{\max}\right], (i, j = 1, 2, \cdots, k; i \ne j)$$

2.2. *Deterministic transformation of objective function*

Because the optimization model is a minimal optimization problem, interval order relation '$\le_{cw}$' is used to determine the interval[3].

For the two interval A^I and B^I, the interval order relation '$\le_{cw}$' is as follows:

$$\begin{cases} A^I \le_{cw} B^I & if\ and\ only\ if\ A^c \ge B^c\ and\ A^w \ge B^w. \\ A^I <_{cw} B^I & if\ and\ only\ if\ A^I \le_{cw} B^I\ and\ A^I \ne B^I. \end{cases} \quad (2)$$

In the formula, A^c, A^w are the mean and the radius of the interval A^I; B^c, B^w are the mean and the radius of the interval B^I.

The order relation of interval number expresses the decision-making preference to the midpoint of the interval and the radius, by (2) known, only when the mean and the radius of interval A^I is less than the mean and the radius of the interval B^I, interval A^I is better than interval B^I.

Formula (1) in the objective function is transformed into:

$$\min_x \left[m\left(f_i\left(Z, x_i, y_i, p_i\right)\right), w\left(f_i\left(Z, x_i, y_i, p_i\right)\right)\right]. \quad (3)$$

In the formula, $m\left(f_i\left(Z, x_i, y_i, p_i\right)\right)$ is the mean value of the objective function, and the $w\left(f_i\left(Z, x_i, y_i, p_i\right)\right)$ is the radius of the objective function:

$$m\left(f_i\left(Z, x_i, y_i, p_i\right)\right) = \frac{1}{2}\left(f_i^L\left(Z, x_i, y_i\right) + f_i^R\left(Z, x_i, y_i\right)\right); w\left(f_i\left(Z, x_i, y_i, p_i\right)\right) = \frac{1}{2}\left(f_i^R\left(Z, x_i, y_i\right) - f_i^L\left(Z, x_i, y_i\right)\right).$$

$$(4)$$

In the formula, A and B are the upper and lower bounds of the objective function:

$$f_i^L(Z,x_i,y_i) = \min_{p \in \Gamma} f_i(Z,x_i,y_i,p_i); f_i^R(Z,x_i,y_i) = \max_{p \in \Gamma} f_i(Z,x_i,y_i,p_i). \tag{5}$$

The double objective function of the formula (3) considers both the mean and the radius of each objective function. The mean value of the objective function $m(f_i(Z,x_i,y_i,p_i))$ and probability theory in the "expected value" similar, the radius of the objective function $w(f_i(Z,x_i,y_i,p_i))$ values similar to the variance in probability theory: the former is the assurance objective function has better average performance, and the latter can reduce sensitivity to uncertainties and ensure the robust[4]. By using the weighted method, the Bi objective optimization problem of formula (3) is transformed into a single objective optimization problem:

$$\min_x f_{qi} = (1-\beta) m(f_i(Z,x_i,y_i,p_i)) + \beta w(f_i(Z,x_i,y_i,p_i)). \tag{6}$$

In the formula, β is the weight coefficient, $0 \le \beta \le 1$, the actual project, the designer can design the β value according to the concrete situation.

2.3. Deterministic transformation of inequality constraints

Possibility degree of interval is a description of whether an interval is greater than the specific degree of another interval. On the axis, interval C^I and interval D^I has a total of six kinds of position relations, literature [5] considers all possible position relations, and puts forward the improved method for possibility degree of interval.

Using Possibility degree of interval method, it can contain uncertain parameters of the inequality constraints $g_i(Z,x_i,y_i,p_i) \le V_i^I$ conversion into a deterministic optimization problem:

$$P(G_i^I \le V_i^I) \ge \lambda_i, i = 1,2,\cdots,k$$
$$G_i^I = \left[g_i^L(Z,x_i,y_i), g_i^R(Z,x_i,y_i) \right] \tag{7}$$

In the formula, λ_i is given in advance the possibility level; G_i^I is uncertain constraint in x_i by the uncertainty of the interval parameters, $g_i^L(Z,x_i,y_i)$ and $g_i^R(Z,x_i,y_i)$ are lower and upper bounds of the constraint function:

$$g_i^L(Z,x_i,y_i) = \min_{p \in \Gamma} g_i(Z,x_i,y_i,p_i); g_i^R(Z,x_i,y_i) = \max_{p \in \Gamma} g_i(Z,x_i,y_i,p_i). \tag{8}$$

2.4. *Solution strategy of equality constraints*

Multi-disciplinary three layer loop nest optimization architecture of the inner loop design variables and the uncertain variables are set as a constant, for solving the state variables; intermediate layer circulation design variables set is constant, the uncertain variables for optimization; the outer loop only to the design variables of optimization[5]. In the mechanical structure design, the design parameters are uncertain factors exist, is not controllable. Therefore the design parameters as the mechanical structure of multi discipline robust optimization of parameter uncertainty, only need to consider the upper and lower bounds, without the optimization. At the same time, interval extension problem exist in the interval arithmetic. So, before solving the equality constraints, the designer should choose the value of uncertain parameters according to the experience of engineering design.

In MDO, there are state variables including subject state variables, system state variables and coupling variables. Multidisciplinary Feasible Method (MDF) is a commonly used single stage optimization method, optimization results of high accuracy. So using MDF to solve the problem of iteration. Gauss - Seidel iterative method (G-S) can save time and computing complexity. Therefore, the G-S method is used to solve the state variables in the MDF method. MDF calculation principle diagram is shown in Figure 1, which is composed of two subsystems.

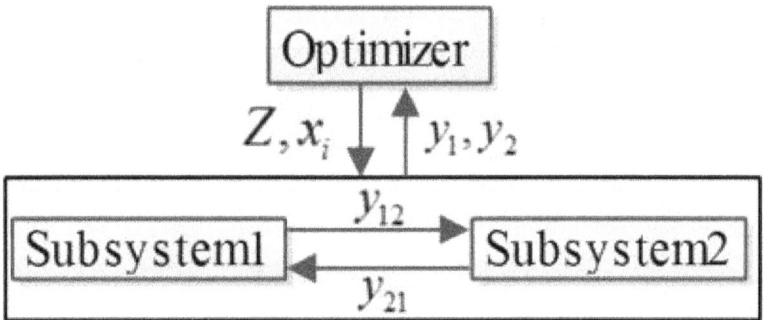

Fig. 1 MDF calculation principle diagram.

2.5. *Unconstrained transformation for constrained optimization problems*

Through the previous processing, get deterministic constrained optimization model:

$$\min_{Z,x_i} f_{qi} = (1-\beta)m\big(f_i(Z,x_i,y_i,p_i)\big) + \beta w\big(f_i(Z,x_i,y_i,p_i)\big); s.t.\ P\big(G_i^I \le V_i^I\big) \ge \lambda_i. \tag{9}$$

Using the penalty function method, the type (10) is transformed into unconstrained objective function:

$$\min_{Z,x_i} f_{ti} = (1-\beta) m\left(f_i\left(Z, x_i, y_i, p_i\right)\right) + \beta w\left(f_i\left(Z, x_i, y_i, p_i\right)\right) + \sigma \sum_{i=1}^{n} \varphi\left(P\left(G_i^I \leq V_i^I\right), \lambda_i\right).$$
$$(10)$$

In the formula, σ is a penalty factor, φ is a penalty function:

$$\varphi\left(P\left(G_i^I \leq V_i^I\right), \lambda_i\right) = \left(\max\left(0, -\left(P\left(G_i^I \leq V_i^I\right) - \lambda_i\right)\right)\right)^2. \qquad (11)$$

3. The Process of Multi-discipline Robust Optimization Design of Mechanical Structure Based on Interval Model

3.1. *Robust optimization design frameworks of multidisciplinary mechanical structure based on interval model*

The mechanical structure multidisciplinary robust optimization design framework is divided into two stages. In the first stage, the calculation process of continuous design variables includes two layers of cycles: In the inner loop, using MDF to solve the state variables and state variables to meet the convergence condition for loop termination criterion; In the outer fixed state variables, using the possibility of interval, the order relation of interval, weighting method, penalty function method will be constrained uncertain optimization problem is transformed into a deterministic constraint optimization problem, optimization design variables. The second stage, considering the effect of tolerance on the size of the structure, using the nearest value method, discrete variable optimization, obtain the discrete solution with tolerance.

3.2. Implementation process of multi-discipline robust optimization design of mechanical structure based on interval model

Continuous variable design flow, as shown in Figure 2.

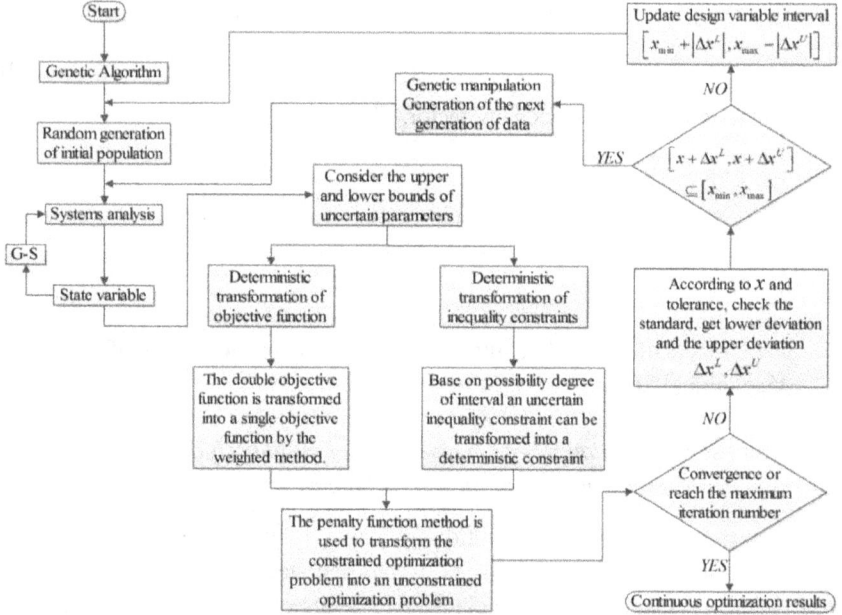

Start

Genetic Algorithm

Random generation of initial population

Systems analysis

G-S

State variable

Genetic manipulation Generation of the next generation of data

Consider the upper and lower bounds of uncertain parameters

Deterministic transformation of objective function

Deterministic transformation of inequality constraints

The double objective function is transformed into a single objective function by the weighted method.

Base on possibility degree of interval an uncertain inequality constraint can be transformed into a deterministic constraint

The penalty function method is used to transform the constrained optimization problem into an unconstrained optimization problem

Update design variable interval

$\left[x_{\min} + |\Delta x^{L}|, x_{\max} - |\Delta x^{U}|\right]$

NO

$\left[x + \Delta x^{L}, x + \Delta x^{U}\right]$

YES

$\subseteq \left[x_{\min}, x_{\max}\right]$

According to x and tolerance, check the standard, get lower deviation and the upper deviation $\Delta x^{L}, \Delta x^{U}$

NO

Convergence or reach the maximum iteration number

YES

Continuous optimization results

Fig. 2 The design process of continuous variable.

Discrete variable design process, as shown in Figure 3.

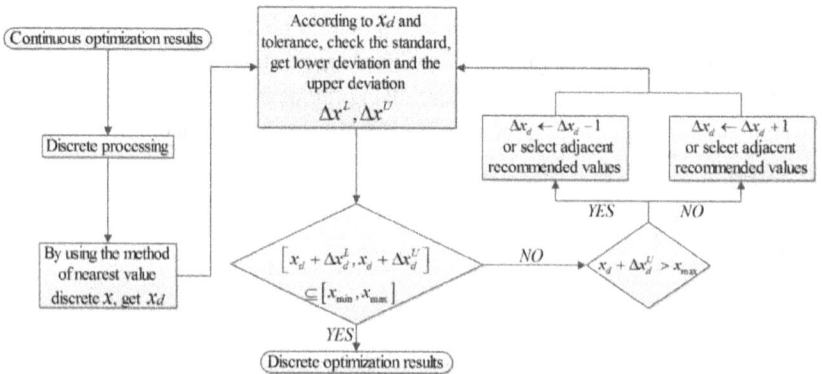

Continuous optimization results

Discrete processing

By using the method of nearest value discrete x, get x_d

According to x_d and tolerance, check the standard, get lower deviation and the upper deviation $\Delta x^{L}, \Delta x^{U}$

$\Delta x_d \leftarrow \Delta x_d - 1$ or select adjacent recommended values

$\Delta x_d \leftarrow \Delta x_d + 1$ or select adjacent recommended values

YES   NO

$\left[x_d + \Delta x_d^{L}, x_d + \Delta x_d^{U}\right]$

$\subseteq \left[x_{\min}, x_{\max}\right]$

NO

$x_d + \Delta x_d^{U} > x_{\max}$

YES

Discrete optimization results

Fig. 3 The design process of discrete variable.

4. Engineering Example

This article uses the literature [6] engineering example, the compound oil cylinder structure diagram as shown in Figure 4. Inside and outside cylinder structure is illustrated in Figure 5.

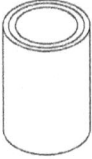

Fig. 4 The compound cylinder structure.   Fig. 5 Inside and outside cylinder structure.

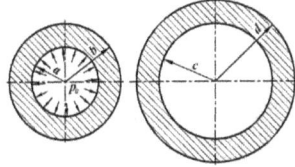

Because of the need to meet certain requirements of the installation, nominal size diameter is d=381mm.The mathematical model is detailed in literature [6].Considering the parameter uncertainty, set the range of uncertain parameters is the experience value of 10%.

Using two kinds of methods to solve the optimization problem in MATLAB[7], respectively method in literature [6] (M1)and the proposed method(M2), continuous variable optimization results are shown in Table 1 below.

Table 1 Continuous variable optimization results.

method	Optimized target value	Value of subject design variables			
	f/mm^2	a/mm	b/mm	c/mm	d/mm
M1	−111856. 5787	188. 6958	262. 0467	262. 0467	381
M2	−125431. 8779	199. 8661	263. 9544	263. 9544	381

Using A, B, C, D to express diameter get the optimization results of discrete variables with tolerance, as shown in Table 2.

Table 2 Discrete variable optimization results.

A/mm	B/mm	C/mm	D/mm
$400^{+0.063}_{0}$	$528^{0}_{-0.067}$	$528^{+0.067}_{0}$	$762^{+0.113}_{+0.031}$

From Table 1, we can see that the method proposed in this paper makes the capacity of the composite cylinder increased by 12%. Table 2 shows, Discrete variable optimization results t is integer with tolerance.

5. Conclusion

A robust optimization design method of two stage mechanical structure model based on multi-disciplinary interval is proposed. Solving the problem is that the continuous optimization results cannot be used in the engineering structure of design directly. The engineering example shows that the method is feasible and effective. It has certain practical value.

Acknowledgment

This work was funded by National Natural Science Foundation of China (51365010).

References

1. Zhong Yi-fang, Chen Bai-hong, Wang Zhou-hong. The multidisciplinary integrated optimization design principle and method [M]. Wuhan: Huazhong University of Science and Technology Press, 2006:1-2.
2. Taguchi G. Quality engineering through design optimization [M] . New York: Krauss International Press,1986:12-15.
3. Jiang Chao. Theories and Algorithms of Uncertain Optimization Based on Interval [D]. Changsha: Hunan University, 2008:15-33.
4. Li Fang-yi. The interval non-probabilisic multi-objective optimization method and its applicaiions of vehicle bady [D]. Changsha: Hunan University,2010:62-79.
5. Jiang C, Han X, Liu G-P. A nonlinear interval number programming method for uncertain optimization problems [J]. European Journal of Operational Research , 2008,188(1):1-13.
6. X-Du, J-Guo, H-Beeram. Sequential optimization and reliability assessment for multidisciplinary systems design [J]. Journal of Structural and Multidisciplinary Optimization, 2008,35(2):117-130.
7. Guo Ren-sheng. Optimization analysis and calculate using matlab [J]. Machinery Design & Manufacture,2004, (02):60-62.

Combustion Simulation of Opposite Axial Piston Engine in Small Scale

Lei Zhang[1,2,†], Hai-jun Xu[1], Cun-yun Pan[1], Fa-liang Zhou[1] and Zhong-bao Qin[2,*]

[1]College of Mechatronics Engineering and Automation, National University
of Defense Technology, Changsha, Hunan, 410073, China

[2]Xi'an Research Institution of High-tech,
Xi'an, Shanxi, 710025, China
Email: * zhongb_qin@163.com;
† cvx1987@163.com

A new kind of Opposite Axial Piston Engine (OAPE) in small scale specially designed for the portable generating system was presented. The working theory of OAPE was studied, based on which the combustion process was simulated in FLUENT. Then the internal flow field and the variation of temperature and pressure in cylinder were recorded. Results show the flame front is a sphere centered at the spark point. The maximal pressure in power cylinder is 2.13Mpa and maximal pressure in charge cylinder is 0.235Mpa. The work done by power cylinder per cycle is 1.24J, while the charge cylinder consumes 0.22J per cycle.

Keywords: Opposite axial piston engine; Combustion; CFD; Numerical simulation.

1. Introduction

With the wide application of portable digital devices and movable robots, the novel electric power sources with high energy density, high power density and fast dynamic response are urgently needed. Up to now, three kinds of novel power sources are invented, novel battery, fuel cells and micro heat power generators [1,2].

Scientists from MIT proposed a micro tubing which could generate power of $10\sim20w$ with a diameter of 10mm and the thickness of $3mm^2$. Researches from Berkley invented a micro rotary engine which is said to owing a combustion chamber of 1 mm^3 [3]. The Sandia state lab combined the free piston engine and linear electric generator together to make a new type of generator, which can be expected to improve the energy transmission efficiency and power density effectively [4,5]. Professor Guo from Inner Mongolia Polytechnic University made a novel micro swing engine, which could change its sizes easily for its 2D structure [1]. However all the novel power generators proposed above are not big enough to fit the power need of robots used in outdoor exploration [5].

In order to invent a portable high-power-density engine with the power range from 50w to 500w, a novel opposite axial piston engine(OAPE)which applies a novel power conservation mechanism is proposed. Benefiting from the novel structure, OAPE has a much bigger power density and a longer endurance. The working process is studied in detail. A CFD model is established to analyze the combustion process in cylinder, based on which the flow field is observed. Results show that the maximal pressure in power cylinder appearing at 22deg after the top dead center(TDC) of piston could reach 2.2 MPa. The maximal pressure in charge cylinder is 0.235Mpa. During one cycle power cylinder generates 1.24J energy, while charge cylinder consumes 0.22J.

2. Working Principle of OAPE

The schematic of OAPE is shown in Fig. 1. OAPE applies a cam-roller mechanism as the main power conservation mechanism, of which two groups of cylinders are oppositely laid on both sides. The rollers pinned to the linkers are in contact with the lateral surface of cam which has a sinusoid profile. Pistons are connected to two bottom of the linker. The main function of power conservation mechanism is to transfer the back and forth movement of pistons to the rotary movement of cam, meanwhile the internal energy of burned gas is converted to the mechanical energy of the outer rotor which is further transferred to the electric energy.

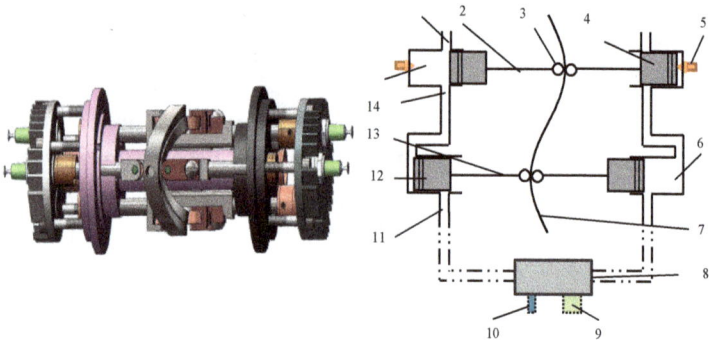

Fig. 1 Outline and schematic of OPAE.

3. Establishment of 3D Combustion Model

The combustion process of internal combustion engine is a complex process which is still under research now. The nonlinear phenomenon of engine combustion is obvious. Any interruption could have a big influence on engine.

3.1. *Fluid model of engine*

Fresh gas is sucked into cylinder through gas pipe. The intaking process needs to follow several conservation laws [6]. The mass conservation law and the momentum conservation law

170

$$\frac{\partial \rho}{\partial t} + \nabla \cdot (\rho \vec{v}) = S_m \tag{1}$$

$$\frac{\partial}{\partial t}(\rho \vec{v}) + \nabla \cdot (\rho \vec{v} \vec{v}) = -\nabla p + \nabla \cdot (\bar{\tau}) + \rho \vec{g} + \vec{F} \tag{2}$$

Where S_m is the source of control volume, F is the control force, τ is the viscous damping coefficience.

3.2. Building of combustion model

OAPE works as a nature-inspiration spark-igniting gasoline engine, combustion process of which could be simulated using pre-mixture model. The fresh gas and fuel is mixed in carburetor, then the mixture is sucked into cylinder. The mixture is ignited by the spark. The flame propagates from the flame kernel. The whole combustion chamber can be divided into two zones by the flame sheet [7]. The location of flame sheet can be deprived from the following equation:

$$\frac{\partial}{\partial t}(\rho \bar{c}) + \nabla \cdot (\rho \vec{v} \bar{c}) = \nabla \cdot (\frac{\mu_t}{Sc_t} \nabla \bar{c}) + \rho S_c \tag{3}$$

Where c is the mean reaction efficiency, S_{ct} is the Schmidt turbulent constant, S_c is the reaction source. The mean reactions efficient can be defined as below [7]:

$$c = \sum_{i=1}^{n} Y_i \bigg/ \sum_{i=1}^{n} Y_{i,eq} \tag{4}$$

Where Y_i is the instant mass fraction of combustion production i, $Y_{i,eq}$ is the mass fraction of combustion production i.

3.3. Kinematical parameters of engine

OAPE applies a cam-roller mechanism as the main power conservation mechanism, the motion of piston is determined by the profile of cam. The piston moves in sinusoidal law when the velocity of output shaft keeps constant [8,9]. The rated speed of OAPR is 3000r/min, so the motion of piston can be determined.

using the following equation:

$$S_i = \frac{s}{2} \sin(wt + \pi i) \tag{5}$$

171

Where s is the stroke of piston, $i\pi$ is the phase angle of liners system i, i=1~4.

3.4. Mesh result of working volume

The simulation model of OAPE can be established after simplifying the 3D model of working volume. Benefiting from the low friction loss of carburetor, the gas pressure at the entrance of power cylinder can be assumed to 1 atm.

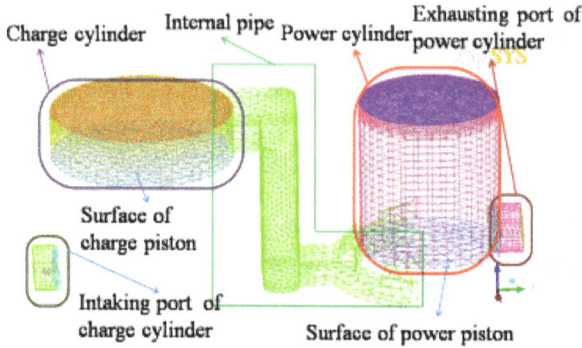

Fig. 2 Mesh result of cylinder system.

As shown in Fig. 2, cylinder system can be divided into five components, intaking port, charge cylinder, internal pipe, power cylinder and the exhausting port. All contact faces between those parts are defined as the interfaces, which mean the fluid can flow through those contacting faces from one part into the next part. The motion of piston surfaces is controlled by the user defined function. The sweep mesh method is applied to the cylinders.

The normal k-epsilon model is used to simulate gas flow process in cylinder. The pre-mixture combustion model is enabled to calculate the combustion process in cylinder. The mixture in power cylinder is ignited by spark, the advance angle of igniting is set to 6 deg. The low heat value is set to 2.73MJ/Kg while the laminar flame speed is set to 0.45m/s. The compression ratio of charge cylinder is set to 3.5, and the compression ratio of power cylinder is set to 9.

4. Analysis of Simulation Results

As shown in Fig. 3. The pressure in charge cylinder decreases with the downward motion of charge piston, which results the mixture in carburetor flows into the charge cylinder through the intaking port under the ambient pressure.

172

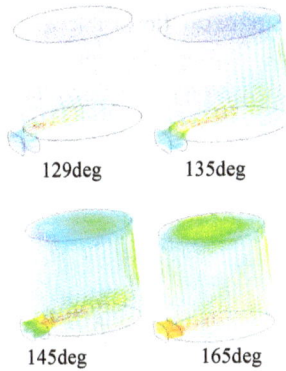

129deg   135deg

145deg   165deg

Fig. 3 Gas fluid in charge cylinder at different crank angle.

As shown in Fig. 4, about 165deg after TDC gas flows into cylinder in too high speed. Mixture could hit the cylinder surface which may result the precipitation of the gasoline. Since the mixture in cylinder is ignited by spark, the combustion process is controlled by the flame sheet.

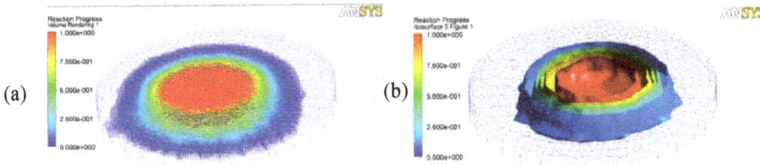

Fig. 4 Reaction process in power cylinder.

As shown in Fig. 4(a), the flame propagates from the center of spark to the surface of cylinder. The iso-surface of reaction fraction is a sphere with a center at spark center as shown in Fig. 4(b). The volume surround by the red surface in Fig. 4(b) is the burnt zone during which all the gas is totally reacted .

Achieving pressure curve in cylinder is primary goal of simulation. Using the CFD post tools the mean pressure in power cylinder and charge cylinder can be deprived, which are shown in Fig. 5. The pressure in power cylinder increases during the compression process. After igniting, pressure in power increases sharply to the maximal value which is 2.127Mpa. The maximal pressure appears at 24deg after TDC. Pressure in power cylinder decreases with expansion of the working cylinder during working stroke. After open of the exhausting port, burnt gas in cylinder flows out of the power cylinder, which results immediately decrease of pressure in power cylinder. As to the charge cylinder, the minimal pressure which is 0.518Bar appears at 135deg after TDC. Once the intaking port opens, the mixture from carburetor flows into charge cylinder, as a result the pressure in power cylinder increases. Then charge piston moves up to the TDC, which will close the intaking port and increases the pressure in charge cylinder. At the moment when the intaking port of power cylinder opens the maximal pressure in charge cylinder can reaches 2.35Bar. Then the burnt gas in power cylinder is exhausted out by the fresh gas from charge cylinder in high pressure.

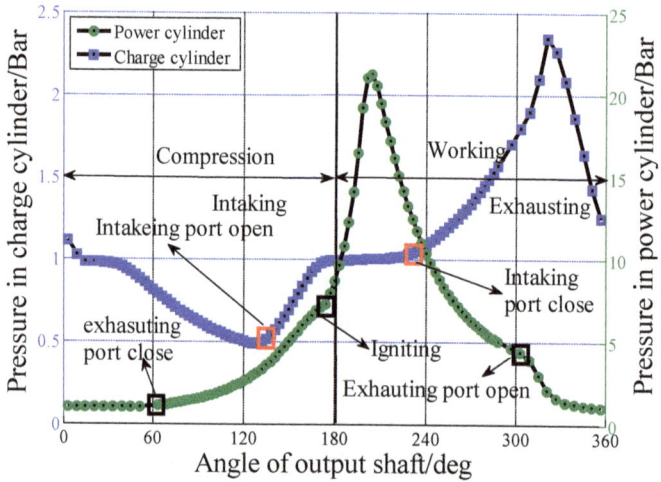

Fig. 5 Pressure in cylinder vs. time.

As shown in Fig. 6, power cylinder always consumes energy before TDC. During the scavenge pressure in power cylinder is approximately equal to atmosphere pressure. However after the exhausting port is closed, the pressure in power cylinder increases sharply, which results the instant power of power piston decreases to the minimal value. After TDC power piston moves down under the push of burnt gas.

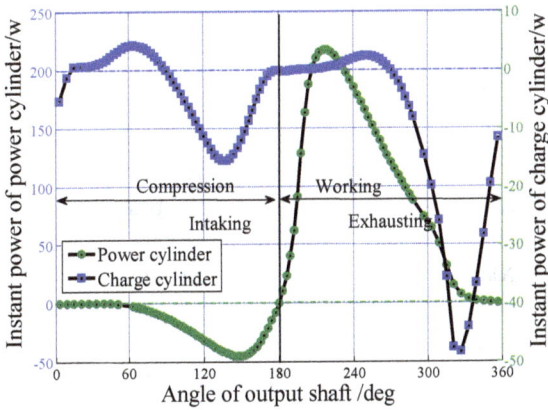

Fig. 6 Instant power vs. angle displacement of outer rotor.

After the integration operation the work done by engine per cycle can be deprived. The power cylinder generates 1.24J per cycle, while the charge cylinder consumes 0.22J per cycle. The pure work done by engine is 1.02J per cycle. Taking the number of cylinders and the speed of engine in to consideration the mean power of OAPE is 204w.

5. Conclusion

1. As a novel engine specially designed for portable power generator, OAPE has the advantages of compacted structure, higher power density, better integration and simplification.

2. After the presentation of working principle of OAPE, the combustion process of OAPE is analyzed. Then simulation of combustion is carried out. The pressure and instant power varies with the angle of output shaft. The power of engine running at a speed of 3000r/min could reach 204w, which could fit the power need of outdoor robots.

Acknowledgment

Work partially supported by Grant No: 51575519 and 51475464 of National Natural Science Foundation of China.

References

1. S M Zhang, Z P Guo. Design methods of novel two stroke swing engine (Beijing: Nation Defense Technology Press 2013).
2. P Chen, Study on the micro power supply based on MEMS technology (Nanjing: Nanjing Technology University 2009).
3. Q. F Li, J. Xiao. Research Status of Free Piston Generator Engine. *Small Internal Combustion Engine and Motorcycle*, 2008, 37, 4:91–96.
4. I A Waitz, G Gauba, Y S Tzeng. Combustors for Micro-gas Turbine Engines. *Journal of Fluids Engineering*, 1998,120, (1):109–117.
5. L Zhang, H J Xu, C Y Pan. 13th International Conference on Control, Automation, Robotics and Vision, on internet.(Singapore, 2014)
6. R Sierens, S Verhelst. A quasi-dimensional model for the power cycle of a hydrogen-fuelled ICE. *International Journal of Hydrogen Energy*, 2007, 32:3545–3554.
7. W Polifke, V Zimont, M Bettelini and W Weisenstein. An Efficient Computational Model for Premixed Turbulent Combustion at High Reynolds Numbers Based on a Turbulent Flame Speed Closure. *Journal of Gas Turbines Power*, 1998, 120:526–532.
8. L Zhang, C Y Pan, H J Xu. Design of air ports for rotary piston engine. *Journal of Zhejiang University*, 2014, 48(12): 2181–2187.
9. L Zhang, H J Xu, C Y Pan. Combustion simulation and key parameter optimization for opposite axial piston engine in small scale. *Journal of Central South University*, 2015, 22(12): 3397–3408.

Mechanical Structural Design of Hot-Line Robot

Wen-hao Wang[†], Cun-yun Pan, Hai-jun Xu and Bo-wen Ni

College of Mechanical Engineering and Automation,
National University of Defense Technology,
Changsha Hunan, 410073, China
[†]E-mail: 1253802049@qq.com

Hot-line robot is a safe, reliable and efficient manner used to repair and inspect high-voltage transmission lines. According to the characteristics of obstacles on line such as damper, strain clamp, insulators, a mechanical structure for hot-line robot is devised. Firstly, the configuration of robot's different parts is fully explained. Then, a motion planning for obstacle climbing and lifting is proposed, using Solidworks COSMOS Motion software to analyze and simulate for the motion. Finally, dynamics analysis is determined for the running gear of robot. The motion simulation verifies the feasibility of mechanical structure and dynamics analysis succeeds in calculating the maximum of climbing angle.

Keywords: Hot-line Robot; Motion Planning; Mechanical Structure; Climbing Angle.

1. Introduction

Periodically repair of high voltage road and replacement of working parts, is an important guarantee for reliable high voltage power lines. Power lines and tower annex long-term exposure in the field, suffering sustained mechanical tension, electrical flashover, material aging which effects cause broken strand, wear, corrosion and other damage, if we don't not repair it and replace the original minor damage in time, defects may expand, then leading to serious accidents eventually, causing widespread power outages, leading to great economic losses and serious social impact [1]. Traditional manual practices are poor efficiency, and by the weather, geography and other natural conditions, workers are always in high voltage, a strong electric field when the job environment, personal safety pose a great threat to workers [2]. In order to ensure operator safety and improve productivity, the researchers at home and abroad have launched a hot-line robot research to improve the manual operation. From the overall structure, the current hot-line robot can be roughly divided into two categories based on Japanese Phase Series as the representative of the control lever, the staff operating in high altitude handle studio, controlling the arm of hot-line robot. One is the United States, represented by remote control; manipulate staff in the face of high altitude remote control robot [3]. In China, the Wuhan University developed a "Dream No. I" hot-line robot prototype in 2014. This robot is a remote control robot, after experiments have been able to achieve charged replace insulators, fastening Clamp and other simple functions, this robot, there are still some obvious drawbacks: the arms are stationary, cannot overcome obstacles yet; few completed job tasks, need artificial assistance, and cannot be online independently.

For the shortcomings, "Dream No. I", in its design based on a can overcome obstacles, autonomous online, multifunctional remote-control hot-line robot "Dream II number." The robot walks online by three walking round, can be online independently with on-line agencies, according to different job tasks, working arm carry two different operating tools to complete a variety of job tasks with each other. You can delete our sample text and replace it with the text of your own contribution to the proceedings. However, we recommend that you keep an initial version of this file for reference.

2. Overall structure design of the robot

Robot works between the two towers with carrying the work tool and damper. During the operation, robot is required to travel across some obstacles, such as spacers, damper, suspension insulator and so on. Robots also need to lift with the on-line agencies, then based on different job tasks, carrying different tools to reach the designated location and working, its tasks include replacing damper, fastening Clamp, replacing insulators with artificial supplementary. hot-line robot structure is the foundation of the whole system, is one of the key technologies of hot-line robot, the mechanical structure of the design requirements of general points: (1) walking stability on the high-altitude wire; (2) climbing ability with a certain angle; (3)overcome obstacles in high-voltage, such as the damper, clamps, insulators and other obstacles; (4) light quality of the robot , to prevent doing damage to wire; (5) online self-locking, to prevent dropping due to on-line fault.

Robot consists of five parts: the walking gear with three walking arm, on-line agencies, modular work platforms, control systems, chassis and battery. The walking gear is a key technology for hot-line robot body structure design. The proposed robot walking gear with three walks arm, walking wheel roll on the same wire, when robot come cross obstacles, three walking arm will cross obstacle in turn, there is always two arms keeping online at the same time ensure the stability of the robot walking posture. Each walking arm has only 2 degree of freedom, structure is simpler and reliable control is also easier. Design quality is 60kg.

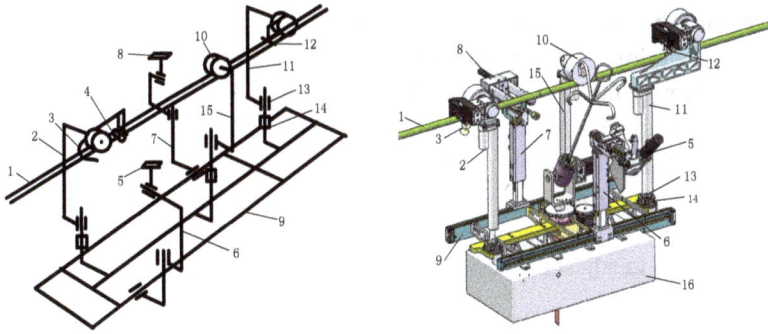

1-line; 2-rear-arm; 3- after pressing wheel; 4-equipotential wheel; 5-working tool 1; 6- working arm 1; 7- working arm 2; 8-working tool 2; 9 transverse rail conductor; 10-wheel; 11-front arm; 12-pressing wheel; 13- prismatic joint; 14- rotating joint; 15- middle arm; 16- case

Fig. 1. Structure of hot-line robot.

2.1. Design of mechanical structure for overcoming obstacle

According to the structure of obstacles, as well as ensure that the process of stabilizing the pose when robot is walking and working, robot structure will be designed into three walking arm which are adjacent and anti-symmetric suspension, the robot suspend on the wire through three or any two walk arm, respectively; three walking arms are called front arm, middle arm and rear-arm. Three arms with two freedom degree, lift and rotate in the plane, each freedom degree is driven by a single motor. Rear-arm install a gripper, forearm install a pressing wheel, played a role in limiting and protection when robot is walking. Front-wheel and rear-wheel are driving wheel; middle wheel isn't drive wheel.

2.2. Design of mechanical structure for lifting

On-line lifting mechanism has three major components: winch, hook, adjustment mechanism of hook's position. Winch is driven to rotate by motor, then winch trip rope, leading to climbing of robot along trope. Winch device consists of a driving wheel and two driven wheels, a worm geared motor, driving wheels and driven wheel fixed on two different bases, two bases connected by a hinge. Rope twine on the winch with 8-shaped, so winch will be twined more tightly, and increasing the friction between the rope and winch. Friction supply tension for robot when it's lifting. Rope is made of insulating material. The control system can control the power output of the platform, and then control the speed of lifting. Pothook connected with cone fixedly, is concentric with cone, when conical tank is lifting, it goes up under the support of conical tank. Stay fork have a turning joint with conical tank, they can relatively rotate to each other. Rope go through the center hole of the cone and the support fork, in this case, both the support fork, cones, rope, conical tank is concentric with each other when robot lift, so as to ensure the accuracy of lifting, increasing flexibility of lifting. Taking into account the robot's swing caused by wind and its own weight effects. A pair of gears, connect on-line agencies to rack, are used to transmit torque to adjust the position relative to the wire in the air.

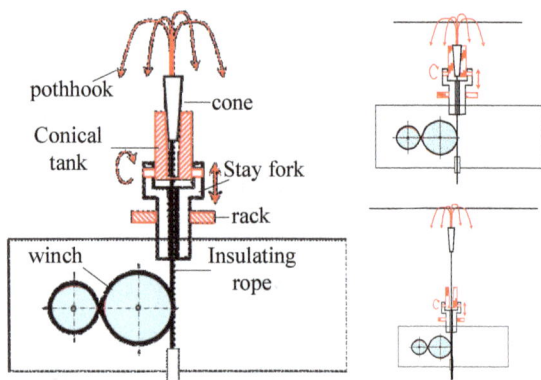

Fig. 2. On-line agencies and two pose of it.

2.3. *Modular design of operating arm*

There is a conflict between the number of completed task and the limit of quality, modular design of operations arm is a way to resolve this conflict, which also can lighten robot quality, and simplify mechanical structure. Aimed at different tasks, different tools are designed for the job. According to the needs of job tasks, corresponding tool is installed in the end of the robotic arm, in the end of the robotic arm; a slide with dovetail groove is equipped so that different tools can be installed easily. Robot design goal is to complete three tasks such as replacing insulator, replacing damper, fastening Clamp, so the design of the three respective working tool is shown in Fig. 3. The bottom of each work tool has the same interface, by this interface dovetail groove can be connected with arm's longitudinal mobile slide; this design is easy for installation and removing of tools. Every arm has three mobile freedom degree; their mobile directions are mutually perpendicular. By mutually coordinated movement in three directions, the tool can reach anywhere in the workspace, and meet the requirements for the position of the tool. Three directions have an accuracy of 0.1mm to move that meet the accuracy requirement of work.

180

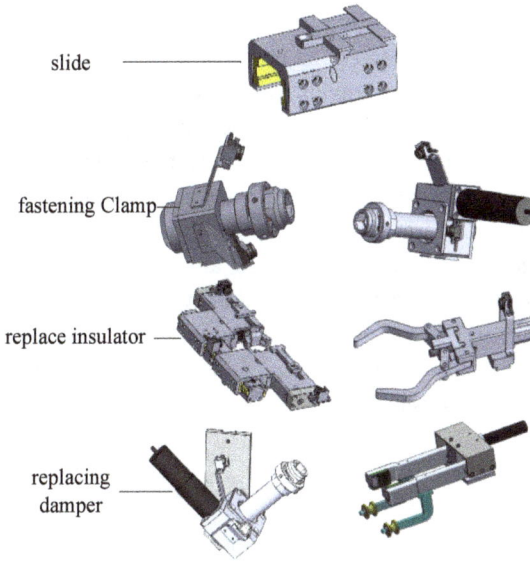

slide

fastening Clamp

replace insulator

replacing
damper

Fig. 3. Working tools.

3. The motion planning and simulation

Firstly, we set up 3D modeling, and then the model is assembled. After that model will be imported to COSMOS Motion in Solidworks environment, finally, adding constraints, force, collision to corresponding position. According to the actual motion, define the direction and angle of rotation in the joint. Before robotics simulating, motion planning need to be finished. After that, robot walks in line and across the barriers in simulation according to motion planning. According to analysis result of simulation, we can adapt the design and size of every individual component. Here a few screenshots of simulation are used to explain the process that robot climb to line and across the insulator.

3.1. *Motion planning and simulation for overcoming obstacle*

Motion planning across the insulator as Fig. 4: When the robot travels obstacles, front arm stop to open the pressing wheel and rise along telescopic joint as shown in (a); then front arm rotation joint rotate to one position as shown in (b); the robot continue walking, the front arm across the insulator, and then front arm return back to the original position to re-contact with the conductor as shown in (c); After front arm return back to working position, middle arm rise along telescopic joint and rotate to reach the position as shown in (d); the middle arm descend along its own telescopic joint until the wheel drop below the level of the line as shown in (e); the robot travels over obstacles arm ,then return back to the working position shown in (f); After middle arm suspend in the line again, two working arm move from the rear to the front, so the center of gravity is changed and falls in between middle arm and front arm, and then open pressing wheel of the rear arm, then rear arm rise along telescopic joint to reach the position as shown in (g); rear arm rotate a certain angle along its own rotary joints, and then descend along its own telescopic joint until the wheel drop below the level of the line as shown in (h); robot continue to move forward, after that rear arm travels over obstacles and back to the original operating position as shown in (i); robot complete overcoming obstacle.

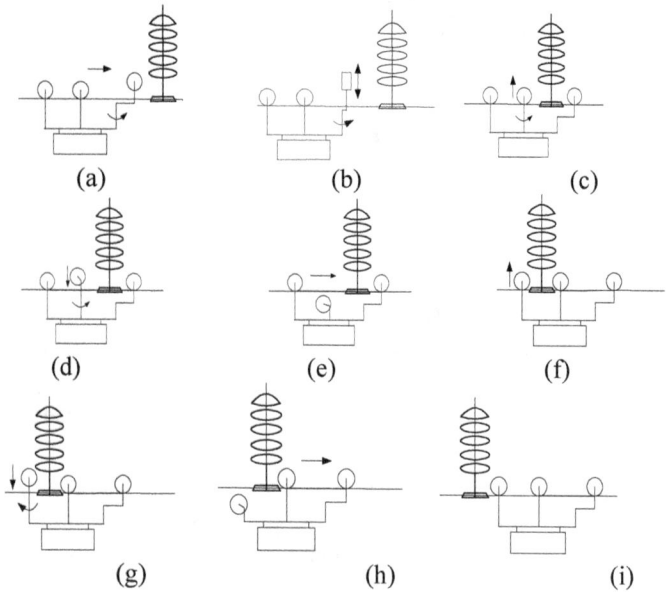

Fig. 4. Motion plan for crossing insulator.

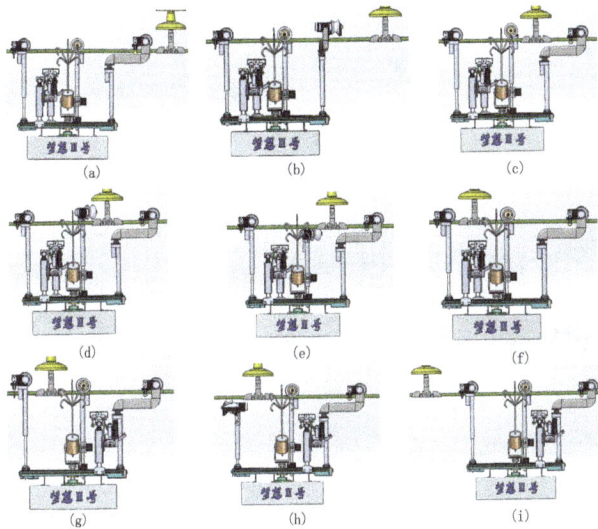

Fig. 5. Simulation for overcoming insulator.

3.2. Motion planning and simulation for lifting

As shown in Fig. 6, the simulation of lifting has a few steps. Winch device rotate in the case and supply force for robot's climbing, so robot can climb along insulating rope as shown in (a); when robot rise to a certain height, cone gradually contact with conical tank as shown in (b) and (c); robot keep rising until cone fit with conical tank fully as shown in (d); the motor installed in rack drive small gear so it can rotate around the big gear, adjust robot to correct position and then stop moving as shown in (e); front arm and rear arm rise to a certain height as shown in (f); front arm and rear arm rotate and then shrink back to contact with line as shown in (g); supported by front arm and rear arm, rack rise by contraction of those two arms, and then conical tank rise along with rack. Because of rising of conical tank, pothook will be pushed out and break away from line as shown in (h); under the driving of conical tank, pothook swing outward as shown in (i), now robot complete lifting.

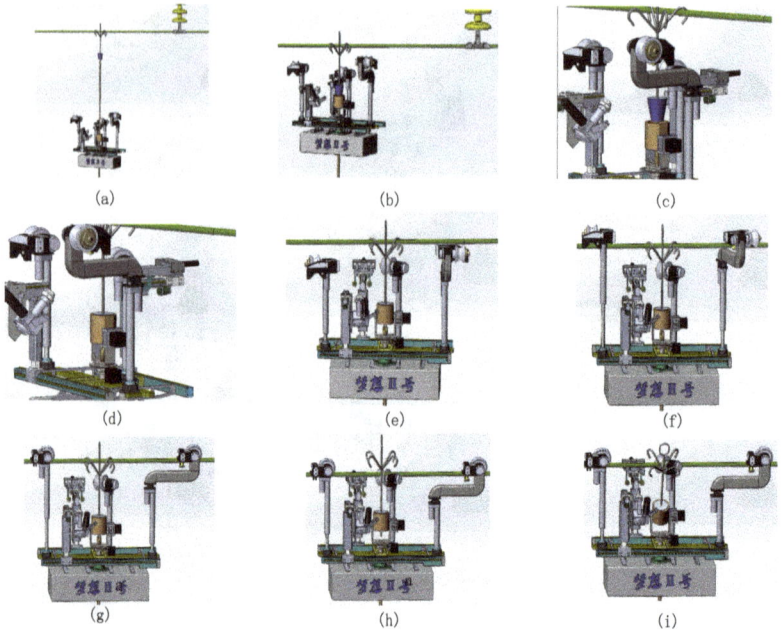

Fig. 6. Simulation of lifting.

4. Principle and Dynamics Analysis for robot's walking

4.1. *Stress Analysis for walking straight*

A couple which consist of driving force F and static friction force F_s, providing power for rolling of wheel. Wheel suffers supportive force that distribute in a plane instead of a line. When the wheel is static, the distribution of supportive force is symmetrical as shown in (a). When the wheel is rolling, the distribution of supportive force wire is changed as shown in (b).

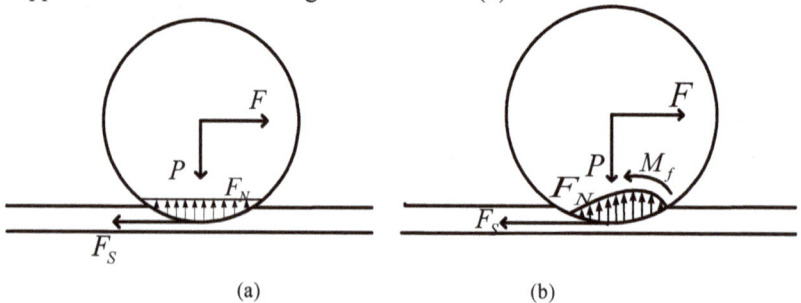

(a)   (b)

F -traction P -gravity F_N -supportive force F_s -sliding friction M_f -rolling resistance moment

Fig. 7. Force analysis of wheel.

184

Rolling resistance moment M_f is in direct proportion to the positive pressure proportional F_N. δ is proportional constant and called rolling resistance coefficient.

$$M_f = \delta F_N$$

Order: $F_r R = M_f = \delta F_N$ so: $F_r = \dfrac{\delta F_N}{R}$

Traction must not exceed the maximum of static friction: $F_{\max} = f_S \cdot F_N$, f_S is friction coefficient

When wheel start: $\quad F = F_S + F_r + F_P + \dfrac{P}{g} a$

When wheel take a uniform motion: $F = F_r + F_P + F_S$

F_P is component of gravity force which is parallel to wire, when robot is walking straight: $F_P = 0$

Sliding friction still can't be measured via theoretical at the present stage, so when making analysis, wheel rolling will be considered as pure rolling when robot is walking, after prototype is made, sliding friction will be measured by test.

4.2. Stress Analysis for climbing

In the process of climbing robot front arms and rear arm are contact with the wire and loaded, middle arm is not in contact with the wire. According to the torque balance principle, ignoring sliding friction, we can obtain:

$$M_1 = \left(m_1 g \sin\alpha + \frac{\delta}{r} m_1 g \cos\alpha \right) r$$

$$M_3 = \left(m_3 g \sin\alpha + \frac{\delta}{r} m_3 g \cos\alpha \right) r$$

Robot weight $60kg$, what m_1, m_3 mean is the load of front arm and rear arm, both of them are taken $30kg$. Wheel center is about 35mm apart from the contact surface, after checking mechanical design manual, rolling resistance coefficient δ is taken 0.18 (cm), the continuous torque M_1 of drive motor is $2.5 N \cdot m$. Taking it into the equation, solution is: $\alpha = 10.8°$.

4.3. Stress analysis for overcoming obstacle

The process of stress analysis for overcoming obstacle is same as climbing, giving the example of walking across for front arm. When front arm departs from wire, according to the balance principle of torque, equation can be obtained:

$$M_1 = \left\{ \left(m_1 + m_2 \right) g \sin \alpha + \frac{\delta}{r} \left(m_1 + m_2 \right) g \cos \alpha \right\} r$$

Middle arm will bear most of the weight of robot, so: $m_1 = 20kg$ $m_2 = 40kg$, the solution is solved: $\alpha = 4.27°$.

Comprehensive analysis of the above situations, when the robot is going to climb obstacle, the maximum angle the robot can climb, meet the requirements for walking normally.

5. Conclusion

This paper designs a new type of self-elevating hot-line robot that owns three mechanical arms. In the future, we will manufacture a prototype and conduct an experiment. On the basis of experimental study, we will make a further improvement of function and mechanical structure until robot can work and walk smoothly in simulative line. There still are many jobs to finish to ensure that robot can work in realistic line. Increasing demand for reliability of supplying power will promote the progress of live working technology. The marketing prospect of hot-line robot is vast with the development of robotics and live working technology.

Acknowledgment

Work supported by the funding of National Science Foundation of China with Grant No: 51575519 and 41475464.

References

1. Zhou Fengyu, Li Yibin, Wu Aiguo, etc. *Design and implementation of high-voltage transmission line inspection robot* [J] Mechanical Science and Technology 2006.25 (5) 623–626.
2. Lu Shouyin. *High voltage live working robot research* [D] Shanghai Jiaotong University. 2003.
3. Wang Kaijun, should Hong, Lu Shouyin. *High voltage live working robot development background and technology trends* [C] Zhejiang Electric Power Development 2005.

4. Yu Xiaoxin, Tian Lianfang, Wang Xiaohong etc. *Based on SolidWorks line patrol robot obstacle body design and motion simulation* [J]. 2010 Mechanical Design and Manufacturing (8): 180–182.

5. Xing Xiangfei. *Inspection robot system design and simulation of Super high-tension line* [D]. 2011 Changchun Science and Engineering.

6. Xiao Xiaohui *study High Voltage Transmission Line Inspection Robot Several dynamical problems* [D] Huazhong University of Science. 2005.

7. Zhang Dehui. *Live working robot from the lift mechanism analysis* [J] 2012.29 (1): 40–43.

8. Yang Dewei, Feng Zuren, Zhang Xiang. *The new three-arm inspection robot Mechanism Design and Motion Analysis* [J] 2012.46 (9): 43–54.

Chapter 3

Electronics Engineering
and Electrical Engineering

Study on Dynamic Contamination Depositing Model of Insulators Using DEM Methods

Hai-zhen Sun[†], Guo-zhi Wang, Wen-hai Wu and Jian Ke

School of Mechanical Engineering, Southwest Jiaotong University, Chengdu 610031,China

[†]*E-mail: sunhaizhen@my.swjtu.edu.cn*

DEM and CFD methods are proposed to study the dynamic contamination depositing characteristics of the insulator surface, established the movement model of particles, the contact and adhesion model, the agglomeration model and the removal model. Through the analysis and simulation of the contamination model, the pollution distribution on the insulator surface, the critical normal initial velocity of particle desorption, the effect of removal by wind are determined. The research result shows that particles with diameter below 20μm primarily deposit on the lower surface through drag force while particles over 50μm mainly deposit on the upper surface through gravity sedimentation. Half fill angel of liquid bridge is the most important factor. The larger the half fill angle is, the better absorbability the insulator gets. The contact angle is the least factor and the particles have the best absorbability when the contact angle is 25°. The critical normal initial velocity and the adhesion ability increase with the decreasing of the particle diameter and it will be easier for bigger particles to be separated from surface. The dynamic removal efficiency of the upper surface is greater than the lower surface and the particles that have adhered to the surface cannot be effectively removed by wind. This paper describes the complete dynamic depositing stages of particles to analyze the contamination on insulator that has never been done before.

Keywords: insulator; Discrete Element *Method (DEM)*; contact and adhesion; critical normal initial velocity; half fill angel; contact angle.

1. Introduction

The contamination being affected with damp on insulators which are the most used devices in the power system would reduce the electrical strength of the external insulation leading to the flashover which will affect people's lives and industrial production seriously. In order to prevent the pollution flashover, the dynamic contamination depositing model of insulators needs to be studied to understand the deposition and adhesion mechanism of particles on insulators which can be based on to design the anti-pollution measures and remove the contamination particles effectively.

Many studies have been done on the contamination characteristics of insulators. The results show that the movement of particles in the flow field outside the insulator and the location of the dust deposition are mainly affected by wind and gravity; and the adhesion relatives to the material property, roughness, and surface charge of the samples.[1] But there was no research which introduced the mathematical model of the whole stage of dynamic contamination in detail.

In order to study the dynamic contamination mechanism of particles on insulator surface, the whole stage of dynamic contamination model was established using both the CFD and DEM methods, which can provide the theory foundation for the anti-pollution.

2. Analysis of the Particle Movement in the Flow field

The flow around cylinder phenomenon will happen when the wind flow through insulator, which will change the wind speed and produce a vertical component as shown in Figure 1 and Figure 2, resulting in a relative velocity between particles and fluid, and then particle trace will be affected by fluid drag force. Through the analysis, the drag force is the main factor in horizontal direction, then the free flow resistance model was used to describe the force.[2, 3]

$$
m_p \frac{d_{v_{px}}}{d_t} = 0.5 C_D \rho_f A_p \left(v_{fx} - v_{px} \right) \left| v_{fx} - v_{px} \right| \tag{1}
$$

Where A_p is the projected area of the particle; C_D is the drag coefficient which is dependent on the local Reynolds Re.[3]

$$
Re = \frac{\alpha \rho L |v|}{\eta} < 1
$$

$$
C_D = \frac{24}{Re} \tag{2}
$$

Particles in the vertical direction mainly affected by gravity, buoyancy and fluid drag force:

$$
\text{Resultant force } F_y = F_{Dy} + F_B - F_G \tag{3}
$$

$$\text{Gravity } F_G = \rho_p \frac{\pi}{6} d_p^3 g \tag{4}$$

$$\text{Buoyancy } F_B = \rho \frac{\pi}{6} d_p^3 g \tag{5}$$

$$\text{Drag force } F_{Dy} = 3\pi\eta d_p \left(v_{fy} - v_{py}\right) \tag{6}$$

Figure 3 shows that the resultant force F_y in the vertical direction becomes downward, the change amplitude increases and gravity sedimentation begins to play a major role with the increasing of particle diameter in the lower region. It can be seen that F_y changes the direction into a vertical upward for the same size of the particles with the increasing of wind speed, then the drag force plays a major role in the particle trace leading to the increases of the pollution on the upper surface. It can be concluded that particles that the diameter is below 20μm primarily deposit on the lower surface through drag forces, and particles that the diameter is over 50μm mainly deposit on the upper surface through gravity sedimentation in conditions of light air and calm.

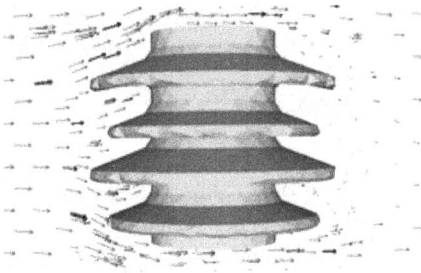

Fig. 1 Insulator outflow field.

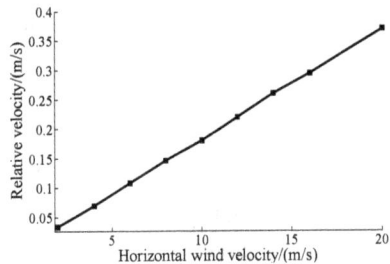

Fig. 2 The relative velocity.

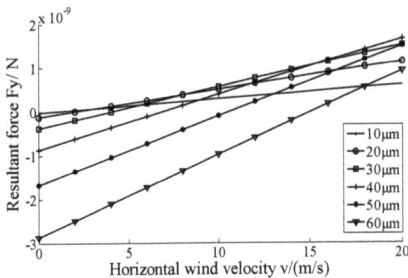

Fig. 3 The resultant force F_y.

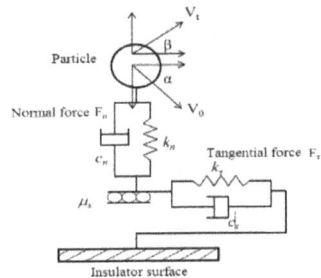

Fig. 4 The contact physical model.

3. Analysis of the Contact and Adhesion Model

Agglomeration is a basic feature of cohesive particles, so both single particle and particle agglomeration coexist in the atmosphere which will be studied at the same time in this paper.

3.1. Force analysis of contact model

The inelastic collision will happen when particles contact with insulator surface, and meanwhile, the particles are squeezed and deformed on the elastic surface which will cause feedback force on particles from surface. A physical model that describes the contact between particle and insulator surface was established according to the Hertz and Mindlin-Deresiewicz theories in this paper as shown in Figure 4. [4] Based on the physical model, the dynamic mathematical model is established using the mathematical balance equations.

In the normal direction, Hertz elastic force and van der Waals force are the universal force, the liquid bridge force between particles and surface also exists in the wet interface, so the normal force expression is given:

$$F_k = \frac{4}{3} E \sqrt{R} \delta_n^{3/2} \tag{7}$$

$$F_c = F_L + F_W \tag{8}$$

$$F_n = F_k - F_c = \frac{4}{3} E \sqrt{R} \delta_n^{3/2} - 4\pi \gamma R \cos\theta - \frac{HR}{6D^2} \tag{9}$$

Where F_k is the Hertz elastic force; F_c is the normal damping force which is mainly affected by the liquid bridge force F_L and the van der Waals force F_W; E is the elastic modulus of particles, E is 7.698e9pa for the material of Calcium sulphate; δ_n is the normal overlap; R is the radius of particles; γ is the surface tension of liquid; θ is the contact wetting angle; and H is the Hamaker constant.

In tangential direction, the particles are squeezed by rough surfaces and relative sliding exists between particles and surface in the tangent motion causing the tangential elastic force and the friction force. The tangential elastic force depends on the tangential overlap and tangential stiffness, and the tangential damping is caused by the sliding friction force as follows:

$$T_k = 8G\sqrt{R\delta_n}\,\delta_\tau \tag{10}$$

$$T_c = \mu F_n \tag{11}$$

$$F_\tau = T_c - T_k \tag{12}$$

194

Where T_k is the tangential elastic force; T_c is the friction force; G is the shear modulus; δ_τ tangential overlap; and μ is the coefficient of sliding friction.

But for the agglomeration, the outermost layer is always in the dynamic balance of adhesion and detachment, so the detachment of particles need to be considered in the collision stage from the agglomeration, the force analysis between agglomeration and surface is shown in Figure 5.

The van der Waals force and the liquid bridge force between agglomeration and insulator depend on the particles of outermost layer which is contact with insulator directly. The expressions of the interaction force can be expressed as follows:[5]

$$F_W' = \frac{A(1-\varepsilon_a)}{\pi R^2} \times \frac{HR}{6D^2} \tag{13}$$

$$F_L' = \frac{A(1-\varepsilon_a)}{\pi R^2} \times 4\pi \gamma R \cos\theta \tag{14}$$

$$\varepsilon_a = 1 - \frac{\rho_a}{\rho_p} \tag{15}$$

Where A is the adhesion area between agglomeration and surface; ε_a is the cluster porosity; and ρ_a is the density of agglomeration.

The adhesion shear force F_s is balanced with the gravity F_g of the outermost layer before the collision which can be expressed as follows:

$$F_s = CA' \tag{16}$$

$$F_g = \frac{2}{3}(1-\varepsilon_a)\rho_p g d_p A' \tag{17}$$

Where A' is the adhesion area of the outermost layer; C is the adhesive shear strength, C is 951.05N/m^2 for Calcium sulphate;

When the normal force in the collision stage between the outermost layer and insulator surface is larger than the adhesion shear force of the outermost layer shown in expression 18, particles of the outermost layer will shed from the agglomeration which will increase the probability of adhesion because of the energy loss.

$$\frac{\frac{4}{3}E\sqrt{R}\delta_n^{3/2} - 4\pi \gamma R \cos\theta - \frac{HR}{6D^2}}{CA'} > 1 \tag{18}$$

195

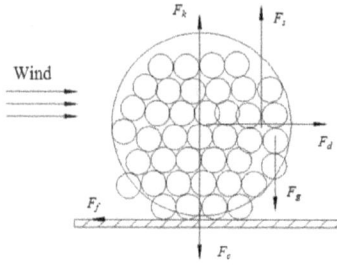

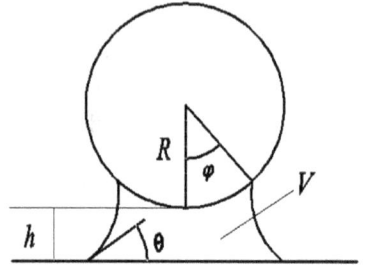

3.2. Collision analysis of contact model

The soft sphere model is used for the inelastic collisions which can be divided into three stages: extrusion deformation stage, recovery stage and particle desorption stage; [6] and both the extrusion deformation and recovery satisfy the theorem of momentum in both normal and tangent direction:

$$v_{nt} = k v_{n0} \tag{19}$$

$$\mu \bar{F}_n (t_1 + t_2) = m_p (v_{\tau t} - v_{\tau 0}) \tag{20}$$

$$t_{1,2} = 1.47 \left(\frac{15(1 - \upsilon^2) m_p}{8E\sqrt{R}} \right)^{2.5} v_{n0,ny}^{-0.2} \tag{21}$$

Where t_1, t_2 are the time of extrusion deformation stage and recovery stage respectively; $v_{nt}, v_{n0}, v_{\tau t}, v_{\tau 0}$ are the velocity of the end of recovery stage and the begin of extrusion deformation respectively; υ is the Poisson's ratio; $\bar{F}_n$ is the average normal force which can be calculated using the mean value theorem with the hypothesis of the maximum normal overlap R as follows:

$$\bar{F}_n = \frac{\int_0^R F(\delta_n) d_{\delta_n}}{R} \tag{22}$$

3.3. Analysis of adhesion stage

Particles are mainly affected by liquid bridge force from the end of recovery stage to detachment naming desorption stage. If the normal moving distance is greater than the maximum liquid bridge distance, particles will be desorbed from surface, otherwise the adhesion would happen which can be expressed as follows:

196

$$\frac{1}{2}m_p v^2{}_{ny} \le F_L l_b \tag{23}$$

Where l_b is the maximum liquid bridge distance which can be calculated by the equation(24) when the contact angle is less than 40 ° as follows: [7]

$$l_b = (1 + 0.5\theta)V^{1/3} \tag{24}$$

$$
\begin{aligned}
V = \pi\{ &(a^2 + r^2)r[\cos(\varphi + \theta) + \cos\theta] - \\
&\frac{1}{3}r^3[\cos^3(\varphi + \theta) + \cos^3\theta] - \\
&ar^2[\sin(\varphi + \theta)\cos(\varphi + \theta) + \sin\theta\cos\theta] + \\
&ar^2(\varphi + 2\theta - \pi) \} - \frac{\pi}{3}[(2 - 3\cos\varphi + \cos^3\varphi)R^3]
\end{aligned}
\tag{25}
$$

$$r = \frac{R(1 - \cos\varphi) + h}{\cos(\varphi + \theta) + \cos\theta} \tag{26}$$

$$a = r\sin(\varphi + \theta) + R\sin\varphi \tag{27}$$

Where V is liquid bridge volume; φ is the half fill angel of liquid bridge; h is the spacing between particles and surface, the general value of spacing is $R/1000$; and r is the radius of curvature of liquid ring. The liquid bridge force diagrams were shown in Figure 6.

Then the critical normal initial velocity of particle desorption v_{n0}^c was calculated and analyzed using the contact and adhesion model with the different particle diameter ranging from 10μm to 100μm. Figure 7 and Figure 8, Figure 9 show the relation between the critical normal initial velocity and the contact angle, the diameter of particles, the half fill angel respectively.

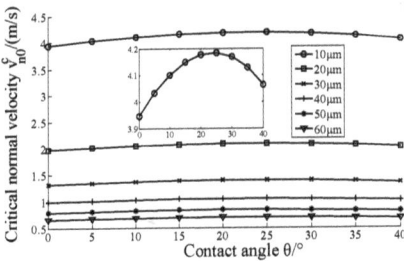

Fig. 7 The relation between v_{n0}^c and θ.

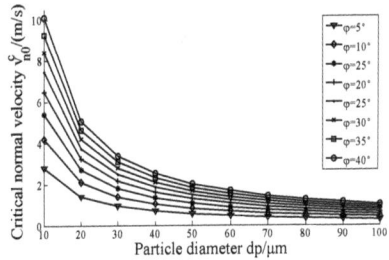

Fig. 8 The relation between v_{n0}^c and d_p.

197

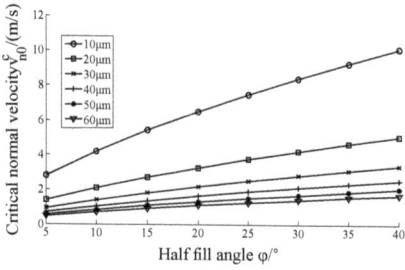

Fig. 9 The relation between V_{n0}^c and φ.

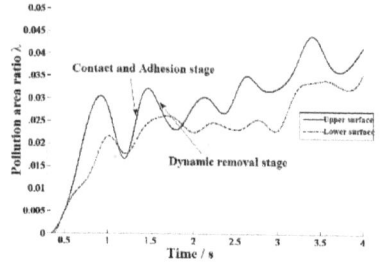

Fig. 10 The polluting area ratio when d_p=20μm.

From Figure 7, it can be observed that the relation between the critical normal initial velocity and the contact angle is a parabola curve, and the critical normal initial velocity reaches the maximum when the contact angle is 25°. The contact angle depends on the wettability of the particles which is tightly related to the type of particles. The weaker the wettability of the particle is, the larger the contact angle will be, which will reduce the force of liquid bridge but increase the maximum distance of the liquid bridge; when the contact angle reaches a certain value, the critical normal initial velocity gets the maximum, however the change of critical velocity is not obvious with the change of contact angle, and the desorption of particles has no remarkable difference.

Figure 8 shows that the critical initial velocity decreases with the increasing of particle diameter, which indicates that it requires a bigger initial velocity to be dislodged from the surface for the smaller particles, then we can concluded that the smaller the particle diameter is, the stronger adhesion ability will be, and then the easier the adhesion happens. Because the influence of inertia and gravity become weaker and the effect of the acceleration generated by the action of a liquid bridge becomes stronger for the smaller particles, which results in the increasing of difficulty to remove the particles from the surface. Also it can be observed that the relation between the critical initial velocity and diameter is in accordance with the equation:

$$v_{n0}^c = Ad_p^{-1} \tag{28}$$

The critical initial velocity shows an increasing trend and changes significantly with the increasing of the half fill angle which can be seen from Figure 9. It can be concluded that particles adhere to the insulator surface more easily and the ability of desorption is reduced with the increasing of the half fill angle because of a greater critical initial velocity. In comparison, the effect of half fill angle on the adhesion property of particles is greater than that of the contact

198

angle. The half fill angle determines the volume of liquid bridge, [7] and the liquid bridge becomes bigger with the increasing of the half fill angle, then the larger normal initial velocity is needed to get away from surface for particles. As we know that the half fill angle depends on the hydrophilicity of the material of the insulator surface, and the better the hydrophilicity is, the larger the half fill angle will be, so the material of the insulator needs a poor hydrophilicity.

4. Analysis of Removal by Wind

There are two kinds of removal ways by wind, one is the dynamic removal stage which can be described as separation of particles from insulator surface after the collision; and the other way is the removal of the adhered particles called the static removal stage.

Particles are desorbed from surface for the dynamic removal stage which should satisfy the equation as follows:

$$\frac{1}{2}m_p\left(v_0\cos\alpha\right)^2 > \mu F_L x_{\max} \tag{29}$$

In order to describe the dynamic removal stage, the model of removal was simulated using the coupled DEM and CFD methods when the diameter is 20μm, getting the pollution area ratio of insulator surface shown in Fig. 10. It can be seen that the dynamic removal stage is obvious and the desorption efficiency of the upper surface is better than the lower surface.

For the static removal stage, it depends on the effect of shear flow in the boundary layer which satisfies the Stokes equation when the airflow passes through the insulator surface. Particles are mainly affected by drag force and static friction in tangential direction, and they are mainly affected by liquid bridge force and van der Waals force in the normal direction. When the air drag force is greater than the static friction force causing a relative movement between particle and surface, particles will be removed.[8] This can be expressed as follows:

$$F_{Dy} - \mu F_n > 0 \tag{30}$$

$$F_{Dy} = 3\pi\eta d_p v_{fy} \tag{31}$$

$$F_n = F_L + F_W = 4\pi\gamma R\cos\theta + \frac{HR}{6D^2} \tag{32}$$

The critical wind velocity which was calculated is 30.8m/s, setting the static friction coefficient μ to be 0.01, cosθ to be 0.8 and D to be 0.01μm. It can be

199

concluded that the particles that have adhered to the surface cannot be effectively removed by wind.

5. Conclusion

1) The motion of particles, the contact and adhesion between particles and insulator, and the removal stage of the dynamic contamination depositing model of the insulator surface were established and analyzed using the coupled CFD and DEM methods to provide the theoretical basis for the design of insulator.

2) Particles which the diameter is below 20μm primarily deposit on the lower surface through drag force, particles that the diameter is over 50μm mainly deposit on the upper surface through gravity sedimentation when the wind speed is smaller than 8m/s.

3) The effect of the half fill angel on dynamic contamination is the most serious factor, however the contact is the least factor. The larger the half fill angle is, the critical normal initial velocity will be needed, then it will be easier for particles to adhere to surface. The critical normal initial velocity reaches the maximum and the adhesion property of the particles is the best when the contact angle is 25°. The follow-up nature and adhesion nature decrease with the increasing of the diameter. It is easier for smaller particles to adhere to surface, and there is more chance for larger particles to rebound to be separated from insulator.

4) The dynamic removal efficiency of the upper surface is greater than the lower surface, and the particles that have adhered to the surface cannot be effectively removed by wind.

References

1. Cher Lin Clara Tan, Shaokai Gao, Boon Siong Wee, et al, Adhesion of Dust Particles to Common Indoor Surfaces in an Air-Conditioned Environment, *Aerosol Sci. Technol.* 48(5),541(2014).
2. Guoming Hu (ed.), *Analysis and Simulation of particle system by discrete element method* (Wuhan University of Technology Press, 2010).
3. B. Oesterlé, T. Bui Dinh. Experiments on the lift of a spinning sphere in a range of intermediate Reynolds numbers. *Exp. Fluids*, , 25(1),16(1998).
4. Qicheng Sun and Guangqian Wang (eds.), *An introduction to the mechanics of granular materials* (The Science Publishing Company, Beijing,2009).
5. Hongzhong Li and Musun Guo(eds.). *Particulation of Gas-Solids Fluidization* (Chemical Industry Press, Beijing, 2002)pp.21-26.
6. Bangguo Ling. The Quantitative Calculation of Impact in the Process of Collision, *J. Journal of Nantong Institute of Technology*. 17(04), 7(2001).

7. Hui Wang, Yang Jiao, Wenyu Xin, Effect of liquid bridge force and critical velocity for the separation of wet granule, *J. College Physics*. 34, 44 (2015).
8. Burdick G M, Berman N S, Beaudoin S P. Hydrodynamic particle removal from surface, *Thin Solid Films*, 488(1), 116 (2005).

Research and Implementation of Key Technologies for Electronic Circuit Board Performance Test

Wei Xu[†], Hui-gang Xu, Qi Xie, Mei Dai and Zheng-yang Wu

School of Electrical and Automation Engineering,
Changshu Institute of Technology,
Changshu 215500, China
[†]E-mail: xu_wei985@163.com

In order to improve the level of performance test for electronic circuit board, some key techniques as analog isolation acquisition, standard test signals generation and test software architecture design were discussed in this paper. The hardware components and its selection principle of the techniques were discussed. The design idea and structure of the technique software developed by the LabVIEW software development platform was introduced in detail. Finally, practical application shows that the key technologies studied in this paper can significantly improve the precision and efficiency of electronic circuit board performance test.

Keywords: Performance test; Electronic circuit board; LabVIEW.

1. Introduction

Electronic circuit board is the core of the electrical products, whose performance is directly related to the quality of the whole electrical appliances, and even causes some safety accidents. Therefore, in accordance with the relevant standards, the performance of the electronic circuit board must be tested before the product assembly, to find the quality problem as soon as possible, and subsequent processing [1].

It has the disadvantages of more wiring, poor accuracy, trival process and inefficient during the traditional electronic circuit board testing, resulting in detection quality can not be guaranteed. In addition, some of the circuit board test system with special test hardware, visualization of application software, has been unable to meet the demand of electronic circuit board performance testing, whose generality is not strong, and the late maintenance upgrade cost is high [2,3]. In order to satisfy the new demand of producers, three key technologies such as analog isolation acquisition, standard test signals generation and test software architecture design for electronic circuit board testing are studied in this paper, through the practical application shows which can significantly improve the accuracy and efficiency of electronic circuit board performance testing.

2. Analog Isolation Acquisition

In allusion to the characteristics of more reference ground, strong and weak electrical signals coexist for electronic circuit boards performance test; the isolation method must be used. The digital input/output isolation technology is mature, such as using optocoupler for photoelectric isolation. There is a difficulty in the design of the isolation acquisition for quick-change analog. In order to solve this problem, an analog isolation acquisition circuit is designed, as shown in figure 1. HCNR201 is AVAGO linear optocoupler production, whose nonlinearity is only 0.01%. It contains two matching photodiode, whose current transfer ratio is consistent, to ensure the accuracy of the analog signal isolation by the use of feedback approach. A high slew rate rail-to-rail precision op-amp combined with a linear optocoupler is used to realize tracking isolation acquisition for high voltage and quick-change signal. By adjusting circuit parameters, optimizing the circuit structure, improving the linear optocoupler drive current, make its work in a suitable linear workspace, to improve the accuracy of measurement.

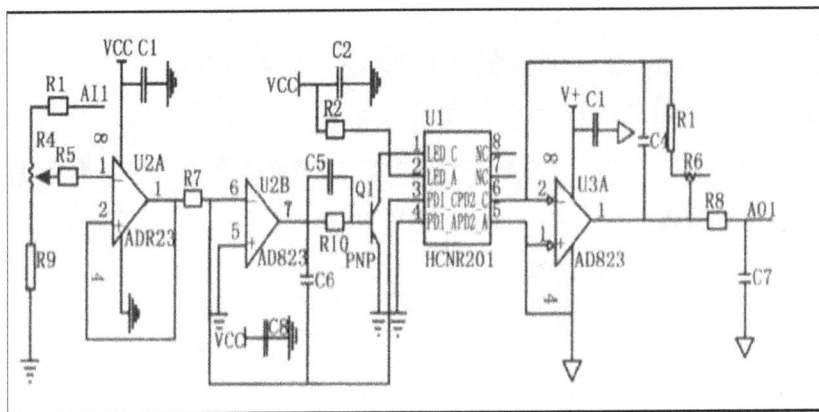

Fig. 1. The analog isolation acquisition circuit.

3. Standard Test Signals Generation

According to the requirements of electronic circuit board inspection, high precision standard sine signal source need to be provided, to simulate the actual

203

running state. Standard signal source can output the effective value from 0 ~ 7 v of A, B, C, N the four phase voltage signal, whose amplitude, phase and frequency are controlled by software.

The analog output channel of the PCI - 6229 data acquisition card is used to output the multi-channel high accuracy sine signal with the continuous output mode.   First the DAQ opens up a circular buffer to hold the sampling values of output waveform, then the data acquisition card continuously outputs the data stored in the buffer. The whole process is made up of giving DAQmx task, creating channels, configuring channels, setting time and buffer, writing the data to the buffer, starting the output task, continuing to write the task, and stop tasks, etc., according to this programming idea, development of the corresponding signal module. Real-time continuous Signal produce software programming process is shown in figure 2; the voltage output program block diagram is shown in figure 3.

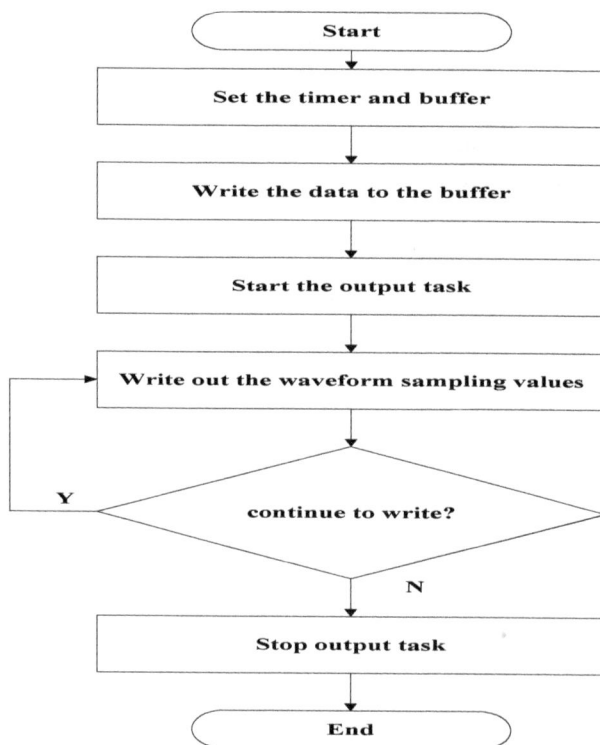

Fig. 2. The real-time continuous signal produce software programming process.

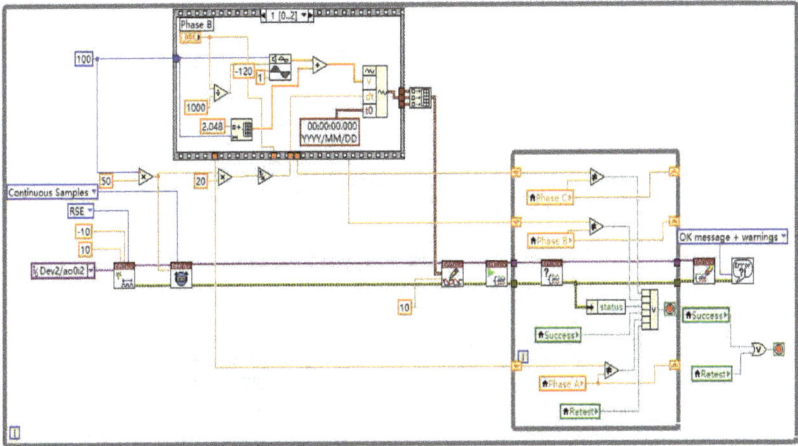

Fig. 3. The voltage output program block diagram.

4. Test software architecture design

The main task of the performance test software for electronic circuit board is testing process control, data acquisition, data analysis, data display and test report generation, etc. [4,5,6]. The architecture design of the software is directly related to the accuracy and efficiency of the electronic circuit board performance test. So the choice of appropriate software architecture becomes particularly important.

In order to improve the flexibility of the test, the test software adopts the modular design idea, what means to make each test project as a module. So that operators can screen the test project according to the actual situation, skip the needless project, which can shorten the test time, and improve the production efficiency.

To make the test software be extensible and easy to maintain, the general framework is chosen. Test software uses the hierarchical structure of the overall framework, divided into two layers, independent, and utility, as shown in Fig. 4. Utility functional layer provides services for independent, and independent functional layers complete itself unique function by calling utility layer in the form of sub V I. Each function module in the independent function layer is independent of each other, and not affected. At the same time, when the detection function needs to expand and update, the impact on the software is not very clear. As long as changing the specific function module not all of the software, it can reduce workload, improve the production efficiency, and also fully reflect the advantage of strong flexibility.

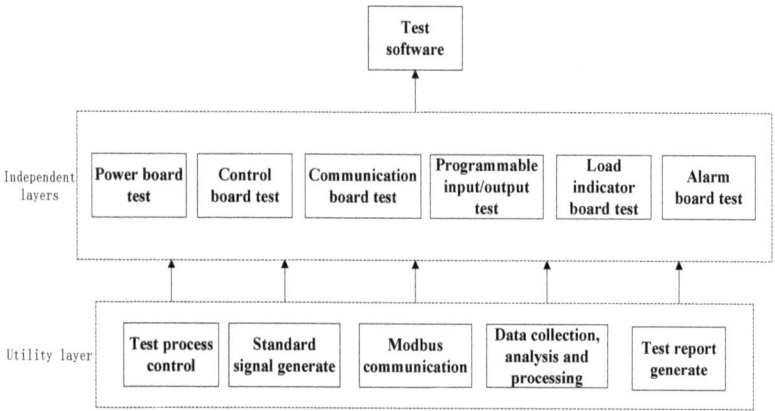

Fig. 4. The test software architecture diagram.

5. Application example

Here take the control and protection switch device circuit board performance test system as an example to shows the superiority of the key technology introduced in this paper.

CPS circuit board performance test system mainly includes two parts of hardware and software. Hardware system consists of industrial computer, printer, signal conditioning circuit, multi-function data acquisition card and PC-6229I and PCI -6133), etc. Software part is realized by the currently widely used software development platform ——Lab VIEW 7. The overall structure of the circuit board test system is shown in figure 5.

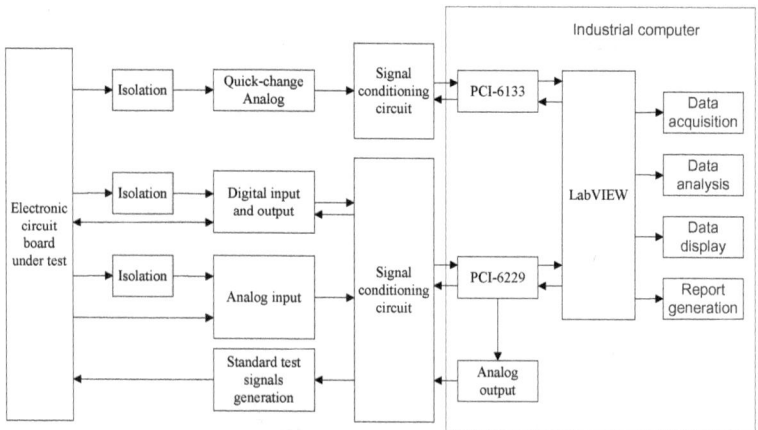

Fig. 5. The overall structure of the circuit board test system.

206

During the test, the standard signal source is controlled by the computer to output the proper test signal which is loaded into the measured circuit board, and then the output value of the test points in the circuit board are acquired back to the computer by the use of data acquisition card and analog isolation acquisition technology. Comparing with standard value, if the error is in the scope of the permit the correspondence performance of the circuit board is qualified. At last the test results are displayed in the interface of the test software. The test interface is shown in figure 6.

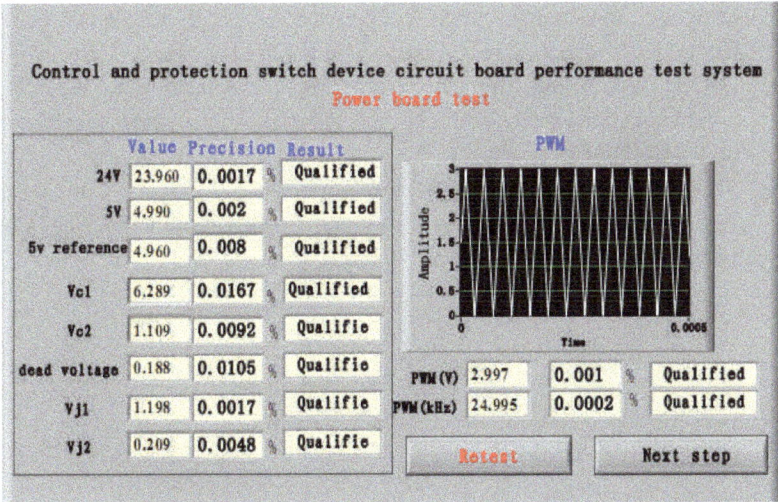

Fig. 6. The test interface of the system.

6. Conclusion

Three key technologies for the electronic circuit board performance testing — analog isolation acquisition, standard signal generation, and test software architecture design are studied in this paper, which are applied to a performance test system of control and protection switch device circuit board. Through the practical application shows that the key technologies not only can satisfy the demand of circuit board performance test also has the benefit of high precision, strong anti-interference ability and the higher efficiency. Therefore, the key technologies have a good prospect and promotional value.

Acknowledgment

This work is funded by grant BY2014075 of Jiangsu Science and Technology Project Foundation.

References

1. Degui Chen, New intelligent technique of low voltage apparatus, Low voltage apparatus 1, pp.1-5 (2008)
2. Bairong Cao and Haihua Yu, Designing automatic test system of electronic circuit board by using VB, Process automation instrumentation 11, pp.40-42 (2007)
3. Qi Xie and Yu Fang, Research on key techniques of developing test system based on LabVIEW, Machine Tool & Hydraulics 10, pp.151-153 (2005)
4. Yi Wang, Qi Xie, Ting Lv and Qimin Gu, Design and realization of high accuracy and multi–channels synchronous sinusoid signal generator, Computer Measurement 11, pp.2663-2665 (2010)
5. Qi Xie, Qimin Gu, ShuiLin Tu, Shaolin Ji. Realization of Modbus RTU communication protocol based on Labview, Coal mine machinery 27, pp.95-97 (2008)
6. Shui Lin Tu, Qimin Gu Qi Xie and Yunfei Yang, Design of Comprehensive Performance Test System for Air Circuit Breaker, Low voltage apparatus 2, pp.45-48 (2010)

A Low Power Double/Single-Leg Cascaded Type AC-AC Switched-Capacitor Converter

Li-ting Bao⁺, Hui Cai, Yin Wang and Zi-xing Zhang

College of Mechanical and Electrical Engineering,
China Jiliang University,
Hangzhou, Zhejiang, China
E-mail: ⁺litingball@hotmail.com
caihui@cjlu.edu, cnsarawy@cjlu.edu.cn

This paper presents a cascaded type switched-capacitor converter in the ac-ac field. The converter topology is described and analyzed. The main advantage of this proposed converter is the absence of magnetic elements which significantly reduces its volume and improves the efficiency. Based on the principle of switched-capacitor, the double/single-leg cascaded converter is able to achieve a fixed conversion ratio of 1/4 for the step-down configuration and a ratio of 4 for the step-up configuration. The theoretical analysis of the equivalent resistance of the converter is given and a prototype of 150w, 220V_{rms} high-side voltage, 55V_{rms} low-side voltage and switching frequency of 50kHz is built to verify the correctness through experimental results.

Keywords: Switched Capacitor; Cascaded Converter; Step-Down; Equivalent Resistance.

1. Introduction

Switched capacitor converters (denoted henceforth as SCC for singular and plural) have been popular for many years as their performances such as power density and efficiency have improved and there is an increasing demand for low power supplies for mobile electronic systems[1]. Various topologies and control methods have been proposed and successfully applied[2,3].

Unlike the conventional power converters implemented using magnetic components which lead to large sizes and low power density, SCC can achieve good performance with simple structures and they are suitable for chip-level power conversions[4,5]. Previous studies on SC dc-dc converters have verified that the use of SC principle provides good results[6]. Recent publications[7,8] for the first time extended SC principle to ac-ac static conversion, whose results are promising and show that the employment of an SC can contribute efficient solutions in ac-ac area.

In this paper, a low power dual/single-arm cascaded ac-ac converter based on the SC principle is proposed and analyzed. The operation principle and analysis of equivalent resistance are described. A prototype of 150w, 220V$_{rms}$ high-side voltage, 55V$_{rms}$ low-side voltage and switching frequency of 50kHz is built to demonstrate the performance of the cascaded converter.

2. The Proposed Dual/Single-leg Cascaded Switched-Capacitor Converter

The proposed cascaded switched-capacitor ac-ac converter is presented in Fig. 1(a) and Fig. 1(b), which operate as step-down converters. The only difference between the step-up and step-down configuration is the points where the source and the loads are connected. In this paper, we focus on the dual/single-arm topology in Fig. 1(a).

Taking the Fig. 1(a) as an example, the first stage of the circuit is a double-leg switched capacitor converter and the second stage is a single-leg switched-capacitor converter. The first stage consists of eight MOSFETs represented as S_1 to S_8 which are connected in series on two legs and six capacitors represented as C_1 to C_6. The second stage consists of four bidirectional switches represented as S_9 to S_{12} and three capacitors represented as C_7 to C_9. Each bidirectional switch is comprised of two MOSFETs.

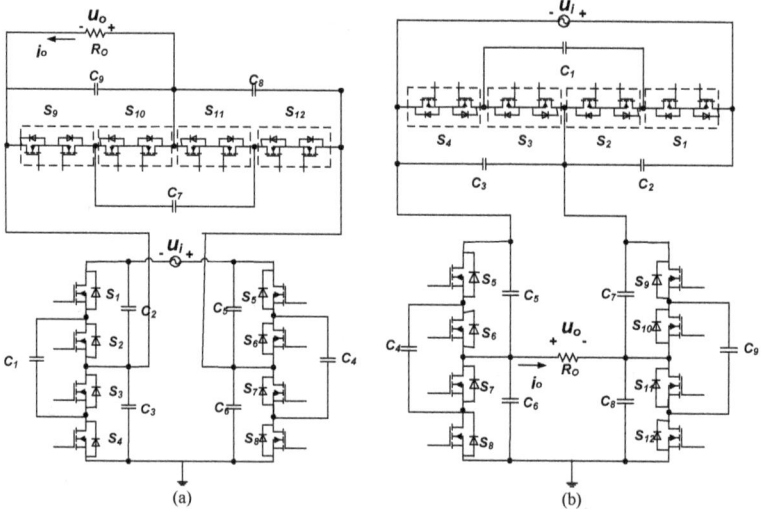

Fig. 1. The proposed cascaded step-down switched-capacitor ac-ac converter: (a) the dual/single-leg cascaded configuration and (b) the single/dual-leg cascaded configuration.

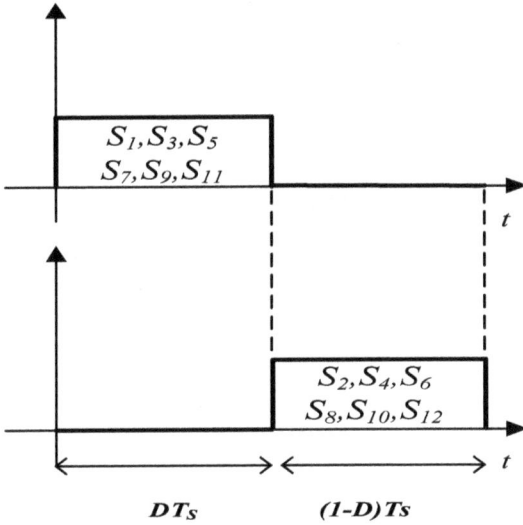

Fig. 2. Gate drive signals of the proposed converter.

The gate drive signals of the converter are shown in Fig. 2. The switching frequency is 50kHz and duty cycle is 0.5 to minimize the losses and consequently increase the efficiency of the converter.

3. Theoretical Analysis

3.1 Operation Principle

The main output characteristic of the proposed cascaded converter is the fixed conversion ratio of 1/4 for the step-down operation and a ratio of 4 for the step-up operation. Because the capacitors in the first stage keep charging and discharging, this stage can be seen as an ideal voltage source which is isolated from the second stage and this noninterference feature ensure the electrical parameters of the two stages independent from each other.

Thanks to the switched-capacitor principle, capacitor C_1 and C_4 ensure the voltage balance between C_2, C_3, C_5 and C_6 and capacitor C_9 ensure the voltage balance between C_7 and C_8 in Fig. 1(a). During the working stages of the converter, capacitors C_1, C_2 and C_3 in the dual-leg converter have an ac component equal to $V_i/2$ in the positive half-cycle of the grid voltage while capacitors C_4, C_5 and C_6 have that in the negative half-cycle of the grid voltage. All the capacitors in the single-leg converter have an ac component equal to $V_i/4$.

211

The theoretical voltage gain is $U_o/U_i=1/4$ for the step-down configuration. The theoretical voltage waveforms are given in Fig. 3.

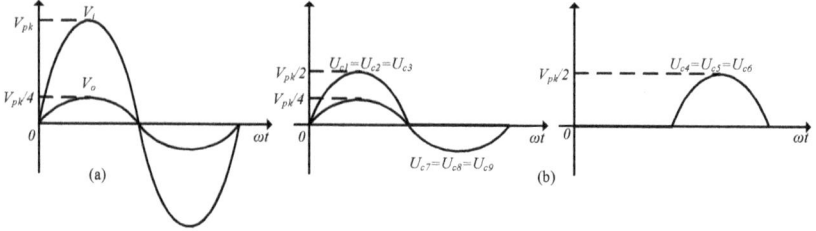

Fig. 3. Theoretical low-frequency waveforms. (a)input and output voltages. (b)voltages across all the capacitors.

The proposed cascaded step-down converter has two operation stages per switching period. The left leg of the dual-leg converter only works in the positive half-cycle and the right leg only works in the negative half-cycle.

Stage I starts when switches S_1, S_3, S_5, S_7, S_9 and S_{11} are turned on and the rest are turned off during the positive half-cycle of the grid voltage. Capacitors C_1 and C_7 are charged during this stage. At first capacitors C_2 and C_8 discharge and C_3 and C_9 charge until their currents reach zero(Δt_{11}). Then capacitors C_2 and C_8 start to charge and capacitors C_3 and C_9 start to discharge until the end of stage one(Δt_{12}). Switches S_1, S_3, S_5, S_7, S_9 and S_{11} are turned off and the rest switches are turned on.

Stage II: Capacitors C_1 and C_7 discharge during this stage. At first, capacitors C_2 and C_8 discharge and C_3 and C_9 charge until their currents reach zero(Δt_{21}). The capacitors C_2 and C_8 start to charge and capacitors C_3 and C_9 start to discharge until the end of the stage. Switches S_2, S_4, S_6, S_8, S_{10} and S_{12} are turned off and the rest switches are turned on to start the Stage I again. These stages are similar during the negative half-cycle of the grid voltage. The theoretical waveforms of capacitors of the converter are presented in Fig. 4(a) and Fig. 4(b).

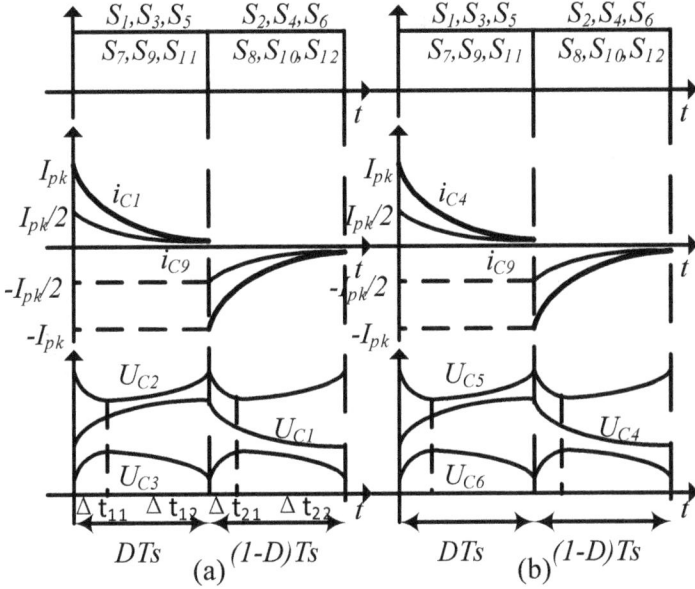

Fig. 4. Theoretical high-frequency waveforms. (a) the positive half-cycle (b) the negative half-cycle.

3.2 Equivalent Resistance Analysis

Generally, SC converters are characterized by a voltage transfer ratio which is determined by the topology. Additionally, they have an equivalent resistance determined by the switching frequency and the choice of components of the converter[9].

The equivalent resistance of the first stage of the ac-ac converter for a duty cycle of 0.5 can be defined as:

$$R_{eq1(D=0.5)} = \frac{1}{2 \cdot f_s \cdot C} \cdot \frac{(1 + e^{-\frac{1}{2 \cdot f_s \cdot R_{on1} \cdot C}})}{(1 - e^{-\frac{1}{2 \cdot f_s \cdot R_{on1} \cdot C}})} \tag{1}$$

The equivalent resistance of the second stage of the ac-ac converter for a duty cycle of 0.5 can be defined as:

$$R_{eq2(D=0.5)} = \frac{1}{f_s \cdot C} \cdot \frac{(1 + e^{-\frac{1}{2 \cdot f_s \cdot 2 \cdot R_{on2} \cdot C}})}{(1 - e^{-\frac{1}{2 \cdot f_s \cdot 2 \cdot R_{on2} \cdot C}})} \tag{2}$$

In Equation(1) and (2), f_s is the switching frequency of all MOSFETs, R_{on1} is the on resistance of a bidirectional switch which is twice the on resistance of a

213

MOSFET, R_{on2} is the on resistance of a MOSFET and C represents the capacitance of all capacitors.

In case that the two stages are electrically isolated, the cascaded converter can be seen as an ideal transformer and the energy transferring efficiency as 100%. A simplified equivalent circuit can be derived as in Fig. 5.

The current i_1 in the first stage is half of that(i_2) in the second stage and the relation between U_{o1} and U_{o2} is just the opposite. These relations lead to Equation(3). The equivalent resistance of the proposed converter is described through the load-side of the converter, which is presented in Equation(4).

$$R_{eq1}/4 = R_{eq1}' \tag{3}$$

$$R_{eq(D=0.5)} = \frac{1}{4 \cdot 2 \cdot f_s \cdot C} \cdot \frac{\left(1 + e^{-\frac{1}{2 \cdot f_s \cdot R_{on1} \cdot C}}\right)}{\left(1 - e^{-\frac{1}{2 \cdot f_s \cdot R_{on1} \cdot C}}\right)} + \frac{1}{f_s \cdot C} \cdot \frac{\left(1 + e^{-\frac{1}{2 \cdot f_s \cdot 2 \cdot R_{on2} \cdot C}}\right)}{\left(1 - e^{-\frac{1}{2 \cdot f_s \cdot 2 \cdot R_{on2} \cdot C}}\right)} \tag{4}$$

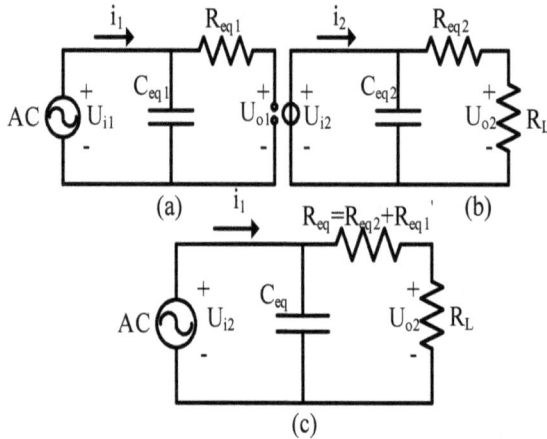

Fig. 5. Equivalent circuits of (a) the first stage, (b) the second stage and (c) the cascaded ac-ac converter seen by load-side.

4. Experimental Verification

In order to validate the correctness of the theoretical analysis as is stated above, a prototype of 150w, 220V$_{rms}$ high-side voltage, 55V$_{rms}$ low-side voltage and switching frequency at 50kHz is built with components listed in Table 1. The input current of the converter is filtered by an LC circuit to reduce the high-frequency components to improve the transfer efficiency.

Table 1 Parameters of components in the prototype

Description	Values
Input Voltage	220V
Output Voltage	55V
Output Power	150w
Switching frequency	50kHz
Capacitors	10μF
MOSFETs	FQA25N62C

The experimental results are shown in Fig. 6.

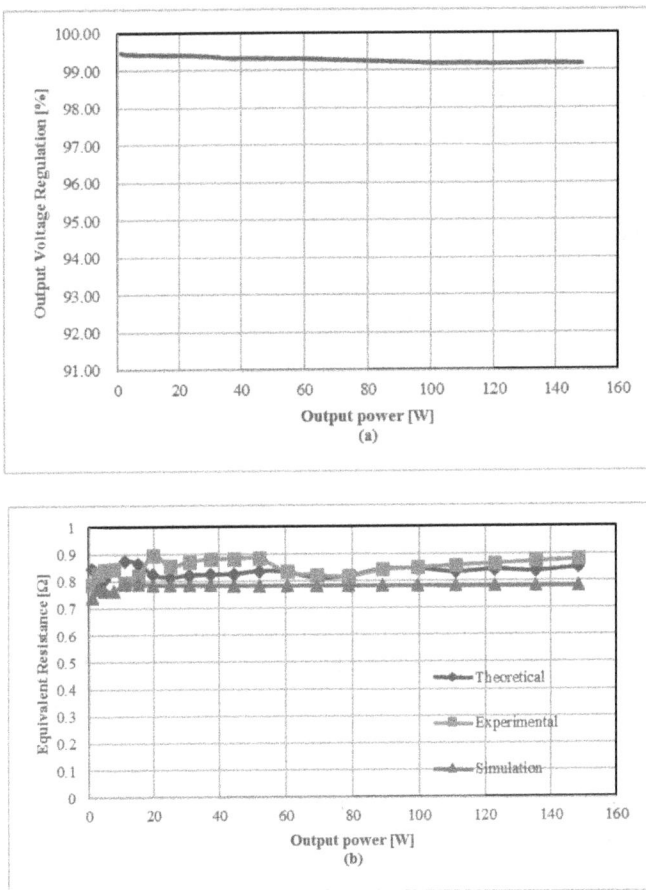

Fig.6. (a) Output voltage Regulation. (b) Equivalent Resistance. Theoretical curve is obtained from Equation (4). Simulation curve is obtained from PSpice A/D.

The voltage conversion ratio is close to 1/4 and remains nearly unchanged when output power increases. The experimental curve of equivalent resistance matches quite well with the theoretical calculation curve as well as the simulation curve. The efficiency is quite good for a cascaded converter. The experimental results are quite consistent with the previous theoretical analysis.

5. Conclusion

The paper presents a cascaded type of ac-ac converter based on the switched capacitor principle without magnetic components. The converter is able to achieve a conversion ratio of 1/4 for the step-down configuration with simple control methods. The theoretical equivalent resistance analysis of the converter is validated by experimental results.

The SCC is a potential candidate for replacing the conventional low-power and high-power autotransformer. There is still much to be extensively researched in the field of SCC, whether in the dc-dc or the ac-ac field.

Acknowledgment

This work is funded by National Natural Science Foundation of China under Grant 51407173, and Zhejiang Provincial Natural Science Foundation of China under Grant LQ13E070001.

References

1. Ioinovici, A., "Switched-capacitor power electronics circuits," in *Circuits and Systems Magazine, IEEE*, Vol.1, No.3, pp.37-42, Third Quarter 2001.
2. Arntzen, B; Maksimovic, D, "Switched-Capacitor DC/DC Converters with Resonant Gate Drive," in *IEEE Trans. on Power Electronics*, Vol.13, No.5, Sept.1998.
3. Sano, K.; Fujita, H., "Voltage-balancing circuit based on a resonant switched-capacitor converter for multilevel inverters," in *IEEE Trans. Ind. Appl.*, Vol. 44, No. 6, pp. 1768–1776, Nov. 2008.
4. Thiele, G.; Bayer, E., "Voltage doubler/tripler current-mode charge pump topology with simple "gear box", in *Proc. Power Electron. Spec. Conf.*, 2007, pp. 2348–2352.
5. C.-H. Hu and L.-K. Chang, "Analysis and modeling of on-chip charge pump designs based on pumping gain increase circuits with a resistive load," in *IEEE Trans. Power Electron.*, Vol. 23, No. 4, pp. 2187–2194, Jul. 2008.

6. Seeman, M.D., Sanders, S.R., "Analysis and Optimization of Switched-Capacitor DC-DC Converters," in *Computers in Power Electronics*, 2006. COMPEL '06. IEEE Workshops on, Vol., No., pp.216-224, 16-19 July 2006.

7. Lazzarin, T.B., R. L. Andersen, Martins, G.B.; Barbi, I., "A 600 W switched-capacitor ac-ac converter for 220 V/110 V and 110 V/220 V applications," in *IEEE Trans. Power Electron.*, Vol. 27, No. 12, pp. 4821–4826, Dec. 2012.

8. Andersen, R.L., Lazzarin, T.B., Barbi, I., "A 1-kW Step-Up/Step-Down Switched-Capacitor AC–AC Converter," in *Power Electronics, IEEE Transactions on*, Vol.28, No.7, pp.3329-3340, July 2013.

9. Kimball, J.W., Krein, P.T., Cahill, K.R., "Modeling of capacitor impedance in switching converters," in *Power Electronics Letters, IEEE*, Vol.3, No.4, pp.136-140, Dec. 2005.

Research on Soft-Switching Equipment and Its Control Method

Jian Chen, Wei Wang, Hui Cai†, Li-tin Bao and Chun-wei Song

College of Mechanical and Electrical Engineering, China Jiliang University,
Hangzhou, Zhejiang, China
†*E-mail: caihui@cjlu.edu.cn*

Soft-switching equipment is used for switching load softly among power phases without powering down the system, whose key component is a single-phase inverter. The equipment is designed to solve the three-phase unbalanced problems of small power supply from load-side. This paper presents a new topology of soft-switching equipment. In order to solve the grid-connected circulation problem at the voltage mode, built-in load is designed. Then the mathematical model of transfer function is built for analyzing the reason circulation is inhibited when the equipment with built-in load is grid-connected. In this case, it would produce smaller magnitude difference. The design method for the parameter of built-in load is given. Finally, it is verified through experiments that built-in load can restrain circulation current and the correctness of theoretical analysis are proved.

Keywords: Soft-switching; Single-phase inverter; Grid-connected; Circulation current.

1. Introduction

There are a lot of solutions for the three-phase imbalance at home and abroad, the reactive compensation devices based on thruster controlled capacitor switching have been widely used in three phase imbalance practical problems. But such methods are not able to solve the problem of three phase imbalance fundamentally, but provide static compensation and dynamic compensation for supply-side.

Zeng et al. [1] proposes a suppressing method of three-phase unbalanced overvoltage based on distribution networks flexible grounding controlling. Fang, et al. [2] proposes a method for real-time online control of three-phase unbalanced load in distribution area. Zheng et al. [3] proposes a method of fast unbalanced three-phase based on single-phase load switching.

Cai et al. [4] presents a soft-switching device of load supply phase for terminal power grid, and its general structure design is shown in Fig. 1. The core component of the device is single-phase grid inverter.

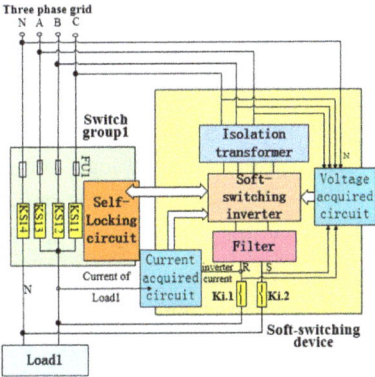

Fig. 1. Structure of soft-switching equipment.

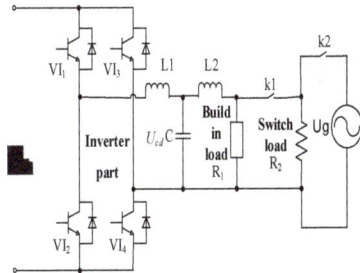

Fig. 2. The main circuit of inverter.

This device allows the loads to work in the normal online operation, and complete the task of soft-switching the position of power supply phase. This device can be used to adjust the position of the three-phase terminal grid load power supply phase, so as to balance three-phase load imbalance and brings new solutions for terminal grid three-phase unbalanced. In this paper, we propose a new soft-switching device topology based on the study of Cai et al. [4], presents research on mathematical model of soft-switching device (i.e., grid-connected inverter) firstly. Then the case of the device with built-in load was analyzed based on the mathematical model. Finally, it is concluded that built-in load has better restrain effect on the circulation generated by soft-switching devices when it is connected with the grid and improves the operational reliability of the device.

2.   The Implementation of Soft-Switching Device

Hui Cai et al. [4], the terminal loads are grouped according to a fixed power, and each load group is designed with a three-phase power supply phase adjustment switch KS, so as to make the load work in any position of power supply phase. Soft-switching device acts as a transition power of position adjustment of power supply phase when power supply phase is switching. The device will first track the current power supply phase voltage signal, connect with the current power grid after synchronization, and turn off the switch of current power supply phase, and the load transitions to the soft-switching device. In this case, the device begins tracking the target power supply phase position, closes grid switches to connect with it after synchronization, and then disconnect switch of device

219

to complete the whole switching task. Fig. 2 is the model of soft-switching inverter, and k1 is grid-connect switch, k2 is grid-disconnect switch, R_1 is build-in load, R_2 is switch load.

Chen [5], Gan Ding et al. [6], Bo Liu et al. [7], and C.H. Wu [8] pointed out that the use of voltage control mode PIP control strategy can reduce instability under the current control mode, the output current distortion and other issues. Soft-switching devices also use voltage mode PIP in this paper.

3. Analysis of Circulation Inhibition

Soft-switching device use voltage source control mode during the operating time, since the load voltage is clamped to U_g, in the case of that inverter works in the form of voltage and in parallel with grid, an instant circulation between the grid and the inverter is easily formed [9].

3.1. Analysis of before connect with grid

In order to analyze influence on circulation when the soft-switching device connects with grid with and without built-in load R_1, there will be a theory analysis through Matlab simulation.

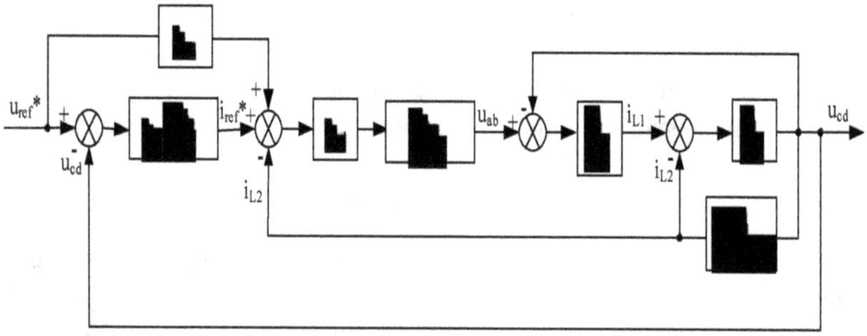

Fig. 3. With a built-in load block diagram of the system before the grid-time.

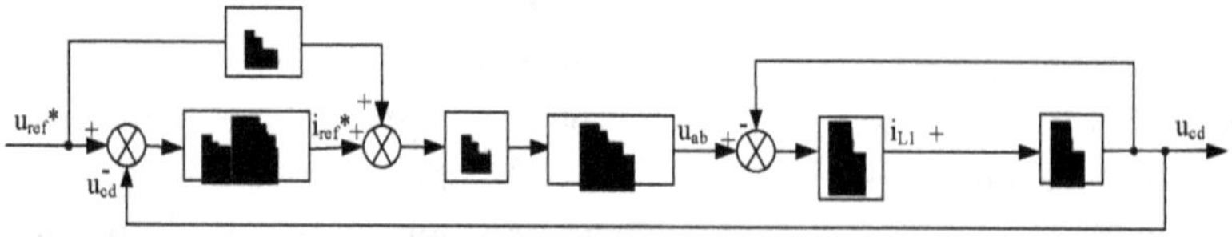

Fig. 4. Without a built-in load block diagram of the system before the grid-time.

220

This device adopts i_{L2} as the feedback current of current inner loop, structure diagram of the device with a built-in load is shown in Fig. 3, and the structure diagram without built-in load is shown in Fig. 4.

Comparing Fig. 3 with Fig. 4, it can be found that before connecting with the grid without built-in load, $i_{L2} = 0$, and the system is a single voltage outer loop closed loop system, but when it's with built-in load, $i_{L2} \neq 0$, and the system is voltage current double closed loop system. At the moment of connecting with grid, the change is not the same under these conditions, so $\Delta i_{L2}' < \Delta i_{L2}''$, and $\Delta i_{L2}'$ is variation of i_{L2} with built-in load, and $\Delta i_{L2}''$ is variation of i_{L2} without built-in load.

3.2. Analysis after connect with grid

In order to analyze the influence to U_{cd} from both variations of change at the moment of connecting with grid, and in case that the current inner loop response speed is faster, so the effect from voltage outer loop does not need to be considered, structure diagrams under two kinds of conditions respectively are built, setting Δi_{L2} as given input of the inner loop, and ΔU_{cd} as a current loop output. Fig. 5(a) is a structure diagram of current inner loop with built-in load, and $R = R_1 // R_2$. Fig. 5(b) is a structure diagram of current inner loop without built-in load, and $R = R_2$.

Below are the parameters of structure diagram: $k_{inv} = 123$, T=0.0001s, $L_1 = 0.01\text{H}$, $C = 20 \times 10^{-6}\,\text{F}$, $L_2 = 2 \times 10^{-3}\,\text{H}$, $R_1 = 10\Omega$, $R_2 = 50\Omega$.

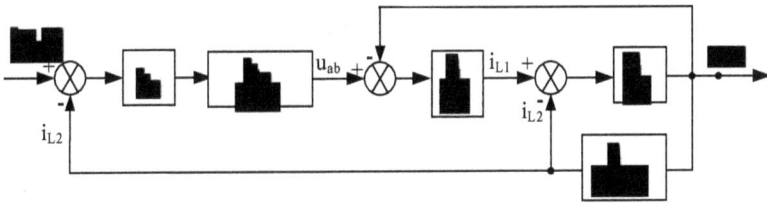

Fig. 5(a). Block diagram of a built-in load current loop.

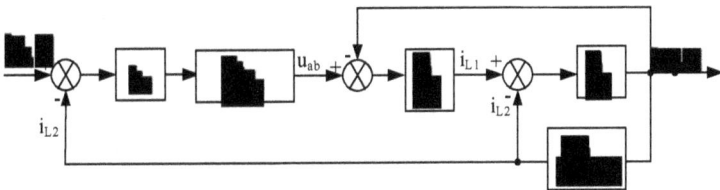

Fig. 5(b). Block diagram of current loop without build-in resistance.

221

The closed-loop transfer function is as follows:

$$G(S) = \frac{a_1 s + a_2}{b_1 s^2 + b_2 s + b_3} \tag{1}$$

In(1) $a_1 = k_{pi}k_{inv}L_2$, $a_2 = k_{pi}k_{inv}R$, $b_1 = T(RCL_1 + L_1 + L_2)$, $b_2 = RT + RCL_1 + L_1 + L_2$, $b_3 = R + k_{pi}k_{inv}$.

Ignoring higher-order terms and the coefficient of smaller items, current inner loop transfer function can be written as:

$$G(s) = \frac{K(As+1)}{Bs+1} \tag{2}$$

K, A, B is

$$K = \frac{k_{pi}k_{inv}R}{k_{pi}k_{inv} + R}, A = \frac{L_2}{R}, B = \frac{RT + RCL_1 + L_1 + L_2}{R + k_{pi}k_{inv}}$$

After substituting the actual values, it is obtained that current inner loop transfer function under condition of with built-in load is:

$$G_1(S) = 3.089 \times \frac{2.40 \times 10^{-4} s + 1}{9.81 \times 10^{-4} s + 1} \tag{3}$$

Current inner loop transfer function under without built-in load is:

$$G_2(S) = 4.48 \times \frac{3.995 \times 10^{-5} s + 1}{2.27 \times 10^{-4} s + 1} \tag{4}$$

Fig. 6 is the comparison chart of Bode diagram of transfer function with a built-in load before and after simplification, it can be obtained from the figure that both waveforms are approximate before the frequency of 100π, but this paper studies the amplitude under the frequency of 100π, so higher order terms is reasonably ignored. Simplify of the transfer function with no built-in load is similar too.

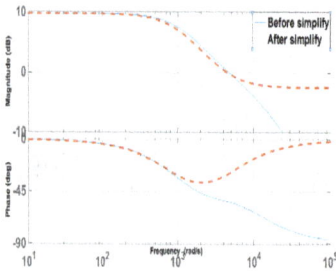

Fig. 6. Comparison chart of simplified transfer function.

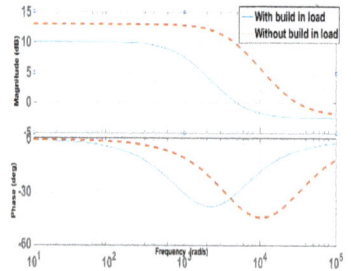

Fig. 7. Amplitude-frequency characteristics of the two conditions and phase-frequency characteristic comparison.

After simplification, the figure of amplitude-frequency characteristics and phase-frequency characteristic of the two conditions are shown in Fig. 7. According to Bode diagram, it can be seen that $\Delta U'_{cd} = 10.075 < \Delta U_{cd}" = 13.025$ under the same input conditions, that is to say comparing with the condition of without built-in load, the output Δi_{L2} of system with built-in load have smaller affects to output ΔU_{cd} of the system. But the reality is $\Delta i_{L2}' < \Delta i_{L2}"$, so the output at the condition of with built-in load is even smaller.

Suppose that before connecting with the grid, the output U_{cd} and U_g of two conditions are the same, so at the moment of connecting with the grid

$$U_{cd} = U_g + \Delta U_{cd} \tag{5}$$

Based on the analysis above, compare the voltage of both cases at the moment of connecting with the grid

$$U_{cd}' < U_{cd}" \tag{6}$$

Because the circulation is related to grid voltage U_g and the output voltage U_{cd} of inverter, so the circulation is

$$I_H = \frac{U_{cd} - U_g}{Z} = \frac{\Delta U_{cd}}{Z} \tag{7}$$

z is the line impedance. Theoretically only when $U_g = U_{cd}$, namely the circulation will not appear when U_g and U_{cd} are equal among both amplitude and phase, the circulation is larger when the difference between U_g and U_{cd} is larger.

So it can be concluded that a built-in load can obtain better circulation inhibition effect.

3.3. Analyze for the ability of circulation inhibition among different resistance values of built-in load control capability analysis

Because the amplitude is related to the K value when the system frequency is 100π, therefore the graph K under the different R value is shown in Fig. 8.

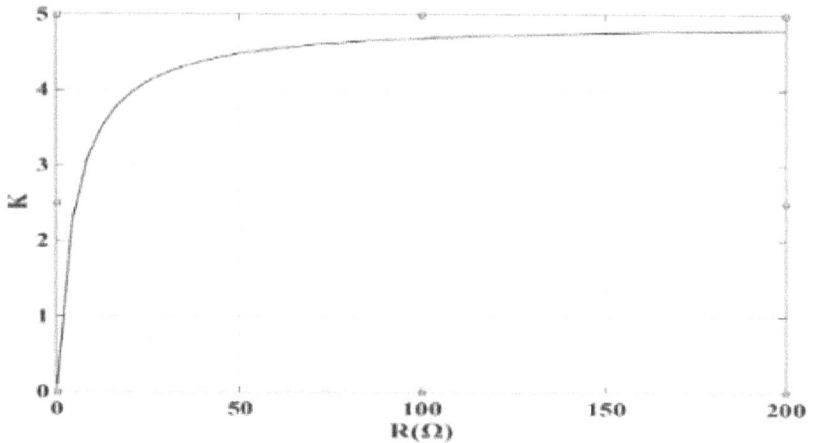

Fig. 8. Plot of K-R.

According to Fig. 8, it can be drawn that with the increases of R, the impact on amplitude is increasingly smaller, and the amplitude rises fast at R = 0 ~ 10, the rising rate of the magnitude slows after R> 10, so it can be concluded that the smaller built-in load is, the greater the ability of inhibit circulation is. But on the other hand, the smaller built-in load is, the greater the power consumption is. Because soft-switching device switches from the A phase to the B phase in a very short time, if the device run time is short, in order to suppress the circulation and network time, a certain degree of power loss is acceptable, so we can choose a smaller resistance load a built-in load.

4. Experimental Verification

Inverter of soft-switching device experimental platform is obtained by the three-phase inverter converted, control chip is DSP (TM320LF2407), and its main parameters are: power is 4kW, grid power line voltage is 380V, bus voltage is 530V, switching frequency is 10kHz, inverter filter L1 is 8mH, C is 1uF, circulation suppression inductor L2 is 1.5mH. Because the output voltage peak exceeds the voltage value of the oscilloscope, so oscilloscope sampling to the actual voltage using resistance in series 3 partial pressure, so the test voltage is 1/3 of the actual.

Soft-switching experimental inverter platform is obtained by the three-phase inverter converted, control chip is DSP (TM320LF2407), and its main parameters are: power is 4kW, grid power line voltage is 380V, bus voltage is 530V, switching frequency is 10kHz, inverter filter L1 is 8mH, C is 1uF, circulation suppression inductor L2 is 1.5mH. The output voltage

224

peak exceeds the voltage value of the oscilloscope, so oscilloscope sampling to the actual voltage using resistance in series 3 partial pressure, so the test voltage is 1/3 of the actual one.

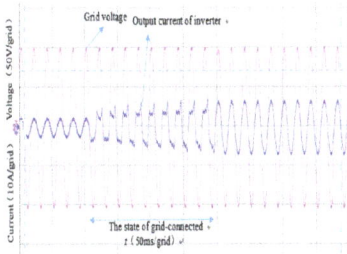

Fig. 9. Waveform of inverter without built-in resistance connected with grid directly.

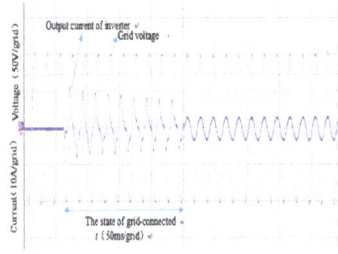

Fig. 10. Waveform of soft-switching equipment with built-in load connect with grid indirectly.

Fig. 9 is output voltage and current experimental waveforms of the soft-switching device without built-in load when connecting with grid and Fig. 10 is output voltage and current experimental waveforms of the soft-switching device with built-in load when connecting with grid. It can be seen more apparently from experimental results that the circulation is suppressed when connecting with grid with built-in load, and at the stage of connecting with the A phase, output current of soft-switching device gradually increases, which is of great benefit for the security of soft-switching device itself.

5. Conclusion

This paper proposes a new soft-switching device topology. A mathematical model of soft-switching device is set up, and theoretical analysis is carried out to explain the reason why built-in load can inhibit circulation, parameter is given for design of built-in load. Finally, the experimental results of the device with and without built-in load are given through experiment, which confirmed and verified soft-switching device with built-in load has better circulation inhibition effect.

Acknowledgment

This paper is funded by Zhejiang Provincial Natural Science Foundation of China under grant LQ16E070001.

References

1. X.J. Zeng, J.Y. Hu, Y.Y. Wang, et al Suppressing method of three-phase unbalanced overvoltage based on distribution networks flexible grounding control. In Proceedings of the CSEE, vol.34, no.4, pp: 678-684, Feb.2014.
2. H.F. Fang, W.X. Sheng, J.L. Wang, et al Research on the method for real-time online control of three-phase unbalanced load in distribution area. In Proceedings of the CSEE, vol.35, no.9, pp: 2185-2193, May.2015.
3. Zheng Y, Zoul, He J, et al. Fast unbalanced three-phase adjustment base on single-phase load switching. In TELKOMNIKA Indonesian Journal of Electrical Engineering, vol.11, no.8, pp: 4327-4334, Aug.2013.
4. Hui Cai, Yu Guo, Hong Yan. Design of soft-switching system on three-phase load balance for IDC electric power. In Journal of Zhejiang University (Engineering Science), vol: 48, no.12, pp: 2210-2215, Dec.2014.
5. W.M. Chen, Research on the Key technologies of PV Inverter Based on Micro Grid Operation, PhD thesis, Shanghai University, (Shanghai, China,2011), pp.ix+60.
6. Gan Ding, Y.S. Li, X.X. Wang. Study on New Applied Dual-Loop Control Technique for PWM Inverters. In Telecom Power Technology, vol: 26, no.5, pp: 22-25, sep.2009.
7. Bo Liu, Xu Yang, F.L. Kong, H.Z. Ye, Yu Hong. Control Strategy Study for Three Phase Photovoltaic Grid-Connected Inverters. In Transactions of China Electrotechnical Society, vol: 27, no.8, pp: 64-70, Aug.2012.
8. C.H. Wu, Zhang Yi, K.Y. Cui, G.C. Chen, W.M. Chen, Peng Xiao. Control Strategy for Single-Stage Three-Phase Boost-Type Grid-Connected Inverter. In Transactions of China Electrotechnical Society, vol: 24, no.2, pp: 108-113, Feb.2009.
9. X.F. Xun, H.R. Gu, L.Q. Wang, W.Y. Wu (eds.), High-frequency-switching inverter and parallel and network technology. (China Machine PRESS, 2011).

Modeling and Simulation of Electromagnetic Radiation of DC Link in Electric Vehicle Drive System

Guo-qiang Chen[†], Ya-hui Su and Jian-li Kang

School of Mechanical and Power Engineering, Henan Polytechnic University, Jiaozuo, 454000, China

[†]E-mail: jz97cgq@sina.com

Aiming at the problem of complicated structure and difficulty of analyzing electromagnetic interference of the drive system in the electric vehicle, a combination simulation method using several software packages is proposed to analyze the electromagnetic radiation distribution law of the Direct Current (DC) link. Firstly, the electric vehicle control model is built in MATLAB. Secondly, the geometric model of the vehicle body is built in CATIA and meshed in Hypermesh. Thirdly, the mesh model is integrated in FEKO and the DC link cable is modeled. Finally, the electromagnetic radiation distribution law is computed based on the DC link current measured from the control model used as the excitation source. The result verifies its efficiency and convenience.

Keywords: Electric Vehicle; Direct Current Link; Electromagnetic Radiation.

1. Introduction

The electric vehicle is regarded as a promising one that can solve the serious problem of environment pollution and the oil energy crisis [1-3]. The electric vehicle substitutes the electric motor and the battery for the conventional internal combustion engine and the fuel tank. The battery provides the power source to drive the electric motor in the drive system. The alternating current (AC) motor driven by an alternating current (AC) has been widely used in the electric drive system, so the direct current (DC) power source from the battery has to be converted to the AC source to feed the AC motor. Therefore, the three-phase two-level power inverter is always used to change DC to AC [4, 5]. The space vector PWM (SVPWM) scheme is one of key modulation schemes to control the three-phase two-level inverter and has been widely used in the electric vehicle drive system [6, 7]. Because SVPWM is based on the volt-second balance principle, the output waveforms of the inverter consist of series of pulses with different widths and the harmonic is inevitable besides the required fundamental or first harmonic [8,9]. For a highly inductive load, there are lots of current ripples in the load current. The steep pulses and current ripples result in serious effects on performance of the application, such as electromagnetic interference (EMI) and audible noise [4, 10, 11]. The repeated high frequency switch operations provide different EMI noises, such as differential and common mode noises and radiated noises, and the EMI noises disturb not only the passenger and the device in the vehicle itself but also the nearby vehicles [4]. The issues on how to analyze, assess and control EMI have gained plenty of research [4, 5, 12-14]. Because the drive system in the electric vehicle is very complex, it has been a challenge for vehicle design and assessment for a long time. It is difficult to analyze and reduce EMI only through experiments and theoretical analysis, so the simulation is a powerful tool here. Therefore, a combination simulation method using several software packages is proposed to analyze the electromagnetic radiation distribution law of the DC link in the paper. And the key simulation models and results are given and discussed.

2. Simulation Model of Drive System

The permanent magnet synchronous motor (PMSM) has been widely used in the electric vehicle and is used in the paper because of its high efficiency, high power density and large startup torque. The direct torque control strategy and the vector control strategy are the two mature PMSM control strategies, and every strategy has advantages and disadvantages. The vector control strategy is adopted because of the low torque ripple. The simulation model of PMSM control of an electric vehicle prototype is shown in Fig. 1, as is built in MATLAB/Simulink.

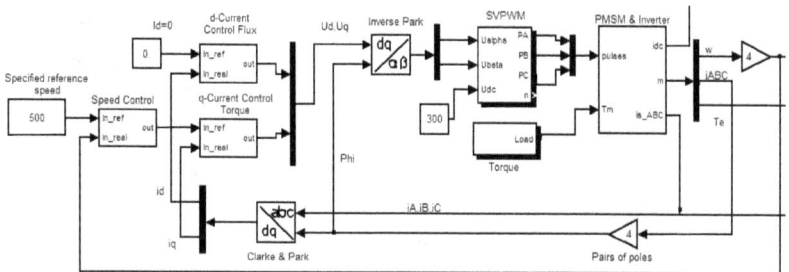

Fig. 1. Simulation model of PMSM control of an electric vehicle prototype.

The vector control strategy uses three reference frames: a stationary reference frame OABC, a stationary reference frame Oαβ and a rotating reference frame Odq.

The three phase currents in the stationary reference frame OABC are measured using two current sensors. For example, the currents of Phase A and B are measured as i_A and i_B, and the current i_C of Phase C is $-(i_A + i_B)$.

The three phase currents are transformed into a two-dimensional vector [i_α , i_β] in the stationary reference frame Oαβ using the Clarke transformation. The position angle φ (Phi in Fig. 1) of the motor is measured using an encoder or a rotary transformer. And the angle φ represents the angle between the rotating frame Odq and the stationary frame Oαβ . The two-dimensional vector [i_α , i_β] is transformed into a two-dimensional vector in the rotating reference frame Odq using the Park transformation.

The three controllers (Speed Control, d-Current Control Flux and q-Current Control Torque) in Fig.1 are based on the proportional–integral (PI) controller. The PI controller with anti-windup based on back-calculation is adopted in the simulation, and the simulation model is shown in Fig. 2.

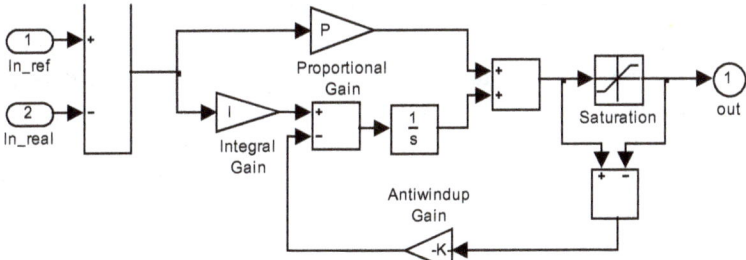

Fig. 2. Simulation model of the PI controller with anti-windup based on back-calculation.

The inputs of the two current PI controllers (d-Current Control Flux and q-Current Control Torque) are the errors between the reference and measurement currents of d-axis and q-axis, respectively. The outputs of the current PI controllers u_d and u_q , are transformed into the stationary reference frame Oαβ using the inverse Park transformation. The SVPWM scheme is used to generate the 6 control signals for the inverter. The simulation model of the inverter and the PMSM is shown in Fig. 3.

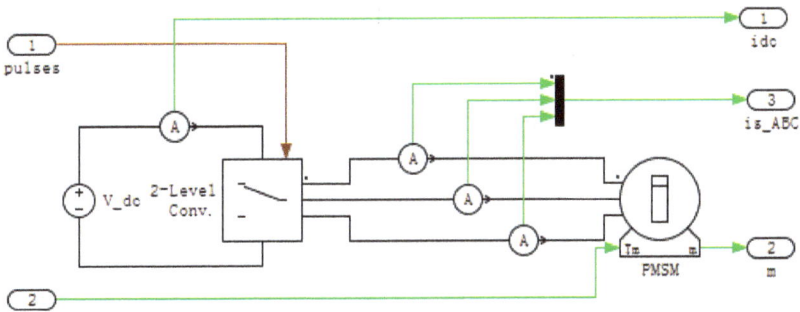

Fig. 3. Simulation model of the inverter and PMSM.

The three-dimensional geometric model of the vehicle body with the size 3570mm×1300mm×1440mm is built in the CATIA software. And then the geometric model is imported into the Hypermesh software and meshed. The mesh model is imported into the FEKO software and the DC link cable is modeled using the transmission line method. In simulation, the DC link current measured from the control model in Fig. 1 is used as the excitation source of electromagnetic radiation.

3. Results and Discussions

The simulation experiment is done using the built simulation model described in Section 2. The DC link voltage is 300V. The number of the pole pairs of PMSM is 4, the stator resistance is $0.0113\,\Omega$, the d-axis and q-axis inductance of the combined stator leakage and magnetizing inductance [Ld, Lq] is [0.000175, 0.000284] H, and the flux induced by magnet is 0.08424 Wb. The three phase and DC link currents, and the radiation current distribution in the body surface are shown in Fig. 4 and Fig. 5 given that the specified reference speed is 1200rad/s, the switching frequency is 10000Hz, and the DC link cable length without shielding is 100mm.

From Fig. 4, the DC link current composes of segments of the three phase currents i_A, i_B, i_C or $-i_A, -i_B, -i_C$, so it has the same period with the phase currents, and the period is determined by the steady speed of the PMSM. From another perspective, the DC current presents a feature of pulses with changing magnitudes as a whole. The time interval of the current pulses has an approximate period that is the switching frequency 10000Hz. Because of the symmetry of the SVPWM control signals fed to the inverter, the current waveform of the DC link current is symmetric in a switching period to some extent. Therefore, the incremental step of the frequencies that are corresponding to the large magnitudes in the spectrum charts of the radiated electric field is the double switching frequencies, which is verified by simulation. The current distributions in the body surface at 300 kHz and 500 kHz are shown in Fig. 5.

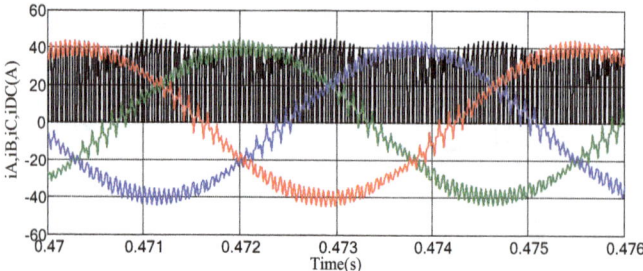

Fig. 4. Three phases and DC link currents

230

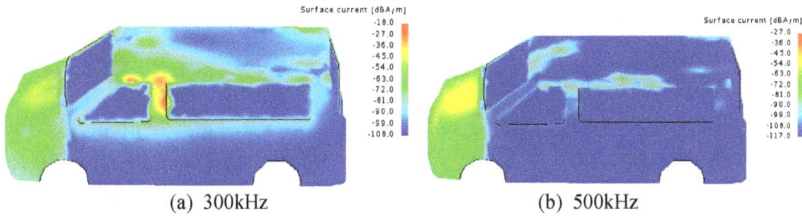

| (a) 300kHz | (b) 500kHz |

Fig. 5. Currents distribution in the body surface at two specified frequencies.

4. Conclusion

The simulation model is built and analyzed in the paper. The modeling procedure can be concluded as follows. Firstly, the PMSM control model is built in MATLAB/Simulink. In addition, the geometric model of the vehicle body is built in the CATIA software and meshed in the Hypermesh software. Thirdly, the mesh model is integrated in the FEKO software and the DC link cable is modelled. Finally, the electromagnetic radiation distribution is computed based on that the DC link current measured from the PMSM control model is used as the excitation source. The analysis and computation results verify the simulation model and show that the simulation method is very efficient and convenient for assessment and analysis of the electric vehicle. However, it should be noticed electromagnetic shielding is the highly effective practice of reducing electromagnetic radiation while the DC link is unshielded in the paper. To simulate electromagnetic radiation with shielding is our future work.

Acknowledgment

This work is funded by National Science Foundation of China (No. U1304525) and Scholarship Fund of China Scholarship Council (File No. 201408410031).

References

1. Ali Emadi, Young Joo Lee and Kaushik Rajashekara, Power Electronics and Motor Drives in Electric, Hybrid Electric, and Plug-in Hybrid Electric Vehicles, IEEE Transactions on Industrial Electronics 55, 2237(2008).
2. Kai-Wei Hu and Chang-Ming Liaw, Incorporated Operation Control of DC Microgrid and Electric Vehicle, IEEE Transactions on Industrial Electronics 63, 202 (2016).
3. Mahran Quraan, Taejung Yeo and Pietro Tricoli, Design and Control of Modular Multilevel Converters for Battery Electric Vehicles, IEEE Transactions on Power Electronics 31, 507 (2016).

4. Nobuyoshi Mutoh, Mikiharu Nakanishi, Masaki Kanesaki and Joji Nakashima, EMI Noise Control Methods Suitable for Electric Vehicle Drive Systems, IEEE Transactions on Electromagnetic Compatibility 47, 930(2005).

5. Hyunwoo Shim, Hongseok Kim, Younghwan Kwack, Minkang Moon, Hyunsuk Lee, Jinwook Song, Joungho Kim, Beomshik Kim and Eulyong Kim, Inverter Modeling Including Non-Ideal IGBT Characteristics in Hybrid Electric Vehicle for Accurate EMI Noise Prediction, in *Proc. IEEE International Symposium on Electromagnetic Compatibility (SEMC2015),* (Dresden, Germany, 2015).

6. Hemza Saidi, Rachid Taleb, Noureddine Mansour and Abdelhamid Midoun, Three Phase Inverter Using SVPWM Method for Solar Electric Vehicle, in *Proc. IEEE 6th International Renewable Energy Congress (IREC2015),* (Sousse, Tunisia, 2015).

7. Gang Zheng, Shengjie Guo, Shibin Cai, Shicai Fan and Yahui Ren, Efficiency Analysis of Two Dead-Time Compensation Methods for SVPWM in Motor Drive System of the Electric Vehicle, in *Proc. IEEE 7th International Power Electronics and Motion Control Conference (IPEMC 2012), (Harbin, China, 2012).*

8. Xiaolin Mao, Jain Amit Kumar and Ayyanar Rajapandian, Hybrid Interleaved Space Vector PWM for Ripple Reduction in Modular Converters, IEEE Transactions on Power Electronics 26, 1954 (2011).

9. D G Holmes, and T A Lipo, *Pulse Width Modulation for Power Converters: Principles and Practice.* (IEEE Press, USA) (2003).

10. Kaboli Shahriyar, Mahdavi Javad and Agah Ali, Application of Random PWM Technique for Reducing the Conducted Electromagnetic Emissions in Active Filters, IEEE Transactions on Industrial Electronics 54, 2333 (2007).

11. S. H. Na, Y. G. Jung, Y. C. Lim and S.H. Yang, Reduction of Audible Switching Noise in Induction Motor Drives Using Random Position Space Vector PWM, IEEE Proceedings Electric Power Applications 149, 195 (2002).

12. Hong-wu Wang, Ji Zhang and Teng-fei Yang, Status of the EMC Research on Drive System in Electric Vehicle, Electronic Measurement Technology 34, 18 (2011).

13. Ling Zheng and Hai-qing Long, Overview of Research on EMC of the Electric Vehicle Motor Drive System, Chinese Journal of Automotive Engineering 4, 319 (2014).

14. Ming-lei Du, Zhong-ming Xu and Liang-xu Ding, Simulation and Experimental Study on the Electromagnetic Radiation Interference of Power Cables in Hybrid Electric Cars, Automotive Vehicle 36, 734 (2014).

Research on the Characteristics of the Grid Icing Erlang Mountain Region

Song-hai Fan, Yi-yu Gong and Zhi-hang Xue

State Grid Sichuan Electric Power Research Institute
Chengdu, Sichuan, China
E-mail: 1051244307@qq.com

Based on grid icing observation data of Erlang Mountain region in the winter during 2013-2014, the characteristics of the grid icing in the regional Erlang Mountain and time variations were researched. The regional power grid and icing intensity classified into mild, moderate, severe icing, and ice characteristics of the growth process were also studied. Air temperature, wind direction, wind speed and other meteorological elements were discussed in the influence of the strength of the power grid ice. The results showed that: (1) in Erlang mountain areas, most icing are mild to moderate icing, severe icing phenomenon is relatively rare, except for the special effects of the weather system. (2) When the temperature is low, roughly -5 °C ~ -8 °C is the most conducive to ice crystals due to the air in the water, constant humidity and the lower the air temperature, the faster the ice formation. Certain temperature and humidity result in longer freezing and thicker ice. Wind speed plays a role in transporting water vapor and water droplets have an important influence on the formation of ice. This study demonstrates when the wind speed is at 2 ~ 6m/s, ice forms the fastest. (3) Temperature, wind speed and quantitative relationship of ice thickness are linear correlation.

Keywords: Grid icing; Temperature; Wind Direction; Wind Speed; Standard Ice Thickness.

1. Introduction

Icing electric wires can cause a wire fault and collapsed pole and so on, under certain weather conditions, such as wind speed and direction when reaching a certain level, will lead to the occurrence of dancing wire and cause wire flogger, line failure.

Research on wire icing foreign started earlier, scholars in Finland, Norway, Britain, Canada and other countries discussed the relationship between meteorological elements and wire icing.[1] Chinese scholar Tengzhong Lin discussed the impact of wind speed, terrain, wire diameter, electric field on the icing, and gives different heights icing changes. [2-4] Tan Guanri further explore ice thickness variation with height from the ground. [5-7] Jiang Xingliang and others are comprehensively analyzed the relationship between the transmission lines icing and meteorological elements, terrain, geographical conditions. [8-9]

Sichuan Basin is a relatively closed, the southeast area of mountainous relatively low, which is in favor of water vapor to enter and evacuation. However, higher ground in northwest mountain of Sichuan, go against moisture loss, high air humidity, perennial rainy foggy little wind, more parts of the grid to form ice created favorable conditions. The northern regions of Western Sichuan Plateau, North Mountain, mount emei, liangshan state are serious power grid ice regions in Sichuan basin. Erlang Mountain Region average altitude of 3000 meters is easier to form a grid ice, which will have a big impact on transmission work. Therefore, it is extremely important to study the characteristics of power grid ice mountain area and the influence of various meteorological elements on ice that can provide a reference and technical support for the design of transmission lines.

2. Relationship between Icing and Meteorological Elements

2.1. *Effect of temperature on the icing*

Li Luping and other people think that the air temperature is -5 °C ~ -8 °C, the vapor in the air most likely to form ice crystals. [10] At the same time, the moisture content of the air will also greatly affect the formation of ice crystals. Under certain conditions of humidity, the lower the temperature, the faster the water vapor freezes; when temperature and humidity is constant, the longer freeze time, the thicker icing.

In figure 1 the blue curve is the temperature change map between February 4 to 13, 2014. Since February 5, the temperature showed a downward trend, on the 10th to reach a minimum temperature -12.0°C, expected to have a more serious form of icing. The black curve is ice thickness curve from February 4 to February 13, 2014. It can be seen from the figure; the ice thickness began to increase from the 4th until the 11th reaches a maximum 45mm. The following figure also reflects the inverse relationship between temperature and ice thickness, the lower the temperature, the larger ice thickness, the faster icing. In addition, although 11 February temperatures higher than the 10th, but the ice thickness has reached the maximum value, can be seen not only icing intensity and temperature-related, and other meteorological factors are also linked.

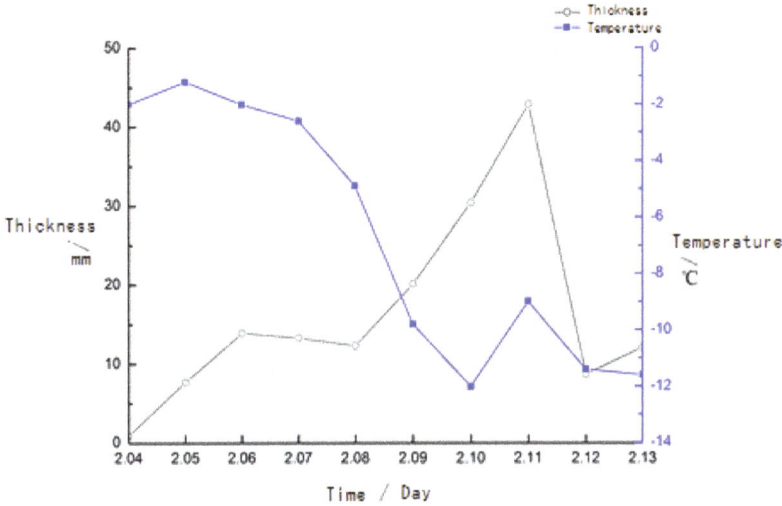

Fig. 1. The curve of the temperature and the ice thickness.

2.2. Influence of wind direction on icing

Table 1. Wind direction Distribution.

Wind direction	S	N	SW	SSW	C
Days	8	10	1	3	2

As can be seen from Table 1, when the angle of wind direction and wire is zero or less than 45°or greater than 150°, less icing; when the angle of wind direction and wire is 90°or greater than 45°or less than 150°, relatively severe icing.

2.3. Influence of wind speed on icing

On the icing process, wind speed also plays an important role. Figure 2 shows when wind speed reaches 2 ~ 6m / s, icing fastest.

Yang Mingli, etc. think the wind speed plays a role in transporting water vapor and droplets, has an important impact on the formation of icing. [11] When there is no wind or breeze, it is conducive to the formation of crystalline rime; When wind speed is large, it is conducive to the formation of granular rime. And near surface the density of wind speed and fog increases with height increases from the ground, thus rime, glaze and other icy objects also increases with increasing height above the ground, the ice thickness will also increase with the height of hang wire point increases.

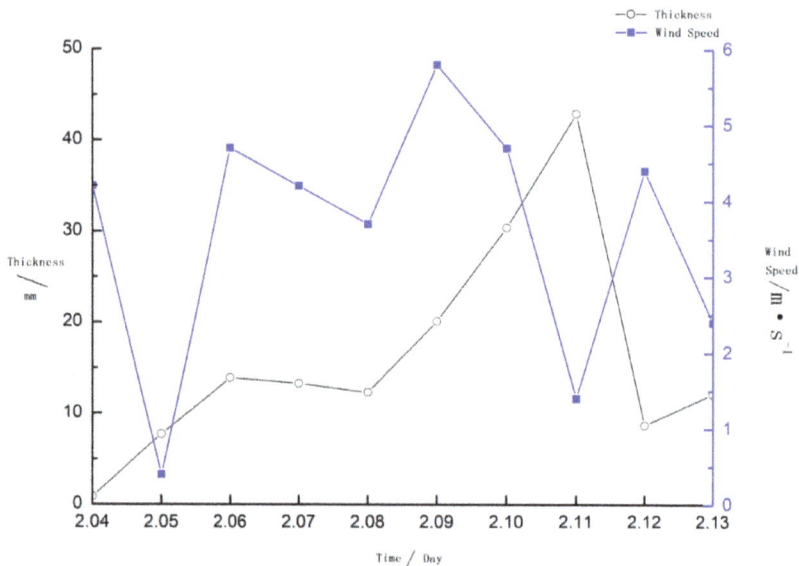

Fig. 2. Relationship between wind speed and ice thickness.

3. Linear Regression Analysis of Ice Thickness, Temperature and Wind Speed

Table 2 below used icing data from February 4 to February 11, 2014, lasting eight days, ice thickness data after treatment using a least squares method to fit linear regression analysis with various meteorological elements, then got the following linear regression equations.

Table 2. Feb. 4 - Feb. 11 ice thickness and wind speed, temperature data.

Date	Temp (°C)	Wind Speed (m/S)	Thickness (mm)
2.4	-2.0	4.2	0.9
2.5	-1.2	0.4	7.7
2.6	-2.0	4.7	13.9
2.7	-2.6	4.2	13.3
2.8	-4.9	3.7	12.3
2.9	-9.8	5.8	20.1
2.10	-12.0	4.7	30.4
2.11	-9.0	1.4	42.9

Thickness data and temperature data by linear regression, the equation is: $Y = -0.25X_1 + 4.08$, wherein, X_1 is the temperature, 4.08 is a constant.

236

Ice thickness and wind speed data were fit linear regression analysis, then got linear regression equation is: $Y = -0.94X_2 + 21.1$, wherein, X_2 is the wind speed, 21.1 is a constant.

The two meteorological elements of air temperature and wind speed were fit linear regression analysis, to achieve the quantitative relationship between temperature, wind speed and ice thickness. Fitting the linear regression analysis of the data above, to obtain a linear regression equation: $Y = -2.84X_1 - 2.77X_2 + 12.36$, wherein X_1 is air temperature, the coefficient -2.84 for the X_1, X_2 as wind speed, the coefficient -2.77 for the X_2, 12.36 is a constant. Wherein the confidence factor of X_1 is 99 percent, the confidence factor of X_2 is 99 percent, 98% confidence is fitting for the linear regression requirements. The icing on behalf of 10 February data into the formula, ice thickness obtained was 33.42mm, with the difference 3mm between the actual thicknesses 30.4mm.

4. Conclusion

In this paper, by using of grid icing data of Erlang Mountain regional during February 4 to February 13, 2014, analyzes the impact of Erlang Mountain area temperature, wind direction and wind speed and other meteorological elements on icing, provides a reference for transmission line design and technical support.

References

1. Makkonen, Ice and construction, Quarterly Journal of the Royal Meteorological Society, 15 DEC 2006.
2. Teng Zhonglin etc, Research of wire icing growth, Science Bulletin, 1983, 15.
3. Xie Yunhua, The research of wire ice time and space distribution rule in southwest of Sichuan, China Power, 2013, 03.
4. Huang Xinbo, The design and application of transmission line on-line monitoring mechanical sensors, Electric Power Construction, 2014, 03.
5. Tan Guanri, the Soviet Union on research progress of towering buildings icing, Meteorological Science and Technology, 1982, 4.
6. Wang Wen, Cai Xiaojun, The reconstruction of the sequence of wire ice thickness in Erlang Mountain, atmospheric sciences, Nanjing University of Information Science& Technology. March 2010, 28 (1).
7. Linli, Xu Juan etc., Mechanism analysis of transmission lines ice formation, Journal of Jiang Xi Vocational & Technology College of Electricity, 2008, 2.

8. Jiang Xingliang, Influence of Chongqing micro topography on power grid ice, Chongqing institute of electrical engineering in 2010 academic conference proceedings.
9. Yin Xianzhi, Ren Yulong etc., Characteristics and meteorological condition analysis of Hua Jia Ling wire icing in 2008, September, 2013, 31 (3).
10. Li Luping, Liu Shengxian etc., the temperature and humidity of the air on blade experimental study of influencing the ice, in October, 2012, 27(5).
11. Yang Mingli, Wei Lin, Wen Zengchuang, Li Ying etc., The influencing factors of guide, ice of ground and the application in the circuit design, in June 2008, 21 (2).

Decentralized Variable Structure Steam Valve Control for Multi-Machine Power System

Xin Kong and Bao-hua Wang[†]

College of Automation, Nanjing University of Science and Technology, Nanjing, Jiangsu Province, China
[†]E-mail: 13951845674@163.com

A decentralized variable structure generator steam valve controller is designed for stability enhancement and dynamic performance of multi-machine power system. Adaptive back stepping method, block control approach and variable structure control are combined to effectively deal with unmatched uncertainties of multi-machine power system and possess good dynamic performance. The simulative results of the two-area four-machine power system demonstrates that the designed controller restrains the power oscillation quickly. It not only improves the system stability effectively, but also possesses good robustness to disturbance.

Keywords: Decentralized Control; Variable Structure Control; Steam Valve Control; Electric Power System; Stability.

1. Introduction

Electric power systems are becoming more complex because they must meet the increasing power demand. These are highly nonlinear systems that include a number of synchronous machines as producers; consequently, it is important to enhance the transient stability and dynamic performance. As power systems are large scale, distributed and highly nonlinear systems with fast transients, their uncertainties mainly embody on the uncertainties of interconnection. Dealing complex large system with decentralized control can reduce the complexity and cost of the control design, therefore increases the realizability [1,2].

In recent years, robust stabilization problem on multi-machine power systems has caused wide public concern [3-8]. Huerta et al. [3] proposed a decentralized sliding mode stabilizing excitation controller to improve power system stability is designed. Karimi et al.[4], put forwarded a decentralized adaptive backstepping excitation controller, tuned using a PSO technique, is designed for stability enhancement of multi-machine power systems. Zhanxun et al. [5], suggested a decentralized variable structure stabilizer based on nonlinear variable structure control theory is presented.

With the rapid development of steam valve control systems, numerous advanced control method have been applied to steam valve controller of synchronous generator systems [9-12]. Xi [9] uses the Hamiltonian structure of the control system, a decentralized steam valve controller is designed. In dealing uncertainties of damping coefficient, Cai[12] studied a multi-machine steam valve controller using backstepping control theory and fuzzy control technique to solve the problem. Although these researches on multi-machine system have obtained excellent dynamic performance and improved robustness of power system, all mentioned studies mostly do not take uncertainties into account or just assume that there is only one uncertainty in the system. Nevertheless, uncertainties may be everywhere.

In this paper, a novel decentralized variable structure generator steam valve controller has been designed by applying adaptive backstepping method block control approach and variable structure control to effectively deal with unmatched uncertainties and improve transient stability performance. This method can be applied to handle power systems with uncertainties and external disturbances and enhance the robustness of the power system. During the design procedures, even if there are some uncertainties, the closed loop stability can be ensured, meanwhile all signals of the system are bounded and exponentially converge to the neighborhood of the origin. The Simulative results show that the designed controller restrains the power oscillation quickly, it not only improves the system stability effectively, but also possesses good robustness to disturbance.

2. Decentralized variable structure control design

2.1. *Mathematical model*

Assuming that only the main adjust steam valve control by high-pressure cylinder is considered, and all the generators are used in rapid excitation mode. In addition, E'_{qi}, the transient EMF in q axis of generator i is considered as a constant during the dynamic process. Then the mathematical model of the power system with N generators can be established as follows[14]:

$$
\begin{cases}
\dot{\delta_i} = \omega_i - \omega_0 \\
\dot{\omega_i} = \dfrac{\omega_0}{H_i} P_{Hi} + \dfrac{\omega_0}{H_i} C_{Mi} P_{m0i} - \dfrac{D_i}{H_i}(\omega_i - \omega_0) - \dfrac{\omega_0}{H_i} P_{ei} + w_{i1}(t) \\
\dot{P}_{Hi} = -\dfrac{1}{T_{H\Sigma i}} P_{Hi} + \dfrac{C_{Hi}}{T_{H\Sigma i}} P_{m0i} + \dfrac{C_{Hi}}{T_{H\Sigma i}} u_{Hi} + w_{i2}(t)
\end{cases}
\tag{1}
$$

240

where, $i=1,2,...,N$, $w_{i1}(t)$ and $w_{i2}(t)$ are both expressed as internal or external disturbances and uncertainties of the multi-machine system.

2.2. Design procedure

For multi-machine valve control system, according to function (1), divide the mathematical model into three pieces. To simplify the controller design process, transform the states x_{i1}, x_{i2}, x_{i3} into: $x_{i1} = P_{Hi} - P_{H0i}$, $x_{i2} = \omega_i - \omega_0$, $x_{i3} = \delta_i - \delta_{i0}$. Thus, function(1) can be transformed into:

$$\begin{cases} \dot{x}_{i1} = f_{i1}(x_i) + g_{i1}(x_i)u_i + \Delta f_{i1} \\ \dot{x}_{i2} = f_{i2}(x_i) + g_{i2}(x_i)x_{i1} + \Delta f_{i2} \\ \dot{x}_{i3} = f_{i3}(x_i) + g_{i3}(x_i)x_{i2} + \Delta f_{i3} \end{cases} \quad (2)$$

where, $\quad f_{i1}(x_i) = -\dfrac{x_{i1}}{T_{H\Sigma i}}, \qquad g_{i1}(x_i) = \dfrac{C_{Hi}}{T_{H\Sigma i}}, \qquad \Delta f_{i1} = w_{i1}(t);$

$f_{i2}(x_i) = -\dfrac{\omega_0}{H_i}P_{ei} + \dfrac{\omega_0}{H_i}P_{m0i}$, $g_{i2}(x_i) = \dfrac{\omega_0}{H_i}$, $\Delta f_{i2} = -\dfrac{D_i}{H_i}x_{i2} + w_{i2}(t)$; $f_{i3}(x_i) = 0$,

$g_{i3}(x_i) = 1$; $\Delta f_{i3} = 0$. Let the damping coefficient D_i and the disturbances $w_{i1}(t)$, $w_{i2}(t)$ be the unknown parameters, so the abovementioned Δf_{i1}, Δf_{i2} and Δf_{i3} are the uncertainties.

In order to deal with the uncertainties Δf_{i1}, Δf_{i2} and Δf_{i3}, we assume that there exists a function $\varsigma_{ij}(\cdot)$ such that the following inequation is satisfied.

$$\|\Delta f_{ik}\| \le \sum_{j=1}^{N} g_{ik}\varsigma_{ijk}(\|x_i\|)\|x_i\| \quad (3)$$

where $\varsigma_{ij}(\|x_i\|)$ is a vector field continuous with t and smooth with x_i, and $\varsigma_{ij}(\|0\|) = 0$, $k = 1,2,...,r$. The equilibrium point of the system is $x_0^{\mathrm{T}} = (0,0,\cdots,0)$.

As we divide the multi-machine valve control system mathematical model into three pieces, then the design process is composed by 3 steps.

Step1: Define $z_{i3} = x_{i3}$, $s_{i3} = c_{i3}z_{i3}$, $g_{i3} = g_{i30}c_{i2}$, $z_{i2} = x_{i2} - \alpha_{i3}$; according to function(4), we obtain as follows:

$$\dot{s}_{i3} = c_{i3}\left[f_{i3}(x_i) + g_{i30}s_{i2} + g_{i3}\alpha_{i3} + \Delta f_{i3}\right] \quad (4)$$

The virtual trajectory is selected as:

$$\alpha_{i3} = -\left(c_{i3}g_{i3}\right)^{-1}\left[c_{i3}f_{i3}\left(x_i\right) + h_{i3}s_{i3}\right] \tag{5}$$

where, h_{i3} is the parameter of the controller, and $h_{i3} > 0$.

Take the Lyapunov function as:

$$V_{i3} = \left\|s_{i3}\right\| \tag{6}$$

$$\dot{V}_{i3} = \frac{s_{i3}^T \dot{s}_{i3}}{\left\|s_{i3}\right\|} \leq \frac{1}{\left\|s_{i3}\right\|}\left\{-h_{i3}s_{i3}^T s_{i3} + s_{i3}^T c_{i3}g_{i30}s_{i2}\right\} + \left\|c_{i2}g_{i2}\right\|\sum_{j=1}^{N}\zeta_{ij3}\left(\left\|x_i\right\|\right)\left\|x_i\right\| \tag{7}$$

Step 2: Define $z_{i1} = x_{i1} - \alpha_{i2}$, $s_{i2} = c_{i2}z_{i2}$, $g_{i2} = g_{i20}c_{i1}$. According to function (4), we obtain as follows:

$$\dot{s}_{i2} = c_{i2}\left[f_{i2}(x_i) + g_{i20}s_{i,r-2} + g_{i2}\alpha_{i2} + \Delta f_{i2} - \dot{\alpha}_{i3}\right] \tag{8}$$

The virtual trajectory is selected as:

$$\alpha_{i2} = -\left(c_{i2}g_{i2}\right)^{-1}\left[c_{i2}f_{i2}\left(x_i\right) + h_{i2}s_{i2} - c_{i2}\dot{\hat{\alpha}}_{i3} + \rho_{i2}\operatorname{sgn}\left(s_{i2}\right)\right] \tag{9}$$

where, h_{i2} is the parameter of the controller, and $h_{i2} > 0$.

Take the Lyapunov function as:

$$V_{i2} = \left\|s_{i2}\right\| \tag{10}$$

$$\dot{V}_{i2} = \frac{s_{i2}^T \dot{s}_{i2}}{\left\|s_{i2}\right\|} \leq \frac{1}{\left\|s_{i2}\right\|}\left\{-h_{i2}s_{i2}^T s_{i2} + s_{i2}^T c_{i2}g_{i20}s_{i1}\right\}$$
$$+ \left\|c_{i2}g_{i2}\right\|\sum_{j=1}^{N}\zeta_{ij2}\left(\left\|x_i\right\|\right)\left\|x_i\right\| + \frac{s_{i2}^T}{\left\|s_{i2}\right\|}\left[c_{i2}\left(\dot{\hat{\alpha}}_{i3} - \dot{\alpha}_{i3}\right) - \rho_{i2}\operatorname{sgn}\left(s_{i2}\right)\right] \tag{11}$$

where, $\dot{\hat{\alpha}}_{i3}$ is the output of the tracking differentiator[10] while the input is $\dot{\alpha}_{i3}$; and $\rho_{i2} = \sup\left(c_{i2}\left\|\dot{\hat{\alpha}}_{i3} - \dot{\alpha}_{i3}\right\|\right)$. $\sup(\bullet)$ is the supremum.

Step 3: According to function (4), we obtain as follows:

$$\dot{s}_{i1} = c_{i1}\left[f_{i1}\left(x_i\right) + g_{i1}\left(x_i\right)u_i + \Delta f_{i1} - \dot{\alpha}_{i2}\right] \tag{12}$$

Select the control law as:

$$u_i = -\left(c_{i1}g_{i1}\right)^{-1}\left[c_{i1}f_{i1}\left(x_i\right) + \left\|c_{i1}g_{i1}\right\|\xi_{i1}\left(\left\|x_i\right\|\right)\left\|x_i\right\| - c_{i1}\dot{\hat{\alpha}}_{i2} + \rho_{i1}\operatorname{sgn}\left(s_{i1}\right)\right.$$
$$\left. + h_{i1}s_{i1} + \sum_{k=2}^{r}s_{fi1}s_{i,k-1}^T\left(c_{ik}g_{ik0}\right)^T s_{fik} + \sum_{k=2}^{r}s_{fi1}\xi_{ik}\left(\left\|x_i\right\|\right)\left\|c_{ik}g_{ik}\right\|\left\|x_i\right\|\right] \tag{13}$$

where, $\hat{\alpha}_{i2}$ is the output of the tracking differentiator while the input is α_{i2},

$$\rho_{i1} = \sup\left(c_{i1}\left\|\hat{\alpha}_{i2} - \dot{\alpha}_{i2}\right\|\right) \ ; \quad s_{fik} = \begin{cases} \dfrac{s_{ik}}{\|s_{ik}\|}, & \|s_{ik}\| \neq 0 \\ 0, & \|s_{ik}\| = 0 \end{cases}, \quad \xi_{ik}\left(\|x_i\|\right) = \sum_{j=1}^{N} \zeta_{jik}\left(\|x_i\|\right),$$

$k = 1, 2, 3$. And h_{i1} is the parameter of the controller, and $h_{i1} > 0$.

Take the Lyapunov function as:

$$V_{i,1} = \left\|s_{i1}\right\| \tag{14}$$

$$\dot{V}_{i1} = \frac{s_{i1}^T \dot{s}_{i1}}{\|s_{i1}\|} \leq \frac{-h_1 s_{i1}^T s_{i1}}{\|s_{i1}\|} - \frac{s_{i1}^T}{\|s_{i1}\|} s_{fi1} \sum_{k=2}^{3} \|c_{ik} g_{ik}\| \xi_{ik}\left(\|x_i\|\right)\|x_i\| - \sum_{k=2}^{3} s_{i,k-1}^T \left(c_{ik} g_{ik0}\right)^T s_{fik}$$

$$+ \frac{s_{i1}^T}{\|s_{i1}\|} [\|c_{i1} g_{i1}\| \sum_{j=1}^{N} \zeta_{ij1}\left(\|x_i\|\right)\|x_i\| - \|c_{i1} g_{i1}\| \xi_{i1}\left(\|x_i\|\right)\|x_i\| + c_{i1}\left(\hat{\alpha}_{i2} - \dot{\alpha}_{i2}\right) - \rho_{i1} \operatorname{sgn}(s_{i1})] \tag{15}$$

2.3. Stability proof

The Lyapunov function of the whole system can be expressed as:

$$V = \sum_{i=1}^{N} \sum_{k=1}^{3} V_{ik} \tag{16}$$

then, According to function(11), function (16), and function (21), we obtain:

$$\dot{V} = \sum_{i=1}^{N} \sum_{k=1}^{3} \frac{s_{ik}^T \dot{s}_{ik}}{\|s_{ik}\|} \leq \sum_{i=1}^{N} \{ -\sum_{k=1}^{3} h_{ik} \|s_{ik}\| + \sum_{k=2}^{3} \frac{s_{i,k-1}^T}{\|s_{i,k-1}\|} [c_{i,k-1}\left(\hat{\alpha}_{ik} - \dot{\alpha}_{ik}\right) - \rho_{i,k-1} \operatorname{sgn}\left(s_{i,k-1}\right)]$$

$$+ \sum_{k=1}^{3} \|c_{ik} g_{ik}\| \sum_{j=1}^{N} \zeta_{ijk}\left(\|x_i\|\right)\|x_i\| - \sum_{k=1}^{3} \frac{s_{i1}^T}{\|s_{i1}\|} s_{fi1} \|c_{ik} g_{ik}\| \xi_{ik}\left(\|x_i\|\right)\|x_i\| \} \tag{17}$$

Note that

$$\|c_{ik} g_{ik}\| \sum_{j=1}^{N} \zeta_{ijk}\left(\|x_i\|\right)\|x_i\| - \frac{s_{i1}^T}{\|s_{i1}\|} s_{fi1} \|c_{ik} g_{ik}\| \xi_{ik}\left(\|x_i\|\right)\|x_i\| = 0 \ , k = 1, 2, 3 \tag{18}$$

$$\frac{s_{i,k-1}^T}{\|s_{i,k-1}\|} \left[c_{i,k-1}\left(\hat{\alpha}_{ik} - \dot{\alpha}_{ik}\right) - \rho_{i,k-1} \operatorname{sgn}\left(s_{i,k-1}\right) \right] \leq 0 \ , k = 2, 3 \tag{19}$$

Thus

$$\dot{V} \leq -\sum_{i=1}^{N} \sum_{k=1}^{3} h_{ik} \|s_{ik}\| \leq -hV \tag{20}$$

243

where, $h = \min(h_{ik}) > 0$, and $i = 1, 2, \cdots, N$, $k = 1, 2, 3$.

As long as $V = \sum\limits_{i=1}^{N} \sum\limits_{k=1}^{3} V_{ik} = \sum\limits_{i=1}^{N} \sum\limits_{k=1}^{3} \|s_{ik}\| \neq 0$ is satisfied, then we can get:

$$\dot{V} < 0 \qquad (21)$$

3. Simulation analysis

This section gives the simulation analysis for the design of decentralized variable structure controller of system (1) when the three-phase short-circuit occurs. The proposed control algorithm was tested on the model of two-area four-machine power system[14] whose schematic is shown in Fig 1. The partial parameters for simulation are selected from [13]. Compare the proposed new control theory with the Terminal sliding mode control proposed by Zou and Wang [8]. During the simulation, there is a limitation : $C_{Mi}P_{mi0} \leq P_{mi} \leq P_{mi0}$, by assuming that only the main adjust steam valve control by high-pressure cylinder is considered.

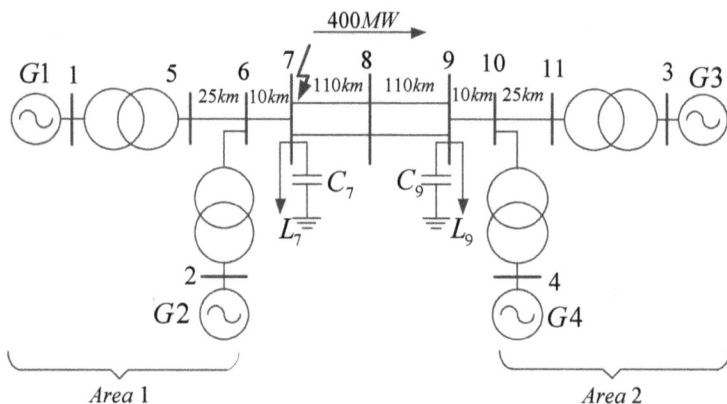

Fig. 1. Schematic diagram of two-area four-machine power system.

When $t=0.2$s, a three phase fault is applied at bus 7 and completely removed at 0.35s. To the designed large-disturbance attenuation controller, the dynamic response is shown in Fig. 2. It can be seen that the dynamic response of the system is very quick and it does not significantly change when the three-phase short-circuit occurs, moreover, the disturbed system rapidly returns to the steady state and there's little system static error. In addition, the dynamic response curve of decentralized variable structure controller with unknown damping coefficients is almost coincided with that of the controller with the known damping coefficients. The simulation results show that the proposed one not only improves the system stability more effectively, but also possesses better robustness to disturbance when compared with Terminal sliding mode control.

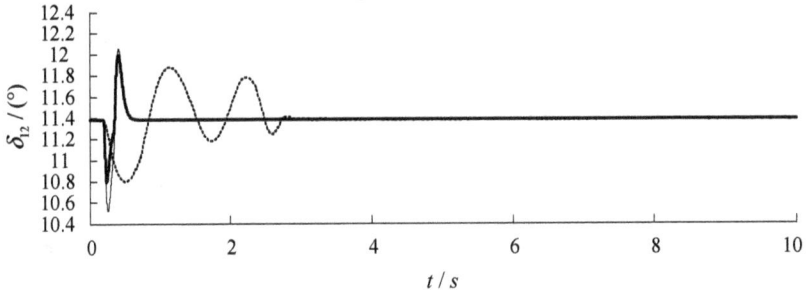

(a). Relative rotor angle δ_{12}.

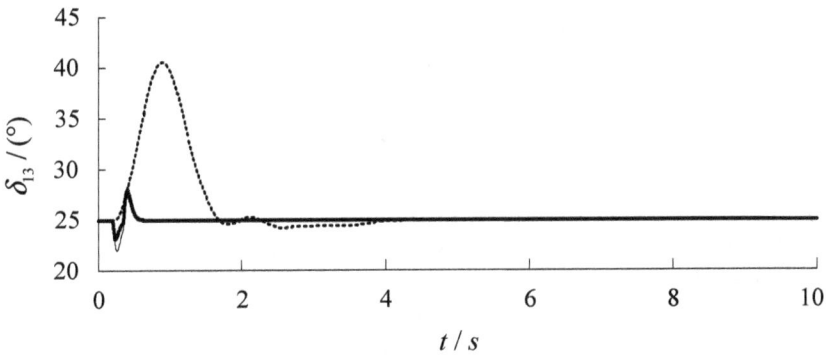

(b). Relative rotor angle δ_{13}.

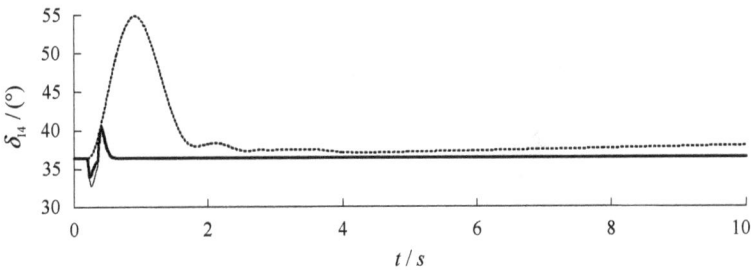

(c). Relative rotor angle δ_{14}.

Fig. 2. Dynamic response to three-phase short-circuit. Proposed control with D_i unknown (thick solid line), proposed control with $D_i = 1$ (fine solid line), Terminal sliding mode control [14] (dotted line) Dynamic response to three-phase short-circuit.

4. Conclusion

A new decentralized variable structure control theory is developed to effectively deal with unmatched uncertainties of multi-machine power system and possess good dynamic performance, which is combined with adaptive backstepping method, block control, and variable structure control. Tests results show the effectiveness of the variable structure generator steam valve control in improving dynamic stability of system under large disturbances in comparison with Terminal sliding mode control. Uncertainties lie in the interconnection between each subsystem, then the backstepping method can exactly deal with the unmatched or matched uncertainties. During the design procedure, the nonlinear tracking differentiator is particularly mentioned to track every virtual trajectory and its differential. Moreover, there is no need to handle the uncertainties in every step, consequently, the design process becomes more simple and convenient. In summary, the proposed controller is able to achieve the improved performance of dynamics stabilization and possess good robustness.

References

1. W. Liu, J. Sarangapani, Neural Network based Decentralized Excitation Control of Large Scale Power Systems, *in Proc. Int. Joint Conference on Neural Networks (IJCNN'2006)*, (Vancouver, Canada, 2006).
2. M. Sun, Nonlinear decentralized control of interconnected large-scale power systems, Control Conference*(CCC'2008)*, (Kunming,China,2008).
3. H. Huerta, Alexander G. Loukianov, Jose M. Cañedo, Decentralized sliding mode block control of multimachine power systems. International Journal of Electrical Power and Energy Systems, **32**, 1(2010).
4. Ali Karimi, Ali Feliachi, Decentralized adaptive backstepping control of electric power systems. Electric Power Systems Research, **78**, 484(2008).
5. YU Zhanxun, CHEN Xueyun, The study of decentralized variable structure stabilizer for multimachine power system, Automation of Electric Power Systems, **05**, 19 (1996).
6. S. Yan, Y. Sun, LI Xiong, Decentralized robust excitation controller for improvement of power system stability, Automation of Electric Power Systems, **22**, 21(2002).
7. M. Ouassaid, M. Maaroufi, M. Cherkaoui, A real-time nonlinear decentralized control of multimachine power systems. Systems Science & Control Engineering, An Open Access Journal, **2**, 135(2014).

8. D. Zou, B Wang, Adaptive and robust excitation control with terminal sliding mode for multi-machine power system, Electric Power Automation Equipment, **30**, 79(2010).

9. Z. Xi, A decentralized steam valving H_∞ controller for nonlinear multi-machine power systems, Automation of Electric Power Systems, **26**,7(2002).

10. G. Liu, X. Lin, Nonlinear Fast Valving Control of Multi-machine systems, Automation of Electric Power Systems, **20**, 10(1996).

11. G. Liu, X. Lin, Application of Lyapunov Theory to Nonlinear Control of Turbine Fast Valving (Multi-machine System), Electric Power Automation Equipment, **3**, 13(1996).

12. Jingwen Cai, Liying Sun, Direct Fuzzy Backstepping Control for Turbine Main Steam Valve of Multi-machine Power System, Journal of Liaoning University of Technology (Natural Science Edition), **34**, 351(2014).

13. Jingqing Han, Wei Wang, Nonlinear tracking-differentiator, Systems Science and Mathematical Science, **14**, 177(1994).

14. Kundur P, *Power system stability and control* (McGraw-Hill, New York, USA,1994).

Loss Optimization of Distribution Network Based on HUPFC

Meng-ze Yu

Grid Planning & Research Center,
Guangdong Power Grid Corporation, Guangzhou, China
Email: yumengze98@163.com

Lei Liu, Wen-li Fei, Bai-chao Chen and Jia-xin Yuan

School of Electrical Engineering, Wuhan University,
Wuhan, China

According to statistics, half of the grid loss occurred in the distribution systems, of which line loss is of considerable portion. Since distribution systems are normally connected by a radial network into a ring network to improve reliability, collectively, impact due to the line loss are mulitpled. Threfore, it is important to invitagate issue to idenify problem to compensate for line loss in loop distribution systems. This paper analyzes the criteria required to minimize line loss, and compensation by HUPFC. It is proved by the simulation results based on MATLAB/ Simulink.

Keywords: Loop Distribution Systems, Line loss, UPFC, SEN Transformer.

1. Introduction

Distribution network faces directly to the user, which is the key to control and ensure the quality of the user power. At present, 80% of the user power outage is caused by the power distribution system, and half of the loss of the power grid is in the distribution network [1]. It can improve the power supply reliability, reduce the power supply recovery time and increase the standby capacity to use ring topology. But compared to the tree network, the ring structure can increase the line loss. FACTS (Flexible AC Transmission Systems) have been gradually used in the distribution network system, especially UPFC (Unified Power Flow Controller), which is more and more popular. [2] "SEN" Transformer (ST) is an improved phase shifted transformer proposed by Dr. K. K. SEN in 2003. [3, 4] Through the switching of the auxiliary winding taps, it can output different series injection voltage, adjust the system voltage, change the transmission power flow and improve the system stability.

HUPFC is composed of ST and traditional UPFC in series. The ST realizes the series voltage that can be discretely adjusted in 360 degrees by the control of the ordinary on-load voltage-regulating switch. The UPFC uses principle of power electronic switch to realize the series voltage that can be continuously adjusted in 360 degrees, working together to realize the function of control in the system flow.

Based on these works, this paper establishes the basic model of looped distribution network loss optimization, and puts forward a method of optimizing distribution network loss and power flow based on Namely, HUPFC adjust the line parameters or compensation voltage to reduce line losses [5] and optimize the power flow by joining HUPFC to the looped distribution network. It has a significant meaning in improving the reliability of power network operation and economic benefit of the power network.

2. Distribution network model and minimum loss conditions [6]

The simplified equivalent distribution ring network is shown in Fig. 1

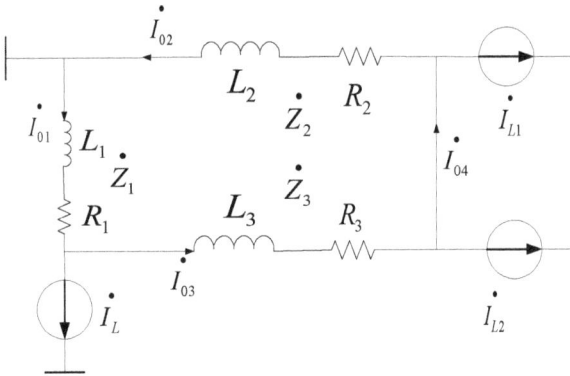

Fig. 1 Simplified equivalent distribution ring network

249

The loss of the ring network is:

$$P_i = \sum_{i=1}^{3} R_i \left| \dot{i}_{0i} \right|^2 = R_{Loop} \left| \dot{i}_{01} - \frac{R_2 \dot{i}_{L1} + R_2 \dot{i}_{L1} + (R_2 + R_3) \dot{i}_L}{R_{Loop}} \right|^2$$

$$- \frac{\left| R_2 \dot{i}_{L1} + R_2 \dot{i}_{L1} + (R_2 + R_3) \dot{i}_L \right|^2}{R_{Loop}} + R_2 \left| \dot{i}_{L1} + \dot{i}_{L2} + \dot{i}_L \right|^2 + R_3 \left| \dot{i}_L \right|^2 \tag{1}$$

$$R_{Loop} = \sum_{i=1}^{3} R_i$$

From the formula (1), it can be seen that the loss of the ring network is greatly increased compared with the loss of the tree network. It is because $\dot{i}_{L1}$, $\dot{i}_{L2}$ and $\dot{i}_L$ all are constant values, to minimize the loss of the line, the first part of the above formula should be zero. At this time, the smallest line loss [6] is:

$$P_{l\,min} = \sum_{i=1}^{3} R_i \left| \dot{I}_{mi} \right|^2 \tag{2}$$

$$\dot{I}_{Loop} = \dot{I}_{01} - \dot{I}_{m1}$$

$$= \frac{j\omega}{R_{Loop} Z_{Loop}} [\{(R_1 + R_3)L_2 - (L_1 + L_3)R_2\} \dot{I}_{L1}$$

$$+ \{(R_1 + R_3)L_2 - (L_1 + L_3)R_2\} \dot{I}_{L2} \tag{3}$$

$$+ \{R_1(L_2 + L_3) - L_1(R_2 + R_3) \dot{I}_L]$$

$$= - \frac{\sum_{i=1}^{3} j\omega L_i \dot{I}_{0i}}{R_{Loop}}$$

It is thus clear that $\dot{I}_{Loop}$ and $\dot{I}_{01}$ are in the same direction (i.e., increase the line loss). To enable the loss at least, make $\dot{I}_{Loop}$ equal to zero. From the above calculation formula to reach this condition, there are two kinds of conditions, namely:

$$\frac{R_1}{L_1} = \frac{R_2}{L_2} = \frac{R_3}{L_3} \tag{4}$$

or:

$$\sum_{i=1}^{3} j\omega L_i \dot{I}_{0i} = 0 \tag{5}$$

Therefore, when reduce line loss by way of compensation, only needs to satisfy the two conditions of equation (4), (5).

3. The Principle of HUPFC

Novel electromagnetic Hybrid Unified Power Flow Controller (HUPFC) is composed of large capacity UPFC and small capacity ST, the topology structure and schematic diagram are shown in Fig. 2.

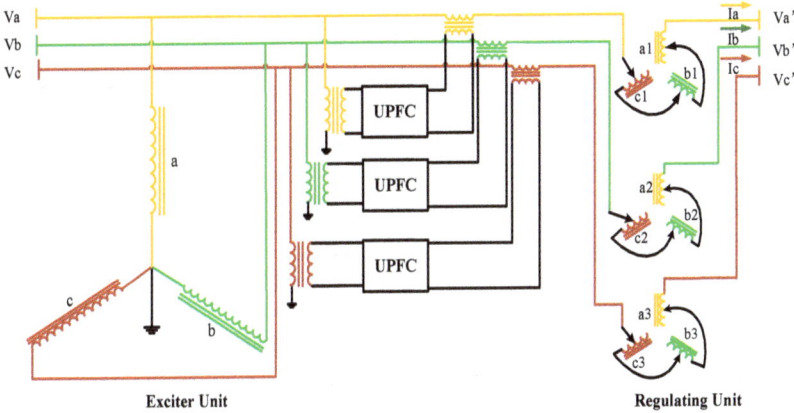

Fig. 2 The topology of HUPFC

ST and UPFC to adjust the system of the power flow together UPFC parallel converter VSC1 compensated system reactive power. ST through the series part of the series connected to an amplitude and phase are discrete adjustable voltage $\dot{V}_{ST}$, UPFC through the series part of the series connected to an amplitude and phase are continuously adjustable voltage $\dot{V}_{upfc}$. When the system power flow needs to be improved, $\dot{V}_{ST}$ and $\dot{V}_{upfc}$ working together can implement. $\dot{V}_{ST}$ provides discrete, larger parts, $\dot{V}_{upfc}$ provide accurate, the rest of the part; the two coordination to achieve a continuous, large capacity of the power flow control. The UPFC parallel converter VSC1 can provide continuous, adjustable, reactive power from the inductive region to the capacitive region. When the system needs to compensate reactive power (or stable point voltage), the UPFC shunt converter VSC1 realizes the reactive power compensation.

4. Simulation

Through the compensation function of HUPFC, the circuit resistance and inductance of the ring network can satisfy the relationship $\frac{R_1}{L_1} = \frac{R_2}{L_2}$, so as to achieve the purpose of power economic distribution and reduce the line loss. Calculated, when the compensation occurs in the line 1 when the network loss.

In order to satisfy the relationship between the resistance and inductance of the circuit, it is necessary to:

$$L_c = \frac{R_1}{R_2}L_2 - L_1 = \frac{0.13}{1.41} \times 1.6 - 6.0 = -5.85(mH).$$ It is the reactance

of HUPFC is -5.85 mH. Table 1 shows the consumption of the line after compensated HUPFC.

Table 1. Simulation results in ring network with HUPFC compensation

	Theoretical value	Simulation value		Theoretical value	Simulation value
i_1(kA)	1.332	1.348	i_2(kA)	0.123	0.109
i_{loop} (kA)	0.605	0.620			
P_{L1}(MW)	15.865	15.910	P_{L2}(MW)	15.865	15.910
P_{loss}(MW)	0.252	0.253			

Table 2 shows the comparison of the ring network in the HUPFC before and after compensation of line loss, the results show that the HUPFC can greatly reduce the line loss, a decrease of more than 80%.

Table 2. The comparison of line loss before and after HUPFC compensation

Items	Without HUPFC	With HUPFC	Reduction
Theoretical value	1.304	0.252	80.7%
Simulation value	1.299	0.253	80.5%

Fig. 3 was compared in radiation type network, ring network without compensation and ring network using HUPFC compensation three cases line loss, compared the ring network using HUPFC before and after compensation line loss (kw), the results show that the HUPFC can greatly reduce the line loss, a decrease of more than 80%.

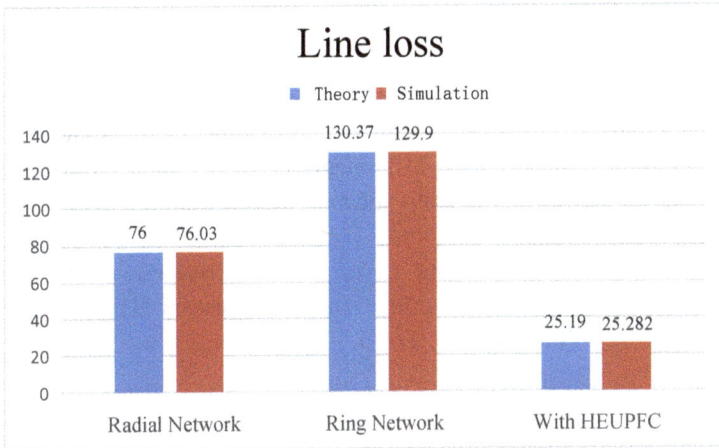

Fig. 3 The line loss in different situations

5. Conclusion

After the radial network is connected to the ring network, the reliability of the system is greatly improved, and the loss of the circuit is greatly increased at the same time. This paper deduces the compensation loss in a ring network, the requirements of the resistance and inductance of all lines in a certain proportion, the HUPFC of this device to compensate for the distribution network, and build a distribution ring network and simulation model based on HUPFC compensation. The simulation results show that HUPFC can play an important role in minimizing line loss in the distribution network; the highest loss reduction can reach more than 80%. The compensation effect of HUPFC on the damage of ring network is verified.

References

1. Mahyar Zarghami, Mariesa L. Crow, Jagannathan Sarangapani, Yilu Liu. A Novel Approach to Interarea Oscillation Damping by Unified Power Flow Controllers Utilizing Ultracapacitors [J]. IEEE Transactions on Power systems, 2010, 25(1): 404-412.

2. Tanaka K, Oshiro M. Decentralized Control of Voltage in Distribution Systems by Distributed Generators [C]//Generation, Transmission & Distribution. [S.l.]: IET, 2010; 1251-1260.

3. M. Omar Faruque, Venkata Dinavahi, A tap-changing algorithm for the implementation of "Sen" Transformer [J], IEEE Trans. Power Del., vol.22, pp.1750-1757, Jul. 2007.

4. K. K. Sen, M. L. Sen, "Comparison of the "SEN" Transformer with the unified power flow controller," *IEEE Trans. on PWRD*, vol. 18, pp. 1523-1533, Oct. 2003.

5. Mahmoud A. Sayed, Takaharu Takeshita, All Nodes Voltage Regulation and Line Loss Minimization in Loop Distribution Systems Using UPFC [J]. IEEE Trans. Power Elec, vol.26, pp.1694-1703, Jun. 2011.

6. Liu Keyan, Sheng Wanxing, Li Yunhua. Reactive Power Optimization Based on Improved Immunity Genetic Algorithm [J]. Power System Technology, 2007, 31(13): 11-16.

The Research on Image Fusion for Infrared and Visible Images in Production Safety

Xing-zheng He[†], Rong-xue Kang and Gui-wen Ren

China Academy of Safety Science and Technology, Beijing 100012, China
Key Laboratory of Major Hazard Control and Accident Emergency Technology,
State Administration of Work Safety, Beijing 100012, China
[†]Email: hygehy@126.com

The background and significance of the image fusion research in production safety are researched in this paper. The hardware configurations are described and the principles of image fusion are analyzed. A suitable method for infrared image and visible image fusion for product safety is proposed. The data processing ways are demonstrated. The effects of the fusion system are displayed and analyzed.

Keywords: Image Processing, Image Fusion, Fusion System, Infrared Image, Visible Image.

1. Introduction

With the rapid development of science and technology, production safety has become more and more important to the society and every enterprise. The fire always brings the most serious harm to the production safety. There is a lot of fire accidents occurred in the world, causing huge losses every year. The two main factors that affect the occurrence of fire are high temperature and combustible gases [1]. Infrared temperature measurement is a kind of radiation temperature measurement, which is always used to measure the temperature of the object.

In nature, all the objects higher than 0 K (-273.15 °C) emit infrared radiation energy to the surrounding space, the radiation energy density and the temperature of the object are ruled with the Kirchhoff's radiation law [2]. The infrared camera is always used to get the distribution images of the temperature. It is to say, that the temperature of the objects can be seen with eyes. But because of the technique limitations of the infrared image sensors, the infrared images are usually with lower resolutions and not very clear. Because of the lower resolution, infrared image have less definition than visible light image. So people usually hope to observe information including infrared and visible light synchronously. Image fusion is a kind of effective way that people can "see" the infrared image information and visible light image information at the same time.

2. Image Fusion

2.1. *System configuration*

The main hardware of the system includes an infrared camera, a visible camera and a data processing card with software embedded. System configuration is shown as Fig. 1.

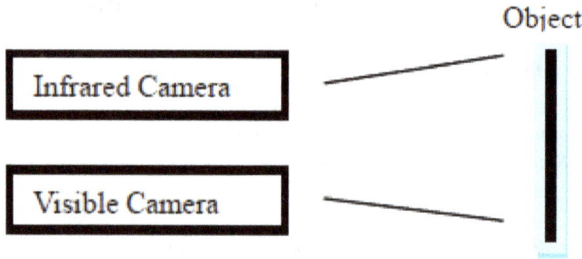

Fig. 1. System configuration

2.2. *Image fusion principle*

The target of image fusion is to make the 2 images come into alignment. There are 4 coordinate systems, which are established to complete the conversions. They are world coordinate Ow, camera coordinate Oc, image coordinate in lens system O1 and pixel coordinate in computer O0 [3] as shown in Fig. 2.

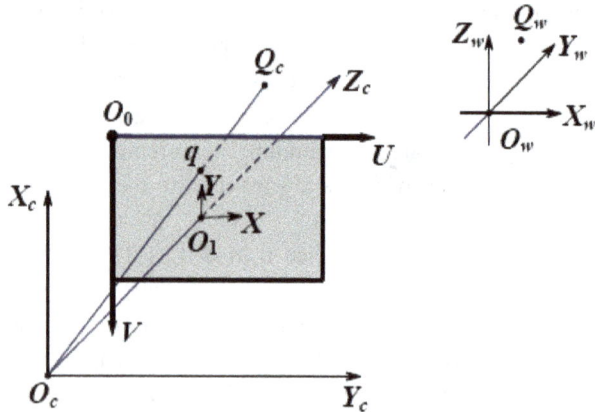

Fig. 2. Relations of coordinates

There are 4 steps to finish the conversion as following.
Step 1. Qw (xw, yw, zw) to Qc(xc, yc, zc).
Formula 1 is used to get this conversion.

$$\begin{bmatrix} x_c \\ y_c \\ z_c \\ 1 \end{bmatrix} = \begin{bmatrix} R & t \\ 0^T & 1 \end{bmatrix} \begin{bmatrix} x_w \\ y_w \\ z_w \\ 1 \end{bmatrix} \qquad (1)$$

In the formula 1 above, R is rotation matrix of 3*3, t is translation matrix of 3*1.

Step 2. Qc(xc, yc, zc) to Q1(x1, y1).

In geometrical optics system, to get a good fusion precision, the optical distortions, including tangential distortion and radial distortion, have to be considered. The following formulas 2 and 3 are used to finish the conversions.

$$\begin{cases} x' = x_c / z_c \\ y' = y_c / z_c \end{cases} \qquad (2)$$

$$\begin{cases} x_1 = x'(1 + k_1 r^2 + k_2 r^4 + k_3 r^6) + 2p_1 x' y' + p_2 (r^2 + 2x'^2) \\ y_1 = y'(1 + k_1 r^2 + k_2 r^4 + k_3 r^6) + 2p_2 x' y' + p_1 (r^2 + 2y'^2) \end{cases} \qquad (3)$$

In the formula 3 above, K1, k2, k3 are tangential distortion coefficients, p1, p2 are radial distortion coefficients, $r^2 = x'^2 + y'^2$.

Step 3. Q1(x1, y1) to Q0(u, v).

The formula 4 is used to get this conversion.

$$\begin{cases} u = f_x x_1 + c_x \\ v = f_y y_1 + c_y \end{cases} \qquad (4)$$

In the formula 4 above, fx is the product of the focal length of the lens and the image pixels number per mm (pixels / mm) in horizontal direction. fy is in vertical direction. Cx and Cy are the coordinates of the image center.

Step 4. Solving relative relation of 2 cameras.

Based on the results of step 1 to step 3, it is easy to calculate the position relations between the two images.

2.3. Calibration images and data processing

A chess grid calibration board is used to calculate the relations of the two cameras [4]. The images of the calibration board are shown as Fig. 3. The left one is the infrared image and the right one is the visible image.

Fig. 3. The images of the calibration board

OPENCV is used to process the images and the data [5], shown as Fig. 4 and Fig. 5.

Fig. 4. Images processing 1

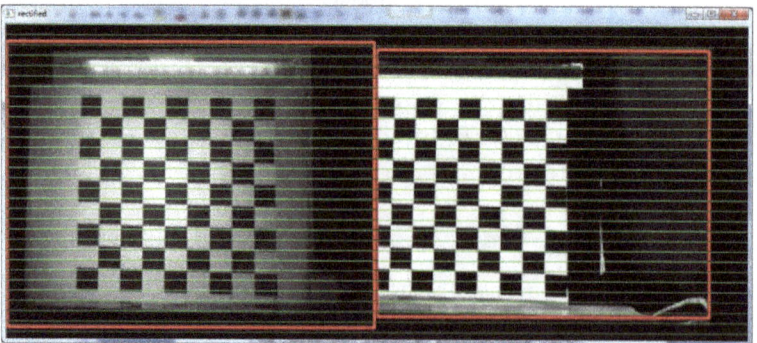

Fig. 5. Images processing 2

3. Fusion Result

Good fusion result was reached, shown as Fig. 6. The left image is the infrared image, the middle one is the visible image, the fused image is the one appear on the right.

Fig. 6. Images fusion result

4. Conclusion

The fusion of infrared image and visible light image is an important branch in image data fusion. Infrared image indicates the temperature distribution of certain object. There are some shortcomings of infrared image, such as lower resolution, less definition than visible light image, etc. Image fusion is a kind of effective way that people can observe the infrared image information and visible light image information at the same time. The paper researched the image fusion technology of infrared image and visible image, discussed the hardware configuration, and got a good fused image result.

Acknowledgments

This work is funded by "National Science-technology Support Program of China" (#2015BAK16B03), "Fundamental Research Funds for the Central Research Institutions of CASST" (#2016JBKY15, #2016JBKY05) and "Key technology project of production safety for major accident prevention and control" (#zhishu-0006-2015AQ, #zhishu-0003-2016AQ).

References

1. Z. Yan, J. Dong: *Low Voltage Apparatus.* Vol. 2 (2000), p. 32
2. Q. Zeng, F. Shu, Q. Li: *Sensor World.* Vol. 7 (2007), p. 32
3. L.Yuan, H. Liu, X. Lu: *Modern Scientific Instruments.* Vol. 2 (2013), p. 78
4. Z. Zhang: *IEEE Transactions on Pattern Analysis and Machine Intelligence.* Vol. 22 (2000), p. 1330
5. http://www.opencv.org.cn/opencvdoc/

PSS Compensator Topology ICPT Transmission Performance Analysis

Xue-qing Yuan[1,†], Kan Wang and Yan-hua Liao

1Shenyang Institute of Automation Chinese Academy of Sciences,
Shenyang, China
2University of Chinese Academy of Science,
Beijing, China
†E-mail: yuanxueqing@sia.cn

In this paper, power transmission characteristics and main circuit phase angle characteristics of PSS type Inductively Coupled Power Transfer (ICPT) system are analyzed. In general, increasing the resonance frequency of the system can improve the power transmission capability of the system, but output power of PSS compensator topology is found to decrease with increasing resonance frequency. The PSS compensation topology has the feature that can transfer rated voltage with high variation of coefficient of coupling. This paper analyzes the system transmission efficiency and phase angle variation in the voltage range with a constant coupling coefficient. The feasibility of the design method is certified by experimental results.

Keywords: Inductive Coupled Power Transfer; PSS topology; Transmission power; Efficiency

1. Introduction

Inductively coupled power transfer is a contact-less method to transmit electrical energy without any mechanical contact between two magnetic coils. As the system does not have connector, it makes the system more flexible, and avoiding the inconvenience caused by regular plug. At the same time the power transmitting and receiving terminals are not electrically connected, thereby eliminating the common connector caused by arcing, sparks and other fundamental. It can be used at underwater, colliery, electricity, etc. [1-7]. In electric vehicle charging, which solve the requirements of the electricity supply for safety and reliability[8-12]. Thereby it becomes a hot research of domestic and foreign scholars in recent years.

There are still some problems in traditional SS, SP, PS, PP and other ICPT system topology. When the coupling distance changes, the core will offset which led to changes in the coupling coefficient load terminal voltage fluctuations, the stability of the output load is unable to guarantee. PSS compensator topology has a good behavior for the load voltage fluctuation because of the changing of coupling distance and core offset, which can ensure that the changing size of the coupling coefficient rate change at 30% while load voltage does not exceed 5%. PSS compensator topology has obtained more and more research in recent years, but there are not literature to study output characteristics, transmission efficiency and phase angle characteristics when the coupling coefficient change.

In this paper, we will study the ICPT system based on PSS [13,14] compensator topology, analyzes the system transmission efficiency and phase angle variation in the voltage range with a constant coupling coefficient. The feasibility of the design method is certified by the experimental results.

2. PSS adaptability analysis of topology load

Fig. 1 shows the structure of the proposed system using PSS topology. Lp represents the primary coil inductance. Ls represents the secondary coil inductance. M is the mutual inductance between the primary coil and the secondary coil. In primary side, the coil is compensated by two coils: Cps in series and Cpp in parallel. The capacitor Cps is adopted to compensate the leakage inductance. The parallel capacitor can reduce the current rating in main circuit. Lf and Cf are filter inductance and filter capacitor. Lf and Cf can eliminate the high order current harmonics. In secondary side, the capacitor Cs and secondary inductance Ls is connected in series. Capacitor Cs is fully resonance with inductance Ls, and then the reactance of secondary side is zero.

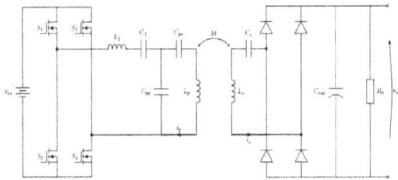

Fig. 1 PSS compensation ICPT system circuit structure

Supposing the resonance angular frequency of the system is ω, secondary pickup coil resistance can be ignored. Under this assumption, in order to achieve maximum power transfer system, the secondary side resonant capacitor selection meets [15,16]:

$$\omega^2 L_s C_s = 1 \qquad (1)$$

Based on paper, the value of capacitor C_{ps} in conventional series-series compensation topology named Cpsis determined as follow:

$$C_{ps} = \frac{1}{L_p \omega_0^2} \tag{2}$$

In the proposed topology, we didn't choose Cps as the primary series compensation capacitor, instead, we chooseseries compensation capacitor using principles which can be called 'partly compensate'.

$$C_{ps} = \frac{1}{(1-a) L_p \omega_0^2} \tag{3}$$

where a is an auxiliary parameter and constrained by inequality $0 < a < 1$. Z_p is the winding impedance after partly series compensation. Z_p can be derivedas:

$$Z_p = j\omega_0 L_p + \frac{1}{j\omega_0 C_{ps}} \tag{4}$$
$$= j\omega_0 a L_p$$

where r_p is the parasitic in primary side. It's obvious that Z_p is inductive when using partly compensation capacitor C_{ps}.

$$Z_r = \frac{\omega_0^2 M^2}{\text{Re}_q} \tag{5}$$

When two coils are exactly in position, the main circuit should be working in ZPA (zero phase angle) condition. Selecting a suitable C_{pp}, we can guarantee that the reactance of Z_{source} is zero. According to (2), (3), (4), (5) C_{pp} will be given by

$$C_{pp} = \frac{a L_p}{(a\omega L_p)^2 + (\frac{\omega^2 M_{max}^2}{R_{eq}})^2} \tag{6}$$

M_{max} is the mutual coupling value (maximum coupling coefficient k). The output voltage is

$$u_{out} = \frac{j\omega_0 M}{\frac{\omega_0^2 M^2}{\text{Re}_q} + j\omega_0 L_p a} v_{in} \tag{7}$$

Using symmetrical coil $L_p = L_s = L$, coupling coefficient is $k = \frac{M}{\sqrt{L_p L_s}}$, system output power and input power ratio is :

$$P_{out} = \frac{V_{in}^2}{\left(\left(\frac{a^2 L_p}{L_s}\right)^2 \left(\frac{1}{k}\right)^2 + \left(\frac{\omega_0 \sqrt{L_p L_s}}{R_{eq}} k\right)^2\right) R} \tag{8}$$

The maximum value of the voltage is

$$G(k)|_{max} = \frac{1}{\sqrt{2bq}} = \sqrt{\frac{R_{eq}}{2a\omega_0 L_p}} \tag{9}$$

The maximum value of the output power is

$$P_{max} = \frac{V_{in}^2 R_{eq}}{2\omega^2 L_s^2 k^2} \tag{10}$$

The output voltage out is proportional to G(k) when vin is fixed. Therefore, the ratio of output voltage to its maximum will be equal to the ratio of G(k) to G(k)|max. Through the ratio, output voltage drop when misalignment happens can be investigated. The ratio is defined as

$$H(k) = \frac{u_{out}}{u_{out}|_{max}} = \frac{G}{G_{max}} = \frac{\sqrt{2k^2}}{\sqrt{k^2 + \frac{k_0^4}{k^2}}} \tag{11}$$

For k < k0, H(k) increases monotonically with increasing in coupling coefficient k. H(k) peaks at k0. H(k) decreases slightly as the coupling coefficient increases continuously. By introducing an auxiliary parameter $0 < \zeta < 1$, output voltage uout can be designed to be always largerthan $\zeta \cdot$ uout|max for a specific coupling range [kmin, kmax]. In this condition, the following inequality should be satisfied

$$\zeta < \sqrt{\frac{2k_0^2}{k^2 + \frac{k_0^4}{k^2}}} \tag{12}$$

$$\delta_k = \frac{k_{mi} - k_{ma}}{k_{ma}} = 1 - \sqrt{\frac{1 - \sqrt{1 - \zeta^4}}{1 + \sqrt{1 - \zeta^4}}} \tag{13}$$

To give a better understanding of this, we take k0 = 0.2, ζ = 0.95 as an example. When this set of parameters is given, kmin and kmax are calculated as 0.1588 and 0.2518. That means the output voltage drop will be less than 5% as k varies in [0.1588, 0.2518]. The variation interval of k can up to 36.9% of the total interval when voltageare guaranteed.

3. Characteristics of the phase angle

In this study, we employ a full-bridge as inverter topology and inductive circuit will result in soft switching inverter and better efficiency. The main circuit should be designed to achieve an inductive impedance which means that the AC voltage should lead the inverter current.

When using the compensation capacitors, the real part of Zsource can be calculated as:

$$\Re_{in} = \frac{\omega_0^2 M^2 / R_{eq}}{\left(1 - \omega_0^2 a L_p C_{pp}\right)^2 + \left(\omega_0 C_{pp}\right)^2 \left(\omega_0^2 M^2 / R_{eq}\right)^2} \tag{14}$$

the imaginary of Zsource is:

$$\chi_{in} = j \frac{\omega_0 a L_p \left(1 - \omega_0 a L_p C_{pp}\right) - \left(\omega_0^2 M^2 / R_{eq}\right)^2 \omega_0 C_{pp}}{\left(1 - \omega_0 a L_p C_{pp}\right)^2 + \left(\omega_0 C_{pp}\right)^2 \left(\omega_0^2 M^2 / R_{eq}\right)^2} \tag{15}$$

The phase of the impedance which is given the symbol _x0012_ is determined through the following relations:

$$\theta = \frac{180^o}{\pi} \arctan \frac{\chi}{\Re} \tag{16}$$

Substituting (14) (15) to (16)

$$\theta = \frac{180^o}{\pi} \arctan\left(\frac{k_0^2}{k_0^4 + k_{max}^4} \frac{k_{max}^4 - k^4}{k^2}\right) \tag{17}$$

When k0 is equal to 0:2 and kmax is equal to 0:25, phase angle _x0012_ will result in Fig. 7. From the picture we can see that the angle of impedance has been always positive, then, providing a soft switching condition.

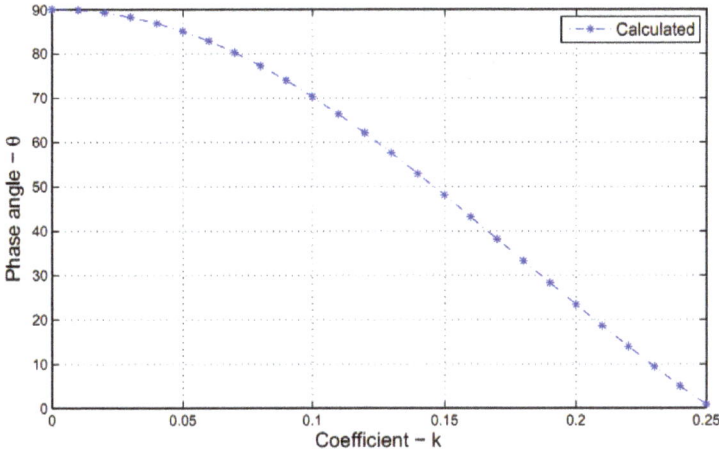

Fig. 2 Phase angle curve with the coupling coefficient

4. Efficiency analysis

System efficiency is a very important aspect of system design. Based on the above analysis, we designed a PSS compensation ICPT system, experimental apparatus are shown in Fig. 3.

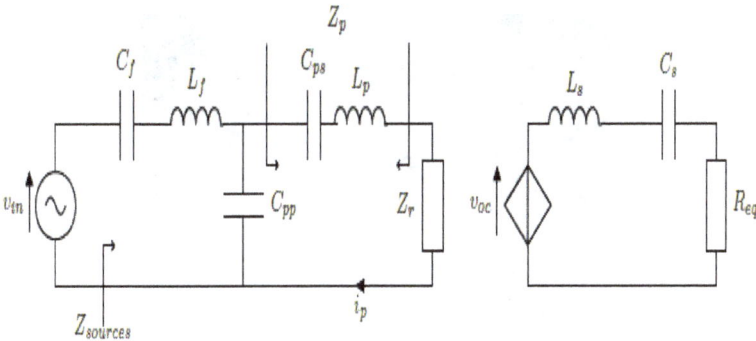

Fig. 3 System equivalent circuit

The total power losses caused by parasitic can be calculated as:

$$P_{loss} = I_0^2 r_0 + I_1^2 r_1 + I_p^2 r_p + I_s^2 r_s \qquad (18)$$

The output power can be expressed as

$$P_{out} = I_s^2 R_{eq} \qquad (19)$$

$$\left|\frac{I_1}{I_p}\right| \approx \left|\frac{Z_r + j\omega_0 aL_p}{\dfrac{1}{j\omega_0 C_{pp}}}\right| = \frac{k_0^2}{k_0^4 + k_{max}^4}\sqrt{k_0^4 + k^4} \tag{20}$$

$$\left|\frac{I_s}{I_p}\right| \approx kq\left|\frac{I_0}{I_p}\right| = \left|\frac{I_1 + I_p}{I_p}\right| \tag{21}$$

Considering only parasitic resistance loss, the efficiency of the system η can be calculated as

$$\eta = \frac{P_{out}}{P_{loss} + P_{out}} \tag{22}$$

$$\eta = \frac{(kq)^2 R_s}{(R_s + r_s)(kq)^2 + (\dfrac{k_0^2}{k_0^4 + k_{max}^4}\sqrt{k_0^4 + k^4})^2 r_1 + r_p + (1 + \dfrac{k_0^2}{k_0^4 + k_{max}^4}\dfrac{k_0^2}{k_0^4 + k_{max}^4})^2 r_0} \tag{23}$$

5. Experimental results and analysis

5.1 Experimental platform

Fig. 4 Experimental Device

5.2 Efficiency analysis

The measured and simulated output voltage when lateral misalignment x varies are illustrated in Fig. 5. The air gap between two coils is maintained at 40mm for different lateral misalignment x. As shown in the experimental curves, the output voltage slightly increases and arrives its maximum value with a lateral misalignment of 30mm. It is worth nothing that the maximum lateral misalignment can be up to 30% of coil diameter D showing a good position tolerance and validating the applicability of the proposed concept.

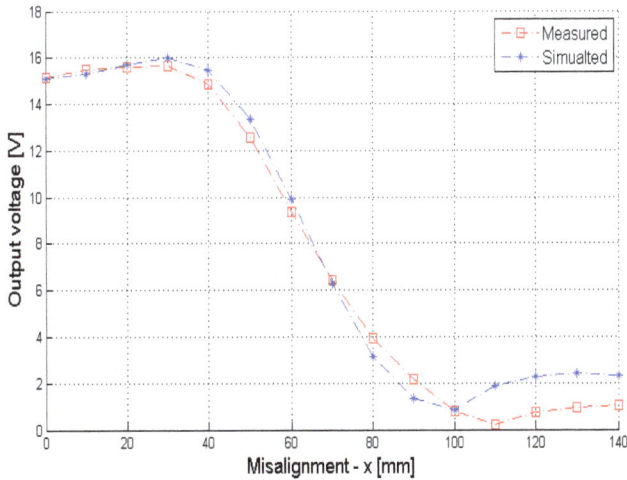

Fig. 5 Output voltage with different lateral misalignment

When the system input voltage is 48V, PSS compensation topology ICPT system efficiency curve under different coil distance is shown in Fig. 6. We can learn from the graph, when the coil distance is small, the system has high efficiency. When the eccentric distance of the coil is less than 20mm, the efficiency is always above 90%. When the coil eccentric distance continues to increase, the efficiency decreases quickly, as magnetic loss, resonant capacitor and the coil resistance loss are increasing.

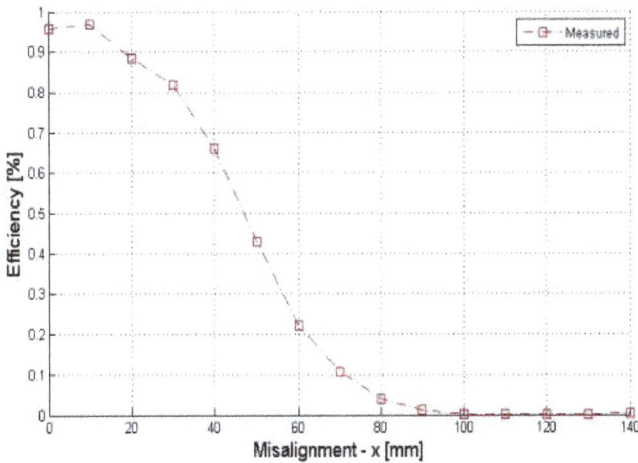

Fig. 6 Transmission efficiency with different lateral misalignment

6. Conclusion

In this paper, we studied the ICPT system based on PSS compensator topology, analyzes the system transmission efficiency and phase angle variation in the voltage range with a constant coupling coefficient. From our invitagation verified by experimental results, we are able to confirmed that

1) The maximum output is determined by compensation factor and resonance frequency;
2) Impedance phase angle is always positive when the coupling coefficientchanges within a certain range, providing a soft switching condition. It can reduce the loss of power devices, the resonance element and transformer coils.
3) Voltage gain increases with the resonant frequency decreasing, the transmission efficiency of the system increases with the resonance frequency increasing, and the system resonance frequency exists an optimal value.

References

1. Liu X, Hui S Y R. Optimal design of a hybrid winding structure for planar contactless battery charging platform [J]. IEEE Transactions on Power Electronics, 2008, 23(1): 455-463.
2. Fu Wenzhen, Zhang Bo, Qiu Dongyuan, Wang Wei. Maximum efficiency analysis and design of self-resonance coupling coils for wireless power transmission system [J]. Proceedings of the CSEE, 2009, 18:21-26.
3. Zhou Hao, YaoGang, ZhaoZiyu, ZhouLidan, JiangDawei, GuoFeng. LCL Resonant Inductively Coupled Power Transfer Systems [J]. Proceedings of the CSEE, 2013, 33:9-16+2.
4. Wu Ying, Yan Luguang, Xu Shangang. Stability analysis of the new contactless power delivery system [J]. Proceedings of the CSEE, 2004, 24 (5): 63-66 (in Chinese).
5. Cao Lingling, Chen Qianhong, Ren Xiaoyong , etal.Review of the efficient wireless power transmission technique for electric vehicles[J].Transactions of China Electro-technical Society, 2012 , 2012, 08:1-13.
6. Zhao Zhibin, SunYue, SuYugang, et al. Primary side constant input voltage control and parameters optimization of ICPT systems by genetic algorithm [J]. Proceedings of the CSEE, 2012, 32(15); 170-176 (in Chinese).
7. ChenKerui, WangZezhong, LiuShengnan, XiaTian, FangZhou, YanLei, ZhaoLili, GuoRuoying. Impacting Factors on Power and Efficiency of Inductively Coupled Power Transfer System [J]. Power System Technology, 2014, 03:807-811.

8. Ma Hao, SunXuan. Design of voltage source inductively coupled power transfer system with series compensation on both sides of transformer [J]. Proceedings of the CSEE, 2010, 30(15); 48-52 (in Chinese).

9. Sallan, J., Villa, J., Llombart, A. Optimal design of ICPT systems applied to electric vehicle battery charge. IEEE Transactions on Industrial Electronics, 2009, 56, (2), pp. 2140-2149.

10. Kacprzak D, Covic G A, Boys J T. An improved magnetic design for inductively coupled power transfer system pickups[C]//The 7th International Power Engineering Conference (IPEC), Singapore, 2005, 2:1133-1106.

11. Zhang Kai, Pan Mengchun, Weng Feibing, et al. The study status quo and application analysis on the inductively coupled power transfer technology [J]. Power Electronics,2009, 43(3): 76-78 (in Chinese).

12. Zhao Zhibin, SunYue, SuYugang, et al. Primary side constant input voltage control and parameters optimization of ICPT systems by genetic algorithm [J]. Proceedings of the CSEE, 2012, 32(15):170-176 (in Chinese).

13. Hu A P, Hussmann S, Improved power flow control for contactlessmoving sensor applications [J]. IEEE Power Electronics Letters, 2004, 2 (4): 135-138.

14. Huang, C.,James, J., Covic, G.: 'Design Considerations for Variable Coupling Lumped Coil Systems', IEEE Transactions on Power Electronics, 2014, 30, (2), pp. 680-689.

15. Hao, H., Covic, G., Boys, J.: 'A parallel topology for Inductive Power Transfer power supplies', IEEE Transactions on Power Electronics,2014, 29, (3), pp. 1140-1151.

16. Chwei-Sen Wang, Grant A Covic. Power transfer capability and bifurcation phenomena of loosely coupled inductive power transfer systems [J]. IEEE Trans. Industrial Electronics, 2004, 51(1):148-157.

Temperature Extraction Based on Infrared Image of High Voltage Power Equipment

Si-yuan Wang

*Jinan Power Supply Company, Shandong Electric Power Corporation,
Jinan, 250012, China
Email: sgh168@263.net*

De-wei Zhang

*Xinjiang Power Supply Company, Xinjiang Electric Power Corporation,
Urumqi, 830002, China*

Ge-hao Sheng and Xiu-chen Jiang

*Department of Electrical Engineering, Shanghai Jiao Tong University,
Shanghai, 200240, China
Email: shenghe@sjtu.edu.cn; xcjiang@sjtu.edu.cn*

An infrared temperature prediction method for power equipment is proposed based on the Radial Basis Function (RBF) network optimized by Quantum Genetic Algorithm (QGA) and Orthogonal Least Squares Algorithm (OLSA). The modified compound algorithm was used to optimize parameters of the RBF network. A temperature prediction model was established through the fitting of pixels and temperatures of the infrared image of an equipment. After image matching, the infrared temperature, at a position can be directly obtained from the visible image. Meanwhile, temperature values of different positions from the infrared image can also be directly read and identify the corresponding positions in the visible image. Experimental results indicate that the algorithm proposed has a better prediction performance than the RBF network optimized by OLSA alone and by Adaptive Genetic Algorithm (AGA) and OLSA. It improves the generalization capacity of RBF network, resulting in a more stable input and a higher prediction accuracy. The algorithm proposed facilitates temperature analysis and condition-based maintenance for substations.

Keywords: Infrared Temperature Prediction Method; Temperature Prediction Model; Prediction Performance.

1. Introduction

Infrared thermography has found wide applications in the diagnosis of thermal faults of power equipment due to multiple advantages such as high efficiency, safety, non-contact temperature measurement, large detection range and rapid detection [1-3]. The operating condition of main power equipment in the monitored substation is considerably useful for the prediction or diagnosis of potential faults and defects in the equipment. However, limited by working principle, external environment and the device itself, infrared images are less clear than normal images and the contrast between target equipment and background is weaker, which are unfavorable for subsequent fault analyses.

Online monitoring systems based on infrared and visible images [4-7] have currently been piloted in developed regions and will be further promoted. But the infrared monitor market is almost occupied by foreign large companies, e.g. the US FLIR Systems. Due to industrial monopoly and blockade on new techniques, power companies have no choice but to purchase and use the built-in analytic software of equipment. Personalized requirements cannot be satisfied and the capability of diagnosing faults of power equipment in the substation can hardly be improved, bringing about potential risks for the safe and stable operation of smart grid.

The key point of infrared thermography research is to determine the general relationship between temperature and image, namely, temperature fitting and prediction. At present, artificial neural network (ANN) theory has attracted great attention in the research of temperature prediction because of strong self-learning ability and fitting capability for complex nonlinear functions. Therein, radial basis function (RBF) network is able to achieve the global optimal approximation and give a better prediction. In this article, we optimized the RBF network-based infrared temperature prediction method for the substation's equipment by using quantum genetic algorithm (QGA) and orthogonal least squares algorithm (OLSA). The obtained infrared images were processed, and the pixels and temperatures of these images were fitted to establish an infrared temperature prediction model for the equipment. This model was then registered to visible images. Thus, the temperature of a position on an infrared image can be known by directly clicking on the corresponding place of the visible image; meanwhile, the infrared temperature can be directing obtained by clicking on the infrared image, which will also help find the corresponding position in visible image. Using experimental data, we made a comparison on the evolution situation of the fitness of QGA and adaptive genetic algorithm (AGA). Subsequently, the modified RBF algorithm was compared with OLSA-RBF and AGA-RBF algorithms, thus verifying the superiority and effectiveness of the former algorithm.

2. Temperature Prediction Algorithm based on Modified RBF Network

2.1. *RBF network*

RBF network is a traditional technique of multi-dimensional spatial interpolation, overcoming the defects of back propagation neural network such as local minimization and slow convergence. Composed of input layer, hidden layer and output layer, RBF network enjoys a favorable capability of global approximation and a self-adaptive structure. And its output is irrelevant to the initial weight [8]. The structure of RBF network is displayed in Fig. 1.

The hidden layer of RBF network has various basis functions of which the most common one is Gaussian kernel function

$$R_j(X - c_j) = \exp(-\| X - c_j \|^2 / 2\sigma_j^2), j = 1, 2, \cdots, p \tag{1}$$

where X is an n-dimensional input vector ($X=[x_1,x_2,...,x_n]$); c_j is the center of the j-th basis function, a vector with the similar dimension as X; σj is a generalized constant of the j-th neuron, i.e., the variance of Gaussian kernel function; n and p denote the number of neurons of input layer and hidden layer, respectively. After determining the function of the hidden layer, the relationship between input and output of the RBF network is expressed as

$$y_i = \sum_{j=1}^{p} w_{j,i} \exp(-\| x - c_j \|^2 / 2\sigma_j^2), i = 1, 2, \cdots, m \tag{2}$$

where m denotes the number of neurons in the output layer; y_i is the output value of the i-th neuron of the output layer; $w_{j,i}$ is the weight of connection between the j-th unit of hidden layer and the i-th unit of output layer. To determine the structure of RBF network, three parameters need to be solved: the basis function center c_j, variance j, and the weight from hidden layer to output layer (wj,i).

The construction of a RBF network depends largely on the optional selection of basis function center, unit number in the hidden layer, and network weight [9]. But in traditional RBF network, the algorithm training is prone to local minimization in the adjustment of each parameter.

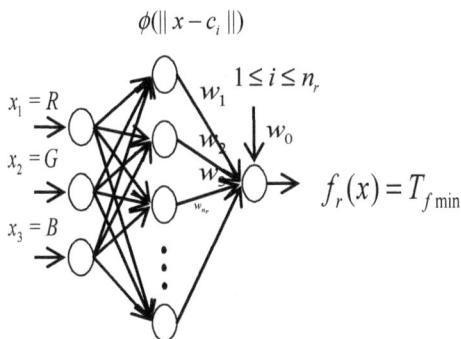

$$\phi(\| x - c_i \|)$$

$$x_1 = R$$

$$x_2 = G$$

$$x_3 = B$$

$$1 \le i \le n_r$$

$$w_1$$

$$w_0$$

$$f_r(x) = T_{f\min}$$

Fig. 1 Structure of RBF network

2.2. OLSA-RBF network

OLSA is a popular algorithm due to small computation amount, less occupation on storage space and rapid convergence [10-11]. OLSA can find the basis function center relatively accurately by introducing an error term:

$$y_i = \sum_{j=1}^{p} w_{j,i} \exp(-\| x - c_j \|^2 / 2\sigma_j^2) + e_k, i = 1,2,\cdots,m \tag{3}$$

It can also be expressed in the form of matrix:

$$Y = BW + E \tag{4}$$

where $Y \in \Re^{m\times 1}$ is the expected output vector of the neural network; $B \in \Re^{m\times q}$ is the regression matrix of each column of vectors; $W \in \Re^{q\times 1}$ is the network weight vector and $E \in \Re^{m\times 1}$ is the vector of errors between actual and predicted values of the network output.

Through Gram-Schmidt orthogonalization, the regression matrix B can be decomposed into a set of orthogonal basis vectors:

$$B = DA = [d_1 \quad d_2 \quad \cdots \quad d_q] \times \begin{bmatrix} 1 & a_{12} & \cdots & a_{1q} \\ 0 & 1 & \cdots & a_{2q} \\ \cdot & \cdot & \cdots & \cdot \\ \cdot & \cdot & \cdots & \cdot \\ 0 & 0 & \cdots & 1 \end{bmatrix} \tag{5}$$

where $A \in \Re^{m\times q}$ is an upper triangular matrix and $D \in \Re^{m\times q}$ is an orthogonal matrix; di can be calculated through Equation (6) and (7).

$$D^T D = H = diag(h_1, h_2, \cdots, h_q) \tag{6}$$

$$h_i = d_i^T d_i = \sum_{k=1}^{q} d_{ik}^2 \tag{7}$$

273

By combining the two equations, the expected network output Y is obtained as follows:

$$Y = DAW + E = DG + E \qquad (8)$$

Gram-Schmidt orthogonalization can ensure that matrix E and DG are orthogonal to each other, so

$$Y^T Y = G^T D^T DG + E^T E = \sum_{k=1}^{q} h_k g_k^2 + E^T E \qquad (9)$$

Thereby, the equal error rate (EER) of the k-th center is defined as

$$ERR_k = \frac{h_k g_k^2}{Y^T T} \qquad (10)$$

During continuous forward regressions of the RBF network, EER provides an effective standard for determining the network center. In each forward regression, when EER reaches the maximum an appropriate network center will be selected, and the regression will terminate at step q1 if the following condition is satisfied:

$$1 - \sum_{k=1}^{q_1} ERR_k < 0 \qquad (11)$$

In the construction of a neural network with OLSA, the selection of the initial σ value has great impact on the unit number of the hidden layer [12]. Hence, the parameter selection for OLSA-RBF network should be further optimized.

2.3. Modified OLSA-RBF network

We introduced QGA and optimized the initial σ value and the unit number of the hidden layer, to improve the efficiency of OLSA-RBF network.

QGA, firstly proposed by Ajit Narayanan and MarkMoore [13], is an optimized probabilistic searching algorithm combining quantum computational theory and evolutionary algorithm. It adopts quantum bit (qubit) to encode chromosomes and achieves evolutionary search by using the effect and updating of quantum gate. Compared with normal genetic algorithms, QGA has higher population diversity, faster convergence, and the ability of global optimization [14-15]. The steps of building a modified RBF network are described as follows.

(1) Qubit encoding

Quantum state is employed to encode information. Besides the two states, 0 and 1, a qubit can also represent any immediate state between 0 and 1. So the state of one qubit can be expressed as

$$|\Psi\rangle = \alpha|0\rangle + \beta|1\rangle \qquad (12)$$

where α and β (possibly complex number) represent the probability amplitude of corresponding states and satisfy the following normalization condition:

$$|\alpha|^2 + |\beta|^2 = 1 \qquad (13)$$

where $|\alpha|$ represents the probability of $|0\rangle$,and $|\beta|$ represents the probability of $|1\rangle$. Hence, a chromosome with m qubits can be expressed as

$$q = \begin{bmatrix} \alpha_1 & \alpha_2 & \cdots & \alpha_m \\ \beta_1 & \beta_2 & & \beta_m \end{bmatrix} \qquad (14)$$

(2) Population initialization

Suppose n is the population size (i.e., the number of chromosomes). In the initial population $Q(t)=\{q1^t, q2^t, ..., qnt\}$, the qubits of all chromosomes were assigned the value $1/\sqrt{2}$. This means that the state of each chromosome is the result of superposition of all possible states at an equal probability.

(3) Adjusting strategy for quantum revolving gate

Individual adjustment is realized through quantum revolving gate. In other words, the revolving gate in QGA is the final actuator of evolution. The working principle of revolving gate is as follows:

$$\begin{bmatrix} \alpha_i' \\ \beta_i' \end{bmatrix} = \begin{bmatrix} \cos(\theta_i) & -\sin(\theta_i) \\ \sin(\theta_i) & \cos(\theta_i) \end{bmatrix} \begin{bmatrix} \alpha_i \\ \beta_i \end{bmatrix} \qquad (15)$$

$$\theta_i = s(\alpha_i \beta_i)\Delta\theta_i \qquad (16)$$

where (αi, βi) is the i-th qubit in chromosome; θ denotes the revolving angle which controls the convergence rate of the algorithm; S(αi, βi) and Dθi denote revolving direction and step size of the revolving angle, respectively. The delta in Table is a coefficient related to the algorithm's convergence rate, to which a reasonable value should be given. By referring to the idea of dynamically adjusting the quantum revolving angle [16], the coefficient can be determined with the following equation.

$$delta = 0.05\pi\left(1 - \frac{k \cdot n}{MAXGEN + 1}\right) \qquad (17)$$

where n denotes the current generation of evolution and MAXGEN is the final generation; k is a constant in the range of [0,1]. The convergence rate of the algorithm is raised in the early operation period because the grid searched is larger. In the late operation period, the searched grid narrows, thus realizing precise searching and facilitating the seeking of optimal solutions.

Table 1 Methods for determining the revolving angle

x_i	α_i	$f(x)\geq f(b)$	Literature [17] $\Delta\theta_i$	The present work $\Delta\theta_i$	$S(\alpha_i, \beta_i)$			
					$\alpha_i\beta_i>0$	$\alpha_i\beta_i<0$	$\alpha_i=0$	$\beta_i=0$
0	0	F	0	0	0	0	0	0
0	0	T	0	0	0	0	0	0
0	1	F	0	0	0	0	0	0
0	1	T	0.05π	delta	-1	+1	±1	0
1	0	F	0.01π	delta	-1	+1	±1	0
1	0	T	0.025π	delta	+1	-1	0	±1
1	1	F	0.005π	delta	+1	-1	0	±1
1	1	T	0.025π	delta	+1	-1	0	±1

In Table 1, x_i and b_i represent the binary bit corresponding to the solution x and the i-th qubit of the current optimal individual b, respectively; f(x) is the function of fitness. Guaranteed by the quantum revolving gate, the algorithm will rapidly converge to obtain the chromosome with a higher fitness.

This study conducted a comparison between the evolution of QGA and AGA fitness, as shown in Fig. 2. The result indicates that QGA has a great improvement in evolutionary efficiency and its best fitness is more ideal (2.3212 vs. 1.5944) compared with the AGA. The average fitness of QGA and AGA is 1.9191 and 1.4369, respectively.

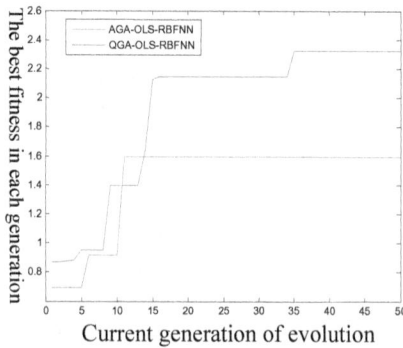

Fig. 2 Comparison of QGA and AGA evolution

3. Infrared Temperature Prediction Model

The model was established for the sake of power equipment analysis by relevant staff. They will be able to know infrared temperature by directing viewing visible images. This progress will reduce the difficulty in positioning thermal anomalies due to vague infrared images, thus improving positioning accuracy. Fig. 3 shows the infrared and visible images which reflect thermal anomalies of the transformer in a substation.

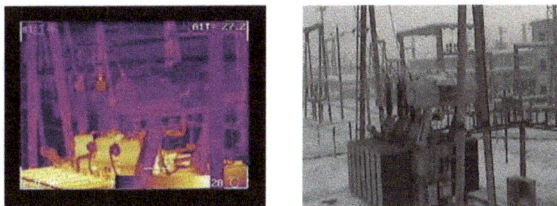

Fig. 3 Images of substation equipment

It can be found from Fig. 4 that the upper and lower limits of each temperature bar of infrared images will adjust automatically due to the configuration of bundled software. And infrared images are generally different from each other, e.g. the infrared temperature range in Fig. 3 is 3°C~28°C and those in Fig. 4 are -6°C~7°C and -13°C~3°C. In this article, the upper and lower limits of temperature bars were manually input tentatively. This step can be improved using digital intelligent recognition in the future.

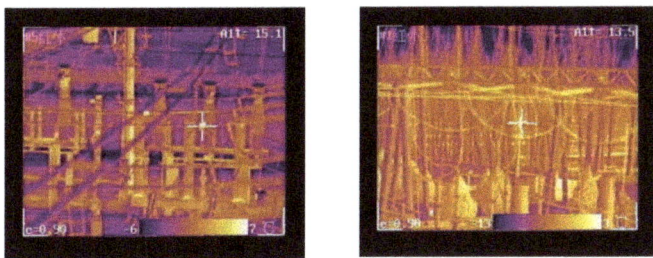

Fig. 4 Infrared images of the substation equipment

Preprocessing, registration and fusion were firstly carried out on visible and infrared images of the same scene. It was supposed that the infrared image was smaller than the visible image; if not, the former would be clipped. The image effect after preprocessing is displayed in Fig. 5.

Fig. 5 Images fused by visible and infrared images

Subsequently, the pixel matrices of visible and infrared images were matched to realize that the location information can be acquired by directly clicking on the visible image, and the temperature on the infrared image can be known through the prediction of the modified RBF network. Similarly, when clicking on the infrared image, we can know the temperature of a target position immediately and be led to the corresponding area on the visible image. The flow chart of the model is shown in Fig. 6: Image preprocessing

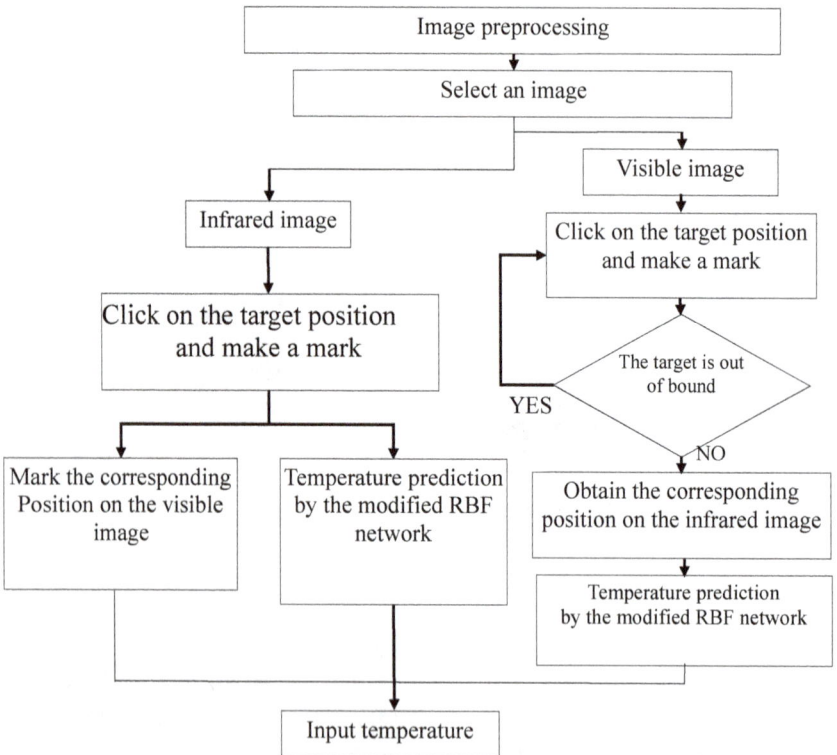

Fig. 6 Flow chart of temperature analysis model

4. Experimental results and analyses

Tests were conducted in the MATLAB environment. By setting points on the visible image, the coordinates of and RGB values at the corresponding positions of these points on the infrared image can be obtained automatically. Fifty sets of data were selected randomly as the input of the modified RBF network while the temperature on infrared image was taken as output. The output value was compared with those of OLSA-RBF network and AGA-OLSA-RBF network. See Table 2 for the comparison result.

Table 2 A comparison between evaluation indices of each algorithm

Algorithm	Average relative error	Maximum relative error
OLS-RBF	0.486425	1.819685
AGA-OLS-RBF	0.141767	0.598378
QGA-OLS-RBF	0.060528	0.240022

Fig. 7 shows the predicted temperatures of the modified RBF network almost completely accord with actual temperatures, with little error.

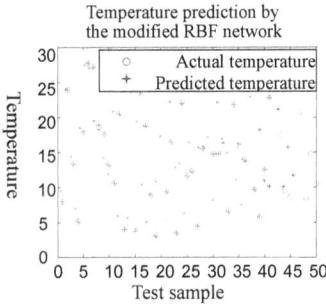

Fig. 7 Temperature prediction by the modified algorithm

The distribution of relative errors of each algorithm is exhibited in Fig. 8, where the green, blue and red curves are the error curve of OLSA-RBF network, AGA-OLSA-RBF network and QGA-OLSA-RBF network, respectively. It clearly reveals that compared with the former two algorithms, our modified algorithm (QGA-OLSA-RBF network) performed better in temperature prediction and enhanced the generalization capacity of RBF network, with a stable input and a high prediction accuracy. The maximum relative error of the OLSA-RBF network was about 1.8, while that of the QGA-OLSA-RBF network was merely 0.24. As for the average relative error, the OLSA-RBF network was eight times that of the QGA-OLSA-RBF network and the AGA-OLSA-RBF network doubled the latter.

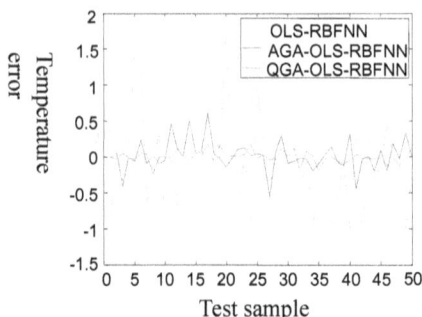

Fig. 8 A comparison of prediction errors of each algorithm

We designed a simple operation interface for the temperature analysis program using Matlab GUI. Taking the image of a transformer's thermal anomalies in a substation as an example, the specific interface is shown in Fig. 9. On the interface, one can separately analyze visible or infrared image, or simultaneously analyze both images. In Fig. 9, we randomly selected a point on the visible image and obtained the infrared temperature value of 21.6°C for the target position, which is marked with a red dot. We also selected a point on the infrared image randomly and obtained the infrared temperature value of 8.3°C for the target position, which is marked with a green dot. Its corresponding position on the visible image is marked with a yellow dot, with the coordinate of (123, 100).

Fig. 9 Interface of infrared temperature analysis result

5. Conclusion

An infrared temperature prediction method for power equipment in the substation was proposed by optimizing the radial basis function network with quantum genetic algorithm and orthogonal least squares algorithm. It overcomes many shortcomings of original infrared images of those power equipment, such as unclear picture, weak contrast between target equipment and the background, and the resultant inconvenience for fault analysis. The parameters of RBF network were optimized with the modified compound algorithm. Experimental results indicate that the algorithm modified in this article had a better prediction performance than OLSA-RBF network and AGA-OLSA-RBF network, whose maximum relative error was more than 7 times and twice that of our algorithm, respectively. Moreover, the average relative error has also been greatly lowered. These suggest that our algorithm has improved the generalization capacity of RBF network, leading to a stable input and an increased prediction accuracy. The algorithm proposed makes it more convenient to analyze the temperature of power equipment in substations and position the area of thermal anomalies, which is favorable for the condition-based maintenance in substations.

References

1. Men Hong, Yu Jiaxue, Qin Lei. Segmentation of electric equipment infrared image based on CA and OTSU[J]. Electric Power Automation Equipment, 2011, 31(9): 92-95.
2. Lin Lihua, Wu Dongmei, Liu Jian, et al. Substation infrared early warning system using object segmentation and image registration[J]. High Voltage Engineering, 2010, 36(7): 1718-1724
3. Li Hongwei. Intelligent vision system of smart substation[J]. Electric Power Automation Equipment, 2012, 32(8): 141-147.
4. Li Ran. Research and realization of convergence between infrared temperature measurement technology and video surveillance system of substation[J]. Power System Technology, 2008, 32(14): 80-84.
5. Zhao Zhenbing, Gao Qiang, Yuan Jinsha, et al. An on-line monitoring system for substation power equipment dual-channel image[J]. Automation of Eletric Power Systems, 2009, 32(10): 76-79.
6. Liu Lihui, Zhang Yongsheng, Sun Yong, et al. Development and application of equipment inspection robot for smart substation[J]. Automation of Eletric Power Systems, 2011, 35(19): 85-88.
7. Lu Ming, Sheng Gehao, Zhang Weidong, et al. A tower and line tracking algorithm for power transmission line inspection based on unmanned aerial vehicles[J]. Automation of Eletric Power Systems, 2012, 36(9): 92-97.
8. Hong Cui, Wen Buying, Lin Weiming. Short-term forcasting of wind power output based on improved OLS-RBF ANN model[J]. Electric Power Automation Equipment, 2012, 32(9): 40-43.

9. Shi Yan, Jiang Xingliang, Yuan Jihe. Flashover voltage forecasting model of iced insulator based on RBF network[J]. High Voltage Engineering, 2009 (3): 591-596.

10. Huang C M, Wang F L. An RBF network with OLS and EPSO algorithms for real-time power dispatch[J]. Power Systems, IEEE Transactions on, 2007, 22(1): 96-104.

11. Liao C C. Enhanced RBF network for recognizing noise-riding power quality events[J]. Instrumentation and Measurement, IEEE Transactions on, 2010, 59(6): 1550-1561.

12. Chang W Y. Estimation of the state of charge for a LFP battery using a hybrid method that combines a RBF neural network, an OLS algorithm and AGA[J]. International Journal of Electrical Power & Energy Systems, 2013, 53: 603-611.

13. Narayanan A, Moore M. Quantum-inspired genetic algorithms[C]// Evolutionary Computation, 1996., Proceedings of IEEE International Conference on. IEEE, 1996: 61-66.

14. Lee J C, Lin W M, Liao G C, et al. Quantum genetic algorithm for dynamic economic dispatch with valve-point effects and including wind power system[J]. International Journal of Electrical Power & Energy Systems, 2011, 33(2): 189-197.

15. Liao G C. Solve environmental economic dispatch of Smart MicroGrid containing distributed generation system–Using chaotic quantum genetic algorithm[J]. International Journal of Electrical Power & Energy Systems, 2012, 43(1): 779-787.

16. Xing Huanlai, Pan Wei, Zou Xihua. A novel improved quantum genetic algorithm for combinatorial optimization problems[J]. Acta Electronica Sinica, 2007, 35(10): 1999-2002.

17. Jiao Songming, Han Pu, Huang Yu, et al. Fuzzy quantum genetic algorithm and its application research in thermal process identification[J] . Proceedings of the CSEE, 2007, 27(5): 87-92.

18. Polat Ö, Yıldırım T. Genetic optimization of GRNN for pattern recognition without feature extraction[J]. Expert Systems with Applications, 2008, 34(4): 2444-2448.

Study of Power Supply Structure for Jilin Province Power Grid

Ping Ren, Tian Dong and Li-shen Chu

Jilin Province Electric Power Research Institute,
Jilin Changchun, China

Qing-sheng Bi and Deng-chao Li[†]

Dept of Energy and Power Engineering, Changchun Institute of Technology,
Jilin Building Energy Supply and Indoor Environment Control Engineering
Research Center, Changchun, China
[†]Email: 18843148689@163.com

In this paper, the reasonable matching mathematical model proposes a multiple power supply structure power grid with a certain capacity based on the analysis of the unreasonable structure of Jilin province power grid. According to the mathematical model, a reasonable matching of power supply in Jilin province is analyzed and the principle and development direction of the future power supply in Jilin province are proposed.

Keywords: Power grid; Power supply structure; Mathematical model.

1. Introduction

Great changes have taken place in Jilin province power grid since beginning in 2003. Under the new system, multiple power generation group are derived from the power generation enterprise base on the no substantial practice of originally bid online mechanism, with the capacity of competition between the power generation group has gradually escalating and growth of Jilin power grid capacity is far higher than the growth in electricity eventually lead to power construction planning in a state of disorder out of control. Hours in order to increase the utilization and reports to the project approval through, there was a group of 300 MW heating units, some of the original condensing steam turbine are also create conditions into a heating unit, lead to the heating units of rising year by year, by heating load of power grid is more and more. At the same time, the construction of wind power, such as a new energy sources, cost greatly reduced, and the wind power installed capacity in the west of Jilin Province is doubled every year, based on the strongly inclined to foster policy of the country to give new energy. As the winter of 2007, the unreasonable structure of Jinlin province power grid has been exposed seriously, and the unreasonable state continues today escalating, has affected each aspect, mainly displays in the following three prominent problems:

- Severe imbalances happen on power grid during the winter heating
- Electric peak-shaving during the holidays more difficult, especially during the holiday season is due to the large proportion of large-scale heating unit and heating unit in power grid load adjustment is limited by the heating load.
- Electric peak-shaving affecting the heating unit

To meet, because of the large heating unit of heating load increased the boot mode, the load rate of province's overall thermal power unit is low, run at night need to be further depth electrical load to satisfy the stability of power grid peak shaving down, heating quality is limited to peak shaving.

- The grid to abandon the wind happen occasionally
- Wind turbines are limited by power grid peak shaving and heating effects, winter heating season, it was the season is suitable for wind power in the province, the wind load of reverse has happened more, to increase the using rate of green energy, the power grid has as much as possible given wind power adjustment ability, but there are still part of wind power capacity is abandoned.

How to solve the above problem, is not only solve the current primary problem of the impact on the safe of Jilin power grid, is also related to the heating quality of Jilin province residents heating, related to the green to improve energy efficiency, reduce the pollutant emission in the process of electric power production, improving the efficiency of the electric power production, reduce the blindness of power construction investment and other problems. Regional power grid in the power structure reasonable matching is the key to solve the problem. This article elaborates the different kinds of power supply in the regional power grid how to match to get the benefits, to make the electric power production, thermal production, and green energy use are safe, stable and reasonable running.

2. A variety of power source structure reasonable matching mathematical model

With a certain scale capacity can achieve the safe and stable and economic operation of power grid operation, besides is closely related to the distribution of grid architecture, power source planning and the generator set to run in adjustment characteristics also affect the power grid security stability, also directly affect the electric power transmission and power generation operation economy. For there is a lot of cold north area with heating of the heat load and industrial heating load of cogeneration units, large power condensing generator set, wind turbine, hydropower and other variety of generating set in terms of regional power grid, rational matching of different power supply combination can not only optimize the grid power source planning and power grid structure, optimizing the allocation of power network capacity, realize the power grid security and stability of economic operation. Can play a maximum cogeneration unit benefits at the same time, improve the safe operation of heat supply network stability and reliability, ensure the quality of heating, also can maximize the utilization of clean energy and renewable energy power generation equipment, and achieve the goal of energy conservation and emissions reduction.

2.1. A variety of power source structure reasonable matching mathematical model

In order to achieve the goals mentioned above, the author puts forward the reasonable matching network in a variety of power structure mathematical model for a certain scale capacity of power grid, as shown in figure 1, and unit of generation units are MW.

$$N_{base} = N_{r\min} + N_{wind} + N_b + N_g + N_{n0.5} + N_{nu} \quad (1)$$

Where, N_{base} denotes daily basic electrical load. $N_{r\min}$ is the minimum load under the heating unit to ensure the heating quality. N_{wind} is the load of wind power. N_b denotes biomass power generation load. N_g illustrates the power generation load of waste incineration. $N_{n0.5}$ expresses the large capacity condenser 50% generating unit load (peak load operation). N_{nu} denotes the load of nuclear power generating units.

$$N_{\min} = N_{base} + N_{ps} \quad (2)$$

Where, $N_{\min}$ stands for electricity trough of the load. N_{ps} denotes a power load of pumped storage.

$$N_{\max} = N_{r\max} + N_{wind} + N_b + N_g + N_{nu} + N_{n(>0.5)} + N_{hy\max} \quad (3)$$

Where, $N_{\max}$ denotes daily peak electrical load. $N_{r\max}$ is one of the largest electricity co-generation unit load. $N_{n(>0.5)}$ expresses the large capacity condensing steam generating units more than 50% load (peak load operation). $N_{hy\max}$ denotes the maximum load of hydropower.

2.2. Operation mode of power supply

List several possible ways：
    This is both energy saving and safe mode, established mathematical model according to above operation when power grid load N meet $N_{max} > N > N_{min} = N_{base}$.

When power grid load N meet $N = N_{min} < N_{base}$, using pumped storage power station, etc have energy storage effect of electric consumption, can improve N_{base} to N_{min} , and peace through power trough.

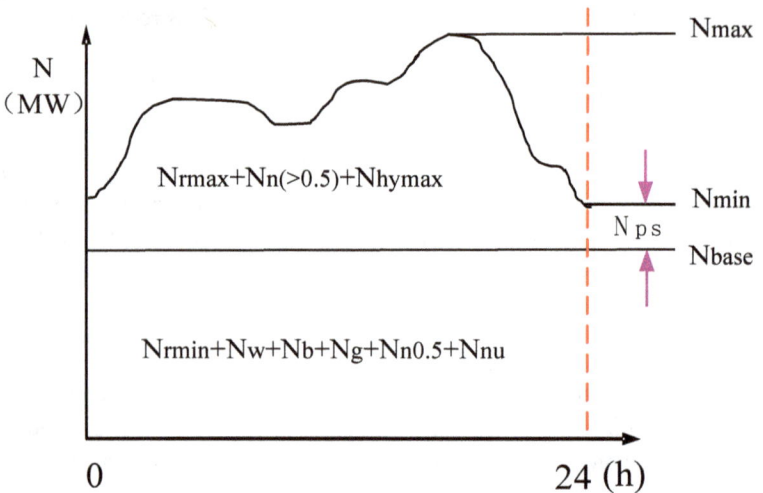

Figure 1. The power daily load of power grid combination curve diagram

When power grid load N meet $N = N_{max} > N_{min} = N_{base}$, the large capacity condensing generators and large heating units and some of the hydropower units bear the load of power grid growth.

When the power grid load N meet $N = N_{min} < N_{base}$, $N_{wind} = 0$ (due to natural conditions such as wind discomfort); thermal power condensing generating set takes capacity exit of wind power.

Known from the analysis of the above, with a certain scale capacity of the grid power supply ideal configuration mode can make sure that the more wind power capacity is absorbed during the trough of power grid, heating units can take enough heat load, nuclear power units can take constant power load. Run at the same time the large condensing generators within their margin of 50% rated capacity adjustment will be able to meet the needs of the power grid peak capacity, this power supply mode configuration are the basic conditions of the economic operation of power grid.

3. The present situation to grid power structure of Jilin province and the reasonable matching analysis

3.1. Jilin province grid power structure of the present situation (by the end of 2014)

By the end of 2014, the power structure of Jilin province is shown in table 1.

Table 1. The power structure of Jilin province (by the end of 2014)

Project		Quantity (MW)		Ratio (%)	
Total installed capacity		20446			
Thermal power unit	Heating unit	15772	13283	77.14	64.97
	Condensing steam turbine		2489		12.17
Hydro power		188		0.92	
Wind power		4080		19.96	
Other (including biomass, litter and photovoltaic power station)		406		2.0	

From table 1, the installed capacity of heating unit occupies 64.97% of the total installed capacity of Jilin province, and wind power accounts for 19.96% of the total capacity of the grid. They are all higher than other provinces, and have formed a unique power grid with a high ratio of wind power and heating unit.

$$N_{base} = N_{r\min} + N_{wind} + N_b + N_g + N_{n0.5} + N_{nu}$$
$$= 13283 \times 5 \div 6 + 4080 \times 0.75 + 406 + 2489 \times 0.5 = 15779.7 \, \text{MW}$$
$$N_{\min} = N_{base} + N_{ps} = 15779.7 \, \text{MW}$$
$$N_{\max} = N_{r\max} + N_{wind} + N_b + N_g + N_{nu} + N_{n(>0.5)} + N_{hy\max}$$
$$= 13283 + 4080 \times 0.75 + 406 + 2489 \times 0.5 = 17993.5 \, \text{MW}$$

3.2. The reasonable matching analysis about grid power structure of Jilin province

We can see the typical daily load in figure 2.

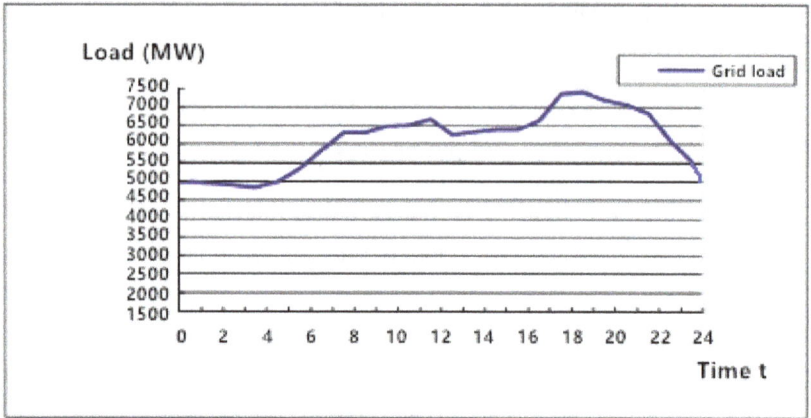

Figure 2. The typical daily load curve in winter heating period of Jilin province grid

The maximum and minimum data, $N_{max} \approx 7500\,\text{MW} \leq 17993.5\,\text{MW}$ and $N_{min} \approx 4800\,\text{MW} \leq 15779.7\,\text{MW}$, it is far do not conform to the reasonable matching mathematical model. There are the following four reasons for analysis:

Power construction growth faster than the growth of the power load, and the new power structure is not reasonable.

- The Proportion of the heating units of thermal power units has increased.
- The capacity of wind power unit increased year by year.

Slow economic development, social power consumption growth is slow, which resulted in increased power grid peak valley load difference, excess capacity at low load of power grid period, and heat supply quantity is growing rapidly, heating is difficult to meet the boot mode of heat-supply unit.

The end results are: the power grid peak shaving difficulties; affected by the power grid peak valley, the residents heating quality affected; abandon the wind have occurred constantly; the investment waste phenomenon of all kinds of power is comparatively serious.

4. Jilin province power arrangement principles and the development direction in the future

According to the heating period statistics for Jilin province calculates the average may be residents heating central heating heat load of 15.83 GW, according to the normal distribution to the installed capacity of 26.38 GW heat load demand, the current heating unit installed with capacity of 12.87 GW, according to the cogeneration device equipped with heat load gap of 13.51 GW. This means that if current configuration scale of thermoelectric generator double again, to meet the needs of heat users in the province, and according to the current power grid electricity growth, or according to the growth rate of electricity in recent 10 years, is likely to have development space of cogeneration unit. Heating demand for cold area in winter, however, is rigid growth, and must be on the basis of overall consideration and proper methods of heating, to meet the needs of the winter heating, blindly on heating reform large condensing turbine unit, and the practice of construction of large heating unit, can make the current situation of heating and power generation conflict worsened, ultimately it is difficult to achieve the desired effect. Therefore, Jilin province in the future development of thermoelectric principle should consider mainly from the following several directions:

New construction or expansion to the north China electric power transmission channel, and given wind power.

Add regenerative electric boiler according to local conditions, on the one hand, supplementary part time heating demand, increase grid electricity, troughs power, on the other hand can reduce the wind volume, and realize effective use of clean energy.

For large heating unit with large heat supply network, according to the requirements of the state of cogeneration policies, or reasonable position, in the end of the heat supply network configuration efficiency and environmental protection indexes meet the requirements of the central heating boiler into peak shaving operation, efforts to reduce the thermalization coefficient of cogeneration, on the one hand is advantageous to the safe and stable operation of power grid peak shaving, on the other hand to ensure the quality of heating, improve heat supply network operation safety and reliability.

In the check expectations with heating way operating conditions can be large condensing generating set heating, strictly prohibited rob large heating unit of electrical energy as heat renovation project put into production. Have been put into operation in the large cogeneration heating unit, in 2 ~ 3 years does not meet the provisions, the thermoelectric than the index of thermal power plant, should be to assess the USES, the heat load under other ways supply is given priority to, create conditions for the given wind power generation and power grid peak shaving operation.

Development of the heating unit and coordinated development of regional electricity; Have to construction of pumped storage power plant; Considering the future electricity consumption growth and wind power given and under the condition of existing capacity for coal-fired power units, should not be built during the twelfth five-year and the thirteenth five-year in large unit.

5. Conclusion

This paper established mathematical model of power grid in a variety of power structure reasonable matching of scale has a certain capacity of power grid power supply has a certain guiding significance to the reasonable matching; representative due to Jilin province grid structure, therefore, the Jilin province future power arrangement principle and the development direction of the northern cold region of significance of reference to the regional power grid power structure.

Acknowledgments

This work is funded by the Department of Science and Technology of Jilin Province (No: 20100638) and the Education Department of Jilin Province (No: 2015305).

References

1. Shui Han, Jin-Dong Zhang. "The northeast power grid power supply structure optimization adjustment problems". Northeast Electric Power Technology, 2012,06:1- 4.
2. Yi-Zi Yuan, Hao Cui. "The northeast power grid power supply structure optimization adjustment problems". East China electric power, 2007,35(1).
3. Jin-Chang Zhang. "Jilin province power grid peak shaving difficulties and solutions". Jilin power grid, 2004,06: 17- 19.
4. Ying-Ling Shi, Yuan-Yuan Liu. "Power supply structure optimization research based on environmental constraints". The China Power, 2009-09:16- 19.
5. Yu-Wei Fu, Li Yang. "Based on the fuzzy pattern recognition method of the power supply structure optimization evaluation". Journal of Statistics and Decision, 2010, 09:53 -55.
6. Li Yang. "Based on the sustainable development of China's power structure optimization research" [D]. Harbin Engineering University, 2010.
7. Alexander G Kozlov. "Optimization of structure and power supply conditions of catalytic gas sensor". Sensors & Actuators: B. Chemical, 2002, Vol.82 (1), pp.24-33.
8. Chun-Hua Wang, Guo-Ying Fan, Lei Guo. "Jilin province the safe and stable operation of power grids in-depth study". Journal of Jilin Electric Power, 2009 01:6-9.
9. Gang Cao, Ming Ding. "Medium and long-term power supply structure optimization research in Anhui province". Journal of China Power, 2000, 09: 40-43.

Impact of Power System Load on Short-Circuit Current

Li-ping Qiu[1], Guang-yao Yu[2], Xiao-jun Tang[1],
Zhen-bin Li[3] and Zhi-gang Huang[3]

[1]*Power Research Institute, Beijing, China, 100192*
[2]*State Grid Tianjin Electric Electric Power Research Institute,
Tianjin, China, 300384*
[3]*State Grid Tianjin Electric Power Corporation Tianjin, China 300010*

Different considerations of the load in the calculation of the short-circuit current may cause different short-circuit current calculation results. In this paper, the influence of the static load and the induction motor on the short-circuit current is analyzed and the mechanism of the effect of various loads on the short-circuit current is demonstrated.

Keywords: Power system load, Short-circuit current, Load model.

1. Introduction

With the construction of power grid, The interconnection of power grid is more and more closely. In some areas, the problem of short circuit current exceeds the standard become more serious. In the short circuit current calculation different considerations to the load may lead to different result, and the result may be too conservative or too aggressive, and then impacts the reasonable selection of switches and other equipment, even the power system safe operation.

How to consider the load in short circuit calculation is a big problem to calculating personnel. In this paper, the impact of static load and different capacity induction motor to the short circuit current is analyzed, and the mechanism is explained. So as to provide a basis for the calculation to practical engineering.

2. Influence of static load to short circuit current

According to the load voltage characteristics, the static load is divided into constant impedance load, constant current load and constant power load.

In this paper, the IEEE-9 system is taken as an example to study the effects of constant impedance, constant current and constant power load on the short circuit current. The IEEE-9 system is shown in Fig. 1. The load character of bus A, bus B and bus C is constant impedance, constant current, constant power load.

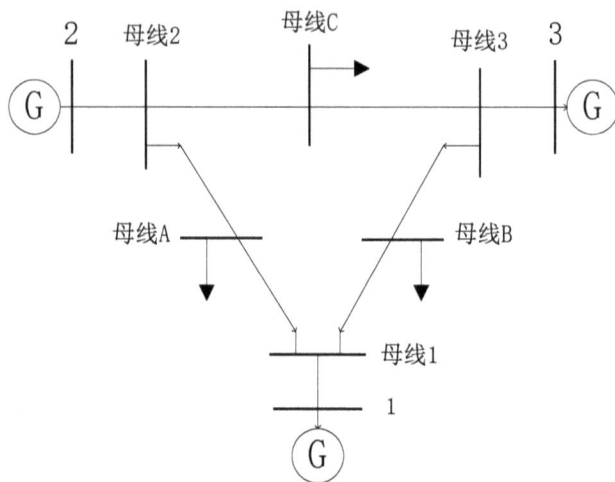

Fig. 1. IEEE-9 SYSTEM

The calculation results of short circuit current of IEEE-9 system are shown in Table 1. From the calculation result we can see: using the three kinds of static load models, the change trend of the short-circuit current is same. Bus 1, bus 2, and bus 3 are closer to the generators, their short current level is higher, more than 6.2kA. When using the constant impedance, the short circuit current level is the highest, when using the constant power, the short circuit current level is the lowest. When using the current model, the short-circuit current level is between the constant resistance and the constant power model.

Table 1. Short-circuit current calculation results (kA)

Load model	Bus1	Bus2	Bus3	BusC	BusB	BusA
Constant resistant	8.91	7.4	7.0	5.84	5.31	5.61
Constant current	8.71	7.13	6.73	5.69	5.16	5.49
Constant power	8.28	6.6	6.21	5.3	4.8	5.17

The static load is non rotating element, which does not produce short circuit current to the short circuit point when short circuit occurs. Its effect on short circuit current is mainly reflected in its shunt effect. When short-circuit fault occurs, the load bus voltage will be reduced, so the load current will change. For constant power load, the load current is bound to increase, and the shunt effect is more obvious. For constant impedance load, the load current is decreased with the decrease of bus voltage, and the shunt capacity is relatively weak. The shunt effect of constant current load is between constant impedance and constant power load.

3. Influence of induction motor load to short circuit current

In actual power system, the proportion of induction motor load is for more than 60%, their function and characteristics are widely different. From practical point of view, induction motor is divided into induction motor with large capacity and small capacity, and their influence on the short circuit current were analyzed in this paper.

Taking the single-machine infinite-bus system as an example, when the load of bus 1 is induction motor or constant impedance load, the short circuit current of bus A is 6.03kA and 4.06kA. It is found from the injection current of bus 1 that the induction motor load provide short circuit current to system during short circuit fault, and the initial value of the periodic component is 5.0kA

Taking the single machine infinite system as an example, when the load of bus 1 is induction motor or constant impedance load, the short circuit current of bus A is 6.03kA and 4.06kA. Analysis the injection current of bus 1, we found that the induction motor load provide short-circuit current to system, and its periodic component initial value is about 5.0kA

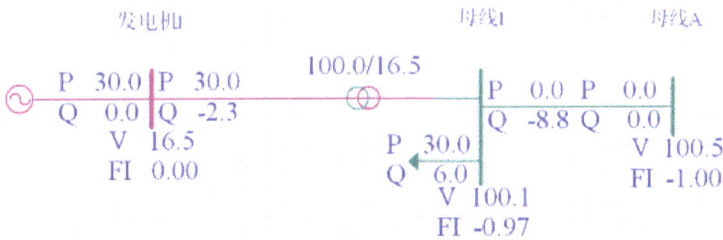

Fig. 2. Single machine infinite system

In order to study the current feedback characteristics of the induction motor load. We firstly determine the analytical expression of induction motor based on the three order electromechanical transient model [1-5]. The analysis is based on the following two assumptions.

(1) The rotor slip of induction motor is close to 0 when the normal operation of induction motor. It is considered that induction motor is an under excited synchronous generator. The normal operation of the motor rotor slip is about 5% ~ 2%, it can be seen as synchronous operation, $\omega^* = 1$. In the early stage of the short circuit, due to the inertia, the speed change of induction motor is very small, which can be approximately considered to be constant. As a result, the induction motor can also be approximated as an under excited synchronous generator[5].

293

(2) Approximate think that the stator and rotor leakage reactance of induction motor are similar. Through coordinate transformation, the induction motor can be studied in the same coordinate system, so the stator winding and rotor winding have no relative motion, and the electromagnetic relationship is similar to that of the transformer. Based on these two simplified conditions, it can be considered that the equivalent circuits of the induction motor is the same when the stator winding end point occurs three-phase short circuit or the induction motor starts. It was shown in figure 3. When the induction motor is connected to the grid at the beginning start stage, the rotor has not yet started rotating and rotor winding is short circuited, the starting current is the short circuit current, and the starting reactance is the equivalent reactance of the stator when the rotor winding is short connected. Therefore, solving the analytical expression of induction motor when the end of the stator winding is short circuit can be considered to solve the starting current of induction motor.

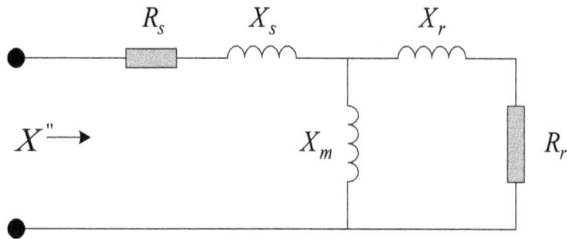

Fig. 3. Equivalent circuit of induction motor starting

For the image above, the circuit can be expressed as follows expression:

$$\begin{cases} L_s \dfrac{di_s}{dt} + R_s i_s + M \dfrac{di_r}{dt} = 0 \\ L_r \dfrac{di_r}{dt} + R_r i_r + M \dfrac{di_s}{dt} = 0 \end{cases}$$

(1)

In expression (1), R_s, R_r, L_s, L_r is stator resistance, rotor resistance, stator inductance, rotor inductance, i_s, i_r is the current of stator and rotor winding; M is coefficient of mutual inductance between stator and rotor windings.

For the equation (1), it includes two parts of current, one part is the current frequency close to zero, called DC component of short circuit current; the other part is the current frequency close to the synchronous frequency, called periodic component of short-circuit current.

Set the closing angle of source is zero, And omit the winding resistance, The analytical expression of the feedback current of the motor is as follows:

$$i = i_d e^{-\frac{t}{Ts}} + I_k^{''} \cos te^{-\frac{t}{Tr}}$$

(2)

In expression (2), T_s is the time constant of the stator when the rotor winding is short. T_r is the time constant of the rotor when the stator winding is short. Based on the assumption (2), the parameters of the stator and rotor windings are considered to be the same; therefore, considering $T_s = T_r = T_D$

For winding type, single mussaurus type, or deep groove type motor, there is only one time constant, above expression can by simplified as:

$$i = (i_d + I_k^{''} \cos t)e^{-\frac{t}{T_D}}$$

(3)

The phasor diagram of the induction motor under normal operation is shown in Fig. 4.

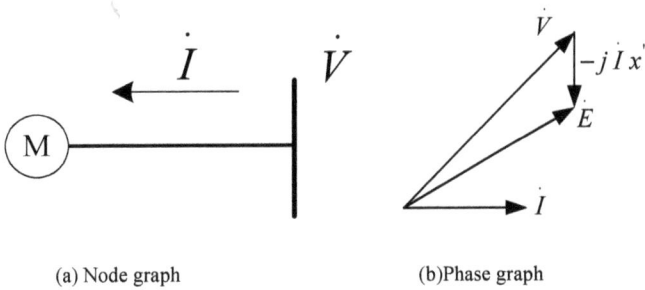

(a) Node graph   (b)Phase graph

Fig. 4. Normal operation state of induction motor

From the phasor diagram, the following expression can be got:

$$\dot{E} = \dot{V} - j\dot{I}x^{''}$$

I is the effective value of the stator current period component, can be expressed as:

$$I = -j\frac{V-E}{x^{''}}$$

(4)

Moment before the short circuit $t = 0$ sec, (denoted by 0^-), Motor stator current can be expressed as

$$i_{0^-} = \frac{V_{0^-} - E_{0^-}^{''}}{X_D^{''}}$$

(5)

295

V_{0^-} and E_{0^-} v are terminal voltage and second transient potential of the induction motor before short circuit.

Moment after the short circuit $t = 0$ sec, (denoted by 0^+), The initial amplitude of the periodic component of the motor is:

$$I_k^{"} = -\frac{E_{0^+}^{"}}{X_D^{"}}$$

(6)

The total short circuit current is:

$$i_{0^+} = i_d - \frac{E_{0^+}^{"}}{X_D^{"}}$$

(7)

According to the principle that the current cannot be mutated before and after short circuit, combining the formula (5) and the formula (7) can get formula (8)

$$i_d = \frac{V_{0^-} - E_{0^-}^{"}}{X_D^{"}} + \frac{E_{0^+}^{"}}{X_D^{"}}$$

(8)

Put the formula (6) and (8) into the equation (2), then the following formula could be got.

$$i = [(\frac{V_{0^-} - E_{0^-}^{"}}{X_D^{"}} + \frac{E_{0^+}^{"}}{X_D^{"}}) - \frac{E_{0^+}^{"}}{X_D^{"}} \cos t]e^{-\frac{t}{T_D}}$$

At the just moment of short circuit occurs, the transient electric potential of the motor does not suddenly change.

That is $E_{0^-}^{"} = E_{0^+}^{"}$. Thus, the formula can be abbreviated as:

$$i = [\frac{V_{0^-}}{X_D^{"}} - \frac{E_{0^+}^{"}}{X_D^{"}} \cos t]e^{-\frac{t}{T_D}}$$

(9)

Formula (9) is the analytic expression of the induction motor of three-phase short circuit at the end of stator winding. The analytical formula is in complete agreement with the physical concept of the induction motor: Due to the law of conservation of flux, when three phase short circuit fault occurs in the system, but the protection device is not yet in operation, the flux of induction motor is constant. The free component of flux and current will appear in the winding, and the stator current will contain DC component and synchronous frequency AC component, show in formula (5).The short circuit current analytic formula includes the DC component $\frac{V_{0^-}}{X_D^{"}}$ and AC component $\frac{E_{0^+}^{"}}{X_D^{"}} \cos t$.

For the feedback current of induction motor, the impulse current, initial value current of periodic component, and DC component current are most cared.

From formula (5) : The feedback current of induction motor is closely related to the sub-transient reactance $X_D^{"}$ and the decay time constant T_D.

Based on the above expressions, the feedback current of induction motor has the following characteristics:

(1) The attenuation speed of motor feedback current is proportional to the motor capacity. The R, X and T_D are different between large capacity induction motor and small capacity induction motor. Typically, the big capacity induction motor has small $X_D^{"}$ and big T_D. Based on the formula (5), for feedback current of large capacity induction motor, its impulse current (i_p)is large and attenuation rate is slow.

For small capacity induction motor, its feedback current peak is small, and will decay after 3~4 cycles. For small capacity induction motor, its feedback current peak is big, and will gradually decay after 6~8 cycles.

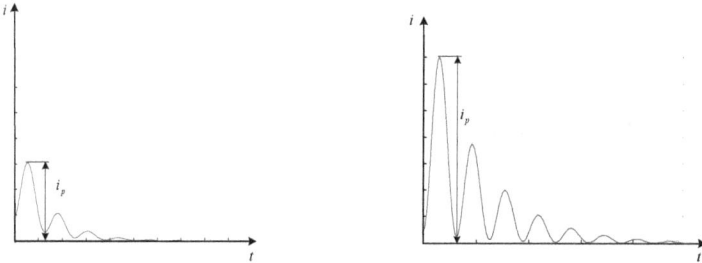

Fig. 5. Feedback current waveform of Small and Big capacity induction motor

(2) Considering the saturation characteristic of the magnetic circuit, the feedback short circuit current is increased. In order to reduce the no-load current of induction motor and to improve the power of induction motor, the air gap should be as small as possible in the manufacturing process of induction motor. The air gap of the medium and small induction motor is 0.2mm ~ 1.5mm, which is less than that of the generator. In the calculation of the feedback current of the induction motor, the influence of the magnetic circuit saturation is generally neglected, such considerations are in line with the actual.

For large capacity induction motor, when the instantaneous current value is larger, some leakage flux is forced through the core part of the magnetic circuit, Which leads to the saturation of magnetic circuit. After saturation, the transient reactance $X_D^{"}$ of induction motor is much smaller than that of the unsaturated state. By type $I_k^{"} = -\frac{E_{0^-}^{'}}{X_D^{"}}$,the initial value of the periodic component of the feedback current of induction motor is inversely proportional to the sub transient reactance. Thus, the decrease of transient reactance will cause the increase of current. For induction motor with large capacity, considering the characteristic of magnetic saturation will cause the increase of current feedback.

4. Conclusion

(1) Static load does not provide short circuit current, but the shunt effect of different static load is different, and the shunt effect of constant power load is the most obvious, the constant current load is the second, and the constant impedance load is the worst.

(2) Induction motor provides short circuit current when the system short circuit fault occurs. And the short-circuit current of the induction motor feedback is proportional to the capacity of the motor. Considering the saturation characteristic of the magnetic circuit, the feedback current of the motor is increased.

References

1.   IEEE Task Force on Load Representation for dynamic Performance. Standard Load Models for Power Flow and Dynamic Performance Simulation [J]. IEEE Trans on Power Systems,Vol.10, No.3, pp 1302–1313, August 1995.

2.   Ju Ping, Ma Daqiang. Power system load modeling [M]. Hydraulic and Electric Power Press, 1995.

3.   Tang Yong, Zhang Hongbin. A Synthesis Load Model with Distribution Network [J]. Power System Technology. 2007, 31(5): 34–38.

4.   Xi'an Jiao Tong University. Power system calculation [M]. Hydraulic and Electric Power Press,1978.

5.   EL-Hawary M E, Dias L G. Incorporation of load models in load-flow studies: form of model effects [J]. IEEE Proceedings C, 1987,134(1): 27–30.

6.   Vaahedi Ebrahim. Dynamic load modeling in large scale stability studies[J]. IEEE Trans on Power Systems, 1988, 3(3):1039–1045

7.   Qiu Liping, Zhao Bing, Zhang Wenchao. Effects of Synthesis Load Models on Dynamic Stabilities of Large-Scale Interconnected Grid, Power System Technology. 2010, 34(10):82–87.

Research on Dual Three-Phase PMSM Series-Connected Drive System

Xiao-tong Lu[†] and Hua-qiang Zhang
Department of Electrical Engineering, Harbin Institute of Technology at Weihai,
Weihai, Shandong, China
E-mail: [†]lulu_5412@163.com; zhq@hit.edu.cn

Ling-shun Liu
Department of Control Engineering, Naval Aeronautical Engineering Institute
Yantai, Shandong, China
E-mail: lingshunliu@sohu.com

Multi-motor variable speed drive system is widely used in electric applications. Since multiphase machines have additional degree of freedom than three-phase machines, it can be utilized to control other machines. Dual Three-phase (DTP) machine can connect in series with another DTP machine or a two-phase (TP) machine or one DTP machine and one TP machine by an appropriate phase transposition, supplied from a single multiphase inverter. Based on the Permanent Magnet Synchronous Motor (PMSM) as the research object, a detailed mathematical modeling of the three series-connected system and the phase transposition rules were proposed. Using Direct Torque Control based on Space Vector pulse width Modulation (SVM-DTC) technology for series-connected system, the method of SVM technology was modified to obtain the full dc bus voltage. The simulation results verified this developed SVM technology.

Keywords: Series-connected drive system; Dual three-phase; Direct torque control; Space vector pulse width modulation; Permanent magnet synchronous motor.

1. Introduction

The researchers have studied the multi-phase machine drive system since 1969. With the rapid development of machine technology, multi-phase machine is widely used in marine electric propulsion, railway traction, electric vehicle traction and more-electric aircraft [1]. Dual three-phase (DTP) permanent magnet synchronous motor (PMSM) has the advantages of both multi-phase machine and PMSM. With the PMSM characteristic of wide speed range and high energy conversion efficiency, DTP PMSM can also eliminate 5th and 7th harmonic magnetic potential which have the biggest influence on the machine performance, while reducing the torque ripple greatly and improving the machine performance [2]. Besides, there are many other significant features of multi-phase machine like reduced torque pulsation, high torque density, higher ripple frequency, less current harmonic of dc bus, potentially higher efficiency, reduced power per phase and greatly improved reliability as a result of higher fault-tolerance [3, 4].

Series-connected multi-motor system is mainly applied to metro, mill, robot, locomotive traction, winding machine, paper machine, marine electric propulsion and more-electric applications. Previous studies indicate multiphase machines can be series connected by an appropriate phase transposition and supplied from a single VSI to achieve decoupling control [5]. The introduction of an appropriate phase transposition in the two-motor five-phase drive system with series connection of stator windings leads to a complete decoupling of the flux/torque-producing currents of one machine from the flux/torque-producing currents of the second machine [6, 7]. A symmetrical six-phase machine can be connected in series with a three-phase machine, the independent and decoupled vector control of this series-connected symmetrical six-phase motor drive system has been studied in [7]. What is more, the modeling of a series-connected seven-phase three-motor drive system has been established in [8], used to validate the high performance control of the drive system. However, the realization of the series-connected system is applicable to not only symmetrical but also asymmetrical multi-phase machine. The research of two asymmetrical six-phase machines series-connected drive system is elaborated in [9]. This paper presents the research on the three series-connected DTP PMSM drive systems, the modeling of the DTP PMSM series-connected system is provided firstly. Then, SVM-DTC control strategy is analyzed and improved. Lastly, simulation results are given.

2. Modeling of DTP PMSM series-connected system

Independent flux and torque control of an AC machine only requires two current components, so DTP machine as a kind of multiphase AC machine has additional degrees of freedom, which can be utilized not only to enhance the torque production by stator harmonic current injection or to improve the fault-tolerance, but also to form a multi-motor drive system with single inverter supply [1]. The stator windings of the machines are series connected by an appropriate phase transposition to make the flux/torque producing (d-q) currents of one machine become the non-flux/torque producing (z_1-z_2) currents for the other machines and the decoupling independent control of the machines is

300

achieved. The normal decoupling transformation matrix for DTP machine with two isolated neutral points is presented in [2].

The relative connection schemes are illustrated in Fig. 1, indicating three possible series-connected drive system. The first one in Fig. 1 has been studied by many researchers. The second one is a novel system, in which the two neutral points of DTP machine are connected with the two windings of TP machine respectively and the voltage of this two points are $(u_a+u_b+u_c)/3$ and $(u_x+u_y+u_z)/3$. As TP machine with spatial displacement of 90 degree between the phases, the current or voltage of neutral point is not zero by applying the Y connection. Thus, add one bridge arm in the inverter, the neutral point of TP machine connected to the midpoint of this bridge arm, in which the duty cycle of the switching device should be 50%. The series-connected DTP and TP machines drive system becomes a system supplied from a single seven-phase VSI. And the third one is the combination of the previous two system.

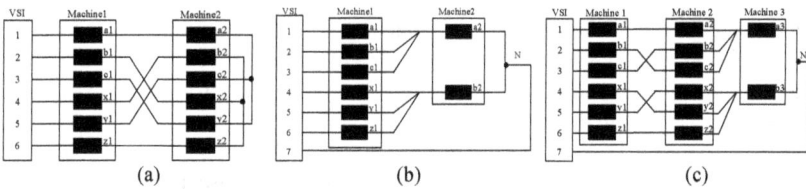

Fig. 1. (a)Series connection of two DTP machines, (b) Series connection of one DTP machine and one TP machine, (c) Series connection of two DTP machines and one TP machine.

3. SVM-DTC control strategy

3.1. Space vector modulation

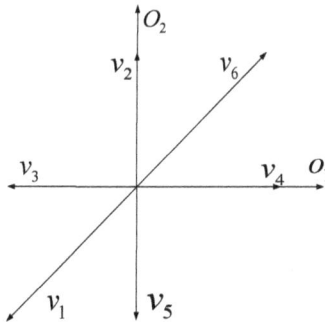

Fig. 2. The space voltage vector distribution map of o1-o2 plane.

Generally, an n-phase two-level VSI is characterized with 2^n voltage space vectors. Thus, for a dual three-phase PMSM there are $2^6=64$ space vectors, four of which are zero vector; for a two-phase PMSM there are 2^3=8 space vectors, two of which are zero vectors. The seven-phase VSI voltage space vectors of the d-q and z_1-z_2 planes are as in [2], the o_1-o_2 plane is shown in Fig. 2.

For the dual three-phase machines, select the largest 12 neighboring non-zero vectors to synthesize the reference voltage vector, the amplitude value of which is $0.644U_{dc}$ (U_{dc} is the DC bus voltage), 4 active space vectors are required to generate sinusoidal voltages. And for two-phase machine, select the total 6 non-zero vectors from Fig. 2, the amplitude of which is U_{dc} or $1.414U_{dc}$. It is assumed that the d, q components of reference voltage vector in d-q plane u_d, u_q are known. According to the projection relationship between the reference voltage vectors and the active space vectors in the coordinate, the four active space vectors acting time are $T_1 = [(2\sqrt{3} - 3)u_d - \sqrt{3}u_q]T_s/2U_{dc}$, $T_2 = [(3 - \sqrt{3})u_d - (3 - \sqrt{3})u_q]T_s/2U_{dc}$, $T_3 = [(3 - \sqrt{3})u_d + (3 - \sqrt{3})u_q]T_s/2U_{dc}$, $T_4 = [(2\sqrt{3} - 3)u_d + \sqrt{3}u_q]T_s/2U_{dc}$, T_s=0.0002s is the switching period. The active space vectors acting time of two DTP machines is same. Similarly, according to the projection relationship between the reference voltage vectors and the active space vectors of TP machine, the application times of active space vectors in each sector are obtained and shown in Table 1.

Table 1. The active space vectors acting time in each sector of two-phase PMSM.

Sector	T_1	T_2
I	$(U_{O1}-U_{O2})T_S/U_{dc}$	$U_{O2}T_S/U_{dc}$
II	$U_{O1}T_S/U_{dc}$	$(U_{O2}-U_{O1})T_S/U_{dc}$
III	$-U_{O1}T_S/U_{dc}$	$U_{O2}T_S/U_{dc}$
IV	$(U_{O2}-U_{O1})T_S/U_{dc}$	$-U_{O2}T_S/U_{dc}$
V	$-U_{O1}T_S/U_{dc}$	$(U_{O1}-U_{O2})T_S/U_{dc}$
VI	$-U_{O2}T_S/U_{dc}$	$U_{O1}T_S/U_{dc}$

Note: T_1, T_2 are two active vectors acting time and zero vector acting time $T_0 = T_S - T_1 - T_2$.

3.2. Improved SVM-DTC method

Not like using the SPWM method, series-connected machines can synchronous operate, the SVPWM method limit the synchronous operation of machines. There is a cutover problem of SVPWM modulators for series-connected drive system as each machine has its independent space vector modulators. Take two-motor system as examples, the traditional way is to control the first machine in the first switching period of two consecutive switching periods and the second machine in the second switching period, illustrated in Fig. 3(a), the effective values of output voltages will be one half of the given reference values because it is only one reference that imposed in every switching periods. Even if one machine is at standstill only half of the whole dc voltage is available for the control of another machine. So, since the active space vectors in one plane can replace the zero space vectors in the other plane, utilizing one switching period to replace two consecutive periods and centering the switching pattern with respect to the midpoint of the switching period, in order to reduce the switching frequency of IGBTs. Then the space vector voltages acting sequence are arranged according to the minimum switch losses as follows.

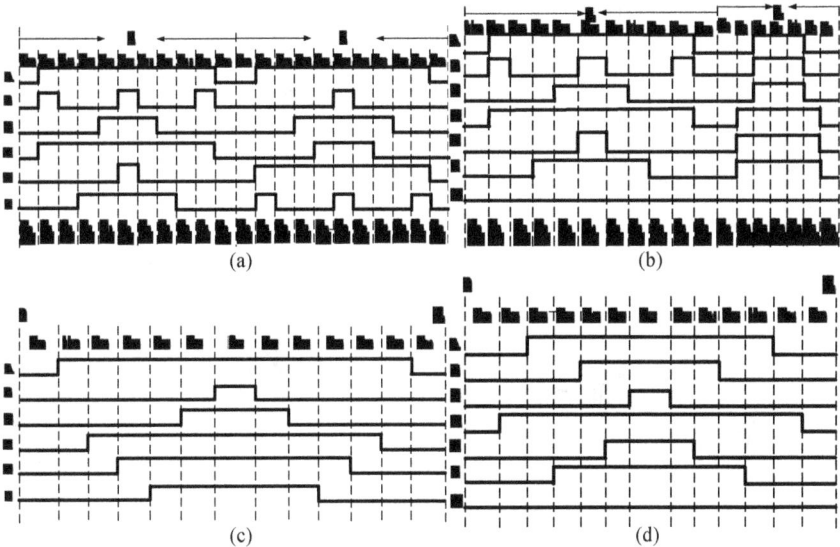

Fig. 3. (a) The switching pattern of DTP PMSM series-connected two-motor drive system, (b) The switching pattern of DTP and TP PMSM series-connected two-motor drive system, (c) The improved switching pattern of DTP PMSM series-connected two-motor drive system, (d) DTP and TP PMSM series-connected two-motor drive system.

The DTP PMSM series-connected two-motor drive system is provided at first. It is assumed that the d-q plane reference is in sector 1, reference in the z_1-z_2 plane is in sector 2 and the reference in the o_1-o_2 plane is in sector 2 the selected space vectors in two planes are v_{55}, v_{45}, v_{44}, v_{64} and v_{63}, v_{43}, v_{42}, v_{52}, which are imposed in a sequential manner as illustrated in Fig. 3(c). In the same way, the switching pattern of the DTP and TP PMSM series-connected two-motor drive system are showed in Figs. 3 (b) and (d).

The simulation model of three series-connected drive system is established in Matlab/Simulink and the results are shown in Figs. 4-6.

303

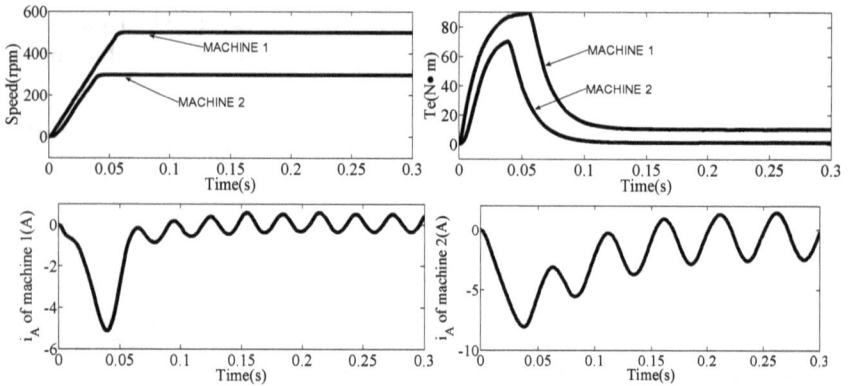

Fig. 4 The waveform of DTP PMSM two-motor drive system.

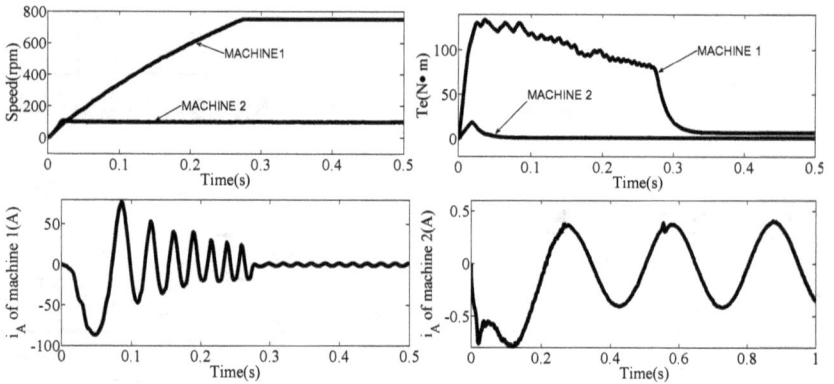

Fig. 5 The waveform of DTP and TP PMSM two-motor drive system.

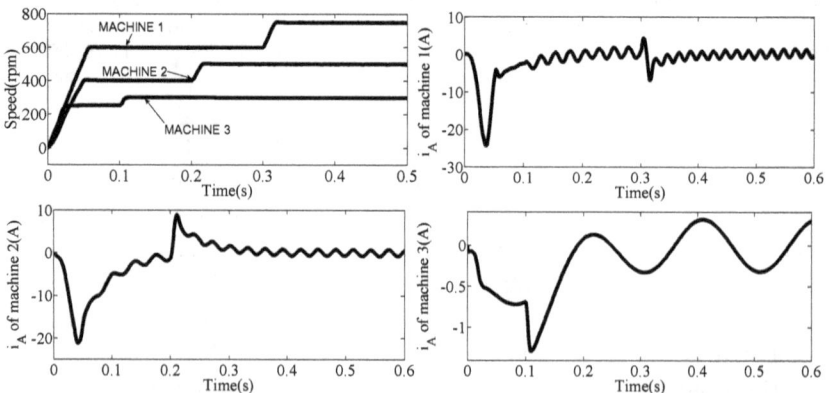

Fig. 6 The waveform of DTP and TP PMSM three-motor drive system.

304

In this way, if one machine is not operate, another one can obtain the whole dc bus voltage; if two machines are operate at different speeds, the faster one will get the more dc bus voltage than another one. The reasonable distribution of dc bus voltage is realized and the utilization is improved.

4. Conclusion

This paper has dealt with the three DTP series-connected drive systems. The mathematical models of DTP PMSM and TP PMSM are established by the space vector decoupling theory. The SVM-DTC control strategy for DTP PMSM series-connected drive system was improved to realize the fully utilization of the dc bus voltage.

Acknowledgment

This work is supported by the Foundation of National Natural Science (51377168).

References

1. E. Levi, R. Bojoi, F. Profumo, H. A. Toliyat and S. Williamson, Multiphase induction motor drives – a technology status review, in *IET Electr. Power Applications*, Vol.1 (2007), pp. 489-516.
2. YF Zhao and T.A.Lipo, Space vector PWM control of dual-three phase induction machine using vector space decomposition, in *IEEE Transactions on Industry Applications*, Vol.31 (1995), pp. 1100-1109.
3. J. Karttunen, S. Kallio, P. Peltoniemi, P. Silventoinen, and O. Pyrhönen,Decoupled vector control scheme for dual three-phase permanent magnet synchronous machines, in *IEEE Transactions on Industrial Electronics*, Vol.61 (2014), pp. 2185-2196.
4. R. Bojoi, M. Lazzari, F. Profumo and A.Tenconi, Digital field-oriented control for dual three-phase induction motor drives, in *IEEE Transactions on Industry Applications*, Vol.39(2003), pp. 752-760.
5. E. Levi, M. Jones, S.N. Vukosavic and H.A. Toliyat, A novel concept of a multiphase, multimotor Vector controlled drive system supplied from a single voltage source inverter, in *IEEE Transactions on Power Electronics*, Vol.19 (2004), pp. 320-335.
6. E. Levi, M. Jones, S.N. Vukosavic, Atif Iqbal, and H. A. Toliyat, Modeling, control, and experimental investigation of a five-phase series-connected two-motor drive with single inverter supply, in *IEEE Transactions on Industrial Electronics*, Vol.54 (2007), pp. 1504-1516.

7. E. Levi, M. Jones, S.N. Vukosavic and H.A. Toliyat, Steady-state modeling of series-connected five-phase and six-phase two-motor drives, in *IEEE Transactions on Industry Applications*, Vol.44 (2008), pp. 1559-1568.

8. M.Thavot, A. Iqbal and Mohammad Saleh, Modeling of a seven-phase series-connected three-motor drive system, in *IEEE GCC Conference and Exhibition (GCC'13)*, (2013).

9. E. Levi, S.N. Vukosavic and M. Jones, Vector control schemes for series-connected sixphase two-motor drive systems, in *IEE Proc-Electr. Power Applications*, Vol.152 (2005), pp. 226-238.

The Development of High Strength-High Conductive Aluminum Conductors

Hong-yun Yu, Rui Li, Miao Qian[†] and Heng Ma

ZheJiang Hua Dian Equipment Testing Institute,
Hangzhou 310015, China.
[†]Email: 153634553@qq.com

Jia-peng Hou and Qiang Wang

Shenyang National Laboratory for Materials Science,
Institute of Metal Research, Chinese
Academy of Sciences, Shenyang 110016, China.

Overhead transmission lines are important carrier transporting the electric energy from the generating end to the receiving end. In this paper, key scientific issues such as structural design, strengthening mechanism and conductive characteristics, which mainly influence the development of aluminum conductors were discussed briefly. In structural design, commercially pure aluminum wire has excellent electrical conductivity, but its strength is very low, and thus combining the high conductive aluminum wire with the high strength steel wire may exhibit outstanding comprehensive properties. In the aspect of strengthening mechanism, solid solution, precipitation hardening, work hardening and fine grain strengthening are important strengthening mechanisms which are commonly applied to strengthen the conductive materials. Sometimes single mechanism can be applied, and also sometimes several mechanisms can be used together when applying these mechanisms. In the aspect of conductivity, boron was usually adopted to make the impurity elements precipitate from the liquid metal by the way of gravity sedimentation in order to purify the matrix of aluminum and its alloys, and finally the electrical conductivity of aluminum wire was improved. In the past ten years, T-81 and HPT techniques were applied to aluminum conductors one after another, and they improved the electrical conductivity of aluminum wires as well as the mechanical properties. However, some of these new techniques still cannot be applied in the conductor field constrained by their disadvantages, and these techniques are still waiting to be transformed into techniques in the preparation of aluminum conductors. In the future, more new techniques are needed to solve the problems from development of high strength-high conductive conductors for the purpose of breaking the bottleneck, which impedes the development of aluminum conductors. Developing some outstanding aluminum conductors is beneficial to the development of human society.

Keywords: Overhead transmission lines; Aluminum and its alloy; Strengthening mechanism; Electrical conductivity; High strength- high conductivity

1. Introduction

Overhead transmission lines are important carrier transporting the electric energy from the generating end to the receiving end, which also play an important role in the economic development. Among structure materials such as gold, silver, copper and aluminum, aluminum is the most suitable one can be used as conductive materials, and thus it got wide application in overhead transmission lines. Aluminum conductor steel reinforced (ACSR) is the oldest overhead conductor, which is firstly designed in 1908 in America. In the 1920s, conductors made of aluminum and its alloys were firstly applied in high voltage transmission lines by America, Switzerland and Germany. Since 1950s, aluminum conductors were applied in more than 50 percent of the transmission lines in Japan and America, and as high as 80 percent in France. In China, the application of ACSR started from 1970s [1]. In recent years, the application of aluminum conductors increased rapidly. Main properties for evaluating the conductive materials are strength, electrical conductivity, elongation and so forth. In these properties, some key scientific and technical issues still need to be investigated systematically, including the structural design, mechanical properties and conductive properties of aluminum conductors so as to improving the operational safety of overhead transmission lines and energy savings. In the future, the application of aluminum conductors may have broad prospects.

2. Structural design of aluminum strand

Aluminum and its alloys are important conductive materials applied in overhead conductors and the factors from two aspects should be considered in its structural design, i.e., mechanical properties (Strength) and conductive properties (Electrical conductivity). Although the commercially pure aluminum gets a good conductive property, it has a lower strength. As a result, it should be applied with other materials which have a higher strength such as steel and high strength aluminum alloys. As is well known, aluminum alloy usually has higher strength than pure aluminum, so it can be used in conductor sindependently. At present, three types of aluminum conductors were used in the world [2]:

(i) Aluminum conductor steel reinforced (ACSR). The core of this type of conductor consists of high strength steel wires, and the outer layer is composed of A6, a commercially pure aluminum, wires. The steel wires and aluminum wires play their respective advantages in ACSR, combining high strength with high electrical conductivity, and simultaneously got good technical effects with safety and energy saving. ACSR is the most widely used overhead conductor in high voltage AC and DC transmission lines. However, the key problem exists in ACSR is steel core, which has high strength with remarkably lower electrical conductivity, bringing eddy current loss.

(ii) All aluminum alloy conductor (AAAC). In AAAC, all strands are constituted by the same aluminum alloy wire. The aluminum alloy wire is very light as compared with the steel core in ACSR due to its high specific strength. The energy saving AAAC has become the main direction of future development. Although this type of conductors has outstanding advantages, there are also some questions which should be considered in the future such as wear, creep and fatigue.

(iii) Aluminum conductor alloy reinforced (ACAR). In ACAR, the core is high strength aluminum alloy wire, and the wire in the outer layer is made of A6. ACAR shows outstanding properties combining the high strength of aluminum alloy with the high electrical conductivity of commercially pure aluminum. The most prominent feature of ACAR is that the core wire cannot bring eddy current loss. However, the ACAR also has weak points, and the strength of its core wires is obviously lower than steel wires. Therefore, the safety problem is the biggest problem restricting its development, and the aluminum alloy with higher strength should be developed in the future to fill the gap.

On the whole, three types of overhead conductors have their own advantages, and all of them have good application foundation. Future, high strength-high conductive aluminum alloy strands will become the most popular direction of development. In addition to the structural design, safety and energy saving are also important factors that influencing the development of overhead conductors. The safety of aluminum conductor is closely related to its strength. Aluminum conductors should bear some loads from its weight, the wind blowing, and icing. Thus, it should own a high strength to protect itself against the damage from instantaneous load and alternating load, and the improvement of its strength is closely related to safety and economy. Aluminum conductor should have a good performance in energy saving in order to decrease the power loss in the transmission process.

3. Strengthening of conductive materials

Strengthening is a key issue in the development of conductive materials. Actually, the safety of structural material usually increases with the improvement of strength. Thus, high strength conductor made of aluminum and its alloy is an objective in the development of overhead transmission conductor. The aluminum rod for ACRS only has a strength of 90-110MPa. Although the strength of aluminum wire increased after the cold drawing, its strength was only 160MPa [3]. Based on the basic theory in material science and engineering, the influencing factors on the strength of conductive materials are mainly from two aspects: alloying elements and microstructure. Manufacturing process is an important link influencing the final properties of aluminum conductor. The production process of aluminum conductor mainly includes: melting, casting, rolling and drawing as well as solid solution and aging treatment [4,5].The state of alloying element and the microstructure in the matrix were significantly influenced by cold drawing in the manufacturing process such as the solution of atoms, the growth of dislocations, and thus influencing the properties of aluminum conductor.

Solid solution strengthening. Solid solution treatment is an important way of strengthening conductive material, because when alloying elements present in the metal matrix in the form of solute atoms will lead to the changes of the local stress field due to the different atomic size between the solvent atoms and the solute atoms or due to the different shear modulus. After solution treatment, the aluminum can be strengthened significantly as shown in Fig. 1, the strength of 6201 and AA6082 aluminum alloy increased significantly after solid solution treatment. This local stress field may react with the dislocation, and impede the movement of dislocation, which cause an increase in yield strength. At present, a widely accepted description on the solid solution is Fleischer equation [6,7]:

$$\Delta\sigma = M \cdot G \cdot b \cdot \varepsilon^{3/2} \cdot c^{1/2} \tag{3}$$

Where, M is the average orientation factor; Gis the shear modulus; ε is the micro-strain; b is the Burgers vector;c is the solubility.

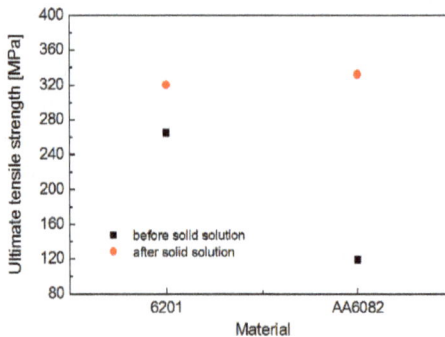

Fig. 1 Strength of 6201 and AA6082 aluminum alloy increases significantly after solid solution [4,5].

Precipitation hardening. Solvent atoms may precipitate from the alloy in the form of inter-metallic compounds which can also impede the movement of dislocation, and thus the alloy is strengthened through the aging treatment, as shown in Fig. 2 and Fig. 3. Orowan strengthening mechanism can be used to describe the role played by the second phase, which can be divided into two kinds of mechanisms includingdislocation bypassing mechanism and dislocation cuting through mechanism. At present, nano-particles areused by some researchers to prepare some high strength-high conductivealuminum alloy and copper [9, 10]. The dislocation bypassing mechanism was commonly described by the Orowan-Ashby equation:

$$\Delta\sigma_{or} = \frac{0.13 \cdot G \cdot b}{L} \cdot \ln\frac{r}{b} \tag{4}$$

Where, G is the shear modulus of the alloy matrix; b is the burgers vector; r is the radius of the precipitation; L is the average spacing between precipitates.

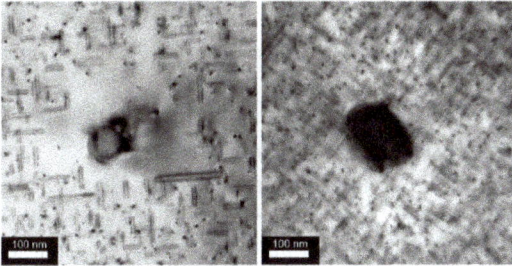

Fig. 2 A number of micro-particles appear in the matrix of 6060 alloy after aging treatment observed under the transmission electron microscopy (TEM) [8].

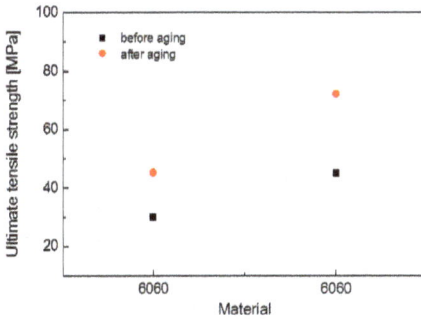

Fig. 3 Strength of 6060 alloy increasesobviously after aging treatment [8], and the particles in the matrix play a strengthening role in the aging treatment.

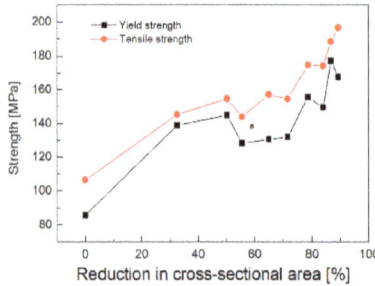

Fig. 4 Variety of strength in the cold-drawn A6 conductor [12].

Work hardening.In Fig. 4, the aluminum rod was drawn from a diameter of 9.5mm to 3.0-3.5mm, and then the aluminum conductor was obtained. In this process, the overall deformation (reduction of cross-sectional area) may achieve 90 percent, and the equivalent strain is in the range of 2.0-3.0 [12]. The dislocation density increased continuously in the drawing process, and the conductor was strengthened with the increase of dislocation. Dislocations interacting with each other in the grain impede the movement of dislocation, and then the yield strength of aluminum alloy can be improved. The strengthening effect of the residual dislocation on the metal can be expressed by Bailey-Hirsch [13,14] equation:

$$\Delta\sigma = M \cdot \alpha \cdot G \cdot b \cdot \rho^{1/2} \tag{2}$$

Where, M is the orientation factor; a is a constant; G is the shear modulus; b is Burgers vector; ρ is the dislocation density.

Cold-drawn process is a conventional strengthening step which can introduce lots of dislocations and grain boundaries into the material, and simultaneously grains in the matrix can be refined. After the serious deformation, the high angle grain boundary (HAGB) increases remarkably in the aluminum and its alloy [15, 16]. The impeding effect of HAGB is more obvious than the low angle grain boundary (LAGB) [17, 18].

Therefore, the increase of the proportion of LAGB is very beneficial to the strength of aluminum conductor, and this relation usually described by H-P(Hall-Petch) equation [19-21]:

$$\sigma_y = \sigma_0 + k_y \cdot d^{-1/2} \tag{1}$$

Where, σ0is the shear stress; ky is the H-P slop; d is the average grain size; σ0 and kyare constants.

Grain refinement strengthening.Nano structures can be used in the design of high strength materials. Serious plastic deformation (SPD) can be introduced to refine the grains up to nano-size. Valiev et al. reported that the grain size of 6201 aluminum alloy was refined to an average size of 130 nm by a technique of high pressure torsion (HPT), and the strength of 6201 alloy was improved up to 510MPa [9]. The SPD technique was also applied to prepare high strength- high conductivity copper. It is a pity that the SPD technique still cannot be used to manufacture conductor snow [10].

In the aspect of strengthening mechanism, solid solution, precipitation hardening, work hardening and fine grain strengthening are important strengthening mechanisms which are commonly applied to strengthen the conductive materials, for example, aluminum and its alloy. Sometimes single mechanism can be applied, and also sometimes several mechanisms can be used together when strengthening materials. The advantages of them were mixed to strengthen conductors in order to get a good strengthening effect and improve the mechanical properties of conductors to a large extent.

4. Conductive properties of aluminum and its alloy

Conductive property is the most important property of the conductive materials, which is closely related to the economy of conductors. In conductive metal materials, many factors such as alloying elements, impurities, intermetallic compounds, grain boundaries and dislocations influence the conductive properties, and the change of above factors can also lead to the variety of electrical conductivity, which indicates that the electrical conductivity can be controlled by the above factors.

Impurities. According to the Nordheim's role, the impurities usually play a negative role in the conductive properties of aluminum and its alloy. Common impurities include: silicon, iron, copper, titanium, vanadium, manganese, chromium, etc., among which titanium, influence vanadium, manganese and chromium influence the conductivity of aluminum conductors seriously [22]. Some alloying elements, for example boron, can purify the aluminum and its alloy and improve their electrical conductivity [23, 24]. In this process, the impurities precipitated from the alloy matrix, which decreases the scattering effects and thus increase the electrical conductivity of aluminum and its alloy. $AlB2$ and $AlB12$ were added into 99.6% Al, and the impurities in this alloy were precipitated through gravity sedimentation in the form of $VB2$, CrB and $TiB2$. As a result, the amount of Ti, Cr and V in the alloy matrix decrease by 60-70%. After the boron treatment, the electrical conductivity of aluminum increases from 60.6-60.8 to 61.5% IACS [25]. Boron treatment has become the most mature technology in the production of conductors made of aluminum and its alloy in the past ten years.

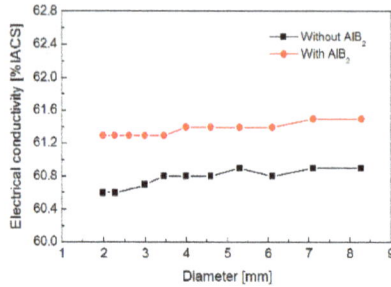

Fig. 5 Average electrical conductivity of 99.6% Al improved up to 1.0-1.5% IACS by the boron treatment [25].

Crystal defects. Point defects, line defects and plane defects are common defects in crystals such as dislocation, vacancies, grain boundaries, phase boundaries. Especially, a lot of crystal defects can be introduced into the alloy matrix during the cold drawing process, and to a large extent these crystal defects play a negative role in the electron transmission. As is well known, in the electron transmission process,crystal defects increase the electron relaxation time and they mainly influence the residual resistivity of metal materials. In recent years, some new researchesshowed that aging treatment can effectively improve the electrical conductivity of aluminum and its alloy. During the aging treatment, dislocations may disappear to some extent, and the lattice distortion may be eliminated.Thus the conductive properties of aluminum can be improved. Alwant et al. found that the peak of electrical conductivity appears during the aging treatment with the variation of aging temperature and time as shown in Fig. 6. At the very beginning, the electrical conductivity of A and B aluminum alloy is 40% IACS and 44% IACS, respectively. After aging treatment under the temperature of 175° C, the electrical conductivity of A and B aluminum alloy reaches 44% IACS and 48% IACS, respecctively. The average electrical conductivity of the tested alloys increased up to 4.0% IACS [24]. Kalabay et al. found that the strength and electrical conductivity of 6201 aluminum alloy conductor increased simultaneously after T-81 treatment, as shown in Fig. 7(a) [4]. The high pressure torsion (HPT) technique was applied to prepare some 6201 aluminum alloy by Valiev et al. with a subsequent aging treatment, and the results show that the electrical conductivity of the 6201 aluminum alloy was promoted as well as its strength [9], as shown in Fig. 7(b). In summary, the conductive property of aluminum and its alloy can be effectively improved by the controlling of microstructure, so the microstructure of conductive materials play an important role in improving its electrical conductivity.

314

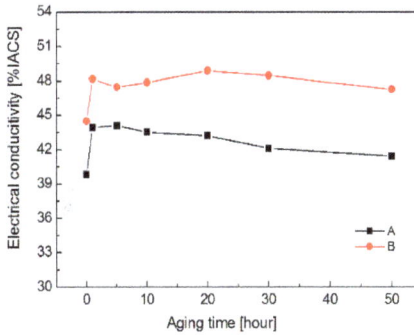

Fig. 6 Electrical conductivity of A and B aluminum alloy varied with aging time [24].

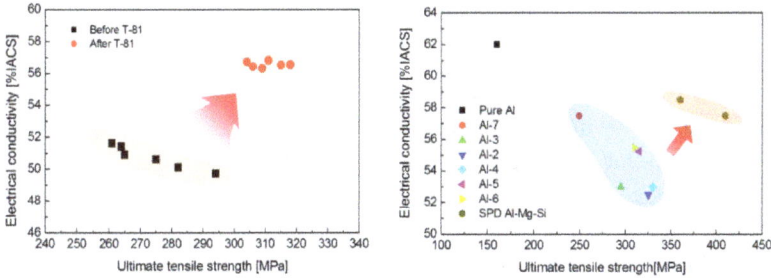

Fig. 7 Variation of the strength and electrical conductivity of 6201 alloy after (a) T-81 treatment and (b) high pressure torsion (HPT) processing [4,9].

5. Conclusion

In the past ten years, boron treatment was widely used to improve the electrical conductivity of aluminum and its alloy. The boron reacts with the impurities and finally precipitating from the alloy matrix by gravity sedimentation. In this process, the impurities in the alloy were driven out of the conductive material and therefore the electrical conductivity of aluminum and its alloys were improved. However, the addition of alloying elements may cause the crystal distortion, and thus few alloying elements were used to modify the aluminum and its alloy to improve the electrical conductivity, so it still need to make a further investigation on the dominant mechanism of electrical conductivity. It is emphasis that the addition of alloying elements strengthens the conductive materials and it is a pity that the electrical conductivity usually decreases in this process. New technology and new design may be very helpful for improving the strength and electrical conductivity of conductors made of aluminum and its alloy. In this aspect, the development of new technology and new design ideas

315

has a good prospect of application, including the optimization of design techniques and the evolution of materials. In this paper, taking the SPD technology for example, the HPT technology has a good application prospect in the field of conductive materials. However, how to transform these new technologies has become a bottle neck. At present, the dimension of sample prepared by HPT is still very small and cannot be used to prepare conductor. The development of high strength-high conductive aluminum conductors calls for new materials, new technologies and new processes.

It is hoped that more and more new materials, new technologies and new processes can be developed based on the understanding of the role of alloying elements and microstructure which may encourage the development of the conductive theory. Accelerating the development of high strength-high conductive materials may very beneficial to mankind.

References

1. B. Liu, Q. Zheng, P. Dang, W. Zeng. Development and Applications of Aluminum Alloy in Overhead Lines. Electric Wire & Cable, 2012,4: 10-16.
2. S. Karabay, F.K. Önder. An approach for analysis in refurbishment of existing conventional HV-ACSR transmission lines with AAAC. Electric Power Systems Research 2004, 72: 179-185.
3. E.N. Yang, X.M. Wu, C.H. Li, Q. Wang, Q.Q. Duan, Z.F. Zhang. Electromechanical properties, current situation and development of aluminum alloys for overhead conductor. Materials Review, 2014, 28:111-116.
4. S. Karabay. Modification of AA-6201 alloy for manufacturing of high conductivity and extra high conductivity wires with property of high tensile stress after artificial aging heat treatment for all-aluminium alloy conductors. Materials & Design, 2006, 27: 821-832.
5. S. Karabay. Influence of AlB$_2$ compound on elimination of incoherent precipitation in artificial aging of wires drawn from redraw rod extruded from billets cast of alloy AA-6101 by vertical direct chill casting. Materials & Design, 2008, 29: 1364-1375.
6. M.Y. Murashkin, I. Sabirov, X. Sauvage, R.Z. Valiev. Nanostructured Al and Cu alloys with superior strength and electrical conductivity. Journal of Materials Science, 2015, 51: 33-49.
7. Z. Zhang, D. Chen. Consideration of Orowan strengthening effect in particulate-reinforced metal matrix nanocomposites: A model for predicting their yield strength. ScriptaMaterialia, 2006, 54: 1321-1326.
8. K. Strobel, M.D.H. Lay, M.A. Easton, L. Sweet, S.M. Zhu, N.C. Parson, A.J. Hill. Effects of quench rate and natural ageing on the age hardening

behaviour of aluminium alloy AA6060. Materials Characterization, 2016,111: 43–52.

9. R.Z. Valiev, M.Y. Murashkin, I. Sabirov. A nanostructural design to produce high-strength Al alloys with enhanced electrical conductivity. Scripta Materialia, 2014, 76: 13-16.

10. M.Y. Murashkin, I. Sabirov, X. Sauvage, R.Z. Valiev. Nanostructured Al and Cu alloys with superior strength and electrical conductivity. Journal of Materials Science, 2015, 51: 33-49.

11. Z. Zhang, D. Chen. Consideration of Orowan strengthening effect in particulate-reinforced metal matrix nanocomposites: A model for predicting their yield strength. Scripta Materialia, 2006, 54: 1321-1326.

12. J.P. Hou, Q. Wang, H.J. Yang, X.M. Wu, C.H. Li, Li XW. Microstructure evolution and strengthening mechanisms of cold-drawn commercially pure aluminum wire. Materials Science and Engineering: A 2015, 639: 103-106.

13. J.E. Bailey. Electron microscope observations on the annealing processes occurring in cold-worked silver. Philosophical Magazine, 1960, 5: 833-842.

14. K. Ma, H. Wen, T. Hu, T.D. Topping, D. Isheim, D.N. Seidman. Mechanical behavior and strengthening mechanisms in ultrafine grain precipitation-strengthened aluminum alloy. Acta Materialia, 2014, 62: 141-155.

15. M. Kawasaki, Z. Horita, T.G. Langdon. Microstructural evolution in high purity aluminum processed by ECAP. Materials Science and Engineering: A, 2009, 524: 143-150.

16. I. Saxl, A. Kalousová, L. Ilucová, V. Sklenička. Grain and subgrain boundaries in ultrafine-grained materials. Materials Characterization, 2009, 60: 1163-1167.

17. Z.F. Zhang, Z.G. Wang. Grain boundary effects on cyclic deformation and fatigue damage. Progress in Materials Science, 2008, 53: 1025-1099.

18. Z.F. Zhang, Z.G. Wang. Dependence of intergranular fatigue cracking on the interactions of persistent slip bands with grain boundaries. Acta Materialia, 2003, 51: 347-364.

19. C.E. Carlton, P.J. Ferreira. What is behind the inverse Hall–Petch effect in nanocrystalline materials? Acta Materialia, 2007, 55: 3749-3756.

20. E.O. Hall. The deformation and aging of mild steel. Proc Phys Soc Sect B 1951, 64: 747.

21. N.J. Petch. The cleavage strength of polycrystals. J Iron Steel Inst 1953, 25: 174.

22. Y.Z. Zeng, S.J. Mu, P. Wu, K.P. Ong, J. Zhang. Relative effects of all chemical elements on the electrical conductivity of metal and alloys: An alternative to Norbury–Linde rule. Journal of Alloys and Compounds, 2009, 478: 345–354.

23. L. Bolzoni, M. Nowak, N. HariBabu. Grain refinement of Al-Si alloys by Nb-B inoculation. Part II: Application to commercial alloys. Materials and Design, 2015, 66 :376–383.

24. H.A. Alwan, N.L. A. Saffar, H.A.H. al-Jubouri. Effect of Adding Boron on the Electrical Conductivity of Alloy AL-Mg-Mn. International Journal of Engineering and Technology, 2013,3: 964-972.

25. S. Karabay, I. Uzman. Inoculation of transition elements by addition of AlB2 and AlB12 to decrease detrimental effect on the conductivity of 99.6% aluminium in CCL for manufacturing of conductor. Journal of Materials Processing Technology, 2005, 160: 174–182.

Design and Optimization of the Wireless Charging System for Running or Motionless Electrical Vehicles Based on Coupled Electric Field Theory

Yu-juan Fang[†], Shu-hong Wang, Quan Zhou, Yu-fei Chen, Qi-hang Huang
and Yi-kai Wang

State Key Laboratory of Electrical Insulation and Power Equipment,
School of Electrical Engineering, Xi'an Jiaotong University,
Xi'an, Shaanxi, China
[†]E-mail: fyj_thu_eea@outlook.com

This paper achieves a new method to charge the vehicle when it is running on the road. Generally, the power is transferred through a capacitor composed of a steel belt covering the tire and a metal pathway attached to the road. First, it demonstrates a new theory. The transmission of electrical energy is mainly due to the capacitor among the vehicle and we seek for the peak of the voltage by LC resonance. Therefore the electricity is transferred in the way of displacement current and electrical field coupling. Then the paper concentrates on the circuit design and simulation, utilizing COMSOL Metaphysics and ADS 2009. The appropriate capacitance and other parameters for the vehicle are attained when reaching the peak voltage, including the resonance frequency, the value of inductance and the transfer functions. In the system, the capacity between the pathways is 12pF, the capacity between the tire and pathway is 23pF, the capacity between the tires is 11pF. For simulation, when a 68μH inductance is added in the circuit, the system achieved a LC resonance at the frequency of 1.9MHz. Taking a LED light as the load for experiment, the charging efficiency can be roughly assessed. The result showed the LED light reached peak current at the frequency of 1.95MHz, which matches well with the simulation result. This proved the feasibility of the method mentioned above for wireless power transmission..

Keywords: Wireless Power Transmission; Coupled Electrical Field; Running Vehicle Charging.

1. Introduction

The electric vehicle, which is the clean and pollution free transportation, has been popular in the recent years. However, the disadvantages of original charging method limit the rapid development of electric vehicle industry. Six or seven charging hours severely diminish individuals' time, and especially in some less developed cities, even the Telsa super charging pile cannot satisfy the customers. In addition, electric vehicle batteries are usually heavy and expensive, and the charging efficiency is a bit low. These kind of disadvantages lower users' expectations. Currently, the charging methods for electric vehicles can be roughly divided into two categories [1]. One is the contact charging, including standard charging, fast charging and battery replacement; another is the contactless charging, which mainly utilize the radio transmission technology, inductive, resonator and microwave radio.

As the original charging methods have high requirements for the structure, location and other parameters of the coil, and the vehicles must be settled in a fixed position, the vehicles cannot move during charging and cause great energy losses [2-3]. Therefore, to promote the new wireless charging technology is of great significance. This paper provides a new wireless charging method and optimizes the charging efficiency for running vehicles innovatively.

This paper is organized as follows: Section II describes the structure of charging system for running vehicles and then establishes the simulation model considering different loads to get the optimal charging frequency. Section III details the simulation process to find the maximum of output voltage and carries out the results in experiments. In Section IV, application prospect and economic assess is analyzed. Section V finally draws conclusions on the obtained results and provides the scope for future work.

2. Design and simulation of the system

2.1. *Structure*

The schematic diagram of the wireless charging system is as follows: The characteristic of this system lies on the metal wheel hub of the electrical vehicle, two metal pathways underground and the circuit included in the vehicles. The two metal pathways connect with the alternating current power supply, and inductance is contacted with the metal pathways to achieve LC resonance with the capacitance between two metal pathways. With the insulated connection fixed on the cover of the axis, the bilateral metal wheel hubs will not be separated. Through the tires, the metal hubs and pathways construct a capacitance. On the metal hubs, there are electrical brushes which can be used to conduct electricity. Then the brushes connect with the input of the charging system. Through the capacitance, the electricity will transfer to charging system and therefore provide power for the running vehicles.

Consider the symmetry of the equivalent circuit, select one side and the circuit diagram is shown below. The tire can be seen as a pure resistor and a capacitor in shunt. Since resistance (Me ohm) is generally much larger than the capacitance (kilo Ohm), the tire can be simplified to a pure capacitance. In order to achieve the reactance resonance, the experimental frequency should satisfy (1),

$$\omega^2 LC = 1 \tag{1}$$

Converts the track of the vehicle to its equivalent two port circuit in figure 2:

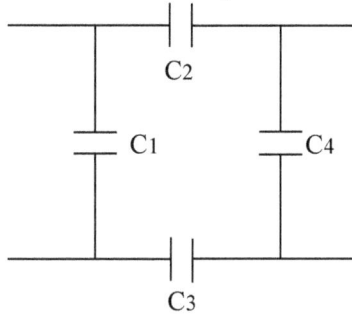

Fig. 1 The equivalent circuit of vehicle model

Where C1 is the capacitance between the metal plate capacitors; C2 and C3 are the capacitances between the left wheel and the left metal plate; C4 is the capacitance between the left and right wheels. Figure 3 details the equivalent circuit of the wireless charging system. The inductances in this circuit are innate in the vehicle model and R represents the input of the charging system.

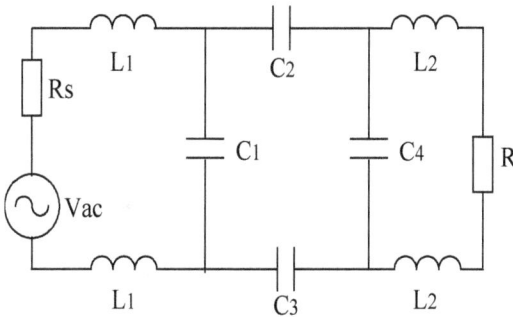

Fig. 2 The charging circuit within the vehicles

2.2. Circuit topology

In order to simplify the analysis, the power supply is considered as an ideal voltage source. Considering the power output voltage is constant, the input impedance of the circuit needs to be minimum to achieve the maximum input power. As the series resonant impedance is zero, the circuit can reach the series resonance by adding the appropriate element in the input and output terminals.

The input and output terminals are coupled with the inductor so that the circuit can reach the resonance state, and the input impedance is minimum at the resonance frequency. The circuit diagram is as follows:

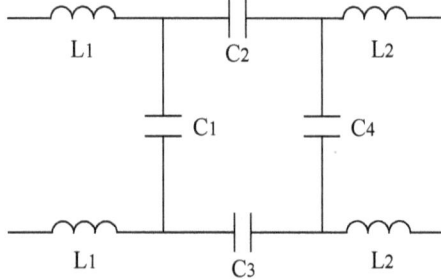

Fig. 3   The original circuit

Where L1 and L2 are the external inductors, the value is limited to hundreds micro Henry level so that the impedance can reach thousands Ohm level with this frequency. As the load impedance is relatively small, the load can be approximated to a short circuit.

In order to achieve the reactance resonance, appropriate L2 is chosen to achieve the shunt resonance with C4, and the experimental frequency should satisfy (2),

$$\omega^2(L_1+L_2)C_4 =1$$

$$(2)$$

At the parallel resonance, the impedance of the impedance is infinite, and it can be considered as the open circuit. The experimental circuit is simplified as follows:

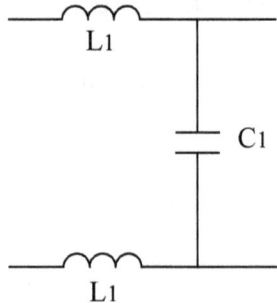

Fig. 4   The simplified circuit

Just choose L1 to satisfy (3),

$$\omega^2(L_1+L_1)C_1 =1$$

$$(3)$$

The series resonance can be fulfilled, and the input impedance is minimal.

3. Simulation and Results

We build a vehicle mini-type model consisting of tires and pathway, and simulate with COMSOL Multiphysics and ADS 2009 to get the parameters that match well with the real model. The result shows that the equivalent resistance and inductance is $3.22 \times 10^{-10} + 2.25 \times 10^{-8} i$.

The tire can be considered as a resistance and a capacitance in shunt, whose equivalent value is $R = 3.11 \times 10^9, C = 4pF$. Under the experimental frequency, which is MHz level, the resistance is far larger than the capacitance, so that it can be approximated to a pure capacitance. Similarly, a comprehensive four tires model can be constructed and simulated. Use a digital meter measuring the capacitance between the plate, the capacitance of the rail and the wheel, and the other parameters, and the tire can be equivalent to the two port model.

Table 1. The circuit parameters

Capacitance between	Value
Pathways	9pF
Tire and pathway	19pF
Tires	10pF
Input inductance	263μH, 264μH

With ADS2009, the relationship between simulated input resistance and frequency shows the load voltage reaches the maximum at the signal frequency, 1.9MHz.

As for the real experiment in laboratory, we switched on the power supply, and gave a 20-volt voltage to the system. The LEDs reach the brightest with the frequency in the vicinity of 1.957MHz (a LED is equivalent to a diode in this paper). According to the experimental results, the device can transfer the displacement current to the load on the condition of a certain gap between tires. The device can transfer the displacement current to the load LEDs. For the experiment scene, when the vehicle separates with the pathways, the LED lights turn off and when the vehicle connects with the pathways, the lights turn on. It means that power transferred successfully from the pathways to the vehicles.

4. Application Prospect and Economic Assess

4.1. Application prospect

The invention has a wide range of applications, and the specific examples are:

- **High ways:** Where there is a heavy traffic, many vehicles can be charged at the same time.
- **Factories:** With wireless charging pathways underground along these specific moving tracks, the problem of robot charging can be efficiently solved.
- **Electrical train rails:** Applying this wireless charging technology, power can be directly transmitted to the train through the rail instead of overhead lines.
- **Bus station:** Applying this wireless charging pathways under each bus station, the bus will be charged with enough electricity to support it to the next station.

4.2. Economic assess

Compared with several existed charging methods, the coupled electric field wireless charging system performs better. Alternative current slow charging is generally under low efficiency and the time of charging ranges from 5h to 8h. As for direct current fast charging, the power efficiency is high, and the charging time ranges from 20min to 2h. However, the lifespan of batteries is greatly shortened. For magnetic resonance charging, if the axis of two coils deviates, the transmission efficiency will decline so dramatically that it's difficult to charge a running vehicle. This paper provides a new charging method based on electric field coupling, which charges for running vehicles. Waiting time is shortened and capacity of batteries is decreased [4]. Moreover, the equipment on the car is light, which not only realizes low energy consumption of cars, but also decrease the construction cost.

5. Conclusion

Based on the coupled electric field theory, this paper proposes a new method to charge the vehicle when it is running. With the capacitances between tires, pathways, displacement current can be achieved under high frequency voltage. According to our simulation and experiment, power can be efficiently transferred to the vehicle. For the minimal model, the transferred power can illuminate 5 LED lights, which proves the feasibility of our charging system.

Acknowledgment

This work was funded by Science and Technology Innovation Foundation of China for The University or College Students.

References

1. Ren Xiofeng, The research on the wireless charging system of electrical vehicles, Master's thesis, University of industry in Harbin, (Harbin, China, 2014).
2. Qiang Xie, Si-zhong Chen. *Key Techniques and Prospects of Developing Electric Vehicles. Bus Technology and Research*, 4th ed, vol. 25 (2003), pp. 1-2.
3. Shi Luofei. *Electrical vehicles: Step into the new era. Research on Automobile Industry*, 3rd ed., vol. 2 (2010), pp. 10-13.
4. Li Bin, Liu Chang, Chen Qichu, et al. *The wireless charging technology of electrical vehicles. Electrical Engineering in Jiangsu*, 1st ed., vol. 32 (2013), pp. 81-84.

Comprehensive Effectiveness Evaluation Based on Entropy Weight Method for Energy Utilization Schemes of Smart Parks

Sheng-jun Zhou

Global Energy Interconnection Research Institute, Beijing, China

Meng Tan and Xu-dong Wang

State Grid Tianjin Electric Power Company, Tianjin, China

Yu-bo Yang[†]

North China Electric Power University, Beijing, China
†E-mail: yangyb3333@163.com

To achieve the comprehensive evaluation on energy utilization scheme of smart parks, an evaluation index system of energy utilization scheme of the parks is established and a comprehensive effectiveness evaluation method based on entropy weight method for energy utilization of smart parks is proposed in the paper. Firstly, the importance of the energy utilization effectiveness index, i.e. the index weight, is quantized by using entropy. Secondly, the relative approach degree from the energy utilization effectiveness index to corresponding ideal optimal index sample is calculated by using the technique for order preference by similarity to an ideal solution (TOPSIS) method, and the comprehensive effectiveness evaluation of energy utilization schemes is achieved by comparing with the relative approach degrees. The indexes, such as N-1 rate of the distribution network, reliability rate of power supply, average failure recovery time, voltage limitation index, new energy power generation proportion, average emission reduction and average energy consumption index, are mainly selected in the index system. Finally, the effectiveness and feasibility of the presented comprehensive evaluation scheme is verified by using comprehensive effectiveness evaluation on three parks.

Keywords: Smart Park; Energy Utilization; Comprehensive Evaluation; Entropy Weight Method.

1. Introduction

The smart park construction is an important part of constructing smart grid in China [1]. It is also an important link of energy structure optimization, energy-saving and emission-reduction. As the high-tech industrial zone, the important manufacturing base and the residents living service area, smart parks gather a large number of energy consumers, such as residents, public facilities, commercial buildings and industrial plants [2]. To improve energy usage efficiency, multiple energy supply and storage devices, as well as intelligent information service network are highly integrated in smart parks. However, considering the huge quantity and variety of energy demands, the energy waste problem of parks has not been settled because the supply reliability is lower and the energy cannot be fully cascaded utilized.

In view of this, park energy utilization schemes are studied by scholars and some valuable results are obtained [3]. However, the research progress on objective comprehensive evaluation of smart park energy utilization schemes is slow. The difference degree of various indexes can be calculated by using the entropy weight method, and then, the weight of each index can be determined [4]. Moreover, the relative approach degree from each index to corresponding ideal reference index can be calculated by the TOPSIS method, and it is used for comprehensive evaluation on different objects. Therefore, the entropy weight method is uesd to quantify the importance of each evaluated index in the smart park energy utilization scheme, and the TOPISIS method is used to calculate the comprehensive effectiveness of different schemes. The results can provide references for objective evaluation and comparison on comprehensive effectiveness of smart park energy utilization schemes.

2. Construction of evaluation index system of the smart park energy utilization scheme

(1) Distribution network N-1 rate of primary grid

After an N-1 fault of the main transformer or the feeder occurs in the distribution network, the load transfer capability is an important manifestation of power supply safety. The index can be calculated as follows:

$$R_{N-1} = n_{N-1} / n \times 100\% \tag{1}$$

Where, R_{N-1} represents the distribution network primary grid N-1 rate, n_{N-1} represents the number of the transformer and feeder that meet N-1, n represents the total number of transformers and feeders. The bigger the index, the higher the reliability of power supply is. So, it is a positive index.

(2) Reliability rate of power supply

Reliability rate of power supply is an index to measure continuous power supply. This index can be calculated by using Formula (2),

$$R_{ps} = T_{ps} / T_{all} \times 100\% \tag{2}$$

Where, R_{ps} is the reliability rate of power supply, T_{ps} is the hours of actual power supply, T_{all} is the total hours for the statistical period. It is a positive index. And the bigger is the value, the higher is the reliability of power supply.

(3) Average fault recovery time

Assume s times faults occurs within a period of time, total power outage time is T_{ft} minutes, the average fault recovery time T_{rec} can be calculated by using the following formula.

$$T_{re} = T_{ft} / s \qquad (3)$$

(4) Voltage limitation index

The bus voltage limitation index is the characterization of a static safety index for the power supply. The voltage limitation index R_{vq} can be calculated by using formula (4). The smaller is the index, the better is the voltage quality.

$$R_{vq} = \frac{1}{m} \sum_{i=1}^{k} \left| \frac{U_i - (U_i^H + U_i^L)/2}{(U_i^H - U_i^L)/2} \right| \qquad (4)$$

Where, m is the bus line, k is the times of the voltage beyond the limit, U_i is the ith voltage amplitude, and U_i^H and U_i^L are the upper limit and the lower limits.

(5) Renewable energy generation proportion

Assume renewable energy power generation is E_{new} in a period of time, the total power consumption is E_{all}, and clean energy generation is accounted for by R_{new}. Thus, the renewable energy generation proportion can be calculated through equation (5). It is positive index.

$$R_{new} = E_{new} / E_{all} \times 100\% \qquad (5)$$

(6) Average emission reduction

The average emission reduction index is represented by C_{ER}, and it can be calculated by using the ratio of average emission reduction to the calculated equivalent emission of total consumption. It is a positive index.

$$C_{ER} = E_{er} / E_{all} \qquad (6)$$

(7) Average energy consumption index

It can be calculated by subtracting total energy consumption E_{all} from total energy generation E_{gen}, and average energy consumption E_{con} can be gotten by the ratio of E_{all} to power generation, which is a negative index.

$$E_{con} = (E_{gen} - E_{all}) / E_{gen} \times 100\% \qquad (7)$$

3. Comprehensive evaluation of energy utilization scheme based on the entropy weight method and the TOPSIS method

3.1. *Entropy weight method*

(1) Establish the evaluation matrix

It is assumed that the index number for each scheme is 7, and Xj=[x1j, x2j, x3j, x4j, x5j, x6j, x7j]T is the evaluation matrix of the jth scheme. Where x1j represents the distribution network N-1 rate, x2j represents the reliability rate of power supply, x3j represents the average recovery time index, x4j represents the voltage limitation index, x5j represents the proportion of renewable energy, x6j represents the Average emission reduction, and x7j represents the average energy consumption index.

If the scheme number is *n*, the evaluation matrix can be expressed as $X=[X_1, X_2, \cdots, X_n]$. It is assumed *w* represents the weight of each evaluation index, the weight vector can be expressed as $w = [w_1, w_2, ..., w_i]^T$. In the vector *w*, *i* represents the evaluation index order number. Then, the evaluation matrix, i.e. $X=(x_{ij})_{m \times n}$ can be established. Where, *m* represents the index number and *n* represents the evaluated scheme number.

(2) Standardization of the original evaluation matrix

If there is a negative index in evaluation indexes, the smaller the value, the better the index. Firstly, the indexes are treated by the same direction method, and they are turned to positive indexes. The original matrix *R* can be obtained.

$$R = (r_{ij})_{m \times n} \tag{8}$$

Where, r_{ij} represents the standard value of the *j*th evaluated object on the *i*th evaluated index, $r_{ij} \in [0,1]$. The negative indexes can be turned into the positive indexes by using Formula (9) and Formula (10).

For the positive and the negative indexes, it is expressed as follows respectively. Where, r_{ij} is an element of the standard evaluation matrix.

$$r_{ij} = (x_{ij} - \min\{x_{ij}\}) / (\max\{x_{ij}\} - \min\{x_{ij}\}) \tag{9}$$

$$r_{ij} = (\max\{x_{ij}\} - x_{ij}) / (\max\{x_{ij}\} - \min\{x_{ij}\}) \tag{10}$$

(3) Definition of entropy

In a case of m indexes and n evaluated scheme, the entropy H_i of the ith index can be defined as follows:

$$H_i = -k \sum_{j=1}^{n} f_{ij} \ln f_{ij}, i = 1,2 \ldots m \tag{11}$$

Where, $f_{ij} = r_{ij} / \sum_{j=1}^{n} r_{ij}$, $k=1/\ln n$, when $f_{ij}=0$, $f_{ij} \ln f_{ij}=0$.

(4) Definition of the entropy weight

After defining the ith index entropy, its entropy weight can be defined as

$$w_i = (1 - H_i) / (m - \sum_{i=1}^{m} H_i) \tag{12}$$

Where, $0 \leq \omega_i \leq 1$, and the sum of all the w is 1.

3.2. The TOPSIS method

(1) Distance between sample point and ideal reference point

Based on the entropy weight method, the weight of the indexes can be calculated according to Formula (13) to determine the weighted data matrix R'

$$R' = R \bullet w_i, i = 1,2 \ldots m \tag{13}$$

Since the indexes are positive, we can use the indexes of the maximum value from the sample to form an ideal reference one R^+, and use the smallest value from the indexes constitute a negative ideal sample R^-, respectively. The distance of the sample point to the best and the worst reference ones can be calculated by using Formula (14) and (15).

$$D_j^+ = \sqrt{\sum_{i=1}^{m} (r_{ij} - r_i^+)^2}, j = 1,2 \ldots n \tag{14}$$

$$D_j^- = \sqrt{\sum_{i=1}^{m} (r_{ij} - r_i^-)^2}, j = 1,2 \ldots n \tag{15}$$

(2) Calculation of relative proximity

The relative approach degree is calculated by using Formula (16). According to the value of the relative approach degree, the evaluation objects can be evaluated and compared. The greater the relative distance between the evaluated object and the ideal sample, the better the evaluation results of the corresponding evaluation object.

$$C_j = D_j^- / (D_j^- + D_j^+) \tag{16}$$

4. Application examples

Three energy utilization schemes are taken as an example, and the comprehensive effectiveness of each scheme is analyzed. As shown in Table 1, the calculation results are listed. Firstly, the entropy weight method is used to generate the evaluation matrix, and it will be standardized. Secondly, the average fault recovery time, the voltage limitation index and the average energy consumption index are translated into positive indexes.

Table 1. Indicators on energy utilization schemes of intelligent parks

Schemes	R_{N-1}	R_{ps}	T_{rec}	R_{vq}	R_{new}	C_{ER}	E_{con}
Scheme1	0.9800	0.9994	6.7000	0.6000	0.9000	0.0030	0.0130
Scheme2	1.0000	0.9999	4.5000	0.5000	0.9500	0.0050	0.0112
Scheme3	0.9900	0.9990	6.8000	0.6000	0.9100	0.0053	0.0156

According to the weight calculation method, the entropy and weight of each index can be calculated. The results are shown in Table 2.

Table 2. Calculation results of parameters

Index name	R_{N-1}	R_{ps}	T_{rec}	R_{vq}	R_{new}	C_{ER}	E_{con}
H_i	0.7925	0.7925	0.7806	0.7899	0.7923	0.7721	0.7858
w_i	0.1389	0.1389	0.1468	0.1406	0.1390	0.1525	0.1433

The distance between the sample points to ideal ones are calculated in D^+ and D^-: Finally, according to Formula (16), the relative closeness from the sample points to the optimal ones can be calculated. The comprehensive evaluation results are shown in Table 3.

$$D^+ = (1.7703 \quad 1.7356 \quad 1.7424) \quad D^- = (1.5391 \quad 1.5882 \quad 1.5297)$$

From the results, scheme 2 is the best, and followed by scheme 1 and scheme 3. The difference in the value of relative approach degree index between scheme 3 and 1 is very small.

Table 3. The comprehensive evaluation results on energy utilization schemes of smart parks

scheme	C_j	sorting
scheme 1	0.4651	3
scheme 2	0.4778	1
scheme 3	0.4675	2

From the indexes in Table 1, the average emission reduction index of scheme 2 is equal to scheme 3, and other indexes are better than scheme 1 and 3. So the comprehensive effectiveness of scheme 2 is the best. In addition, the results are in good agreement with the analytic results of schemes 1 and 3.

Scheme 3 not only has the same limit of voltage with scheme 1, the N-1 rate of the distribution network is better, the renewable energy power generation ratio and the average emission reduction index are all better than scheme 1. However, the remaining three indexes are inferior to scheme 1. From results, the two schemes' effectiveness should be basically the same. Besides, the results are similar to actual situations. Furthermore, some other smart park energy schemes are also evaluated, and the results show that the presented comprehensive effectiveness evaluation method is feasible and effective.

5. Conclusion

In this paper, the entropy weight method and the technique for order preference by similarity to an ideal solution (TOPSIS) method are proposed to evaluate the comprehensive effectiveness of smart park energy utilization schemes. It shows that the presented method based on the entropy weight method and the TOPSIS method is effective and feasible to evaluate the comprehensive effectiveness of smart park energy utilization schemes.

Acknowledgment

This work is partially funded by grant SGTJ0000KXJS1400087 of the State Grid Corporation of China.

References

1. Fangchen, Ling Ping, Bao Hailong, et al. Smart grid architecture with high proportion of renewable energy utilization. East China Electric Power, 2013, vol.41, no.12, pp: 2468-2471.
2. Wang Dong, Yang Yongbiao, Huang Li, et al. Advanced pinch algorithm geared to smart grid area used for multi-energy recycling. Automation of Electric Power System, 2015, vol.39, no.21, pp: 122-137.
3. Cao Zhigang. Design and implementation of user interaction in smart power distribution and utilization park. Automation of Electric Power System, 2013, vol.37, no.9, pp: 79-83.
4. Ouyang Sen, Shi Yili. A new improved entropy method and its application in power quality evaluation. Automation of Electric Power System, 2013, vol.37, no.21, pp: 157-159.

Phase-to-Phase Fault Detection Method for De-Energized Distribution Lines Based on Active Disturbance Technique

Ke Zhu and De-qiang Li[†]

School of Electrical Engineering, Shandong University
Jinan, Shandong, China
[†]E-mail: lee237074@163.com

A phase-to-phase fault detection method is proposed in this paper to prevent breaker to close to a de-energized distribution line with fault. An inverter source connected to the low-side of service transformer is controlled by Thyristor Bridge to momentarily inject a transient high voltage to the de-energized feeders. Fault existence can be judged by detecting and analyzing the voltage waveform near the downstream side of the breaker. The method does not need communication and can be used for both adaptive reclosure and safely closing to a feeder de-energized for an extended period.

Keywords: Distribution Lines; Permanent Fault; Transient Fault; Thyristor; Wavelet Analysis.

1. Introduction

Neutral ineffectively grounded system is widely used in China and some eastern European countries. When a single-phase to ground fault happens, distribution system can keep running for 1~2 hours without trip. For phase to phase fault, the breaker will act to clear fault and then reclose.

With the wide application of cable in distribution networks, the success rate of reclosing is declining. Reclosing to a faulted line will not only have negative impacts on breaker and upstream transformers but also deteriorate power quality. So it is meaningful to detect if the phase to phase fault still exists in a de-energized distribution line before reclosing.

There are two kinds of methods to determine whether the de-energized feeder still experiences short circuits: passive and active method. The former is based on the characteristics of some electrical quantities during discharging of residual electromagnetic energy after tripping [1-2]. This method is applied to adaptive reclosure of transmission lines rather than distribution feeders, which will lose electromagnetic energy immediately after tripping. The active method exerts an external excitation source on de-energized lines. A special mechanical breaker has been put forwarded by Wilson and Hedman [3] to connect upstream source to the de-energized lines momentarily. But it needs to cost lots of money to change breaker. Long and et al. [4] proposed a two transformers connected by a thyristor are in parallel with breaker to impose upstream source to de-energized lines, while it is difficult to be put into effect.

An active detection method for phase-to-phase fault in de-energized distribution lines, which exerts the excitation source rom the downstream low voltage side, is proposed in this paper. It can be used in not only adaptive reclosure but also energizing feeder de-energized for an extended period.

2. Fault Detection Solution

2.1 *Principle*

An inverter power supply is connected to the low voltage side of service transformer through a thyristor bridge and normally on floating charge. After breaker trips out and deionization near fault point disappears, which normally lasts for about 0.5s, inverter power is controlled by Thyristor Bridge to momentarily impose a voltage boosted by service transformer between the faulted phases. Whether the fault exists is judged by detecting and analyzing the transient voltage waveform near the downstream side of breaker and then used to determine closing or not. The single-phase equivalent diagram of the proposed scheme is shown in Fig. 1, in which Thyristor Bridge is represented as a thyristor.

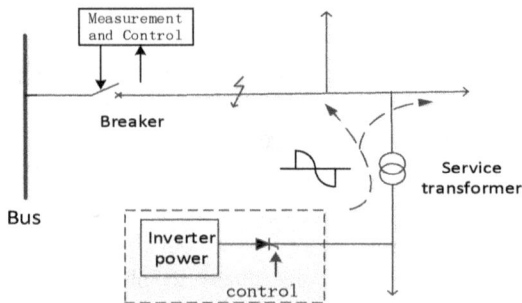

Fig. 1 Single-phase equivalent diagram of the proposed scheme

The amplitude of transient voltage exerted between faulted phases of transformers low voltage side can be controlled by firing angle of thyristor. Controlled voltage and current waveforms under different firing angles are shown in Fig. 2.

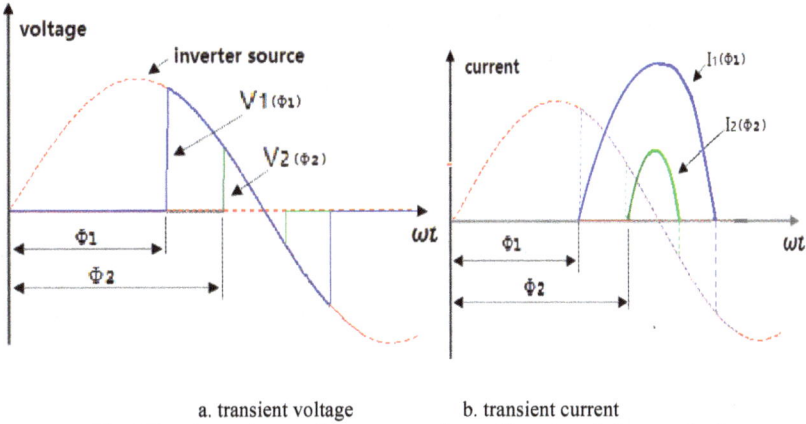

a. transient voltage   b. transient current

Fig. 2 Transient voltage and current waveform with resistor-inductance load

2.2 Theory analysis

The simplified equivalent circuit of Fig. 1 is shown as Fig. 3. $Z_1 = R_1 + j\omega L_1$ is equivalent fundamental load of low voltage side of service transformer. $R_2 + j\omega L_2$ is equivalent load of high voltage side. L_T is leakage reactance of service transformer, R_F is fault resistance. U_s is the voltage of inverter power supply.

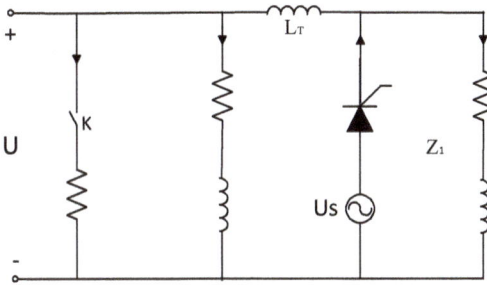

Fig. 3 Simplified equivalent circuit

K switches off (no fault)

Thyristor is triggered to turn on before its voltage crosses zero and turns off at t=0 when its current passes through zero. As the sum of i_1 and i_2 are forced to be zero after thyristor turns off, there is an abrupt change in the waveform of U at the moment when thyristor turns off and can be expressed as:

$$\Delta U = |U_{0+} - U_{0-}| \approx \frac{L_1 U_s}{|Z_1|}\left[\omega\cos(\delta + \phi_u - \varphi_1) + \frac{1}{\tau}\sin(\delta + \phi_u - \varphi_1)\right] \quad (1)$$

335

Where φ_1 impedance is angle of Z_1, ϕ_u is firing angle of thyristor, δ is its conducting angle, τ is the time constant. Equation (1) can be simplified when firing angle ϕ_u is controlled between 90° and 150°:

$$\Delta U \approx \frac{L_1 U_s \omega}{|Z_1|} = U_s \sin \varphi_1 \tag{2}$$

Equation (2) reveals that when thyristor turns off there will be an abrupt change in waveform of U.This is because the variation of i_1 and i_2 with time, i.e. di_1/dt and di_2/dt, change suddenly during the turn-off, which causes the voltage near the downstream side of breaker to change abruptly.

The relation between $\Delta U(p.u)$ and $\cos\varphi_1$ under different ϕ_u is shown in Fig. 4. From the figure we can see that ΔU increases with the decrease of $\cos\varphi_1$ and ϕ_u.When $\cos\varphi_1 = 0.99$ ΔU is still large enough (about 0.1) to be detected if ϕ_u is controlled to 90°. Actually capacitors used for enhancing power factor normally withdraw immediately after de-energization, which will cause power factor to drop.

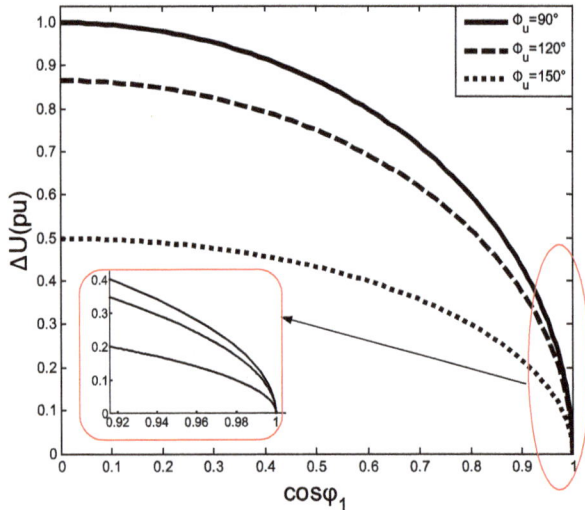

Fig. 4 Relationship between ΔU and $\cos\varphi_1$ under different firing angle Φ_u

K switches on(fault)

As the existence of bypass R_F, di_1/dt and di_2/dt don't need to change abruptly during the turn-off of thyristor. So the voltageU during the turn-off, which mainly consists of the voltage on inductance, will not change abruptly, i.e.

$$\Delta U = |U_{0+} - U_{0-}| \approx 0 \tag{3}$$

It can be seen from (2) and (3), whether fault exists can be determined by detecting the existence of abrupt change in voltage waveform near downstream side of breaker during the moment thyristor turns off.

2.3 Voltage abrupt detection

Db5 wavelet is selected to detect the existence of abrupt change in voltage waveform. According to the analysis in 2.2, an extreme condition such as $\cos\varphi_1 = 0.999$ and $\phi_u = 150^o$ is used to calculate the threshold as [5]:

$$A = \sigma\sqrt{2\ln(n)} \qquad (4)$$

Where n is the number of sampling points; σ is the standard deviation; A is the estimated threshold.

3. Computer Simulation

The simulation circuit is based on Fig. 1 and with total 0.4MVA load remained after feeder de-energization, which includes 0.2MVA parallel RLC load and 0.2MVA motor load. $\cos\varphi_1 = 0.95$, $\phi_u = 120^o$, sampling rate is 256 points/cycle.

The typical transient voltages imposed between faulted phases of low voltage side of service transformer, the upper, and phase to phase voltage near downstream side of breaker, the lower, are shown in Fig. 5.

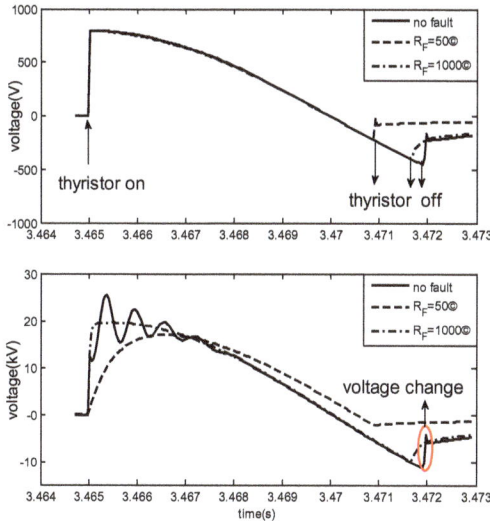

Fig. 5 Voltage waveform on load side of breaker in different fault types

It can be seen that when the thyristor turns off there is a distinct abrupt change in voltage waveform near downstream side of breaker if fault doesn't exist between faulted phases of feeder.

Voltage waveforms near downstream side of breaker during thyristor turn-off and the results of wavelet transform on them are shown in Fig. 6. The abrupt change in the voltage waveform when there is no fault can be effectively identified by the proposed wavelet transform.

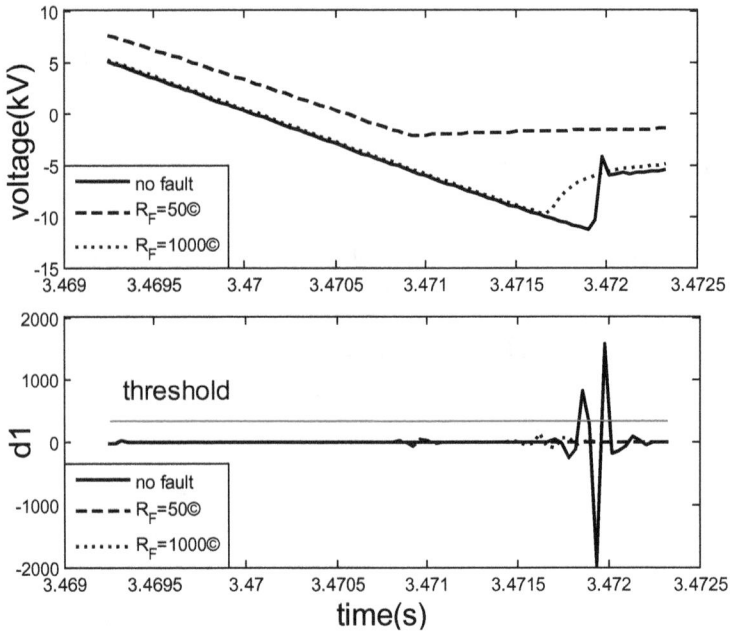

Fig. 6 Results of wavelet transform

4. Conclusion

A phase-to-phase fault detection technique for de-energized distribution lines, which can avoid the blindness of closing, is proposed in this paper. With high feasibility, the method can be used in not only adaptive reclosure but also tentative feeding de-energized distribution lines, and provides a novel idea for fault detection of de-energized feeders.

Acknowledgment

Project Supported by Key Sciences and Technologies of Shandong Province (2015GGX104010); NSFC of Shandong Province (ZR2013EEM026). The author is with the Key Laboratory of Power System Intelligent Dispatch and Control of Ministry of Education (Shandong University), Jinan, China.

References

1. Liu Hongshun, Li Qingmin, Zou Liang, Lou Jie. "Secondary Arc Characteristics and Single-phase Autoreclosure Scheme of EHV Transmission Line with Fault Current Limiter" Proceedings of the CSEE. vol.28, no.31, pp.62-67. 2008.
2. Johns, A.T. and R.K. Aggarwal "Digital simulation of fault autoreclosure sequences. With particular reference to the performance evaluation of protection for EHV transmission lines. Generation, Transmission and Distribution" IEE Proceedings C, vol.4, no.128, pp. 183-195. 1981.
3. D. D. Wilson, Hedman, "Fault isolator for electrical utility distribution systems", US Patent 4,370,609, Jan.25, 1983.
4. Xun Long, Wilsun Xu, Yunwei Li. A New Technique to Detect Faults in De-Energized Distribution Feeders–Part I: Scheme and Asymmetrical Fault Detection [J]. IEEE Transactions on Power Delivery, 2011, 26(3), pp.1893-1901.
5. Surya Santoso, Edward J. Powers, W. Mack Grady, Peter Hofmann. Power Quality Assessment via Wavelet Transform Analysis [J]. IEEE Transactions on Power Delivery，Vol. 11, No 2, pp.924-930. 1996.

An Overview on Piezoelectric Power Generation System for Electricity Generation

Xiao-ming Sun[†], Wei-feng Peng, Li Zhang and Ming-qian Zhang

Department of Electrical Engineering, Chongqing Water Resources and Electric Engineering College,
Chongqing, Yongchuan 402160, China
[†]E-mail: xmsun.whu@163.com

Coal, petroleum and natural gas will remaim the basis power house in drving economic of the world frowarded for the foreseeable future whether we like it of not. However, the impac of green house effect, the limit supply of fossil fuesl, it is wise to work towards energy saving, emission reduction, energy recovery, and exploration of new renewable energy sources. Currently, the electricity generation technology using piezoelectric material to recover the compressional or vibrational energy begins to attract attention, in particular, in areas of designing small self-powered devices. This paper presents an overview of the feasibility of piezoelectric power generation for electric power system, in which the fundamentals of piezoelectric power generation and the feasible structure of the system are discussed.

Keywords: Piezoelectric Power Generation; Renewable Energy; Electricity Generation.

1. Introduction

Coal, petroleum and natural gas will still be the basis of economic development for a long time. However, with a rapider consumption speed, these fossil fuels will be exhausted in the near future. In addition, the usage of these fossil fuels can also cause environmental pollution and greenhouse effect [1]. To deal with energy security and environmental crisis, it is wise to work towards three directions: energy saving and emission reduction, energy recovery, exploration of new renewable energy.

Although wind power generation and solar power generation are the eminent achievements of the above work, their development is restricted for the following reasons. First, for China, except western areas, the resources of wind power and solar power in most of the mainland are not abundant, the exploitation and utilization of which are not economic. Second, the electrical distance from the western wind farms and solar farms to the central and eastern load centers is very large, therefore, with the limitations of transient stability or dynamic stability, the power of the AC transmission lines cannot reach the natural transmission capacity, which may produce "abandoned wind" and "abandoned sunlight"; if the energy is transmitted by DC transmission lines instead, considering the high cost of DC transmission lines, the method can not be used widely. Third, when the penetration rate increases, in order to balance the power fluctuation caused by the intermittency of wind and sunlight, the peak load regulation capacity distributed to thermal power units will increase as well, and this increases the extra cost of coal and thus reduces the low carbon benefit of wind power and solar power [2]. Fourth, large-scale wind farms and solar farms tend to occupy a very large area and even urban-type equipment need also enough spaces to install; moreover, wind power generator may induce noise pollution [3] and solar power generation equipment may induce light pollution; therefore, it is not very appropriate to install wind power generator and solar power generation equipment in cities with high concentration of population and construction and the suburb areas.

Recently, the electricity generation technology using piezoelectric material to recover the compressional or vibrational energy begins to draw attention. The principle is based on the positive piezoelectric effect of piezoelectric material: when piezoelectric material deforms under external force (pressure or stress), inside polarization phenomenon occurs, and charges of different polarity accumulate on two opposite surfaces; when external force disappears, charges disappear accordingly. If intermittent external force is continuously exerted on piezoelectric material, the charges appearing on the surfaces are simultaneously collected by charge collector and stored in energy storage equipment, then the transition from mechanical energy to electric energy is realized. During this process, there are no emission, no electromagnetic conversion, no heat and no big mechanical vibration; therefore, piezoelectric power generation is a new type of green power generation without pollution, electromagnetic interference (EMI), thermal radiation and noise.

In addition to the advantages above, piezoelectric power generation equipment has many other advantages – simple structure, light weight, low cost, large energy density, long service life and easy integration. All these advantages overcome the shortcomings of wind power generation and solar power generation. Piezoelectric power generation is more suitable for using in cities, and because it is almost unaffected by climate and weather, it can be applied throughout a whole nation without any regional limitations. Piezoelectric power generation equipment can be integrated with buildings or facilities, not affecting landscape and not occupying extra area and space. For example, if integrated with floor tile or pavement [4], piezoelectric material can generate electricity when pedestrians or vehicles pass by; if integrated with high-rise building, billboard or traffic light, piezoelectric material can recover the energy of wind or noise [5] to

341

generate electricity. Provided that these distributed piezoelectric power generation equip-ment can be integrated into piezoelectric power generation system and then connected to electric power grid, in cities the closely packed buildings, crowded streets and countless facilities can all be turned into huge, intangible electric power plants. Not only can these electric power plants recover the biomass energy of men, but also recover the kinetic energies of vehicles and wind. As a result, it not only generates electricity, but it also damps vibration, reduces nose and thus improve city environment indirectly. Piezoelectric power generation system is actually distributed throughout the city, going deep into the load center and shortening the electrical distance, therefore, there almost exist no transient stability and dynamic stability problems, avoiding the problem of competing with important power plants for transmission channels.

In short, piezoelectric power generation has a wide range of application prospects in electricity generation area, and there are considerable social and economic benefits in the industrial chain from electrical equipment manufacture to electricity generation. The purpose of this paper is to present an overview of the feasibility of piezoelectric power generation system for electricity generation, in which the fundamentals of piezoelectric power generation and the feasible system structure are discussed. This paper serves as an important basis of the subsequent researches.

2. Choice of Piezoelectric Material

Piezoelectric power generation system is required to have high power generation capacity, high reliability and high stability. Only high power generation capacity can bring economical efficiency, only high reliability can bring low maintenance cost, and only high stability can bring easy control and wide application. All these requirements firstly depend on the piezoelectric material adopted. At present, piezoelectric materials include piezoelectric ceramic, glass ceramic, piezoelectric crystal, piezoelectric polymer, piezoelectric composite, ferroelectric piezoelectric crystal, non-ferroelectric piezoelectric crystal, relaxation electric crystal and inorganic piezoelectric ferroelectric thin films, etc. Their power generation capacity is assessed by piezoelectric strain constant, reliability is assessed by flexibility or mechanical strength, and stability is assessed by the stabilities of performance parameters (e.g. relative dielectric constant) including temperature stability, humidity stability, chemical stability and time stability. References [6], [7] and other design manuals provide the design parameters of various types of piezoelectric materials, and they serve as important evidences at the early stage of design. It is required that the piezoelectric material to be used is of high piezoelectric strain constant, fine flexibility or mechanical strength and stable performance parameters, and Pbʐro3-Pbʈio3-5 (PZT-5) piezoelectric ceramic is a good example.

3. Assembling of Piezoelectric Power Generation Equipment

In addition to piezoelectric material, power generation capacity, reliability and stability also depend on the assembling of piezoelectric power generation equipment. The power generation capacity of piezoelectric power generation is assessed by electromechanical coupling coefficient, which indirectly reflects the transition efficiency from mechanical energy to electric energy. Piezoelectric power generation equipment is required to have high electromechanical coupling coefficient; thus, some efforts have to be made to improve the vibration mode, support mode, motivation mode and interconnection mode of piezoelectric vibrator.

As to vibration mode, 31-mode (the stretching vibration along length direction) is easy to assemble and has low system natural frequency and relatively high vibration [8], and 33-mode (the stretching vibration along thickness direction) has high electromechanical coupling coefficient but is not easy to produce strain. From the principles of physics, the mathematical models of 31-mode and 33-mode can be constructed [9], [10], and then the mathematical formula of electromechanical coupling coefficient, quality factor, input energy, output energy and efficiency can be derived. Hereinto, high quality factor means low extra heat loss and more generation of electricity.

As to support mode, cantilever support and simple support are convenient to install and easy to realize. The deformation of rectangle-section cantilever support concentrates on the root, which greatly shortens the effective use length of piezoelectric material. Therefore, to promote power generation capacity, the section of rectangle-section cantilever support can be replaced by other sections of different shapes, e.g. triangle and trapezoid [11]. The deformation of triangle-section or trapezoid-section cantilever support distributes more widely and more uniformly, which can generate more electricity energy using the piezoelectric material of the same volume. Reference [12] presents a new structure of simple support for circular piezoelectric vibrator, making the deformation of piezoelectric material more uniformly than that of traditional staked configuration and thus expanding the electricity generation area.

As to motivation mode, inertial free vibration mode can recover weak environmental vibration energy (e.g. wind power and noise energy), and forced vibration mode can recover great mechanical pressure energy (e.g. vehicle crushing energy). However, only with the aid of percussion hammer (e.g. metal ball) can shock free vibration mode generate electricity; therefore, this mode may not only reduce the service life of piezoelectric power generation equipment, but also produce noise, and can only be used in situations where transient voltage or current is needed. Therefore, only inertial free vibration mode and forced vibration mode can be adopted by piezoelectric power generation system. After motivation mode is determined, the next task is to determine the resonant frequency under this motivation mode, because only when piezoelectric vibrator resonates with external force can electricity generation be maximized. To maintain the maximum electricity generation, resonant frequency should be able to be adjusted online so as to adapt to operation mode changes and various disturbances. According to whether the extra energy is needed or not, the adjusting method of resonant frequency is divided into active self-adjusting method and passive self-adjusting method. Active self-adjusting method that needs extra energy may increase electricity generation by about 30% [13], [14], however, the additional adjustment circuit may consume much more electricity energy than the increased electricity energy generated by resonant frequency adjustment. Passive self-adjusting method that does not need extra energy is no better than active self-adjusting method in increasing electricity generation [15]; on the contrary, volume is increased and structure is complicated. Therefore, the self-adjusting method of resonant frequency requires in-depth study.

As to interconnection mode, by series mode, parallel mode or series-parallel mixed mode, several piezoelectric vibrators can be interconnected to form multilayer piezoelectric vibrator so as to enable small piezoelectric material to have relatively high power generation capacity under weak motivation and in wide frequency band. n pieces of paralleled piezoelectric vibrators can increase the output current by n times, and n pieces of serialized piezoelectric vibrators can increase the output voltage by n times. Paralleled piezoelectric vibratos have great output current and great equivalent capacitors, and are applicable to small load impedance situation; serialized piezoelectric vibratos have high output voltage and small equivalent capacitors, and are applicable to great load impedance situation. Therefore, through the optimization of the series-parallel mixed mode of piezoelectric vibrators, the voltage and current output characteristics of piezoelectric power generation equipment can be optimized, which is beneficial for electric power grid connection. From analytical calculations, reference [16] demonstrates that the multilayer piezoelectric vibrator consisting of 145 piece series-parallel piezoelectric vibrators (height 1.8cm, section area 1cm²) has 1~10 μF equivalent capacitance and about 30V open circuit voltage, and its matched load resistance is several kilohms. Reference [17], by changing the direction of the traditional interconnection mode, promotes the electricity energy output quantity and the efficiency by 1.5~1.8 times.

4. Charge Collector and Energy Storage Equipment

Considering that the charges generated by piezoelectric vibrator during each vibration period are finite, the design of high efficient charge collector and energy storage equipment is also very important for promoting efficiency and reducing energy loss. In addition, because the output of piezoelectric vibrator is an alternating current with small current and high voltage, in order to ensure the high efficient transmission of electricity energy from piezoelectric vibrator to energy storage equipment, charge collector should not only decrease the output voltage and increase the output current, but also match the equivalent load of energy storage equipment. AC-DC converter (rectifier) or AC-DC-DC converter (rectifier+chopper) can be adopted to realize the charge collector having the two functions above. The synchronous charge collecting technology [18], [19], which synchronizes the charge collecting period with the vibration period of piezoelectric vibrator, can reduce the damping brought by inverse piezoelectric effect and promote the efficiency of piezoelectric power generation equipment by 4~9 times. It is verified by experiment that the output power of AC-DC-DC converter is almost 3 times greater than AC-DC converter [20].

Electrolytic capacitor, supercapacitor, memory-free rechargeable battery [21] and many other devices can all serve as energy storage equipment. It is necessary to comprehensively assess these energy storage devices from experiments, determining their advantages and disadvantages.

5. Conclusion

The current researches related to piezoelectric power generation mainly concentrate in the fields of material science, mechanical science and microelectronic science, and certain limited and special application areas, e.g. self-powered wireless sensor network, piezoelectric road signs, piezoelectric power generation shoes, passive safety-belt detecting device and piezoelectric micro-electro-mechanical (MEM) system. This paper actually presents a prospect of the possible application of piezoelectric power generation in electricity production field. Except the important topics discussed in this paper, as the subsequent researches start, there will generate many other interesting problems worthy of researching.

Acknowledgments

This work is funded by Talent Introduction Foundation and High Level Talent Research Foundation of Chongqing Water Resources and Electric Engineering College.

References

1. International Energy Agency (IEA), in *World Energy Outlook* 2014 (London, Nov. 2014).
2. Y. Qiao, Z.X. Lu and F. Xu, et al, *Performance Evaluation Method of Wind-Coal Coordinating Operation*, in *Automation of Electric Power Systems* (Vol. 37, No. 17, Sept. 10, 2013).
3. Y.M. Park, N.H. Lee, T.R. Choung, *Effect of Noise and Low Frequency Noise Generated by Wind Power Plant (Wind Farm)*, in *the 39th International Congress on Noise Control Engineering* (INTER-NOISE 2010, Lisbon, Portugal, 2010, Vol. 3, pp. 2061-2067).
4. H. Zhang, X.R. Song and J. Feng, *Road Power Generation System Based on Piezoelectric Effect*, in *Applied Mechanics and Materials* (2013, Vol. 329, pp. 229-233).
5. Y. Jia and A.A. Seshia, *White Noise Responsiveness of an AIN Piezoelectric MEMS Cantilever Vibration Energy Harvester*, in *Journal of Physics* (2014, Vol. 557, No. 1, pp. 25-33).
6. T. Ashutosh and U. Lokman, *Advanced Functional Materials* (Wiley Scrivener Publishing, 2014).
7. K. Tomoyuki, *Progress in Advanced Structural and Functional Materials Design* (Springer Verlag, 2012).
8. C. Lu, C.Y. Tsui and W.H. Ki, *Vibration Energy Scavenging System with Maximum Power Tracking for Micropower Applications*, in *IEEE Transactions on Very Large Scale Integration (VLSI) Systems* (Nov. 2013, Vol. 19, No. 11, 2109-2119).
9. P. Li, F. Jin and J.S. Yang, *Effects of Semiconduction on Electromechanical Energy Conversion in Piezoelectrics*, in *Smart Materials and Structures* (Feb. 2015, Vol. 24, No. 2, pp. 210-218).
10. O. Doaré and S. Michelin, *Piezoelectric Coupling in Energy-harvesting Fluttering Flexible Plates: Linear Stability Analysis and Conversion Efficiency*, in *Journal of Fluids and Structures* (Nov. 2011, Vol. 27, No. 8, pp. 1357-1375).
11. Y.S. Li, W.J. Feng and Z.Y. Cai, *Bending and Free Vibration of Functionally Graded Piezoelectric Beam based on Modified Strain Gradient Theory*, in *Composite Structures* (Aug. 2014, Vol. 115, No. 1, pp. 39-47).
12. S. Mohammadi and M.J. Abdalbeigi, *Analytical Optimization of Piezoelectric Circular Diaphragm Generator*, in *Advances in Materials Science and Engineering* (2014, Vol. 16, pp. 46-60).
13. C. Eichhorn, N. Tchagsim and N. Wilhelm, et al, *An Energy-autonomous Self-tunable Piezoelectric Vibration Energy Harvesting System*, in

Proceedings of the IEEE International Conference on Micro Electro Mechanical Systems (MEMS 2010, Cancun, Mexico, Jan. 2011, Vol. 1, pp. 1293-1296).

14. V.R. Challa, M.G. Prasad and F.T. Fisher, *Towards an Autonomous Self-tuning Vibration Energy Harvesting Device for Wireless Sensor Network Applications*, in *Smart Materials and Structures* (Feb. 2011, Vol. 20, No. 2, pp. 81-91).

15. C.G. Gregg, P. Pillatsch and P.K. Wright, *Passively Self-tuning Piezoelectric Energy Harvesting System*, in *the 14th International Conference on Micro- and Nano-Technology for Power Generation and Energy Conversion Applications* (PowerMEMS 2014, Awaji Island, Hyogo, Japan, Nov. 2014, Vol. 557, No. 1, pp. 52-59).

16. J.J. Wang, Z.F. Shi and Z.J. Han, *Analytical Solution of Piezoelectric Composite Stack Transducers*, in *Journal of Intelligent Material Systems and Structures* (Sep. 2013, Vol. 24, No. 13, pp. 1626-1636).

17. H.Q. Ye and R.M. Pan, *Experimental Analysis of Electro-mechanical Characterization of Piezoelectric Stack Actuators*, in *Advanced Materials Research* (2011, pp. 293-297).

18. Y.P. Wu, A. Badel and F. Formosa, et al, *Piezoelectric Vibration Energy Harvesting by Optimized Synchronous Electric Charge Extraction*, in *Journal of Intelligent Material Systems and Structures* (Aug. 2013, Vol. 24, No. 12, pp. 1445-1458).

19. Y.P. Wu, A. Badel and F. Formosa, et al, *Self-powered Optimized Synchronous Electric Charge Extraction Circuit for Piezoelectric Energy Harvesting*, in *Journal of Intelligent Material Systems and Structures* (Nov. 2014, Vol. 25, No. 17, pp. 2165-2176).

20. Y.H. Su, Y.P. Liu and D. Vasic, et al, *Design of Piezoelectric Transformer-based DC/DC Converter to Improve Power by Using Heat Transfer Equipment*, in *the 23rd International Conference on Adaptive Structures and Technologies* (ICAST 2012, Nanjing, China, Oct. 2012, Vol. 1, pp. 90-95).

21. Y. Levron, D. Shmilovitz and S.L. Martinez, *A Power Management Strategy for Minimization of Energy Storage Reservoirs in Wireless Systems with Energy Harvesting*, in *IEEE Transactions on Circuits and Systems I: Regular Papers* (2011, Vol. 58, No. 3, pp. 633-643).

Improved Mental Workload Classification by Kernel Spectral Regression and Transferable Discriminative Dimensionality Reduction Approaches

Jian-hua Zhang[†] and Yong-cun Wang

Department of Automation, East China University of Science and Technology
Shanghai, 200237, China
†E-mail: zhangjh@ecust.edu.cn

Ru-bin Wang

Institute of Cognitive Neurodynamics, East China University of
Science and Technology
Shanghai 200237, China

Considering the poor generalization ability of the subject-specific Mental Workload (MWL) classifier, a cross-subject MWL recognition framework was proposed in this paper. In the proposed framework, Kernel Spectral Regression (KSR) and Transferable Discriminative Dimensionality Reduction (TDDR) methods were employed to reduce the feature dimensionality and transfer the classification knowledge from the source domain to the target one. The data analysis results showed that the proposed framework and the related methods can effectively improve the accuracy of MWL recognition.

*Keyword*s: Human-Machine System; Mental Workload; Dimensionality Reduction.

1. Introduction

In complex human-machine (HM) systems with humans as system operators, human operators must monitor or manually control the system based on real-time auditory and visual information concerning the state of the system[1]. Some researchers have started to consider the problem of how to maintain the operator functional state (OFS) appropriate or optimal during the operation of a HM system. Mental workload (MWL)measurement has become a hot research topic in recent years since MWL is an essential dimension of the OFS[2,3]. Our previous work focused largely on designing subject-specific classifier of MWL[3]. In addition, the features extracted in our recent work [3] were based on power spectral density computed by fast Fourier transform(FFT). Nevertheless, the measured electrophysiological signals from human subjects tend to be non-stationary in nature [4]. In this paper, KSR is adopted to reduce the EEG feature dimensionality as well as to enhance the inter-class discrimination capacity of the MWL classifiers. TDDR technique proposed by Tu and Sun [5] transfer the MWL classification knowledge from one subject to another. Finally, the classification performance of different combinations of classification and dimensionality reduction methods is compared and discussed.

2. Data acquisition experiments

An experimental procedure similar to that developed in [3] is adopted in our experiments. The number of subsystems manually controlled (NOS) and two-level (i.e., standard level (SL) and high level (HL)) actuator sensitivity (AS) are designed to manipulate the level of task difficulty and thus can be used to determine the target (or desired)MWL level of each sample data. Each of six young healthy male subjects(A, B, C, D, E, and F) participated in two sessions of a CAMS (automation-enhanced Cabin Air Management System) software based computer process control experiments arranged on two different days. An experimental session consisted of 10 task-load conditions, each lasting 5 min.

3. Methods

The EEG-based MWL classification procedure is described as follows. Firstly, the normalized logarithmic energy of the coefficients of the 4-th level wavelet-packet decomposition was used as the features. Then the dimensionality of the extracted features was reduced by a combination of KSR[6]and TDDR[5] algorithms.

3.1. An introduction to TDDR algorithm

Suppose a large training set X_{tr}^S in the source domain $\square$and a very small training set X_{tr}^T in the target domain T. The classifier will be evaluated on the testing set X_{te}^T in the target domain T.

The training data is assigned with different weights. Then, the combined training set $X_{ctr}^{ST} = \{X_{tr}^S, X_{wtr}^T\}$ is formed, where X_{tr}^S and X_{wtr}^T denote the source training set and weighted target training set respectively. Then define weight as $W_{tr} = 1 + n_t/n_s$, where n_s and n_t are the number of training samples in the source and target domains respectively. Then, the inter-and intra-class scatter metrics are computed on the combined training set X_{ctr}^{ST} in lower-dimensional latent space as follows:

$$S_{inter}^{ST} = \sum_{i=1}^{n} \sum_{j=i}^{n} \left(\mu_i - \mu_j \right) \left(\mu_i - \mu_j \right)^T \tag{1}$$

$$S_{intra}^{ST} = \sum_{i=1}^{n} \sum_{k=1}^{m_i} \left(x_k^{(i)} - \mu_i \right) \left(x_k^{(i)} - \mu_i \right)^T \tag{2}$$

where μ_i is the i-th class mean of the combined training set, n is the number of classes, $x_k^{(i)}$ denotes the kth sample that belongs to class i, m_i is the number of samples in class i. To reveal the separation between two domains, an inter-domain scatter matrix was defined by:

$$S_L^{ST} = \sum_{i=1}^{n} \left(\mu_i^S - \mu_i^T \right) \left(\mu_i^S - \mu_i^T \right)^T \tag{3}$$

where μ_i^S and μ_i^T are class means of source and target training sets respectively.

The solution of transferable discriminative dimensionality reduction can be found such that the following objective function is maximized:

$$J(\Phi) = \frac{\Phi^T S_{inter}^{ST} \Phi}{\Phi^T (S_{intra}^{ST} + \alpha S_L^{ST}) \Phi} \tag{4}$$

where parameter α is used to control the trade-off between the desired level of discrimination and transferability.

4. MWL classification results and discussion

The parameters set in each of the four primary dimensionality reduction algorithms are given in Table 1. Specifically, in KNN, k = 3. One Against One SVM (OAO-SVM) was employed for multi-class classification. In TDDR the parameter α was set to be 1.

Table 1. Parameters in dimensionality reduction algorithms

Method	Parameter	Kernel Type
PCA	Ratio=0.95	/
KPCA	Ratio=0.95	Gaussian
KSR	α_1=1	Gaussian
KLPP	α_2=1	Gaussian

4.1. Feature dimensionality reduction

Dimensionality reduction is a transformation of higher-dimensional data into meaningful lower-dimensional data. In this paper, seven different dimensionality methods, namely TDDR, PCA+TDDR, KPCA+PCA, LPP+TDDR, KLPP+TDDR, SR+TDDR, and KSR+TDDR, are examined and compared. On the other hand, four types of classifiers, viz. SVM, KNN, NB and LDA, are examined in combination with the seven dimensionality reduction algorithms. Take four-class MWL classification of subject A as an example, the transformed data in the reduced 3-D feature space distribution is shown in Fig. 1. It can be seen that after applying KSR+TDDR the original data are grouped into four distinguishable clusters in the lower-dimensional feature space, which will improve the classification accuracy.

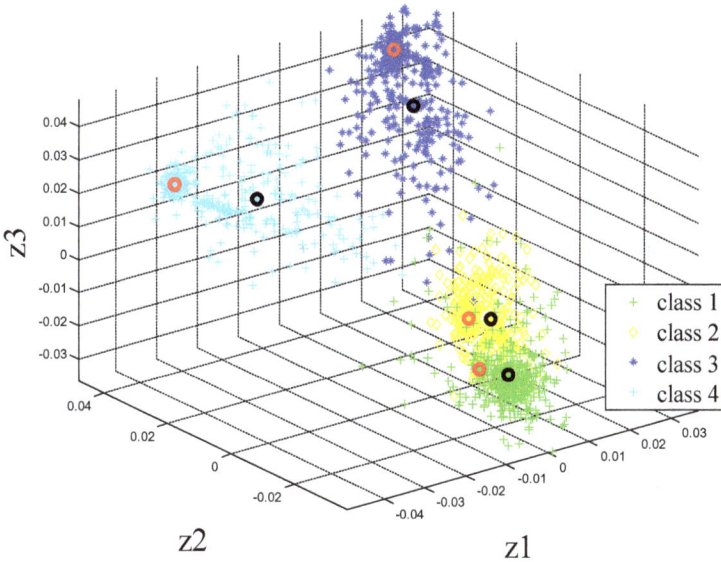

Fig. 1 The transformed data in the reduced 3-D feature space after applying KSR and TDDR: An example of 4-class case for subject A.

4.2. Performance comparison of different classification methods

In Table 2, the 3-class classification accuracy of combinations of KSR+TDDR and four classifiers is compared. Compared with LDA and NB, SVM and KNN achieved higher accuracy. KSR+TDDR+SVM resulted in the highest accuracy of 81.06% on subject C in the case of three-class classification. In Table 3, the 4-class classification accuracy of combinations of KSR+TDDR and four classifiers is compared. It can be seen that SVM and KNN obtained higher 4-

class classification accuracy and that KSR+TDDR+SVM resulted in the highest accuracy of 78.63% on subject C.

Table 2 Performance comparison of different classifiers: 3-class classification case

ACC(%)	A	B	C	D	E	F
KSR+TDDR +KNN	**71.01±5.86**	**80.35±3.87**	81.05±3.14	**64.39±4.69**	69.78±4.48	64.39±4.29
KSR+TDDR +SVM	70.94±5.93	80.30±3.79	**81.06±3.21**	64.30±4.33	**69.81±4.54**	**64.42±4.32**
KSR+TDDR +NB	68.13±7.91	78.89±4.44	78.42±5.47	60.54±5.55	67.48±7.09	60.54±5.55
KSR+TDDR +LDA	69.53±7.32	75.88±5.29	77.12±6.43	62.79±4.82	71.46±6.08	62.79±4.82

Table 3 Performance comparison of different classifiers: 4-class classification case

ACC(%)	A	B	C	D	E	F
KSR+TDDR +KNN	73.11±5.86	77.30±3.87	78.59±3.14	65.35±4.29	71.02±4.48	70.29±4.29
KSR+TDDR +SVM	**73.15±5.93**	**77.32±3.79**	**78.63±3.21**	**65.40±4.33**	**71.08±4.54**	**70.36±4.32**
KSR+TDDR +NB	70.22±7.91	75.97±4.44	75.82±5.47	59.29±5.55	68.07±7.09	66.42±5.55
KSR+TDDR +LDA	70.00±7.32	75.23±5.29	76.80±6.43	62.53±4.82	68.53±6.80	69.29±4.82

Fig. 2 compared the subject-averaged classification accuracy. One-way analysis of variance (i.e., ANOVA with significance level set at 0.05)was performed on the classification testing accuracy between KSR+TDDR and other six dimensionality reduction methods when combined with four classifiers.

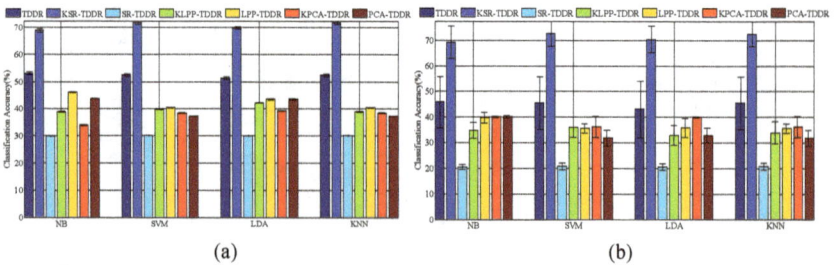

(a)   (b)

Fig. 2 The subject-averaged classification accuracy of different combinations of dimensionality reduction and classification methods: (a)3-class case; (b)4-class case

In the case of three-class classification as shown in Fig. 2(a), when using NB as the classifier, the classification accuracy of KSR+TDDR outperforms TDDR, PCA+TDDR, KPCA+TDDR, LPP+TDDR, KLPP+TDDR, and SR+TDDR by 15.91% (p=0.004), 25.30 % (p=2.29E-05), 35.08% (p=3.10E-06), 22.93% (p=1.19E-04), 30.15% (p=1.25E-05),and 39.01% (p=3.73E-07)respectively. When using KNN as the classifier, the classification accuracy of KSR+TDDR outperforms the other six feature reduction algorithms by 19.36% (p=8.66E-04), 34.54% (p=6.57E-07), 33.37% (p=6.91E-07), 31.47%

(p=1.41E-04), 32.92% (p=3.59E-03),and 41.61% (p=7.90E-06)respectively. When using LDA as the classifier, the classification accuracy of KSR+TDDR outperforms the other six feature reduction algorithms by 18.56% (p=7.17E-04), 26.45% (p=8.19E-06), 30.63% (p=4.14E-07), 26.48% (p=5.87E-06),27.70% (p=2.25E-06), and 39.86% (p=2.15E-08) respectively. When using SVM as the classifier, compared with other six feature dimensionality reduction algorithms the testing classification accuracy of KSR+TDDR is improved by 19.30% (p=8.87E-04), 34.49% (p=6.73E-07), 33.31% (p=7.09E-07), 31.42% (p=1.45E-06), 31.99% (p=4.46E-06), and 41.56% (p=8.08E-08) respectively. In the case of four-class classification as shown in Fig.2(b), when using NB as the classifier, the classification accuracy of KSR+TDDR outperforms TDDR, PCA+TDDR, KPCA+TDDR, LPP+TDDR, KLPP+TDDR, and SR+TDDR by 23.45% (p=6.69E-04), 29.16% (p=5.00E-07), 29.23% (p=4.81E-07), 29.50% (p=7.12E-07), 34.46% (p=2.77E-07), and 48.38% (p=7.26E-09) respectively. When using KNN as the classifier, in comparison with other six algorithms the accuracy of KSR+TDDR is increased by 27.03% (p=1.72E-04), 40.72% (p=8.87E-09), 36.36% (p=6.57E-08), 37.00% (p=8.04E-09), 38.61% (p=4.55E-08), and 48.75% (p=2.26E-10) respectively. When using LDA as the classifier, the accuracy of KSR+TDDR outperforms the other six algorithms by 27.32% (p=2.70E-04), 37.44% (p=2.30E-08), 30.44% (p=4.74E-08), 34.39% (p=1.05E-07), 37.41% (p=5.59E-08), and 49.78% (p=5.00E-10) respectively. When using SVM as the classifier, the accuracy of KSR+TDDR outperforms other six alternatives by 27.08% (p=1.69E-04), 40.77% (p=8.64E-09), 36.41% (p=6.41E-08), 37.04% (p=7.80E-09), 36.61% (p=5.14E-08), and 51.77% (p=2.20E-10) respectively.

5. Conclusion

In this paper across-subject MWL classification framework was proposed. Among seven different dimensionality reduction algorithms examined, KSR+TDDR exhibited higher classification accuracy when combined with four different classifiers. Furthermore, the KSR+TDDR+SVM and KSR+TDDR+KNN demonstrated superior classification accuracy to other combinations of dimensionality reduction and classification methods. Nonetheless, certain aspects of this work may be further improved in the future. For instance, more comprehensive and effective MWL classification validation scheme should be considered, i.e., the first and second session data can be used as the training and testing data, respectively, so as to further validate the prediction (or generalization) performance of the MWL classifiers developed.

Acknowledgments

This work was supported in part by the National Natural Science Foundation of China under Grant No. 61075070 and Key Grant No. 11232005.

References

1. Wilson G F, Russell C A. Performance enhancement in an uninhabited air vehicle task using psychophysiologically determined adaptive aiding. Human Factors, 2007, 49(6): 1005-1018.
2. Borghini G, Astolfi L, Vecchiato G, et al. Measuring neurophysiological signals in aircraft pilots and car drivers for the assessment of mental workload, fatigue and drowsiness, Neuroscience & Biobehavioral Reviews, 2014, 44: 58-75.
3. Yin Z, Zhang J. Identification of temporal variations in mental workload using locally-linear-embedding-based EEG feature reduction and support-vector-machine-based clustering and classification techniques, Computer Methods and Programs in Biomedicine, 2014, 115(3): 119-134.
4. Khushaba R N, Kodagoda S, Lal S, et al. Driver drowsiness classification using fuzzy wavelet-packet-based feature-extraction algorithm, IEEE Trans. on Biomedical Engineering, 2011, 58(1): 121-131.
5. Tu W, Sun S. Transferable discriminative dimensionality reduction, in *Proc. of 23rdIEEE Int. Conf. on Tools with Artificial Intelligence*, 2011: 865-868.
6. Cai D. He X, Han J. SRDA: An efficient algorithm for large-scale discriminant analysis, IEEE Trans. on Knowledge and Data Engineering, Jan. 2008, 20(1): 1-12.

Energy Storage Capacity Optimization for Wind Farms Considering the Reliability and Economy

Hao-chuan Niu, Chun-juan Jia[†], Pei Zhang and Xiao Yang

School of Electrical Engineering, Shandong University, China
[†]E-mail: jiachunjuan@sdu.edu.cn

The reliability of the power grid would change when the wind-storage system connects to the power grid. In the meantime, the economy for the wind farm would also be changed. This paper proposes a new objective function to optimize the wind farm's storage capacity. The objective function built on the reliability and economy for the wind farm. The paper demonstrated a test system named IEEE RTS79 to detect the new objective function's practicability.

Keywords: Grid-Connected Wind Farm, Energy-Storage Capacity, Reliability of the Power Grid, Economy for the Wind Farm.

1. Introduction

As the ecological environment problem has become more and more seriously in recent years, renewable energy usage has been more important. Being one of the renewable energy, wind energy has gained reputation of fast growth and best economic. The instability of wind energy has affected power system much more than before.

When the wind power [1] connects to the power grid only, the contribution to the reliability of the system is much less than conventional power unit, at the meantime, the volatility of the wind has a great impact on the stability of the power system. As a consequence, wind power need energy storage's efforts to regulate and smooth its output power. The energy storage systems can contribute larger reliability and reduce the impact on the stability of the system [2].

Wind-Storage can effectively improve the randomness and volatility of wind power and smooth the output power, thus improve the quality of power supply. But the cost of the wind farm will increase with the increase of the wind power storage capacity. Therefore, optimize the wind farm's storage capacity reasonable has great significance for the wind farm's economy and the power grid's reliability, on the other hand, wind farm has to consider the reliability for the power grid and economy for the wind far together to optimize the wind farm's storage capacity [3].

2. Power System Reliability Assessment and Evaluation Index

Power system reliability assessment requires some quantitative data to evaluate power grid, the reliability index is adopted. Reliability index can be divided into deterministic index and probabilistic index.

Deterministic index is extremum which is related to frequency of a processor an event, not including the information of the whole process. Deterministic index mainly has transmission network load supply ability, power utilization coefficient and coefficient of transmission reserves, the minimum supply ability, etc.

Probabilistic index can reflect the whole simulation process and the characteristics of the process. Probabilistic index mainly has loss of load probability, loss of load frequency, loss of load expectation, loss of load duration, expected energy not served.

EENS (expected energy not served.) would be used in this paper to evaluate the power system reliability. EENS can express the power system's loss power which is caused by load curtailing in the evaluating period. Computational formula for EENS as follows:

$$EENS = T * \int_{\Omega} F_{EDNS}(x)dx \tag{1}$$

In the computational formula, $F_{EDNS}(x)$ is power system's loss power when the system is in state X.T is the evaluation period.

3. Economy index

Energy storage capacity's selection of wind farm has important effects on the technical and economic indicators of the wind farm. When it is small capacity selection, excess capacity of the wind generating set can't be stockpiled fully. When it is large capacity selection, it will increase the investment at first.

Comprehensive cost will be used in this paper to evaluate the wind farm's economy. Comprehensive cost which is synthesize by one-time investment cost of the power system and the operation cost of the power system and the generating revenue of the power system. Computational formula for comprehensive cost as follows:

$$C = C_1 + C_2 - C_3 \tag{2}$$

In the computational formula, C_1 refers to the one-time investment cost of the power system, C_2 refers to the operation cost of the power system, C_3 refers to the generating revenue of the power system.

4. Evaluation Index Which Consider the Reliability and Economy

Power system economy becomes more and more important, but at the same time we can't ignore the reliability of the power system but the reliability for the power grid and economy for the wind farm [4-6] are the two different directions. So this paper introduce an economics commonly used indicators — marginal revenue to establish objective function.

In economics, the marginal revenue refers to the increase of the total return values which is caused by the increase of producers' selling. In this article [7], the reliability of the power system's increase can be seen as the total revenue increment, at the same time, the increase of the energy storage [9] capacity's cost can be seen as the increase of producers' selling to calculate the marginal revenue.

In this article, the new objective function is called marginal EENS MEENS. Due to the reduce of EENS represents the improvement of reliability, the computational formula has to be absolute values. Computational formula for MEENS as follows:

$$MEENS(n) = \left| \frac{EENS(n) - EENS(n-1)}{C(n) - C(n-1)} \right| \tag{3}$$

In the computational formula, EENS（0）refer to the EENS when the wind power storage capacity is zero，C（0）refer to the comprehensive cost when the wind power storage capacity is zero，n refer to the wind power storage capacity [10, 11].

Due to the EENS's dimension and the comprehensive cost's dimension is different, so the data of the EENS and the comprehensive cost have to be changed.

The changed computational formula for MEENS as follows:

$$MEENS(n) = \left| \frac{[EENS(n) - EENS(n-1)] / EENS(0)}{[C(n) - C(n-1)] / C(0)} \right| \tag{4}$$

5. Analysis of Examples

This article is based on RTS79 platform. The installed capacity of original RTS79 system for 3405 MW, the load is 2850 MW. Because Original reliability is at a lower level, this paper has to reduce the installed capacity to 2430 MW to make the EENS have obvious change. And this paper will increase 300MW wind turbines for analysis of examples.

357

When installed capacity is 2430MW, the EENS of the power system is 11.8537 MW. And the EENS of the power system becomes 6.9596MW when 300MW wind turbine is included. Next let's add energy storage systems to the wind farm [9]. 40 kinds of energy storage capacity are adopted in this paper

5.1. Power system reliability assessment and evaluation index

The tendency of the EENS with the energy storage capacity's increase is shown in figure 1.

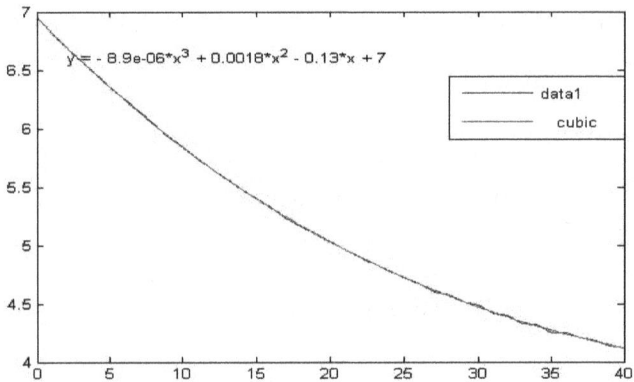

Figure 1. The tendency of the EENS with the energy storage capacity's increase

In the picture we can see that the reducing of the EENS is faster when the energy storage capacity is lower, and the reducing of the EENS is slower when the energy storage capacity is higher, even becomes gentle when the energy storage capacity is very high.

5.2. Economy index

Let's see the change of the comprehensive cost which is caused by the change of energy storage capacity. It is shown in figure 2.

In figure 2, we can see the comprehensive cost's increase is proportional to the ascension of the energy storage capacity.

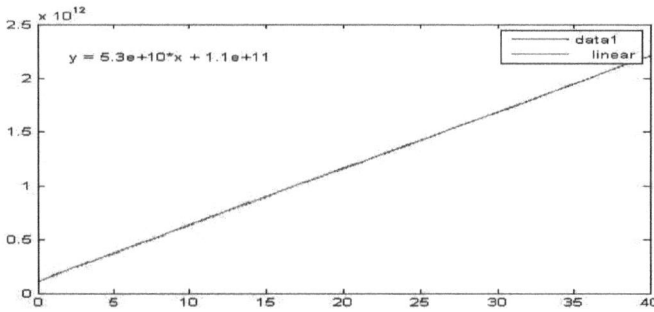

Figure 2. The change of the comprehensive cost

5.3. Evaluation index which consider the reliability and economy

The simulation result of the tendency of the change of MEENS is shown in figure 3.

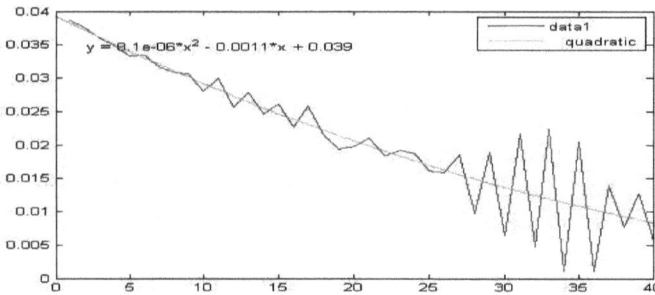

Figure 3. The tendency of the change of MEENS

As we can see in figure 3, the tendency of the change of MEENS is more and more small, finally close to zero. But there are some fluctuations in the picture. Due to RTS79 system reliability analysis is built on the large-scale random number, there is certain randomness on result which is inevitable.

There is an extension curve of MEENS which is shown in figure 4. The figure is drawn by the polynomial fitting result of MEENS.

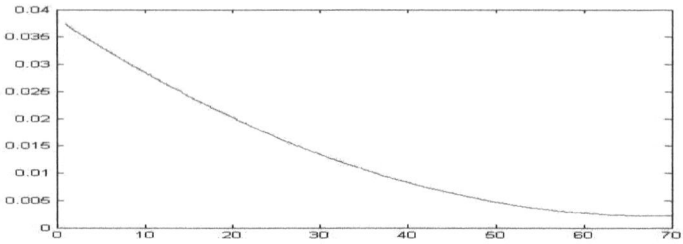

Figure 4. Extension curve of the quantitative relation between the MEENS and the corresponding serial number of energy storage capacity.

From figure 4, we can conclude that when the energy storage capacity increased to 1050 mw, the reliability [12] index would ascend with the increase of the energy storage capacity, but the increasing range is a little small, at the same time, the increasing range of comprehensive cost is invariable, so the MEENS is close to zero. When the energy storage capacity increased to 2100 mw, the MEENS would be a constant which is very close to zero.

6. Conclusion

This paper put forward a new objective function to optimize the wind farm's storage capacity which is built on the reliability of the power grid and economy for the wind farm. After we verify in RTS79 system, the objective function can be effectively combine he reliability of the power grid and economy for the wind farm together. The objective function can help the wind farm to determine the wind energy storage system capacity and it is practical [8].

From the conclusion we can know, simply increase the energy storage system capacity will lead to ascension of the reliability of the power grid. But take into account the economy of the wind farm, it is unwise to increase the energy storage system capacity without thinking about the cost of wind farm. Wind farms could choose the appropriate energy storage capacity in accordance with the requirements of their reliability and economy.

References

1. Le, Ha Thu, and Surya Santoso, "Analysis of voltage stability and optimal wind power penetration limits for a non-radial network with an energy storage system," Power Engineering Society General Meeting, 2007, IEEE. IEEE, 2007.
2. Chi, Yongning, et al, "Voltage stability analysis of wind farm integration into transmission network," Power System Technology, 2006. PowerCon 2006. International Conference on. IEEE, 2006.

3. Chi, Yongning, Studies on the Stabiliy Issues about Large Scale Wind Farm Grid Integration, A Thesis Submitted to the Graduate school of China Eleertie Power Researeh Institute 2006.

4. Zhou, Fengquan, Geza Joos, and Chad Abbey, "Voltage stability in weak connection wind farms," Power Engineering Society General Meeting, 2005. IEEE. IEEE, 2005.

5. Horne, Jonathan, Damian Flynn, and Tim Littler, "Frequency stability issues for islanded power systems," Power Systems Conference and Exposition, 2004. IEEE PES. IEEE, 2004.

6. Meng Zhaonian, Xie Kaigui, "Wind Farm Reliability Evaluation Considering Operation Characteristics of Battery Energy Storage Devices, " Power System Technology, vol. 36, pp.214-219, June 2012.

7. Lou Suhua, et al, "Operation Strategy of Battery Energy Storage System for Smoothing Short-term Wind Power Fluctuation,"Automation of Electric Power Systems, vol. 38, No.2 pp.17-22, Jan. 2014

8. Tesfahunegn, S. G., et al, "Optimal shifting of Photovoltaic and load fluctuations from fuel cell and electrolyzer to lead acid battery in a Photovoltaic/hydrogen standalone power system for improved performance and life time," Journal of Power Sources vol.196, pp. 10401-10414, 2011.

9. Teleke, Sercan, et al, "Control strategies for battery energy storage for wind farm dispatching," Energy Conversion, IEEE Transactions on 24.3 pp. 725-732,2009.

10. Jiang, Zhenhua, and Xunwei Yu, "Modeling and control of an integrated wind power generation and energy storage system," Power & Energy Society General Meeting, 2009. PES'09. IEEE. IEEE, 2009.

11. Yoshimoto, Kayo, Toshiya Nanahara, and Gentaro Koshimizu, "New control method for regulating state-of-charge of a battery in hybrid wind power/battery energy storage system," Power Systems Conference and Exposition, 2006. PSCE'06. 2006 IEEE PES. IEEE, 2006.

12. Subcommittee, P. M, "IEEE reliability test system," IEEE Transactions on Power Apparatus and Systems, pp.2047-2054, 1979.

EAST Cryogenic Instrumentation and Control— Present Status and Reconstruction Plan Based on EPICS

Ming Zhuang, Zhi-wei Zhou, Liang-bing Hu, Zhen-shan Ji and Gen-hai Xia

Institute of Plasma Physics, Chinese Academy of Sciences
Hefei, Anhui 230031, China
E-mail: zzw@ipp.ac.cn

The cryogenic control system for the Experimental Advanced Superconductive Tokamak (EAST) was designed and constructed based on DeltaV distributed control system from Emerson Corporation, which has been operated for eleven cryogenic experimental campaigns since its first commissioning in 2005 and demonstrated stable and reliable over these years. With the reconstruction of EAST cryogenic system, its control system has also been extended and upgraded to ensure its reliability and availability in continuously uninterrupted operation. In the meantime, for breaking the limitation of commercial control software, the cryogenic control system is being redesigned based on the open source distributed control software architecture EPICS. This paper will introduce the present status of EAST cryogenic instrumentation and control system. Also, the reconstruction plan based on EPICS will be presented in detail.

Keywords: Instrumentation and Control; EPICS; Cryogenic System; EAST Tokamak.

1. Introduction

EAST cryogenic system is one of the critical sub-systems in the Experimental Advanced Superconductive Tokamak (EAST) device. [1] It comprises a large scale compressor station with a helium recovery and storage system to supply the high pressure of 2.0 MPa for refrigeration cycle, a 2 kW/4.5 K helium refrigerator and the cryogenic distribution system which provides the proper cold power for the cold components of EAST, including the superconducting coils, thermal shields, high temperature superconductor current leads (HTS CL) with buslines, and built-in cryopumps. Figure 1 is the simplified process flow of the updated EAST cryogenic system. The new MYCOM compressors were paralleled into the compressor station, also did the new turbine expanders in the refrigerator.

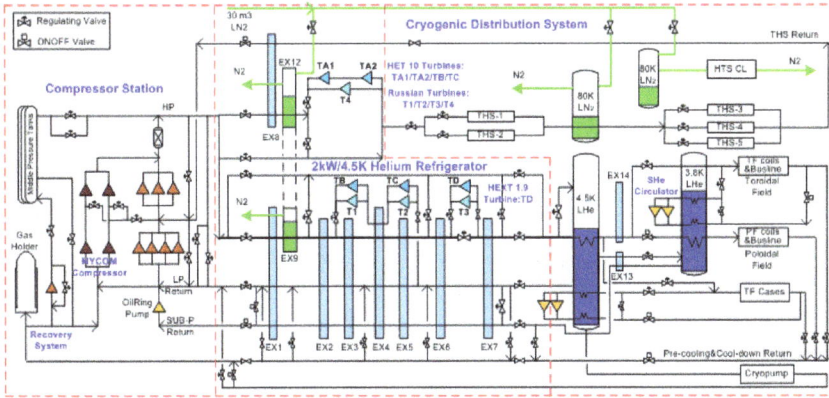

Fig. 1. The simplified process flow of EAST cryogenic system.

The EAST cryogenic control system (ECCS) was designed based on DeltaV distributed control system (DCS) from Emerson Corporation [2], which has been operated over ten years and demonstrated stable and reliable over these years. However, many electrical and control components have been running beyond their expected lifetime gradually. Also the old version of operating system and control software limit the update of control computers. Therefore, with the reconstruction of EAST cryogenic system, the control system were extended and upgraded to ensure its reliability and availability in the future continuously uninterrupted operation.

2. Present Status of EAST Cryogenic Control System

2.1. Latest control network architecture

The EAST cryogenic control system was designed and constructed with a three-layer communication network structure of standard DCS and expanded its function through OPC (OLE for Process Control) protocol. Figure 2 shows the latest architecture of the ECCS, which has three different types of networks. They are Cryogenic Redundant Control Network, Data Exchange Local Area Network (LAN) and Main Control Data LAN. Several OPC clients form the control auxiliary system including the database server/client, FTP server/client, web server and the data acquisition and processing system used for acquiring the complicated cryogenic process parameters such as helium mass flow rate and cryogenic temperature. Some important parameters of the magnets are acquired through TCP/IP communication from other systems. There is a firewall between the cryogenic intranet and extranet to ensure the security of ECCS.

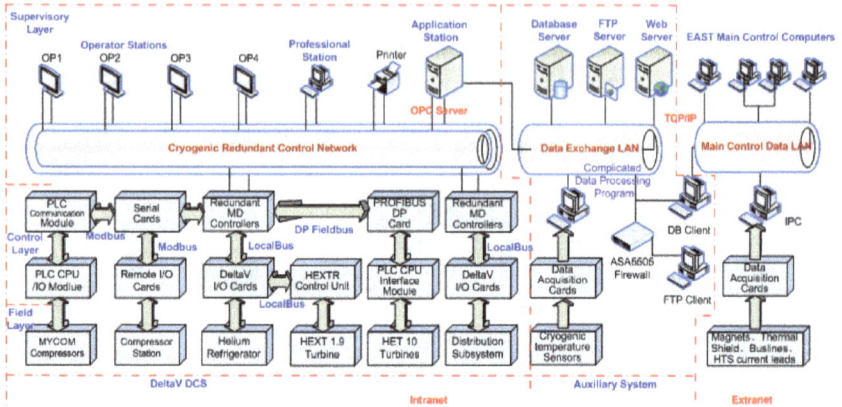

Fig. 2. Network architecture of latest EAST cryogenic control system.

DeltaV DCS has two pairs of redundant MD controllers for refrigerator and distribution subsystem respectively, where the process parameters such as pressure and normal temperature were acquired directly by DeltaV I/O cards. While, because of the long distance the compressor station were supervised by the remote I/O cards, which connects to the redundant controllers of refrigerator for control by serial cards through MODBUS protocol. The new MYCOM compressors are controlled by a set of Programmable Logic Controller (PLC) in field, which connects to DCS by optical fiber and communicates as a sub-station of DCS through Modbus protocol for a global supervisory. The turbine TD controlled by a field control unit based on Digital Signal Processor (DSP) was integrated with DCS through its localbus. Four new HET 10 turbines controlled by the local PLC communicate with DCS through PROFIBUS DP fieldbus.

2.2. Cryogenic instrumentation and control

In the field layer, there are many varieties of cryogenic parameters to be measured and controlled, including pressure, temperature, mass flow rate, liquid level, rotation speed, vacuum degree and purity. The control loops amount to about 160 many of which are coupled multi-variable control loops. Table 1 shows the I/O list of EAST cryogenic system.

More than 177 resistance temperature sensors, of four types, are used in EAST cryogenic system, including the platinum resistance temperature detectors Pt100 used in high temperature range from 52.5 K to 300 K, and the sensors of Negative Temperature Coefficients used in low temperature range from 2 K to 300 K, which are Cernox, Carbon sensors from American LakeShore Company and Russian Carbon sensors. The pt100 resistance whose maximum precision is ±0.5 % are directly measured by RTD cards in DeltaV DCS as a four-wire resistor, without any other signal conditioning. While other cryogenic temperature sensors are acquired by external data acquisition cards and Keithley digital multi-meter and processed in the programs, and then transfer data through OPC protocol to DCS. The precision is ±3 mK at 4.2 K temperature area.

Table 1. I/O list of the latest EAST cryogenic system.

Area ——— I/O	MYCOM Compressor	Compressor Station	Helium Refrigerator	Turbine System	Distribution System	Magnets THS HTS CL	Total
AI	66	89	122	60	108	341	786
AO	2	10	48	20	46	30	156
DI	16	109	17	52	24	/	218
DO	8	69	22	18	68	/	185
Total	92	277	209	150	246	371	1345

About 140 pressure on pipes and vessels are measured by intelligent pressure transducers which have 4-20 mA output connected directly to analog input (AI) cards in DCS. The precision of pressure measurement can be reach ±0.1 %. There are 16 liquid level meters inside the vessels, of three types, which are differential pressure type gauges, buoyancy gauges for measuring liquid nitrogen, and superconducting-wire gauges for measuring liquid helium. There are 4 Orifice plate, 2 vortex and 24 Venturi mass flow-meters to measure the mass flow rate, which is calculated by the parameters such as density, differential pressure and geometric dimensioning in OPC client programs.

The cryogenic control are executed by these actuators including the 135 pneumatic control valves with intelligent positioners driven by the analog output (AO) signals of 4-20 mA, 52 pneumatic ON/OFF valves driven by 0-24 V DC digital output (DO) signals, solenoid valves, slide valves for compressor control, frequency converters for helium circulating pumps control, switches and heaters, and so on.

In the supervisory layer, there are one professional station used to develop control programs and supervisory interface, one application station as the OPC server and the gateway of control network, and four operator stations providing real-time process monitoring and control interface. Sequence Function Charts (SFC) and Function Block Diagrams (FBD) were used to develop the automatic control programs as well as alarms and interlocks for the cool-down and warm-up operation of EAST cryogenic system. It has been demonstrated the good performance which reduced the operation burden and ensured a safe cryogenic status.[3] Figure 3 is the new supervisory and control interface on DeltaV DCS.

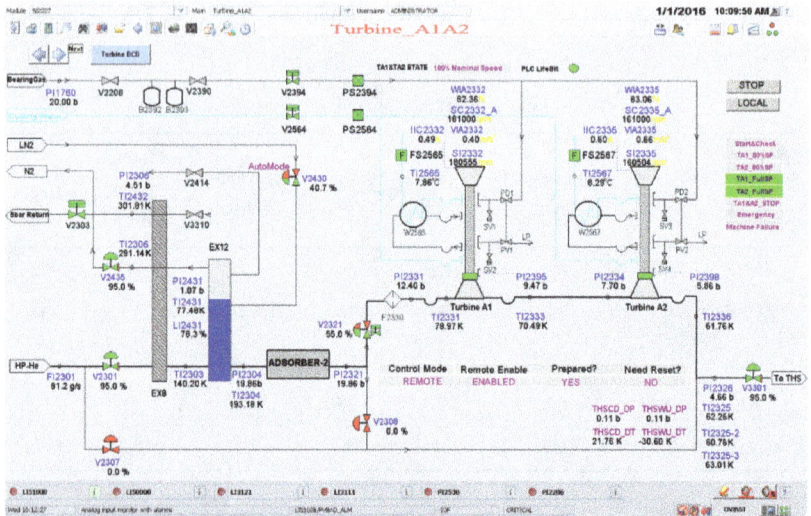

Fig. 3. New supervisory interface of DCS in EAST cryogenic control system.

3. Reconstruction Plan based on EPICS

The control software of DeltaV DCS has been upgraded to the latest version V12.3.1 which operates in Windows Sever 2008 and Windows 7, and some hardware also been updated, including all monitor computers, the redundant network switches, and some instrumentation and transducers. It will take the DCS into a new lifetime for another 4 to 6 years. However, the software would encounter another upgrade demand several years later. So an open source distributed control software architecture EPICS (Experimental Physics and Industrial Control System), which was developed collaboratively and used worldwide from high energy physics field to fusion field, was considered to reconstructed the ECCS.

Figure 4 is the control network architecture based on EPICS for the ECCS reconstruction. It will be a distributed and open system integrated with Control, Data Access and Communication (CODAC).[4] Considering the existing control structure and the slow control of cryogenic system, a pair of redundant Programmable Logic Controllers (PLC) as the master station will be used to realize the global control and interlock as well as the coordination control of other sub-systems. While each sub-system as a Profibus DP sub-station has its own PLC to implement its own control and adopts the remote I/Os or filed I/Os to connect with the filed instrumentation. Compared to DeltaV DCS, the new control structure is distributed control based on field-bus and multi PLCs will relieve the loads of MD controllers, which will promote the system's reliability greatly. PLC is the mature industrial controller which is suitable to develop the slow control process system like cryogenic system and its control software is relatively open.

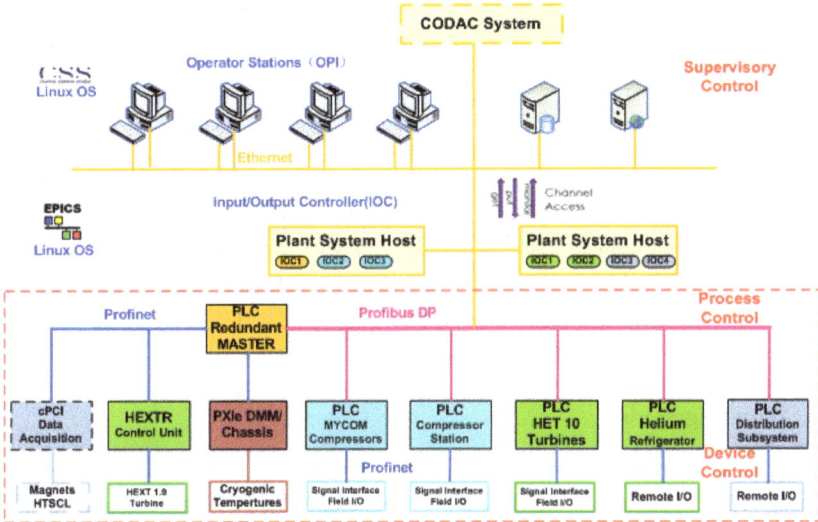

Fig. 4. Control network architecture based on EPICS for the ECCS reconstruction.

EPICS has typical Client/Server mode, which has three basic components, including Operator Interface (OPI), Input/Output Controller (IOC), and Channel Access (CA). The IOCs chose industrial computers with cPCI bus, on which Linux Operating System (OS) running and CA module installed to provide network transparent communication between OPIs and IOCs. There is a distributed real-time database system on IOC to manage the data from each sub-system. The OPIs chose workstation on which runs Linux OS and adopted the open source software Control System Studio (CSS) provided by EPICS to develop the human machine interfaces, alarms, archives and supervision.

The international fusion research and engineering project ITER adopted the CODAC Core System which is also based on EPICS for its instrumentation and Control, including for the slow control system.[4] In view of its successful commissioning, our reconstruction plan based on EPICS is feasible. However, we plan to reconstruct the ECCS according to the EPICS architecture gradually. The device control of each sub-system implemented by PLC wouldn't reconstruct for the time being. Instead, we will design the OPIs and IOCs for

367

supervisory control first. The data from device control layer could be accessed from DeltaV DCS through MODBUS protocol. But at last, we would reconstruct the ECCS thoroughly based on the new EPICS architecture.

4. Conclusion

DeltaV DCS has been demonstrated reliable and flexible for the cryogenic instrumentation and control through over ten years' operation, and it also has a good expandability for various application development. After the extension of control structure and upgrade of control software, the ECCS has become an integrated distributed control system together with varieties of industrial field-bus and communication protocol. However, some functions still have the license limitation such as historical data storage and alarm help, and so on. Therefore, the open source control architecture EPICS was considered to reconstruct the ECCS gradually. The reconstruction plan has been proved feasible and it will provide a reference for better development of more large scale cryogenic control system for the future China Fusion Engineering Test reactor.

Acknowledgments

The funding from the National Natural Science Foundation of China (Grant No. 11505237) and the Application & Development Project of the Institute of Plasma Physics, Chinese Academy of Sciences (Y35ETY130G) are gratefully acknowledged.

References

1. H. Bai, Y. Bi, P. Zhu, Q. Zhang, K. Wu, M. Zhuang and Y. Jin, Cryogenics in EAST, *Fusion Engineering and Design.* 81(2006).
2. Y. Jin, X. Shao, M. Zhuang and H. Bai, EAST Cryogenic Supervisory and Control System Based on Delta-V DCS, *Plasma Science & Technology.*7, 5(2005).
3. Z. Zhou, Q. Zhang, M. Zhuang, L. Hu and G. Xia, Performance Analysis of EAST Cryogenic Control System under Various Operational Modes, *The 13th Cryogenics 2014 IIR International Conference*, (Prague, Czech Republic , 2014).
4. M. Panella, C. Centioli, F. Maio, M. Napolitano, M. Rojo, M. Vellucci, V. Vitale and A. Wallander, ITER CODAC Core System at FTU: State of the art and new perspectives, *Fusion Engineering and Design.* 88 (2013).

Self-Extinguishing in UHV Transmission Lines with Hybrid Reactive Compensation

Hong-shun Liu[†], Zhen Wang, Run-chang Li and Bin Li

School of Electrical Engineering, Shandong University,
Jinan, Shandong Province, China
[†]E-mail: lhs@sdu.edu.cn

Ting-ting Lv and Song-yan Li

State Grid Shandong Electric Power Maintenance Company
Jinan, Shandong Province, China

UHV single-phase grounding fault presents unique physical features under the action of hybrid reactive compensation with the combination of the series compensation and the stepped controlled high resistance. The influence mechanism is more complex than the conventional UHV transmission line. Basing on the introduction of the key elements and structure of the hybrid reactive compensation, this paper studies the arc ignition and self extinguishing characteristics of secondary arc. Firstly, the works undertaken both at home and abroad were presetned, before, the unique physical features of secondary arc self-extinguishing with the UHV hybrid reactive compensation were put forwarded. With regard to single-phase autoreclosure and secondary arc self-extinguishing, our investigation have uncovered the key problems of UHV power grid with hybrid reactive compensation, by which might pave the way for future application.

Keywords: Hybrid reactive compensation; Series compensation; Controlled shunt reactor; Secondary arc

1. Introduction

Long-distance, large-capacity and low-loss power transmission is the inevitable choice to optimize resources distribution and control smog. The large scale construction of UHV projects will become the new development mode of the electric power industry in China [1]. After the establishment of UHV synchronous power grid, the interconnection of regional power grids and the optimization of resources distribution will be realized. But the transmission power will realize a significant improvement and the reactive power will change frequently, so the safety and stability of the system would face both opportunities and challenges [2]. It mainly included follow aspects:

1) owing to the restriction of the system impedance characteristics and stability limits, the demand of the transmission power rise cannot be satisfied;

2) The increase of changing times and amplitude of the system voltage and reactive power. It is an ideal solution to use hybrid reactive compensation with series compensation and controlled shunt reactor, which gives consideration to both the increase of transmission power and the frequent adjustment of reactive power, while overcome the shortages of traditional simple reactive compensation mode (only use fixed reactor).

The application of series compensation and shunt controlled reactor will make the UHV reactive compensation more complex, and several frontier topics in the field of UHV hybrid reactive compensation are supported by the State Grid Corporation of China for further study at present. Compared with normal transmission lines, the process of secondary arc self-extinguishing with UHV hybrid reactive compensation has unique physical features. Under the influence of short-circuit currents and low frequency oscillations of secondary-arc currents, the change of secondary-arc current amplitudes and the decrease of zero-crossing times directly affect the physical features of secondary arc duration and extinction, even leads to the failure of single-phase auto reclosure. The available basic theories and analysis methods of secondary arcs cannot be suited to hybrid reactive compensation condition and guide the application in future projects efficiently, neither. Since series compensation and controlled shunt reactors are the key devices of reactive compensation, their application in UHV transmission lines has broad prospect. With regard to renovation research in relative science and techniques, indicating the influence mechanism of UHV hybrid reactive compensation to the characteristics of secondary arcs offers theoretical basis and technical supports; meanwhile, it has outstanding scientific significance and theoretical value.

2. Structure of Hybrid Reactive Compensation

The hybrid reactive compensation contains the series compensation and stepped controlled reactor, the structure composition of which is demonstrated in Fig. 1. Where, series compensation consists of capacitor banks, Zinc Oxide arresters, spark gaps, bypass circuit breakers and damping devices. The stepped controlled shunt reactor consists of high impedance transformers, series reactance, mechanical switches and thyristors.

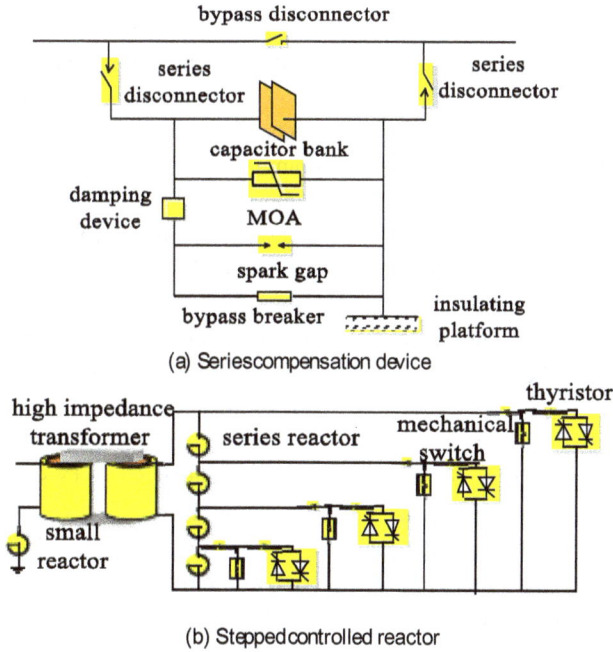

(a) Seriescompensation device

(b) Steppedcontrolled reactor

Fig. 1.   Structure of hybrid reactive compensation.

3. Secondary Arc Self-Extinguishing Physical Properties

Successful self-extinction of secondary arc influences the single-phase autoreclosure. In recent years, researchers from home and abroad studied on the arcing and self-extinguishing characteristics of secondary arc in transmission line by physics experiments, mathematical modeling, numerical simulation and other methods, while most of the researches concentrate on EHV/UHV network and pay little attention to hybrid reactive compensation.

In China, Japan, Brazil, Russia and other countries, some field experiments of secondary arc aim at several EHV/UHV transmission lines has been carried out, which reflect the physical process of secondary arc [3]. The experiment the CEPEL laboratory in Brazil carried out in 500kV transmission line is demonstrated in Fig. 2. While field tests has limitations in experimental scheme, times, conditions and other aspects, simulation experiments in little current and short gap arc can simulate field test of large current and long gap arc, and gradually become the most direct effective method and means to study secondary arc characteristics and its physical nature. By fitting the data of simulation experiments, approximate function between arc extinction time and relative influence factors can be obtained. For all that, present simulation experiments lack of thorough analysis in the motion characteristics of secondary arc, extinguish rekindle characteristic, arcing characteristics under different compensation schemes and other aspects. Despite from this, UHV hybrid reactive compensation contains series compensation and controlled reactors, its influence to secondary arc is different from the fixed reactors. If for some reason makes the circuit breaker on the line of both sides tripped without bypass the series compensation, the short-circuit current and secondary arc current show low frequency oscillation with high amplitude, which affects the physical characteristics of arc burning and self-extinguishing and go against single-phase autoreclosure. The circuits of present simulation experiments only aims at normal transmission lines installed fixed reactors without any consideration of hybrid reactive compensation, so it cannot meet the requirement of low frequency oscillation to the influence mechanism of secondary arc characteristics. As a result, it is necessary to further perfect the equivalent analysis circuit of secondary arc and equivalently design the topological structure of physical simulation experimental circuit in UHV transmission line with hybrid reactive compensation. Then the simulation experiments related to circuit secondary arc is developed to gain thorough understanding of the influence mechanism of the secondary arc current physical characteristics.

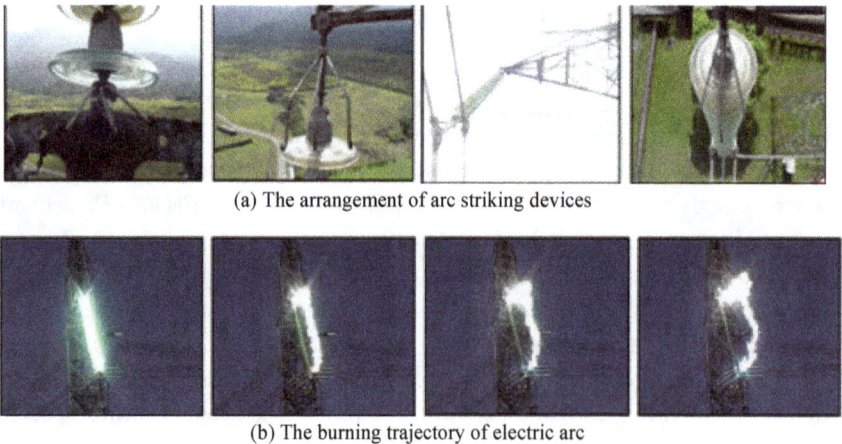

(a) The arrangement of arc striking devices

(b) The burning trajectory of electric arc

Fig. 2. The simulation experiments of secondary arc.

At present, the mathematical models of secondary arc are all built on the basis of normal line simulation experiments, and the physical parameters of the model is determined by fitting the data from secondary arc experiments. Based on the classical black box theory proposed by Cassie and Mayr, A. T. Johns built the EMTP calculation model of secondary arc, and Russian scholars simulated the secondary arc in UHV transmission lines [4]. Chen Weijiang, Sima Wenxia, Li Qingmin and other scholars did related researches on the motion characteristics of secondary arc by using chained arc theory. No matter the black box model or the chained model, present modeling methods of secondary arc in normal transmission lines have deficiencies in the length of arc and the criterion on whether the arc strikes or not. The reason is that they are always restricted to macroscopic external characteristics, cannot describe the distribution of arc discharge temperature and the microscopic changes of internal plasma parameters (electron energy, density distribution of each particles, electric field distribution), and it is difficult to reflect the influence the macroscopic factors have on the microscopic developing mechanism of secondary arc, while it is exactly where the physical essence of electric arc is. MHD analysis has obvious superiority, since it can reflect the physical characteristics of secondary arc like temperature distribution, magnetic field distribution and motion condition or so. It can analysis not only the macroscopic characteristics of electrical arc, but also the distribution of its internal fields. However, such research works are seldom [5]. As a result, based on MHD analysis, describe the arcing and self-extinguishing characteristics of the low frequency oscillation secondary arc from a temporal microscopic/macroscopic physical aspect, which takes external electromagnetic field as well as flow field and the coupling effect between the environment and the electromagnetic transient process of secondary arc burning and self-extinction, then present a more accurate criterion on the extinction and strike of electric arc, build the simulation model of secondary arc and calculate the self-extinguishing characteristics of secondary arc, all these researches has guiding significance to the setting of autoreclosure in UHV transmission lines with hybrid reactive compensation.

4. Conclusion

The electromagnetic transients of single-phase grounding faults in UHV transmission lines with hybrid reactive compensation differ from normal transmission lines, and there are many key technical problems which need to be solved by carrying out basic research. Aiming at the influence mechanism of secondary arc, the author holds that it needs to be studied from following aspects.

The temporal microscopic/macroscopic physical characteristics of low frequency oscillation secondary current under the consideration of the space electromagnetic field, flow field as well as the special environmental coupling effect and MHD equivalent resistance modeling method. The microscopic and macroscopic physical characteristics of low frequency oscillation secondary arc plasmas under the influence of space electromagnetic field would be revealed. Flow field and spatial environment multi-field coupling, the MHD equivalent resistance model used in the simulation of electromagnetic transients would be established. Finally, the optimal cooperation sequence between controlled shunt reactors, series compensation and autoreclosure will be proposed.

References

1. Liu Zhenya. *Innovation of UHVAC transmission technology in China, Power System Technology*, Vol.37(3), pp. 1-8, (2013).
2. Tang Xiaojun, Wang Tieqiang, JiaJinghua, et al. *Function and corresponding strategy for AVC system under interconnection of Ultra-high Voltage Power Grid, Automation of Electric Power Systems*, Vol.37(21), pp. 106-111, (2013).
3. Chen Weijian, Yan Xianglian, He Ziming, et al. *Simulation for secondary arc caused by single-phase grounding in UHVAC transmission line, High Voltage Engineering*, Vol.1(31), pp. 1-6, (2010).
4. A. T. Johns, R. K. Aggarwal, Y. H. Song. *Improved techniques for modeling fault arcs on faulted EHV transmission system, IEE Proceeding-Generation, Transmission, Distribution*, Vol. 141(2), pp. 148-154, (1994).
5. Chen Lunjiang, Tang Deli, Cheng Changming, et al. *Magneto hydrodynamics analysis of a dual anode arc plasma torch. High Voltage Engineering*, Vol.39(7), pp. 1723-1729, (2013).

Application of a Novel Cell Membrane Algorithm in Optimizing Reactive Power

Wei Huang, Ning-kun Li[†], Yi Song and Wen-xue Che

North China Electric Power University, Beijing, Changping District, China
State Power Economic Research Institute Changping District, Beijing, China
[†]E-mail: lnk_0910@126.com

To solve the multi-variable, non-linear, and non-convex reactive power optimization problem, this study proposes a new intelligent optimization algorithm called Cell Membrane Optimization (CMO). This algorithm is based on the diffusion processes of the substance in cell. The substance (solutions) moves through free diffusion, facilitated diffusion and active transport according to its type which is classified by concentration, thereby increasing the speed of searching for the global optimal solution. In the process of reactive power optimization, based on the constraint conditions establish objective function of the minimum net loss using CMO. The algorithm is tested with IEEE 14-bus and IEEE 30-bus systems. Results show that the optimization ability of the algorithm is acceptable. This algorithm has parallelism naturally, which can avoid the local optimal solution.

Keywords: Power System, Cell Membrane Algorithm, Global Optimum, Reactive Power Optimization.

1. Introduction

Reactive power system optimization can reduce network power losses under the premise of voltage quality assurance. However, the reactive power optimization problem is multi-variable. The voltage magnitude is continuous, whereas the transformer winding ratio is discrete. Multi-constraints are added because the active and reactive power must be balanced and the voltage magnitude should be acceptable. Representations by the interior point methods, which have high requirements for the initial value and easy to fall into a local optimal solution, so that the optimal solution cannot be guaranteed. Drawing on the wisdom of nature [1]. Such intelligent algorithms provide new solutions to complex issues [2]. Neural network system, ant colony algorithm [3], and bacterial chemo taxis optimization [4] have been applied. The global optimization capability of these algorithms is relatively strong. However, disadvantages also exist, such as slow calculation and the search area may be limited, which weakens a global search capability [5][6]. The CMO algorithm introduces different ways and directions of movement for different types of material (solution) to improve the speed of the optimization and retains the optimal solution to guarantee that the solution is global. This study proposes a reactive power optimization model using the CMO algorithm based on the abovementioned theory. This algorithm is tested in IEEE 14 and IEEE 30 node systems, shows faster convergence, and achieves a good solution.

2. Reactive Power Optimization Model

2.1 *Objective functions*

The main purpose of reactive power optimization is to reduce active power losses.

$$\min F = \sum_{k \in N_i} g_k (u_i^2 + u_j^2 - 2u_i u_j \cos \theta_{ij}) \tag{1}$$

where P_{LOSS} is the total active power loss; g_k is the conductance of the of kth branch; θ_{ij} is the admittance angle of the transmission line; N_i is the number of transmission lines; and u_i are the voltage magnitudes of the ith bus.

2.2 *System constraints*

(i) Equality constraints

$$P_i = V_i \sum_{j \in N, j \neq s} V_j (G_{ij} \cos \theta_{ij} + B_{ij} \sin \theta_{ij})$$

$$Q_i = V_i \sum_{j \in N, j \in N_{PQ}} V_j (G_{ij} \sin \theta_{ij} - B_{ij} \cos \theta_{ij}) \tag{2}$$

(ii) Inequality constraints

$$V_{Gmin} \leq V_G \leq V_{Gmax}$$
$$K_{Tmin} \leq K_T \leq K_{Tmax}$$
$$Q_{Cmin} \leq Q_C \leq Q_{Cmax}$$
$$V_{Lmin} \leq V_L \leq V_{Lmax} \tag{4}$$
$$Q_{Gmin} \leq Q_G \leq Q_{Gmax}$$
$$S_T \leq S_{Lmax}$$

The active and reactive powers [7] must be balanced in the system. where G_{ij} is the real part of the bus admittance matrix of the $(i,j)^{th}$ entry and B_{ij} is the

376

imaginary part, Pi and Q_i are the active and reactive powers. s is the number of the slack bus, and N_{PQ} is the number of PQ nodes. where V_G is the voltage of the generator, K_T is the transformer ratio, Q_c is the reactive power compensation capacity, V_T is the voltage of the load node, Q_G is the reactive power the generator provides, and S_L is the power that goes across the branch[8][9].

3. CMO Algorithm

3.1 *Principle of the cell membrane optimal algorithm*

The membrane (cell membrane, CM) can control substances to go into and out of cells by free diffusion, facilitated diffusion, active transport. CMO divides the substances into fat-soluble substances, non-fat-soluble substance, and ions or small-molecule substances. Free diffusion refers to the process, by which FS diffuse from the side of the membrane with high concentration to that with low concentration, which requires neither energy nor carrier. Facilitated diffusion refers to the process by which NS diffuse from the higher to lower concentration with the help of membrane proteins, but energy is still unnecessary. Active transport can diffuse from the lower to the higher concentration with the help of both energy and carrier. They must also carry on active transport [10][11].

3.2 *CMO algorithm*

The basic steps of the CMO algorithm are as follows:

Step 1: Substance initialization
Randomly generate m n-dimensional materials in the solution space. Calculate their function values, and choose the X^{best}, which has the smallest function.

Step 2: Classify the substances
Calculate the concentration of each solution based on Equation (10). Sort the top 50% of the substances into the low-density material (LS). The others are high-density material (HS). The substance in HS, whose order is odd classified fat-soluble, the others are non-fat-soluble.

$$con = \frac{s}{m} \qquad (5)$$

where s represents the number of the substances whose distance from the set substance is smaller than a set length; m is the number of all the substances.

Step 3: Free diffusion of fat-soluble HS

Each fat-soluble HS diffuses toward the direction of the LS group lcon times. Choose any substance from the LS group to determine the direction vector of the movement, and unitize the vector.

$$\vec{F} = \frac{LS^{randi} - FS^i}{\left\| LS^{randi} - FS^i \right\|} = \frac{LS^{randi} - FS^i}{\sqrt{\sum_{k=1}^{n} (LS_k^{randi} - FS_k^{\ i})^2}} \tag{6}$$

The fat-soluble HS then moves to the LS group based on this vector to obtain a new material.

$$newFS_k^i = \begin{cases} FS_k^i + rand() * F_k * (u_k - LS_k^{randi}) & F_k \geq 0 \\ FS_k^i + rand() * F_k * (LS_k^{randi} - 1_k) & F_k < 0 \end{cases} \quad k = 1, \cdots, n \tag{7}$$

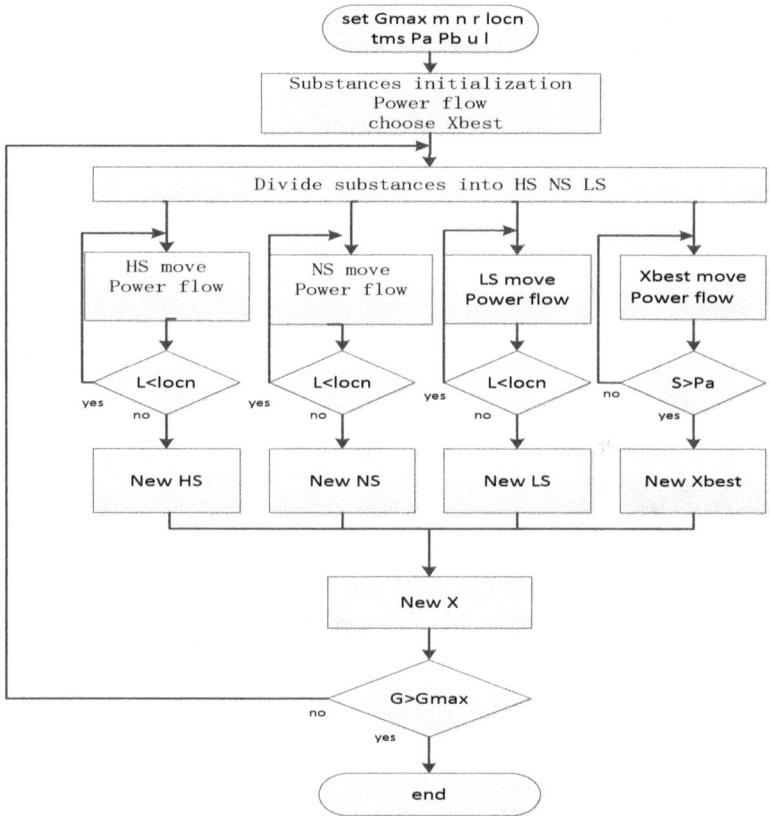

Fig. 1. Flow chart of the CMO.

Step 4: Movement of non-fat-soluble HS

Non-fat-soluble HS moves toward the global optimum material when no carrier is required. The movement of non-fat-soluble HS can be described into three steps:

(i) Distinguish whether a carrier exists.

(ii) The non-fat-soluble FS that has a carrier moves in support of the carrier, just as free diffusion.

$$\bar{F} = \frac{LS^{randi} - NS^i}{\left\| LS^{randi} - NS^i \right\|} = \frac{LS^{randi} - NS^i}{\sqrt{\sum_{k=1}^{n}(LS_k^{randi} - NS_k^{\ i})^2}} \tag{8}$$

$$newNS_k^i = \begin{cases} NS_k^i + rand()*F_k*(u_k - NS_k^{randi}) & F_k \geq 0 \\ NS_k^i + rand()*F_k*(NS_k^{randi} - 1_k) & F_k < 0 \end{cases} \quad k = 1,\cdots,n \tag{9}$$

(iii) The non-fat-soluble FS without a carrier moves toward the global optimum material.

$$newNS_k^i = NS_k^i + rand()*(X_k^{best} - NS_k^i) \quad k = 1,\cdots,n \tag{10}$$

Step 5: Movement of LS

The movement of LS requires a carrier and energy so that the two judgments should be conducted. The LS that cannot meet the conditions of energy moves in the searching area at random Icon times, and the best position is chosen as the final result.

$$newLS_k^i = l_k + rand()*(u_k - l_k) \quad k = 1,\cdots,n \tag{11}$$

The LS that meets the energy condition but not the carrier condition moves toward the current optimum material.

$$newLS_k^i = LS_k^i + rand()*(X_k^{best} - LS_k^i) \quad k = 1,\cdots,n \tag{12}$$

The LS that has both energy and carrier can move toward the HS group.

$$\bar{F} = \frac{HS^{randi} - LS^i}{\left\| HS^{randi} - LS^i \right\|} = \frac{HS^{randi} - LS^i}{\sqrt{\sum_{k=1}^{n}(HS_k^{randi} - LS_k^{\ i})^2}} \tag{13}$$

$$newLS_k^i = \begin{cases} LS_k^i + rand()*F_k*(u_k - LS_k^{randi}) & F_k \geq 0 \\ LS_k^i + rand()*F_k*(L\,S_k^{randi} - 1_k) & F_k < 0 \end{cases} \quad k = 1, \cdots, n \qquad (14)$$

Step 6: Update substance
Update X^{best}, fat-soluble FS, non- fat-soluble HS, and LS.

4. Simulation and Discussion

4.1 *Simulation of the IEEE 14-bus system*

The IEEE 14-bus system as shown in Fig. 2 [12], includes 20 branches, 5 generator nodes. Node 9 can compensate for reactive power; whose compensation capacity is from 0 MVar to 30 MVar. The voltage of every node is supposed to be between 0.95 and 1.05. The eight variables include a reactive power compensation capacity, four generator voltages, and three transformer ratios.

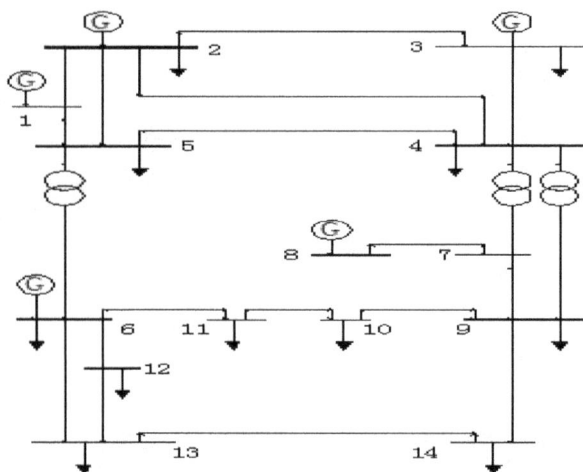

Fig. 2. IEEE 14-bus system.

The results of the simulation on the IEEE 14-bus system are shown in Tables 1 and 2 and Fig. 3.

The power loss of the CMO algorithm is the smallest, as shown in Table 1. Meanwhile, Table 2 shows that the number of iterations is smaller than that of the other algorithms.

Table 1. Values of the variables before and after optimization (14-bus).

Variable	Reference	CMO	CPSO	IMGA
Q_{c9}	0.19	0.3	0.18	0.18
V_{g2}	1.045	0.9866	1.0463	1.0443
V_{g3}	1.01	0.9972	1.0165	1.0138
V_{g6}	1.07	0.9606	1.1	1.1
V_{g8}	1.09	0.9994	1.1	1.0882
T_{4-7}	0.978	1.0625	0.94	1.07
V_{4-9}	0.969	0.9056	0.93	1.04
V_{5-6}	0.932	0.9498	0.97	1
P_{loss}	13.7325	13.6628	13.6934	13.7123

Table 2. Power loss and number of iteration using CMO, CPSO, and IMGA
algorithms for the IEEE 14-bus power system.

Algorithm	Ploss (MW)	Number of iteration	Optimization time (min)
Reference	13.7325	--	--
MO	13.6628	50	1.65
PSO	13.6934	70	2.04
IMGA	13.7123	60	2.45

However, Fig. 3 shows that the advantage is slight.

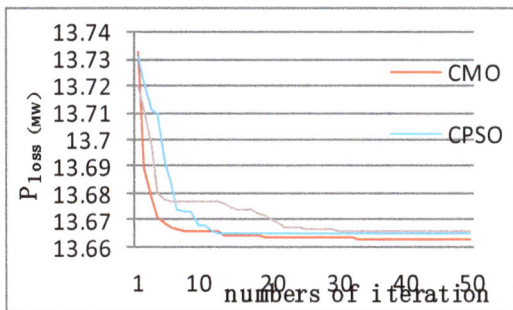

Fig. 3. Convergence characteristics of power loss using CMO, CPSO, and IMGA algorithms for the
IEEE 14-bus power system.

4.2 Simulation of the IEEE 30-bus system

The IEEE 30-bus system, showns in Fig. 4 [13, 14] consists of 41 branches, 6
generators, and 24 load nodes. Each of the branches 6–9, 6–10, 4–12, and 27–28
contains a set of transformers, whose tap is adjustable. Nodes 10 and 24 can
compensate for reactive power whose capacity range is 0 MVAr to 30 MVAr.

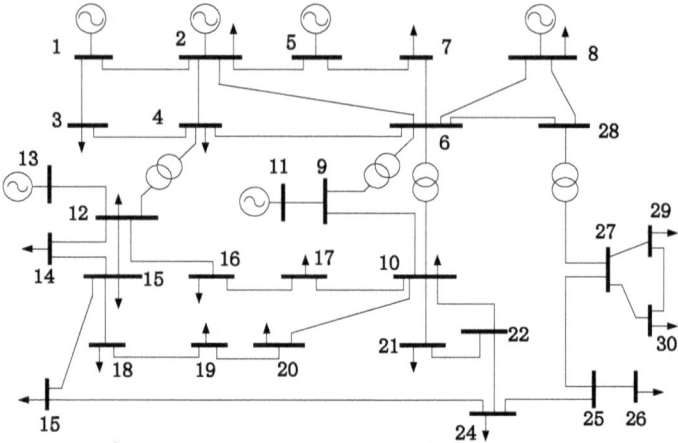

Fig. 4. IEEE 30-bus system.

The results of the simulation of the IEEE 30-bus system, summaried in Tables 3, and 4, show the power loss are dramatically decreased. The CMO algorithm almost finishes its work before 10 times, when the CPSO and IMGA are just beginning to converge as shown in Fig. 5.

Table 3. Values of the variables before and after optimization.

Variables	Q_{C10}	Q_{C24}	V_{G1}	V_{G2}	V_{G5}	V_{G8}
Before	0.19	0.04	1.06	1.04	1.01	1.01
After	0.2765	0.1027	1.0584	1.0664	0.9761	1.0577
Variables	V_{G11}	V_{G13}	T_{9-6}	T_{10-6}	T_{12-4}	T_{28-27}
Before	1.082	1.071	1.016	1.0385	1.013	1.044
After	1.0362	1.0380	1.0138	0.9216	0.9343	0.9277

Table 4. Power loss and numbers of iteration using CMO, CPSO, and IMGA algorithms for the IEEE 30-bus power system.

Algorithm	Ploss (MW)	Numbers of iteration	Optimization time (min)
Reference	17.9464	--	--
CMO	17.3804	50	2.15
CPSO	17.5365	70	2.54
IMGA	17.4702	66	2.96

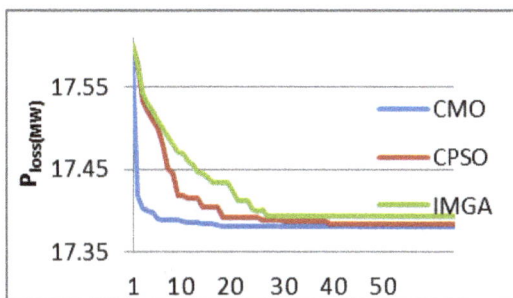

Fig. 5. Convergence characteristics of power loss using CMO, CPSO, and IMGA algorithms for the IEEE 30-bus power system.

4.3 Discussion

Table 5. Time for the solution by CPSO, IMGA, and CMO algorithms

Algorithm	CPSO	IMGA	CMO
14-bus	2.1	2.4	1.6
30-bus	2.5	3	2.3

The CMO algorithm shows a better ability to converge either in the IEEE 14-bus or 30-bus system. The CMO algorithm speeds up by dividing the substance and selecting the appropriate direction of the movement based on their different characteristics. It also guarantees that the global optimal solution achieves good simulation results in the reactive power optimization problem. Thus, this algorithm shows superior speed over the other methods in searching for the best solution. The optimization time shows the advantage of CMO over the other two algorithms.

After the exploration, the iteration time Icon has an effect on the optimization process. The time would be long if it is too large, but the convergence would be slow if it is too small. Future research should determine the suitable value of this algorithm.

5. Conclusion

This study proposes a novel CMO algorithm in solving the reactive power optimization problem; the proposed algorithm achieves better results than CPSO and IMGA. The results show that this algorithm can still perform better if combined with other algorithms. It also still requires further exploration. More studies can be conducted to prove the advantages of this algorithm.

References

1. Robert R G. An Introduction to Cultural Algorithms[C]. Proceedings of the 3th Annual Conference Evolution Programming. Singapore: World Scientific Publishing, 1994: 131-139.
2. Kennedy J, Eberhart R. Particle swarm optimization[C]. Proceedings of IEEE International Conference on Neural Networks. Perth: IEEE, 1995: 1942-1948.
3. Shi Y, Eberhart R. A modified particle swarm optimizer [C]. Proceedings of IEEE World Congress on Computational Intelligence. Honolulu: IEEE, 1998:69-73.
4. A. Likas, M. Vlassis, J. Vetheek. The global k-means clustering algorithm [J]. PatternReeognition, 2003, (36): 451- 461.
5. Deeb N, Shahidehpour S M. Linear reactive power optimization in a large power network using the decomposition approach. IEEE Trans on Power Systems, 1990, 5(2): 428- 438.
6. Ni Wei, Shan Yuanda. A refined genetic algorithm with optimal searching path used in power system reactive power optimization. Automation of Electric Power Systems, 2000, 24(21): 40-44.
7. Paton R, Gregory R, Vlachos C, et al. Evolvable social agents for bacterial systems modeling. IEEE Trans on Nano bioscience, 2004, 3(3): 208-216.
8. Huang Wei, Zhang Jianhua, Zhang Cong, et al. Reactive power optimization in power system based on bacterial colony chemotaxis algorithm. Automation of Electric Power Systems, 2007, 31(7): 29-33.
9. Chuangxin, Zhu Chengzhi, Zhao Bo, et al. Power system reactive power optimization based on an improved immune algorithm. Automation of Electric Power Systems.
10. Zhelong, Zhang Boming, Sun Hongbin, et al. A distributed genetic algorithm for reactive power optimization. Automation of Electric Power Systems, 2001, 25(12): 37-41.
11. Paun G. Computing with Membrane [J]. Journal of Computer and System Science, 2000, 61(l): 108-143.
12. Huang Liang. Research on Membrane Computing Optimization Methods [D]. Zhejing, China: Zhejiang University, 2007, 3-22.
13. Paun G. Menbrane Computing [M]. Germany: Springer-Verlag Berlin and Heidelberg GmbH & Co.K, 2010.
14. Tan Shiheng. A New Swarm Intelligent Optimization: Cell Membrane Optimization and Its Applications [D]. Guangzhou, China: South China University of Technology, 2011. 21-65.

The Out-of-Phase Fault Interruption Characteristics of Circuit Breakers for UHV Transmission Line

Bin Li[†], Qing-quan Li, Hong-shun Liu, Run-chang Li and Zhen Wang
*School of Electrical Engineering, Shandong University,
Jinan, Shandong Province, China
[†]Email: lbsdu2014@163.com*

Zhen Yang
*China Electric Power Research Institute,
Beijing, China*

With the extension of Ultrahigh Voltage (UHV) transmission line, hybrid reactive compensation will be widely used in the future. On the basis of the introduction of key components and the structure of hybrid reactive compensation, simulation analysis was carried out on the characteristics of circuit breaker when out-of-phase fault occurs in the UHV transmission line with hybrid reactive compensation. With regard to the worst condition of the out-of-phase fault, the variation rules of the out-of-phase fault transient recovery voltage of circuit breakers were obtained. Using equivalent lumped parameter circuit, the computational formula of the out-of-phase fault transient recovery voltage rising rate are deduced and computed. Based on the above method, the variation rules of the out-of-phase fault transient recovery voltage in UHV transmission line were studied under the condition with hybrid reactive compensation. In view of the out-of-phase fault interruption characteristics, the research results provided analysis foundation and theoretical basis for parameter optimization of the hybrid reactive compensation and arrangement selection of circuit breaker in UHV transmission line.

Keywords: Hybrid Reactive Compensation; Out-of-Phase Fault Circuit; Transient Recovery Voltage.

1. Introduction

Long-distance, large-capacity and low-loss power transmission is the inevitable choice to optimize resources distribution and UHV transmission technology has become the development orientation of the power industry [1]. UHV network makes it possible to realize the interconnection of regional power network and the optimal distribution of resources in a larger range, while the transmission power will experience a significant improvement and the reactive power will change frequently, so the safety and stability of the system will face both opportunities and challenges [2-3]. Meanwhile, it is an ideal solution to use hybrid reactive compensation with series compensation and shunt controlled reactor, which gives consideration to the increase of transmission power and the adjustment of reactive power. Since both of the inductance and capacitance are the energy storage components in power systems, transient process will occur when the working states change because of operations or faults [4]. In this paper, based on the status of the fault of the transmission line with hybrid reactive compensation, the electromagnetic transient program is used for modeling and simulation. On the basis of theoretical analysis and derivation, the mathematical expression of the relationship between the circuit breaker transient recovery voltage rise rate and the controllable high impedance and the series compensation capacitor is presented.

2. System Model

The hybrid reactive compensation contains the series compensation and graded controlled shunt reactor, its structure composition is demonstrated in Fig. 1, where series compensation consists of capacitor banks, Zinc Oxide arresters, spark gaps, bypass circuit breakers and damping devices. Only the capacitor banks are put into operation during normal operation [5].

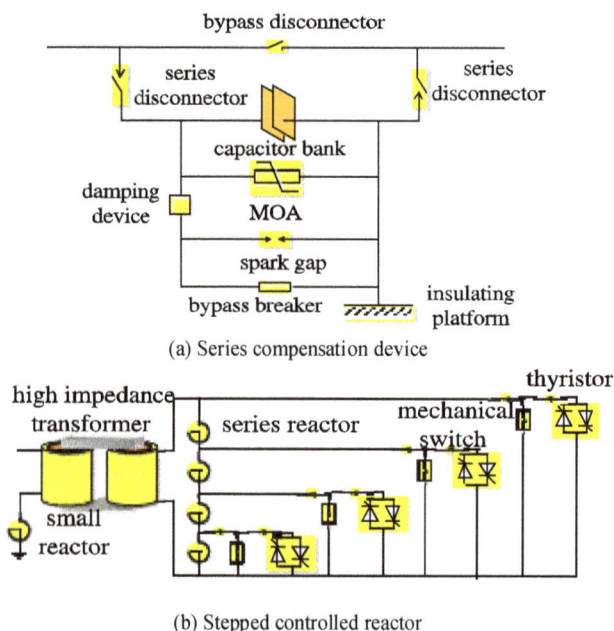

(a) Series compensation device

(b) Stepped controlled reactor

Fig. 1. Structure of hybrid reactive compensation.

The arrangement method of the hybrid reactive compensation in system model can be illustrated as follows: shunt reactors are installed at both circuit ends respectively, and their compensation degrees are assumed as 44%, while two series compensation devices with the compensation degree of 20% are installed at the both circuit ends. Supposed that the length of the circuit is 600km, the power bus voltage of the system is 1 087kV, and the three-phase short-circuit capacity is 50GVA. The main parameters of the transmission lines are showed in Table 1. To demonstrate the effects of the interrupting characteristics in the out-of-phase failure of high-voltage circuit breakers after installing the hybrid reactive compensation, no voltage-restriction measure is taken in calculation.

Table 1. Main parameters of 1 000 kV transmission line

Sequence	$R/(\Omega \cdot km^{-1})$	$L/(mH \cdot km^{-1})$	$C/(\mu F \cdot km^{-1})$
Positive Sequence	0.007 83	0.834 08	0.013 79
Zero Sequence	0.196 37	2.673 66	0.008 46

3. The Analysis Of The Characteristics Of Transient Recovery Voltage In Circuit Breaker Out-Of-Phase Failure

When out-of-phase fault occurs in power system, oscillation happens and the circuit breaker breaks the fault to get the system disconnected. When the circuit breaker breaks the out-of-phase fault, though the out-of-phase fault current is relatively low, the transient recovery voltage is extremely high, so the electric arc is difficult to extinct and using circuit breaker to break the out-of-phase is not easy. The most severe out-of-phase fault is that the voltage phases of two power system are opposite, which means the phase difference is 180 electrical angles. Under this circumstance, the circuit breakers installed in the tie lines should be reliable to break the tie lines, in case that the system collapses. So the reliability to break the out-of-phase faults is an important function to circuit breakers.

Aiming at the parameters listed in Table 1, the UHV with hybrid reactive compensation transmission line is simulated. When the out-of-phase fault current crosses zero, the circuit breakers break the line. The transient recovery voltage is determined by the value and the distribution of the inductors, capacitors and resistance in the circuit. The simulated waveforms of the transient recovery voltage after the installation of hybrid reactive compensation are demonstrated in Fig. 2.

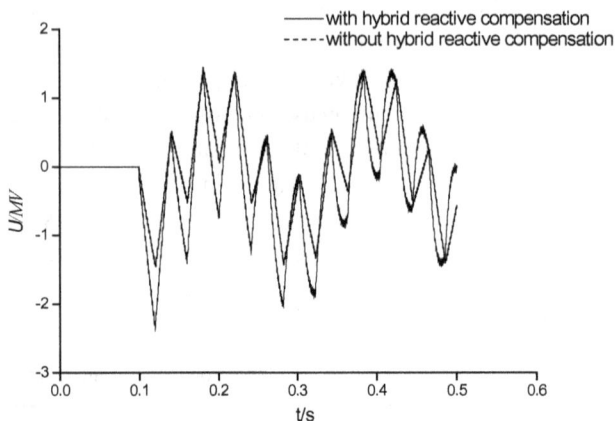

Fig. 2. Transient recovery voltage waveform of circuit breakers.

As can be seen from Fig. 2, in the breaking process of UHV transmission line out-of-phase fault, the transient recovery voltage peak value of break circuit rose significantly after installing hybrid reactive compensation. From the analysis and calculation of the waveform of the transient recovery voltage, the rising rate of recovery voltage increased obviously. As the results showed, after the installation of hybrid reactive compensation, the circuit transient recovery voltage and its rise rate rose significantly, which adds difficulties to circuit breakers to break the out-of-step fault, it also puts forward higher requirements to the insulation and stability of the system. To tell the concrete difference between the interrupting characteristics of circuit breaker out-of-phase fault before and after the installation of hybrid reactive compensation, it is necessary to analyze the transient recovery voltage in out-of-phase fault.

The installation of hybrid reactive compensation changes the structure of the UHV power system, and affects the interrupting characteristics of circuit breakers.

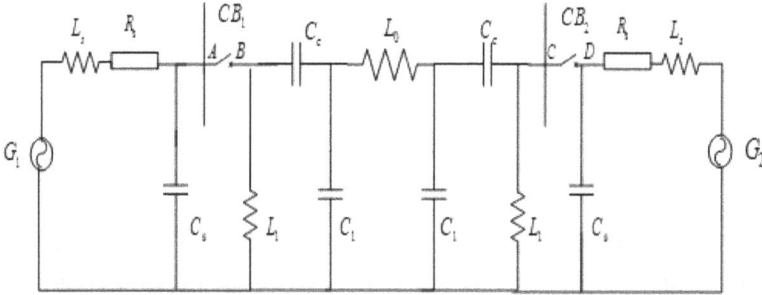

Fig. 3 Equivalent calculation circuit for transient recovery voltage of the out-of-phase fault.

The equivalent calculation circuit for transient recovery voltage of the out of phase fault is showed in Fig. 3, among which G_1 and G_2 are power supply, U_m is the amplitude of power supply phase voltage, CB_1 and CB_2 are circuit breakers, L_s, R_s, C_s are inductor, resistance, and shunt capacitance at the sending end separately, and C_c, L_1 are the series compensation capacitance and shunt high resistance. Assumed that the inductor and shunt capacitance per unit length of lines are represented by l_1 and c_1 respectively and the length of line is s, so the inductance of line and shunt capacitance are $L_0 = l_1 s$, $C_1 = c_1 s$ separately, The transient recovery voltage rising rate of fracture of circuit breaker $RRRV$ can be showed as:

$$RRRV = V_p / t_p \qquad (1)$$

$$V_p = u_{tr}\big|_{t=t_p} - u_{tr}\big|_{t=0} \qquad (2)$$

$$RRRV = \frac{\Delta u_A + \Delta u_C + \Delta u_L}{t_p} \qquad (3)$$

$$\Delta u_A = 2U_m\left[cos(100\pi \cdot s\sqrt{l_1 c_1}) - 1\right]$$

$$+2\frac{U_m L_s[cos\left(s\sqrt{\dfrac{l_1 c_1}{L_s C_s}}\right) - 1]}{2L_s + sl_1(1-\beta)}$$

$$\Delta u_C = -\frac{2U_m \beta l_1 s\left[cos\left(\omega^2 s^2 l_1 c_1\sqrt{\dfrac{\beta \cdot \gamma}{2}}\right) + 1\right]}{2L_s + sl_1(1-\beta)}$$

389

$$\Delta u_L = -\frac{2U_m l_1 s}{\left(2L_s + sl_1(1-\beta)\right)}$$

According to the situation of the special high voltage system, the failure of the step fault is analyzed. The simulation results are consistent with the calculation results.

Considering the tendency of transient recovery voltage rising rate changing with the compensation degree of capacitor in left of Fig. 4, series compensation capacitors with different compensation degrees have obvious effect on the transient recovery voltage rising rate of circuit breakers. As can be seen, the compensation degree of controlled shunt reactor is set at 0.44, after the compensation degree of series compensation exceed 0.5, the rise speed of the rising rate of transient recovery voltage accelerates, and the value exceeds the tolerance peak value established by 1100kV circuit breaker national standard.

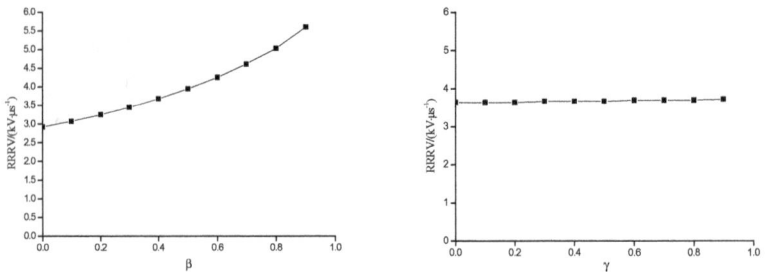

Fig. 4 Relationship between rising rate of transient recovery voltage and different compensation degrees of series compensation and controlled shunt reactor.

4. Conclusion

It is not advantageous to limit the value of the fault current when the series compensation installed in the UHV transmission line, which increase the difficulty to the circuit breaker open.

The peak value of transient recovery voltage and its rising rate increases with the increase of the compensation degree of series compensation in the hybrid reactive compensation. However, the change of the controllable high compensation degree has little effect on the peak value and the rise rate of the transient recovery voltage. The installed of hybrid reactive compensation will increase the difficulty of the circuit breaker, and put forward higher requirements on the breaking performance of the circuit breaker.

Acknowledgments

This work is funded by National Science Foundation of China (51507095), Shandong Provincial Science Foundation of China (BS2015NJ011) and China Postdoctoral Science Foundation (2015M572032) are hereby acknowledged.

References

1. Liu Zhenya. *Innovation of UHVAC transmission technology in China, Power System Technology*, Vol.37(3), pp. 1-8, (2013).
2. Zheng, Xiang, Zutao Ban, Liangeng, et al. *Analysis on Circuit Breaker Transient Recovery Voltage UHV AC Transmission Lines Fixed with Series Capacitors, High Voltage Engineering*,2013, 03:605-611.
3. Na Hu, Li Ya-mei, Wei Zhi-yong, et al. *Analysis and Study on Out-Of-Phase Fault of Circuit Breaker's Break-Off with Opening Resistance[J]. Electrotechnics Electric*, 2011, 12:13-18.
4. Zhou Pei-hong, Gu Din-xie, Dai Min, et al. *Research on Transient Recovery Voltage of UHV Circuit Breakers, High Voltage Engineering*, 2009, 02: 211-217.
5. Liu Hong-shun, Lü Ting-ting, Han Ming-ming, et al. *Overvoltage Amplitude and Frequency Characteristics of Energizing Off-load UHV Transmission Line with Hybrid Reactive Compensation, Power System Technology*, 2015, 07: 1970-1976.

APF Harmonic Detection Method Based on the Average Algorithm

You-hua Jiang, Jian Chang[†] and Shu-jin Tian

School of Electrical and Information Engineering, Shanghai University of Electric Power, Yangpu District, Shanghai 200090, China

[†]Email: 1353371903@qq.com

The application of APF in power transformer is analyzed. At the same time, the harmonic detect methods of instantaneous reactive power theory have some disadvantages. A novel detect method based on the average value is proposed, which can reduce the detecting delay and calculation time, so that the harmonic can be real-time and dynamic compensated. The correctness of arithmetic and the feasibility of converter design are verified by simulation results and experiments.

Keywords: Power Transformer; Active Power Filter; Average Value Algorithm.

1. Introduction

With the development of modern industrial technology, a large number of nonlinear loads in power system increases, the negative effect is increasingly obvious, a lot of harmonic and sub-harmonic components are poured into to the power grid, resulting in a serious AC voltage and current waveform severe distortion, increasing the degree of deterioration of power quality, Which not only affects the load's Safe and economic operation of power supply transformers, but also increased power transformers their loss [1, 2]. Therefore, it is necessary to take appropriate technical measures to reduce the impact of harmonics on the transformer, and enhance power supply reliability and energy-saving operation of power transformers.

2. Based on the Theoretical Basis of Power Transformers Eliminate Harmonic Energy-Saving Technology

Generally, transformer load loss is mainly composed of the hysteresis loss, eddy current loss of the core material. Hysteresis loss is a loss which repeatedly alternating ferromagnetic material magnetization process due to hysteresis generated. Area is proportional to the size and the hysteresis loop hysteresis loss, in other words, the smaller the area of the hysteresis loop, hysteresis loss also smaller. Furthermore, hysteresis loss is generated by the alternating magnetization, so its size is also related to alternating frequency f, f greater, the greater the loss. Specifically, the hysteresis loss can be calculated as the size of the PC [2].

$$P_c = C_1 B_m^2 \bullet f \bullet V \qquad (1)$$

The eddy current loss is due to the core itself is a metal conductor, so the metal induction phenomenon induced potential will produce circulation in the core, is the eddy. Due to eddy currents flowing in the core, and the core itself and its resistance, it caused the eddy current loss. Because the core of the eddy is a circulation EMF generated, and because the square is proportional to the eddy current losses and eddy current, so it is proportional to the square of EMF, and EMF depends on the rate of change of magnetic flux. Thus the eddy current loss is proportional to the square of the maximum magnetic flux density Bm and the square of the frequency f. In addition, the classical eddy current losses are related to the thickness and resistivity silicon. Specifically, the size of the classical eddy current loss can be calculated. [2]

$$P_w = C_2 B_m^2 \bullet f^2 \bullet V \qquad (2)$$

As can be seen from equation (1), (2), hysteresis loss and eddy current losses are related to their own material and alternating frequency, high harmonic currents will produce harmonics leakage magnetic field in the power transformer, and produces high harmonic losses in the winding and metal structure, making the internal power transformer produce local overheating. At the same time, computing power transformer winding leakage magnetic field necessary to consider the non-linear ferromagnetic material, but also consider the harmonic current (considering the 49th) the synthesis of non-sinusoidal variation with time. Therefore, the calculation of the power transformer leakage magnetic field is calculated the actual non-sinusoidal transient field, the research winding leakage magnetic and eddy current losses in the distribution of high harmonic currents of great significance. As the restul, the higher harmonic current frequency, the greater the loss. Thus reducing the harmonic current transformer, particularly high harmonics, you can greatly reduce the loss of the transformer, which is the theoretical basis of the proposed elimination of power transformers based on harmonic energy-saving technologies.

3. Algorithms commonly used in Active Filters

To make effective compensation active filter, it must accurately detect harmonic components of the load [3]. The most commonly used harmonic detection algorithm is the instantaneous reactive power [4]. Its block diagram as shown, it mainly consists of a phase locked loop, 3/2 transform, a rotation transformation, a low-pass filter, rotate counter-transform modules [5].

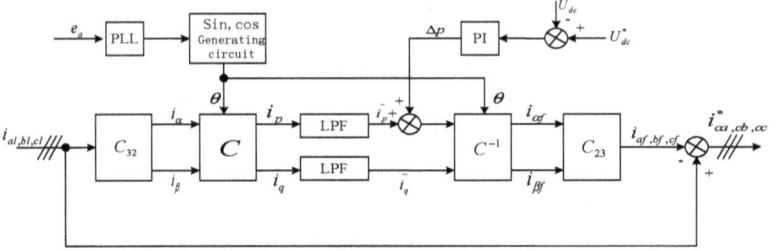

Figure 1. Active filter harmonic detection block diagram

4. Improved Detection Algorithm

The traditional method of instantaneous reactive power during harmonic extraction process, the need for a low-pass filter, calculates the volume is very large, resulting in a delay. Therefore, this article intends to use the improved detection algorithm. Set symmetrical three-phase load current

$$
\begin{pmatrix} i_{la} \\ i_{lb} \\ i_{lc} \end{pmatrix} = \begin{pmatrix} i_l(t) \\ i_l(t - 2\pi/3) \\ i_l(t + 2\pi/3) \end{pmatrix} \tag{3}
$$

After Fourier decomposition, there is formula

$$
\begin{pmatrix} i_{la} \\ i_{lb} \\ i_{lc} \end{pmatrix} = \sqrt{2} \bullet \begin{pmatrix} \sum_{k=0}^{\infty}\{I_{l(6k+1)}\sin[(6k+1)\omega t + \varphi_{6k+1}] + I_{l(6k+3)}\sin[(6k+3)\omega t + \varphi_{6k+3}] + I_{l(6k+5)}\sin[(6k+5)\omega t + \varphi_{6k+5}]\} \\ \sum_{k=0}^{\infty}\{I_{l(6k+1)}\sin[(6k+1)\omega t + \varphi_{6k+1} - 2\pi/3] + I_{l(6k+3)}\sin[(6k+3)\omega t + \varphi_{6k+3} - 2\pi/3] + I_{l(6k+5)}\sin[(6k+5)\omega t + \varphi_{6k+5} - 2\pi/3]\} \\ \sum_{k=0}^{\infty}\{I_{l(6k+1)}\sin[(6k+1)\omega t + \varphi_{6k+1} + 2\pi/3] + I_{l(6k+3)}\sin[(6k+3)\omega t + \varphi_{6k+3} + 2\pi/3] + I_{l(6k+5)}\sin[(6k+5)\omega t + \varphi_{6k+5} + 2\pi/3]\} \end{pmatrix} \tag{4}
$$

The three-phase currents are converted to the $\alpha\beta$ coordinate system can be expressed as

$$
\begin{pmatrix} i_\alpha \\ i_\beta \end{pmatrix} = \sqrt{3} \bullet \begin{pmatrix} \sum_{k=0}^{\infty}\{I_{l(6k+1)}\sin[(6k+1)\omega t + \varphi_{6k+1}] + I_{l(6k+5)}\sin[(6k+5)\omega t + \varphi_{6k+5}]\} \\ \sum_{k=0}^{\infty}\{-I_{l(6k+1)}\cos[(6k+1)\omega t + \varphi_{6k+1}] + I_{l(6k+5)}\cos[(6k+5)\omega t + \varphi_{6k+5}]\} \end{pmatrix} \tag{5}
$$

Then after rotation transformation can be obtained

$$
\begin{pmatrix} i_p \\ i_q \end{pmatrix} = \sqrt{3} \bullet \begin{pmatrix} \sum_{k=0}^{\infty}\{I_{l(6k+1)}\cos(6k\omega t + \varphi_{6k+1}) - I_{l(6k+5)}\sin[6(k+1)\omega t + \varphi_{6k+5}]\} \\ -\sum_{k=0}^{\infty}\{I_{l(6k+1)}\sin(6k\omega t + \varphi_{6k+1}) + I_{l(6k+5)}\sin[6(k+1)\omega t + \varphi_{6k+5}]\} \end{pmatrix} \tag{6}
$$

394

Formula: ω The power supply angular frequency; $I_{I(6k+1)}$, $I_{I(6k+3)}$, $I_{I(6k+5)}$, φ_{6k+1}, φ_{6k+3}, φ_{6k+5}, respectively, for the corresponding rms current times and phase angle, Where k is the non-negative integer. It should be noted, formula(3)-(6) only requires three-phase current I_{la}, I_{lb}, I_{lc} are symmetric, Whether of them contain zero-sequence component are not required.

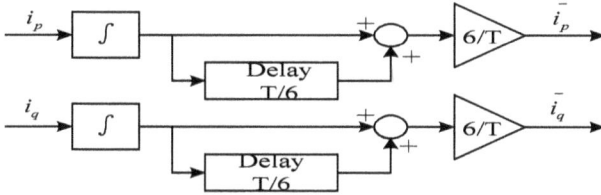

Figure 2. A schematic diagram of the average algorithm

From equation (6) can be seen, in addition to the DC component in i_p, i_q, the cycle period of AC component is 1/6 of the power supply, i.e., the AC component in the 1/6 line cycle average value is zero. This can be obtained by averaging algorithm, the DC component i_p, i_q in i_p, i_q, as shown in figure. Figure 2, T is a power cycle, this detection algorithm with a simple integral, delay plus gain links to replace the traditional low-pass filter to reduce the delay to the harmonic current detection method 1/6 a power cycle, but also reduces the amount of calculation, so that the operation speed of the controller can reduce the requirements.

The average of algorithm analysis and presentation is presented above, but the actual digital processing is within 1/6 of a power cycle which is sampled, accumulate, and then take the average, so you need more storage data variables. In addition, according to the traditional FIFO serial shift accumulates DSP also take a lot of time, increasing the burden on the control chip DSP, the following will introduce a parallel shifted to calculate the average covering method, you can omit a lot of DSP computation time.

Assuming 15K Hz sampling period, you need to save power cycle within 50 1/6 months of data, the need to open up 50 data variables, namely $D_0, D_1, D_2, D_3, D_4, D_5 .... D_{45}, D_{46}, D_{47}, D_{48}, D_{49}$.

The traditional way is a first-in first-out serial processing, as shown in Figure 3 (a) shows, a new sampling data, put the top of the discard data, so each time data processing, you need to shift 50 times, take a lot of DSP processing time; improved process is shifted in parallel overlay, 3 (b), the new data as shown in a sample, put the top data coverage, then a data sample is then covered

395

with the back of the data, this treatment does not need to be shift, and other data covering the last one, the count is cleared, the new data overwrites the old data.

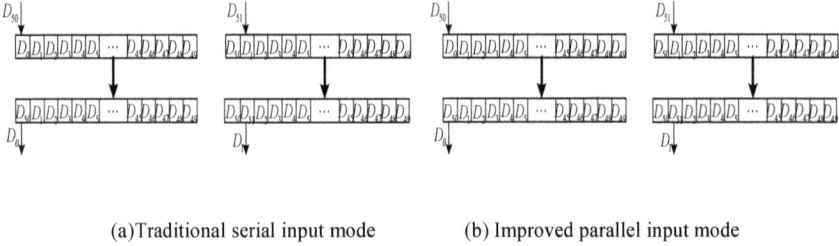

(a)Traditional serial input mode   (b) Improved parallel input mode

Figure 3. Contrast operations of two data

Specific algorithm structure can be seen, the active filter shown in Figure 4. Fundamental wave, as extracted; the figure i_{al}, i_{bl}, i_{cl}, for the detected load current; for after active after coordinate transformation, i_p, i_q are the reactive component (including DC, AC component);after averaging algorithm for the DC component i_p, i_q after component; i_{af}, i_{bf}, i_{cf} is the fundamental wave component extracted from; the i_{ca}, i_{cb}, i_{cc} are instruction current component; They are U_{dc}, U_{dc}^* the DC bus detection component and a given reference component to compensate for the power loss, and the DC bus voltage is usually 800V voltage control; e_a is the A-phase grid voltage, phase-locked loop through, get precise grid frequency and phase of the signal; C_{32} represents a three-phase to two-phase coordinate system coordinate transformation, C_{32} is a two-phase to three-phase transformation, C is the rotation transformation, C^{-1} is the inverse transform rotation.

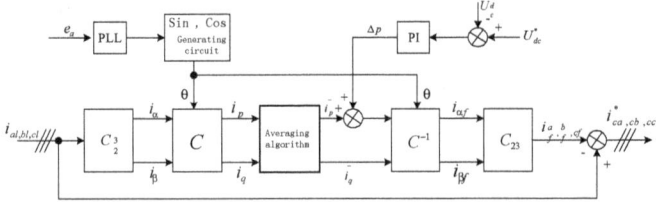

Figure 4. Active filter algorithm block diagram based on the average method

5. Simulation and Experiment

In this paper, the structure of PSIM are established by simulation software system as shown in Fig. 5, the system parameters for the power transformer capacity is 200KVA, active filter capacity is 50KVA, load for three-phase bridge rectifying means, an inductance of 0.3 mH, resistance 0.4 ohms.

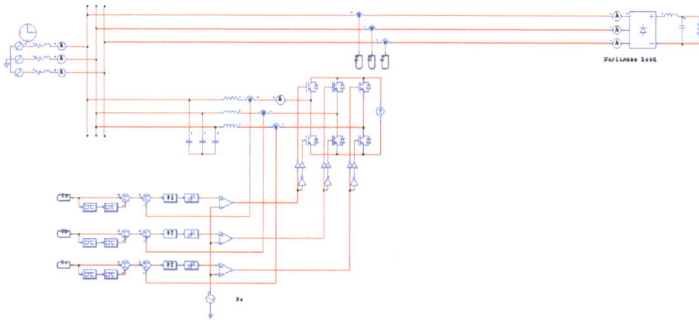

Figure 5. Power transformer active filter device system simulation block diagram

According to the system the simulation model built in Fig. 6, and then based on the formula (3) to (6) of the averaging algorithm simulation, the simulation results are shown in Fig. 6. Fig. 6 (a) for the three phase load current, comprising 6k ± 1 (k is a positive integer) harmonic; Fig. 6 (b) is active, both containing a DC component, also contains an AC component, and 1/6 of a period of the AC component of the grid cycle; Fig. 6 (c) for the active power DC component, which is active through income after averaging algorithm, there are about 3.3 millisecond delay, and the AC component is effectively filtered out, which, after the inverse transform, can extract the fundamental RMS load current. It can be seen, the average of the algorithm not only reduce the amount of computation, but also less delay, you can achieve dynamic compensation active filter. Figure 8 is a system simulation rendering, it can be seen from Figure 7, although the load current harmonic content is relatively large, but through active filter harmonic filter, the power transformer near sinusoidal output current, less harmonic content so no additional load harmonic power transformer losses.

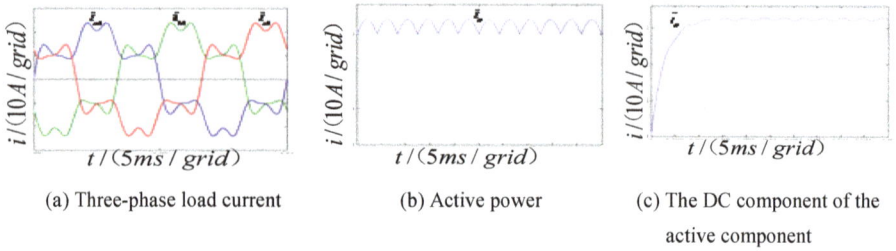

(a) Three-phase load current　　　　(b) Active power　　　　(c) The DC component of the
active component

Figure 6. Improved algorithm of active filter simulation

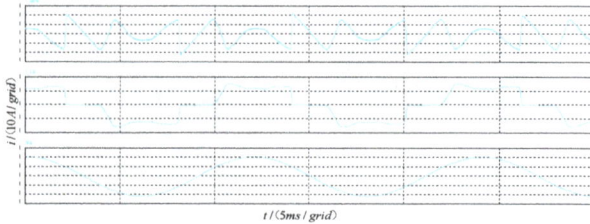

Figure 7. Simulation results

Figure 8 shows the measured waveform 10A prototype, in which Figure 8 (a) for the PWM pulse drive waveform can be seen from the waveform, PWM pulses using hysteresis control, so the PWM cycle is not fixed, but variable. Figure 8 (b) to compensate the harmonic waveform in Figure 8 (c) is linked to network operation waveform, timely, fast load harmonic compensation.

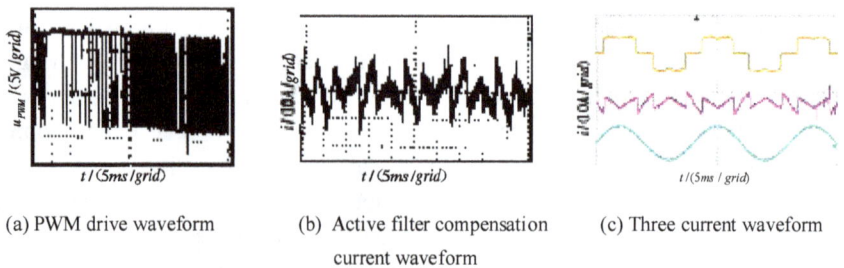

(a) PWM drive waveform　　　　(b) Active filter compensation　　　　(c) Three current waveform
current waveform

Figure 8. Active filter experimental waveform

6. Conclusion

With the rapid development of the national economy, the increase of non-linear loads will affect the safety and economic operation of power transformers. If the

398

active power filter device for power transformers for harmonic elimination, not only can increase the reliability of electricity supply transformer, but each transformer can reduce energy consumption by about 10%, and which is of great significance for advocating energy-saving and environment-friendly society.

Acknowledgement

Project is funded by construction of local capacity in Shanghai City (14110500900).

References

1. Fang Wen. EHV load tap reactive power flow regulator. Energy Technology, 2002, 23(1): 37-39.
2. Yi Ke Ning. Transformer Design Principles [M]. Beijing: China Electric Power Press, 2006.
3. Xia Xiang Yang, Zhang Yi Bin. Novel Type Hybrid Active Power Filter and Its Application in High Voltage Substations [J]. Grid technology, 2006, 30(1): 75-79.
4. Dong Mi, Cheng Jan, Luo An. Parallel hybrid active filter current tracking iterative learning control [J]. China Electrical Engineering, 2006, 26(9): 104-107.
5. Liu Feng Jun. Mains power quality compensation technology [M]. Beijing: Science Press, 2005.

Research of No-Contact Tap-Changer On-Load Voltage Regulator in 10 KV Distribution Power Network

You-hua Jiang, Shu-jin Tian† and Jian Chang

College of Electronic and Information Engineering, Shanghai University
of Electric Power, Shanghai, China
†E-mail: sophia0103@163.com

There are some disadvantages in mechanism tap-changers, and this paper proposes a new 10KV no-contact tap-changer on-load voltage regulator. The compensated principle is used to regulate output voltage and the switch process is completed by SCR, so the dynamic response is fast and the harmonic is little. In addition, the electric arc is not produced. At the same time, the advanced combination technology and optimum topology are adopted. The compensated voltage precision is high and have more grades. The relationship between compensated voltage and switch stage is also given. Based on the principle and topology, the power flow is analyzed and introduced to reveal the inner running essence. The correctness of power flow analysis and the feasibility of circuit design are verified by experiment results.

Keywords: Voltage Regulator, Compensated Principle, Power Flow Analysis.

1. Introduction

No-contact tap-changer on-load voltage regulator [1] has the advantages of both traditional on-load voltage regulator and ac voltage-stabilizer, and is an ideal voltage regulator device for distribution network [2].

2. The design of no-contract tap-changer on-load voltage regulator

2.1. *Compensation principle of no-contact tap-changer on-load voltage regulator*

Figure 1 illustrates that:

$$U_o = U_i \pm \Delta U \tag{1}$$

According to the cosine theorem, we can get :

$$U_o = (U_i^2 + \Delta U^2 + 2U_i \Delta U \cos \alpha)^{\frac{1}{2}} \tag{2}$$

Where α is the angle between Ui and ΔU.

Eq. (1) shows that we can adjust the magnitude and phase of output voltage U_o by changing the size of the compensation voltage ΔU or the input voltage U_i angle. Generally, we adjust compensation voltage ΔU, and the phase is in phase with the input voltage U_i, or anti-phase, so as to remain the voltage U_o unchanged.

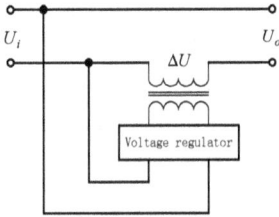

Fig. 1. Block diagram of a voltage regulator   Fig. 2. Voltage vector superposition

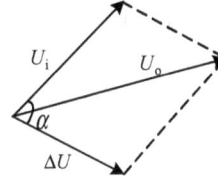

Eq. (2) shows that we can adjust the magnitude and phase of output voltage Uo by changing the size of the compensation voltage ΔU or the input voltage Ui angle. Generally, we adjust compensation voltage, and the phase is in phase with the input voltage Ui, or anti-phase, so voltage Uo remains unchanged.

As shown in Figure 3(a), when the input voltage Ui decreases by ΔU, to maintain output voltage Uo, we can add $+\Delta U$ (positive phase compensation voltage) through the compensation transformer. As shown in Figure 3(b), when the input voltage Ui rise ΔU, we can compensate with reverse voltage $-\Delta U$ (reversal phase compensation voltage) through the compensation transformer to maintain output voltage Uo .

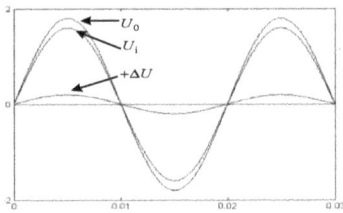

(a) Cophase compensation   (b) Anti-phase compensation

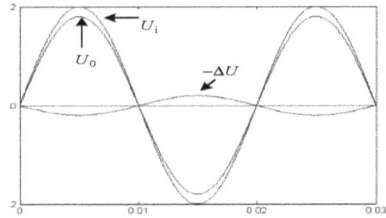

Fig. 3. Schematic diagrams of the voltage compensation

2.2. No-contact Tap-changer On-load Voltage Regulator's Main Circuit Topology and Work Principle[3]

As can be seen from Figure 4, the voltage regulator is composed of regulating transformer T1, compensating transformer T2, TRIAC regulation circuit and shoring circuit. The regulating transformer T1's primary side connects to 10KV distribution network, and its secondary side has multiple taps, respectively, 50V, 150V, 100V, with the dotted terminals marked with " "; the compensating transformer T2's turns ratio is 5:1; TRIAC regulation circuit contains SCR1--SCR8, FUSE1--FUSE8, of which TRIAC is used to switch the taps to provide different compensation voltages, and absorption circuits are in parallel to prevent the breakdown voltage destroying TRIAC; fast acting fuse FUSE1--FUSE8 protect components against overcurrent fault. And shorting circuit is composed of SCR9 and current limiting reactor L9, whose function is to provide a loop for T2's secondary side when the TRIAC switching circuit is in transient state, to prevent the compensating transformer open. In Figure 4, U_i represents the input voltage, i_1 represents input current; U_o represents the output voltage, i_2 represents output current; U_1 represents compensating transformer T2 primary voltage, and U_2 represents the compensation transformer T2 secondary voltage, i_3 is compensating transformer T2's primary current.

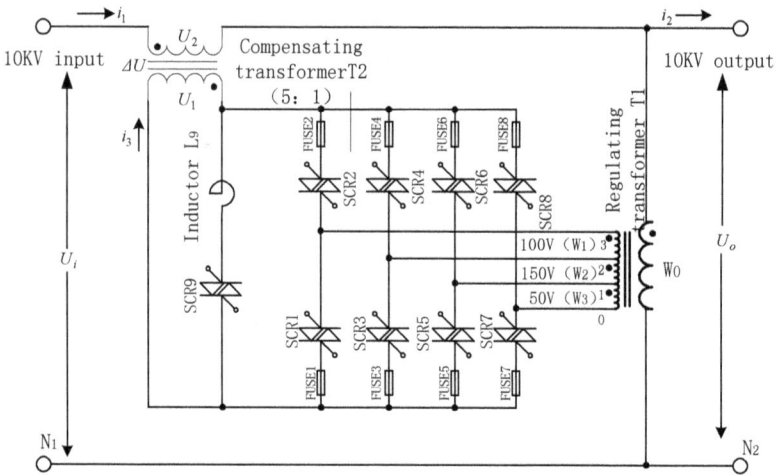

Fig. 4. Main circuit topology of no-contact tap-changer on-load voltage regulator for 10KV distribution network

2.3. No-contact Tap-changer On-load Voltage Regulator's Main Circuit Topology and Work Principle

When the output voltage is higher than the rated value (equivalent to the input voltage increases or load decreases), such as 11000V input voltage, it requires compensating transformer T2 to produce reverse polarity voltage 1000V to offset the increased voltage, then we can trigger SCR8, SCR3, on AC positive half cycle, the current path is: regulating transformer T1 tap 0---TRIAC SCR8--- compensating transformer T2--- TRIAC SCR3--- regulating transformer T1 tap 2; on AC negative half cycle, the current flows along the pathway back (due to the use of Triac).

Table 1. Triac switch status and compensation voltage

Voltage range of U_i (V)	Corresponding trigger TRIAC number	Offset voltage (Unit: V)	Voltage range of U_o (V)
$11500 < U_i \leq 11750$	SCR7, SCR2	1 Level(-1500)	$10000 < U_o \leq 10250$
$11250 < U_i \leq 11500$	SCR5, SCR2	2 Level(-1250)	$10000 < U_o \leq 10250$
$11000 < U_i \leq 11250$	SCR7, SCR4	3 Level(-1000)	$10000 < U_o \leq 10250$
$10750 < U_i \leq 11000$	SCR5, SCR4	4 Level(-750)	$10000 < U_o \leq 10250$
$10500 < U_i \leq 10750$	SCR3, SCR2	5 Level(-500)	$10000 < U_o \leq 10250$
$10250 < U_i \leq 10500$	SCR7, SCR6	6 Level(-250)	$10000 < U_o \leq 10250$
$10000 < U_i \leq 10250$	SCR$_k$, SCR$_{k+1}$ (k=1,2...7)	7 Level(0)	$10000 < U_o \leq 10250$
$9750 < U_i \leq 10000$	SCR6, SCR7	8 Level(+250)	$10000 < U_o \leq 10250$
$9500 < U_i \leq 9750$	SCR2, SCR3	9 Level(+500)	$10000 < U_o \leq 10250$
$9250 < U_i \leq 9500$	SCR4, SCR5	10 Level(+750)	$10000 < U_o \leq 10250$
$9000 < U_i \leq 9250$	SCR4, SCR7	11 Level(+1000)	$10000 < U_o \leq 10250$
$8750 < U_i \leq 9000$	SCR4, SCR7	12 Level(+1250)	$10000 < U_o \leq 10250$
$8500 < U_i \leq 8750$	SCR2, SCR7	13 Level(+1500)	$10000 < U_o \leq 10250$

As can be seen from Table 1, the input voltage fluctuation range ± 15%, after voltage regulator's compensation, the output voltage range from 0 to 2.5%; In addition, voltage regulator's compensation is divided into 13 stalls, 2.5% per file, so the accuracy of offset voltage is relatively high.

3. Power Flow Analysis

10KV distribution network no-contact tap-changer on-load voltage regulator operating status is complex, but it is still comply with the input and output power balance principle. Assuming that the loss is zero, then the input power of voltage regulator must be equal to the output power, namely

$$U_i \bullet i_1 = U_o \bullet i_2 \qquad (3)$$

Since there is voltage regulator, so that $U_o = U_r$ (is the reference voltage $U_r = 10KV$), so we can get: $U_i \bullet i_1 = Ur \bullet i_2$ OR $i_1 = Ur \bullet i_2 / U_i$
When $U_i > U_o = U_r$, for example, $U_i = U_r(1+15\%)$.

403

$$i_1 = \frac{U_r \bullet i_2}{U_r(1+15\%)} = \frac{i_2}{115\%} = i_2(1-13.1\%) \qquad (4)$$

By voltage regulator principle we can know that when the input voltage increases, it needs to compensate with negative input voltage. Compensating transformer T2's secondary side produces a negative offset voltage, as shown in Figure 5(a). We can get $U_i - \Delta U = U_r(1+15\%) - U_r(15\%) = U_r$. The power that compensating transformer absorbed from the distribution network is:

$$\Delta U \bullet i_1 = \Delta U \bullet i_2(1-13.1\%) = U_r(15\%) \bullet i_2(1-13.1\%) \approx U_r i_2(13.1\%) \qquad (5)$$

Convert it into current:

$$\frac{\Delta U i_1}{U_r} = \frac{U_r(15\%) \bullet i_2(1-13.1\%)}{U_r} \approx i_2(13.1\%) \qquad (6)$$

The output current is:

$$i_1 + i_2(13.1\%) = i_2(1-13.1\%) + i_2(13.1\%) = i_2 \qquad (7)$$

By the Eq. (4) – Eq. (7), we find that when we maintain the balance of output power and input power (assuming that there is no loss), since the distribution net input voltage U_i rises $U_r(15\%)$ leaving the distribution network to reduce the input current $i_2(13.1\%)$. Distribution network receives the compensation voltage from voltage regulator and then makes output voltage, become $U_o = U_i - \Delta U = U_r(1+15\%) - U_r(15\%) = U_r$, and transform the power absorbed by transformer T_2 from distribution network into putout current $\Delta U \bullet i_1 = U_r(15\%) \bullet i_2(1-13.1\%) = U_r i_2(13.1\%)$. So that ensure that the input power is equal to the output power. When $U_i < U_o = U_r$, for example, $U_i = U_r(1-15\%)$, then $i_1 = \frac{U_r \bullet i_2}{U_r(1-15\%)} = \frac{i_2}{85\%} = i_2(1+17.6\%)$.

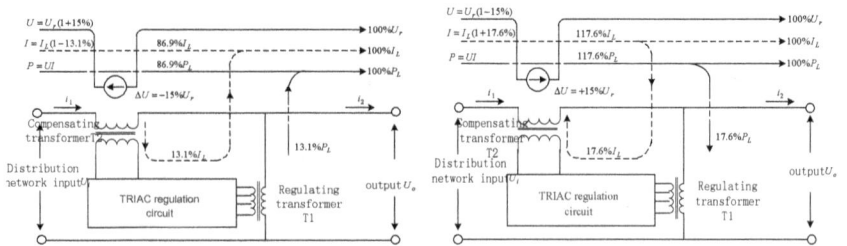

(a) Distribution network voltage increases 15%   (b) Distribution network voltage reduced by 15%

Fig. 5. Power flow diagram when distribution network input voltage fluctuates 15%

As mentioned above, transformer T2's secondary side produces positive compensation $\Delta U = +U_r (15\%)$, as showed in Figure 5(b). At the same time $U_i + \Delta U = U_r(1-15\%)+U_r(15\%) = U_r$, the power compensation from transformer T2 for distribution network is:

$$\Delta U \bullet i_1 = \Delta U \bullet i_2(1+17.6\%) = U_r(15\%) \bullet i_2(1+17.6\%) \approx U_r i_2(17.6\%) \quad (9)$$

Which can be inverted into current:

$$\frac{\Delta U i_1}{U_r} = \frac{U_r \bullet i_2(17.6\%)}{U_r} = i_2(17.6\%) \quad (10)$$

so that the output current is:

$$i_1 - i_2(17.6\%) = i_2(1+17.6\%) - i_2(17.6\%) = i_2 \quad (11)$$

4. Conclusion

Taking the investment, the safety and reliability factors of 10KV distribution system of the experiment into account, the unit experimented on 220V single-phase prototype. Experimental device capacity is 10KVA, and there are 7 compensation grades which offset voltage levels are + 30, + 20, + 10, 0, -10, -20, -30. Regulating transformer T1 and compensating transformerT2 capacity are both 1.8KVA, thyristor SCR1-SCR8 select 50A, 1000V solid-state relays. Figure 6(a) shows a waveform with phase compensation, when the input voltage is 200 volts, the positive phase compensation voltage is 19V; Figure 6(b) is a waveform with phase reversal compensation, the input voltage is 239 volts, the offset voltage is 20V.

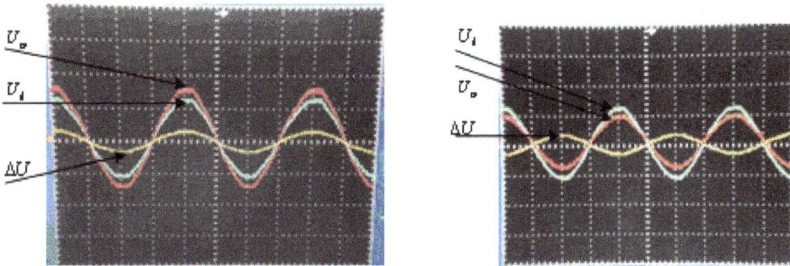

    (a)   Cophase compensation   (b) Anti-phase compensation

Fig. 6. Experimental waveforms (Ordinate: 200 volts / division; abscissa: 5 ms / div)

Based on the principle and topology, the power flow is analyzed and introduced to reveal the inner running essence. The correctness of power flow analysis and the feasibility of circuit design are verified by experiment results.

Acknowledgment

This project is funded by Shanghai local capacity building project (Project number: 14110500900).

References

1. N. Yorino, M. Danyoshi, M. Kitagawa. Interaction among multiple controls in tap change under load transformers [J]. *IEEE Transactions on Power Systems*, 1997, 12(1):430-436.
2. Ming Tsung Tsai. Design of a Compact Series-Connected AC Voltage Regulator With an Improved Control Algorithm [J]. *Transaction on industrial Electronics*, 2004, 51(4): 933-936.
3. Xiaoming Li, Qingfen Liao, Xianggen Yin, et al. A new on-load tap changing system with power electronic elements for power transformers [C]. *Power System Technology, 2002. International Conference on Power Con* 2002: 556-559.

Study on Feature Extraction Method of Mechanical Vibration Signal for Transformer Winding

Xue-bin Li[†], Tong Luo, Bin Li, Tie Guo and Bin Zhang
Electric Power Research Institute of State Grid Liaoning Electric Power Co., Ltd. Shenyang, Liaoning 110006, China
[†]*E-mail: lixuebin1985@163.com*

Chong Li
SIASUN Robot & Automation Co., Ltd.
Shenyang, Liaoning 110168, China
E-mail: 286402065@qq.com

The deformation failure and mechanical vibration is closely related to power transformer windings. The problem of a transformer is poor accuracy in the diagnosis of mechanical deformation and difficult to accurately judge the winding fault type. In this paper, as the research object of the S11-M-500/35 type distribution transformer, to obtain the vibration information through the obstructing of transformer winding insulation, low voltage winding compression and loose winding deformation failure. Two signal processing methods, wavelet packet energy spectrum entropy and the short-time Fourier transform are adopted respectively to extract the feature of different deformation information. Using the method of fuzzy c-means clustering analysis to compare the classification of the two methods of feature extraction to determine the one with more effectiveness. The results demonstrate that the wavelet packet energy spectrum entropy feature extraction method can achieve the best classification results for the winding failure deformation type. Its membership degree is above 0.94, providing the basis for the later mechanical failure diagnosis of winding type.

Keywords: Transformer winding; Vibration signal; Wavelet packet energy spectrum entropy; Short-time Fourier transform; Fuzzy C clustering analysis.

1. Introduction

In the recent years, mechanical vibration method has become a major concern of the winding detection in the world. The extract result of transformer winding vibration signal has effect on diagnosis model design and performance, choosing the best extraction method is important to the fault diagnosis [1-2].

In this paper, as the research object of the S11-M-500/35 type distribution transformer, through the obstructing of transformer winding insulation, low voltage winding compression and loose winding deformation failure to obtain the vibration information. Two signal processing methods that wavelet packet energy spectrum entropy and the short-time Fourier transform are adopted respectively to extract the feature of different deformation information, using method of fuzzy c-means clustering analysis to compare the classification of the two methods of feature extraction which one is the more effectiveness, and then to provide a theoretical basis for the winding faults type diagnosis.

2. Transformer Winding Fault Set and Vibration Signal Collection

2.1. *Transformer winding fault set*

To obtain the power transformer different types winding fault information, The typical flaw is designed to simulate the different faults of winding. Figure 1 show a physical picture in the different winding fault setting.

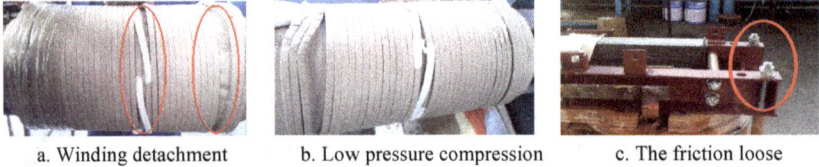

a. Winding detachment   b. Low pressure compression   c. The friction loose

Fig. 1. The different deformation fault maps of Transformer winding.

2.2. *Transformer winding fault vibration signal collection*

The transformer vibration test system is composed of vibration acceleration sensor, dynamic Data Acquisition Instrument, signal test analysis software and other parts.

The transformer winding insulation off, low voltage winding compression and winding loose deformation fault to obtain the vibration information. Figure 2 shows two groups of measured vibration data in the three states.

3. Transformer Winding Vibration Signal Feature Extraction

Transformer winding vibration signal belong to varying signal of non-stationary cycle that contains a wealth of information feature. To choose the best feature extraction method, this paper compares the two methods that wavelet packet energy entropy and STFT to obtain vibration characteristics by using typical time-frequency domain analysis.

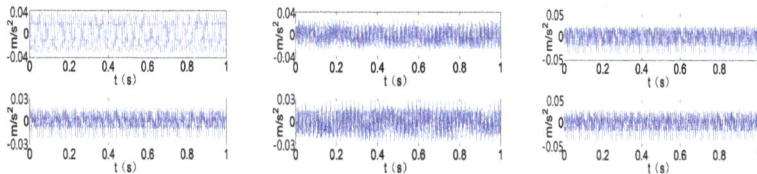

a. Winding detachment   b. Low pressure compression   c. The friction loose

Fig. 2. Transformer winding vibration signal pictures in the three states.

3.1. Wavelet packet energy entropy extract vibration characteristics

Wavelet packet energy entropy belong to time-frequency analysis method of nonlinear, it has a good time-frequency localization characteristics and the signal adaptive ability, and thus capable of varying signal decomposition effectively. In the same frequency band, vibration signal energy will be a corresponding change in different states, according to the wavelet energy entropy feature vectors to reflect different states of vibration characteristics of transformer windings [3].

The winding vibration information obtained mainly in the fundamental frequency at 100Hz by using the time-frequency analysis method of nonlinear, accompanied by low-frequency harmonic multiples, 500Hz attenuated to zero. Therefore, a septal point sampling method is used to sample frequency envelope signal that decomposed down to 2kHz.

The extracted energy entropy in Figure 2, with the wavelet packet method, the and the result is showed in Figure 3.

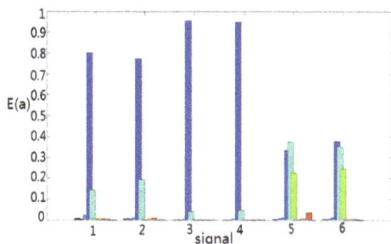

Fig. 3. Vibration signal extraction energy entropy in different winding states.

3.2. Short-time Fourier transform extract vibration characteristics

Short-time Fourier Transform (STFT) is the essence of non-stationary time-varying signal windowing truncation processing, this approach can be viewed as

409

local smoothing process, and then Fourier transform. It will discrete Fourier transform in practical applications.

The power spectrum of the signal is usually used for classification in the character extraction. Every peak value in the power spectrum reflects the energy measure of corresponding frequency. The fault type can be diagnosed and classified, because changes of peak value can be result in energy transformation of the faulted frequency component.

Shown in figure 3, the power spectrum of vibration signal of the windings in various statuses is extracted through the short-time Fourier transform. The length of characteristic is set 256, shown in Figure 4.

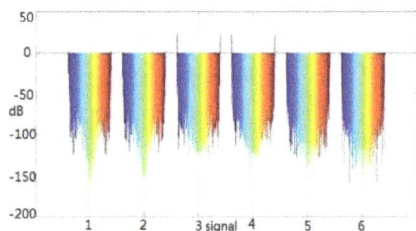

Fig. 4. Vibration signal extraction STFT in different winding states.

4. Comparison of The Effectiveness of Feature Extraction

In this paper, cluster analysis are using to compare the two methods, using the best features of the winding vibration signal feature extraction classification.

Clustering analysis is the process that refers to a collection of different objects grouped by similar objects of multiple classes, is an unsupervised self-learning method. Fuzzy C-means clustering (FCM) is the Clustering Algorithm that refers to the membership to determine the extent of each of the data belonging to a cluster, is an improvement of C-means clustering (HCM) method. The FCM uses the given data point membership interval (0,1) to determine the extent of each of the groups. The membership degree of Sparse Matrix U values allowed in (0,1), which is the total is equal to 1.

$$\sum_{i=1}^{c} u_{ij} = 1, \quad \forall j = 1,...,n \tag{1}$$

410

Wherein: u_{ij} shows data points the membership in the cluster center. The expression of the objective function of FCM:

$$J(U, c_1, ..., c_c) = \sum_{i=1}^{c} J_i = \sum_{i=1}^{c} \sum_{j}^{n} u_{ij}^m d_{ij}^2 \qquad (2)$$

Wherein: $c_1, \cdots, c_c$ is cluster center of the fuzzy group; U is membership Sparse Matrix, $d_{ij} = \|c_i - x_j\|$ is the euclidean distance between cluster center of the i-th and data points of j-th; $m \in [1, \infty)$ is weighted index.

The input parameters derivation, the objective function achieves to the minimum requirement, computing cluster center and membership matrix equation:

$$c_i = \frac{\sum_{j=1}^{n} u_{ij}^m x_j}{\sum_{j=1}^{n} u_{ij}^m} \qquad u_{ij} = \frac{1}{\sum_{k=1}^{c} \left(\dfrac{d_{ij}}{d_{kj}}\right)^{2/(m-1)}} \qquad (3)$$

According to the fuzzy C-means clustering analysis, the two methods of transformer winding vibration signal extraction are compared under different conditions; transformer winding vibration data are compared to judge the accuracy of clustering under different conditions. The clustering analysis of results that uses wavelet packet energy entropy and STFT transform as shown in Table 1 to Table 2.

Table 1. The result of wavelet packet energy entropy characteristic extract FCM.

The actual state	Fuzzy membership matrix			Classification
1 Winding detachment	0.982	0.012	0.006	1
1 Winding detachment	0.921	0.016	0.063	1
1 Winding detachment	0.976	0.023	0.001	1
2 low pressure compression	0.013	0.978	0.009	2
2 low pressure compression	0.032	0.941	0.005	2
2 low pressure compression	0.008	0.978	0.014	2
3 The friction loose	0.003	0.007	0.99	3
3 The friction loose	0.011	0.002	0.987	3
3 The friction loose	0.028	0.007	0.965	3

By using the fuzzy C clustering analysis method, the results of the membership matrix of three feature extraction can be obtained: STFT feature

411

value is lower than wavelet packet energy entropy what make difference of the actual status, brings to the mistake the judgement in the research. Power spectrum entropy of little wave packe is more accurate in classification than other method of character extraction. The membership result is higher than 0.94, and the result of cluster is accordance to the actual state of the winding type.

Table 2. The analytical result of STFT characteristic extract FCM.

The actual state	Fuzzy membership matrix			Classification
1 Winding detachment	0.270	0.527	0.203	2
1 Winding detachment	0.972	0.017	0.011	1
1 Winding detachment	0.268	0.519	0.213	2
2 low pressure compression	0.037	0.637	0.326	2
2 low pressure compression	0.061	0.409	0.53	3
2 low pressure compression	0.320	0.581	0.099	2
3 The friction loose	0.153	0.692	0.155	2
3 The friction loose	0.118	0.592	0.29	2
3 The friction loose	0.054	0.327	0.619	3

5. Conclusion

(1) By building the winding vibration signal acquisition platform, the obtained vibration signal of windings in different deformation is mainly concentrated in the fundamental frequency of 100Hz, associated a high harmonic of multiple fundamental frequencies, attenuated to zero at 500Hz.

(2) By using the fuzzy C clustering analysis method, the results of the membership matrix of three feature extraction can be obtained: Power spectrum entropy of little wave packe is more accurate in classification than other method of character extraction. The membership result is higher than 0.94, and the result of cluster is accordance to the actual state of the winding type.

Acknowledgment

This work is funded by Science and Technology Project of State Grid Corporation (2016YF-26).

References

1. Li Jun, Dong Liwen, Zhao Hong. Aging and life assess-ment of oil-immersed transformer [J]. High Voltage Engineering, 2007, 33(3): 186-189.
2. Liao Ruijin, Meng Fanjing, Zhou Nianrong, et al. Assessment strategy for inner insulation condition of power transformer based on Set-pair analysis

and evidential reasoning decision-making [J]. High Voltage Engineering, 2014, 40(2): 474-481.

3. Xu Jianyuan, Zhang Bin, Lin Xin, et al. Application of Energy Spectrum Entropy Vector Method and RBF Neural Networks Optimized by the Particle Swarm in High-voltage Circuit Breaker Mechanical Fault Diagnosis [J]. High Voltage Engineering, 2012, 38(6): 1299-1306.

Influence of Splashing Water on Insulator Hot Washing

Guo-zhi Wang, Hao-qi Peng[†], Jian Ke and Lan-ying Yu

School of Mechanical Engineering, Southwest Jiaotong University,
Chendu Sichuan, 610031, China
[†]E-mail: haoqipeng@my.swjtu.edu.cn

Hot washing has been applied with more and more frequency on power systems and splashing water has a great influence on hot washing cleaning effect, but the corresponding research was rare. Simulation analysis and theoretical research were carried out to analyze the influence of splashing water on insulator hot washing cleaning effect. The results demonstrated that splashing water will increase the cleaning range and improve the cleaning efficiency, but may also cause flashover occurrence, affecting the security of hot washing; leading to the change of cleaning mechanism. The main reason of cleaning had changed from internal water pressure to directly shear pressure. It also can improve the adhesion force as it caused high relative humidity around the contamination, but having little influence on dynamic pressure.

Keywords: Splashing Water; Hot Washing; Cleaning Mechanism; Adhesion Force.

1. Introduction

Hot washing is a kind of live working that is used to clean the electric power equipment with pressure water, because of the advantages of simple device, high efficiency, good effects on prevent flashover, and significant economic benefits, and it is wildly applied on Power Systems. The mechanism of hot washing is that the kinetic energy of water column will be converted into the force on contamination, making contamination fall off and finish the cleaning work. After the injection of water from the nozzle, the jet fluid exchange energy and mass with air, makes the jet cross-section expanding with the increase of downstream distance, and eventually the entire cross-section is isolated by air into water droplets. In the process of hot washing, the cleaning work is mainly completed by the main body of the jet, but the splashing water that produced by jet impact with air and insulator surface still has an important influence on cleaning effects and efficiency. However, the research about the influence of splashing water on insulator hot washing was rarely, and usually just alluded to, never discussed in detail. Simulation theoretical studies were carried out in this paper, and analyze the influence of splashing water on insulator hot washing cleaning effect on three aspects.

2. Sources, Composition and Structure of Insulator Contamination

The components of insulator contamination related to many factors, the difference of geographical location, climatic conditions and different surroundings will lead to different ingredients of insulator contamination. Insulator contamination sources are mainly divided into two categories: one is produced under natural conditions, called natural-type contamination, such as dust and salt, another is caused by the pollution substance that emissions on industrial process, known as industrial-type contamination, such as chemical fumes, cement dust and other dirt.

Numerous research scholars had studied the components of insulator contamination from different areas, although the various parts of its composition is slightly different, the mainly components are $CaSO_4$ and SiO_2 [1].

Affected by the collision mechanism and the turbulence on insulator surface, small particles can't collide with insulator, and it is hard to fall on the insulator surface, which is usually gathered together with a larger particle mass can it finish the pollution process. The analysis of insulator surface contamination particle size shows that the majority diameters of particles are less than $50\mu m$, and more than 50% particles diameters are less than $15\mu m$ [2]. The SEM (scanning electron microscope) analysis found that the insulator contamination was packed by messy arrangement crystalline, it has many gap on the inside but closely combined with insulator surface.

Therefore, the target of insulator hot washing is an aggregate that build up mainly by $CaSO_4$ and SiO_2 particles, which the diameters is about $20\mu m$.

3. Influences on Washing Work

Numerical simulation was conducted through the fluid simulation software FLUENT, three-dimensional analysis was carried out and obtained the velocity vector of jet flow. Analyzing the flow characteristics of splashing water and its influence on washing work.

The vof-model and the standard k–ε turbulence model with standard wall functions were used to capture the flow physics, water was treated as the secondary phase, the nozzle used in the simulation is a common conical nozzle that its outlet diameter is 3mm, and the inlet pressure is 6MPa, with a 4.5m target distance. The velocity vector of jet flow is shown in figure 1.

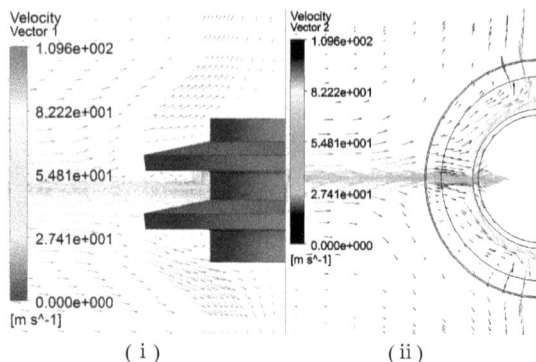

(i) Velocity vector on front surface of the flow field
(ii) Velocity vector on plan surface of the flow field

Fig. 1. Velocity vector of jet flow field

Figure 1 shows that before the jet flow impact with insulator, the cross-sectional of jet flow has increased a certain extent because of the friction between water column and air, and after the impact, the water column spread around, producing a large amount of splashing water which has different velocity. Part of those splashing water is detaching from the surface of insulator and splashing into the air, others is adhering on the insulator and flowing along the surface, cleaning those sites that jet can't directly impact.

3.1. Influences on effective cleaning area

The cleaning work depends on whether the kinetic energy carried by jet flow is greater than the critical cleaning pressure of the contamination, and only those jet flow that satisfied this condition can it finish the cleaning work.

In the simulation conditions that setting above, the radial dynamic pressure distribution of the jet cross-section at the insulator surface is shown in Figure 2.

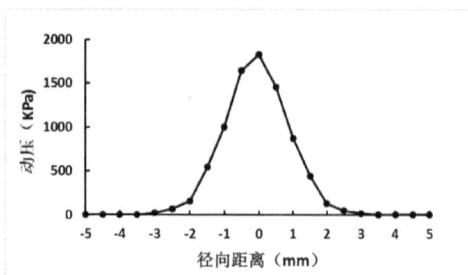

Fig. 2. Radial dynamic pressure distribution

416

The dynamic pressure on the radial cross-section spread of radiation to the surrounding uniformly, and with increasing radial distance, the dynamic pressure of water gradually reduced to ambient pressure.

Assuming the critical cleaning pressure of contamination is 0.2MPa, the effective cleaning radius is 2mm on this simulation, while the radius of the outlet is only 1.5mm. This is because the splashing water expended in the air. With the expanding of splashing water, the effective cleaning areas was increasing from $S=7.068mm^2$ to $S'=12.566mm^2$. All nozzles with different outlet diameters are in line with the above law, that the splashing water can increase the effective cleaning area, and has a great effect on the improve of cleaning efficiency.

3.2. Influences on washing safety

When the splashing water has a sufficient velocity that enables it to detach from the insulator surface, the splashing water will fly away after collision. Those splashing water forming water mist around the insulator, while the insulator is distributed densely. It can easily wet the near insulator, increasing their surface conductivity. As the wetted area is large enough, can even lead to flashover, known as splashing flashover, increasing the risk of cleaning work.

When the splashing water has a low velocity, it can only adhere to the surface. Under the action of gravity, the splashing water gradually gathered on the surface, and flowing from the above umbrella to the bottom umbrella. With some contamination dissolved in, it becomes sewage, which can reduce the electrical resistivity. When those sewage forming connection on the surface of insulator, it can cause arcing, even flashover may occur. Therefore, it is important to choosing a right cleaning way and retrace timely, to avoid the sewage connection.

4. Influences on Flushing Mechanism

High pressure water jet cleaning is a complex process that includes a series of phenomena, such as squeeze, stretch, shear, corrosion, broken, stress wave propagation, cavitation, wear and so on [3, 4]. Due to the diversity of contamination components and structure, the cleaning process also varies.

Scholars have made a lot of researches on the mechanical mechanism of water jet impact broken objects, and put forward a lot of hypotheses, including quasi-static elastic breaking, impact crushing stress wave breaking, cavitation effect crushing, crack extending breaking and damage breaking, seepage damage coupling crushing. Because of the different emphasis of the research, most hypothesis are limited to describe the breaking phenomenon, does not involve in the essence of breaking process, and can't form a system theory. At present, the generally accepted view was put forward by Foreman and Secor, they conclude the main reasons of the damage on water jet impingement process for two points, one is the stress wave and continuous water jet impact stress; another is the internal stress caused by the penetration of water on the gap and crack.

For the hard and loose insulator contamination, the cleaning process also contains two aspects: on the one hand, the jet carrying a large kinetic energy, and can constantly impact to the contamination surface, through fatigue effect and shear action, breaking and cracking the contamination; on the other hand, the pressure water enters the inner space of contamination through penetration, cracks are produced by the effect of the water wedge effect and the squeezing action, which promoted the propagation of crack and the detachment of contamination. For this cleaning process, the main effect is the water pressure on particles that caused by water penetration. The pore in the contamination and the crack caused by jet impact makes sure that the pressure water can penetrate into the interior of the contamination continuously. When the internal water pressure is greater than the adhesion force between particles, the contamination broken, and with the action of subsequent jet, the crack extended, then finally be cleaned.

Splashing water is a product of entrainment between water and air at the jet boundary, as it is spray-like and speed is very slow, the kinetic energy it carried is not enough to finish the clean work through direct impact. As the jet cross-section continues to expand in the air, the coverage range of splashing water is far greater than the water jet effective cleaning area. With the cleaning process goes on, splashing water can land on the contamination surface continuously before the nozzle moved to the region for cleaning, wetting the contamination and makes it soft. It will change the stress and deformation characteristics of contamination and finally decrease the intensity in a certain extent. At the same time, the splashing water fallen to the contamination will penetrate, decreasing the penetration rate of pressure water when cleaning. Because the contamination has softened, and the permeation rate of water was declined, the main reason of cleaning is no longer the internal pressure caused by penetration of pressure water, and changed to the extrusion and shearing action from direct impact of jet. When the jet hitting force is greater than the ultimate strength of contamination, it will be crushing, expansion and finally be cleaned.

5. Influences on the Adhesion Strength of Contamination

Splashing water also can affect the adhesion force between particles by changing the relative humidity around the contamination.

The adhesion force between solid particles is mainly composed of four forces, including capillary force, van der Waals force, electrostatic force and gravity. Because of the small particle size and the charge neutralization after bonding, the influence of gravity and electrostatic force on the adhesion strength is very small, in the analysis of the adhesion of contaminant and insulator surfaces, only the effects of Van der Waals force and capillary forces was considered.

As the particle size of contamination is d=20μm, and the diameter of insulators umbrella is between 10 ~ 30cm, the umbrella surface is an infinite plate for particles. So, it is reasonable to assume that the adhesion between the contamination particles and the insulator can be equivalent to the adhesion force between spherical particles and an infinite plate, known as sphere/plane model.

Capillary force is the additional pressure generated by the curved surface. As there are many gaps between the contaminations, when water contacted with it, can easily form a liquid bridge, as shown in Figure 3.

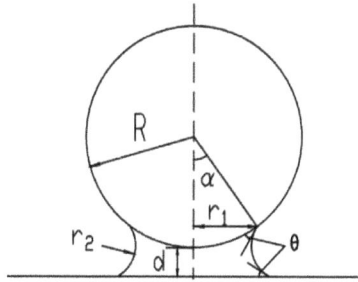

Fig. 3. Liquid bridge model between particles and plate

The liquid bridge makes an additional force between the two surfaces, which is composed of two parts, one is the surface tension of the liquid curved surface, and the other part is the interface pressure generated by the capillary action. The additional force is the capillary force, can be expressed by Eq. (1).

$$F_c = F_s + F_p . \tag{1}$$

Where F_c, F_s and F_p are capillary force, surface tension and capillary pressure, respectively.

Regarding the boundary of liquid bridge as a section of circular arc. Surface tension and capillary pressure can be obtained through the geometric relations and the Young-Laplace equation [5]

$$F_p = -\pi r_1^2 \Delta p = \pi \gamma R(-\sin\alpha + \frac{\cos(\theta+\alpha)+\cos\theta}{d/R+1-\cos\alpha}\sin^2\alpha) . \tag{2}$$

$$F_s = 2\pi \gamma r_1 \sin(\alpha+\theta) = 2\pi \gamma R \sin\alpha \sin(\alpha+\theta) . \tag{3}$$

Where α is the filling angle, γ is the surface tension of water, d is the separation between the particle and the plate, for most solid contacts can be taken as 2×10^{-10}m, θ is the contact angle.

The main components of insulator contamination $CaSO_4$ and SiO_2 are hydrophilic substances, which means the contact angle is very small after wetting. Taking the contact angle as $0°$ to simplify the analysis. At thermal equilibrium, the filling angle α is determined by the Kelvin equation [6].

$$\frac{KT}{\gamma v_0}\ln\frac{p}{p_s} = \frac{\Delta p}{\gamma} = \left(\frac{1}{r_1} - \frac{1}{r_2}\right) = \frac{1}{R\sin\alpha} - \frac{\cos(\theta+\alpha)+\cos\theta}{d+R(1-\cos\alpha)} . \quad (4)$$

Where v_0 is the molecular volume and p/p_s is the relative vapor pressure (relative humidity for water).

From Eqs. (1-4), the capillary force can be calculated, and obtained the trend of capillary force with relative humidity, as shown in Figure 4.

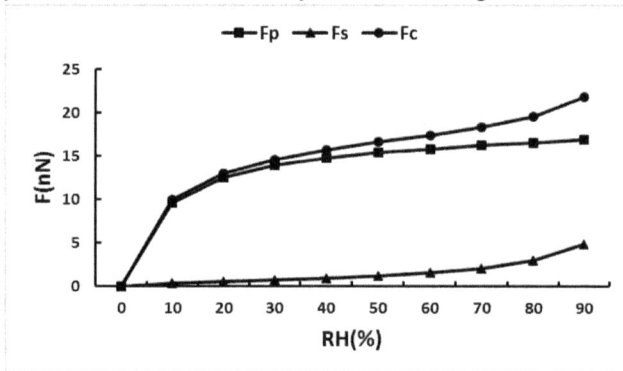

Fig. 4. The trend of capillary force with relative humidity

When the relative humidity is 0%, the capillary force is also zero, and the capillary force increases with the increase of relative humidity. The reason is that when the relative humidity is 0%, the particles is in a dry state, there is no chance to produce capillary force. In the low relative humidity, the water around the particles is few, and the contact area between the water and the particles is small, it can't form enough liquid bridge. With the increase of relative humidity, the effective contact area increased, generating a large number of liquid bridge and capillary forces between particles is also increased.

Van der Waals force is generated by the interaction of two objects, and the interaction distance is relatively short, usually for one or several molecular diameters. For sphere/plane geometry, the van der Waals force is given as

$$F_{vdw} = \frac{AR}{6d^2} .$$ (5)

Where A is Hamaker constant, which depends on the medium that the two objects are in. For $CaSO_4$/SiO_2 contact, if the medium is air, $A=8.86\times10^{-20}J$, if the medium is water, $A=1.9\times10^{-20}J$. As the humidity increases from 0% to 100%, it is expected that the van der Waals force will take a value between $F_{vdw.w}$ and $F_{vdw.a}$. An approximation formula to account for the intermediate humidity is given [7]:

$$F_{vdw} = F_{vdw.w}\left\{1-\frac{1}{[1+R(1-COS\alpha)/d]^2}\right\}+F_{vdw.a}\left\{\frac{1}{[1+R(1-COS\alpha)/d]^2}\right\}$$
(6)

From Eqs. 4-6, the van der Waals force can be calculated, and obtained the trend of van der Waals force with relative humidity, as shown in Figure 5.

With the increase of relative humidity, the van der Waals force first appears to be significantly reduced, and then the amplitude of decreases gradually to be gentle. The reason is that when the relative humidity is 0%, particles and insulator surface is in the dry state, the medium between the two part is air, and their contact radius is very small, with the increase of relative humidity, the medium between the two part change to the mixture of liquid and air, the occur of liquid bridge makes contact radius nasty increased, then reached saturation and tend to be gentle.

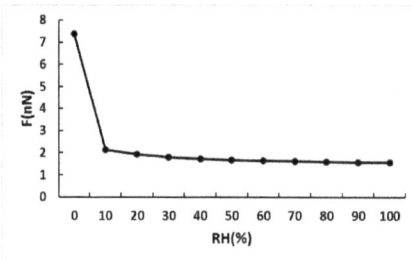

Fig. 5. The trend of van der Waals force with relative humidity

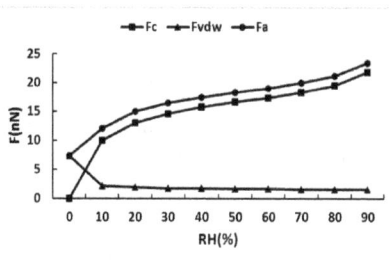

Fig. 6. The trend of adhesion force with relative humidity

The trend of adhesion force between particles and insulator surface with the relative humidity can be obtained, as shown in Figure 6.

The adhesion force between particles and the surface of insulator increased with the increase of relative humidity. At the low humidity, it's hardly to form a liquid bridge, the adhesion force is mainly dominated by the van der Waals force, with the increase of relative humidity, capillary forces began to dominate, the adhesion force between particles is mainly determined by capillary forces.

High relative humidity will lead to higher adhesion force between particles and insulator surface, but the liquid bridge will increase the contact radius at the same time, thus, the change of adhesion strength is little, and the dynamic pressure that needed for cleaning will not be affected too much.

6. Conclusion

Through theoretical analysis and simulation research, studied the influence of splashing water on insulator hot washing process, and get the following conclusion:

(1) Splashing water will increase the cleaning area and improve the cleaning efficiency, but also may cause splashing flashover and sewage connection, which can affect the safety of washing operation.

(2) Splashing water will wet the contamination and makes it soft before cleaning work, leads to the change of cleaning mechanism, the main reason of cleaning has changed from internal water pressure to directly shear pressure.

(3) As the splashing water can increase the relative humidity around the contamination, it will lead to the increase of adhesion force between particles and insulator surface, but will not affect the dynamic pressure of cleaning too much.

References

1. Bin Fen, et al. "Study on structure and chemical components of the deposits of natural polluted insulators." *Hunan Electric Power* 27.1(2007):9-10.
2. Ying-yan Liu, et al. "Adhesion Force and Long-range Attractive Force between Contamination Particles and Insulator Surface." *High Voltage Engineering* 40.4(2014):1010-1016.
3. Obara, T., N. K. Bourne, and J. E. Field. "Liquid-jet impact on liquid and solid surfaces." *Wear* 186 (1995): 388-394.
4. Hashish, M. "Theoretical and experimental investigation of continuous jet penetration of solids." *Journal of Engineering for Industry* 100.1 (1978): 88-94.
5. Zarate, Nyah V., et al. "Effect of relative humidity on onset of capillary forces for rough surfaces." *Journal of colloid and interface science* 411 (2013): 265-272.

6. Chen, Sheng Chao, and Jen Fin Lin. "Detailed modeling of the adhesion force between an AFM tip and a smooth flat surface under different humidity levels." *Journal of Micromechanics and Microengineering* 18.11 (2008): 115006.
7. Xiao, Xudong, and Linmao Qian. "Investigation of humidity-dependent capillary force." *Langmuir* 16.21 (2000): 8153-8158.

Distribution Network Vulnerability Analysis
Based on the Worm Algorithm

Jie Zhang[†] and Min-fang Peng

Hunan University,
Changsha, Hunan, China
[†]E-mail: successzhangjie@163.com

Liang Zhu, Hong-wei Che and Jing-ying Hou

State Grid Hunan Electric Power Company,
Changsha, Hunan, China

With the continuous development of distributed power and micro grid, distribution network structure becomes increasingly complex. Being able to estimate vulnerability of the distribution network becomes an important part in modern society. A simple margin analysis could not clearly reveal inherited characteristics of the power system, so complex network theory begins to be applied to analyze structural vulnerability of power systems. Therefore, from the basic concept of complex networks departure, it defines the electrical characterization of the nature of the distribution network topology improved correlation matrix. Taking into account the evolving smart grid, distribution network and the internet gradually reflect a great similarity. In this paper, the worm simulation model determines the vulnerability of the distribution network nodes, combined with improved incidence matrix squaring algorithm to get the distribution network node vulnerability. It provides a theoretical basis for the future with vulnerability research in grid.

Keywords: Worm; Vulnerability; Correlation Matrix; Linear Fitting.

1. Introduction

With the development of the distribution network, the complexity of the distribution network is also increasing. The problem in transmission network now becomes more prominent and severe in the distribution network. With the further development of smart grid, traditional power grid emphasizing on power equipment physical network gradually changed into large-scale complex systems including power equipment and computer communication network. Under stable operation condition of the gird, the complex structure of the system itself will still exist many big security risks. The demand of electric accelerate the pace of development, the grid focus on the construction of the transmission network, it brings about a corresponding decrease in the investment and construction of the distribution network. This has resulted in an imbalance between distribution network user demand and electricity distribution capacity. Distribution network is always in the limit operation state in the most of time, therefore, how to protect the distribution network security has become an important issue.

With the implement of the power industry market reform and the scale of power system expanding, security and stability issues are more and more concerned [1]. Distribution network node vulnerability assessment is based on the whole distribution network; it includes reliability, economy and security, in which we can calculate the degree of vulnerability of the distribution network in every node. Then we find the biggest degree of the node, and focus on the node for its protection, and it provide theoretical support for the safe and stable operation of the power distribution networks.

2. Distribution Network Vulnerability Assessment

According to the existing literature, there is different angle of vulnerability studies, it can be roughly divided into two categories of vulnerability research: structural vulnerability studies and the status of vulnerability studies. In the structural vulnerability studies, research based on complex network theory is a hot research, which mainly based on complex network theory and the topology characteristics of the grid and fault simulation to explore the internal mechanism of the power grid cascading failures [2, 3]. From the perspective of network topology model and vulnerability assessment indicators, this paper proposes that the next focus of the study is to improve the accuracy of the assessment; but it only takes into account to calculate static index, and the power system is a system of self-organization. Although accuracy can be further optimized to improve the assessment of static index, but the dynamic changes of the power system and make this method has some limitations [4-7].

State (refer to the operating state of elements or units) vulnerability refers to the amount of component status changes (such as voltage drops or downward trend) after suffering a fault, and may be approaching the threshold [8]. Paper [9] based on transient energy function, combined with high-precision numerical analysis method and the direct method gives the advantage of energy margin, which use hybrid power system transient methods to assess vulnerability; but those methods no matter the time or the energy margin only show the security level of the system when the fault is cleared after the accident, therefore the results does not reflect the overall level of vulnerability [10-11].

Although there are a lot of researches for transmission vulnerability analysis, but the study for the distribution network vulnerability analysis is still shallow, the above method is suitable for high-voltage transmission network, taking into account the similarity of the grid structure and the state can be appropriately reference grid vulnerability studies theory.

3. Improved Correlation Matrix Distribution Network

A simplified model of distribution network to take this test as follows:

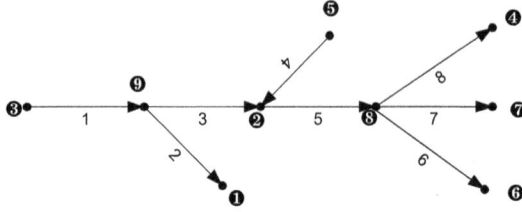

Fig. 1. Nine-node distribution network experimental model

It is defined as follows:

$$a_{ij} = \begin{cases} 1 & \text{Node i and node j is associated} \quad \text{inflow from node i node j} \\ 0 & \text{Node i and node j is independent} \\ -1 & \text{Node i and node j is associated} \quad \text{inflow from node i node j} \end{cases}$$

According to the above definition, we can obtain the following matrix A.

Dealing with the problem of distribution network, the number of nodes is the changing all the time, after several tests we confirmed the obtained correlation matrix which is raised to improve the number of the following formula:

$$\kappa = round(\sqrt{n} - 0.5) \tag{1}$$

Based on the above resulting matrix A and the number of iterations of the algorithm, we can characterize the distribution network node vulnerability matrix C:

$$C = A^{\kappa} \tag{2}$$

Now we find the maximum value of each column, observed that:

According to the column of the matrix, we found that contact with the node two relatively large proportion node is node two and node eight and node night, which are in line with figure.

We can draw the structure of the radiation network by matrix c, but its specific direction is still to be verified.

1, if $C_{ij} < 0$, then the current flow from j to i.

2, if $|C_{ij}| = 1$ or $|C_{ij}| = 0$, then there is no direct link between i and j.

3, if $C_{ij} > 0$, the current flow from i to j.

The node arrangement by degree of importance is $B_{impt} = [8\ 2\ 9\ 4\ 6\ 7\ 5\ 3\ 1]$.

4. Worm Algorithm and its Mathematical Model

Currently the main method for power system in Vulnerability Analysis is to analyze the network topology and system operation. In parallel with the analysis of the topology of the network, we consider the power flow changes in the system, it makes simulation are closer to the actual operating state, the current research in this area has achieved some results.

Paper [12] proposed an index of node degree on importance as follows:

$$K_i = \omega_1 D_i + \omega_2 P_i \qquad (3)$$

K_i, D_i, P_i respectively are node important degree, the degree of distribution, the node power, ω_1, ω_2 are weighted value, the value is 0.5.

The evaluation index takes into account of the importance of the node characteristics including instantaneous power and topological analysis, but for some of the load, which is edge node, although load's power is large, but after its removal, there is not much impact on the power supply capacity of the power distribution network.

The diagram shows the network transmission performance IEEE 114 nodes after the attack.

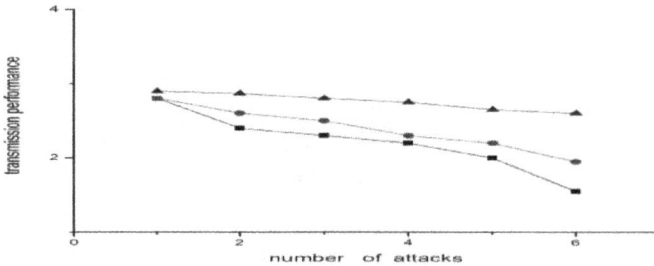

Fig. 2. Node network transmission performance curve after attack

From this figure we can find that under random attack the impact on network transmission performance is relatively small, but under node important degree attack and node degree attack, the impact is more obvious, but the impact between the node important degree attack and node degree attack are very close, therefore it does not reflect well the degree of importance of the superiority of the node degree.

However, in the power system distribution network, the faults propagation mode and speed, as well as the mechanism is changing due to the recent spread of smart devices access, leading to the distribution network fault propagation characteristics of the computer network worm propagation has some similarities, we plan to use worm algorithms to evaluate the distribution network node important degree.

We define that the load and line intersections are the network nodes, we define transmission line as edge of the network, then distribution network will be entitled to a weighted and directed network which is simplified $G = (V, E)$, where V is the set of nodes, E is the set of edges, and the number of nodes is n, the edge the number is m.

To assess structural vulnerability of distribution network, we need to calculate the following parameters:

1. The weight of each side of the network: ω_{ij} the edges are represented as nodes directly connected between the nodes i and j, {i, j} the right, the value decided by the line impedance, namely:

$$\omega_{ij} = X_{ij} \tag{4}$$

2. The right value of each node in the network: node i itself indicates the trend of the impact parameter, namely

$$\varepsilon_i = \phi_1 P_i + \phi_2 Q_i \tag{5}$$

3. Select the worm k value from the current node i to the probability of the nodes connected to j:

$$p_{ki}^j(t) = \begin{cases} \dfrac{\tau_i^j(t)}{\sum\limits_{s=0}^{n} \tau_i^s(t)}, q > q_0 \\ \operatorname{argmax}(\tau_i^j(t)), q \le q_0 \end{cases} \tag{6}$$

4. In the worm propagation process, worm find path by the amount of information, and during its propagation, mortality rate and update evolutionary rates all need to be concerned, node information is:

$$\tau_i^j(t+1) = \rho \tau_0 + (1-\rho)\tau_i^j(t) \tag{7}$$

In the formula, ρ represents the coefficient update worms, $1-\rho$ represents mortality coefficient.

5. Numerical Example

We plan to use examples of IEEE 114 nodes 5kv weakly meshed network, as shown in figure 3. For each node, the number in the figure are the node number, Number one is the head node of the power, nodes which number bigger than 114 are the three-phase switching nodes, it out of our consideration. Linear fitting algorithm was obtained to find out the vulnerability of distribution network, we obtain squaring algorithm vulnerability assessment programs and improved worm algorithm vulnerability assessment program, and introduces actual operation of vulnerability results to fit the results.

5.1. Squaring algorithm

The distribution network in accordance with the formula (1) is defined to give 114 nodes improved correlation matrix. After calculation, the result is shown in Figure 3.

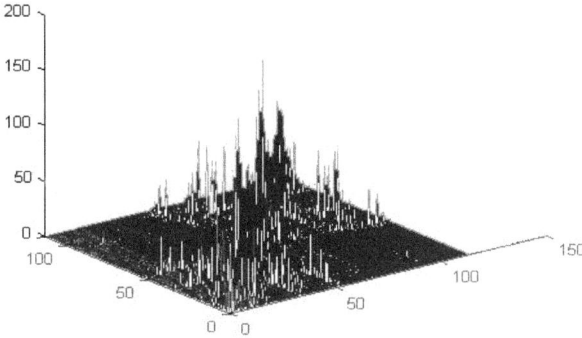

Fig. 3. Squaring algorithm results

5.2. Worm algorithm

For improved correlation matrix which is proposed in squaring algorithm, through the network reconstruction algorithm, the program distribute the coordinates of each point randomly to within a specific range, and then we use the program describe it as directed topology.

Due to the nature of the topology of the network, after we get the random network topology whose network node location will be the location of nodes and edges as worm algorithm, finally get the fitness global optimal solution.

According to the resulting, we use two algorithms sort to linear fit, since the node vulnerability worm sorting algorithm is in accordance with the optimal path length in ascending order, the results from worm algorithm need to be processed before it was fitted with squaring algorithm.

Since the large numbers optimal path length cardinal number, discrimination on the result is not dominant; we deal with optimal path as follows:

$$\tilde{L}_i = 100 / (L_i - 1000) \tag{8}$$

Assume that node vulnerability sorting algorithm obtained is B_{1*N}, results from node vulnerability worm sorting algorithm is L_{1*N}. The final order of the node vulnerability is G_{1*N}. The proportion of the two methods were k and p. The formula can be obtained as follows:

$$G_{(l,i)} = k * B_{(l,i)} + p * L_{(l,i)} \tag{9}$$

Where, k takes 0.6, p takes 0.4 and will be arranged in order according to the final results obtained results are shown as follows:

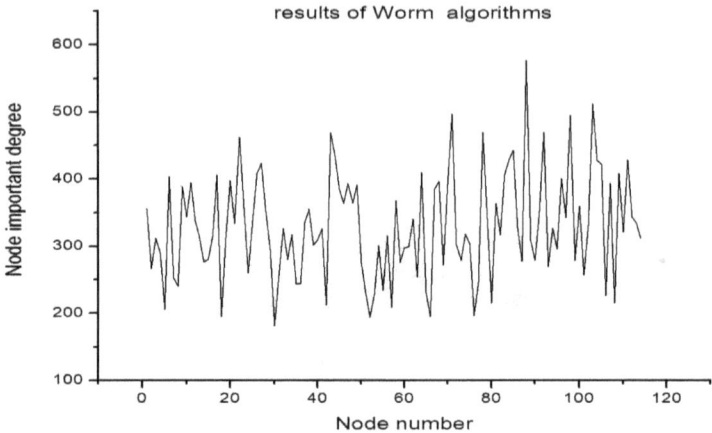

Fig. 4. Result of worm algorithms

6. Conclusion

Based on a complex distribution network, we proposed a new method for vulnerability assessment of network nodes which combine electric power system characteristic parameters and consider the line reactance and power transmission, and use this new method on IEEE 114 bus system. From the analysis calculation it shows that the method considered network characteristics and power characteristics of the power system can calculate the vulnerability of the distribution network, which is closer to the actual operation condition.

Acknowledgments

This work is funded by National Natural Science Foundation of China under Grant No. 61472128 and 61173108, and Hunan Provincial Natural Science Foundation of China No. 14JJ2150.

References

1. Cai Ming, Xie Xiaoling, Wang Xue Chang, et al. Smart grid vulnerability analysis and counter measures [J]. Information Engineering University, 2013 (03): 376-379.

2. Zhao Bo; Wang Cai Sheng; Zhou Jin Hui; Zhao Jun Hui; Yang Ye Qing; Yu Jin Long, Active distribution network status and future development [J]. Automation of Electric Power Systems, 2014 (18): 125-135.

3. Zhang Haixiang; Lv Fei Peng, Vulnerability assessment grid under weighted topological model vulnerability protection [J]. Chinese Electrical Engineering Science, 2014 (4): 613-619.

4. Dai Tingting; Jun Yong; Wei Zhen Bo; Chen Ye. Vulnerability Analysis Based on Complex Network Theory of Power System [J]. Modern Electric Power, 2010 (1): 56-60.

5. Liu Yao Nian; Shu Qian; Yu Bing; Gong Wei Guo; Zhang Wei Min. Vulnerability analysis based on the number of referrals electrical power system [J]. Automation of Electric Power Systems, 2011 (7): 61-64.

6. Guo Hua; Zhang Jian Hua; Yang Jing Yan; Wang Ce; Zhang Yin. Vulnerability Assessment Based on weighted directed graph and the theory of complex networks of large power systems [J]. Electric Power Automation Equipment, 2009 (4): 21-26.

7. Xiao Sheng; Zhang Jian Hua. Power Grid Vulnerability Assessment Based on Small World Topological Model [J]. Power System Technology, 2010 (8): 64-68.

8. Wei Zhen Bo, Jun Yong, Zhu Guo Jun, etc. Comprehensive vulnerability assessment model grid status and structure [J]. Automation of Electric Power Systems, 2009 (08): 11-14 + 55.

9. Li Qian; Li Huaqiang; Huang Zhao Meng; Li Yanqing. Vulnerability analysis based on transient energy function mixing power system [J]. Power system protection and control, 2013 (20): 1-6.

10. Wang Yong; Pengmao Jun; Renjiang Bo. Multi-area interconnected power system state estimation based on sensitivity analysis [J]. Automation of Electric Power Systems, 2007 (19): 27-31.

11. Lu Jinling; Ji Qun Xing; Zhu Yong Li. Power Grid Vulnerability Assessment Based on energy function method [J]. Power System Technology, 2008 (7): 30-33 + 45.

12. Bao Zhe Jing, Cao Yi Jia. Cascading Failures in Local-world Evolving Networks [J]. Journal of Zhejiang University Science, 2008, 9(10); 1336-1340.

Effective Load Carrying Capacity of Wind System
for Electric Vehicles

Pei-dong Lu, Wei-xing Zhang and Chun-juan Jia†

School of Electrical Engineering, Shandong University,
Jinan 250061, China
†E-mail: jiachunjuan@sdu.edu.cn

With wind power and other renewable energy access to power grid, its influence on the reliability of power network is a concern. As a new type of vehicle, electric vehicles are becoming more popular. Firstly, this paper establishes the model of electric vehicle charging load and the output of wind power. Then the impact of EVs and wind power on system reliability are analyzed by sequential Monte Carlo approach in the RTS79 system. Finally, the load carrying capacity of wind system for EVs is calculated, in order to research the relationship between wind power and EV load under a certain reliability index.

Keywords: Electric vehicles; credible capacity; sequential Monte Carlo simulation; matching capacity.

1. Introduction

At present new energy, such as wind resource is becoming widely used today. Meanwhile, as alternative vehicles, electric vehicles have a great contribution to reducing vehicle exhaust emissions, to getting rid of dependence on oil and to alleviating pressure on resources. Wind power and EVs have an impact on the reliability of the grid.

To build the model of the electric vehicles charging load, there are three methods to accomplish the program. The model of the load of charging is built by using Monte Carlo simulation [1]. The impact of PHEV on the grid is reported by Edwin Haesen and Johan Driesen [2], which is illustrated by computing the power losses and the maximum voltage deviation. The relationship of the renewable energy and the PHEV was put forwarded by Soumyo et al. [3].

This paper mainly discusses the relationship between EV load and wind power generation at a reliability level in grid. The main index of power system reliability is the loss of load probability (LOLP) and the expected demand not supply (EDNS). By the Monte Carlo approach, we build the model of the EV load. Based on IEEE reliability test system RTS79, we used sequential Monte Carlo approach in the MATLAB software to calculate reliability index. The impact of wind power and EVs on the reliability of the power system is analyzed. Finally, the carrying capacity of wind power generation for electric vehicle charging load is researched based on the secant method.

2. The Charging Load Model of Electric Vehicles

In this paper, we use Monte Carlo sampling method to model the charging power load of EVs. Monte Carlo sampling method solves random problems through random experiment according to the probability distribution.

The time, when the EVs start to be charged, obeys the normal distribution. According to Liting Tian et al. [1], the statistical data of the starting time of electric vehicle charged in an area can be fitted, and the probability density function of the starting time is depicted as formula (1).

$$f(x) = \begin{cases} \dfrac{1}{\sigma_s \sqrt{2\pi}} \exp[-\dfrac{(x-\mu_s)^2}{2\sigma_s^2}], & (\mu_s - 12) < x < 24 \\ \dfrac{1}{\sigma_s \sqrt{2\pi}} \exp[-\dfrac{(x+24-\mu_s)^2}{2\sigma_s^2}], & 0 < x < (\mu_s - 12) \end{cases} \tag{1}$$

Where $\mu_s = 17.6$; $\sigma_s = 3.4$.

The probability density function of travel mileage can be written as the formula (2).

$$f_D(x) = \dfrac{1}{x\sigma_D \sqrt{2\pi}} \exp[-\dfrac{(\ln x - \mu_D)^2}{2\sigma_D^2}] \tag{2}$$

Where $\mu_D = 3.20$; $\sigma_D = 0.88$.

According to the probability density functions, we use the Monte Carlo approach to calculate the EV load. We can observe the model of electric vehicle load by Fig. 1.

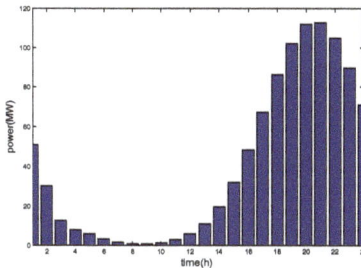

Fig. 1. The model of electric vehicle load

As shown in Fig. 1, in a day, the EV load increases from 4 pm, and the load began to enter the peak at 9 pm. Electric car charging load reached a peak value of 112.738MW. The charging load gradually reduced, at 6 in the morning to 12 points, maintained at a low period.

3. The Model of Wind Output Power

The wind power generation will affect power system stability. In this chapter, the wind power generation system is analyzed, and its output power is calculated and analyzed.

The output power of wind turbine is mainly determined by the parameters of the wind turbine, and output power of wind turbine is generally derived as:

$$P = \begin{cases} 0, & (V \leq V_{ci} \text{ or } V \geq V_{co}) \\ P_r \dfrac{V - V_{ci}}{V_r - V_{ci}}, & (V_{ci} \leq V \leq V_{co}) \\ P_r, & (V_r \leq V < V_{co}) \end{cases} \tag{3}$$

Where V_{ci}, V_r, V_{co} represent the cut-in, rated and cut-out wind speeds respectively, and P_r is the rated capacity of the wind turbine.

The installed capacity of wind turbines is 500MW respectively, and Fig. 2 shows the 24-hour average output power of the wind turbines.

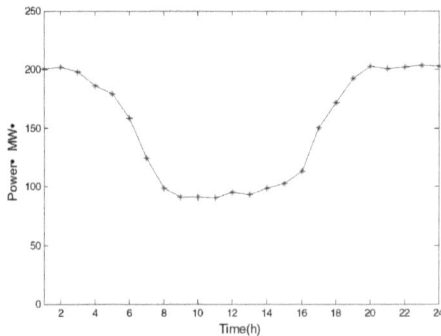

Fig. 2. 24-hour average output power

As shown in Fig. 2, the output power of wind turbines is lower in the day of the 6-point to 16-point and higher on the other time.

4. The impact of wind power generation and electric vehicles on the reliability of power grid

Tests in this section are based on the modified IEEE-RTS79. Considering the higher reliability of the original system, the total installed capacity of the original system is reduced to 2613MW. The impact of EVs and wind power generation on the reliability of power grid is illustrated by the loss of load probability (LOLP). The LOLP of 5.12% is taken as the reference value in the initial RTS79 system. When the wind generation is connected with the system, the LOLP are shown in Table 1.

Table 1. The reliability index in case of wind power

Installed capacity of wind turbines	LOLP	EDNS(MW)
300 MW	0.03251	5.51922
400 MW	0.03011	5.10777
500 MW	0.02656	4.36022
600 MW	0.02473	4.09923
700 MW	0.02257	3.68053
800 MW	0.02206	3.66325

Table 2 depicts the LOLP of the power system with different EV penetration. The estimates show that EVs leads to a significant increase in LOLP. It illustrates that the reliability level of the power system will be decreased, with the increase of the EV penetration.

Table 2. The reliability index with different EV penetration

Penetration level	LOLP	EDNS(MW)
0%	0.0512	9.14129
10%	0.0607	11.60857
20%	0.0771	16.00365
30%	0.0921	21.31351

Table 3 depicts, with electric vehicle penetration of 10%, wind power lead to increased reliability of power systems and LOLP is lower than the initial system.

Table 3. The reliability index of wind power with EV penetration of 10%

Installed capacity of wind turbines	LOLP	EDNS (MW)
300 MW	0.04084	7.10647
400 MW	0.03631	6.36102
500 MW	0.03384	5.68856
600 MW	0.03185	5.50394
700 MW	0.02817	4.83686
800 MW	0.02689	4.53143

5. The Relationship between Wind Power Generation and Electric Vehicle Charging Load

ELCC of an energy source is defined as its ability to support additional load without decreasing the grid's reliability. Alternatively, ELCC of a source can be thought as the amount of existing supply capacity that can be replaced by the given source while serving the same load without increasing the grid's LOLP.

Credible capacity can be expressed as the formula (4):

$$F(C,L) = F(C+\Delta C, L+\Delta L) \qquad (4)$$

R_0 represents the reliability index of the initial system when the installed capacity is C and the load is L. C_{ad} is the capacity of wind turbines. R_2 is the reliability index as adding renewable power C_{ad} in original system and electric vehicle load increase by C_{ad}. As shown in Fig. 3, mark the reliability index in the i th iteration as R_i.

We used the secant method to adjust the peak load of electric vehicles to corresponding with the installed capacity of wind generation, and then the reliability index of the system gradually approaches R₀. Finally, we can get the load carrying capacity of wind power for electric vehicle load by the calculation. Tests in this section are still based on the modified IEEE-RTS79. Table 4 shows the peak value of EV load corresponding to different wind farm installed capacity when the reliability of system is set at 5.12%.

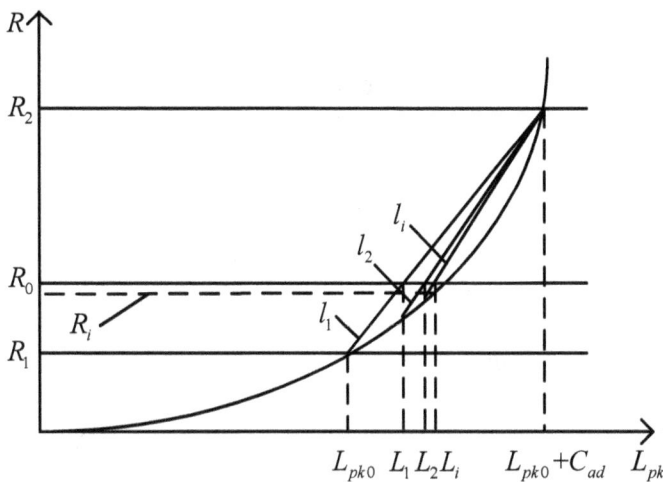

Fig. 3. The principle of the secant method

As shown in Table 4, when the reliability index is maintained at a given value, the peak load of electric vehicle will gradually increase as the installed capacity of wind power generation rises. Meanwhile the ratio of the EV load and wind power installed capacity is reduced with the capacity of wind power increasing. The calculation result illustrate the efficiency of the wind power cannot be improved by the capacity.

Table 4. The relationship between EV load and wind power

Capacity of wind power	EV charging load(MW)	Ratio of EV and wind generation
300 MW	231.8163	0.7727
400 MW	259.4350	0.6486
500 MW	313.4914	0.6270
600 MW	349.7903	0.5830
700 MW	387.8187	0.5540
800 MW	436.4343	0.5455

6. Conclusion

This paper presents the effective load carrying capacity of wind power for the electric vehicle load. We established the model of EV load by the Monte Carlo sampling method. Afterwards we use the sequential Monte Carlo approach to analysis the reliability of the power system, when the EVs are charged. The system reliability is reduced when the EVs are charged without control.

Wind power can significantly improve the reliability of the power system, and eliminate the negative impact of uncoordinated charging electric vehicles on the grid. With the increase of the installed capacity, the number of electric vehicles is increasing. However, the ratio of the EV load and wind power installed capacity is decreasing when the number of wind turbines is increasing. The cost of the wind generations will be not valuable due to expend the scale of the wind power plants. The next assignment is to improve the utilization of renewable resources by controlling the EV charging load.

References

1. Liting Tian, Shuanglong Shi, Zhuo Jia. A statistical modeling method for the charging power demand of electric vehicles [J]. Technology of Power Grid, 2010, 34(11): 126-130
2. Edwin Haesen, Johan Driesen. The Impact of Charging Plug-In Hybrid Electric Vehicles on a Residential Distribution Grid [J]. IEEE Transactions on Power Systems. 2010, 25(1): 371-380
3. Soumyo V. Chakraborty, James Thorp. Computing Optimal Solar Penetration in the Presence of Plug-in Electric Vehicles [Online]. http://www.researchgate.net/publication/261046097
4. Garver, LL. Effective load carrying capability of generating units [J]. IEEE Transon Power Apparatus and Systems, 1996, 85(8):910-919
5. Claudine D A, Surya Santoso. Non-iterative method to approximate the effective load carrying capability of a wind plant [J]. IEEE Transon Energy Conversion, 2008, 23(2): 544-550

Bottom-Up Harmonic Analysis Modeling for Residential Loads

Kun Zhang[†] and Yuan-yuan Sun

School of Electric Engineering, Shandong University,
Ji'nan, 250061, China
[†]Email: zk_sdu@163.com

This paper proposes a bottom-up, probabilistic harmonic assessment technique for residential loads. The modeling, divided into two levels, uses the Markov-Chain Monte Carlo (MCMC) method to build harmonic profiles of 24h. At the top, Monte Carlo controls the household size, the ownership of electric devices and other factors that result in different use patterns of home appliances between families. At the bottom, Markov-Chain is used to determine the starting time of each appliance in individual household. A varying harmonic equivalent circuit based on human behavior representing a residential house, combining user activity model with corresponding electric harmonic model, is thus derived. The total harmonic produced by residential loads is carried out by accumulating multiple residential houses supplied with residential feeders. Field measurements have confirmed the validity of the proposed technique.

Keywords: Harmonic Analysis; Residential Loads; Markov-Chain; Monte Carlo.

1. Introduction

As power electronic equipment propagates in residential loads, their harmonic impact on the power distribution system becomes increasingly serious. The previous survey did in Japan demonstrated that 40 percent of harmonics in the power system came from residential loads. There is an urgent need for techniques that determine the collective harmonic impact of modern residential loads. The common method to access harmonic impact on power system is the harmonic power flow. Amplitude and phase angle of the harmonic currents, which can be available directly by installing metering devices at the point of common coupling (PCC), have a significant influence on the outcome of the harmonic power flow. However, the quantity and geographical distribution of residential loads bring difficulties for data available with field measurements. Moreover, there is no obvious evidence to indicate where the harmonics should be detected. Harmonic analysis modeling for residential loads, therefore, is necessary to solve the problem. The objective of this paper is to develop a model to predict the total harmonic currents produced by multiple houses.

Compared with industrial harmonic source loads which previous studies mainly focused on, the main characteristic of residential ones is its random nature presented in two fields. One is the switch states of every home appliance. The other is which phase providing power energy for residents due to residential power supply mode in China. Thence, the method to describe the randomness of residential loads is crucial to establish the model. Probability statistics, a classical method to solving the randomness problem, has been applied to the summation of random harmonic phasors [1], the algorithms of stochastic harmonic power flows [2], etc. All of these works help to understand harmonic analysis and modeling of systems with random harmonic source loads. However, the techniques available cannot meet requirements of the model to be developed. A novel method to predict the harmonic distortions caused by residential loads was proposed in [3]. The author estimated residential load harmonic distortions with bottom-up modeling, which has been used to predict load profiles. In this paper, we use a different bottom-up method to model the harmonics produced by residential loads. An important feature of the new model is in its approach to represent time-correlated appliance use. Section 2 presents the methodology of the model. Section 3 presents the simulation of this method extended to multiple households and the verification results. Section 4 summarizes the conclusions.

2. Harmonic Model Development Methodology

Single household is used as an example to illustrate the model developed here. Each home has many electric appliances, which can be divided into linear loads and non-linear loads referred to harmonic source loads. Suppose that these loads are denoted as current source model [7], we can get the circuit representing the household electric model in Fig. 1(a). In order to solve this circuit, the states of switches and the electric parameters of components involved is required. In other words, the harmonic currents can be calculated out at any given time, as long as kinds of home devices connected to the power network are definite. In the process of calculation, the circuit is generally simplified into the equivalent circuit shown in Fig. 1(b).

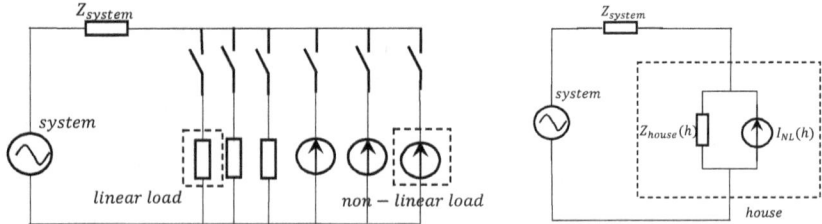

Fig. 1. (a) circuit representing individual household   (b) equivalent circuit

And multiple households supplied by the same residential feeder can be finally converted into the circuit in Fig. 1(b), which simplifies the calculations.

When the approach is extended to other families, the diversity, discussed in detail in the following section, should be considered.

In summary, the model uses the appliance as the basic building block, where appliance refers to any individual home electricity load, such as a television, refrigerator or microwave oven. It is therefore a bottom-up model, which is split into two parts as shown in Fig. 2. Other households are represented by the outer square block. One is switch functions which determine the starting time of every appliance involved in a household. The other is electric harmonic source model which provides corresponding electric parameters. Combining the circuit topology and component parameters involved, a definite circuit representing one house at a given time is thus derived. Further, the harmonic currents injected into the distribution network can be figured out.

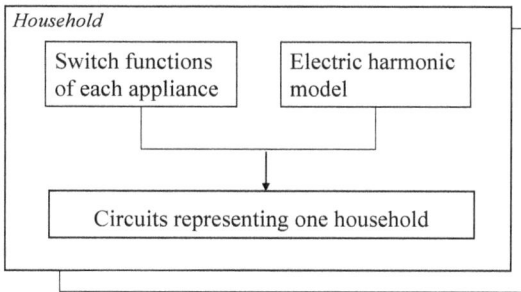

Fig. 2. Harmonic model development for one household

The two models have been studied in the field of load forecasting and harmonic analysis respectively. The former contains three main models [6], statistical random model, probabilistic empirical model and time of use based model, determining the switch function of each home appliance. In view of the accuracy of the model and the data available, time of use based model is used to determine the starting time of every home device during one day. Time of use data, empirical sequences of activities in households, is normally collected with time diaries where household members write down their daily activity sequences. The methodology of time of use based model is that: on/off states of residential loads is determined by user activities. For example, if the cooking activity occurs, it has a great chance to use related electric appliances. This treatment is reasonable, since most residential loads are driven by user behavior which has resulted in the random nature of home electric appliances.

Harmonic current source model is selected to represent the electric appliance based on the following reasons. Once the residential load is on, the harmonics produced by it follow a deterministic spectrum. The attenuation effect can be neglected, though some research works demonstrated the attenuation and

441

diversity effects [4] have much influence in the summation of multiple harmonic sources. The attenuation effect becomes significant only when the voltage total harmonic distortion exceeds 10%. This situation occurs rarely. And this selection can reduce the complexity of the complete model. Therefore, Harmonic current source model is used to represent the residential loads.

3. Harmonic model for multiple households

The diversity between households should be considered when the method is extended to other houses. The number of the household occupants, the building type, the time of year and other factors have influence on user habitual patterns, which are closely related with electric appliance use. And the types and the brands of electric appliances, determining the electric parameters, are different in each household. The method processing the diversity as shown in Fig. 3 is briefly presented. The details of transition matrices, sharing probabilities and user activity definition required in the model, could be found in [5].

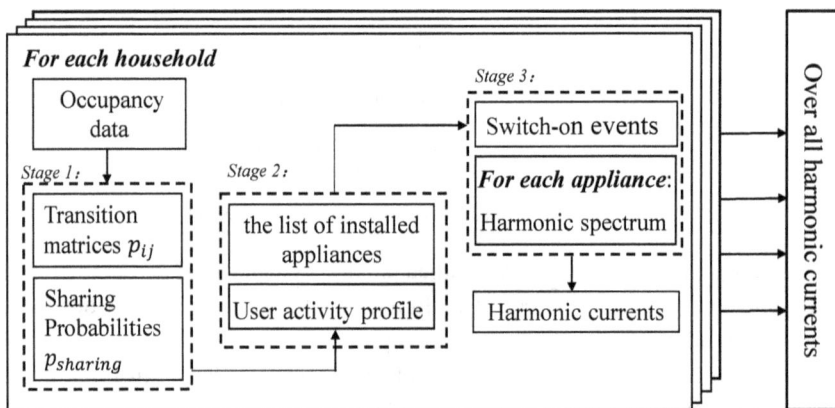

Fig. 3. harmonic model development for multiple households

The simulation time step for user activity modelling is 10 min, due to the available input data, and is reduced to 1 min during the conversion to switch-on events. Each household is defined by the number of occupants, and the list of installed electric appliances. In stage 1, the unmodified user activity profile for each occupancy in one household is derived with comparing transition probability with a random figure first. Transition probabilities indicate the likelihood of the transfer from the current activity state to another including itself. There are 13 user activity states defined in the model. And transition matrices contain 13×13 elements for each household. For multiple occupancy households, sharing probabilities that more than one occupant will use certain appliances at any given time, are used to modify other user activity profiles

442

based on one of occupants in the household. For example, the current activity state of other users is transferred into non-electrical activity state if sharing probabilities in the algorithm suggest the situation of sharing the same electric appliance occurs.

The other factor, the ownership of electric appliances as presented as the list of installed appliances for each household in stage 2, is used to convert user activity profiles into switch-on events. Conversion from user activity states to switch-on events for residential loads is based on two principles. Whether user activity states is electrical activity and related electric appliances are available in one household, should be checked. The two points are determined by the user activity definition mentioned and the list of installed appliances respectively. The harmonic spectrum is available simultaneously since the list of installed appliances for one household is determined. The calculation process of stage 3 has been mentioned in Section 2. Residential loads are divided into linear loads and non-linear loads. As the harmonic current source model, linear loads are modeled as constant power loads at the fundamental frequency (50 Hz) and as impedance at harmonic frequencies. The nonlinear residential loads are modeled as constant power loads at the fundamental frequency and as current sources at harmonic frequencies [7]. Monte Carol simulation distribute different number of occupants and electric appliances for each household in order to simulate the diversity between households. For each household, Markov-chain simulation generates synthetic activity patterns for each household member. Overall harmonic currents produced by summing up multiple houses. The 24h curve for 3rd harmonic current of one household is presented in Fig. 4. Obvious variation occurs at 6:00 to 8:00 and night probably because occupants at home is relatively active during this time.

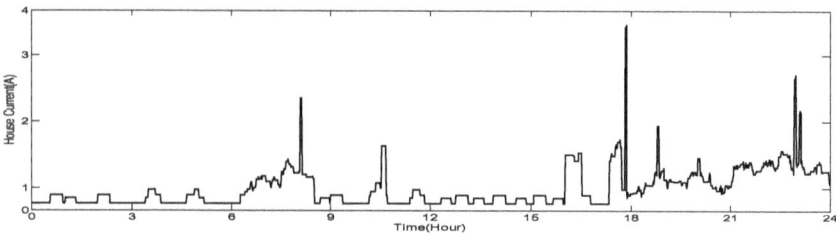

Fig. 4. 3rd harmonic current magnitude of one household varying during 24h

In the statistics perspective, large number of samples makes sense. Field measurements are used to valid against the model developed. Whether the harmonic current change trend during one day is consistent with field measurements should be check firstly. The next step is to calculate the mean and standard deviation of harmonic currents obtained from both the simulation and field measurements. Through comparisons, the validity of this model is proved.

443

The variations of user habitual pattern and home appliances both have effect on the final results of the model due to the method adopted. The model could be also used to evaluate the influence on power system which technological advance and electricity use policy change have.

4. Conclusion

This paper uses a probabilistic method to determine the harmonic impact of residential loads and houses. At the individual household level, Markov-chain method is used to determine the starting time of each home appliance. At upper level, Monte Carol simulation controls the household size, the ownership of residential device and other factors that have influence on the use of home appliances. Thus a bottom-up based harmonic analysis model is derived. We can assess the harmonic impact of residential loads on power system accurately and economically with it. This harmonic analysis technique is also ideally suited for studying the consequences of consumer behavior or regulatory policy changes.

References

1. M. Lehtonen, "A general solution to the harmonics summation problem," Eur. Trans. Elect. Power Eng., vol. 3, no. 4, pp. 293-297, Jul./Aug. 1993
2. A. Cavallini and G. C. Montanari, "A deterministic/stochastic framework for power system harmonics modeling," in *IEEE Transactions on Power Systems*, vol. 12, no. 1, pp. 407-415, Feb 1997.
3. D. Salles, C. Jiang, W. Xu, W. Freitas, "Assessing the Collective Harmonic Impact of Modern Residential Loads—Part I: Methodology," in IEEE Transactions on Power Delivery, vol. 27, no. 4, pp. 1937-1946, Oct. 2012.
4. B. Nassif and W. Xu, "Characterizing the Harmonic Attenuation Effect of Compact Fluorescent Lamps," in IEEE Transactions on Power Delivery, vol. 24, no. 3, pp. 1748-1749, July 2009.
5. J. Collin, G. Tsagarakis, A. E. Kiprakis and S. McLaughlin, "Development of Low-Voltage Load Models for the Residential Load Sector," in IEEE Transactions on Power Systems, vol. 29, no. 5, pp. 2180-2188, Sept. 2014.
6. Joakim Widén, A. Grandjean, J. Adnot, "A review and an analysis of the residential electric load curve models," Renewable and Sustainable Energy Reviews, Volume 16, Issue 9, Pages 6539-6565, December 2012.
7. "Task force on harmonics modeling and simulation, modeling and simulation of the propagation of harmonics in electric power networks–part I: concepts, models, and simulation techniques," IEEETrans. Power Del., vol. 11, no. 1, pp. 452–465, Jan. 1996.

The Layout Constraint Research of Base Stations in Multistation and Time-Sharing Measurement Method to Measure Machine Tool Geometric Error

Yu-fen Deng[*,§], Jun-jie Guo[*], Hai-tao Li[*], Jun-feng Lu[†] and Jin-dong Wang[‡]

*State Key Laboratory for Manufacturing Systems Engineering,
Xi'an Jiaotong University, Xi'an, China
§E-mail: kmdyf@126.com*

*Qinghai Heavy-duty Machine Tool Co.Ltd, Qinghai, China
E-mail: fengliwuxia @163.com*

*‡Southwest Jiaotong University, Chengdu, China
E-mail: wangjindong198205 @163.com*

It is very important in multi-station and time-sharing measurement method. Through computer simulations and experiments, in the paper, the layout constraint of laser tracker is studied by analyzing the Jacobian matrix of equations to verifies the layout constraint to improve the measurement accuracy.

Keywords: System Self-Calibration, Geometric Error, Layout Constraint.

1. Introduction

Mathematical method of self-calibration, arrangement method of the laser tracker and the motion trajectory of the targets have considerable influence on the measurement accuracy of machine geometric error. In the paper, we analyze the Jacobin matrix of the equations and derive the layout constraint of each station and target at the self-calibration.

2. The Mathematical Model of Self-calibration

Assuming coordinates of base station centers as $P_j(x_{pj}, y_{pj}, z_{pj})$ and the coordinates of the measuring points as $A_i(x_i, y_i, z_i)$ and the distances between base station centers and initial target A_0 as L_0, the following equation could be established according to the distance formula between two points.

$$\left(x_i - x_{pj}\right)^2 + \left(y_i - y_{pj}\right)^2 + \left(z_i - z_{pj}\right)^2 - \left(L_{0j} + \Delta l_{ji}\right)^2 = 0 \ (i = 1,2,3,\cdots,n, j = 1,2,\cdots,m) \quad (1)$$

Wherein, the m represents the number of base stations, n represents the number of measurement points. During the system self-calibration, only Δl_{ji} are known parameters, the others are unknowns. The self-calibration method is to solve the overdetermination equation (1) by the least squares method. There are $m \times n$ equations in formula (1) and their left side are represented by $f_k(x)$, $k=1,2,...,m \times n$. Let $k = m \cdot (i - 1) + j$ and x represents a vector composed of all the unknowns. The evaluation function is defined as:

$$\emptyset(x) = \sum_{i=1}^{n} \sum_{j=1}^{m} f_{m(i-1)+j}^2 (x) \tag{2}$$

3. Analyzing Jacobian Matrix

In establishing the coordinate system, the first station center is the original point and the second station center is in the x-axis and the third station center is in the plane of XY and the forth station center is outside the XY plane. Such the unknowns are reduced to $3n+10$, self-calibration could be completed.
When $m=4$, the $4n$ equations is:

$$f_k(x) = \left(x_i - x_{pj}\right)^2 + \left(y_i - y_{pj}\right)^2 + \left(z_i - z_{pj}\right)^2 - \left(L_{oj} + \Delta l_{ji}\right)^2$$
$$(i = 1,2,3,\cdots, n, j = 1,2,3,4) \tag{3}$$

Wherein, $(x_i, y_i, z_i)(i = 1,2,\cdots, n)$ is target point coordinates and $(x_{pj}, y_{pj}, z_{pj})(j = 1,2,3,4)$ is base station coordinates and L_{oj} is the initial distances and Δl_{ji} is the incremental distance of laser trackers. Only Δl_{ji} are known parameters, the other are unknown parameters. The vector constituted with $3n+10$ unknown parameters is:

$$x = \left[x_1, y_1, z_1, x_2, y_2, z_2, \ldots, x_n, y_n, z_n, x_{p2}, x_{p3}, y_{p3}, x_{p4}, y_{p4}, z_{p4}, L_{01}, L_{02}, L_{03}, L_{04}\right]^T \tag{4}$$

Six system parameters have been defined as zero in the coordination system. So the Jacobian matrix $J(x)$ is:

$$J(x) = \begin{bmatrix}
b_{1,1}b_{1,2}b_{1,3} & 0 & 0 & 0 & \cdots & 0 & 0 & 0 & 0 & 0 & 0 & b_{1,3n+7} & 0 & 0 & 0 \\
b_{2,1}b_{2,2}b_{2,3} & 0 & 0 & 0 & \cdots b_{2,3n+1} & 0 & 0 & 0 & 0 & 0 & 0 & b_{2,3n+8} & 0 & 0 \\
b_{3,1}b_{3,2}b_{3,3} & 0 & 0 & 0 & \cdots & 0 & b_{3,3n+2} & b_{3,3n+3} & 0 & 0 & 0 & 0 & 0 & b_{3,3n+9} & 0 \\
b_{4,1}b_{4,2}b_{4,3} & 0 & 0 & 0 & \cdots & 0 & 0 & 0 & b_{4,3n+4} & b_{4,3n+5} & b_{4,3n+6} & 0 & 0 & 0 & b_{4,3n+10} \\
0 & 0 & 0 & b_{5,4}b_{5,5}b_{5,6} & \cdots & 0 & 0 & 0 & 0 & 0 & 0 & b_{5,3n+7} & 0 & 0 & 0 \\
0 & 0 & 0 & b_{6,4}b_{6,5}b_{6,6} & \cdots b_{6,3n+1} & 0 & 0 & 0 & 0 & 0 & 0 & b_{6,3n+8} & 0 & 0 \\
0 & 0 & 0 & b_{7,4}b_{7,5}b_{7,6} & \cdots & 0 & b_{7,3n+2} & b_{7,3n+3} & 0 & 0 & 0 & 0 & 0 & b_{7,3n+9} & 0 \\
0 & 0 & 0 & b_{8,4}b_{8,5}b_{8,6} & \cdots & 0 & 0 & 0 & b_{8,3n+4} & b_{8,3n+5} & b_{8,3n+6} & 0 & 0 & 0 & b_{8,3n+10} \\
\cdots & \cdots & \cdots & \cdots & \cdots & \cdots & \cdots & \cdots & \cdots & \cdots & \cdots & \cdots & \cdots & \cdots & \cdots
\end{bmatrix} \tag{5}$$

Let $k=4(i-1) +j$, each element of the matrix:

$$b_{k,h} = \frac{\partial f_k(x)}{\partial x_h}$$

Following elements are: $b_{4(i-1)+j,3i-2} = 2(x_i - x_{pj})$

$b_{4(i-1)+j,3i-1} = 2(y_i - y_{pj})$ $\quad b_{4(i-1)+j,3i} = 2(z_i - z_{pj})$

$b_{4i-2,3n+1} = -2(x_i - x_{p2})$ $\quad b_{4i-1,3n+2} = -2(x_i - x_{p3})$

$b_{4i-1,3n+3} = -2(y_i - y_{p3})$ $\quad b_{4i,3n+4} = -2(x_i - x_{p4})$

$b_{4i,3n+5} = -2(y_i - y_{p4})$ $\quad b_{4i,3n+6} = -2(z_i - z_{p4})$

$b_{4(i-1)+j,3n+6+j} = -2(L_{0j} + \Delta\, l_{ji})$ $\qquad (i = 1,2,3,\cdots,n, j = 1,2,3,4)$ $\qquad$ (6)

Other elements are zero. To ensure $J(x)$ full rank, now we analyzed the linear independence between any columns in formula.

3.1. Each column vector is not zero

When the $3i$-2 column vector is not zero, the following formula could not simultaneously satisfy: $b_{4(i-1)+j,3i-2} = 2(x_i - x_{pj}) = 0$

Namely $x_i = x_{pj}$ $\qquad (i = 1,2,\cdots,n, j = 1,2,3,4)$ $\qquad$ (7)

This shows that the abscissa of any target could not be equal to the one of the four stations simultaneously. In other words, the four stations could not be coplanar and parallel to the YZ flat. Similarly when the $3i$-1 or the $3i$ column vector is not zero, the four stations could not be coplanar and the flat is parallel to the XZ or XY flat.

When the $3n+6+j$ column vector is not zero, the following formula could not simultaneously satisfy:

$b_{4(i-1)+j,3n+6+j} = -2(L_{0j} + \Delta\, l_{ji}) = 0$ $\quad (i = 1,2,\cdots,n, j = 1,2,3,4)$ $\qquad$ (8)

That is, the distance between stations and targets is not zero constantly. This is obviously always tenable, the system unrestricted.

When the $3n+1$ column vector is not zero, the following formula could not simultaneously satisfy: $b_{4i-2,3n+1} = -2(x_i - x_{p2}) = 0$

That is, $x_i = x_{p2}$ $\qquad (i = 1,2,\cdots,n)$ $\qquad$ (9)

Namely, the horizontal ordinate of any target could not be equal to the one of station *P2*. That is, the target trajectory is not the flat through the station *P2* and parallel to the *YZ* flat. Similarly, When the *3n+2*(or *3n+3*) column vector is not zero, the target trajectory is not the flat through the station *P3* and parallel to the *YZ* or *XZ* flat. When the *3n+4*(or *3n+5* or *3n+6*) column vector is not zero, the target trajectory is not the flat through the station *P4* and parallel to the *XY* or *XZ* or *YZ* flat.

3.2. *Linear independence between any two columns*

When the *3i-2* column is linearly independent on the *3i-1* column, the following formula could not simultaneously satisfy:$2(x_i - x_{pj}) = k \cdot 2(y_i - y_{pj})$

That is, $\dfrac{y_i - y_{pj}}{x_i - x_{pj}} = \dfrac{1}{k}$ $\quad (i = 1,2,\cdots,n, j = 1,2,3,4)$ (10)

Wherein, the k is a constant that is irrelevant on the i and j. Namely, the projection of the ligature from the target to the four stations could not be collinear in the *XY* flat. That is, the four stations could not be coplanar and parallel to *Z* axis and through any measuring point. Similarly, when the *3i-2* column is linearly independent on the *3i* column, the four stations could not be coplanar and parallel to *Y* axis and through any measuring point. When the *3i-1* column is linearly independent on the *3i* column, the four stations could not be coplanar and parallel to *X* axis and through any measuring point.

When the *3n+2* column is linearly independent on the *3n+3* column, the following formula could not simultaneously satisfy:$-2(x_i - x_{p3}) = k[-2(y_i - y_{p3})]$

That is, $\dfrac{y_i - y_{p3}}{x_i - x_{p3}} = \dfrac{1}{k}$ $\quad (i = 1,2,\cdots,n)$ (11)

Wherein, the k is a constant that is irrelevant on the i. Namely, target trajectory could not be a flat through the station *P3* and parallel to *Z* axis. Similarly, when the *3n+2* column is linearly independent on the *3n+9* column (or the *3n+3* column is linearly independent on the *3n+6* column), the same requirements are needed.

When the *3n+4* column is linearly independent on the *3n+5* column, the following formula could not simultaneously satisfy:$-2(x_i - x_{p4}) = k[-2(y_i - y_{p4})]$

That is, $\dfrac{y_i - y_{p4}}{x_i - x_{p4}} = \dfrac{1}{k}$ $\quad (i = 1,2,\cdots,n)$ (12)

Wherein, the k is a constant that is irrelevant on the i. Namely, target trajectory could not be a flat through the station $P4$ and parallel to Z axis. Similarly, when the $3n+4$ column is linearly independent on the $3n+6$ column (or the $3n+5$ column is linearly independent on the $3n+6$ column or the $3n+4$ column is linearly independent on the $3n+10$ column or the $3n+5$ column is linearly independent on the $3n+10$ or the $3n+6$ column is linearly independent on the $3n+10$), the same requirements are needed.

When the $3n+1$ column is linearly independent on the $3n+8$ column, the following formula could not simultaneously satisfy: $-2(x_i - x_{p2}) = k[-2(L_{02} + \Delta l_{2i})]$

$$\text{That is, } k = \frac{x_i - x_{p2}}{L_{02} + \Delta l_{2i}} = cos\,\theta_{2i} \qquad (i = 1,2,\cdots,n) \tag{13}$$

Wherein, the k is a constant that is irrelevant on the i. θ_{2i} are included angles between the connection of the station $P2$ and the target Ai and x positive axis. Namely, the target trajectory could not be a flat through the station $P2$.

3.3. Linear independence among any three columns

When the $3n+2$ column is linearly independent on the $3n+3$ column and $3n+9$ column, the following formula could not simultaneously satisfy:

$$-2(L_{03} + \Delta l_{3i}) = k_1[-2(x_i - x_{p3})] + k_2[-2(y_i - y_{p3})] \tag{14}$$

$$\text{That is,} 1 = k_1 \frac{x_i - x_{p3}}{L_{03} + \Delta l_{3i}} + k_2 \frac{y_i - y_{p3}}{L_{03} + \Delta l_{3i}} = k_1 cos\,\theta_{3i} + k_2 sin\,\theta_{3i} \qquad (i = 1,2,\cdots,n)$$

θ_{3i} are included angles between the connection of the station $P3$ and the target A_i and x positive axis. The $k1$ and $k2$ are two constants that are irrelevant on the i. If the formula is always tenable for all i, θ_{3i} could only be two values. Namely, if the three columns are linearly independent, the target trajectory could not be one flat or two flats through the station $P3$ and parallel to Z axis.

3.4. Anyone among the last ten columns is linearly independent on the previous $3n$ columns

When the $3n+1$ column is linearly independent on from the first column to the $3n$ column, the following formula could not simultaneously satisfy:

$$\begin{cases} 0 = k_{3i-2} \cdot 2(x_i - x_{p1}) + k_{3i-1} \cdot 2(y_i - y_{p1}) + k_{3i} \cdot 2(z_i - z_{p1}) \\ -2(x_i - x_{p2}) = k_{3i-2} \cdot 2(x_i - x_{p2}) + k_{3i-1} \cdot 2(y_i - y_{p2}) + k_{3i} \cdot 2(z_i - z_{p2}) \\ 0 = k_{3i-2} \cdot 2(x_i - x_{p3}) + k_{3i-1} \cdot 2(y_i - y_{p3}) + k_{3i} \cdot 2(z_i - z_{p3}) \\ 0 = k_{3i-2} \cdot 2(x_i - x_{p4}) + k_{3i-1} \cdot 2(y_i - y_{p4}) + k_{3i} \cdot 2(z_i - z_{p4}) \end{cases} \tag{15}$$

Wherein, k_{3i-2}、 k_{3i-1} and k_{3i} represent respectively the coefficient of the *3i-2* and the *3i-1* and the *3i* column in linear independence.

From equation (14), there must be three appropriate values k_{3i-2}、 k_{3i-1} and k_{3i} could make the equation (14) tenable. That is, target trajectory could not be a flat through *P1* and *P3* and *P4*.

Similarly, when the column from the *3n+2* to the *3n+10* is respectively linearly independent on the previous *3n* column, the system's requirements are that the target trajectory could not be a flat through any three stations. The connection of three stations among the nine groups is respectively *P1P2P4*, *P1P2P4*, *P1P2P3*, *P1P2P3*, *P1P2P3*, *P2P3P4*, *P1P3P4*, *P1P2P4*, *P1P2P3*.

3.5. *Any two columns among the last ten columns are linearly independent on the previous 3n column*

When the *3n+1* column and *3n+8* column are linearly independent on previous *3n* column, the relevant equation is:

$$\begin{cases} 0 = k_{3i-2} \cdot 2(x_i - x_{p1}) + k_{3i-1} \cdot 2(y_i - y_{p1}) + k_{3i} \cdot 2(z_i - z_{p1}) \\ -2(L_{02} + \Delta l_{2i}) = k_{3i-2} \cdot 2(x_i - x_{p2}) + k_{3i-1} \cdot 2(y_i - y_{p2}) + k_{3i} \cdot 2(z_i - z_{p2}) + k_{3n+1}[-2(x_i - x_{p2})] \\ 0 = k_{3i-2} \cdot 2(x_i - x_{p3}) + k_{3i-1} \cdot 2(y_i - y_{p3}) + k_{3i} \cdot 2(z_i - z_{p3}) \end{cases} \quad (16)$$

where must be $k_{3i-2}, k_{3i-1}, k_{3i}$ and k_{3n+1} making the formula (16) tenable. So target trajectory could not be a flat through *P1* and *P3* and *P4*.

Similarly, when any two columns among the *3n+2* and *3n+3* and *3n+9* or all column are linearly independent on the previous *3n* columns, the target trajectory could not be a flat through the station *P1* and *P2* and *P4*.

There are many other possibilities will not make Jacobin full rank. We will no longer analyze. So the analysis conditions in the paper are necessary, but not sufficient.

4. Computer Simulation

In order to validate the analysis of the Jacobian matrix, we had the computer simulation about the aforementioned results. Let the four stations coordination are *A(0,0,0)mm*, *B(1200,0,0)mm*, *C(600,300,0)mm* and *D(800,400,300)mm* respectively. When the four stations are coplanar, the coordination of the station *D* is *(800,400,0)mm*. The number of target points was *n=35*. The maximum deviation of the iterative initial value . We had the simulation in two conditions: (1) Without the error of laser trackers E=0; (2) The error of laser trackers E=1um.

Eight layouts were set up in simulation:
(1) General condition.
(2) Four stations are coplanar.
(3) Target trajectory was a flat through the station B and parallel to YZ flat.
(4) Target trajectory was a flat through the station C and parallel to Z axis.
(5) Target trajectory was two flats through the station C and parallel to Z axis.
(6) Four stations are coplanar and the flat was through a measuring point.
(7) Target trajectory was a flat through the station A.
(8) Four stations were coplanar but no pass any measurement points. The simulation results are shown in Table 1.

Table 1. The data sheet of simulation about the self-calibration matrix analysis

Layout	E=0		E=1	
	Final deviation (mm)	Iteration deviation(mm4)	Final deviation (mm)	Iteration deviation (mm4)
(1)	1.011102e-10	7.78569e-18	0.165823	38.4896
(2)	2.78512	5.010211e-05	4.10824	95.5486
(3)	6.95612	4.26715e-03	8.34012	91.5087
(4)	4.41539	1.34586e-10	3.84668	67.6052
(5)	0.037715	1.54012e-08	0.395812	39.8692
(6)	0.543612	3.75246e-09	13.8623	23.1574
(7)	0.714532	2.19945e-11	1.21495	55.4793
(8)	1.16945	1.35486e-06	1.38432	44.5263

As shown in table 1, the final deviation and the iteration deviation from the second layout to the eighth layout are larger than the first layout without considering the laser tracker error. In addition, the final deviation and the iteration deviation from the second layout to the eighth layout are also shown to be larger than the first layout considering the laser tracker error. Which show that the aforementioned conclusion is correct.

5. Experiment

To verify the correctness of the inference, the experiment was conducted in a blade measurement instrument. A laser tracker was placed in various locations

around the blade measurement instrument. The experiment was conducted in accordance with the simulation layout. Due to the limited experiment conditions, only four layouts was experimented:
 (1) General condition.
 (2) Four stations are coplanar.
 (3) Four stations are coplanar and the flat was through a measuring point.
 (4) Four stations were coplanar but no pass any measurement points.

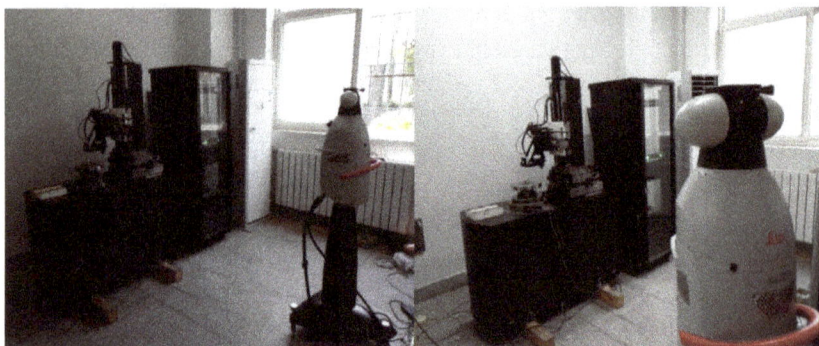

Figure 1.The photo of the experiment

The experiment results are shown in Table 2.

Table 2. The result of experiment

Layout	Final deviation (mm)	Iteration deviation (mm4)
(1)	1.020156e-10	7.79541e-18
(2)	2.80632	5.15896e-05
(3)	0.554693	3.76325e-09
(4)	0.726589	2.20489e-11

6. Conclusion

The station layout and the target trajectory have a great impact on the measurement accuracy. In the paper, the conditions of space self-calibration are derived by analyzing the Jacobin matrix.
 (1) The four stations are not coplanar.
 (2) The target trajectory could not be the plane through $P2$ or $P3$ or $P4$ and parallel to axis.
 (3) The target trajectory could not be the plane through any three stations. The layout constraint must be satisfied in self-calibration, otherwise the correct result will not be obtained.

It is very important in multi-station and time-sharing measurement method. Through computer simulations and experiments, in the paper, the layout constraint of laser tracker is studied by analyzing the Jacobian matrix of equations to verifies the layout constraint to improve the measurement accuracy.

Acknowledgments

This work is supported by China Technology Major Project (no. 2015ZX 04005001) and National Natural Science Foundation of China (no. 51305370).

References

1. Wang J D and Guo J J 2012 The technical method of geometric error measurement for multi-axis NC machine tool by laser tracker *Measurement Science and Technology*.23 1-11
2. Sartoria S and Zhang G X 1995 Geometric error measurement and compensation of machines *CIRP Ann*.44 599–609
3. Lin Y B and Zhang G X 2003 The optimal arrangement of four laser tracking interferometers in 3D coordinate measuring system based on multi-lateration *Int. Symp. on Virtual Environments, Human–Computer Interfaces, and Measurement Systems* pp 138–43

Investigation of the Stabilization Method of DC Microgrid by Passive Damping

Zhong-lin Yang[†] and Xiao-ming Zha

School of Electrical Engineering, Wuhan University,
Wuhan City, Hubei, China
[†]E-mail: blueduny@sina.com

There are a lot of converters in the DC microgrid. This tightly regulated closed-loop converters may cause stability problem of bus voltage when used as a load since constant power is drawn. In this paper, a novel passive damping method was introduced to enhance the stability of the DC microgrid. This method adds the passive damper to the DC bus, which enlarges the system's damping and insures the stability of DC microgrid. The feasibility and effectiveness of this method are verified by the simulation results.

Keywords: DC Microgrid; Stabilization; Passive Damping; Constant-Power Load.

1. Introduction

In recent years, in order to solve the energy shortage, the microgrid has gained more and more attention and application [1]. Compared to AC microgrid, DC microgrid has the advantages of simple control, high efficiency, high reliability, good power quality and so on. It is gradually becoming more and more concerned about [2]. However, the DC microgrid contains a large number of load converters, which are characterized by the constant power loads and cause the instability of the DC bus voltage of the DC microgrid [3].

Some of the literature on the stability of the DC microgrid is analyzed, and some methods to improve the stability of the DC microgrid are proposed. By increasing the equivalent impedance of the active damping signal, the [4] can improve the stability of the DC micro grid. The method for improving the stability of the system with the introduction of virtual capacitors in the load point is proposed [5]. In this paper, a new method is proposed to stabilize the multi-load DC microgrid. The method is simple and requires no modification to the converter. It can ensure the stability of DC microgrid.

2. System Model and its Stability Analysis

Typical DC microgrid structure is shown in Figure 1, which contains a large number of power electronic converters. The source side AC/DC or DC/DC

454

converter is connected to the DC microgrid, and the bus voltage is maintained. When the load point converter works in the constant voltage mode and the control performance is good, the load point converter and its load relative to the DC micro grid as the constant power load [5]. Under normal weather conditions, the distributed power, such as the photovoltaic and wind power, is generally working in the maximum power tracking (MPPT) mode. At this time, the load converters can be seen as the constant power loads [2].

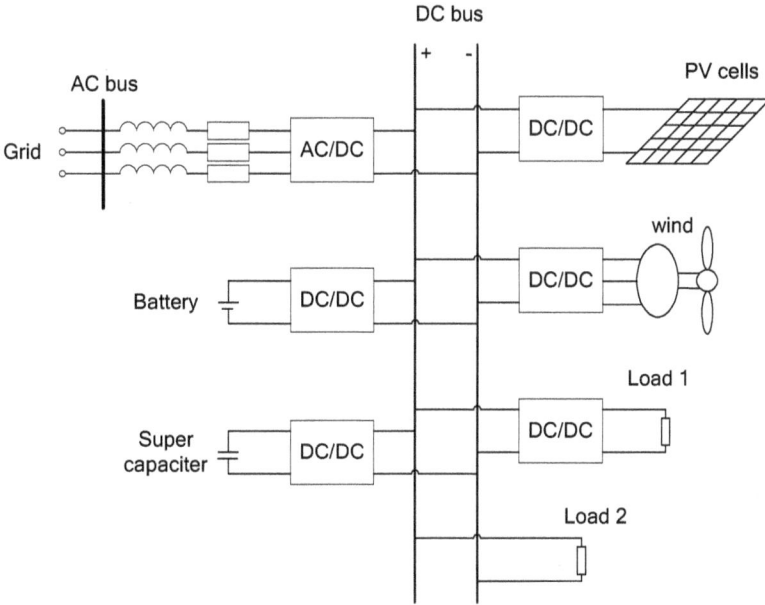

Fig. 1. The typical structure of a DC microgrid

In the following, the situations in which the Boost converter maintains the DC bus voltage of the microgrid were analyzed.

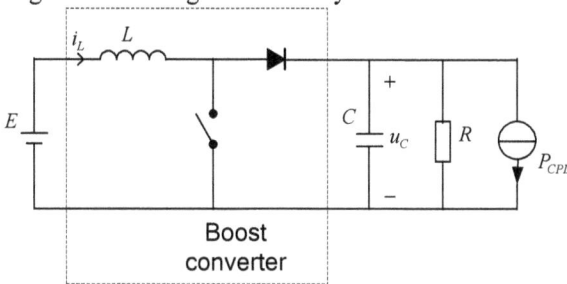

Fig. 2. The simplified model of a dc microgrid which supports bus voltage by the Boost converter

The simplified model of the DC microgrid is shown in Figure 2, in which the distributed power source is connected to the DC bus voltage by the Boost converter.

The circuit equation in Figure 3 is written

$$\begin{cases} L\dfrac{di_L}{dt} = E - (1-d)u_C \\ C\dfrac{du_C}{dt} = (1-d)i_L - \dfrac{u_C}{R} - \dfrac{P_{CPL}}{u_C} \end{cases} \tag{1}$$

The characteristic equation (1) is linearized at the equilibrium point of the system, and the condition of the system can be obtained [6]

$$P_{CPL} < \frac{U_C^2}{R} \tag{2}$$

The power of the constant power load in the stable system must be less than that of the resistive load. However, the typical DC microgrid contains about 80%-85% of the constant power load, the 15%-20% of the resistive load [4], so the system is difficult to be stable. So some measures must be taken to improve the stability of the DC microgrid.

3. Stabilization of DC microgrid based on passive damping

By (1), the DC microgrid can be equivalent to the model shown in Figure 3.

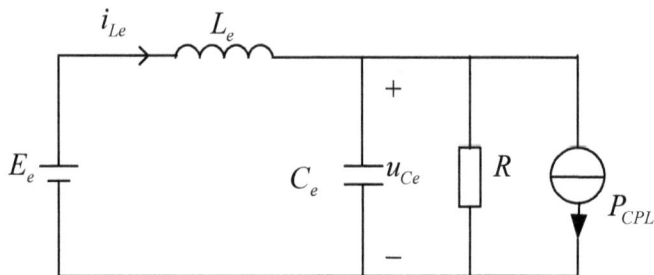

Fig. 3. The equivalent model of a dc microgrid

Where $E_e = \dfrac{E}{1-d}$, $L_e = \dfrac{L}{(1-d)^2}$ and $C_e = C$.

Due to the negative impedance of the constant power load, the system's damping is negative, resulting in the system instability [5]. In order to improve the stability of the system, the passive damping circuit was add to the system. After adding passive damping circuit, the system's simplified model is shown in

456

Figure 4. The method of determining the damping resistance and the damping capacity of the damping circuit are derived.

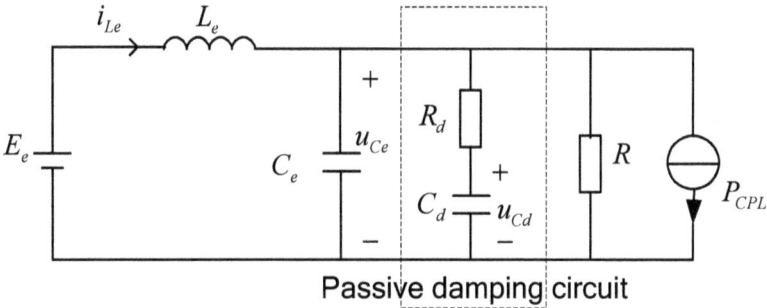

Fig. 4. The simplified model of a dc microgrid with passive damper

Hypothesis

$$\begin{cases} x_1 = i_{Le} \\ x_2 = u_{Ce} \\ x_3 = u_{Cd} \end{cases} \tag{3}$$

The state equation of the system is available

$$\begin{cases} L_e \dfrac{dx_1}{dt} = E_e - x_2 \\ C_e \dfrac{dx_2}{dt} = x_1 - \dfrac{x_2 - x_3}{R_d} - \dfrac{x_2}{R} - \dfrac{P_{CPL}}{x_2} \\ C_d \dfrac{dx_3}{dt} = \dfrac{x_2 - x_3}{R_d} \end{cases} \tag{4}$$

From the above equations, the equilibrium point of the system can be obtained

$$\begin{cases} X_1 = I_{Le} = \dfrac{U_{Ce}}{R} + \dfrac{P_{CPL}}{U_C} \\ X_2 = U_{Ce} = E_e \\ X_3 = U_{Cd} = E_e \end{cases} \tag{5}$$

From the above equation, it can be seen that the adding of the passive damping circuit does not change the system's equilibrium point.

The equation (4) is linearized at the equilibrium point, and is organized in the matrix form.

$$\begin{bmatrix} \hat{x}_1' \\ \hat{x}_2' \\ \hat{x}_3' \end{bmatrix} = \begin{bmatrix} 0 & -\dfrac{1}{L_e} & 0 \\[2mm] \dfrac{1}{C_e} & -\dfrac{1}{C_e}\left(\dfrac{1}{R_d}+\dfrac{1}{R}-\dfrac{P_{CPL}}{U_{Ce}^{2}}\right) & \dfrac{1}{C_e R_d} \\[2mm] 0 & \dfrac{1}{C_d R_d} & -\dfrac{1}{C_d R_d} \end{bmatrix} \begin{bmatrix} \hat{x}_1 \\ \hat{x}_2 \\ \hat{x}_3 \end{bmatrix} \tag{6}$$

Hypothesis

$$\frac{1}{R}-\frac{P_{CPL}}{U_{Ce}^{2}}=-\frac{1}{R_L} \tag{7}$$

Where $-R_L$ is the small signal equivalent load impedance.

From (6) and (7), the system's characteristic equation can get

$$s^{3}+\left[\frac{1}{C_e}\left(\frac{1}{R_d}-\frac{1}{R_L}\right)+\frac{1}{C_d R_d}\right]s^{2}+\left[\frac{1}{L_e C_e}-\frac{1}{C_e R_L C_d R_d}\right]s+\frac{1}{C_e L_e C_d R_d}=0 \tag{8}$$

By the Ross-Holzer stability criterion, the necessary and sufficient conditions for stable system can be deduced

$$\begin{cases} C_d R_d < (C_d+C_e)R_L \\[2mm] C_d R_d > \dfrac{L_e}{R_L} \\[2mm] \dfrac{C_d+C_e}{C_e}\left[\dfrac{1}{C_d R_d}-\dfrac{1}{(C_d+C_e)R_L}\right](C_d R_d - \dfrac{L_e}{R_L})>1 \end{cases} \tag{9}$$

Hypothesis

$$R_0=\sqrt{\frac{L_e}{C_e}} \tag{10}$$

$$n=\frac{C_d}{C_e} \tag{11}$$

When C_d is fixed, the damping of the system changes with the change of R_d. The optimal damping of the system is obtained when Rd satisfy following equation [7]

$$R_d=\frac{R_0}{n}\sqrt{\frac{(n+2)(3n+4)}{2(n+4)}} \tag{12}$$

From (10), (11) and (12), the system stability condition can be determined

$$\frac{R_o}{R_L}<n\sqrt{\frac{2(n+2)}{(n+4)(3n+4)}} \tag{13}$$

4. The Simulation Experiment

In order to verify the effectiveness of the proposed method, the Boost converter was used to maintain the voltage of the micro grid. The simulation model of the

DC micro grid is built by Matlab/Simulink. The system structure is shown in Figure 5.

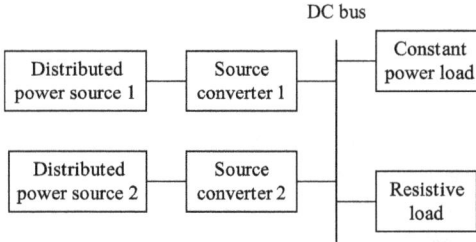

DC bus

Fig. 5. The DC microgrid with two generators and two loads

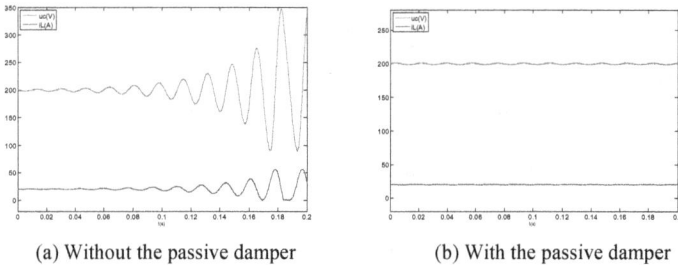

(a) Without the passive damper   (b) With the passive damper

Fig. 6. The bus voltage and inductance current of DC microgrid supported by Boost converter

When the passive damping circuit does not added to the system, the characteristic value of the system is 25 + j374.17. As the real part of the characteristic value is greater than zero, the DC bus voltage and inductor current diverge until the inductor current drops to zero, as shown in Figure 6(a).

In order to maintain the stability of the DC microgrid, the passive damping circuit is added to the DC bus. The passive damping circuit parameters are as follows: $R_d= 20.97\Omega$, $C_d= 0.145mF$. By (8) the characteristic value of the system in the equilibrium point is $-0.43\pm j351.93$, -373.45. As the real part of the eigenvalues is less than zero, the system is stable. DC microgrid bus voltage u_C and the Buck converter inductor current i_L waveform as shown in Figure 6(b).

5. Conclusion

Aimed to the problem that the constant power load may cause the instability of the DC bus voltage in the DC microgrid, this paper increases the damping of the system by adding the passive damping circuit on the DC bus, thus ensuring the stability of the system. This method does not need to change the control mode of the original converter, and the realization is simple. The simulation results verify the validity of the proposed method.

References

1. Hiroaki Kakigano, Yushi Miura, Toshifumi Ise. Distribution voltage control for DC microgrids using fuzzy control and gain-scheduling technique[J], IEEE Transaction on Power Electronics, 2013, 28(5):2246-2257.
2. Andre Pires Nobrega Tahim, Daniel J. Pagano, Eduardo Lenz, etc. Modeling and stability analysis of islanded DC microgrids under droop control[J], IEEE Transaction on Power Electronics, 2015, 30(8):4597-4607.
3. Weijing Du, Junming Zhang, Yang Zhang. Stability criterion for cascaded system with constant power load[J], IEEE Transactions on Power Electronics, 2013, 28(4): 1843-1851.
4. Amr Ahmed A. Radwan, Yasser Abdel-Rady I. Mohamed. Linear active stabilization of converter-dominated DC microgrid[J]. IEEE Transaction on Smart Grid, 2012, 3(1):203-216.
5. Active damping in DC/DC power electronic converters: a novel method to overcome the problems of constant power loads, IEEE Transactions on Industrial Electronics. 2009, 56(5): 1428-1439.
6. A. Emadi, B. Fahimi, M. Ehsani. On the concept of negative impedance instability in advanced aircraft power systems with constant power loads[C], The 34th Intersociety Energy Convers. Eng. Conference, Vancouver, BC, Canada, Aug, 1999:1-11.
7. Mauricio Cespedes, Lei Xing, Jian Sun. Constant-power load system stabilization by passive damping[J], IEEE Transaction on Power Electronics, 2011, 26(7):1832-1836.

Passive Location of Anomaly Target Based on Space-Borne ADS-B

Sun-quan Yu[†], Li-hu Chen and Yan Li

College of Aerospace Science and Engineering,
National University of Defense Technology,
Changsha, 410073, China
[†]E-mail: 15574857554@163.com

For the purpose of national security and defense, it is of great importance to detect and locate spacecraft invaded into our country airspace. A satellite with ADS-B receivers can realize universal air traffic surveillance services by providing global ADS-B coverage to oceans, the poles and remote areas[1]. But the ADS-B system cannot work properly if the GNSS fails or the data have been tampered before transmission. On the other hand, the data transmitted by non ICAO targets will be encrypted. In this paper, we put forwarded a passive location of anomaly target based on space-borne ADS-B signal, which includes positioning with single satellite; twin satellites, and the one with formation flying and the one with satellites constellation. In particular, the basic concepts, characteristics, critical technologies and the performance for these applications are discussed, before finally proposed a space-borne ASD-B, which adopted from the modification of above methods.

Keywords: Space-based ADS-B; Passive Location; Anomaly Target.

1. Introduction

ADS-B (Automatic Dependent Surveillance-Broadcast) will be the global standard for ATM and will bring significant benefits to commercial airlines through the world before 2020. Space-based ADS-B is the idea to place sensitive receivers on board satellite (low earth) orbit, which can receive ADS-B packages and relay them to the relevant stakeholders. It can improve safety, efficiency and reliability of the aircraft[2]. And the space-borne ADS-B has already attracted much attention from many countries and organizations owing to its great advantages (e.g. Danish GomX-1 satellite[3], German Proba-V satellite[4], FAA's NextGen plan[5]).

However, on the one hand, ADS-B system relies on GNSS (Global Navigation Satellite-Based Systems), and the system itself cannot verify the location information transmitted by the target. So the ADS-B system cannot work properly the GNSS fails or the data have been tampered before transmission. On the other hand, the data transmitted by non ICAO targets will

461

be encrypted. It is of great significance if we can locate this type of aircraft, for the purpose of national defense and society security.

The position can be estimated by measuring one or more location-dependent signal parameters such as angle of arrival, time of arrival, received signal strength or Doppler frequency shift. Passive location of anomaly target based on Space borne ADS-B signal will not be affected by geography and climatic conditions, and it can guarantee an accurate, rapid, and timely detection of targets.

2. Localization with Single Satellite

There are two methods that are designed for passive location of single satellite: Doppler frequency shift or arrival angle measurements. Localization by arrival angle measurements includes many methods (e.g. amplitude-comparison direction-finding method, time-difference direction-finding method, and spatial spectrum estimation method). Japan has developed an experimental system using S-band multi-beam phased array antenna on the experimental Engineering Test Satellite VI. And phase interferometer is also important in the direction-finding system. But all the direction-finding and position systems require high-performance satellites (e.g. big antenna, large spacing elements). ADS-B receivers are implemented on LEO satellites using carrier frequency of 1090 *MHz*, and the Doppler frequency shift can reach 30 *kHz*. This characteristic of space borne ADS-B signal is appropriate for passive location by frequency measurements of single satellite.

2.1. *The principle*

It is assumed that the receiver location and velocity are known and therefore the emitter location can be estimated using the Doppler frequency shift between receiver and emitter. The relationship between observed frequency f_i and emitted frequency f_T at a certain time t_i is given by:

$$f_i = f(1 + |\vec{V_i}| \cos(\theta_i) / c) \tag{1}$$

where c is the velocity of light; θ_i is the angle between the satellite's forward velocity and the line of sight from the object to the satellite; V_i is the velocity of the satellite.

Assume that V_i and f_T is known, θ_i will be estimated if we can measure f_T. In 3-D space, we can get a cone with the satellite position as its vertex and angle with θ_i. If we can get two cones in this way and considering that the emitter is deployed on the surface of the earth, the intersection of these three surfaces, defines two points. One of the points is the location of emitter, and the other one is called "fuzzy point". And we shall get rid of the "fuzzy point" in practical application.

Since the phased-array antenna is introduced and it would be an important technology of space-borne ADS-B, we can eliminate "fuzzy point" by amplitude-comparison using two channels (antennas). By comparing the relative amplitude of the signals in the two beams, its position can be determined.

2.2. Critical technologies

Currently, the measurements of satellite position and attitude can be estimated precisely (e.g. the measuring accuracy can be reached $10m$ for position and 10^{-2} degree for attitude). So the measurement of frequency is more critical than that of position and attitude. For ADS-B signal, the frequency estimation accuracy should achieve several hertz, which satisfies the request of the localization accuracy of several kilometers.

Passive location using the Doppler frequency shift with single satellite relies on a group of data which is obtained from repeated measurements, so we cannot locate the target in real time. Actually, this method is originally designed for stationary target. Motion compensation must be introduced to accommodate moving targets.

3. Double-star Positioning Using TDOA/FDOA

3.1. The principle

Double-star positioning system uses a method of cross locating. A hyperbolic and an ellipsoid surface is determined by TDOA and FDOA, respectively. The earth's surface is introduced as the third surface. And two of them intersect in a curved line which in turn intersects with the third hyperboloid in a point corresponding to the unknown three-dimensional target position.

Let $\vec{v}_i$ be the velocity of satellites $i,(i=1,2)$, r_i is the distance between satellites and ground at a certain time. The time difference τ and frequency f_d between the two satellites are given by:

$$\tau = (r_2 - r_1)/c \qquad (1)$$

$$f_d = \Delta f_1 - \Delta f_2 \qquad (2)$$

Where Δf_1 and Δf_2 are the Doppler frequency shift of two satellites, respectively; c is the speed of light. Let $(x_1,y_1,z_1),(x_2,y_2,z_2),(x_3,y_3,z_3)$ be the coordinates of the two receivers and emitter in the earth coordinate system. So we have:

$$r_i = \sqrt{(x_i - x)^2 + (y_i - y)^2 + (z_i - z)^2}, i = 1,2 \qquad (3)$$

463

$$\Delta f_i = \frac{1}{\lambda} \frac{\vec{v}_i \cdot \vec{r}_i}{\|r_i\|} \frac{f_0}{c \cdot r_i} [v_{ix}(x_i - x) + v_{iy}(y_i - y) + v_{iz}(z_i - z)], i = 1,2 \qquad (4)$$

where λ is the wavelength of carrier frequency; f_0 is the carrier frequency; $(v_{ix}, v_{iy}, v_{iz}), (i = 1,2)$ are the components of two satellites' velocity along x,y,z axes. Considering that emitter is located at the earth's surface with radius of R, so we have:

$$R = \sqrt{x^2 + y^2 + z^2} \qquad (5)$$

A hyperbolic, an ellipsoid and a sphere surface are obtained by equation (1), (2) and (5), respectively. The intersection of these three surfaces is the location of the emitter.

Location equations of double-star positioning system are nonlinear equations, and parameters of them are too many to solve directly. Therefore, we use numerical approximation to solve these equations. And there is also a "fuzzy point" of the solution.

3.2. Error analysis

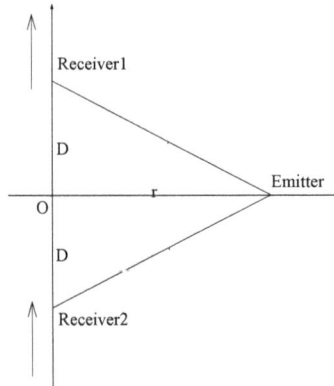

Figure 1. Error analysis of double-star positioning system.

Positioning error of double-star positioning system is composed of two parts: the measuring error and geometric factor. The measuring error is determined by satellites' self-localization error and measuring errors of parameters. Geometric factor is determined by the geometrical relationship between emitter and receivers. Keeping receivers synchronized in velocity and the same in motion direction by controlling the attitude of satellites, we can get the minimum positioning error if the emitter is located just below the midpoint of the line connecting two receivers.

where 2D is the distance between two satellites; r is the distance between emitter and midpoint of base line; $(x_1, y_1), (x_2, y_2)$ are coordinates of two receivers; $\vec{v}_1 = \vec{v}_2 = \vec{v}$ is the speed of two receivers. We can get the GDOP (Geometric Dilution of Precision) by minimum mean square error estimation:

$$GDOP = \sqrt{\frac{\lambda^2 r^4}{4v^2 D^2} \delta_{fd}^2 + \frac{r^2}{4D^2} c^2 \delta_\tau^2} \tag{6}$$

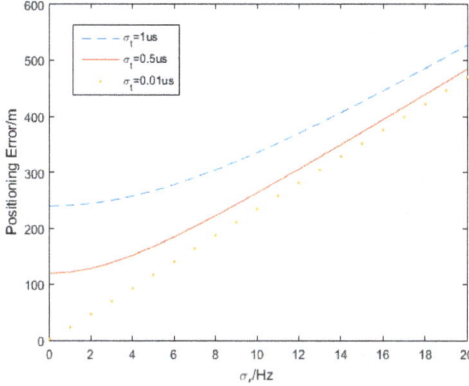

Figure 2. The relationship between measurements error and positioning accuracy.

In order to get high positioning accuracy, the value of GDOP should be as small as possible. For $D=500km$, $r=800km$, $v=7.5km/s$, $1/\lambda = f = 1090MHz$. From this figure we can see that in order to achieve positioning accuracy of $300m$, FDOA with 10 Hz and TDOA with 1 μs are required. Therefore, double-star positioning system relies on not only modern satellite technology, but also high-precision of FDOA/TDOA estimation.

4. Three-satellites Location System

4.1. The principle

This method uses multiple distributions of receivers at different positions in space receiving emitter signal, and measuring signal TDOA (time difference of arrival). The time difference corresponds to the space stations of a group focused on two surfaces. The multiple set of double-surface intersection is the location of emitter. TDOA location technology system has high positioning accuracy compared to the method with single or double satellite. The system includes one master satellite and two concomitant satellites. Receiving system on each satellites is composed of four parts: signal detection, TDOA extraction, position solving and data processing.

After the TDOA is estimated, the equation for determining the emitter location is established:

$$r_i = \sqrt{(x-x_i)^2 + (y-y_i)^2 + (z-z_i)^2}$$
$$r_0 = \sqrt{(x-x_0)^2 + (y-y_0)^2 + (z-z_0)^2} \tag{7}$$
$$r_i - r_0 = cD_i, i = 1,2$$

Where (x, y, z) is the coordinate of the emitter; $(x_i, y_i, z_i), (i = 1,2,3)$ are coordinates of three satellites;

And we also have the earth's surface model as ellipsoid:

$$x^2 / N + y^2 / N + z^2 / [N(1-e^2)] = N \tag{8}$$

The height of flights is about 8000~10000m, which is relatively small comparing to the radius of the earth (about 6400km). We are more interested in flights' latitude and longitude rather than their height in aviation surveillance.

4.2. *Accurate estimation of TDOA*

Accurate estimation of TDOA in three-star positioning system is important. Studying on TDOA is relatively comprehensive and mature. There are many solutions for accurate estimation of TDOA aiming at different characteristics of signal and transmission environment. The following table summarizes feature and application range of each method.

Table 1. Algorithms for TDOA estimation.

Algorithm	Feature and application range	Disadvantages
Generalized correlation algorithm	Simple and easy to implement	Sensitive to noise
Method for parameter	Simple and easy to implement	Need a prior knowledge
High order statistics	Fit for incoherent noise	Great computation
Cross-ambiguity function	Taking the Doppler effect into account	Great computation
Adaptive method	Without prior knowledge	Slow rate of convergence
MMSE	Simple and easy to implement	Execution
Wavelet transform method	Good anti-noise performance	Need a prior knowledge

5. Constellation Positioning

5.1. *The principle*

Distance intersection method can be used to solve the emitter position when more than four satellites are in sight of the emitter. Let $\rho_1, \rho_2, \rho_3, \rho_4$ be pseudo-range between an emitter and four satellites. The position of satellites can be

466

estimated by ephemeris computation. Equations of distance intersection method are given by:

$$\rho_i = \sqrt{(x-x_i)^2+(y-y_i)^2+(z-z_i)^2}+c\cdot\delta t, i=1,2,3,4 \qquad (9)$$

where c is the speed of light; δt is the receiver clock error. That's how GPS locate a target.

5.2. *Accurate estimation of TOA*

Constellation positioning system requires accurate estimation of TOA. The ADS-B signals are ASK signals at a carrier frequency of 1090 MHz with a data rate of about 1Mbits/s[6]. Fig. 3 shows the ADS-B signal data format. The former four 8 μs signal pulse is preamble. The follow-up a total of 112 μs transmit information contains the data block. The preamble has good recognition ability, and can be used as a standard test location for TOA testing.

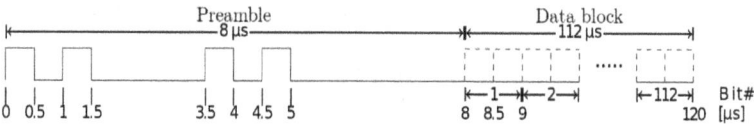
Figure 3. ADS-B signal waveform.

Each bit lasts 1 μs and the time of rising edge and falling edge of pulse are less than 10 ns according to the standard given by ICAO.

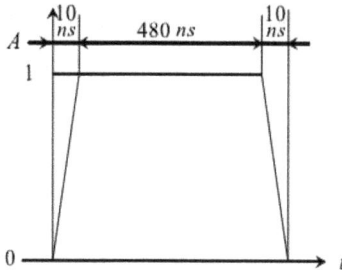
Figure 4. Pulse structure.

We will discuss the precision of measuring value of TOA by detecting the rising edge (the first pulse usually) [7]. Let T_0 be the time of arrival over base time, where signal power is equal to the threshold at this point. Let A be amplitude of pulse; t_r be rise time; ΔT_0 be measuring error caused by noise; $n(T_0)$ is the amplitude of noise at T_0. It is assumed that slope of rise edge with noise is the same as that without noise.

$$A/t_r = n(T_0)/\Delta T_0 \qquad (10)$$

467

Measuring accuracy of TOA is the root mean square value of measuring error of TOA:

$$\delta T_0 = t_r \left[\left(\Delta T_0 \right)^2 \right]^{1/2} / \left(A^2 / n^2 \right)^{1/2} = t_r / \left(2S / N \right)^{1/2} \tag{11}$$

where S/N is the SNR of intermediate frequency part in receiver. Signal power is proportional to effective value. Time of rising edge is restricted by IF band width, which can be given by $t_r \approx 1/B$. Let $S = E / \tau$ and $N = N_0 B$, then we have:

$$\delta T_0 \approx \left(\tau / \left(2BE / N_0 \right) \right)^{1/2} \tag{12}$$

where τ is the pulse width; E is the power of signal; N_0 is the noise per unit bandwidth. We can get the theoretical precision of TOA from equation(12), and this conclusion is based on geometrical relations. So it's not the optimum estimation. Using maximum likelihood method, inverse probability method or other statistics methods, the root mean square error of TOA is given by:

$$\beta^2 = \frac{\int_{-\infty}^{\infty} \left(2\pi f \right)^2 \left| S(f) \right|^2 df}{\int_{-\infty}^{\infty} \left| S(f) \right|^2 df} \approx \frac{2B}{\tau} \approx \frac{2}{\tau t_r} \tag{13}$$

Where β is the effective band widths of signal; $2E/N_0$ is the matched filter's peak signal to noise ratio. When $B_\tau \gg 1$ for rectangular pulse:

$$\delta T_0 = \left[\tau / \left(4BE / N_0 \right) \right]^{1/2} \approx 1 / \left[\beta \left(2E / N_0 \right)^{1/2} \right] \tag{14}$$

So we have the optimum measuring accuracy of TOA:

$$\delta T_0 = \left[\tau / \left(4BE / N_0 \right) \right]^{1/2} \approx 1 / \left[\beta \left(2E / N_0 \right)^{1/2} \right] \tag{15}$$

The main factor of measuring error is noise, and all algorithms are interested in reducing the effect of noise. Only one pulse has been used in the algorithm mentioned above. Considering that the noise is independent identically distributed in M bits (M=52/116, there is no pulse during the last half of the preamble, the time of which is 4 μs), and then they will be cancelled out. The accumulated SNR will increase M-fold if we use all M pulse to measure TOA. So the accuracy of TOA is given by:

$$\delta T_0 = t_r / \left(2MS / N \right)^{1/2} = 1 / \left[\beta \left(2ME / N_0 \right)^{1/2} \right] \tag{16}$$

We can see that the accuracy of TOA will be reduced to $1/\sqrt{M}$ of that measured by single pulse from equation(16).

6. Conclusion

Passive location of anomaly target based on Space-borne ADS-B provides a new tools for aviation surveillance. Passive location with single satellite cannot locate the target in real time and has low accuracy. Double-star positioning system with TDOA/FDOA can locate the target in real time, but requires high accuracy measurements of TDOA & FDOA. When there are more than three satellites in sight, we can locate them in real time and with high accuracy, but the location system is rather complex. Finally, We improved the measurement accuracy of TOA by using multiple pulses of ADS-B signals.

Acknowledgments

This work is funded by the Natural Science Foundation of China under Grant 61302092. Work partially supported by the program ZDYYJCYJ20140701 granted by National University of Defense Technology. Work partially supported by the program ZDYYJCYJ20140701 granted by National University of Defense Technology.

References

1. Trillingsgaard, K.F., et al., Space Based ADS-B. Aalborg University, (2011).
2. Werner, K., J. Bredemeyer, and T. Delovski, ADS-B over Satellite. 2014).
3. Alminde, L., et al., Gomx-1: A nano-satellite mission to demonstrate improved situational awareness for air traffic control. 2012).
4. Delovski, T., et al., ADS-B over satellite the world's first ADS-B receiver in Space. 2014).
5. Noschese, P., S. Porfili, and S.D. Girolamo. Ads-b via iridium next satellites. in Digital Communications-Enhanced Surveillance of Aircraft and Vehicles (TIWDC/ESAV), 2011 Tyrrhenian International Workshop on. 2011. IEEE.
6. Zhang, C. and Y. Wang. 1090ES signal generator simulation design. in Multimedia Technology (ICMT), 2011 International Conference on. 2011. IEEE.
7. Hong, W., et al., Accurate estimation of TOA and calibration of synchronization error for multilateration. Systems Engineering and Electronics, (2013)(04): p. 835-839.

On-Orbit Sensor Calibration Method for Navigation Satellite

Hao-guang Wang[†], Guo-qiang Wu and Tao Bai
The Shanghai Engineering Center for Microsatellites,
Shanghai, China
[†]E-mail: sacrrot@163.com

In this paper, an autonomous calibration method for sensor of navigation satellite is presented. Based on the requirement of pointing to earth, the sensors are calibrated by use of the ephemeris, and a calibration scheme designed to prevent random noise influence. The on-orbit data verification indicates the availability and feasibility of the method, and it is easy to realize in onboard software.

Keywords: Calibration; Navigation Satellite; Sensor; Ephemeris.

1. Introduction

With the development of aerospace science and technology, the navigation satellite system to supply positioning and guiding service has become an indispensable tool in people's daily life, such as global position system (GPS), Beidou system of China. The prime requirements for these satellites are long lifetime, high reliability, and full autonomy. Therefore, several kinds of attitude sensor are deployed on the satellite, and variable attitude determining method are developed by combining different sensors. The sensors or methods can be switched by onboard software. But the sensors' measure will deviate the real attitude in the presence of changes of sensor's inner parameters and mounting matrix error, which are generated by the space physical environment (such as the mechanical environment, thermal environment, magnetic environment, etc.). In addition, the switch between different attitude determining methods will cause a slight fluctuation of attitude. All these cases will influence the navigation task. Therefore, it is very important to design calibration method so that keep the accuracy and consistency of attitude.

In view of above requirements, some sensors' calibration method are designed. An on-orbit calibration for star sensor and gyroscope, which uses extended Kalman filter (EKF) to estimate the mounting error, constant bias and drift based on the error model of sensors [1,2,3]; Based on these algorithm [1,2,3], a new error model of sensor is provided, and misalignment is calibrated [4,5]. The algorithms provide above need an accuracy error model of sensor, and the

470

adopted EKF has 9 dimension of state equation at least. It is difficult to realize in onboard software on the basis of existing computing capacity.

The purpose of this paper is to design an autonomous sensor calibration method for the navigation satellite system, which require less computing power. The main work includes the algorithm of correcting misalignment of sensors (star sensor, static infrared earth sensor, sun sensor) based on the ephemeris of navigation payload, and bias error of static infrared earth sensor.

2. The Analysis of Sensor Error Model and Effect

The errors calibrated include two types: one is the system error generated by the inner noise of sensor; the other is the mounting misalignment.

Because of the low dynamic of navigation satellite, the random noise of sensor has small influence to the attitude. In addition, the constant bias for star sensor is negligible. In this paper, the mounting error and constant bias are considered only. Based on the above analysis, the error models for star sensor is built as following.

$$\mathbf{q}_i = \mathbf{q}_\sigma^s \circ \mathbf{q}_s^b \circ \hat{\mathbf{q}}_i \ . \tag{1}$$

Where, $\mathbf{q}_i$ is real quaternion under nominal mounting matrix; $\mathbf{q}_\sigma^s$ is the quaternion of misalignment matrix; $\mathbf{q}_s^b$ is the quaternion of nominal mounting matrix; $\hat{\mathbf{q}}_i$ is the quaternion under the real mounting matrix.

The error model for other sensors is described as following.

$$\mathbf{r} = \mathbf{R}_e \mathbf{R}_s (\hat{\mathbf{r}} + \mathbf{r}_0) \ . \tag{2}$$

Where, $\mathbf{r}$ is the real attitude under nominal mounting matrix; $\mathbf{R}_e$ is the misalignment matrix; $\mathbf{R}_s$ is the nominal mounting matrix; $\hat{\mathbf{r}}$ is the attitude of sensor; $\mathbf{r}_0$ is the constant bias of sensor.

3. The Method of Calibration

The prime requirement for attitude of navigation satellite is to keep pointing to the earth (the +Z axis in body frame), so that payload can normally broadcast the ephemeris used by ground receiver. Furthermore, the attitude must satisfy the power supply requirement, which means sunlight is parallel to normal of solar panel. In this way, the output of earth sensor is constant, and the output of other sensors are time-varying. Based on the character of navigation satellite, an effective calibration method is proposed for sensors as shown as following.

471

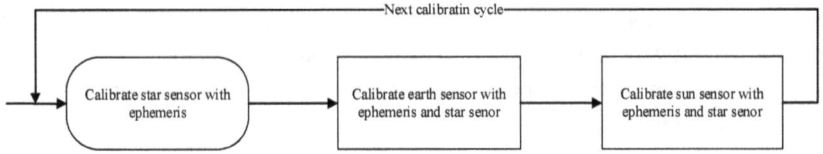

Fig. 1. The method of calibration for sensor

3.1. The algorithm for star sensor

For the navigation satellite, the equation between the quaternion of star sensor and the earth vector in J2000 inertial frame is given as following.

$$\mathbf{E}_i = -\mathbf{R}_{Err}\mathbf{R}_{sb}\mathbf{R}_{ss}\begin{bmatrix} 0 & 0 & 1 \end{bmatrix}^T \tag{3}$$

Where, $\mathbf{R}_{Err}$ is the rotation matrix produced by misalignment; $\mathbf{R}_{sb}$ is the nominal mounting matrix; $\mathbf{R}_{ss}$ the rotation matrix between star sensor coordinate frame and J2000 frame; $\mathbf{E}_i$ is the earth vector in J2000 inertial frame. Because the misalignment angle is very small, $\mathbf{R}_{Err}$ can be simplified as following formula.

$$\mathbf{R}_{Err} \approx \begin{bmatrix} 1-\varphi\theta\psi & \psi+\varphi\theta & -\theta \\ -\psi & 1 & \varphi \\ \theta+\varphi\psi & \theta\psi-\varphi & 1 \end{bmatrix}. \tag{4}$$

Where, φ, θ, ψ are the Euler angle. By substituting Eq. (4) into Eq. (3), φ, θ, ψ can be solved. Finally, the correcting mounting matrix can be described as $\hat{\mathbf{R}}_{sb} = \mathbf{R}_{Err}\mathbf{R}_{sb}$.

3.2. The algorithm for earth sensor

Based on the above analysis, the earth sensor's attitude is constant for navigation satellite. So, the equation between the earth vector in earth sensor body frame and the earth vector in J2000 inertial frame is given as following.

$$\mathbf{E}_i = -\mathbf{R}_{is}\mathbf{R}_{Err}\mathbf{R}_{sb}\mathbf{E}_e \tag{5}$$

Where, $\mathbf{R}_{is}$ the rotation matrix between satellite body frame and J2000 frame, given by the measurement of star sensor; $\mathbf{R}_{Err}$ is the rotation matrix produced by misalignment; $\mathbf{R}_{sb}$ is the nominal mounting matrix; $\mathbf{E}_e$ is the real earth vector in earth sensor body frame. Base on the error model of the earth

472

sensor, the constant bias can be present as a vector rotated from $\hat{\mathbf{E}}_e$. So, $\mathbf{E}_e$ can be described as $\mathbf{E}_e = \hat{\mathbf{E}}_e + \mathbf{R}\hat{\mathbf{E}}_e$, where $\hat{\mathbf{E}}_e$ is the measured attitude by earth sensor, and $\mathbf{R}$ is the rotation matrix between bias and $\hat{\mathbf{E}}_e$. Furthermore, the following equation is given.

$$\mathbf{E}_i = -\mathbf{R}_{is}\mathbf{R}_{err}\mathbf{R}_{sb}(\mathbf{I}+\mathbf{R})\hat{\mathbf{E}}_e = -\mathbf{R}_{is}\hat{\mathbf{R}}_{err}\hat{\mathbf{E}}_e \qquad (6)$$

The same as the algorithm for star senor, the error angle can be calculated.

3.3. The algorithm for sun sensor

By use of the quaternion of star sensor and the accurate solar model, the sun vector in satellite body frame can be obtained as the following equation.

$$\mathbf{S}_b = \mathbf{R}_{si}\mathbf{S}_i = \mathbf{R}_{Err}\mathbf{R}_{sb}\mathbf{S}_s \qquad (7)$$

Based on the above analysis, the error angle can be calculated. However, due to the satellite attitude change, resulting in the measurement of the solar vector in one track is changed, the constant error is not accuracy.

Because the ephemeris data is updated once every hour, for the purpose of preventing influence of random noise, the calibration period is 24 hours. Hourly, according to the above mentioned algorithm, satellite onboard software calculates the new mounting matrix, and records error angle between vector after calibrating and theory vector. Finally, the mounting matrix with the minimum error angle is selected as a new matrix for the next day.

4. The Result Of Calibration

For the purpose of verifying the method, a three-day on-orbit sensor data of a navigation satellite is processing. The proximity of attitude of sensor to the real satellite attitude is chosen as the evaluation criterion. The result of calibration is shown as Table 1, Figs. 2~4.

Table 1. The proximity of attitude of sensor to the real satellite attitude

Item	Before calibration	After calibration
The angle between the earth vector calculated by star sensor and $\mathbf{E}_i$	0.45°	0.015°
The angle between the earth vector calculated by earth sensor and $\mathbf{E}_i$	0.65°	0.05°
The angle between the sun vector calculated by sun sensor and the one calculated by sun model	0.58°	0.1°

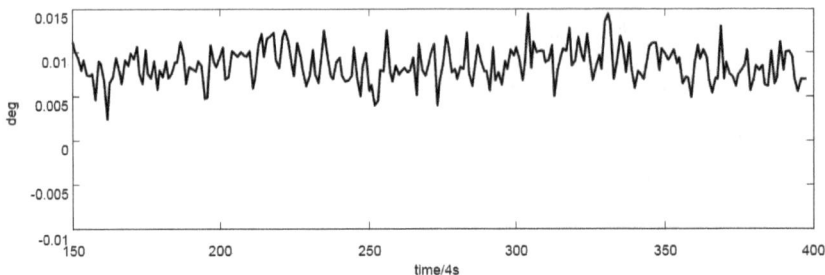

Fig. 2. The angle between the earth vector calculated by star sensor and E_i after calibration

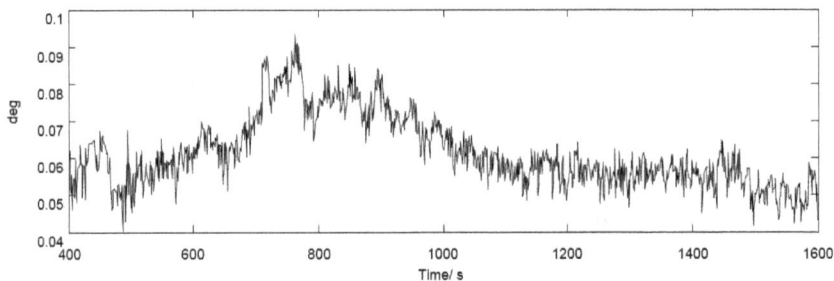

Fig. 3. The angle between the earth vector calculated by earth sensor and Ei after calibration

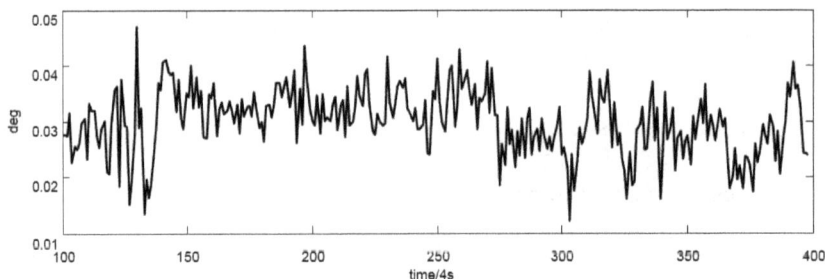

Fig. 4. The angle between the sun vector calculated by sun sensor and the one calculated by sun model

Figs. 2 ~ 4 illustrate that the calibration method can make the sensor's output more proximity to the real attitude, and result is better for star sensor and earth sensor. The reason is that the sun sensor has a bigger constant bias than star sensor, and the bias cannot be correct by the mounting matrix.

5. Conclusion

This paper presents a method to calibrate the system error and misalignment of sensors for navigation satellite autonomously, which is low-computing-consume and able to realize for onboard software. It is verified with the on-orbit data, the results show the method is credible, and it has guiding significance and engineering practical value.

References

1. Murrell J W. Precision attitude determination for multimission spacecraft[C] *Proceedings of the AIAA Guidance and Control Conference. AIAA-78-1248*, 1978:70-87
2. Mark E Pittelkau. Kalman filter for spacecraft system alignment calibration[J]. *Journal of Guidance, Control and Dynamics*, 2001, 24(6):1187-1195
3. Chen Xue-qin, Geng Yun-hai. On-orbit calibration algorithm for gyros/star sensor. *Journal of Harbin Institute of Technology*, 2006,38(8):1369-1373
4. Pal M, Bhat M S. Star Camera Calibration Combined with Independent Spacecraft Attitude Determination. *American Control Conference. St. Louis: AIAA*, 2009:4836–4841
5. Liu Lei, Liu Ye, Cao Jianfeng. A real time on-orbit calibration method for gyro combined calibration of star sensor. *Journal of Spacecraft TT&C Technology*, 2014, 33 (2): 152-157.

Research on Response of Memcapacitive and Meminductive Circuit Under Periodic and Aperiodic Input Signals

Yu Hang[†], Li-li Zhang and Zhi-yun Chen
*School of Electronic Science and Engineering,
Nanjing University of Posts and Telecommunications,
Nanjin, Jiangsu, China
[†]E-mail: b13020228@njupt.edu.cn*

As the fourth fundamental circuit element, Memristor has been achieving lots of attention since its realization. This paper presents the analysis of memcapacitive and meminductive circuit and investigate the response of the circuit to periodic and aperiodic input voltage signals. The simulation results show that the periodic input would change the memcapacitance sharply while the aperiodic input will not change the state of the memcapacitor. Similar results have been observed in meminductive circuit and we have also analysed this phenomenon.

Keywords: Memristor; Memcapacitor; Meminductor; Nonlinear Circuits.

1. Introduction

Major Headings Memristor is the fourth fundamental circuit element which was predicted by Leon O. Chua theoretically in 1971 [1]. Until 2008, memristor did not achieve much attention when memristor was realized by Strukov *et al* [2]. After the realization of this element, memristor has attracted more and more attentions in different fields [3-5] etc.

Pershin *et al* investigated the amoeba's learning behaviour by memristive circuit with LC contour, and the proposed circuit can change the state under the periodic external voltage which has the same feature with the amoeba's locomotion under the periodic variations of temperature or humidity [6]. Furtherly, using a set of LC contours or a single memcapacitor-based adaptive contour with memristive damping, models which can be responding to different frequencies have been presented [7]. It is suggested that the proposed memory models can adapt the internal frequency to the external stimulus and change to the expected state subsequently.

In this paper, we present the similar analysis of memcapacitive and meminductive circuit. In section 2, we describe the models and circuit, and in section 3 we investigate the response of the circuit to periodic and aperiodic input voltage signals. Before we presented the conclusion in Section 4.

476

2. Models and Circuits

In this paper, the capacitive and inductive models we use are as similar as the resistive model used in ref [6]. As shown in Fig. 1, C_M is the capacitance of memcapacitor. The change of C_M is described by the Eq. (1):

$$\frac{dC_M}{dt} = f(V_{CM})[\theta(V_{CM})\theta(C_M - C_{M1}) + \theta(-V_{CM})\theta(C_{M2} - C_M)]. \tag{1}$$

where $\theta()$ is a step function, C_{M1} and C_{M2} is the minimum and maximum value of memcapcitor, and V_{CM} is the voltage drop across the memcapacitor. The characteristic of $f(V)$ is described by several linear segments as shown in Fig. 1(a), and is defined as:

$$f(V) = -\beta V + 0.5(\beta - \alpha)(|V + V_T| - |V - V_T|). \tag{2}$$

where α and β are the slopes of the linear segments.

The change of inductance of meminductor we adopt is similar with the memcapacitor, which is described by :

$$\frac{dL_M}{dt} = f(V_{LM})[\theta(V_{LM})\theta(L_M - L_{M1}) + \theta(-V_{LM})\theta(L_{M2} - L_M)]. \tag{3}$$

where the value of L_M was limited from L_{M1} to L_{M2} ($L_{M1} < L_{M2}$). $\theta()$ and $f(V)$ are the same functions in the memcapacitive model.

In this work, the circuits we investigate are described in Figs. 1(b) and (c). Based on these circuits, we investigate the response of the memcapacitor and meminductor to the input periodic and aperiodic voltage signals.

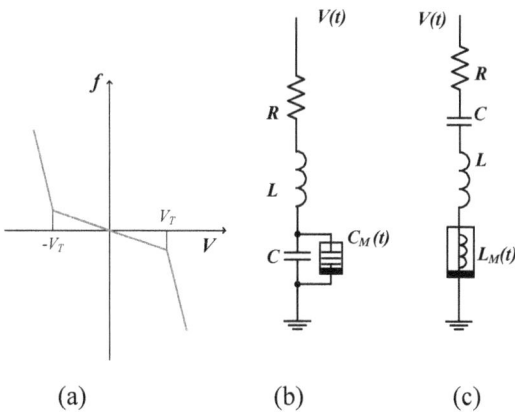

(a)   (b)   (c)

Fig. 1. The graph of $f(V)$ and the schematic of LC contours with mem-elements. (a) Sketch of the function $f(V)$. The slope is $-\alpha$ and $-\beta$ separately, and $|\beta| >> |\alpha|$. V_T are threshold voltages. (b) The memcapacitive model. (c) The meminductive model

3. Results and Discussion

The simulated results of the response of two circuits above are given in Fig. 2. The input voltage impulse is a triangular wave, which minimum is $-2V$ and width (t_w) is $0.5s$, as shown in Fig. 2. Under the periodic and aperiodic impulses, the change of C_M and L_M has been investigated.

The simulations for circuit with memcapacitor were made using the following parameters: $R=0.17\Omega$, $L=0.1H$, $C=0.2533F$, $C_{M1}=30nF$, $C_{M2}=200nF$, $C_M(0)=30nF$, $\alpha=10nF/(Vs)$, $\beta=3\mu F/(Vs)$ and $V_T=3V$. The simulations for circuit with meminductor were made using the following parameters: $R=0.17\Omega$, $L=0.01H$, $C=0.2533F$, $L_{M1}=0.001H$, $L_{M2}=1H$, $L_M(0)=0.09H$, $\alpha=0.0001H/(Vs)$, $\beta=0.1H/(Vs)$ and $V_T=2.8V$.

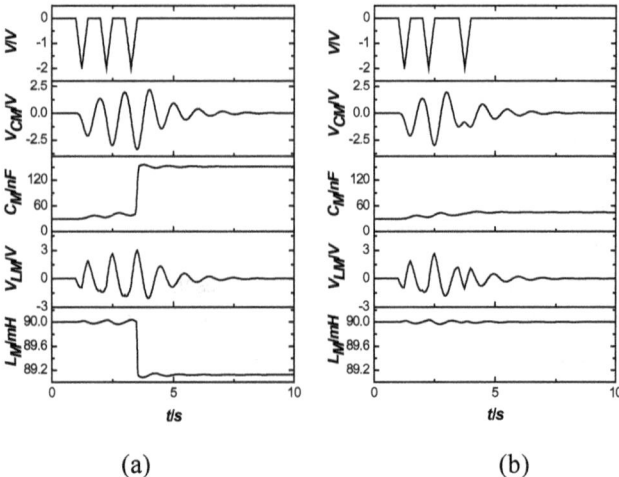

(a)　　　　　　　　　　　　　　　(b)

Fig. 2. Simulated results of two circuits. (a) Under periodic voltage. (b) Under aperiodic voltage

As shown in Fig. 2, the input signal $V(t)$ is composed of three negative impulses and if its frequency is equal to the resonant frequency of the LC contours, C_M or L_M will have a sharp step after the coming of the third impulse. But if the input signal was not periodic or the frequency was not equal to the resonant frequency, the phenomenon will not happen.

Based on the results above and the results of Ref. 6, we can see that, these three models have the similar phenomenon that the state of the mem-elements will change apparently under the stimulating of periodic voltage signals. The following part of this paper is trying to explain this phenomenon analytically.

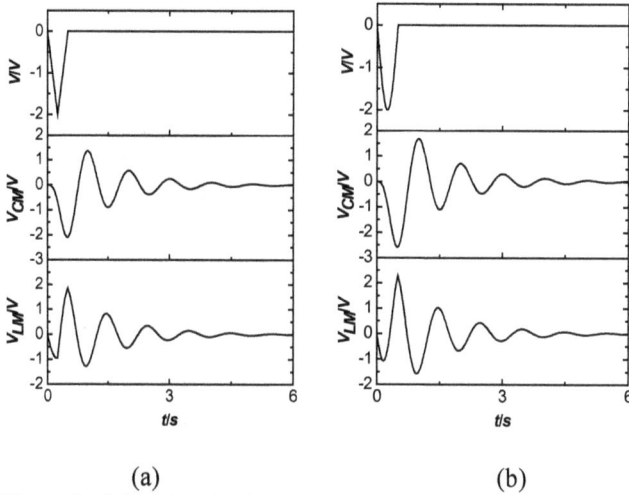

(a)   (b)

Fig. 3. The result of single impulse simulation. (a) Triangular impulse. (b) Sinusoidal impulse

As we all know, memcapacitor is nonlinear device, and the circuit is nonlinear circuit. Compared with linear circuit, nonlinear circuit is hard to analyze. In our work, C_{M2} we chosen is far less than C which is in parallel with memcapacitive, so the circuit can be considered as linear approximately. According to Fig. 1(b), following equation can be obtained:

$$\frac{d^2 V_{CM}}{dt^2} + \frac{R}{L}\frac{dV_{CM}}{dt} + \frac{1}{L(C+C_M)}V_{CM} = \frac{1}{L(C+C_M)}V. \tag{4}$$

Because C_M is far less than C, then it can be ignored and the equation above reads:

$$\frac{d^2 V_{CM}}{dt^2} + \frac{R}{L}\frac{dV_{CM}}{dt} + \frac{1}{LC}V_{CM} = \frac{1}{LC}V. \tag{5}$$

Dealt with Laplace transform, the following equation can be obtained:

$$V_{CM}(s) = \frac{1}{LCs^2 + RCs + 1}V(s). \tag{6}$$

Because Eq. (4) is a linear differential equation, it must satisfy superposition theorem, so the input signal $V(t)$ can be split into three single impulses. Eq. (5) should be solved when input signal $V(t)$ is a single impulse firstly, which can help us analyze the circuit easily. Fig. 4 shows the simulation of single impulse of triangular wave and sinusoidal wave. And $v(t)$ represents

479

single impulse and $v_{CM}(t)$ is the response of $v(t)$. $v(t)$ and $v_{CM}(t)$ can be seen in Fig. 3. $v(s)$ is the Laplace transform of $v(t)$, which is:

$$v(s) = \int_0^\infty v(t)e^{-st}\,dt. \tag{7}$$

According to Eqs. (7) and (8), Eq. (9) can be obtained:

$$v_{CM}(s) = \frac{1}{LC[(s+a)^2 + \omega^2]} \int_0^\infty v(t)e^{-st}\,dt. \tag{8}$$

In Eq. (9), we introduce parameters:

$$a = \frac{R}{2L}. \tag{9}$$

$$\omega = \frac{1}{\sqrt{LC}}. \tag{10}$$

It's very hard to get the inverse Laplace transformation of Eq. (8), but it is considered that it is not necessary to solve it directly. By the analysis of form of $v_{CM}(s)$, we get the approximate simulated formula of $v_{CM}(t)$:

$$v_{CM}(t) = |v_{CM\,min}|\,e^{-a(t-t_w)}\cos\omega t. \tag{11}$$

v_{CMmin} is the minimum of v_{CM}, and according to Fig. 3(a), it's about -2.1V. v_{CMmin} is hard to be calculated directly, and it can be obtained from the simulation. If parameters of LC contours change, v_{CMmin} will also change. So when changing parameters, we need to make the simulation again. Fig. 3(b) shows the changing of v_{CMmin} according to the different impulse. If $V(t)$ is composed of three impulses with different time-delay, according to superposition theorem, the V_{CM} can be as:

$$V_{CM}(t) = v_{CM}(t-t_1) + v_{CM}(t-t_2) + v_{CM}(t-t_3). \tag{12}$$

Because $v_{CM}(t)$ is periodic exponential decay signal, and if intervals between t_1, t_2 and t_3 are equal to the period in Eq. (11), resonance phenomenon will happen according to Eq. (12), so $V_{CM}(t)$ will achieve the minimum at (t_3+t_w).

$$V_{CM\,min} = v_{CM\,min}(1 + e^{-aT} + e^{-2aT}). \tag{13}$$

where T is the period in Eq. (11).

After superposition of two times, V_{CMmin} can be pretty small. From Fig. 1(a), we can see that, if V_{CM} exceeds threshold voltage V_T, $f(V_{CM})$ will change sharply, because $\beta >> \alpha$, and C_M will also change sharply. So if C_M have a sharp step, it means:

$$V_{CM\,min} = (1 + e^{-aT} + e^{-2aT}) < -V_T. \qquad (14)$$

If intervals between t_1, t_2 and t_3 are not equal to the period in Eq. (11), resonance phenomenon won't happen. And $|V_{CM}|$ won't be greater than $|-V_T|$, so C_M won't change sharply.

As the same as the analysis of memcapacitive model, the process for the analysis of meminductive model is similar. But, the $L_M(0)$ we have chosen is big enough which can be ignored. However, α is small enough, so that L_M can be seen as a constant $L_M(0)$ before exceeding threshold voltage. The inductance of the circuit can be seen as the sum of L and $L_M(0)$:

$$L_{sum} = L + L_M(0). \qquad (15)$$

Fig. 3 shows the simulation of the response of meminductive model to single impulse. Obviously, we have $|v_{LMmin}| < |v_{LMmax}|$, which means the condition of exceeding threshold voltage is:

$$V_{LM\,max} > V_T. \qquad (16)$$

As the same with the analysis of the circuit with memcapacitor, Eq. (17) will be:

$$v_{LM\,max} = (1 + e^{-aT} + e^{-2aT}) > V_T. \qquad (17)$$

In Eq. (17), we introduce parameters:

$$a = \frac{R}{2L_{sum}}. \qquad (18)$$

$$\omega = \frac{1}{\sqrt{L_{sum}C}}. \qquad (19)$$

Based on the analysis above, we can see that, Eq. (14) is the condition for state-changing of memcapative model, and Eq. (17) is the condition of meminductive model.

The key point of analyzing such memcapacitive circuit is C_M is far less than C in the circuit, so that C_M can be ignored. But if C_M could not be ignored, make sure α is small enough, so that when $|V_{CM}| < V_T$, C_M can be seen as a constant $C_M(0)$. These two situations can help us simplify the nonlinear circuit as linear circuit which can help us analyze the circuit more easily. Also, it is suggested that, meminductive and memristive model are under the same analyzing method.

4. Conclusion

In this paper, we investigate the response of memcapacitive circuit and meminductive circuit to input periodic and aperiodic voltage signals. The simulation results show that the periodic input would change the memcapacitance sharply while the aperiodic input will not change the state of the memcapacitor. Similar results have been observed in meminductive circuit and we have also analyzed this phenomenon.

References

1. L. Chua, Memristor-the missing circuit element, IEEE Trans. Circuit Theory **18**, 507 (1971).
2. D. Struckov, G. Snider, D. Stewart and R. Williams, The missing memristor found, Nature **453**, 80 (2008).
3. T.A. Wey and W.D. Jemison, Variable gain amplifier circuit using titanium dioxide memristors, IET Circuits Devices Syst. **5**, 59 (2010).
4. S. Duan, X. Hu and Z. Dong, Memristor-based cellular nonlinear/neural network: design, analysis, and applications, IEEE Trans. Neural Netw. Learn. Syst. **26**, 1202 (2015)
5. M.F. Chang, S.M. Yang and C.C. Kuo, Set-triggered-parallel-reset memristor logic for high-density heterogeneous-integration friendly normally off applications, IEEE Trans. Circuits Syst. II, Exp. Briefs **62**, 80 (2015)
6. Y.V. Pershin, S.L. Fontaine and M.D. Ventra, Memristive model of amoeba learning, Phys. Rev. E **80**, 021926-1 (2009)
7. F.L. Traversa, S.L. Fontaine and M.D. Ventra, Memory models of adaptive behavior, IEEE Trans. Neural Netw. Learn. Syst. **24**, 1437 (2013)

Fault Probability Model for Calculating Limiting Capacity of Transmission Line

Yong Xu[†], Min-fang Peng and Guang-ming Li

*College of Electrical and Information Engineering, Hunan University,
Changsha, China
†E-mail: xuyong57002126.com*

Liang Zhu, Hong-wei Che and Zheng-yi Liu

*State Grid Hunan Electric Power Company,
Changsha, Yuelu Zone, China*

Currently, the methods of calculation limiting capacity of transmission line are lack of economy and security. In this paper, based on the theory of the power circles and the conservative utility function, the proposed model can calculate limiting transmission capacity and meeting the security status of running transmission lines. Furthermore, on the condition that transmission lines meet economic and security policy, this paper utilizes the fault probability model that was improved by the conservative utility function to calculate limiting transmission capacity. Through analyzing the result of a calculation example, the paper can improve transmission capacity of existing transmission lines, and has great significance for promoting the future of grid transmission capacity.

Keywords: Transmission Lines; The Fault Probability Model; Conservative Utility Function; Transfer Capability.

1. Introduction

Economy and security of the transmission line are two important aspects considered of power system operation [1]. Under the premise of ensuring safety, making full use of the transmission capacity of the transmission line and improving transmission capacity have become a hot topic of current research [2]. Limiting transmission capacity of transmission line plays a decisive role in transmission grid capacity. the power system to transfer more capacity securely and stably, security and stability study of the limiting transmission of the transmission line is crucial.

Currently, common indicators reflecting lines include load factor, capacity-load ratio benchmark, overloaded rate, which are too conservative. For example, the reference of capacity-load ratio is the average load power of supply area; however the limiting transmission capacity of the transmission line cannot be truly reflected [3]. The article [4]considers the influence of reactive power on the limiting transmission capacity of the transmission line, voltage limit and safe operating. It can be calculated tolerable the limiting theoretical thermal, but this case will accelerate the aging of line and heat loss of line, which will increase maintenance cost of the transmission line.

Based on the literature [4] and the conservative fault probability model, this paper calculates the limiting transmission capacity of transmission lines and reclassifies the operating states of the power grid. Not only can this paper exploit the potential of transmission of transmission lines, but also gives full consideration to the safety of transmission lines and stable operation. It is important for the improvement of transmission capacity of the future grid.

2. The state of grid operation

Accurately determining the state of the power grid will play a crucial role in the stable operation of the power system. The grid in different operating states will affect the resistance of a number of factors, for example the line disturbances, random failure, the appliance fault and so on [5]. After the transmission line to be disturbed, according to the different resistance of fault and the probability of occurrence of different disturbances, the grid will be divided into non-operational state, unsafe operational state and security state.

(1) The non-operational state

Under that limit operation state, even small disturbances may also make the system crash. In this state small disturbances can make the grid caused widespread power outages. Its fault probability closes to 100%. The grid cannot in the state.

(2) The unsafe operational state

The short-running state can make grid resist small disturbances. When the grid is in this state, it cannot resist a massive fault line. Although you can run a short time, there is a higher probability of security risks.

(3) The security state

The grid having a strong ability can resist the most dominant disturbance and some general types of faults, with good economy and security. The transmission line should be in the long-running state. The probability of its faults should meet the security principle of economy within an acceptable range [6].

3. Safe and Feasible Calculating Model

3.1. Calculating model of limiting capacity basing on power circle

Transmission lines Π type equivalent model [7] is shown in Figure 1.

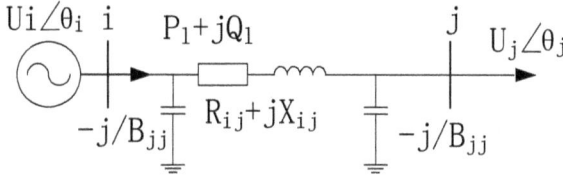

Figure 1. Π type equivalent circuit

For i and j being nodes in a line, the line L flowing into the apparent power can be expressed as:

$$S_{ij} = P_{ij} + jQ_{ij} = \dot{U}_i[(\dot{U}_i - \dot{U}_j)Y_{ij}]^* - jB_{jj}U_i^2 = U_i^2 Y_{ij}^* - \dot{U}_i \dot{U}_j^* Y_{ij}^* - jB_{jj}U_i^2$$
$$= U_i^2 G_{ij} - U_i U_j Y_{ij} \cos(\theta_i - \theta_j - \alpha_{ij}) + j[-U_i^2 B_{ij} - U_i^2 B_{jj} - U_i U_j Y_{ij} \sin(\theta_i - \theta_j - \alpha_{ij})] \quad (1)$$

Wherein: U_i, θ_i, U_j, θ_j respectively represent endpoint, i j corresponding voltage amplitude and phase angle. The above formula can be equaled to the following equation:

$$(P_{ij} - U_i^2 G_{ij})^2 + [Q_{ij} - (-U_i^2 B_{ij} - U_i^2 B_{jj})]^2 = (U_i U_j Y_{ij})^2 \quad (2)$$

The Complex power track of transmission line was shown in Figure 2.

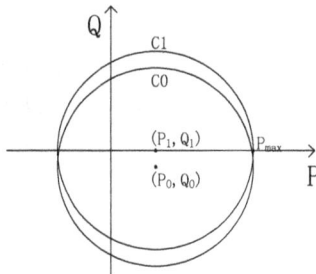

Figure 2. Running track of transmission line

When transmission line absorbing longitudinal inductance of reactive power is equal to the lateral capacitance of reactive power, the power circle will be located C1 position. When the circle moves to the C1 position, maximum active power P_{max} with the x-axis are intersecting. That is, operation state reaches the maximum theoretical thermal limit.

3.2. Calculating model of limiting capacity basing on utility function and the probability of fault

Working state of transmission line is decided by the state of the grid and Transmission Towers. We considered the faults probability that were caused by Transmission Towers. Analyzing the faults probability with increasing transmission capacity, we can fully consider the relationship between risk and benefits. Based on the probability model of initial faults caused by transmission line outage formula is as follows [8].

$$p_{ri} = \begin{cases} p_0 & 0 < P_i \le P_N \\ \dfrac{(1-p_0)(P_i-P_N)}{P_{max}-P_N} & P_N < P_i \le P_{max} \\ 1 & P_i > P_{max} \end{cases} \tag{3}$$

Wherein: p_{ri} is the probability of line fault. P_N is the rated power. p_0 is the probability of faults for the line operation at rated power.

Random faults and hidden faults existing in the grid will also have a huge impact on the operating state of the grid. For the pursuit of economic efficiency and in the premise of more sensitive to safety, the model uses the fault probability model that was improved by the conservative utility function. The math expression is as follows [9].

$$p = i \begin{cases} p_0 & 0 < P_i \le P_N \\ \dfrac{1}{C_1}(1-e^{-a_1(P_i-P_N)}) + p_0 & P_N < P_i \le P_{max} \\ 1 & P_i > P_{max} \end{cases} \tag{4}$$

a_1 and C_1 are positive. Moreover, The following mathematical expresses C_1 [11]

$$C_1 = 1 - e^{-a_1(P_{max}-P_N)} \tag{5}$$

When the transmission line running under the conservative fault probability model, which overestimates the probability of fault. We hope to transmission lines as much as possible to maintain stable operation. In the premise of the pursuit of economic efficiency, this model achieved better security. When transmission lines reach the limiting capacity, the probability of its faults will reach 1.

4. Example analyzing

In order to verify the reasonableness of the proposed model, we use IEEE14 bus system example to analyze. We calculated the results based on the model proposed in this paper. Taking into account the IEEE14 node system as a standard test system, we might first initialization parameters [10]. When the faults probability of transmission lines reaches 1, it will reach the limiting capacity. Calculation results are shown in Table 1.

Table1 Results of limiting capacity

Transmiss ion lines	Pmax	Conservative Pmax	Neutral probability of faults
1-2	1055.969	731.6	0.2829
2-4	329.1335	283.54	0.5642
2-5	333.888	286.79	0.5613
3-4	421.4819	333.82	0.4934
4-5	1428.282	883.15	0.1981
6-11	432.5925	351.47	0.5032
6-12	335.936	282.26	0.549
6-13	675.942	510.78	0.3997
9-10	853.365	601.43	0.3272
9-14	300.1773	256.3	0.5676
10-11	423.0982	337.76	0.4968
12-13	551.6714	424.09	0.4407
13-14	241.8528	208.89	0.5901

From Table 1, in the conservative limiting capacity, the probability of faults occurring within the line is still in an acceptable range. And when the power grid to run certain routes in the case of conservative limiting capacity, each other transmission line capacity will be improved largely. Therefore, in conservative limiting capacity, the lines still have a very high safety and stability, and still meet the long run, so when the transmission capacity of the line is not higher than the conservative limiting capacity, the grid is still in a safe and feasible state. When capacity of the transmission line than conservative limiting capacity, the probability of fault of the power grid is greatly increased. Small disturbances can be resisted, but there is a big security risk probability, which will lead an unsafe but feasible state to the grid.

5. Conclusion

This paper presented a model to calculate limiting capacity of transmission lines based on the conservative probability of faults, and on this basis, the operating status of the power grid are re-divided. The model is the use of transmission capacity to the limit, the probability of faults is 1. Combined with regard to the power circle considering of voltage limit and safe operating limits, the model reverse calculated of limiting capacity of transmission line. In addition, when the system is running in safe runnable circumstances, the model has greatly improved the degree of utilization of transmission line and fully excavates the transmission potential of transmission lines which has great significance on improving the future grid transmission capacity.

6. Acknowledgments

This work is supported by National Natural Science Foundation of China under Grant No. 61472128 and 61173108, and Hunan Provincial Natural Science Foundation of China No. 4JJ2150

References

1. Lei Xuejiao, Pan Shijuan, Guan Xiaohong, et al. Transmission Safety Margin Constrained Unit Commitment In Power Systems [J]. Proceedings of the CSEE, 2014, 34(13).
2. Wang Yanling, Han Xueshan, Kong Lingyuan, et al. Value On Loadability Of Transmission Line Under Operation [J]. Proceedings of the CSU-EPSA, 2012,24(5).
3. Sun Yanlong, Kang Chongqing, Chen Songsong, et al. Low Carbon Network Evaluation Index System and Method [J]. Automation of Electric Power Systems, 2014, 38(17): 157-162.
4. Yang Bangyu, Peng Jianchun, He Yuqing. A Novel Algorithm for Calculating Available Transfer Capability Using Power-circle Determine Stability Constraints [J]. Proceedings of the CSU-EPSA, 2008, 28(34): 66-71.
5. Liu Ruiye, Li Weixing, Li Feng, et al. State Features and Tendency Indices Embodying Abnormal Operation of Power Systems [J]. Automation of Electric Power Systems, 2013, 37(20): 48-53
6. Ding Ping, Zhou Xiaoxin, Yan Jianfeng, et al. Calculation of Online Total Transfer Capability in Bulk Interconnected Grid Integrating Rationality and Security Principle [J]. Proceedings of the CSU-EPSA, 2010, 30(22):1-6.
7. Hu Z, Wang Xi-fan. A probabilistic load flow method considering branch outages[J]. IEEE Transactions on Power Systems,2006, 21(2): 507-514.
8. Jiang Le, Liu Junyong, Wei Zhenbo, et al. Running State and Its Risk Evaluation of Transmission Line Based on Markov Chain Model [J]. Automation of Electric Power Systems, 2015, 35(13): 52-57.
9. Liu Ruoxi. Research on Several Fundamental Theory and Algorithm of Power System Operational Security [D]. Beijing: School of Electrical and electronic Engineering, 2012.
10. Qi Xianjun, Ding Ming. Risk analysis and Utility Decision-making of Spinning Reserve Scheme in Power Generating System [J]. Automation of Electric Power Systems, 2008, 32(13):9-13.

Feasibility Study of Utilizing UPFC to Reduce the Interaction in Multi-Infeed HVDC Transmission System

Ding-wen Hu[†], Xian-wei Wang, Jun Zhang and Jin-long Wu

Xi'an XJ Power Electronics Technology Co. Ltd.
Xi'an, 710075, China
[†]E-mail: hudw890@163.com

Interaction between inverter stations is the facing challenge in multi-infeed direct current (MIDC) transmission system. Unified power flow controller (UPFC) is a new generation equipment and using UPFC to solve the facing problem in MIDC system is a new idea. UPFC is the combination of static synchronous compensation (STATCOM) and static synchronous series compensator (SSSC). SSSC is equivalent to impedance connected in series in line. This paper reviewed the achievement of STATCOM in MIDC system and analyzed the relationship between multi-infeed interaction factor (MIIF) and series impedance. The result shows that it's effective to reduce interaction by changing series equivalent impedance. Considering the effect of STATCOM, it's feasible to reduce MIIF index and enhance MIDC system performance through UPFC.

Keywords: HVDC; MIIF; MIDC; STATCOM; SSSC; UPFC.

1. Introduction

High Voltage Direct Current (HVDC) transmission systems, which are fit for transmission electric power in long distance and large capability, play an important role in the policy of "West power to East" and "National Network"[1]. So far, there are seven HVDC transmission lines feed into East China Power Grid and eight HVDC transmission lines feed into Guangdong Power Grid. In the future, there are more HVDC lines plan to feed into these two regions. Undoubtedly, it will be the most complicated multi-infeed direct current (MIDC) transmission system in the world[2-3]. Compared with single infeed system, the interactions between subsystems in MIDC system have a great influence on the stability of the whole system[4]. So exploring the way to reduce the interactions in MIDC system is meaningful.

Many works on MIDC system have been done and many valuable results have been achieved. Multi-infeed interaction factor (MIIF) is the most important achievement defined by CIGRE WG B4-41 to measure the degree of interactions between subsystems in MIDC system[5]. On the basis of MIIF, some worthy results have been achieved. Denis[6] and Guo et al.[7] give out the theoretical computing method which overcomes the disadvantage of MIIF that it is achieved through simulation earlier. Critical MIIF was proposed by Shao and Tang[8] to estimate the probability of commutation failure and secondary commutation failure. The influence factors of MIIF were reported by Chen et al. [9], Xiao et al. [10], Shao, and Tang [11]. It can be concluded from these articles that the main influence factors of MIIF are electrical distance, DC control model, active load and the strength of AC system. Also, there are some researchers focus on enhancing system stability. Some results show that static synchronous compensation (STATCOM) can enhance the stability of MIDC system and reduce the fault recovery time.

However, no achievements give out the method of reducing MIIF. Just under such a background, this paper focus on how to reduce MIIF and enhance system stability. Unified power flow controller (UPFC) is proposed in this paper to reduce MIIF. The research result shows that it is feasible of using UPFC to reduce MIIF.

2. Define and Computing method of MIIF

An indicator based on the observed ac voltage change at one inverter ac bus for a small ac voltage change at another inverter bus provides a first level indication of the degree of interaction between two HVDC systems. This interaction factor is called MIIF and is defined mathematically as:

$$MIIF_{ji} = \frac{\Delta U_j}{\Delta U_i} \tag{1}$$

where ΔU_j is the observed voltage change at bus j for a small induced voltage change at bus i. It is suggested that the voltage change at bus i should be 1% [5]. In references [6], the calculation method of MIIF was proposed. It's based on the matrix Z_{bus}, the $N \times N$ bus impedance matrix of the AC system. It's given out as:

$$MIIF_{ji} = \frac{Z_{ji}}{Z_{ii}} \tag{2}$$

Where Z_{ii} is the i-th row, i-th column element of the impedance matrix, also is called self or driving-point impedance. Z_{ji} is the j-th row, i-th column element of the impedance matrix, also is called the transfer impedance.

3. Brief introduction of UPFC

UPFC could be seen as the combination of STATCOM and static synchronous series compensator (SSSC) through parallel connection at their DC side, shown in Fig. 1. It possesses the advantage of STATCOM and SSSC both. What's more,

490

UPFC has the capacity of four-quadrant operation. That means UPFC can absorb or issue active power and reactive power. In this way, it's possible to control the power flow by utilizing UPFC.

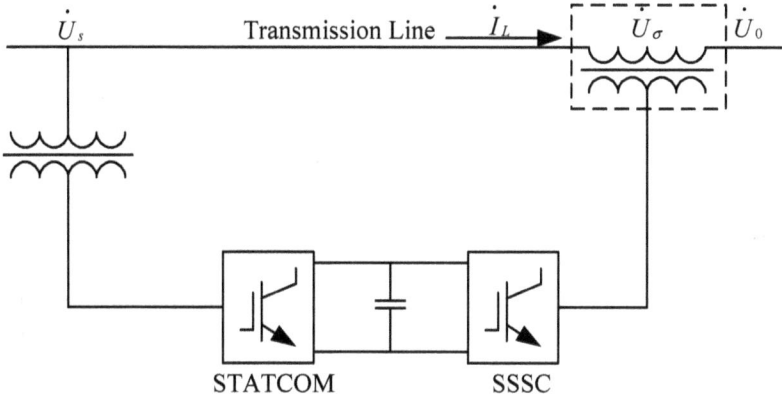

Fig. 1. Schematic diagram of UPFC

UPFC can realize many function, such as voltage regulation, series compensation, phase-angle regulation and multi-function power flow control, by control the voltage $\dot{U}_\sigma$, which is the series compensation voltage controlled by SSSC. Thus, the part of dashed box in Fig. 1 can be seen as an adjustable impedance which can be shown as:

$$Z_{eq} = \frac{\dot{U}_\sigma}{\dot{I}_L} \tag{3}$$

where $\dot{I}_L$ is the line current.

4. Review of the effect of STATCOM in MIDC system

Since UPFC could be seen as the combination of STATCOM and SSSC, thus we can analysis the effect of STATCOM and SSSC respectively. Many work have been done on the application of STATCOM in MIDC system. The researches focus on using STATCOM to improve the system stability, reducing the risk of communication failure, the control strategy of STATCOM in MIDC system and the optimal distribution of STATCOM in the network.

References [12] concluded that STATCOM can reduce the critical multi-infeed short circuit ratio (CMISCR) in MIDC system. That means STATCOM can enhance the system stability. The concept of CMISCR was proposed by Lin, et al. [1] and is the evaluation criterion of the strength of AC system. Lin, et al.

also given out the definition of multi-infeed short circuit ratio (MISCR) which takes into consideration of the interaction between DC links on the basis of short circuit ratio (SCR) which is the most important index to reflect the system stability. If MISCR is greater than CMISCR, then the MIDC system is strong. Otherwise, the system is weak.

STATCOM can restrain the commutation failure caused by voltage fluctuation [13]. In most cases, the fluctuation of voltage is caused by the variation of reactive power. STATCOM can absorb or generate reactive power to keep the bus voltage stable. The major reason for commutation failure is voltage sag which makes the voltage-time integral area decrease and the extinction angle less than the critical extinction angle for a recommended value $7°$ [8]. Thus STATCOM can reduce the interaction of DC links.

Other study points focus on control strategy and the distribution of STATCOM in network. Saichand and Padiyar, [14] had proposed two controls methods: constant reactive power and constant AC voltage. Constant AC voltage control method was used widely to restrain commutation failure. The choosing of installation node of STATCOM is determined by the sensitivity of voltage and reactive power [15].

In conclusion, STATCOM can improve the performance of MIDC system. It have been used in China Southern Power Grid and located at three stations: Mumian station, Beijiao station and Shuixiang station with the capacity 200Mvar for each. This project completed and went into operation at 2013.

5. The effect of series impedance to MIIF

So far, no paper shows the relationship between series impedance in grid and MIIF. While SSSC could be seen as series adjustable impedance, it's necessary to research the effect of series impedance. This section gives out a case study on the basis of simplified real regional power grid in Eastern China.

5.1. Geographical Structure of power grid

The geographical structure of the regional power grid is shown as Fig. 2. There are eight 500kV nodes are reserved and all the 220kV and below power grids are equivalent to load. Node 1, 2 and 3 are the DC line terminals, the AC buses of the inverter stations.

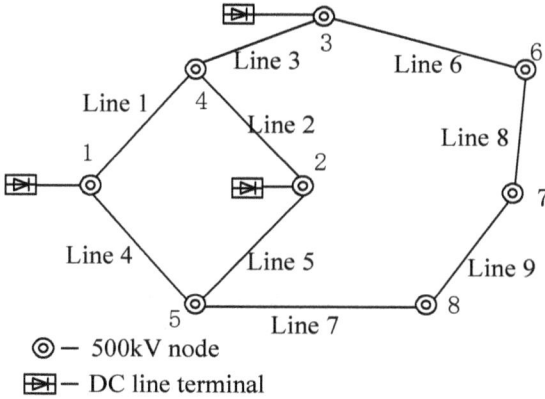

○ — 500kV node

⊞ — DC line terminal

Fig. 2. Geographical structure of the regional power grid

5.2. MIIF value before series compensation

According to the equation (2), the MIIF values between inverter stations could be achieved. Since the self-impedance at each bus is unequal, that is $Z_{ii} \neq Z_{jj}$, the MIIF value of i to j is differ to the value of j to i, which lead to a result of $MIIF_{ji} \neq MIIF_{ij}$. Table 1 gives out the MIIF values between each stations. The self-impedance of inverter station shown in the row of table 1 is the denominator of equation (2).

Table 1. MIIF matrix

	Inverter Station 1	Inverter Station 2	Inverter Station 3
Inverter Station 1	1	0.6	0.36
Inverter Station 2	0.36	1	0.3
Inverter Station 3	0.29	0.41	1

5.3. The effect of series impedance

In general, PI equivalent model is selected as the transmission line model if the line length is less than 241.39km [16]. Since the line length in regional power grid is relatively short, the model with series impedance could be simplified as Fig. 3. Z is the series impedance, Z_0 is the equivalent series impedance of transmission line and Y is the equivalent parallel admittance.

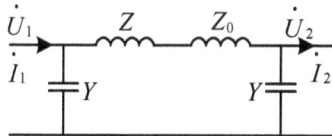

Fig. 3. Transmission line model with series impedance

493

In this paper, the series equivalent impedance of UPFC is considered as reactance or capacitance only. Choosing the equivalent series reactance of 50km length transmission line as the reference impedance, that is:

$$Z_b = 0.285 \times 50\Omega = 14.25\Omega \tag{4}$$

There are nine transmission line in this regional power grid. The series impedance inserted into each line will lead to a different results. Fig. 4~ Fig. 12 show the research results. Z varies from -0.5pu to 1pu.

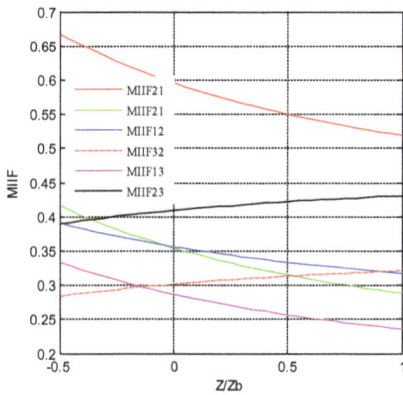

Fig. 4. Inserted in line 1

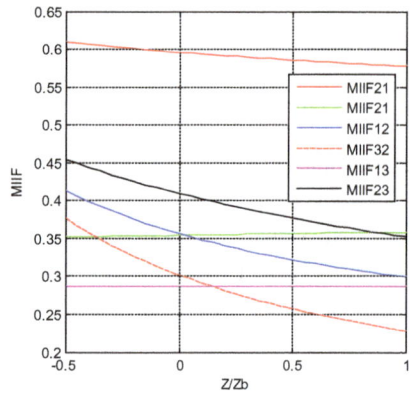

Fig. 5. Inserted in line 2

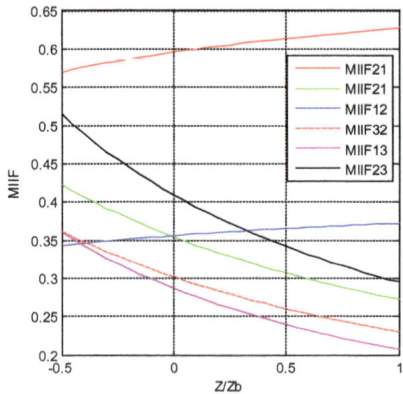

Fig. 6. Inserted in line 3

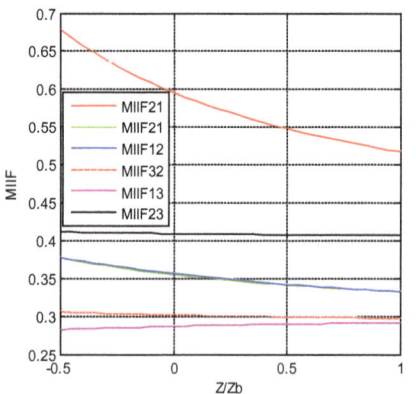

Fig. 7. Inserted in line 4

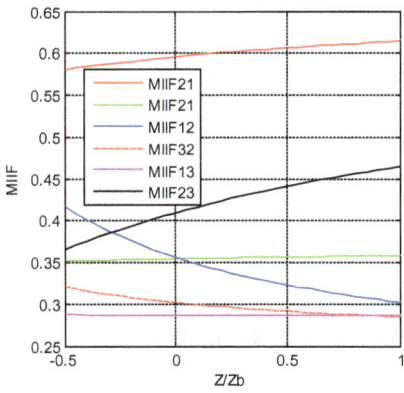

Fig. 8. Inserted in line 5

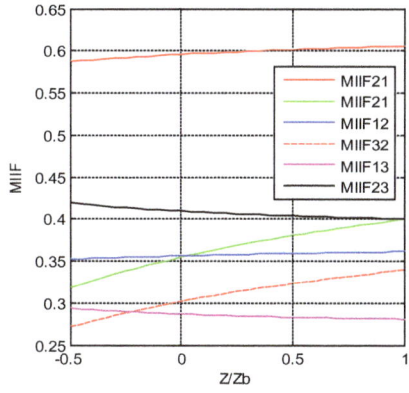

Fig. 9. Inserted in line 6

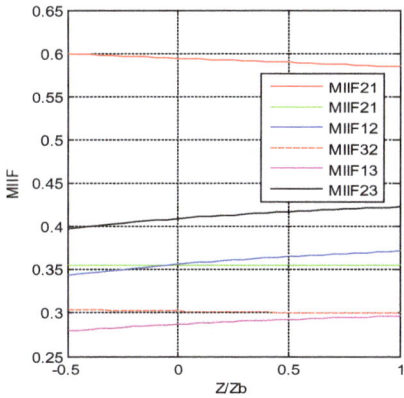

Fig. 10. Inserted in line 7

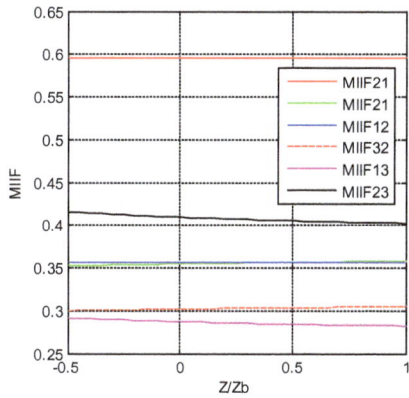

Fig. 11. Inserted in line 8

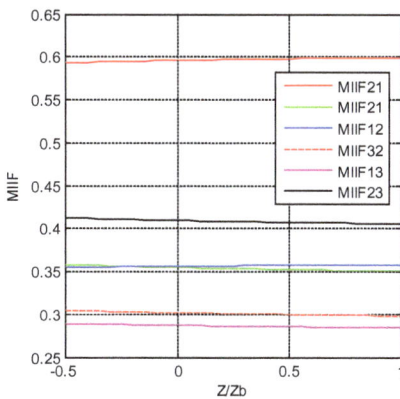

Fig. 12. Inserted in line 9

495

Fig. 4~ Fig. 12 show the relationships of MIIF and series impedance while the series impedance inserted into different line. When the impedance inserted in line1~line4, some of MIIF indexes drop evidently with the series impedance vary from capacitive to inductive. Also a little index varies hardly, or even increase slightly. In Fig. 8, only $MIIF_{12}$ decreases very clearly and $MIIF_{23}$ increases very clearly. When the impedance is tandem in line 6, see Fig. 9 $MIIF_{21}$ and $MIIF_{32}$ vary obviously. $MIIF_{21}$ and $MIIF_{32}$ are lower if the equivalent impedance is capacitive. From Fig. 10~ Fig. 12, we could find that MIIF indexes hardly vary with the series impedance when inserted in line 7, line 8 and line 9.

From all above, we achieve that series impedance can reduce MIIF index when it is inserted in special line. In this case, when the impedance is tandem in line1~line4, the effect is significant. Thus researchers could take advantage of this character to reduce the MIIF indexes. So from this aspect, the series side of UPFC, that is SSSC, have the capability to adjust the MIIF index. Considering the parallel side, STATCOM, UPFC may contribute a lot to MIDC system.

6. Conclusion

UPFC is a new kind of power electronic equipment which is powerful in power system control, such as power flow control, voltage control, phase angle control and power factor control. Utilizing UPFC to reduce the interaction in MIDC system and enhance the system stability is a new idea. STATCOM have been used in MIDC system, and proved to be effective. While SSSC or UPFC has no engineering project in the field of MIDC. The analysis in section 5 shows that UPFC can reduce the MIIF index through adjust the equivalent impedance. However, to solve the facing problem in MIDC system through UPFC need a deeper and comprehensive research.

References

1. W. F. Lin, Y. Tang, and G. Q. Bu, "Definition and application of short circuit ratio for multi-infeed AC/DC power systems," *Proceedings of the CSEE*, vol. 28, no. 31, pp. 1-8, 2008.
2. Y. Shao, Y. Tang, and X. J. Guo, "Analysis on commutation failures in multi-infeed HVDC transmission systems in North China and East China Power Grids Planned for UHV Power Grids in 2015," *Power System Technology*, vol. 35, no. 10, pp. 9-15, 2011.
3. Y. Shao, Y. Tang, and X. J. Guo, "Analysis of influencing factors of short circuit ratio of multi-infeed AC/DC power sys-tems," *Power System Technology*, vol. 35, no. 8, pp. 64-68, 2011.
4. Y. Shao, and Y. Tang, "Current situation of research on multi-infeed AC/DC power systems," *Power System Technology*, vol. 33, no. 17, pp. 24-30, 2009.

5. CIGRE Working Group B4.41, "Systems with multiple DC infeed," CIGRE, Paris, 2008.
6. L. H. A. Denis. (2007). Voltage and power interactions in multi-infeed HVDC systems. [online]. Available:http://www.eeh.ee.ethz.ch/uploads/tx_ethpublications/Voltage_an d_Power_Interaction_Report.pdf
7. X. J. Guo, J. B. Guo, and C. S. Wang, "Practical calculation method for multi-infeed short circuit ratio influenced by characteristics of external characteristics of DC system," *Proceedings of the CSEE*, vol. 35, no. 9, pp. 2143-2151, 2015.
8. Y. Shao, and Y. Tang, "A commutation failure detection method for HVDC systems based on multi-infeed interaction factors," *Proceedings of the CSEE*, vol. 32, no. 4, pp. 108-114, 2012.
9. X. Y. Chen, M. X. Han, and C. R. Liu, "Impact of control modes on voltage interaction between multi-infeed AC-DC system," *Automation of Electric Power Systems*, vol. 36, no. 2, pp. 58-63, 2012.
10. J. Xiao, and X. Y Li, "Analysis of multi-infeed interaction factor of multi-infeed AC/DC system and multi-terminal AC/DC system and its influencing factor," *Power System Technology*, vol. 38, no. 1, pp. 1-7, 2014.
11. Y. Shao, and Y. Tang, "Analysis of influencing factors of multi-infeed HVDC system interaction factor," *Power System Technology*, vol. 37, no. 3, pp. 794-799, 2013.
12. C. Y. Guo, D. Li, and M. L. Peng, "Influences of STATCOM on critical operating limits of HVDC transmission system," *High Voltage Engineering*, vol. 41, no. 7, pp. 231-239, 2015.
13. Q. Kang, X. C. Ma, and M. Yuan, "Application of cascaded STATCOM in prevention of HVDC in Commutation Failure," Academic Annual Meeting of CES, Shan Dong, China, Oct19-20, 2013.
14. K. Saichand, K. R. Padiyar, "Analysis of voltage stability in multi-infeed HVDC systems with STATCOM," IEEE International Conference on Power Electronic, Drives and Energy System, Bengaluru, India, Dec16-19, 2012.
15. Y. Zhao, C. Hong, and Y. G. Zeng, "The voltage support strength factor(VSF) for multi-infeed HVDC systems and its application in CSG's STATCOM allocation scheming," *Southern Power System Technology*, vol. 8, no. 4, pp. 22-26, 2014.
16. T.T. Liu, J. Wen, and G. Y. Qiao, "Research on the influence of harmonic in multi-infeed direct circuit system," *Advanced Technology of Electrical Engineering and Energy*, vol. 35, no. 1, pp. 42-47, 2016.

Analysis and Optimization of Power Output Characteristics of CT Power Supply Based on Starting Current

Chang-tao Chen, Jie Lou, Yu-ying Zhang and Ruo-chen Guo

School of Electrical Engineering, Shandong University,
Jinan, 250061, China
E-mail: cct_1989@163.com

Lei Duan

State Grid Weifang Power Supply Company
Weifang, 261000, China
E-mail: duanl2008@yeah.net

The actual CT model is established under the Saber software by using the MCT (Magnetic Component Tool) in this paper. The research demonstrates that the output power of CT first increases and then decreases with the increase of the filter capacitor voltage. The output power has a maximum point and the maximum power is independent on the turn number when the voltage drop of the rectifier is ignored. But when considering the voltage drop of the rectifier, the maximum power increases with the increase of turn number, and becomes saturated in the end. This has important significance for the selection of turn number. In the output power characteristic curve, there are two operating points corresponding to the load power consumption and the point with higher voltage is the stable operating point. In order to characterize the stability margin, the half power width is defined. Taking the minimum output 0.5W as an example, the turn number is selected by simulation and the hardware circuit is designed. The correctness of the analysis is verified by experiments.

Keywords: Current Transformer; Power Output Characteristic; Starting Current; Saber; Turn Number.

1. Introduction

The online monitoring device is widely used on high-voltage transmission lines with the development of smart grid. But due to the installation position and operating environment, it is difficult to provide stable and reliable power supply. At present, the common methods used are solar cell [1], CT (current transformer)[2,3] and so on. But the output power of the solar cell can be influenced by the weather, and its size is large, installation is not convenient, battery life is short. The using of CT as the power supply has the characteristics of small size, convenient installation and large outputting power. As a result, it has a good application prospect and is widely concerned by scholars.

Because of the fluctuation of the line current, the CT has the problems such as the iron core saturation, low power output when line current is low. In order to solve the problems, many scholars have done some research on it. For example, by adding air gap to suppress the magnetic saturation in the case of large line current [4]; Using battery as the auxiliary power supply when the output power of CT is low [5]; By using nanocrystalline material which has high permeability as the core, to improve the output power in the case of small current and reduce the starting current [6]. These studies have solved some of the problems.

However, in the previous studies, the output characteristic of CT which connects to the actual voltage conversion circuit is less studied. This is different from the characteristics of the CT which directly connects to the resistive load. This paper will discuss the power output characteristics of the CT when it connects the rectifier, filter and the DC-DC circuit. Then give the turn number selection method based on starting current and design corresponding circuit.

2. Power output characteristics of CT when connected to rectifier circuit

When the CT is directly connected to the resistance load, the theory proves that there is a maximum power output point. When the core loss is ignored, the maximum output power is determined by the following formula [7]:

$$P_{max} = \frac{\pi \mu f S I_1^2}{l} \tag{1}$$

Where μ is magnetic permeability, S is core area, l is average length of magnetic path, f is current frequency, I_l is line current (RMS).

It can be known from the formula (1) that the maximum power is only related to the primary side current, core material and size, and is independent of the turn number. When the secondary side coil of CT is connected with the rectifier and filter circuit, the output characteristics are different from that of the directly connecting resistive load. In order to study the power output characteristics of this kind of situation, the model is established and the simulation is carried out under the Saber software.

In this paper, chose the ring type silicon steel core as the material, and its size is $40 \times 95 \times 30$mm (inner diameter, outer diameter, height). The model of the CT is established under the Saber software with its MCT (Magnetic Component Tool). The B-H curve of the core is described by Jiles-Atherton model [8]. Its primary side coil is 1 turn and flows through a constant sinusoidal current. The simulation circuit is shown in Fig. 1. When the circuit reaches to steady state, the filter capacitor voltage is stable, at this time, the output power of CT is equal to the power consumption of resistive load. The output power of the CT is obtained by calculating the power consumption of the resistor. Then the relationship between the output power of CT and the filter capacitor voltage is obtained by changing the load resistance.

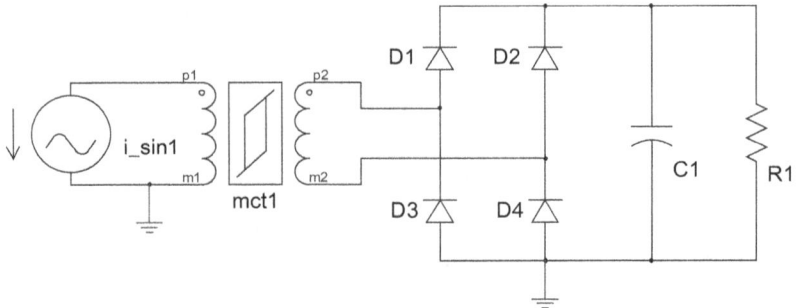

Fig. 1 Simulation circuit of power output characteristic

First simulate the power output characteristics when ignoring the voltage drop of rectifier. The primary side current is 8A RMS. The results are shown in Fig. 2 (a), and from left to right, turns are 50, 100, 150, 200. It can be seen from the figure that, the output power of CT increases first and then decreases with the increase of filter capacitor voltage. There is a maximum power output point, and the maximum output power is independent of the turn number. The voltage corresponding the maximum power output point increases with increase of turn number.

When the voltage drop of the rectifier is considered, the simulation results are shown in Fig. 2 (b). In the figure, from left to right, the turn number is 20, 40, 60, 80, 100, 120, 160, 200. As can be seen from the figure, the change trend of the output power is the same. But when the turn number is low, the output power is low. The reason is that when the turn number is less, the output voltage of CT is lower, the influence of the rectifier loss is larger, so the output power is low. With the increase of the turn number, the influence of the rectifier loss becomes smaller, and the output power is increased. But the output power does not increase infinitely with the increase of turn number. In the end, it is close to the case when the voltage drop of the rectifier is ignored in Fig. 2 (a).

The relationship between the maximum output power of CT and turn number when considering voltage drop of rectifier is shown in Fig. 3. It can be known that the maximum output power of CT increases with the increase of turn number. This phenomenon is more obvious when the line current is small and the output voltage of CT is low. In addition, the curve also has the characteristics of saturation. The increase trend becomes slow. This is important to reduce the starting current of the power supply. It is needed to choose a reasonable turn number, so that the maximum output power is close to the ideal value, while the turn numbers are not too many.

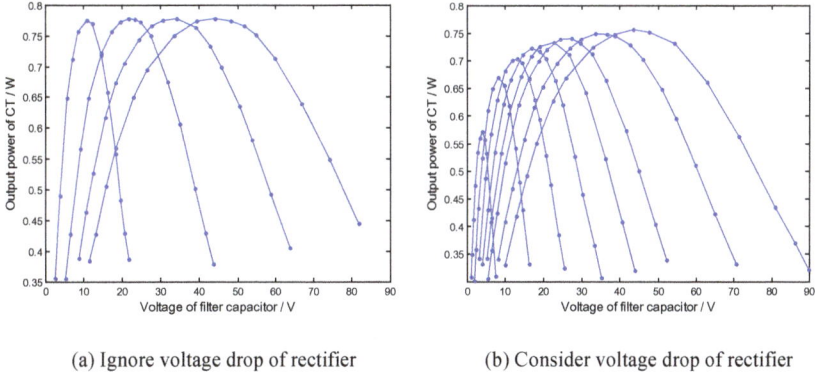

| (a) Ignore voltage drop of rectifier | (b) Consider voltage drop of rectifier |

Fig. 2 Relationship between output power of CT and filter capacitor voltage

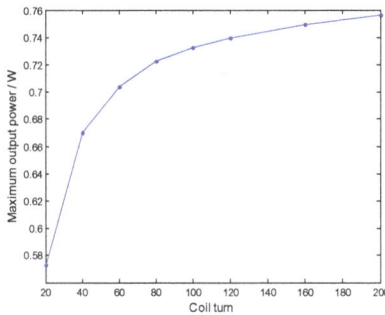

Fig. 3 Relationship between maximum output power and coil turn

3. Analysis of the relationship between output power of CT and load consumption

Considering the output power of CT increases first then decreases with the increase of filter capacitor voltage. For the same power output point, the output voltage may be different, such as point A, B in Fig. 4. It can be known by analyzing that point A is unstable, point B is stable. For fixed power consumption, when the voltage of point B appears small fluctuations. For

501

example, if the voltage drops a little, the output power of CT will increase, then the voltage will rise to point B. If the voltage increases a little, the output power of CT will decrease, then the voltage will decrease to point B. However, point A is on the contrary, when the voltage of point A appears small fluctuations, it will deviate from the point A. The voltage will possibly drop to a lower value, or increase to point B. Then it can be known that the area I in Fig. 4 is an unstable area, and the area II is a stable area.

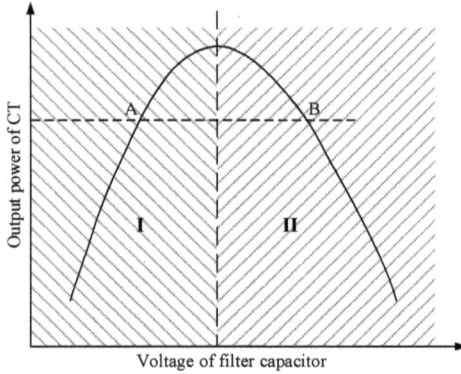

Fig. 4 Output power characteristic curve of CT

In order to reduce starting current of CT power supply, the working point should be in the area II of Fig. 4. Considering the efficiency and stability margin, the stability margin is different when turn number is different. In order to characterize the stability margin, the half power width is defined, which means the voltage width corresponding to half power in the power characteristic curve. As can be seen from Fig. 4, when the maximum output power is certain, the wider the half power width is, the larger the stability margin is.

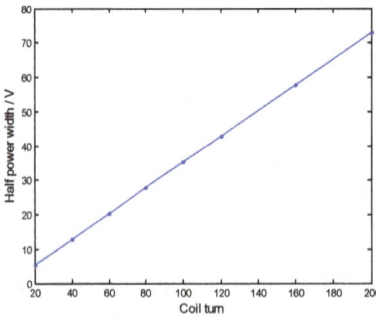

Fig. 5 Relationship between half power width and coil turn

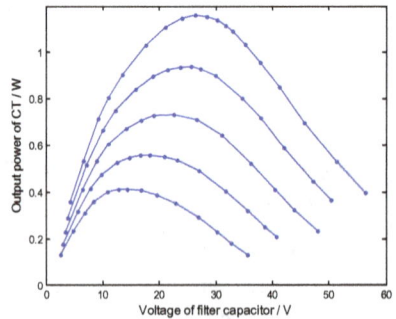

Fig. 6 Relationship between output power and primary side current

502

At the same time, the curves in Fig. 2 show that the turn number is different, the half power width is different. The relationship between half power width and turn number is shown in Fig. 5. The change law is that the half power width increases with the increase of turn number.

The half power width can also be used to analyze the case when the line current is changed. Fig. 6 shows the power output characteristic of CT power supply when primary side current is changed. The turn number is 100 and keeps constant. The curves from bottom to top, the primary current is 6A, 7A, 8A, 9A, 10A RMS. It can be known that with the increase of the primary side current, the power output characteristic curve is shifted upward, and the half power width is enlarged at the same time. When the primary side current decreases, the power curve decreases too, and the half power width becomes small. Finally when the current drops below the starting current, the working point is located in the area I. The CT power supply cannot be stable and is unable to work.

4. CT power supply design

4.1 *Turn number selection*

The simulation results show that the output power of the CT is related to the filter capacitor voltage, and the maximum output power is independent of the turn number when the voltage drop of the rectifier is ignored. However, when considering the voltage drop of rectifier, the maximum output power has the saturation characteristic as is shown in Fig. 3. Therefore, in order to reduce the starting current, the turning point in Fig. 3 should be selected. It can not only guarantee the power required at the starting current, but also ensure a certain stability margin. At the same time, it can reduce the turn number of secondary side, and reduce the volume and cost of the CT device.

Taking the temperature monitoring device as an example, its power consumption is generally not more than 0.5W. This paper chose the ring type silicon steel core and its size is $40 \times 95 \times 30$mm (inner diameter, outer diameter, height) as mentioned before. The design goal is that the output power is more than 0.5W, and at this time, the starting current is as small as possible.

Considering the maximum power output point is the critical stable point, in fact, the circuit cannot be stable for a long time at the maximum power output point. And taking into account the conversion efficiency of the actual circuit. The simulation results should be kept in a certain margin. From Fig. 6 we can see that, when the primary side current is 8A, the maximum output power is more than 0.7W. According to the results of Fig. 3 and Fig. 5, select 80 turns as the turn number. When the turn number is more than 80, the increase trend of maximum power is not obvious, and the half power width is large enough at this turn number.

4.2 *Design of hardware circuit*

Based on the analysis above, the circuit is designed as shown in Fig. 7. In this circuit, in addition to including the main rectifier filter and BUCK form DC-DC circuit. It also includes two parts of the control circuit: overvoltage protection circuit and DC-DC start control circuit.

The role of DC-DC start control circuit is to make the power supply directly cross the area I and working in area II (reasons see below). Through the control of hysteresis comparator composed by amplifier Amp2, at the rising stage of the voltage of capacitor C1, control the DC-DC circuit in off state. At this time, the DC-DC circuit does not output voltage, and its own standby current is very low. The output current of CT quickly charges C1 to a high voltage (higher than the voltage corresponding to maximum power output point). At this time, start the DC-DC circuit and then the circuit can automatically reach to the power balance. And the output power of the circuit can reach to the maximum power in present line current. When the power is no longer balanced and the voltage of capacitor C1 drops to the voltage under the minimum input voltage of DC-DC circuit, switch the DC-DC circuit off. This can avoid outputting voltage to the load below its rated value, so as not to cause abnormal operation of the load.

In the circuit, operational amplifier Amp1 is used for overvoltage protection. When the line current is large or the power consumption is relatively small, in order to prevent the voltage of C1 from exceeding limit and damage the circuit, and also prevent the magnetic saturation of the iron core, turn on Q1 to discharge the output current of CT. The power supply of amplifier and other devices is provided by capacitor C1 through voltage regulator LM317. The DC-DC circuit is composed of LM2596, and its operating voltage range is 4.5-40V. It can be turn on or off by controlling the voltage of pin 5.

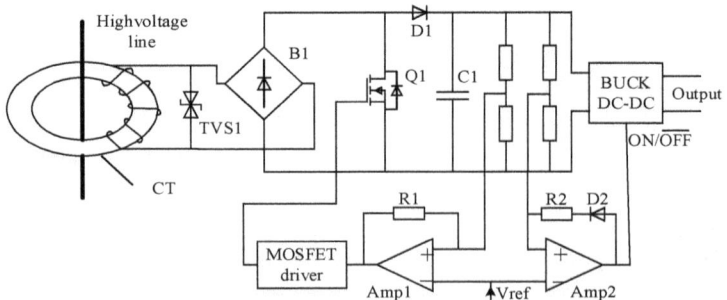

Fig. 7 Schematic diagram of hardware circuit

504

5. Experimental results

Use the CT designed above, the primary side current is 8A RMS. The relationship between the output power of CT and the filter capacitor voltage is obtained by experiments, and the compare with simulation results is shown in Fig. 8. In the figure, from left to right, the turn number is 40, 60, 80, 100, 120. The maximum output power is shown in Table 1.

It can be seen from Fig. 8 and Table 1 that the simulation result is close to the experimental data. And it can be seen from the experimental data that, the maximum output power increases slowly with the increase of the turn number when the turn number is more than 80. The CT is able to output more than 0.7W and also has a large half power width when turn number is 80. So it is reasonable to select turn number to be 80.

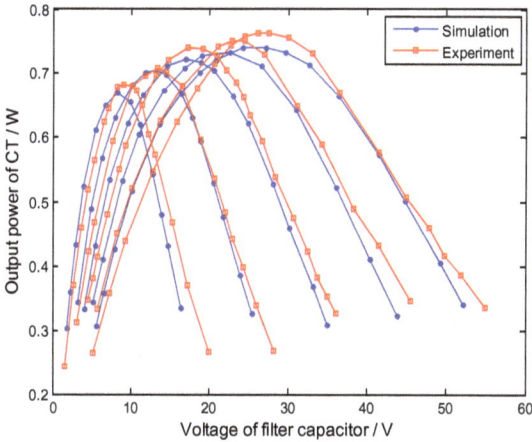

Fig. 8 Contrast of experiment and simulation

Table 1. Effect of turns on the maximum output power

Turn number	Maximum output power (W)	
	Experiment	Simulation
40	0.68	0.67
60	0.71	0.71
80	0.74	0.72
100	0.75	0.73
120	0.76	0.74

When connecting the DC-DC circuit, the minimum required line current is measured when the circuit can output 0.5W, 1W, 2W, 5W and 10W. The turn number is 80 and the DC-DC circuit fixed output 5V. The results are shown in

Table 2. It can be known from the table that the DC-DC circuit is able to output more than 0.5W power when the line current is more than 8A RMS.

Table 2. Output power experiment

Power (W)	Minimum line current required (A)
0.5	7.5
1	11.5
2	17.8
5	34.6
10	63

In order to verify the effect of the DC-DC start control circuit, first the result when this part circuit is not added is tested. The DC-DC circuit works when there is an input voltage. The output voltage of DC-DC circuit is 5V and the load is 50Ω. It is equivalent to a constant power (0.5W) load when it is working.

The measured waveform is shown in Fig. 9 (a) when the primary side current is 10A RMS. It can be seen that the filter capacitor voltage is lower than the minimum input voltage of DC-DC circuit, the DC-DC circuit cannot output the normal voltage. Until the primary side current increases to 16A RMS, the filter capacitor voltage can rise to high enough and the DC-DC circuit can output normal voltage, as shown in Fig. 9 (b).

When the DC-DC start control circuit is added, the measured waveform is shown in Fig. 10. The line current is 8A RMS. It can be seen that, the filter capacitor voltage has crossed the value corresponding to maximum power point when the DC-DC circuit starts. The filter capacitor voltage drops slightly and finally be stable. The DC-DC circuit can output normal voltage.

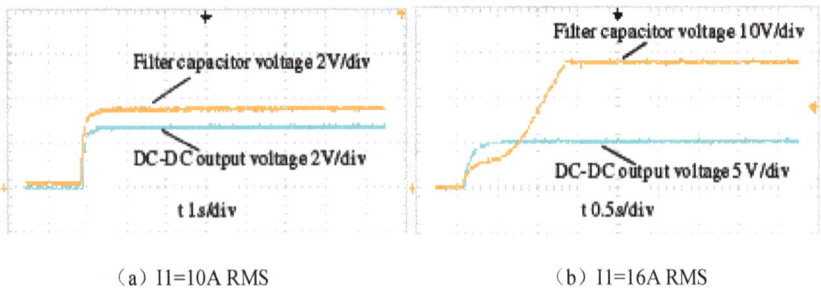

(a) I1=10A RMS   (b) I1=16A RMS

Fig. 9 Filter capacitor and DC-DC output voltage (DC-DC no control)

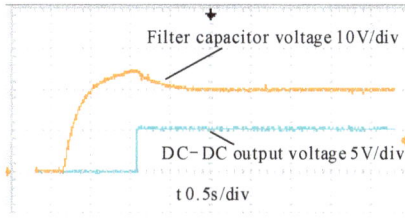

Fig. 10 Filter capacitor and DC-DC output voltage (DC-DC under control)

By contrast it can be known that, it is necessary to add the DC-DC start control circuit. Because there is the circumstance that the line current is too low and the CT power supply cannot work. When the line current starts to rise, if the DC-DC circuit is in working state, the condition is the same as in Fig. 9. This is equivalent to increase the starting current. The problem can be solved by adding this part of circuit.

6. Conclusion

(1) When the primary side current is constant, the output power of CT first increases and then decreases with the increase of the filter capacitor voltage. There is a maximum power output point, and the maximum power is independent of the turn number when the voltage drop of the rectifier is ignored.

(2) In the output power characteristic curve, there are two operating points corresponding to the load power consumption, in which the point with higher voltage is the stable operating point. In order to characterize the stability margin, the half power width is defined. The more the turn number is, the greater the half power width is.

(3) When the voltage drop of rectifier is considered, with the increase of turn number, the maximum power increases and is close to the value when voltage drop of rectifier is ignored in the end. Therefore, the secondary side coil turns can be determined based on the turning point of the maximum power curve and the stability margin described by the half power width.

References

1. Bojie Sheng, Wenjun Zhou, "Ultra-low Power Wireless-Online-Monitoring Platform for Transmission Line in Smart Grid, " High Voltage Engineering and Application (ICHVE), pp. 244-247, October 2010.

2. Zhimin He, Shiguang Nie, Guangyu Qu, Yadong Liu, Gehao Sheng, and Xiuchen Jiang, "The design of CT energy harvesting power supply based

on phase-controlled method" Power and Energy Engineering Conference (APPEEC), pp. 1-5, March 2012.

3. Shouqiao Xin, Liye Xiao, Guomin Zhang, Jiaxing Zhai. "Power Supply Based on Small Current Transducer for Wireless Sensor in Smart Grid, "Electrical and Control Engineering (ICECE), pp. 390-3993, June 2010.

4. Jie Chen, Rongge Yan, Wenrong Yang, Hongyu Zhao, "A Power-line Energizing System for Active Power Electronic Current Transformer," Electrical Machines and Systems (ICEMS), pp. 4431-4435, October 2008.

5. Lan Xiong, MinJie Xu, YouZhong He, DaoJun Song, YanLong Zhao, Wei He. "A novel energy obtaining system for on-line monitoring devices of transmission line," Electrical and Control Engineering (ICECE), pp. 4300-4304, September 2011.

6. Liming Wang, Haidong Li, Changlong Chen, et al. A Novel Online Energy Extracting Device with Low Lower Limit Deadband on Transmission Line [J]. High Voltage Engineering, 2014, 40(2):344-352.

7. Tianchen Yue, Yadong Liu, Zhimin He, et al. Power Output Characteristics of Magnetic Core in CT Energy Harvesting Devices [J]. High Voltage Apparatus, 2015, 51(1):18-23.

8. Jiles D C, Atherton D L. Theory of ferromagnetic hysteresis [J]. Journal of Magnetism and Magnetic Materials, 1986, 61(1-2): 48-60.

Theoretical and Experimental Research of Non-Contact Overvoltage Warning Device of Distribution Network on the Needle Plate Electrode Structure

Jie Lou[†], Ruo-chen Guo, Chang-tao Chen and Yu-ying Zhang
*School of Electrical Engineering, Shandong University,
Jinan, 250061, China
[†]Email: loujie00@sdu.edu.cn*

There are many tremendous risks of the secure and reliable operation of the power distribution network due to the fault, resonance oscillation, operation and lightning over-voltage in the power grid. The present measurement methods about over-voltage are based on the voltage transformer installed in the substation or its improved equipment and need complex data if just for over-voltage warning. This paper researches to use mechanism of production about corona to offer a simple method which can measure over-voltage qualitatively as a supplement of the over-voltage measurements. Experimental results show that, the needle plate electrode structure can be used as a supplemental measure about over-voltage measurement and warning of distribution network which is simple and practicable. In addition, in the normal operating voltage range the needle plate electrode structure can detect voltage waveform as non-contact voltage transformer to some extent. Capacitance of the needle plate electrode structure is tiny, so, it can detect transient voltage of charged conductor effectively.

Keywords: Device; Voltage Measurement; Needle Plate Electrode; Intelligent Online Monitoring[1]; Non-Contact.

1. Introduction

The requirement about power distribution reliability and power quality is higher and higher in 10-35kV distribution network. Statistics show that there are many tremendous risks of the secure and reliable operation of the power distribution network due to the fault, resonance oscillation, operation and lightning over-voltage in the power grid.

The present measurement methods about over-voltage are based on the voltage transformer installed in the substation and extract voltage signal as the data of over-voltage though its secondary port. Under the effect of high-frequency voltage, voltage signal extracted from the PT secondary port distorts badly that can't be used directly. For 10kV and 35kV distribution network, capacitive voltage divider is directly installed on substation bus to measure over-voltage. For 110kV and above voltage level of the high-voltage grid, capacitive voltage divider is composed by the end of the casing screen connected with capacitance of high voltage capacitive device, which can get the voltage signal. This is currently a mature over-voltage data acquisition method. The divider is directly connected to a bus of this method, which will lead to a relatively large change at the scene[2-5]. Therefore, over-voltage warning needs sophisticated data acquisition methods which must be integrated in online monitoring system for the present. This paper attempts to provide a simple method for the qualitative detection of over-voltage from the principle, which can be a supplement of over-voltage measurements.

2. Principle of the needle plate electrode over-voltage warning

The Basic principle of the needle plate electrode over-voltage warning device is that two electrodes are formed by using a needle plate structure. The plates are connected to the bus of the measured circuit while the needle detects if the resistance is grounded by making them series, as shown in figure 1. When there is AC in the circuit, a small capacitive current through the needle plate electrode and a possible corona current will cause a voltage drop on the series resistance. It can be determined whether the circuit is in the status of voltage presence by taking out the voltage signal.

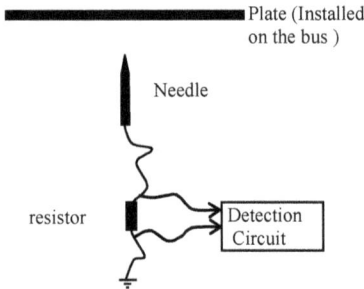

Fig. 1 Schematic diagram of needle plate electrode live displaying device

3. Principle Verification Experiment

3.1. Experimental circuit

The plate is made of stainless steel plate with area of 300×300mm and thickness of 1.5mm. The needle is made of copper with diameter of 0.3mm.The spacing between the plate and the needle is 10mm.The detecting resistance is 2.7MΩ.To improve the ability to resist interference, the connecting line is coaxial cable with shielding net.

In the experimental circuit, the step-up transformer type is YD-JZ, a single phase light high voltage test transformer, which output voltage is AC50/DC70kV and secondary current is 100mA. Experiment uses capacitive voltage divider to measure the voltage, which type is SGB-200, voltage ratio is 10000:1 and capacitance is 150pF. Current limiting resistance in DC experiment is 150Ω. Detecting resistance is 2.7MΩ. Oscilloscope is Agilent DSO5034A, which bandwidth is 300MHz and sampling frequency is 2GSa/s.

In the experiment, we mainly measure and analyze the AD voltage. Figure 2 shows the schematic diagram of the circuit. Figure 3 shows the wiring of experimental circuit.

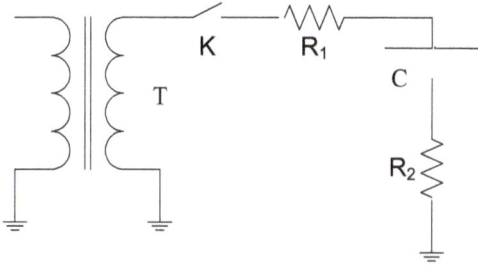

T-step-up transformer, K-charge switch, R_1- current-limiting resistance, R_2-sense resistance, C-the needle plate structure.

Fig. 2 Schematic diagram of ac experimental circuit

Fig. 3 The photo of experimental circuit

3.2. Experimental Result

To be convenient in the experiment, the oscilloscope probe connects to the detecting resistance by inversed grafting. Therefore, the signal of the oscilloscope is the waveform after reversing the phase. Before the boost, interference signal measured by oscilloscope shows in figure 4.

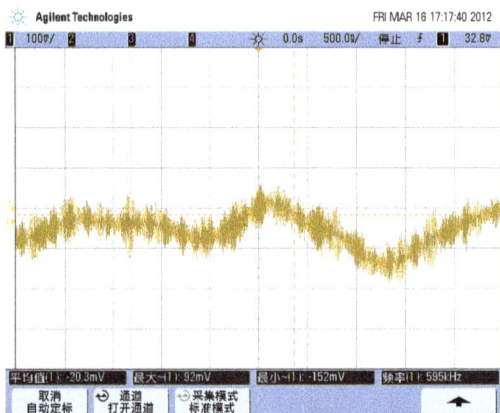

Fig. 4 The waveform of interference signal

The amplitude is less than 150mV basically. After the boost, 4 groups of signals show in figures 5 to 8 as 720V, 1220V, 2100V, 3200V.

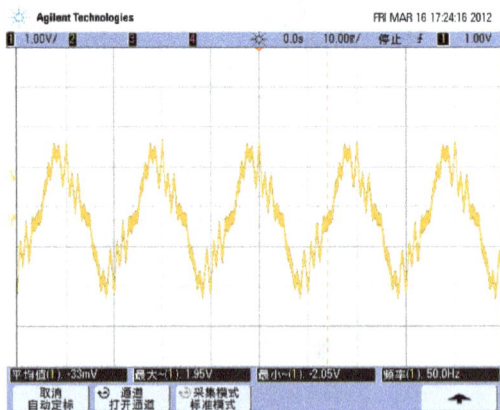

Fig. 5 The waveform of ac 720V

Due to the output waveform of the light high voltage test transformer has high harmonics, a certain extent of distortion will exist of the waveform. As is shown in figures 5 and 6, the waveform is essentially unchanged before 1220V and the signal amplitude is proportional to the applied voltage.

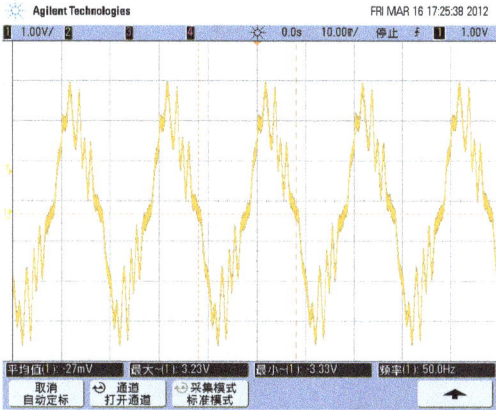

Fig. 6 The waveform of ac 1220V

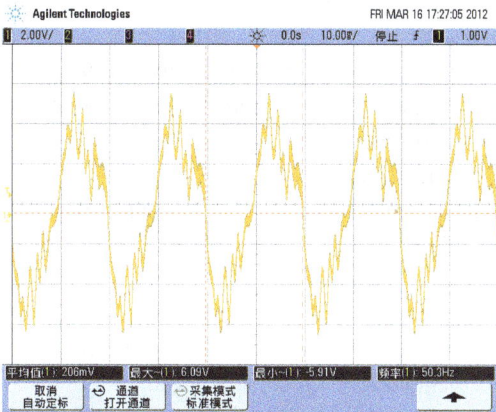

Fig. 7 The waveform of ac 2100V

The result shows that, in addition to the harmonic wave caused the non-sinusoidal waveform, there is no other factors leading to distortion. But voltage waveform distortion is more serious in figures 7 and 8, which is significantly different from figures 5 and 6.

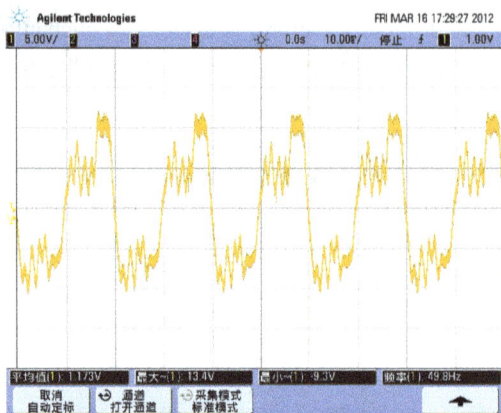

Fig. 8 The waveform of ac 3200V

4. Experimental analysis

This paper found the reason why over-voltage warning function can be realized by this structure is mainly that small capacitor formed by the needle plate flows through the small capacitive current at the normal operating voltage by experiment. When over-voltage happens, capacity corona pulse current occurs in the capacitive current due to the corona phenomenon in needle electrode, which results in waveform distortion. By detecting the change, it can be determined whether the over-voltage occurs, which is shown in figures 7 and 8. According to the actual circuit parameters, the needle plate capacitance value of the experiment is approximately 2.2pF combined with simulation.

From the perspective of gas discharge, when inception voltage is reached, the occurrence of electron avalanche, streamer and other gas discharge will lead to space charge at the tip. The superposition of corona discharge current and capacitive displacement current which flow through the sense resistor, will lead the detected current to distortion. In table 1, the peak voltage of each experiment is compared. As can be seen from the table, the occurrence of corona current leads to waveform distortion, which improves current value through the sense resistor. For this reason, the detected waveform peaks becomes larger and the negative half-wave corona current is large, which also consistent with the positive and negative corona discharge characteristics [5].

However, this will not lead to a breakdown of the needle plate gap directly. On the contrary, according to the corona stability theory, the similar structure may have a stable breakdown voltage, which is generally much higher than the corona inception voltage [6].

Table 1 The comparison of voltage peaks

Voltage (V)	The peak value without considering interference	The peak value considering interference	Reference ratio	The ratio without considering interference	The ratio considering interference
720	1.95	1.858	1	1	1
	-2.05	-1.898		1	1
1210	3.23	3.138	1.68	1.66	1.69
	-3.33	-3.178		1.62	1.67
2100	6.09	5.998	2.79	3.12	3.23
	-5.91	-5.758		2.88	3.03
3200	13.4	13.308	4.44	6.87	7.16
	-9.3	-9.148		4.54	4.82

Corona is extremely uniform electric field's self maintaining discharge, it can calculate the corona inception voltage by self maintaining discharge conditions in theory, but the calculation is very complex and the results are not very accurate, practical calculation often gets results by the empirical formula. Empirical formulas of the needle plate electrode as follows [7, 8]:

$$E_0 = 27.7\delta(1+\frac{0.337}{\sqrt{r\delta}}) \tag{1}$$

$$U_c = E_0\frac{dr}{0.9(d+r)} \tag{2}$$

In formula above, E_0 is corona inception field strength, U_c is corona inception voltage, δ is correction factor, d is clearance distance, r is the radius of the needle plate, length unit is cm, electric field unit is kV/cm, voltage unit is kV. Put parameters of the needle plate in the experiment into formula above:

$$E_0 = 27.7\times1\times(1+\frac{0.337}{\sqrt{0.015\times1}}) = 103.9(\text{kV/cm})$$

$$U_c = 103.9\times\frac{1\times0.015}{0.9(1+0.015)} = 1.706(\text{kV})$$

It will begin to arise corona when the RMS is about 1200V. Since then due to the emergence of space charge, the measurement waveform will distort.

Therefore, when there was no corona, the needle plate structure can measure the voltage in some way. When Corona occurs, it can determine if there would be over-voltage or not according to the waveform distortion. Meanwhile, because of the breakdown voltage of the needle plate structure is relatively stable, so it can withstand larger over-voltage. What's more, under the situation that could satisfy the requirement of insulation, we can narrow needle plate electrode's gap to decrease interference.

5. Conclusion

(1) Through the experiment and analysis, the needle plate electrode structure can be used as a supplemental measure about over-voltage measurement and warning of distribution network, which is simple and practicable.

(2) In the normal operating voltage range the needle plate electrode structure can detect voltage waveform as non-contact voltage transformer to some extent if corona will not occur. Capacitance of the needle plate electrode structure is tiny, so, it can detect transient voltage of charged conductor effectively. Compared with other voltage measurement and early warning device, the structure is non-contact, has less insulation problems. It's easy to adjust to achieve a variety of functions, and it also can be combined with weak current and communication system into the smart grid monitoring system, to develop the smart on-line monitoring function of distribution network power equipment.

References

1. Liu Qiang, Zhang Yuan-fang. Current Status of over-voltage On-line Monitoring Technology and its Developing Propect [J]. East China Electric Power,2002,(8):5-8.

2. Zhao Jun, Lü Yan-ping, Wang Han-guang. New Scheme to Identify Lightning Disturbance for the UHV Transmission Lines Based on Multi-scale Morphology Decomposition[J]. High Voltage Engineering, 2009, 35(5):994-998.

3. He Zheng-you, Chen Xiao-qin. A Study of Electric Power System Transient Signals Identification Method Based on Mylti-scales Energy Statistic and Wavelet Energy Entropy[J]. Proceedings of the CSEE, 2006,26(10):33-39.

4. Chen Xiao-qin, He Zheng-you, Fu Ling. Electric Power Transient Signals Classification and Recognition Method Based on Wavelet Energy Spectrum[J]. Power System Technology,2006,30(17):59-63.

5. Li Er-ning, Liu Yan-bing, Shen Yue. Research of corona discharge characteristics[J]. High Voltage Electrical Apparatus, 1998(6):16-21.

6. Qiu Yun, Jin Xiao, Song Falun, etc. Stabilization performance of rod-plane corona stabilized switch[J]. High Power Laser and Particle Beams, 2010, 22(3): 545-549.

7. Liang Xi-dong, Chen Chang-yu, Zhou Yuan-xiang, etc. High voltage engineering[M], Tsinghua University Press, Beijing, 2003.
8. Qiu Yun. The Research of corona stabilized switch [D]. China Academy of Engineering Physics, 2010.

Analysis of DC Pole-to-Pole Short Circuit Fault Behavior in MMC-HVDC Transmission Systems with Arm Damper

Huan Ma[†], Jin-long Wu, Kun Han, Xin-he Liu, Dao-yang Li and Wei-zheng Yao

Xuji Group Corporation
Xuchang, 461000, Henan, China
[†]*E-mail: mahuan1224@126.com*

The conventional method of clearing the DC pole-to-pole short circuit fault of the half-bridge sub-module based Modular Multilevel Converter (H-MMC) is to trip the ac circuit breakers, which takes a long time to restart the system. The H-MMC with arm damper can help the fault current decay quickly, and realize fast system recovery. In this paper, an analytical model is proposed to analyze the DC pole-to-pole short circuit fault behavior in MMC based high voltage direct current (MMC-HVDC) transmission systems with arm damper. The fault mechanism is analyzed in detail including three stages: before blocking the converter, after blocking the converter, and after tripping the ac circuit breakers, and the derived analytical equations are given at each stage. Using the analytical model, the fault current can be accurately and quickly predicted. Finally, the proposed analytical model is verified by Matlab/Simulink, and the calculated results agree well with the simulation results.

Keywords: Modular Multilevel Converter (MMC); Arm Damper; DC Pole-to-Pole Short Circuit Fault; Analytical Model.

1. Introduction

With the development of power electronic technology, the voltage-source converter based HVDC (VSC-HVDC) system has attracted widespread attentions [1]. Compared with the conventional two-level or three-level converters, the MMC-HVDC as shown in Fig. 1 is considered to a competitive candidate, which presents many advantages such as very low harmonics, no necessity of series connection of power semiconductors, and DC bus capacitor elimination [2-4]. In commercial projects, most of the converters employ the half-bridge sub-modules (HBSMs). However, the conventional H-MMC lacks the DC-fault clearance capability. When a pole-to-pole short circuit fault occurs, the H-MMC operates like a three phase diode rectifier and a inrush current flows in both AC and DC sides. The conventional method of clearing the pole-to-pole short circuit fault is to trip the ac circuit breakers. However, this method takes a long time for fault current decaying and system restarting [5-6].

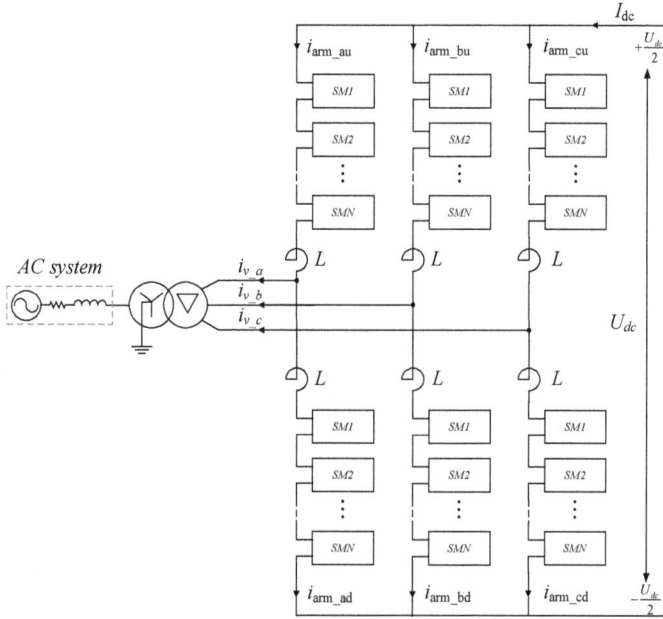

Fig. 1. Simplified schematic of a MMC-HVDC transmission system.

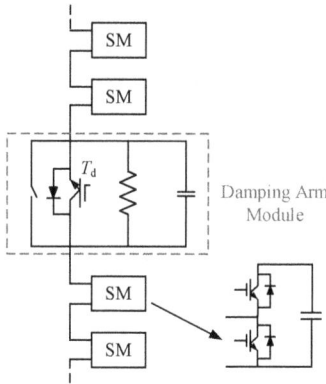

Fig. 2 The topology of MMC-HVDC with arm damper.

An improved H-MMC topology is called H-MMC with arm damper, which adds several arm dampers in series with HBSMs in each arm, as shown in Fig. 2. The damping arm module consists of a IGBT with an anti-parallel diode, a damping resistor, a capacitor, and a high speed bypass switch. When the MMC operates at normal mode, the IGBT keeps conducting state so that the damping resistor is bypassed. When the DC pole-to-pole short circuit fault occurs, the

IGBTs are turned off, so the fault current flows through the damping resistor. The arm dampers can accelerate the fault current attenuation, so that the DC pole-to-pole short circuit fault can be cleared quickly, and the system can be restarted in a very short time.

The fault mechanism of MMC-HVDC have been broadly studied in different technologies such as full-bridge sub-modules, clamp double sub-modules and circuit breakers [7-8]. DC pole-to-pole short circuit fault transient behavior of MMC in HVDC application based on the current stress analytic equation was analyzed [9]. However, there is lack of a comprehensive study of DC pole-to-pole short circuit fault characteristics in H-MMC with arm damper. This paper aims to propose an accurate analytical model to analyze the DC pole-to-pole short circuit fault mechanism of H-MMC with arm damper. The fault mechanism is analyzed in detail and the analytical equations are derived at each stage. The fault currents are calculated by the proposed analytical model, which agree well with the Matlab/Simulink simulation results.

The paper is organized as follows. In Section 2, the DC pole-to-pole short circuit fault mechanism of H-MMC with arm damper is analyzed. In Section 3, the theoretical analyses are verified by simulation. Section 4 concludes the paper.

2. The Analysis of DC Pole-to-Pole Short Circuit Fault Mechanism

When the DC pole-to-pole short circuit fault occurs, the DC bus voltage drops to zero instantaneously, while the DC current and the arm currents rises sharply due to the discharge of the capacitors in HBSMs. All IGBTs including the HBSMs and the damping arm modules are blocked as soon as the fault is detected. Then the fault current starts to decay. After the ac circuit breakers are tripped, the residual dc fault currents gradually decay to zero. According to the above process, the analysis of DC pole-to-pole short circuit fault mechanism are divided into three stages: before blocking the converter; after blocking the converter; after tripping the ac circuit breakers.

2.1. Fault mechanism analysis before blocking the converter

At this stage, the arm overcurrent mainly contains two parts: the discharge current of the capacitors in HBSMs and the ac grid feeding current.

Before the converter is blocked, the main reason lead to overcurrent is the discharge of the capacitors in the input HBSMs. For the input HBSMs, the capacitors are discharged through the upper IGBT; for the being cut HBSMs, the capacitors are not discharged and the fault current flows through the anti-parallel diode of the lower IGBT; for the damping arm modules, the fault current flows through the IGBTs. The equivalent circuit at this stage is a second-order circuit as shown in Fig. 3:

$$\begin{cases} 2L\dfrac{di_1}{dt} + R_{stray}i_1 - u_c = 0 \\ -\dfrac{2C}{n}\dfrac{du_c}{dt} = i_1 \end{cases} \tag{1}$$

where L and C are the arm inductance and the capacitance of HBSMs respectively, n is the number of HBSMs in each arm, and R_{stray} is the stray resistance of circuit, including the reactors' parasitic resistance, capacitors' ESR and the switches' resistance.

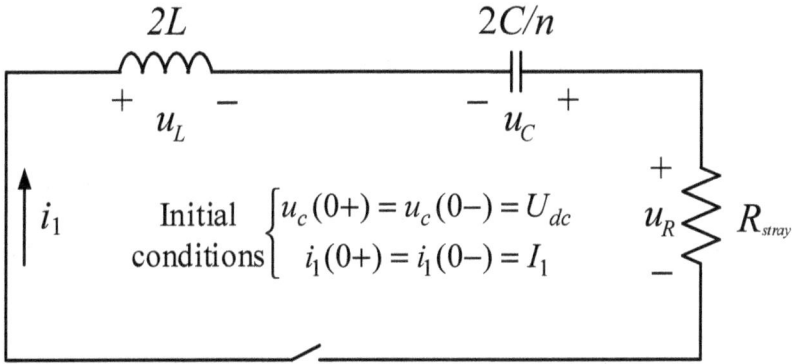

Fig. 3 The equivalent circuit when the capacitors of HBSMs are discharged.

According to Eq.(1), the voltage of the capacitor can be derived as follows:

$$u_C = e^{-\frac{t}{\tau}}\left[\frac{U_{dc}\omega_0}{\omega}\sin(\omega t + \alpha) - \frac{nI_1}{2\omega C}\sin(\omega t)\right] \tag{2}$$

where $\tau = 4L/R_{stray}$, $\alpha = \arctan\left(\sqrt{4nL/(CR_{stray}^2)-1}\right)$, $\omega_0 = \sqrt{n/(LC)}/2$,

and $\omega = \sqrt{n/(LC)-\left(R_{stray}/2L\right)^2}$.

The discharging current of the capacitors can be calculated as follows:

$$i_1 = e^{-\frac{t}{\tau}}\left[\sqrt{\frac{C}{nL}U_{dc}^2 + I_1^2}\sin(\omega t + \beta)\right] \tag{3}$$

where $\beta = \arctan\left(U_{dc}/I_1\sqrt{C/nL}\right)$. Since the impedance of the discharge circuit is very small, the arm currents rise rapidly. Once the fault current is detected, all the IGBTs will be blocked.

The main contributor of the arm overcurrent is the discharge of capacitors in HBSMs, and the detection time of DC pole-to-pole short circuit fault is very short, so the ac feeding current at this stage are approximately equal to the current at normal.

The total arm currents are the sum of the discharge current of capacitors and the current fed by AC grid:

$$i_u = i_1 + \frac{i_s}{2} = e^{-\frac{t}{\tau}}\left[\sqrt{\frac{C}{nL}U_{dc}^2 + I_1^2}\cos\left(\omega t + \beta\right)\right] + \frac{I_m}{2}\sin(\omega_n t + \varphi_i) \qquad (4)$$

$$i_d = i_1 - \frac{i_s}{2} = e^{-\frac{t}{\tau}}\left[\sqrt{\frac{C}{nL}U_{dc}^2 + I_1^2}\cos\left(\omega t + \beta\right)\right] - \frac{I_m}{2}\sin(\omega_n t + \varphi_i) \qquad (5)$$

where $i_s = I_m\sin(\omega_n t + \varphi_i)$ is the ac grid current.

2.2. Fault mechanism analysis after blocking the converter

After the converter is blocked, the capacitors in HBSMs will not discharge any more and the fault current starts to decay. Due to the unidirectional conductivity of diode, the arm current will be interrupted when the current decays to zero. After a short transition period, the system enters the steady state. So the DC pole-to-pole short circuit fault mechanism analysis of this stage is divided into three stages:

Transition I: both the upper and lower arm currents continuously decay;

Transition II: both the upper and lower arm currents decay but sometimes are interrupted;

Steady state: both the upper and lower arm currents reach the steady state.

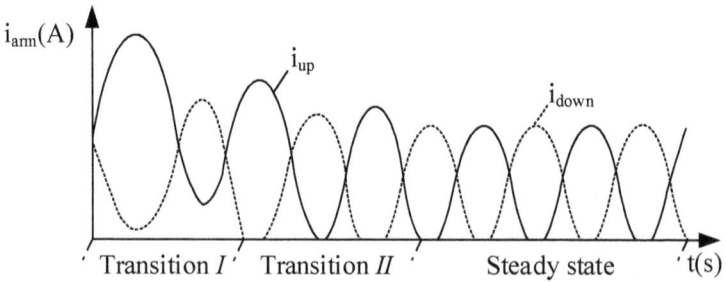

(a) The arm current waveforms

Fig. 4 The arm current waveforms and the equivalent circuits after blocking the converter.

522

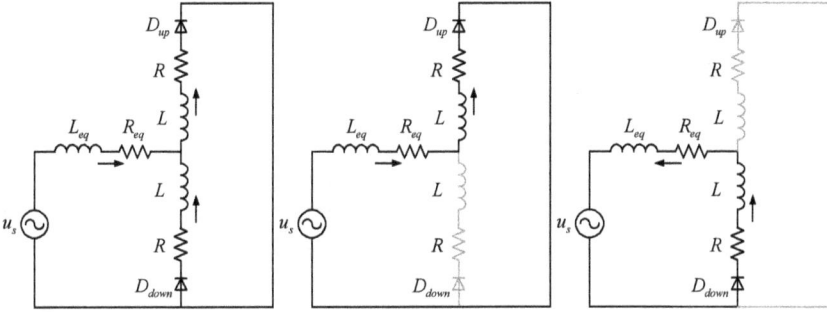

(b) Both upper and lower arms are in conducting state (c) Only upper arm is in conducting state (d) Only lower arm is in conducting state

Fig. 4 The arm current waveforms and the equivalent circuits after blocking the converter. (*continued*)

It can be assumed that the three legs are symmetrical at this stage, thus the arm overcurrent of each arm can be analyzed independently. The arm currents and the equivalent circuits after the converter is blocked are shown in Fig. 4, where $R=R_{damp}+R_{stray}/2$, and R_{damp} is the damping resistance.

2.2.1. Transition I

The equivalent circuit of this stage is shown in Fig. 4(b). The arm current can be calculated as follows:

$$\begin{cases} i_{u1} = i_{dc1} + \dfrac{i_{s1}}{2} \\[2mm] 2L\dfrac{di_{dc1}}{dt} + 2Ri_{dc1} = 0 \\[2mm] (L_{eq} + \dfrac{L}{2})\dfrac{di_{s1}}{dt} + (R_{eq} + \dfrac{R}{2})i_{s1} = U_m \sin(\omega_n t + \varphi) \end{cases} \quad (6)$$

The initial conditions are:

$$\begin{cases} i_{dc1}(0+) = i_{dc1}(0-) = I_{dc0} \\ i_{s1}(0+) = i_{s1}(0-) = I_{s0} \end{cases} \quad (7)$$

where i_{dc1} is the DC component of the upper arm current, i_{s1} is the ac grid current, and $i_{s1}/2$ is the AC component of the upper arm current.

The upper and lower arm currents can be derived as follows:

$$i_{u1} = I_{dc0}e^{-\frac{t}{\tau_1}} + \frac{\sqrt{2}U_s}{2|Z_1|}\sin(\omega_n t + \gamma + \varphi) + (\frac{I_{s0}}{2} - \frac{\sqrt{2}U_s}{2|Z_1|}\sin(\gamma + \varphi))e^{-\frac{t}{\tau_{10}}} \quad (8)$$

$$i_{d1} = I_{dc0}e^{-\frac{t}{\tau_1}} - \frac{\sqrt{2}U_s}{2|Z_1|}\sin(\omega_n t + \gamma + \varphi) - (\frac{I_{s0}}{2} - \frac{\sqrt{2}U_s}{2|Z_1|}\sin(\gamma + \varphi))e^{-\frac{t}{\tau_{10}}} \quad (9)$$

523

where $\tau_1 = L/R$, $\tau_{10} = (L_{eq} + L/2)/(R_{eq} + R/2)$, $\gamma_1 = -\arctan(\omega_n \tau_{10})$, and $|Z_1| = \sqrt{(R_{eq} + R/2)^2 + \omega_n^2 (L_{eq} + L/2)^2}$.

Benefiting of the damping resistor, the arm currents decay quickly. This stage ends when the upper or lower arm current decays to zero.

$$i_{d1}(t_1) = 0 \quad \text{or} \quad i_{u1}(t_1) = 0 \tag{10}$$

where I_{s1} is the non-zero arm current at t_1.

2.2.2. Transition II

In this stage, the upper or lower arm currents has dropped to zero. If the lower arm drops to zero first, the lower arm current will be interrupted due to the unidirectional conductivity of diode. When the voltage of the lower arm reverses, the lower arm will conduct current again. According to Fig. 4(a), the equivalent circuit at this stage alternates between the Figs. 4(b)~(d) and then enters the following cycle: Figs. 4(c)~(b)~(d)~(b).

1) *Case I*: when the lower arm current decays to zero firstly, the equivalent circuit is shown in Fig.4(c), and the upper and lower arm currents can be calculated as follows:

$$\begin{cases} R_{eq} i_{u21} + L_{eq} \dfrac{di_{u21}}{dt} + L \dfrac{di_{u21}}{dt} + R i_{u21} = U_m \sin(\omega t + \varphi_{21}) \\ i_{d21} = 0 \end{cases} \tag{11}$$

The initial conditions are:

$$i_{u21}(0+) = i_{u21}(0-) = I_{s1} \tag{12}$$

According to Eq.(11), the upper arm current can be derived as follows:

$$i_{u21} = \left[I_{s1} - \frac{U_m}{|Z_{21}|} \sin(\varphi_{21} + \gamma_{21}) \right] e^{-\frac{t}{\tau_{21}}} + \frac{U_m}{|Z_{21}|} \sin[\omega t + \varphi_{21} + \gamma_{21})] \tag{13}$$

where $\tau_{21} = (L + L_{eq})/(R + R_{eq})$, $Z_{21} = \sqrt{(R + R_{eq})^2 + \omega_n^2 (L + L_{eq})^2}$, and $\gamma_{21} = -\arctan(\omega_n \tau_{21})$.

Once the upper arm voltage drops to zero, the lower arm voltage reverses. And then the lower arm current begins to rise, and proceeds to case II. The end conditions of this stage are:

$$u_u(t_{21}) = 0 \tag{14}$$

where I_{s2} is the end current of upper arm.

2) *Case II*: when the voltage of the lower arm reverses, the current will flow through this arm again. The equivalent circuit is shown in Fig. 4(b), and the

upper and lower arm currents can be calculated as follows:

$$\begin{cases} Ri_{u22} + L\dfrac{di_{u22}}{dt} + Ri_{d22} + L\dfrac{di_{d22}}{dt} = 0 \\[4mm] R_{eq}\left(i_{u22} - i_{d22}\right) + L_{eq}\dfrac{d(i_{u22} - i_{d22})}{dt} + L\dfrac{di_{u22}}{dt} + Ri_{u22} = U_m \sin(\omega t + \varphi_{22}) \end{cases} \quad (15)$$

where i_{u22}, i_{d22}, φ_{22} are the upper and lower arm current and the initial phase of phase voltage respectively.

The initial conditions are:

$$\begin{cases} i_{u22}(0+) = i_{u22}(0-) = I_{s2} \\ i_{d22}(0+) = i_{d22}(0-) = 0 \end{cases} \quad (16)$$

The upper arm current can be derived as follows:

$$i_{u22} = \frac{1}{2}I_{s2}e^{-\frac{t}{\tau_{220}}} + \left[\frac{U_m}{|Z_{22}|}\sin(\varphi_{22} + \gamma_{22}) - \frac{1}{2}I_{s2}\right]e^{-\frac{t}{\tau_{22}}} + \frac{U_m}{|Z_{22}|}\sin(\omega_n t + \varphi_{22} + \gamma_{22}) \quad (17)$$

$$i_{d22} = \frac{1}{2}I_{s2}e^{-\frac{t}{\tau_{220}}} - \left[\frac{U_m}{|Z_{22}|}\sin(\varphi_{22} + \gamma_{22}) - \frac{1}{2}I_{s2}\right]e^{-\frac{t}{\tau_{22}}} - \frac{U_m}{|Z_{22}|}\sin(\omega_n t + \varphi_{22} + \gamma_{22}) \quad (18)$$

where $Z_{22} = \sqrt{(R+2R_{eq})^2 + \omega_n^2(L+2L_{eq})^2}$, $\gamma_{22} = -\arctan\left(\omega_n \tau_{22}\right)$, $\tau_{220} = L/R$, and $\tau_{22} = (L+2L_{eq})/(R+2R_{eq})$.

The end conditions of this stage are:

$$i_{u22}(t_{22}) = 0 \quad \text{or} \quad i_{d22}(t_{22}) = 0 \quad (19)$$

The analyses of the other cases are similar to case I and cases II. The arm currents can be obtained using the equations. After several periods, the upper and lower arm currents tends to be the same, but have a phase difference of $180°$. Then the system is considered to enter a steady state. The end condition of stage II is:

$$i_{u\,\text{max}} \approx i_{d\,\text{max}} \quad (20)$$

2.2.3. Steady State

This stage has four different cases, as shown in Fig. 4(a). The analysis of each case is similar to that in transition II and has been solved above. It should be emphasized that only half period is needed to analyze due to the symmetry of the upper and lower arm currents. The initial current value also can be calculated by solving Eq.(21).

$$i_{u3}(t_0) = i_{d3}(t_0 + T/2) \quad (21)$$

So DC current value I_{dc3} at this stage can be predicted:

$$I_{dc3} = \frac{3}{T} \int_{t_0}^{t_0+T/2} \left(i_{u3} + i_{d3} \right) dt \tag{22}$$

where i_{u3}, i_{d3} are the upper and lower arm currents respectively, t_0 is the initial time of integration, and T is the period. This stage will continue until the ac circuit breakers are tripped.

2.3. Fault mechanism analysis after tripping the ac circuit breakers

After the ac circuit breakers are tripped, the ac grid is disconnected with the converter. The AC current will not feed into the DC side anymore, so the residual DC fault current starts to decay again. The equivalent circuit of this stage is shown in Fig. 5.

The arm currents are calculated as follows:

$$i_{u4} = i_{d4} = I_{4d} e^{-\frac{t}{\tau_4}} \tag{23}$$

where I_{4d} is the initial arm current, $\tau_4 = 2L/(R_{stray} + 2R_{damp})$.

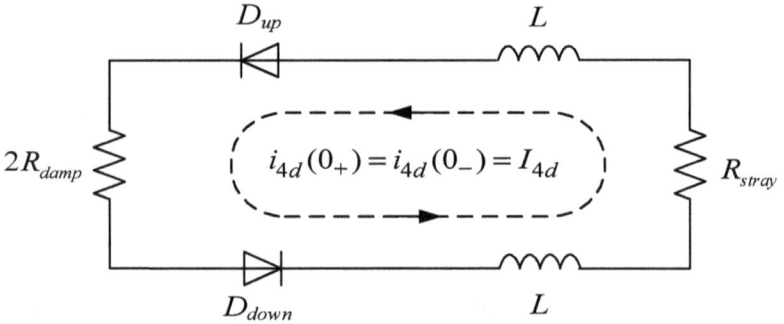

Fig. 5 The equivalent circuit after tripping the ac circuit breakers.

3. Simulation Verification

Table 1. Main circuit parameters of the MMC transmission system.

Parameter	Value
Rated DC bus voltage	±200kV
Rated Power	100MVA
AC grid line-to-line voltage(RMS)	115kV
Transformer MVA	120MVA
Transformer ratio	115kV/208.2kV(Δ/Yn)
Transformer leakage inductance	0.07pu
Arm reactance	350mH
Capacitance of HBSM	2mF
Damping resistance of arm damper	15Ω
Number of HBSMs per arm	250

To verify the proposed analytical model, a simulation model of single-terminal monopolar H-MMC transmission system with arm damper is built in Matlab/Simulink. A Thevenin equivalent circuit method is employed in the simulation to reduce simulation time without sacrificing accuracy [10]. Parameters of the system are given in Table 1.

Fig. 6 and Fig. 7 show the simulation results of the DC bus voltage and current and the capacitor voltages in HBSMs. The MMC operates at the steady state before 0.2s. When a DC pole-to-pole short circuit fault occurs at t=0.2s, the DC bus voltage drops to zero, and the DC current increases rapidly due to the capacitors discharging instantaneously. The fault is detected at t=0.202s, and the IGBTs of all HBSMs and arm dampers are blocked. After the converter is blocked, and the DC current decays about 0.05s. After t=0.25s, the DC current keeps unchanged. The calculated value is 1137.6A obtained from Eq.(22), which is approximately equal to the simulation value 1175A. At t=0.3s, the ac circuit breakers are tripped, and the DC current decays to zero at about 0.4s.

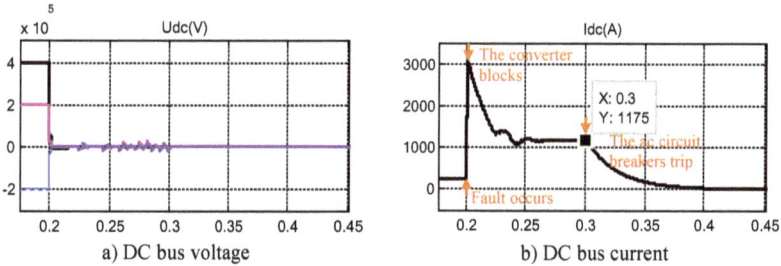

a) DC bus voltage

b) DC bus current

Fig. 6 The simulated waveforms of the DC bus voltage and DC bus current.

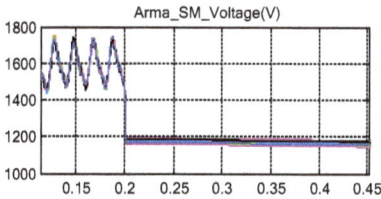

Fig. 7 The simulated waveforms of the capacitor voltages in HBSMs.

In addition, the calculated and simulated upper and lower arm currents of phase A are shown in Fig. 8. The calculated arm currents are consistent with the simulation results. The discharge time of capacitors, the peak values of fault current of each arm are almost the same, which indicates the proposed analytical model are feasible.

527

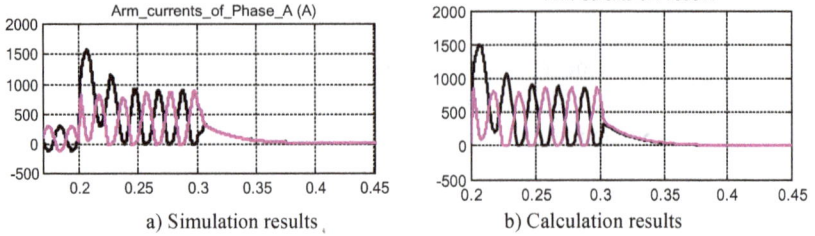

a) Simulation results   b) Calculation results

Fig. 8 Comparison of arm current of phase A between the simulation results and calculation results.

4. Conclusion

This paper proposes an analytical model to analyze the DC pole-to-pole short circuit fault current in MMC-HVDC transmission systems with arm damper. The short-fault process is divided into three stages, and each stage is described by analytical equations. The proposed analytical model gives a deep insight into the fault mechanism and a fast prediction of the fault current. The fault current are calculated by the proposed analytical model, which agree well with the Matlab/Simulink simulation results.

Acknowledgments

This work is funded by Science and Technology Project of SGCC (5292C0160024).

References

1. N. Flourentzou, V. G. Agelidis, and G. D. Demetriades, "VSC-Based HVDC Power Transmission Systems: An Overview," *IEEE Transactions on Power Electronics,* vol. 24, pp. 592-602, 2009.
2. B. Gemmell, J. Dorn, and D. Retzmann. etc. "Prospects of multilevel VSC technologies for power transmission," in *Proc. IEEE/PES TDCE,* 2008, pp. 1-16.
3. A. Nami, J. Liang, and F. Dijkhuizen. etc. "Modular Multilevel Converters for HVDC Applications: Review on Converter Cells and Functionalities," *IEEE Transactions on Power Electronics,* vol. 30, pp. 18-36, 2015.
4. M. A. Perez, S. Bernet, and J. Rodriguez. etc. "Circuit Topologies, Modeling, Control Schemes, and Applications of Modular Multilevel Converters," *IEEE Transactions on Power Electronics,* vol. 30, pp. 4-17, 2015.
5. G. Tang, Z. Xu, and Y. Zhou, "Impacts of Three MMC-HVDC Configurations on AC System Stability Under DC Line Faults," *IEEE Transactions on Power Systems,* vol. 29, pp. 3030-3040, 2014.

6. X. Li, Q. Song, and W. Liu. etc. "Protection of Nonpermanent Faults on DC Overhead Lines in MMC-Based HVDC Systems," *IEEE Transactions on Power Delivery,* vol. 28, pp. 483-490, 2013.

7. S. Cui and S. K. Sul, "A Comprehensive DC Short Circuit Fault Ride Through Strategy of Hybrid Modular Multilevel Converters (MMCs) for Overhead Line Transmission," *IEEE Transactions on Power Electronics,* 2016.

8. G. Liu, F. Xu, and Z. Xu. etc. "Assembly HVDC Breaker for HVDC Grids with Modular Multilevel Converters," *IEEE Transactions on Power Electronics,* 2016.

9. S. Wang, X. Zhou, and G. Tang. etc. "Analysis of submodule Overcurrent Caused by DC Pole-to-pole Fault in Modular Multilevel Converter HVDC System." in *Proc. CSEE,* vol. 31, no. 1, pp.1-7, Jan. 2011.

10. U. N. Gnanarathna, A. M. Gole, and R. P. Jayasinghe, "Efficient Modeling of Modular Multilevel HVDC Converters (MMC) on Electromagnetic Transient Simulation Programs," *IEEE Transactions on Power Delivery,* vol. 26, pp. 316-324, 2011.

Chapter 4

Modeling and Simulation

Research on the Generation of Product Assembly Sequence Algorithm

De-lin Qin[1], Chuang-jian Zhang[2,*] and Kai Zhang[3]

[1]Beijing University of Technology, Chaoyang District, Beijing, China
qin@bjut.edu.cn

[2]Beijing University of Technology, Chaoyang District, Beijing, China
**heze8952@163.com*

[3]Beijing University of Technology, Chaoyang District, Beijing, China
1353370866@qq.com

Assembly planning generally required generating the detailed plans of each of steps in the entired work flow sequence, in terms of solutions. However, with the complexity of the program become more complex, and the number of plans for every conisitional parts will needed to expand and grow in the solution space, so much so that algorithms to understand the process, and generate the sequence of the workflow have becoming essential in order to avoid combination explosion. In order to effectively reduce the number of parts in the product sequence planning, the product assembly hierarchy model is established, and the interference matrix and connection matrix are analyzed, based on assembly mating features to identify sub assembly, the assembly sequence of the product can be automatically generated efficiently.

Keywords: Assembly sequence planning, Interference inspection, Interference matrix, connection matrix.

1. Introduction

In the assembly sequence plan, the number of parts forming the sub-assembly, a sub-assembly is treated as a part treatment, the number of parts can be further reduced when the plan under consideration. it will help to further improve the efficiency of the algorithm. In this paper, a assembly feature information of parts extracted based on 3D assembly model [1], a assembly hierarchy model that greatly reducing the number of parts considered in the planning is constructed. The disassembly sequence of the assembly is analyzed by interference matrix and connection matrix, to identify possible sub assemblies. And further it will reduce the number of parts of the planning, so as to achieve efficient product assembly sequence generation.

2. Assembly Information Model

The assembly process is to establish hierarchical coupling between components in an assembly[2], The relation is implicit in the order of the assembly sequence.Assembly information can be accessed through the two development of the CAD system, which can directly access the internal data of the geometric model.

Taking advantage of mating feature in the feature modeling or mating information between the elements of in solid modeling, mating surfaces, fit direction, joint relation and hinder relations between mating parts is generated. To export hierarchical structure, parts information, contacting information, as well as the relationship between information and assembly constraint information of assembly model, and then build the model required information for assembly planning.

3. Product Assembly Information Model

Product assembly feature is a information set of reactions assembly type, Coordination relation, mutual restraint and assembly operation between related parts. At the same time according to the list of connections between parts of the CAD system, obtainning contact constraints and exposure matrix [3].

Assembly Information Model [4] including parts attribute information, parts hierarchical information, assembly relationship information.

Parts attribute information:

1) Parts of the basic attributes of information, mainly refers to the description of product design, parts assembly and other aspects of the property information;

2) The dimensional accuracy of the information, mainly is the description of the dimensional accuracy information, geometric precision, surface coarse degree information and relevant attribute information;

3) Geometric topology information, mainly refers to the description of the geometric dimensions of information, geometric features information, the topology of geometric features and related information.

   I. Hierarchy relation information, that is, the relationship information between parts, that can be divided into product layer, component layer, part layer.

II. Assembly relation information, mainly recording the feature elements information, constraints type information in the process of assembly. Including

    a. constraint relation, planar alignment (joint or in the same direction), linear alignment relations, angle relations, parallel relations, tangency relations.

    b. the geometric constraint relationship, refers to the assembly interference information along +X, +Y and +Z direction in Cartesian coordinate system.

    c. connection information, including: welding, riveting, screw connection, and other fastening connection; key and spline connection, pin connection, coupling and other functions.

4. Assembly Hierarchy

Mechanical products are made up of components, which have hierarchical relationship.

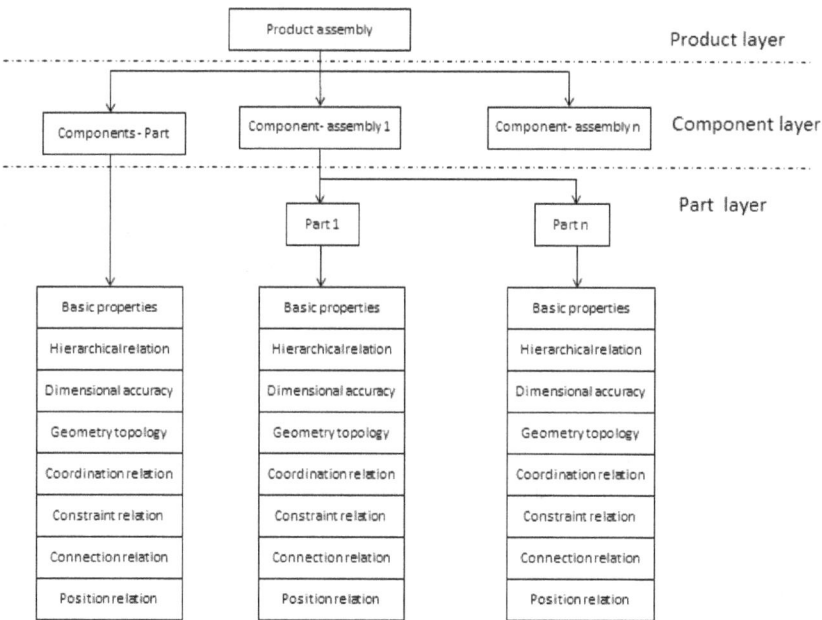

Figure 1. Product assembly hierarchy diagram

A component can be broken down into a number of parts or components, a sub component can be broken down into a number of parts or parts of the lower layer. The relationship between product components can be directly expressed as assembly tree structure. The top node of the tree is an assembly, a child node is a sub assembly or part, and the end node of the tree is a part. Tree data structure can be used to express the assembly structure tree. Hierarchical information for assembly sequence planning in assembly information model can be used to generate prefabricated structure tree. At the same time, the engineers can modify the assembly tree structure according to their own experience knowledge, so as to set up a hierarchical assembly hierarchy model tree. The product assembly hierarchy is shown in Figure 1.

Assembly hierarchy model clearly reflects the close degree of physical and geometric connection between the components in the assembly. When assembly sequence planning for the assembly, the method of bottom-up hierarchical programming can be used, that is, first of all, the assembly sequence planning of parts layer in any part of the component layer; then the assembly sequence reasoning is carried out for the sub assembly of the sub assembly in the component layer, and the assembly sequence of each component unit in the whole assembly is finally carried out.

5. Assembly Reasoning Method

In order to facilitate the computer recognition of the relationship between the parts of the assembly for the automatic generation of assembly sequence. In this paper, the interference matrix is introduced to describe the interference relations between each part.

6. Removing Interference Matrix

The interference matrix [5] can be gained by moving increment interference detection for the position and orientation of the parts in the assembly.

The definition of the disassembly interference matrix is as follows: if the assembly A is assembled by N parts, then $A = (P_1, P_2, ... P_n)$. The disassembly interference matrix established is a n * n matrix, Each element in the matrix represents the interference relationship between the i and the j parts. If $a_{ij} = 0$, description of parts i to the parts j does not cause interference. If $a_{ij} = 1$, description of part i to the part j does cause interference, and that is, part j is removed after the removal of part i. The disassembly interference matrix is expressed as : $\left[a_{ij}\right]_{n\times n}$

$$\left[a_{ij}\right]_{n\times n} = \begin{pmatrix} a_{11} & \cdots & a_{1n} \\ \vdots & \ddots & \vdots \\ a_{m1} & \cdots & a_{mn} \end{pmatrix}$$

It is assumed that the assembly object is a rigid body, and the inverse of the product disassembly sequence is taken as the assembly sequence.

This disassembly reasoning process is simple, but it is difficult to divide components with complex relationships, and solve combinatorial explosion problem when the number of parts is more. Therefore, the assembly characteristic relation information is introduced int this paper to divide the product into several sub parts, and the sequence of the assembly by layer decomposition is divided.

7. Connection Matrix

The connection relationship is based on the contact between parts. Connection parts are generally used as fasteners, and most of them are standard parts. In the process of assembly process planning and analysis, if there is a connection in the relationship (whether it is a fastener connection or a function connection), a stable connection assembly is formed. The components that have a Stable connection are usually combined as a sub assembly or a node to handle, to simplify the analysis of the model. At the same time, the parts of the connection relation are put together in the same process. The connection matrix C is used to show the connection between the parts, if the two parts are tight joint (such as screw connection, interference fit, bonding, welding, riveting, etc.), Then $c_{ij} = 2$, if the contact relationship is general between parts, then $c_{ij} = 1$, if the parts i, j does not contact, then $c_{ij} = 0$, because the part itself does not exist with their own connection,

For an assembly with N parts $p = (P_1, P_2, \ldots P_n)$, the connection matrix C is defined as follows: $C = (c_{ij})_{n\times n}$. The matrix element represents the connection between the parts i and the part j, and the criteria are as follows:

$$c_{ij} = \begin{cases} 0 & \text{There is no connection relation between the parts i and parts j} \\ 1 & \text{Part i and part j have connection relation} \\ 2 & \text{Part i and part j have stable connection relation} \end{cases}$$

8. Sub Assembly Identification

The sub assemblies should be stable and tight, so the parts that are connected to each other can form a sub assembly. In the process of assembly, the parts that have been assembled are stable, so there should be a standard part, other parts or sub assemblies to be assembled with the reference parts. Usually when we remove parts, the connected with the parts is needed to be removed. Common connectors are:bolts, screws, studs, nuts, washers, etc. Therefore, according to the common axis constraint and the common plane constraint rule [2] in the planning of assembly sequence, we select the component nodes which meet the rules to constitute the sub assembly, and then carry out geometric interference check [6]. If no interference, it can be identified as sub assembly parts for reducing the number of parts. Thus, according to the geometry and location information in product design domain, we can directlycall to the CAD of interference check function or constructing motion interference function, to access to assemble domain constraint direction feature, thus feature map each other and interference matrixf each sub assembly is realized. After the simplification, the product is divided into several parts and sub assemblies, and the disassembly sequence is determined according to the established interference matrix, which simplifies the calculation of the workload.

9. Assembly Sequence Generation Algorithm

9.1 *Assembly Sequence Algorithm Process*

Referring to other similar algorithms, a new algorithm is formed by the improvement of the related algorithm. The assembly sequence algorithm of the component assembly is shown in the following steps.

1) We select any part node P as the pre assembly parts and generate component of the connection matrix C, according to the relationship information to participate in the assembly of the feature element, assembly constraint type information;

2) According to the row (column), we search C in the line of all elements and calculate the contact element value, if the first I row (column) in contact with the maximum value, then parts of the I parts within the foundation;

3) For parts of the i line, we find out the whole parts which are contacted with the part i, and make the components of the sub component j, this time the component is not executable sub assembly parts;

4) Traversing the value of the contact relationship between the parts if the value=2, the part can be directly added to the new component. If the value =1,

538

the part can be considered to join the component, the following steps are to check the interference and constitute a new assembly after the determination;

5) According to the interference function, the geometric feasibility of the sub assembly parts is tested, if the direction is no interference, it is sub assembly;

6) Components is removed if the contact value is 1, then go to step 5 for interference detection, until you find through the detection of a new component, the component is part of assembly; otherwise the component no subassemblies;

7) At the end, the component parts and other parts of the node P are confirmed that can be assembled into a new assembly part Q;

8) The sub assembly parts as a separate part, according to the function, the interference matrix of the Q is generated;

9) searching for stem related matrix Q in accordance with the line (column), and if the Q in the first K line (column) in all the elements are zero, then the order of the part k of the record is 1;

10) Removing the elements of the first k row (column) in the interference matrix R, and forming a new interference matrix Q;

11) In the interference matrix R repeat steps 9 operation in order to find out 0; each repeat, part of the sequence of the sequence number plus 1;

12) Until all the elements have been removed, all the feasible disassembly sequence of the pre assembly parts P can be listed at the end of the assembly;

13) Repeat steps 1 to 12 of the operation, in turn traverse each part of the node assembly body, you can calculate the assembly sequence of each assembly in the component layer.

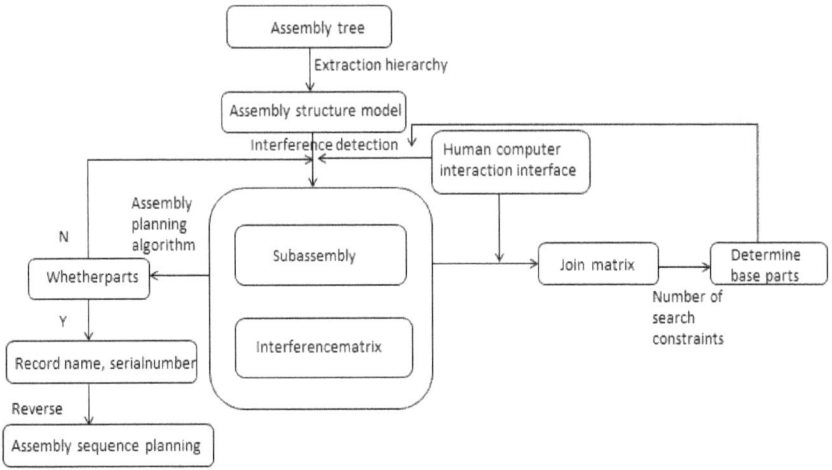

Figure 2. Assembly flow chart

The assembly process of the whole product is as follows:

Step 1: Gainning the level, property and assembly relationship information by two development technology to establish assembly model.

Step 2: According to the product BOM structure tree in accordance with the hierarchical relationship refabricated structure tree is established. According to the rule of the common axis and the common plane, it can be used to verify whether the individual components constitute the sub assembly, and improve the assembly tree and return the level information to the database.

Step 3: Determining the basis, combined with the above information, firstly basic pieces of information is produced , the main basis for according to the attribute information of the parts in the "weight", "volume", "symmetry" property listing possible as the basis of the parts. Search for the relationship between the contact and the number of constraints, the largest number is basic parts.

Step 4: (Two sub nodes for a single part and the base parts) The node of components is treated as the pretreatment of the assembly node.

Step 5: Performing the assembly sequence planning algorithm of the single assembly, to calculate the assembly sequence of the sub assemblies and parts of each assembly. Storing the corresponding assembly sequence information.

Step 6: Each component is regarded as a separate part, and the interference information is calculated by the movement interference function to generate the new interference matrix;

Step 7: To executed component assembly planning algorithm flow, and in order to traverse all the parts to determine the assembly sequence of parts level and storage parts assembly sequence number;

Step 8: From the bottom of the recursive calculation, the parts assembly sequence number and the level of assembly parts assembly sequence number combination, you can get the whole product assembly sequence.

10. Product Assembly Example

According to the assembly sequence planning algorithm, to a certain horizontal machining machine tool as an example of assembly. As shown in Figure 3, the three-dimensional model of the machine tool, power distribution cabinets, water tanks, lubrication systems, and other peripheral parts of the knife and other components are simplified.

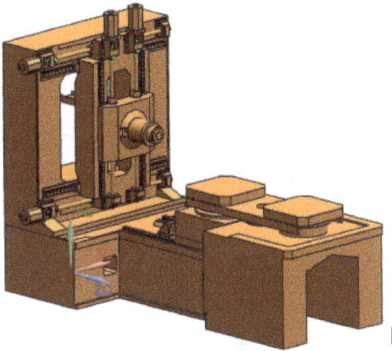

Figure 3. Model of machine tool assembly structure   Figure 4. Column parts assembly

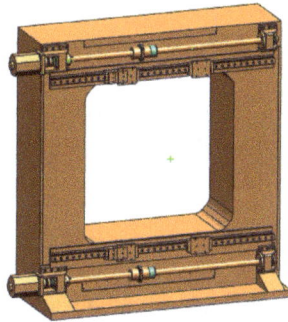

Firstly, the assembly hierarchy structure of assembly model information is extracted, and the assembly structure model tree of assembly sequence planning is generated according to the level information of assembly information model. Component layer is divided into bed, column, a sliding seat, spindle box, electric spindle assembly, workbench, table, 7 assembly parts of the body. According to the assembly sequence algorithm of component assembly, the first part is to take one part, to carry out the assembly planning, to determine whether there is a sub assembly in the component. In the end, the sub assemblies and parts of the parts are ordered according to the planning method.

Below the column parts assembly as an example of assembly, as shown in Figure 4.

1) The first generation of the connection matrix of the component C;
2) All the contact element numerical calculation and search C, connection number of column is 10, the basis of a judgment.
3) Search the column where the line, to find out the various parts of the contact. According to the coupling type bolt, screw, screw, determined on rail and rail 4 components.
4) The interference test, finally determine the screw, screw, rail and rail for sub assembly.
5) The interference matrix of each sub assembly is generated, and the sequence of the internal parts assembly of the sub assembly body is taken up.

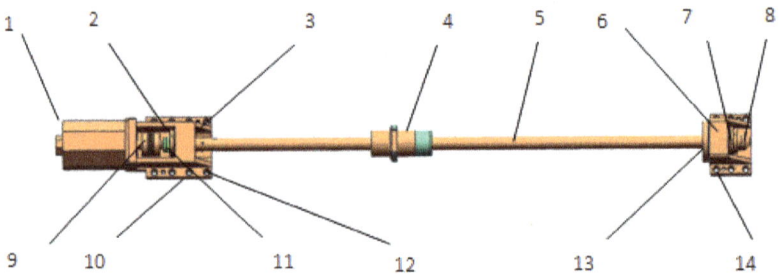

1. Motor 2. Flange plate 3. Inner spacer (top wire) 4. Slider 5. Screw 6. Bearing seat 7. Adjusting nut 8. Fixed sleeve 9. Coupling 10. Motor seat 11. Adjusting pad 12. Bolt (PIN) 13. Axle sleeve 14. Bolt (PIN)

Figure 5. Wire bar assembly diagram

6) Take the lead screw assembly (Figure 5) calculation, according to the X axis direction of the demolition of the matrix can be generated by the assembly sequence:

$$1)5-3-10-12-4-13-6-14-2-11-7-8-9-1$$

$$2)5-3-10-12-4-13-6-14-7-8-2-11-9-1$$

$$3)5-13-6-14-4-3-10-12-7-8-2-11-9-1$$

$$4)5-13-6-14-4-3-10-12-2-11-7-8-9-1$$

7) Each component can traverse, assembly sequence generation on the rail. Under the rail and screw and the assembly sequence, assembly sequence can determine the column of the assembly.

8) Repeat the 1 to 7 steps, bed, slide, spindle box, electric spindle components, working table and the working turntable layers within the six component parts of the assembly sequence in order to generate.

9) Based on the component layer component disassembly matrix to determine the component layer assembly sequence for bed - column - slide - electric headstock spindle components - workbench - work table.

10) Read the sequence number of each component and the component in the product, calculate and determine the order of each part in the product. The assembly sequence of the whole machine tool can be obtained by sorting.

11. Conclusion

A complex assembly after slicing constructed product component parts of hierarchical assembly model, the assembly planning for each processing only the sub components in several parts, in part layer further recognition of subassembly, the need to deal with the number of parts will further reduce. In this way, a complex assembly product can effectively improve the efficiency of assembly planning through the simplified treatment of the method.

References

1. Zhang Youliang, Dai Guohong, You Fei. The assembly constraint relation information of virtual assembly is dynamically generated [J]. Computer integrated manufacturing system, 2007,13 (7): 1406-1411.

2. Zhang Yingzhong, Luo Xiaofang, Fan Chao. Semantic representation of assembly design intent [J]. Computer integrated manufacturing system, 2011,17 (2): 248-254.

3. Yu Jiapeng, Wang Chengen, Zhang Wenlei. Automatic generation method of complex product assembly relation matrix [J]. Computer integrated manufacturing system, 2010,16 (2): 249-255.

4. Zhang Gang, Yin Guofu, Deng Kewen, Li Huosheng. Research on the hierarchical modeling method of assembly oriented [J]. Computer integrated manufacturing system, 2005,11 (7): 916-920.

5. Wang Junfeng, Fu Yan, Peng Tao et al. Automatic generation technology of assembly / disassembly interference matrix [J]. Mechanical science and technology, 2008,27 (7): 955-962.

6. Wu Meiping, Liao Wenhe. Research on layered interference detection algorithm for digital pre assembly [J]. China Mechanical Engineering, 2007,18 (18): 2205-2209.

An Error Separation Algorithm in Radar Astronomic Calibration Based on Singular Value Decomposition (SVD)

Xiao-yong Feng, Heng Yang, Xin-ming Liu, Li-jian Zhao
and Qian-xue Wang

*China Satellite Maritime TT&C Department,
Jiangyin, Jiangsu, China*

To solve the engineering application problems currently exist in radar astronomic calibration and implement the regularly calibration of ship-borne radar while in maritime dynamic condition, through analysis to equipment principles and error model. The calculation method of error equation based on SVD is proposed by trial and error, and a set of typical data from the test results is provided. The test indicates that this calibration method can separate 8 individual errors invalidly, whose precision is better than normal calibration method. By means of SVD, the least square solution of individual error in radar equipment can be obtained. The problem of data rationality, data repeat-ability and data stability in original astronomic calibration has been solved successfully.

Keywords: Radar equipment; Astronomic calibration; SVD; Error model; Separation algorithm.

1. Introduction

Ship-borne radar equipment (such as pulse radar and unified S-band radar) are main tracking and controlling equipment in ocean-going TT&C system, their error calibration and detection accuracy are extremely important in daily use, and the result of calibration has great influence on orbit determination.

Although common method used in individual error calibration is mature, the environment condition required is relatively strict. Astronomic calibration can be easily implemented in maritime dynamic condition automatically and regularly, but it also relies on error model and calculation method. Some negative aspects are listed as following:

The result of calibration is lack of repeat-ability. We can hardly obtain similar result under the same test condition (the same equipment, the same time), even some data are in opposite sign while their absolute value are near to each other. [1-4]

Comparing with common method, the result of calibration is obviously far from the true value, not to mention the empirical data and the design index, data range also over-proof.

In order to solve the problem of error calibration for radar equipment in maritime dynamic condition, it is necessary to study on astronomic calibration. For example, the implementation method, error model and calculation method, especially the single error separation arithmetic of this calibration method.

Instead of regular equations, an error separation algorithm, based on singular value decomposition, is proposed, and relative calibration tests on different kinds of radar equipment are carried out. [5-9]

2. Theory of Constitution

Both the star's theoretical position and the error model of radar equipment are known in astronomic calibration. By real-time stellar observation, measurement error, in fact, brought on by radar's single error, was found between measured value and theoretical value. According to the observation result from different stars and the error model, equations of error are built. Calculation result and precision for each individual error are obtained from the least square solution of error equations, then the calibration result of single error can be used to correct the measurement precision of radar equipment in the light of error model.

The astronomic calibration methods for ship-borne radar consist of 2 portions, software and hardware.

3. Hardware Structure and Theory

Hardware structure including three portions, radar pending to calibrate, microcomputer system for calculation & control, other accessory equipment. Interface circuit is configured according to the specific requirements of radar equipment.

4. Software Composition and Software Principle

There are three functional modules in software part of astronomic calibration method; they are real-time star observation, calculation for single error, measurement error statistics. Guided by ship-borne optical device, star observation in maritime dynamic condition can be carried out simultaneously, and the whole process runs in real time or quasi real-time, as shown in Fig. 1.

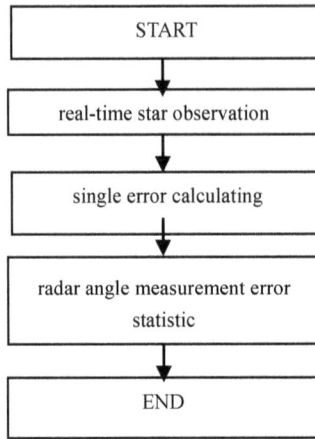

Fig. 1 Astronomic calibration software composition block diagram

According to date, time and station location, theoretical position of star is calculated by system computer in real-time. Guidance information is sent to radar equipment through interface circuit, then the radar equipment point to the star, driven by the servo system. Star observation is accomplished and the measurement information received by system computer through interface circuit as well. Meanwhile, relevant date including star number, star brightness, absolute time, star theoretical position, encoder angle value and TV undershooting value are saved for further use, single error calculation and statics also finished after that.

5. Error Calculation Method

The least square method is a mathematical optimization technique. It can fittingly minimize the squares sum of error-of-fit in predefined measures, but also has its shortage and limitations.

Astronomic calibration of radar depends on its error model, error square based on error model and result of star observation should be built firstly, then solve the least square solution to implement error separation, the core issue is calculation method of error model and error square.

6. Error Model

Measurement error of radar is a result of comprehensive effect of each single error, based on analysis to radar structure and error origination the error model of radar is proposed as following, with the linear error of angle measurement encoder ignored.

$$\begin{cases} \Delta A = a_1 + a_2 SecE_{ri} + a_3 TanE_{ri} + a_4 SinA_{ri} TanE_{ri} + a_5 CosA_{ri} TanE_{ri} \\ \Delta E = e_1 + e_2 CosE_{ri} + e_3 CotE_{ri} + e_4 SinA_{ri} + e_5 CosA_{ri} \end{cases} \tag{1}$$

Therein:

A_{ri}, E_{ri}: measured value for azimuth angle and pitch angle of No. i star;

ΔA_i, ΔE_i: measurement error for azimuth angle and pitch angle of No. i star;

a_1, a_2, a_3, a_4, a_5: azimuth zero value, lateral axis error, non-orthogonality error between transverse & vertical axis, vertical axis tilt error, azimuth tilt error.

e_1, e_2, e_3, e_4, e_5: pitch zero value, gravity sinking, refringence residual error, vertical axis tilt error, azimuth tilt error. (a4=+e5, a5=-e4)

7. Error Equations

According to error model and observation result, the error equations of radar equipment can be built, matrix expression are shown as following:

$$\begin{cases} \Delta A = A \cdot a \\ \Delta E = E \cdot e \end{cases} \tag{2}$$

In the following expressions figure ΔA, ΔE are error vectors based on the measured value of star observation:

$$\Delta A = \begin{bmatrix} \Delta A_1 \\ \Delta A_2 \\ \cdots \cdots \\ \cdots \cdots \\ \Delta A_n \end{bmatrix} \qquad \Delta E = \begin{bmatrix} \Delta E_1 \\ \Delta E_2 \\ \cdots \cdots \\ \cdots \cdots \\ \Delta E_n \end{bmatrix}$$

In the following expressions figure A, E stand for measurement matrices based on the error model of radar and star's space distribution:

$$A = \begin{bmatrix} 1 & SecE_{r1} & TanE_{r1} & SinA_{r1}TanE_{r1} & CosA_{r1}TanE_{r1} & \cdots \\ 1 & SecE_{r2} & TanE_{r2} & SinA_{r2}TanE_{r2} & CosA_{r2}TanE_{r2} & \cdots \\ \cdots & \cdots & \cdots & \cdots & \cdots & \cdots \\ \cdots & \cdots & \cdots & \cdots & \cdots & \cdots \\ 1 & SecE_{rn} & TanE_{rn} & SinA_{rn}TanE_{rn} & CosA_{rn}TanE_{rn} & \cdots \end{bmatrix}$$

$$E = \begin{bmatrix} 1 & CosE_{r1} & CotE_{r1} & SinA_{r1} & CosA_{r1} & \cdots \\ 1 & CosE_{r2} & CotE_{r2} & SinA_{r2} & CosA_{r2} & \cdots \\ \cdots & \cdots & \cdots & \cdots & \cdots & \cdots \\ \cdots & \cdots & \cdots & \cdots & \cdots & \cdots \\ 1 & CosE_{rn} & CotE_{rn} & SinA_{rn} & CosA_{rn} & \cdots \end{bmatrix}$$

In the following expressions figure a,e are coefficient vectors of single error:

$$a = \begin{bmatrix} \Delta a_1 \\ \cdots \\ \cdots \\ \Delta a_5 \\ \cdots \end{bmatrix} \qquad e = \begin{bmatrix} \Delta e_1 \\ \cdots \\ \cdots \\ \Delta e_5 \\ \cdots \end{bmatrix}$$

8. Error Calculation

The astronomic calibration result of radar equipment can be obtained by the least square solution of error equations. The normal equation method is often used, which has implemented the astronomic calibration in principle, cannot meet the need of precision. Thus, an error separation algorithm based on SVD is proposed.

For purposes of comparison, calculation process of single error by means of 2 different methods (normal equation method and SVD) is listed as following:

Achieve the least square solution of error equation by normal equation method

Single error coefficient of radar equipment is given as following:

$$\begin{cases} a = (A^TA)^{-1}A^T\Delta A \\ e = (E^TE)^{-1}E^T\Delta E \end{cases} \tag{3}$$

Co-variance matrix of single error is given as following:

$$\begin{cases} D_a = (A^TA)^{-1}\sigma^2_{\Delta A} \\ D_e = (E^TE)^{-1}\sigma^2_{\Delta E} \end{cases} \tag{4}$$

Achieve the least square solution of error equation by SVD.

Matrix A and E can be decomposed in following form:

$$
\begin{bmatrix} A \end{bmatrix} = \begin{bmatrix} U_a \end{bmatrix} \cdot \begin{bmatrix} W_{a1} & & \\ & \cdots & \\ & & W_{a5} \end{bmatrix} \cdot \begin{bmatrix} V_a^T \end{bmatrix} \tag{5}
$$

$$
\begin{bmatrix} E \end{bmatrix} = \begin{bmatrix} U_e \end{bmatrix} \cdot \begin{bmatrix} W_{e1} & & \\ & \cdots & \\ & & W_{e5} \end{bmatrix} \cdot \begin{bmatrix} V_e^T \end{bmatrix} \tag{6}
$$

U_A, U_E: column orthogonal matrix of nX5, W_A, W_E: diagonal matrix of 5X5, V_A, V_E: orthogonal matrix of 5X5.

Single error coefficient of radar equipment is given as following:

$$
\begin{cases}
a = \sum_{i=1}^{5} \left[\dfrac{U_a(i)\Delta A}{W_a(i)} \right] V_a(i) \\[4mm]
e = \sum_{i=1}^{5} \left[\dfrac{U_e(i)\Delta E}{W_e(i)} \right] V_e(i)
\end{cases} \tag{7}
$$

Co-variance matrix of single error is given as following:

$$
\begin{cases}
D(a_j, a_k) = \sum_{i=1}^{5} \left[\dfrac{Va_{ji}Va_{ki}}{W_a(i)^2} \right] \\[4mm]
D(e_j, e_k) = \sum_{i=1}^{5} \left[\dfrac{Ve_{ji}Ve_{ki}}{W_e(i)^2} \right]
\end{cases} \tag{8}
$$

Error decomposition results from the least square solution are shown in formulas 3 and 4 show, using the normal equation method, formula 7 and 8 are the result of VSD. Single error coefficient represents the size of the radar equipment error components of each system, and the co-variance matrix reflects the solution accuracy and relevance of individual errors coefficients.

9. Calibration Test

To compare the verification accuracy, data stability and reliability of different calibration method. Three different methods, conventional calibration method, astronomical calibration based on normal equations, astronomical calibration based on SVD, are under analysis and comparison, with the statistic results of measurement error given accordingly.

10. Calibration Method and Result

The test is carried out on two different types of radar equipment, total valid data obtained from 202 stars, divided into nine groups. A set of typical calibration result is shown in Table 1:

Table 1. Measurement comparison between astronomic and conventional calibration

individual error calibration results						
single error	azimuth zero value	pitch zero value	lateral axis error	non-orthogonality error	vertical axis tilt error	azimuth tilt error
	$A_0(")$	$E_0(")$	$S_b(")$	$\delta\ (")$	$\beta\ m(")$	$A\ m(°)$
Conventional	-3581.90	+172.60	-39.50	-10.20	271.30	93.414
Astronomic	-3620.02	+134.75	-13.40	-34.21	255.15	90.955
	radar angle measurement error results					
Angle measurement precision	azimuth error statistic results			pitch error statistic results		
	σ_{as}	σ_{ar}	σ_{at}	σ_{es}	σ_{er}	σ_{et}
conventional correction	23.29	11.85	26.13	26.11	12.69	30.58
star-observe correction	7.76	10.85	13.91	22.44	12.61	27.40

11. Calculation Method and Result

For purposes of comparison, two different error separation methods were used during data processing; a set of typical calibration result is also shown in Table 2.

Table 2. Measured value comparison between normal equation method and SVD

Device Model		azimuth zero value (")	pitch zero value (")	lateral axis Error (")	non-orth ogonality error (")	vertical axis tilt error (")	azimuth zero value (°)	gravity sinking (")	refringence residual error (")
Normal Equation method	①	+10.62	+3.39	-17.45	+18.54	2.03	-4.48	\
	②	+24.16	+0.80	-13.96	+27.73	4.63	-1.56	\
SVD	①	+0.32	-0.76	+1.18	-1.55	232.76	+5.28	-0.91
	②	+0.26	+0.69	+1.16	-1.54	233.01	+2.59	-0.34

The calibration results show that, individual error separation algorithm based on SVD can improve the data reliability and accuracy significantly.

12. Conclusion

On a certain space tracking ship, due to its special conditions of use, ship-borne devices are often stayed within dynamic environment, in order to calibrate regularly under dynamic conditions, research for star observation calibration method is in urgent need.

In the process of astronomical calibration for radar, the core issue of which is the use of error separation method through solving the least squares solution. Test proved that this method could help to improve the accuracy of angle-measured value. Therefore, it can be used as a complete, effective and independent calibration means for solving the problem of dynamic maritime calibration.

References

1. Feng Xiao Yong, Zhu Wei Kang. Single Error Separation Arithmetic for Astronomical Calibration of Optical Measurement Equipment. [J]. Opto-Electronic Engineering, 2008, 35(12): 28-33 (in Chinese).
2. Wang Yong hua. The Angel Calibration Method for Vehicle Borne Radar. [J]. Modern Electronics Technique, 2005, (10):97-98.
3. Jiang Da Zhi, Zhao Man Qing, Hu Kui. Research on error correction of ship-borne radar. [J]. Tactical Missile Technology, 2014, 00(03):00-00 (in Chinese).
4. Zhao Xin, Wang Shi-feng, Tong Shou-feng, Song Hong-fei. New method of zero position calibration for ship-borne radar. [J]. Fire Control & Command Control, 2010,00(02) :00-00 (in Chinese).
5. Hou Hong-lu, Zhou De-yun, Wang We etc. Accuracy assessment of photo-electric tracking system based on star calibration. [J]. Photoelectric Engineering, 2006, 33(3):5-10.
6. Che Shuang-liang, Zhang Yao-ming. Theory and Practice of Optical-electronic Theodolite Calibration Method with Star in Range. [J]. Acta Photonica Sinica, 2004, 33(10): 1255-1260.
7. Yao jing-shun, Yang Shi-xing. Dynamic calibration of ship-borne radar. [J]. Fire Control & Command Control, 2008, 00(03): 00-00 (in Chinese).
8. Wang Qi, Zhang Ji-xu. Method for radar calibration and data processing of theodolite base on CCD. [J]. Tactical Missile Technology, 2009, 00(01) :00-00 (in Chinese).
9. Li Xiang-rong, Guo Ping-ping, Qiao Yan-feng, Zhang Yao-yu. On-Sea Precision Calibration Methods of Azimuth Aiming System on Mobile Bedding [J]. Journal of Test and Measurement Technology, 2004, 18(Z3): 173-176.

Simulation Analysis of Flow Field in Saddle Zone for a Tubular Pumping System

Jia Chen, Ji Pei[+], Shou-qi Yuan, Yan-jun Li and Fan Meng

National Research Centre of Pumps, Jiangsu University
Jiangsu, Zhenjiang, 212013, China
Email: chenjia_cj@126.com, [+]jpei@ujs.edu.cn, shouqiy@ujs.edu.cn

To study the flow field in saddle zone of a tubular pumping station, the three dimensional unsteady turbulent flow was numerically simulated with CFD software under different operational conditions. Some characteristics of flow field in saddle zone were obtained, and the results show that there is good agreement between experiment and simulation data and both of them illustrate the flow features of saddle zone under flow of $0.45Q_{DES}$ to $0.75Q_{DES}$. With the decreasing of flow rate, the flow becomes more turbulent, and more vortex and backflow can be found with strong energy consumption, which is the main reason of saddle zone. Besides, the pressure fluctuation intensity with no obvious regularity of impeller inlet in saddle is much stronger than that of impeller inlet under designed flow rate.

Keywords: Tubular pumping system; Flow field; Saddle zone; Numerical simulation; Streamline; Pressure fluctuation intensity.

1. Introduction

With the construction of the East-Route of South-to-North Water Transfer Project and other water resource projects in riverside area of China, large axial-flow pumps with low head have been rapidly applied. As a new axial-flow pumps with low head, shaft tubular pumping systems, with advantages of simple construction, reliable running and maintenance friendly as well as high efficiency [1], have a critical influence on flood and waterlogged. During the practical running of pumping station, strong vibration and noise can be found after a few seconds or more than ten seconds of starting period in station [2], and this is the consequence of unstable saddle zone when axial-flow pumps are under the small flow rate operation. In performance curve there will be two or more flow rate points under the same head, which will significantly reduce the safe and stable operation range of the whole pumping stations. As a result, 3D unsteady turbulent numerical simulation on saddle zone of a shaft tubular pumping system not only has important theoretical significance, but also provides some reference for security and stability in pumping system.

At present, many scholars at home and abroad have some research on saddle zone in axial-flow pumps. Masahiro et al.[3] studied the flow instabilities in a mixed flow pump with a vane diffuser by PIV techniques and found stall phenomenon in the internal flow field of the pump. Kosyna G et al.[4] employed different experimental ways to treat spiral-type vortex when the impeller rotating stall exists in the axial-flow pump, and their study gave a simple but effective type of casing treatment, which has a great effect on performance stability of axial-flow pump, and it was early used to expand the stable operation range of compressors[5,6]. Liu Chao et al.[7] discussed the performance curves of pump and pumping system respectively, and studied the similitude problems of pipeline characteristics and pumping system. They also redefine the unstable operation areas in axial-flow pumps. Zheng Yuan et al. [8] researched the internal flow field of a large axial-flow pumping system with combination way of simulation and experiment, and they found obvious saddle zone under the designed flow rate of 50% ~65%. Yang Hua et al. [9] designed a baffle at the inlet of axial-flow pump impeller to improve the performance of axial-flow pump under unstable condition and decrease the vibration of pump.

On the basis of these studies, unsteady turbulent flow was numerically simulated with SST k-ω turbulent model in saddle zone for a tubular pumping system. This research compared the external characteristics value between simulation and experiment, and analyzed the internal streamlines in pumping system under different unstable operational conditions. Furthermore, the pressure fluctuation intensity was also discussed in saddle as performance curve changing.

2. Numerical Model and Method

2.1 3D Model and Mesh Generation

The tubular pumping system investigated is consisting of an inlet conduit, an axial-flow impeller, a vaned diffuser and an outlet conduit. The main geometrical and hydraulic data is given in table 1. The commercial software Creo was applied on 3D model of the pumping system. Figure 1 is the physical diagram of this tubular pumping system.

Table 1. Main parameters of the investigated pump

Main geometric data		
Diameter of impeller	D	300mm
Blades number	Z1	3
Vane number	Z2	3
Nominal rotational speed	n	1104 r/min
Design flow rate	Q_{DES}	0.28 m³/s
Design head	H_{DES}	1.55m

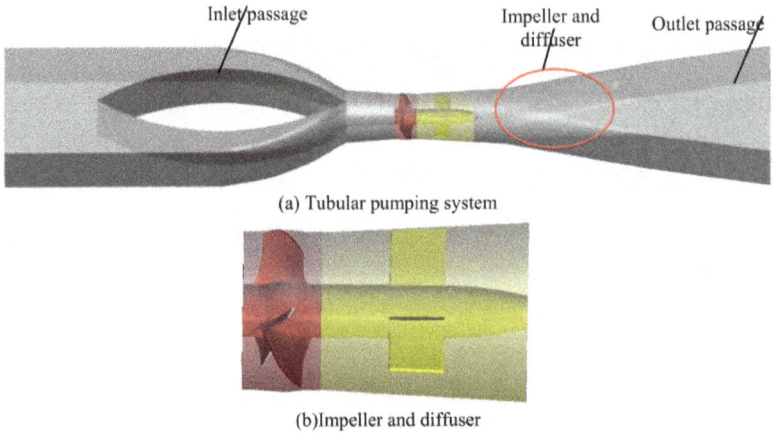

(a) Tubular pumping system

(b)Impeller and diffuser

Figure 1. Three-dimensional diagram of tubular pumping system

The numerical calculation domains include inlet and outlet passages, impeller and diffuser, and structured and unstructured hybrid grids were used for simulation. As the core parts in the pump, the mesh quality and distribution of impeller and diffuser has a direction impact on performance of pumping system. The impeller and guide vane were divided into hexahedron structure mesh, and a localized refinement of mesh is employed at regions close to boundary layers of impeller and diffuser. Besides, the type of inlet and outlet passages mesh was tetrahedral unstructured. The mesh of axial-flow pump is as shown in Figure 2. Tetrahedral grids, which have a good adaptability to different shapes were used in inlet and outlet passages, and the total number of mesh is 4848921.

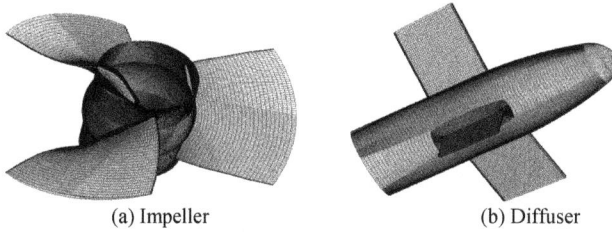

(a) Impeller   (b) Diffuser

Figure 2. Mesh assembly of axial-flow pump

3. Numerical Model Calculation

The flow in pump is considered to be three dimensional, incompressible, viscosity, and the unsteady flow is numerically simulated with ANSYS CFX14.5 in the calculation domains. The SST k-ω turbulent model is used to solve the Reynolds equation while multiple rotating coordinate is used to set rotating and stationary domains. The mass flow rate is given at inlet boundary with opening boundary condition at outlet. The interface between impeller and static components is set to "Transient Rotor stator" and the wall boundary is no sliding. The physical time is$(1/\omega)*0.1$ in which ω is angular velocity of rotation of impeller and the initial flow field of unsteady calculation is the result of steady RANS simulation. Every 360 time step is set to be an impeller rotation period, and the total simulation time is 0.597826s with 8 calculation periods, the volute of convergence residual error (RMS) is 10^{-5}.

4. Results and Discussions

4.1 *Performance Analysis*

After steady and unsteady simulation, performance curve (*H-Q* curve) of the tubular pumping system was obtained, and to verify the calculated results, the comparison of the numerical and experimental data is shown in Figure 3. Besides, the calculated value of the pumping system is the time-averaged head in last period under unsteady numerical simulation.

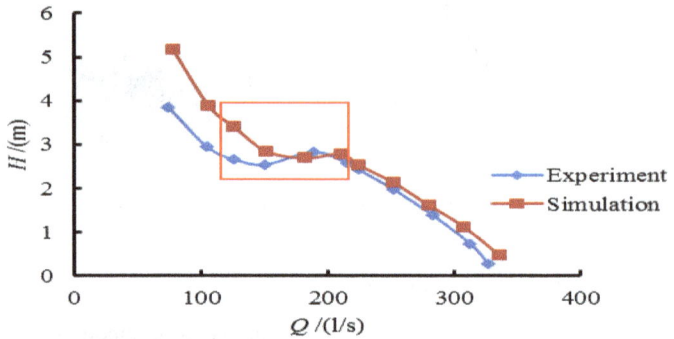

Figure 3. Performance curve

It can be seen from Figure 3 that whether simulation or experiment, there is obvious saddle zone under flow of $0.45Q_{DES}$ to $0.75Q_{DES}$. From the simulation results, when flow decreases to $0.75Q_{DES}$, the head starts to drop and then increases again, and in this area, the pumping system will begin to vibrate with strong noise. Moreover, the calculated head follows the trend very well with experimental data, and simulation values are a little higher than the experimental ones with a bit larger deviation under small flow rate because of unstable flow, and under designed condition, the error is about 3.73%, which is in acceptable range. The agreement indicates that the calculation is reasonable for external characteristics of this tubular pumping system and can be used to perform detail analysis.

5.   Flow Field Distributions in Saddle Zone

To study the flow change in saddle zone of internal flow field, streamlines under six conditions (Q/Q_{DES}=1.0, 0.75, 0.65, 0.54, 0.45, 0.38) on axial plane in tubular pumping system are shown in Figure 4.

Figure 4 shows that the velocity in passages declines with flow rate falling down. The flow is uniform and there are no vortexes in impeller and diffuser field with only few ones in diffuser outlet under designed condition (Q/Q_{DES}=1.0). When the flow jumps from $0.75Q_{DES}$ to $0.65Q_{DES}$, a little of backflow starts to exist at impeller edge with others in diffuser inlet, and the backflow area expands to impeller hub which will plug the impeller passage and decreases the effective width of impeller and finally leads to head dropping. If the flow rate keeps falling down, as can be seen in Figure 4 under conditions of $0.54Q_{DES}$, $0.45Q_{DES}$, there will be more and bigger vortexes both in diffuser inlet and outlet and it makes secondary backflow. Until the flow of $0.38Q_{DES}$, the diffuser passages are almost full of vortexes, and under this circumstance, the flow out of impeller may get secondary pressurization after secondary backflow, therefore, the head of pumping station will increase again. However, under these unstable conditions, the backflow will consume much energy and its instability will make noise and vibration, so the efficiency is quite low and the pump should not operate under these conditions or last running.

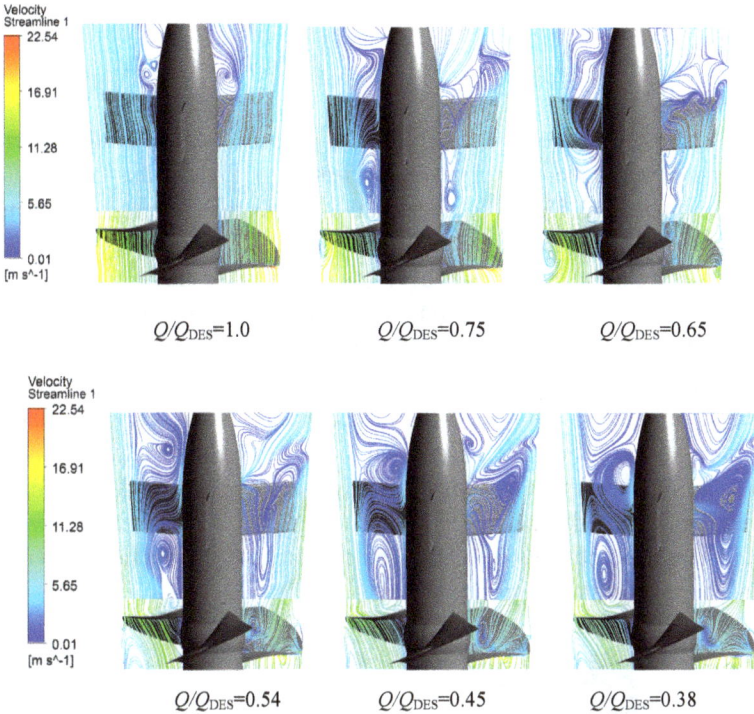

Q/Q_{DES}=1.0 Q/Q_{DES}=0.75 Q/Q_{DES}=0.65

Q/Q_{DES}=0.54 Q/Q_{DES}=0.45 Q/Q_{DES}=0.38

Figure 4. Streamlines of axial plane

6. Pressure Fluctuation Intensity Distribution

In order to analyze the internal unsteady flow phenomenon under saddle zone of pump accurately, the statistical analysis method is adopted to define the pressure fluctuation intensity. Periodically unsteady pressure $P(x,y,z,t)$ at a grid node at some time can be defined and to evaluate periodic pressure as the function of time and impeller rotation position, a time-dependent non dimensional pressure coefficient C_p is also defined $C_p = p/0.5\rho u_2^{\,2}$, here, u_2 is the outlet speed of impeller. Assuming that T is the period of impeller rotation, therefore, the average pressure coefficient (1) and pressure fluctuation intensity (2) in a period are defined respectively.

$$\overline{C_p} = \frac{1}{N}\sum_{n=1}^{N} C_p(x, y, z, \frac{n}{360}T)$$

(1)

$$C_p^* = \sqrt{\frac{1}{N}\sum_{n=1}^{N}\left(C_p(x, y, z, \frac{n}{360}T) - \overline{C_p}\right)^2}$$

(2)

Where N=360.

The flow of impeller inlet from inlet passage will be firstly affected by rotor-stator interaction, so a plane is created to analyse the flow change in impeller inlet, and Figure 5 presents the distribution of pressure fluctuation intensity of impeller inlet in this plane under five conditions (Q/Q_{DES}= 0.75, 0.65, 0.54, 0.45, 0.38, and the intensity of $1.0Q_{DES}$ is far lower than that in saddle zone). It shows that the strong fluctuation intensity is mainly distributed in the out edge of impeller inlet and the intensity of $0.75Q_{DES}$ is much lower and more uniform than that of others which illustrates that pressure change of impeller inlet is going to be more turbulent with the flow rate decreasing. When it comes to $0.65Q_{DES}$, the intensity gradient is the biggest and there is obvious strong intensity area in out edge of impeller inlet, which has a good agreement with head dropping. Under the last conditions, the intensity starts to decrease and then increase, all over, there is no obvious regularity and the intensity distribution presents three petals because of three impeller blades.

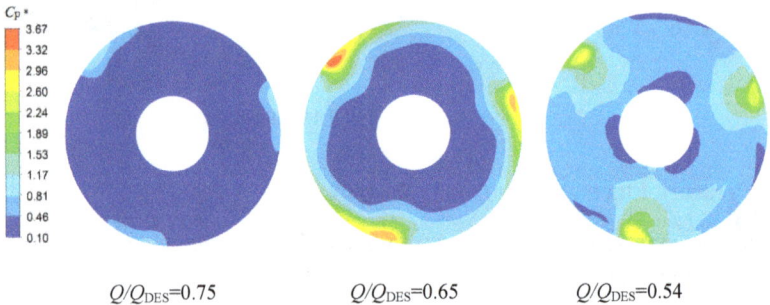

Figure 5. Pressure fluctuation intensity of impeller inlet

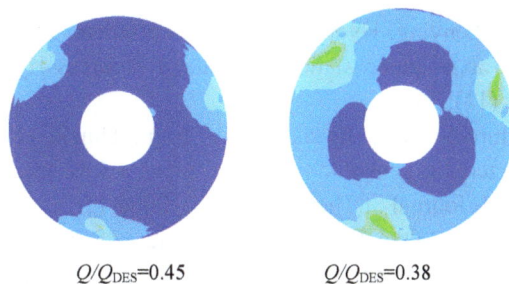

$Q/Q_{DES}=0.45$ $Q/Q_{DES}=0.38$

Figure 5. Pressure fluctuation intensity of impeller inlet (*continued*)

7. Conclusion

In this study, the unsteady numerical simulation in whole passage has been applied to a tubular pumping system, and the analysis of the results shows the following conclusions:

1) Both the experiment and simulation results reveal the flow features of saddle zone, and their agreement which error is about 3.73% shows the further simulation is reasonable and accurate.

2) The streamlines analysis of flow field under $0.75Q_{DES}$, $0.65Q_{DES}$, $0.54Q_{DES}$, $0.45Q_{DES}$, $0.38Q_{DES}$ declares that vortexes or backflow is the reason of saddle zone where there is strong energy consumption.

3) There is no obvious regularity of pressure fluctuation intensity of impeller inlet in saddle, and its distribution is affected by blade number. Besides, the intensity is much stronger than that of impeller inlet under $1.0Q_{DES}$.

Acknowledgement

This research is funded by the Priority Academic Program Development of Jiangsu Higher Education Institutions (PAPD), National Natural Science Foundation of China (Grant No. 51409123), Natural Science Foundation of Jiangsu Province Youth Fund (Grant No. BK20140554), China Postdoctoral Science Foundation (Grant No. 2014M560402) Postdoctoral Science Foundation of Jiangsu Province (Grant No. 1401069B).

References

1. Zhang Rentian. Features of various tubular pumps and its application south-to-north water diversion project, J. China Water Resources, 2005, 4:42-44.

2.   Yuan Zhenying. Hydraulic vibration problem in starting process of large siphon axial pump station, J. Water Resources Hydropower Engineering, 1982, (11): 38-44.

3.   Masahiro Miyabe, Hideaki Maeda, Isamu Umeki, et al. Unstable head-flow characteristic generation mechanism of a low specific speed mixed flow pump, J. Journal of Thermal Science, 2006,15(2): 115-120.

4.   Kosyna G, Goltz I, Stark U. Flow structure of an axial-flow pump from stable operation to deep stall, C// Proceedings of the 2005 ASME Fluids Engineering Division Summer Conference. Houston Texas, USA, 2005: 19-23.

5.   Chu Wuli, Zhang Haoguang, Wu Yanhui, et al. Impact of grooved width of grooved casing on stall margin improvement, J. Acta Aeronautica et Astronautica Sinica, 2008, 29(4) : 866-872.

6.   Zhu Jianhong, Piao Ying, Zhou Jianxing, et al. Investigation of off design performance of a transonic compressor with circumferential grooves, J. Acta Aeronautica et Astronautica Sinica, 2008, 23(4) : 687- 692.

7.   Liu Chao, Tang Fangping, Zhou Jiren et al. Performance and stability analysis for a large vertical axial flow pumping system, J. China Water &Waste Water, 2003,03:69-71.

8.   Zheng Yuan, Mao Yuanyuan, Zhou Daqin et al. Flow characteristics of low-lift and large flow rate pump installation in saddle zone, J. Journal of Drainage and Irrigation Machinery Engineering, 2011,05:369-373.

9.   Yang Hua, Sun Dandan, Tang Fangping et al. Research on the performance improvement of axial-flow pump under unstable condition using CFD, J. Transactions of the Chinese Society for Agricultural, 2012, 11:138-151.

Modeling FCC Risers Using a Three Parameter Kinetic Model Based on Special Pseudo-Components

Li-jia Guo*, Yong Li, Ji-zheng Chu, Ping Wu and Jia-rui Zhang

Department of Automation, Beijing University of Chemical Technology, Beijing 100029, China

2015200721@mail.buct.edu.cn

A three parameter kinetic cracking model has been established using the concept of special pseudo-components and has been used successfully to predict the operational behavior of fluidized catalytic cracking (FCC) risers. Compared with the six parameter kinetic model previously proposed by the authors, the cracking model of this paper is much simplified in calculation and enjoys improved identifiability. Tests with the production data of three commercial risers show that the proposed model has the capability of predicting a riser and catalyst, the kinetic parameters are independent of stock oils.

Keywords: Fluidized catalytic cracking, six parameter kinetic model, Kinetic model, Special pseudo-component.

1. Introduction

As a critical process of petroleum refineries, fluidized catalytic cracking (FCC) produces liquefied petroleum gas, gasoline, diesel and propylene from heavy distillates such as vacuum gas oils or even residues. By the year 2014, the FCC units in China have a total processing capacity of about 150Mt/a of which 40% is residual oil, and provide about 70% gasoline, 30% diesel and 40% propylene of the domestic market [1]. Industrial FCC units are characterized by their huge throughputs, which means great significance of design and operational optimization. Therefore, modeling and simulation of FCC processes have been being an important topic of research since about 1940s.

One of the challenging problems in modeling a FCC unit is to describe the chemical reactions of feed and product oils, which are complicated mixtures of numerous hydrocarbons and non-hydrocarbons. In order to obtain a predictive model for the complex cracking reactions, the authors [2] proposed the concept of special pseudo-components (SPCs), which have unchanged properties and are used to express the stock and product oils. It is shown by the authors that the generic reaction scheme of Gupta et al. [3] is predictive as coupled with SPCs in the sense that for a given riser and given catalyst, the kinetic parameters are independent of stock oils.

However, the identifiability of the six kinetic parameters in the generic reaction scheme of Gupta et al. [3] is poor. In order to improve identifiability of the model, some parameters in the reaction rate equation are fixed, and the reduced model is tested with the production data of three commercial risers.

2. Characterization Using Special Pseudo-components

The oil gas involved in the riser is characterized using special pseudo-components, and the description of the procedure is detailed by Zhang et al. [2]. Special pseudo-components (SPCs) are a set of component entities of invariant properties for expressing or characterizing the complex mixture of the stock and product oils involved in the cracking process. SPCs appear in pairs. The two SPCs in each pair have the same normal boiling temperature or normal boiling temperature range, but different Watson characterization factors. For light (L) and heavy (H) SPCs, the Watson characterization factors are taken to be Kw, L = 12.6 and Kw, H = 10.0, respectively. With the normal boiling temperature and Watson characterization factor known, the properties of a SPC such as density, molecular weight, combustion heat, carbon-to-hydrogen weight ratio and liquid and vapor heat capacities can be fully determined by well established correlations.

SPCs are distributed by their normal boiling temperatures in a range covering the true boiling point (TBP) distillation temperature ranges of all possible stock oils, and the normal boiling temperature interval between neighboring pairs of SPCs is suggested to be 20~30 °C.

With SPCs defined, stock oils are characterized from their TBP distillation curves. The weight recovery and density of a TBP distillation fraction of each stock oil are firstly determined in the normal boiling temperature range of each pair of SPCs. Then, the concentrations of SPCs in the feed oil as a whole are calculated

$$ w_{L,i} = \sum_j \Delta w_{L,ij} $$

(1)

$$ w_{H,i} = \sum_j \Delta w_{H,ij} $$

(2)

with

$$ \Delta w_{L,ij} + \Delta w_{H,ij} = w_{ij} $$

(3)

$$\frac{\Delta w_{L,ij}}{\rho_{L,i}} + \frac{\Delta w_{H,ij}}{\rho_{H,i}} = \frac{w_{ij}}{\rho_{ij}} \tag{4}$$

where w is the concentration in weight fraction, Δw the increment of concentration, ρ the density in kg/m3, subscripts i and j stand for SPC i and stock oil j, and L and H for light and heavy SPC, respectively.

In addition to the SPCs, the characterization procedure also includes light definite components hydrogen, methane, ethylene, ethane, propylene, propane, butene, butane and pentane.

3. Kinetic Reactions of Catalytic Cracking

The cracking scheme of Gupta et al. [3] is used in this study. In this scheme, pseudo-component i (PCi, i=1, 2, ...c) is cracked by the following generic reaction

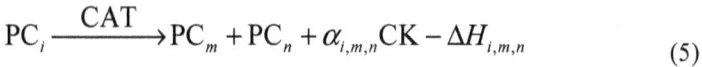

$$PC_i \xrightarrow{\quad CAT \quad} PC_m + PC_n + \alpha_{i,m,n}CK - \Delta H_{i,m,n} \tag{5}$$

where 1 kmol of PCi molecules, with the help of catalyst (CAT) and high temperature, are cracked into 1 kmol of PCm molecules and 1 kmol of PCn molecules as well as $a_{i,m,n}$ coke (CK). The heat consumed by the reaction is $\Delta H_{i,m,n}$ in kJ per kmol of PCi

$$\Delta H_{i,m,n} = \alpha_{i,m,n}\Delta H_{CK}^{comb} + MW_m\Delta H_m^{comb} + MW_n\Delta H_n^{comb} - MW_i\Delta H_i^{comb} \tag{6}$$

with ΔHcomb CK=32950KJ/kg being the combustion heat of coke.
The reaction rate is

$$r_{i,m,n} = \phi k_{i,m,n} C_i C_{CAT} \tag{7}$$

where the catalyst deactivation constant ϕ and the reaction rate constant $k_{i,m,n}$ is calculated by

$$\phi = \frac{B+1}{A + \exp(Ax_{CK} \times 100)} \tag{8}$$

$$k_{i,m,n} = k_0 \times e^{-\frac{E_0 MW_i^{\nu}}{RT}} \times \frac{\exp\left(-\alpha_{i,m,n}/\tau_2 - e^{-MW_i}\right)}{1 - e^{-MW_i}} \tag{9}$$

with A, B, k0, E0, v and τ2 being the kinetic parameter to be determined with experiment or production data.

4. Riser Model

As depicted in Fig. 1, the riser has two feed entrances at the bottom and in the middle, respectively, and plug flow is assumed in the vertical direction. Since the residence time of oil gas and catalyst particles in the riser is short (2~5s), dynamics of the riser is neglected. In addition, instantaneous vaporization of stock oils is assumed at their entrance into the riser, and no heat loss of the riser to the environment is considered. Under these assumptions, the authors have developed a detailed mathematical model [2], which is outlined as Fig. 1.

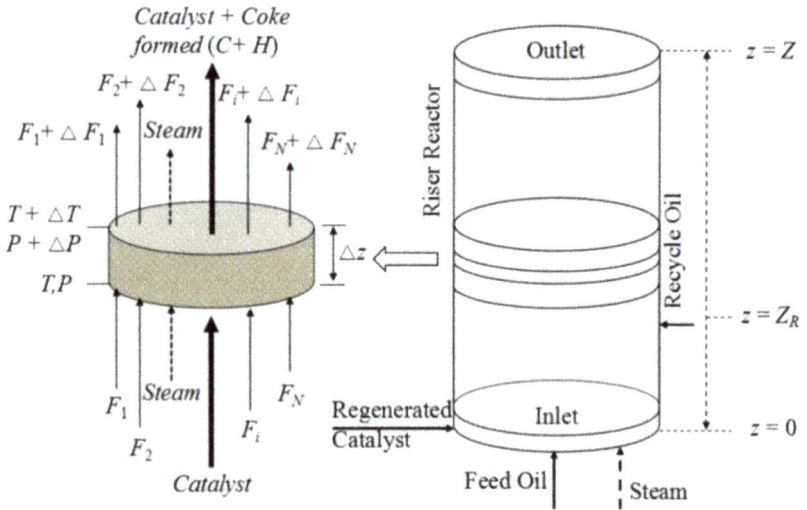

Fig. 1. Schematic diagram of the riser reactor

From material and heat balance, the distribution of special pseudo-component (SPC) i, coke (CK) and temperature along the riser can be derived

$$\frac{dF_i}{dz} = \left[MW_i \sum_{m=i+1}^{c} \left(MW_m^{-1} \sum_{n=1}^{m-i-1} r_{m,i,n} \right) - \sum_{m=1}^{i-1} \sum_{n=1}^{i-m-1} r_{i,m,n} \right] \Omega_{RS}$$

(10)

$$\frac{dF_{CK}}{dz} = \sum_{i=1}^{c}\left(MW_i^{-1} \sum_{m=1}^{i-1}\sum_{n=1}^{m-1} r_{i,m,n}\alpha_{i,m,n} \right)\Omega_{RS} \qquad (11)$$

$$\frac{dT}{dz} = -\frac{\sum_{i=1}^{c}\left(MW_i^{-1} \sum_{m=1}^{i-1}\sum_{n=1}^{m-1} r_{i,m,n}\Delta H_{i,m,n} \right)\Omega_{RS}}{F_{CAT}C_{p,CAT} + F_{CK}C_{p,CK} + F_{STM}C_{p,STM} + \sum_{i=1}^{c}(F_i + F_{F,i})C_{p,i}^{v}} \qquad (12)$$

where Cp, CK = 1.15, Cp, CAT = 1.15, and Cp, STM = 1.79 kg/(kg·K) are the heat capacity of coke, catalyst and stream, respectively.

In evaluating reaction rate ri, m, n with Eq. (7), concentrations of PCi and catalyst particles in the riser are calculated by definition

$$C_i = \frac{F_i}{v_g \delta_g \Omega_{RS}} \qquad (13)$$

$$C_{CAT} = \frac{F_{CAT}}{v_c \Omega_{RS}} \qquad (14)$$

and the weight fraction of coke on the catalyst is

$$x_{CK} = \frac{F_{CK}}{F_c} \qquad (15)$$

$$F_c = F_{CAT} + F_{CK} \qquad (16)$$

As summarized by Gupta et al. [3], Fernandes et al. [4] and Han and Chung [5], the solid phase velocity is determined by the momentum equation of Tsuo and Gidaspow[6].

$$\frac{dv_c}{dz} = \frac{C_f(v_g - v_c)\Omega_{RS}}{F_c} - \frac{2f_c v_c}{D_{RS}} - \frac{g}{v_c} \qquad (17)$$

and the pressure is evaluated by the equation of Pugsley and Berruti [7]

$$\frac{dP}{dz} = -\left(\delta_c \rho_c g + \frac{2f_c \delta_c \rho_c v_c^2}{D_{RS}} + \frac{f_g \rho_g v_g^2}{D_{RS}} \right)\times 10^{-3} \qquad (18)$$

with

$$\delta_g = 1 - \delta_c \qquad (19)$$

$$\delta_c = \frac{F_c}{\rho_c v_c \Omega_{RS}} \qquad (20)$$

$$\rho_c = (1 - \varepsilon_c)\rho_p \qquad (21)$$

$$v_g = \frac{F_g}{\rho_g \delta_g \Omega_{RS}} \qquad (22)$$

$$\rho_g = \frac{F_g}{F_{STM} MW_{STM}^{-1} + \sum_{i=1}^{c} F_i MW_i^{-1}} \frac{P}{RT} \qquad (23)$$

$$F_g = F_{STM} + \sum_{i=1}^{c} F_i \qquad (24)$$

where g = 9.8 m/s2 is the gravity constant, dc = 6.0×10^{-3} m the cluster diameter, and Cf, fc and fg are friction factors.

5. Parameter Estimation

In the above model, there are six parameters A, B, k0, E0, v and τ2 to be identified from production or experimental data. It has shown by the authors [2] that the identifiability of these parameters is poor with respect to the following error function

$$f(\mathbf{X}) = \sum_{p} (F_p - F_p')^2 + (T_{RS} - T_{RS}')^2 \qquad (25)$$

where X = {A, B, k0, E0, v, τ2}, and Fp, F_p', TRS and T_{RS}' are the predicted and measured values of the flowrate of product p and the temperature of the riser's outlet. The product flowrates are calculated through lumping the definite components as dry gas (GS, C1 + C2) and liquefied petroleum gas (LPG, C3 + C4) and the special pseudo-components by their normal boiling temperatures (b.p.) as gasoline (GSL, b.p. 38.5−221 °C), diesel oil (LCO, b.p. 221−410 °C) and recycle oil (RO, b.p. 410−531 °C).

On the other hand, the riser's performance is much less sensitive to parameters A, B and E0 than to the other three parameters k0, v and τ2, as illustrated in Fig. 2. Therefore, only parameters X = {k0, v, τ2} are adjustable in this study, while parameters A = 4.29, B = 10.4 and E0 = 1510 are taken to be constant. Such a selection on tuning parameters are also practiced partially by Pitault et al. [8] and Hernández-Barajas et al. [9]. It should be noted that parameters A, B and E0 may be riser and catalyst dependent.

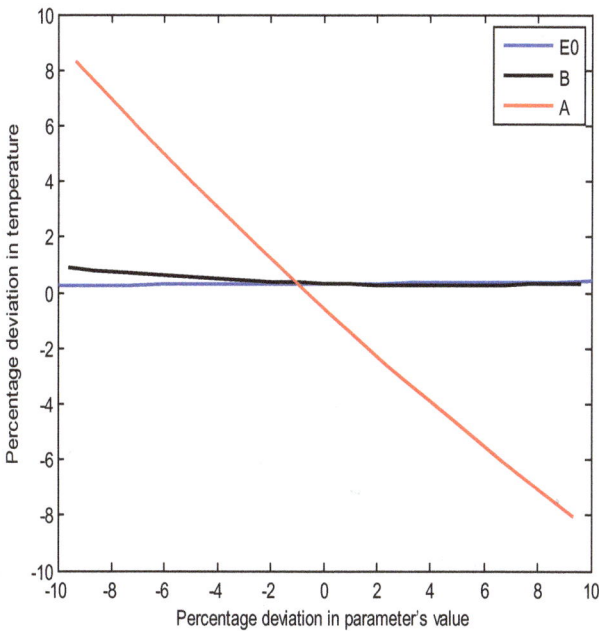

Fig. 2. Change of the riser's outlet temperature with respect to parameters A, B and E0

6. Test Result

The production data of riser I as reported by the authors [2] are used to test the above three parameter model. This riser is from a commercial fluid catalytic cracking unit of SINOPEC, China. For this riser, DRS = 0.98 m, Z = 34.74 m, ZR = 12.38 m, and recycle oil is introduced as the side feed if needed. Two sets of production data representing two different operational conditions were taken from this riser in 2008 (case 1) and 2009 (case 2). In both cases, the catalyst had the same properties and the catalyst particles had an average diameter of 7×10^{-5}

m, density of 1450 kg/m3, and average sphericity of 0.72. In case 1, only the stock oil was fed to the riser, whereas both the stock oil and recycle oil from the main fractionator were introduced into the riser in case 2. The properties of the feed oils and the operational conditions are listed in Tables 1 and 2, respectively.

Table 1. Properties of the three feed oils for the riser

	case 1		case 2	
	stock	recycle	stock	recycle
density in 60 F, kg/m³	914.6	/	944.8	992.3
ASTM D86 IBP, °C	304	/	/	217
ASTM D86 10%, °C	365	/	380	357
ASTM D86 30%, °C	410.41*	/	442	385.34*
ASTM D86 50%, °C	445	/	468	399
ASTM D86 70%, °C	498.56*	/	524	411.54*
ASTM D86 90%, °C	550	/	570	430
ASTM D86 FBP, °C	568	/	/	468
t_{MB}, °C	441.18	/	463.96	392.85
K_w	11.89	/	11.63	10.71

* interpolated value

Table 2. Operational conditions of the riser

	case 1	case 2
flowrate of the stock oil, t/h	146.49	131.2
flowrate of the recycle oil, t/h	0	15
flowrate of the steam, t/h	9	9.3
catalyst/oil ratio (COR)	6.5	7.99
temperature of the stock oil, °C	220	220
temperature of the recycle oil, °C	/	300
temperature of the catalyst, °C	680	710
temperature of the steam, °C	220	220
temperature at the riser's outlet, °C	515	506
pressure at the riser's bottom, kPa	240	220
F_{H2}, kg/s	0.0277	0.48
F_{C1}, kg/s	0.4169	0.4199
F_{C2}, kg/s	0.4293	0.4292
F_{GS}, kg/s	1.5889	1.6111
F_{LPG}, kg/s	6.1972	4.4167
F_{GSL}, kg/s	18.8667	16.8056
F_{LCO}, kg/s	10.425	9.1111
F_{RO}, kg/s	/	4.1667
F_{RES}, kg/s	1.2694	1.8889
F_{CK}, kg/s	2.3444	2.8333

In all the calculation of this study, SPCs of a boiling temperature range of 30 °C are used. The procedures for solving the riser model and identifying the kinetic parameters suggested by the authors [2] were adopted.

Table 3 lists 20 sets of the three parameters and the error in Eq. (25) as a result of regression using the production data in case 1 and 20 different initial values for the parameters taken uniformly in space of $k0 \in [0.003,0.3]$, $v \in [0.01,0.5]$ and $\tau2 \in [5,40]$. It is clear from this table that parameters X = {k0, v, $\tau2$} are fully identifiable. Actually, the reciprocal condition value $(1/\lambda)$, or the ratio of the smallest to largest eigenvalues of the Hessian matrix (H)[10] at k0 = 0.216, $\tau2 = 0.01$ and v = 12.16 is 0.0126, whereas the absolute value of $1/\lambda$ is about $10-11$ if all the six parameters were used [2].

Table 3. Identified parameters and the corresponding errors

| No. | parameters | | | $f(\mathbf{X})$, % |
	k_0	τ_2	v	
1	0.2164	0.010	12.1636	0.166012
2	0.2161	0.010	12.1661	0.165983
3	0.2161	0.010	12.1661	0.165984
4	0.2163	0.010	12.1629	0.166009
5	0.2167	0.012	12.1669	0.165869
6	0.2164	0.011	12.1632	0.166010
7	0.2162	0.010	12.1650	0.165997
8	0.2175	0.012	12.1570	0.166061
9	0.2166	0.010	12.1618	0.166037
10	0.2163	0.010	12.1639	0.166006
11	0.2159	0.010	12.1654	0.165960
12	0.2183	0.013	12.1587	0.165898
13	0.2160	0.010	12.1641	0.165976
14	0.2160	0.010	12.1654	0.165967
15	0.2161	0.010	12.1648	0.165986
16	0.2178	0.012	12.1621	0.165859
17	0.2164	0.010	12.1630	0.166015
18	0.21661	0.010	12.1622	0.166037
19	0.21594	0.010	12.1658	0.165956
20	0.21664	0.010	12.1628	0.166039

In Table 4, the production (experimental) data at the outlet of the riser are compared with the values calculated by the model of this study, with k0 = 0.216, $\tau2 = 0.01$ and v = 12.16. It is noted that, the data in case 2 are predicted by the model using the three parameters identified with the production data in case 1. This table shows that the three-parameter model has the capability of predicting the behavior of the riser at different operational conditions and taking different feed stocks. Furthermore, the regression and prediction accuracy of the model is comparable with that of the six parameter model of the authors [2].

Table 4. Experimental and calculated product compositions and outlet temperatures (ROT) of the riser

	Case 1			Case 2		
	Exptl.	Calc.	Err.%	Exptl.	Pred.	Err. %
RO, wt%	10.204	12.859	2.655	2.780	5.905	3.125
LCO, wt%	22.313	23.938	1.625	24.520	22.787	−1.733
GSL, wt%	41.156	38.643	−2.513	44.370	45.326	0.956
LPG, wt%	10.816	11.934	1.118	14.570	12.797	−1.773
GS, wt%	3.946	2.556	−1.390	3.740	2.769	−0.971
CK, wt%	6.939	7.673	0.734	5.510	6.923	1.413
ROT, °C	506.000	504.66	−1.340	515.000	510.717	−4.283

The predicted distribution of gasoline rate, coke rate and temperature along the riser for case 2 are shown in Fig. 3, where the jump happens at the feed entrance in the middle of the riser.

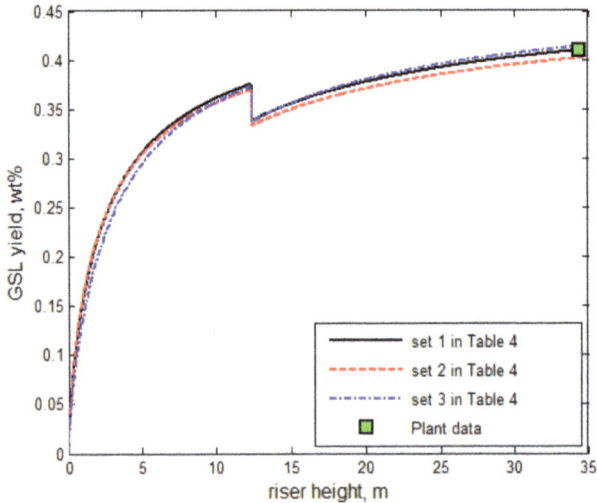

Fig. 3. Predicted distribution of the GSL and coke rates and temperature along the riser in case 2.

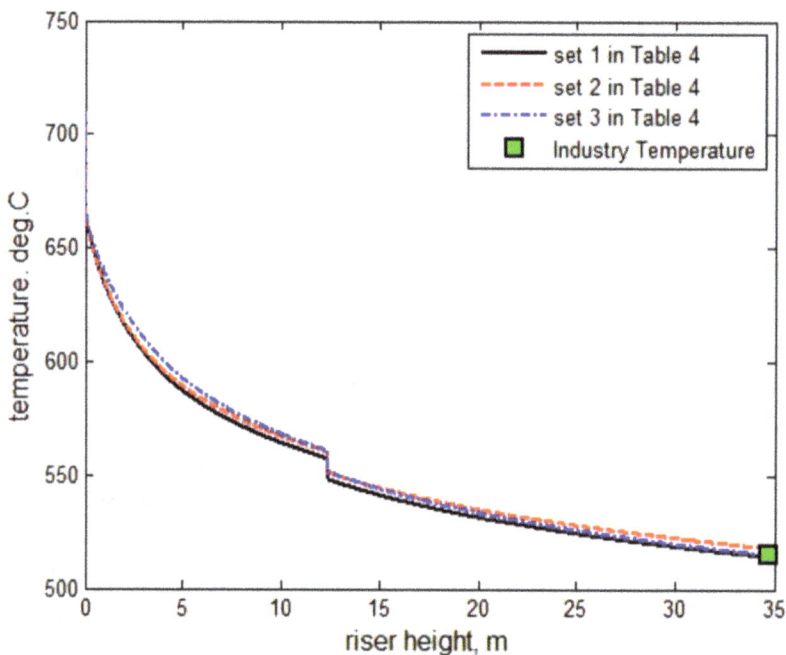

Fig. 3. (*Continued*)

7. Conclusion

1) The behavior of a fluidized catalytic cracking riser can be well correlated and predicted by a model based on the three parameter kinetic cracking reaction scheme with special pseudo-components being the component entity.

2) Compared with the six parameter scheme of cracking reactions, the current scheme enjoys good parameter identifiability.

Nomenclature

A, B	inactivation coefficient	Subscript:	
c	number of pseudo component	*B*	light component or heavy component
C	Concentration, kg/m³	C	carbon
Cₚ	heat capacity, kJ/(kg·K)	CAT	catalytic
COR	catalytic oil ratio	CK	coke

D_{RS}	diameter of riser, m	g	gas phase
E_0	activation energy, kJ/kmol	GS	dry gas
F	quality flow, kg/s	GSL	gas oil
g	gravitational constant, m/s^2	I, j, k	pseudo component
ΔH	reaction heat, kJ/kg	LCO	light circulation oil
k_0	frequency factor, m^3/(kg Cat s)	LPG	liquid petroleum gas
k	the reaction rate constant	RO	return oil
MW	quality percent, kg/kmol	RS	riser
P	Pressure, atm	STM	stream
r	crack reation rate, kmol/s	Superscript:	
R	gas constant, atm $\cdot m^3$/(kmol$\cdot$ K)	$comb$	combustion
ROT	temperature of outlet riser, °C	α	coke yield, kg/kmol
T	Temperature, K	v	kinetic constant
v	Velocity, m/s	ρ	Density, kg/m^3
w	quality percent	τ_2	kinetic parameter
Z	length of riser, m	χ	quality percent
Z_R	position of siding raw material feed, m	ϕ	catalytic inactivation function
		Ω_{RS}	Cross-sectional area of riser, m^2

Acknowledgement

The work of this paper is funded by the Fundamental Research Funds for the Central Universities (Project YS1404), Ministry of Education, China, and by the Hi-tech (863) program (Project 2007AA04Z191), Ministry of Science and Technology, China.

References

1. You-hao Xu. Advance in China fluid catalytic cracking (FCC) process, J. Science China: Chemical, 44.1(2014): 13-24.

2. Jia-rui Zhang, Zhi-qing Wang, Hao Jiang, Ji-zheng Chu, et al. Modeling fluid catalytic cracking risers with special pseudo-components, J. Chemical Engineering Science, 102.15(2013):87-98.

3. Gupta R K, Kumar V, Srivastava V K. A New Generic Approach for the Modeling of Fluid Catalytic Cracking (FCC) Riser Reactor, J. Chemical Engineering Science, 62.17(2007):4510-4528.

4. Fernandes J L, Pinheiro C I C, Oliveira N M C, et al. Steady state multiplicity in an UOP FCC unit with high-efficiency regenerator, J. Chemical Engineering Science, 62.6(2007):6308-6322.

5. In-su Han, Chang-bock Chung. Dynamic modeling and simulation of a fluidized catalytic cracking process. Part I: Process modeling, J. Chemical Engineering Science, 56.5(2001):1951-1971.

6. Tsuo, Y P, Gidaspow, D. Computation of flow patterns in circulating fluidized beds, J. A.I.Ch.E. Journal, 36.6(1990), 885–896.

7. Pugsley T S, Berruti F. A predictive Hydrodynamic Model for Circulating Fluidized Bed, J. Powder Technology, 89.1(1996):57-69.

8. Pitault I, Nevicato D, Forissier M, et al. Kinetic model based on a molecular description for catalytic cracking of vacuum gas oil, J. Chemical Engineering Science, 49.24(1994):4249-4262.

9. Hernández-Barajas, J. R., Vázquez-Román, R. and Félix-Flores, Ma.G. A comprehensive estimation of kinetic parameters in lumped catalytic cracking reaction models, J. Fuel, 88.1(2009): 169–178.

10. Belsley D A. Conditioning Diagnostics, Collinearity and Weak Data in Regression. Wiley, New York, J. Journal of the American Statistical Association, 88(1993).

Data Encryption Algorithm for Intelligent Nodes in the Building Equipment Internet of Things

Guiqing Zhang, Hang Zhao, Ming Wang, Tao Liang,
Xue Wei and Shu-liang Ma

School of Information & Electrical Engineering, Shandong Jianzhu University

Shandong Provincial Key Laboratory of Intelligent Buildings Technology

Jinan, China

E-mail: zhaohang2012@163.com

With the development of Internet of Things industry, network security has become more and more important. The intelligent node in the building equipment Internet of Things realizes the Ethernet communication protocol and wireless network communication protocol. This paper is concerned with the problem of data security between the intelligent node and the server. The study shows that, among different data encryption algorithms, DES encryption algorithm is suitable for encrypting data by embedded system. Firstly, the paper introduces the function of intelligent node and its hardware environment. Then, the data encryption software design is presented. Lastly, DES encryption programs and the encryption effect are tested in Internet of things intelligent node application. The test results show that the algorithm can be implemented stably in Internet of things intelligent node hardware platform. The encryption and decryption efficiency is better than the general algorithm. DES encryption algorithm improves security of the Internet of things intelligent node in network communication.

Keywords: Intelligent node, DES encryption algorithm, Embedded system.

1. Introduction

Among the building equipment Internet of Things, the equipment is interconnected to a unified network by intelligent node, enabling to obtain and share massive data possible, such as, the status and information of the equipment, environmental parameters etc. [1]. With the comprehensive development of the networking industry, people pay much more attention to the safety of data transmission. The protection of network privacy and data security among the communication process has become an important issue to be solved during the construction of building equipment network [2]. However, the current research on the security of building equipment is still at the framework of the macro theoretical research stage. The effective and reliable practical application is relatively less [3]. For example: Ding, Li and Feng analyzed the deficiency of RF data encryption security protocols, and had a research on the model of RF security protocol [4]. European Commission proposes an evaluation framework of privacy and data protection, based on the proposal of the EU [5].

2. The Principle and Process of DES Algorithm

2.2 *Lightweight Encryption Algorithm*

The information security and privacy protection, nowadays, have undergone rapid changing to live up with the market development, in particularly, in aras of cryptography. According to the different key, the password algorithm is divided into: Symmetric key system and public key cryptosystem. The current widely used symmetric key system mainly includes AES, IDEA RC2, RC4 and RC5. Public key system is mainly composed of Elliptic Curve Cryptosystem (ECC), Diffie-Hellman and discrete algorithm cryptosystem, and now cryptography has become a comprehensive science combined with physics, quantum mechanics, electronic information, social linguistics and so on. Advanced research theories such as "quantum cryptography", "chaos" appeared [6-7].

The lightweight encryption algorithm is a kind of encryption algorithm which is high safety, stable, reliable, easy to achieve and suitable for the use in the complex information environment [8]. In this paper, the intelligent node and the terminal equipment whose processing capacity is relatively weak and storage space is also smaller are researched. Owing to the limitation of space resources, the technology of encrypted communication for the intelligent node adopts the lightweight encryption technology

DES encryption algorithm is the most widely used and the most popular symmetric encryption algorithm. Because of fast operation, simple implementation and small calculation, it is suitable for data encryption of the intelligent node and the terminal with limited resources space. Information security and privacy protection of the Internet of Things is realized by DES encryption algorithm. The core of the algorithm is the method of several rounds iterative. The essence is a complex combination of substitution and transposition system. Therefore, this paper uses the DES encryption algorithm as the solution to realize the intelligent node data encryption.

3. Principle of DES Encryption Algorithm

DES encryption algorithm is the effective key 56 bit, 64 bit of plaintext, 64 bit of cipher text and 16 rounds of iterative calculation of the block symmetric cipher algorithm [9]. The structure block diagram of the encryption principle is shown in Fig. 1.

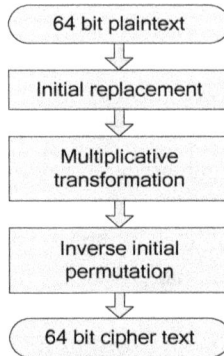

Fig. 1 Structure block diagram of encryption principle

When the data is encrypted by DES, the first 64 bit plaintext of the data packet is rearranged through the initial permutation matrix, and divided into the two parts, L0 and R0.Then the function F uses the encryption key to do the 16 rounds iteration operation and outputs a 64 bit array. Finally, the 64 bit cipher text is obtained through the inverse initial permutation matrix .After that encrypt the next 64 bit until all data packets are completed.

4. The Realization of DES Algorithm in the Intelligent Node

4.1 Function of Internet of Things Intelligent Node

The main function of the intelligent node in the building equipment Internet of Things is to realize the transformation between the communication protocol of the Ethernet communication protocol and wireless network communication protocol. The communication between the intelligent node and the server is the Ethernet communication protocol, and the one between the intelligent node and the terminal equipment module is the wireless communication protocol [10]. The intelligent node can receive the operation instructions from the server and correspondingly control the terminal equipment module. Meanwhile, the intelligent node can transfer the data from the terminal module to the server, and

576

process the data received from the terminal module [11]. Therefore, the intelligent node is the core equipment in building equipment Internet of Things system.

In the premise of ensuring data security, in order to have a good reaction speed and performance, in this paper, not all of the communication data is encryption by DES, instead, only the key data is encrypted based on the priority of the data.

5. The Software Design of Intelligent Node Data Encryption

In this paper, communication protocol which needs encryption is register command packet, control command packet and threshold alarm packet. The flow chart of DES encryption algorithm encrypting data and generating cipher text in intelligent node is shown in Fig. 2.

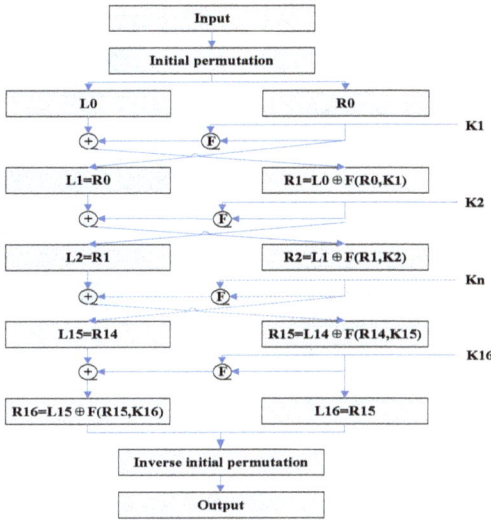

Fig. 2 The flow char of DES encryption algorithm

The steps of encryption are as follows:

Set m as the initial value to represent 64 bits.

a. Each position of m goes through initial permutation matrix, then the rearrangement of the plain text group obtain m0=IP(m),for m0=L0R0, L for the top 32, m for the after 32.

b. For 1≤i≤16, perform the following operations: Li=Ri-1, Ri= Li-1 ⊕f(Ri-1,Ki).

c. The 64 bit array is left and right exchanged to get R16 L16, then obtained 64 bit cipher text through the inverse initial permutation matrix IP, writes as c= IP-1(R16 L16).

Take the register command packet as an example, the data of the register command packet in terminal module is encrypted from the frame class .when the encryption is completed, the cipher text is transmitted in the form of broadcast. The intelligent node receives a packet and judges whether the destinations address of the node, and then determine the category of the frame. If AAH, it is a register command packet, then the data of the Frame class is decrypted. The intelligent node carries on the corresponding operation processing to the data. The flow chart of encryption and decryption is shown in Fig. 3.

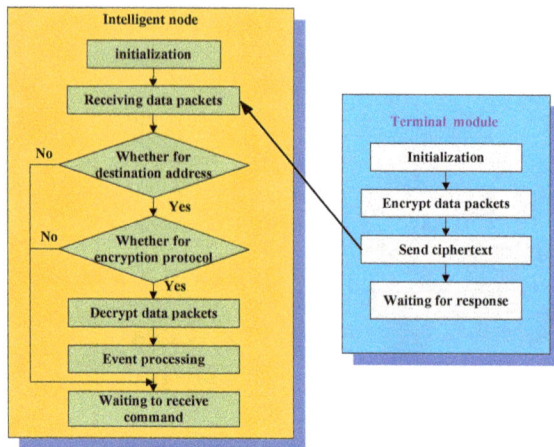

Fig. 3 The flow chart of encryption and decryption

6. DES Encryption Algorithm Testing and Application

6.1 *The Hardware Environment of Data Encryption Test for Intelligent Node*

In this paper, the reliability test and system application test between intelligent node and server communication are carried out. The hardware test environment is built by the computer, wireless router, intelligent node, the terminal module and other related experimental equipment. The physical figure of test equipment is shown in Fig. 4.

578

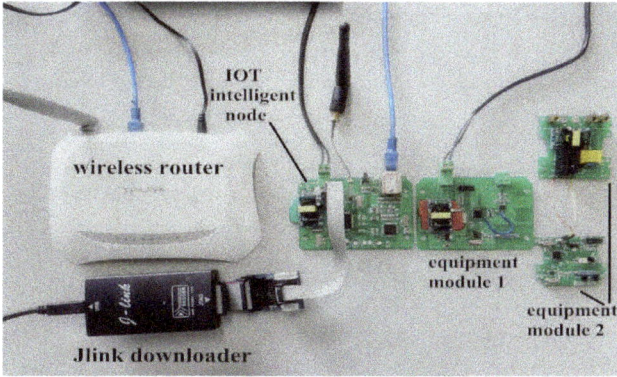

Fig. 4 The physical figure of test equipment

The intelligent node uses the chip of STM32F103xx as the controller. The operating frequency of the system is 36MHz; through the frequency doubling can be promoted to 72MHz [12]. STM32 minimum system is shown in Fig. 5.

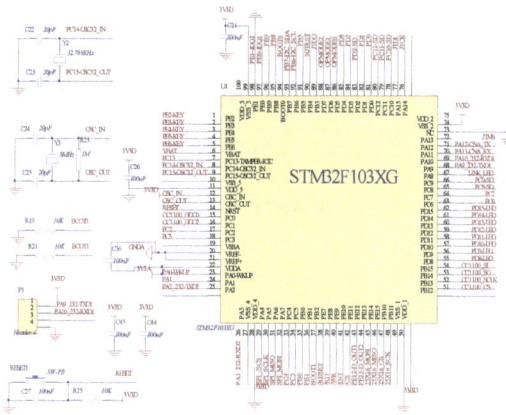

Fig. 5 STM32 minimum system schematic

7. The Test Plan of Data Encryption for Intelligent Node

Encryption program porting and reliability test content of this paper: A. The correctness of the data encryption test, B. The execution test of data encryption program. The correctness of the data encryption test flow chart is shown in

Fig. 6. And the execution time test of data encryption program flow chart is shown in Fig. 7.

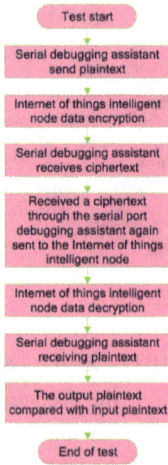

Fig. 6 The correctness of data encryption test

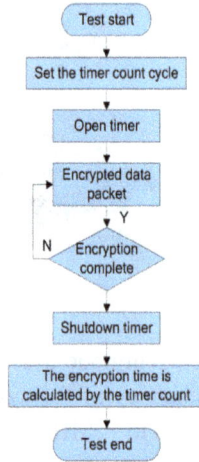

Fig. 7 Execution time test of data encryption program

A. The correctness of the data encryption test

Test procedure:

a. The intelligent node is connected to the computer and downloads the program;

b. Plaintext data is sent to the intelligent node through computer. After encryption, the encrypted message is sent back and displayed in the serial port debugging tool.

c. The encrypted cipher text of the intelligent node is sent through the serial port debugging tool.

B. The Execution time test of data encryption program

DES encryption algorithm performance is reflected through the speed of encrypt and decrypt .The basic expressions are [13]:

$$Encryption\,/\,Decryption\;speed = \frac{block\;length}{number\;of\;cycles\;required\;for\;encryption\,/\,decryption} \times System\;clock\;frequency \qquad (1)$$

So the speed of encryption / decryption in the intelligent node is 144Mbit/s.

The GPTM of intelligent node controller is the TIM2 timer which is mounted in APB1, and the work frequency is 2 times of the system clock frequency, so the clock source of TIM2 is 72MHz.

This test sets the pre frequency division of the number of 719, so the TIM2 single count cycle is:

$$T_{CNT} = (719+1) / 72MHz = 10\mu s \tag{2}$$

Setting the comparative matching count increment value of the TIM2 is 10, therefore the TIM2 interrupt request event interval is:

$$T_{CHN} = 10 \times 10\mu s = 100\mu s \tag{3}$$

According to the relevant requirements about the software design of the intelligent node data encryption,

Test procedure:

a. The intelligent node downloads the program;

b. The intelligent node starts timer and enables TIM2;

c. The intelligent node performs data encryption procedures, requests the TIM2 interrupt and counts the interrupt number.

d. The intelligent node close the timer, TIM2 is disabled. According to TIM2 count and clock frequency, the execution time of e data encryption program is calculated.

8. The test result of data encryption for the intelligent node

A. The result of data encryption correctness test

The test result of DES encryption algorithm for intelligent node transplantation program is shown in Table 1. Using the same main key, encryption and decryption test for different data is showing in test 1 and 2; Using the different main key, encryption and decryption of the same data is showing in test 3and 4; Using the different main key, encryption and decryption of the different data is showing in test 5 and 6. After obtained, the result is comparing with the result of DES standard encryption program. And the conclusion is that DES data encryption algorithm can achieve reliable encryption and decryption functions when transplanted to the intelligent node.

Table. 1 The test result of DES encryption algorithm for intelligent node transplantation program

number	master key	Input plaintext	Output plaintext (Decryption input)	Output plaintext (Decryption output)	Result
1	01030507090B0D0F	0001020304050607	0C06B5504B5CFE5C	0001020304050607	Consistent
2	01030507090B0D0F	050A0F14191E2328	30988F3C1FD4399E	050A0F14191E2328	Consistent
3	70A1B2CD3E0F719	0001020304050607	36BC6E8A 39146770	0001020304050607	Consistent
4	020307091AB3C245	0001020304050607	0CA43D62A1AD171D	0001020304050607	Consistent
5	AAEAD09C51847FB3	2015111010199926	727A8EFBDDDA554B	2015111010199926	Consistent
6	B0EAC19251847FB3	1415926535897932	874EE7AFF5AD5797	1415926535897932	Consistent

B. Data encryption delay test results

TIM2 interrupt count is the number of 47. Therefore,when the intelligent node uses the chip of STM32F103xx as a controller, the data encryption program execution time is:

$$T_1 = 47 \times T_{CHN} = 47 \times 100\mu s = 4.7ms \tag{4}$$

So the intelligent node data encryption program execution time T0=4.7ms, the intelligent node data encryption program will not affect the overall performance of the Internet of Things system because of the execution time.

9. Conclusion

Through the intelligent node in the Building equipment Internet of Things, data from the terminal equipment module in buildings can be easily linked into the Internet to achieve the collection of information. The DES encryption algorithm is used to encrypt the key data in the communication process between the intelligent node and the server. The test results are accurate and reliable. Encryption program execution time is 4.7ms and meets the overall performance of building equipment networking system.

The intelligent node uses the chip of STM32F103xx as controller. In order to reduce the execution time of the data encryption program in the intelligent node, we can take measures to improve the data rate of the intelligent node, such as using more advanced processing chip and optimizing encryption algorithm.

Acknowledgement

This work is partially funded by NSFC Grants (61573225, 61273326), the Science and Technology Development Program of Shandong Province of

China(2012GGX10120); the Excellent Young and Middle-Aged Scientist Award Grant of Shandong Province of China(BS2013DX018); the Science and Technology Development Program of Higher Education of Shandong Province (J11LG16), the Science and Technology Program of Ministry of Housing and Urban-Rural Development of China (2013-K8-069).

References

1. Guiqing Zhang, Xiaolong Wu, Chengdong Li, Liang Tao. Building electrical equipment Internet of Things with applications to energy saving [C]. in Proc. of the International Conference on Mechatronics and Control, Jinzhou, China, 2014.
2. Chuncheng Niu. Security and Privacy Issues of the Internet of Things [D].Jilin: Changchun University of Technology, 2013.
3. Zhengqiang Wu, Yanwei Zhou, Jianfeng Ma. A Security Transmission Model for Internet of things [J].Chinese Journal of Computers, 2011,34(8): 1352-1364.
4. Zhenghua Ding, Jintao Li, Bo Feng. Research on Hash Based RFID Security Authentication Protocol [J]. Journal of Computer Research and Development,2009,46(4):583-592.
5. GSI, European Commission. Privacy and Data Protection Impact Assessment Framework for RFID Applications. Janu-ary 12, 2011, http;//ec,europa.eu/information_society/policy/rfid/doc uments/infso-2011-00068.pdf
6. Bo Yang. Modern Cryptography [M]. Beijing: Tsinghua University Press, 2003.
7. Po Bo. Design and Implement Ation of 3DES Encryption Algorithm Based on the ARM [D]. Beijing: Beijing University of Technology, 2012.
8. Wen Zhao, Chuanwei Wu, Min Luo. Study on Encryption Technology in Internet of Things [J]. Information Security and Communication Privacy, 2012, 7:103-104.
9. Randall K. Nichols. ICSA Guide To Cryptography [M]. Asia:McGraw-Hill Education Co., 2004.
10. Wei Peng. The intelligent terminal research in the building equipment Internet of Things [D].Shandong: Shandong Jianzhu University, 2013.
11. Guiqing Zhang, Zhaohai Wang, Ming Wang, Xingchao Duan, Wei Peng. Design of UHF Reader for building equipment Cyber Physical Systems [J]: 2012, 31(8): 1-4.

12. Xianghe Ji.Research and Realization of Household Energy-Saving Controller in the Internet of Things [D].Shandong: Shandong Jianzhu University, 2014.
13. Liang Fang. DES Encryption Algorithm IP Module [D]. Sichuan: University of Electronic Science and Technology of China, 2011.

A Phytoplankton-Zooplankton Model with Holling-I Function Response and State Feedback Control

Yuan Tian*, Zhan-tao Zhang and Hai-ting Sun

School of Information Engineering, Dalian University,

People's Republic of China

*Email: tianyuan1981@163.com

In this paper, a phytoplankton-zooplankton system with state feedback control strategies is proposed and analyzed. Firstly, the dynamic behaviour of the system including the existence and stability of equilibria are investigated, the global asymptotical stability of equilibria is analyzed by Lyapunov method. And then, the existence and stability of the semi-trivial periodic solution and order-1 periodic solution for $P_r \leq P_c$ and $P_r > P_c$ are discussed. The analytical results provide a possibility of obtain a stable output in a phytoplankton-zooplankton system.

Keywords: Phytoplankton-zooplankton, Holling I functional response, State feedback control, Positive periodic solution, Semi-trivial periodic solution, Orbital stability.

1. Introduction

Plankton are microscopic organisms that float freely with oceanic currents and in other bodies of water which are made of phytoplankton and zooplankton [1]. Over recent years, many scholars have pay attention to the ocean food chain and thus set up a series of phytoplankton-zooplankton models [2-8] with in-depth research on the sustainable use of ocean resources. For example, Li et al. [2] analyzed global stability of phytoplankton-zooplankton model of plankton, Janga et al. [3] considered nutrient phytoplankton-zooplankton models with a toxin substance that inhibits either growth rate of phytoplankton, zooplankton or both thophic levels, Mukhopadhyay et al. [4] dealt with a nutrient-plankton model in an quatic environment in the context of phytoplankton bloom with spatial heterogeneity. In the established models, few scholars have studied it by state feedback control strategies. In this article, we present a toxin producing phytoplankton-zooplankton model with Holling I functional response and state feedback control strategies which can be described by the following differential equations:

$$\begin{cases} \left.\begin{array}{l} \dfrac{dP}{dt} = rP\left(1-\dfrac{P}{K}\right) - \beta f(P)Z \\ \dfrac{dZ}{dt} = \beta_1 f(P)Z - \rho g(P)Z - dZ \end{array}\right\} P < P_r \\ \left.\begin{array}{l} \Delta P = -c_1 EP \\ \Delta Z = -c_2 EZ \end{array}\right\} P = P_r \end{cases}$$

(1)

with

$$f(P) = \begin{cases} P, P \le P_C \\ P_C, P > P_C \end{cases} \quad g(P) = f(P)_.$$

Where P and Z denote the density of toxin producing phytoplankton population (TPP) and zooplankton population at time t subject to the non-negative initial condition $P(0) = P_0$ and $Z(0) = Z_0$, respectively. $0 < \beta_1 < \beta$, $\rho < \beta_1$ and $\gamma > 0$ is a real constant; the constants c_1 and c_2 ($0 < c_1, c_2 < 1$) are the catch ability coefficients of the two species, constant $E > 0$ is the harvesting effort, P_r describes the toxin producing phytoplankton population threshold.

2. Existence and Stability without Impulsive Effect of System (1)

In system (1), when $c_1 = c_2 = 0$, we obtain the following system without impulsive effect:

$$\begin{cases} \dfrac{dP}{dt} = rP\left(1-\dfrac{P}{K}\right) - \beta f(P)Z \\ \dfrac{dZ}{dt} = \beta_1 f(P)Z - \rho g(P)Z - dZ \end{cases}$$

(2)

System (2) has three equilibria $E_0(0,0)$, $E_1(K,0)$ and $E_2(P^*,Z^*)$ where $P^* = d/(\beta_1 - \rho)$, $Z^* = r\left(1 - P^*/K\right)/\beta$. It is easily obtained that $E_2(P^*,Z^*)$ is locally stable.

Theorem 1. The interior equilibrium $E_2(P^*,Z^*)$ is globally asymptotically stable.

Proof. Define $V(P,Z) = P - P^* - P^* \ln(P/P^*) + \beta(Z - Z^* - Z^* \ln(Z/Z^*))/(\beta_1 - \rho)$. Then

$$\frac{dV}{dt} = \frac{P - P^*}{P}\frac{dP}{dt} + \frac{\beta}{\beta_1 - \rho}\frac{Z - Z^*}{Z}\frac{dZ}{dt} = (P - P^*)\left[r\left(1-\frac{P}{K}\right) - r\left(1-\frac{P^*}{K}\right)\right]$$

$$+ (P - P^*)\left[\frac{f(P)}{P}\beta Z^* - \frac{f(P)}{P}\beta Z\right] + \beta(Z - Z^*)\left[\frac{f(P)}{P}(P - P^*)\right] = -\frac{r}{K}(P - P^*)^2$$

which implies that $\dot{V}<0$. The Lyapunov-Lasalle theorem implies that all solutions ultimately approach the equilibrium $E_2(P^*,Z^*)$. □

By the biological background, we consider system (1) in the region $\mathbb{R}_+^2 = \{(P,Z):P>0,Z>0\}$. Obviously, $\mathbb{R}_+^2$ can be divided into four domains as follows:

$$I \triangleq \left\{(P,Z)\in R_+^2 : \frac{dP}{dt}<0, \frac{dZ}{dt}<0\right\}, II \triangleq \left\{(P,Z)\in R_+^2 : \frac{dP}{dt}>0, \frac{dZ}{dt}<0\right\}, \tag{3}$$

$$III \triangleq \left\{(P,Z)\in R_+^2 : \frac{dP}{dt}>0, \frac{dZ}{dt}>0\right\}, IV \triangleq \left\{(P,Z)\in R_+^2 : \frac{dP}{dt}<0, \frac{dZ}{dt}>0\right\}.$$

Let $\Gamma_1 = \{(P,Z):P=(1-c_1E)P_r,Z>0\}, \Gamma_2 = \{(P,Z):P=P_r,Z>0\}$. Then any trajectory $O^+(E_0,t_0)$ starting from $E_0(P_r,Z_0)\in\Gamma_2$ at time t_0 will jump to $E_0^+\left((1-c_1E)P_r,(1-c_2E)Z_0\right)\in\Gamma_1$ at $t=t_0^+$ due to impulse effects $\Delta P = -c_1EP$ and $\Delta Z = -c_2EZ$. Then, it intersects Γ_2 at $E_1(P_r,Z_1)$ again. Therefore, we define a Poincaré map on Γ_2 as follows:

$$Z_1 = F(Z_0,c_1,c_2,P_r) \triangleq F(Z_0) \tag{4}$$

3. Existence of Periodic Solution with $P_r \le P_C$

When $P_r \le P_C$, then from system (1), we have

$$\begin{cases} \left.\begin{aligned} \frac{dP}{dt} &= rP\left(1-\frac{P}{K}\right)-\beta PZ \\ \frac{dZ}{dt} &= \beta_1 PZ - \rho PZ - dZ \end{aligned}\right\} P < P_r \\ \left.\begin{aligned} \Delta P &= -c_1EP \\ \Delta Z &= -c_2EZ \end{aligned}\right\} P = P_r \end{cases} \tag{5}$$

Case of $P_r \le P^*$

Theorem 2. For any solution $(P(t), Z(t))$ of system (5) with initial condition $(P_0, Z_0) \in \mathbb{R}_+^2$, if $P_0 \le P_r$, then $\lim\limits_{t \to \infty} Z(t) = 0$.

Proof. Suppose $P_0 \le P_r$, since any trajectory of system (5) starting from domain I will enter into II, solution $(P(t), Z(t))$ starts from points $E_0((1 - c_1 E) P_r, Z_0)$ must intersect with Γ_2 at $E_1(P_r, Z_1)$, and then jump to Γ_1 at point $E_1^+((1 - c_1 E) P_r, (1 - c_2 E) Z_1)$. Obviously, in domain II, $\dot{P} > 0$, then we have $Z_1 < r(1 - (1 - c_1 E) P_r / K)/(1 - c_2 E)\beta \triangleq \omega$. So, we only consider trajectories start from points on Γ_1 with $Z_0 \le \omega$. Suppose that trajectory then reaches to Γ_2 at $E_2(P_r, Z_2)$ again, $Z_1, Z_2 \in (0, r(1 - P_r/K)/\beta)$. Repeating the process above, we can get two point sequences $\{E_n(P_r, Z_n)\}$ and $\{E_n^+((1 - c_1 E) P_r, Z_n^+)\}$, where $Z_n^+ = (1 - c_2 E) Z_n$. The corresponding impulsive time sequences are marked by $\{t_n\}$. Integrating both sides of the two equations of system (5) from the orbit $E_n^+ E_{n+1}$, we have

$$t_{n+1} - t_n = \int_{(1 - c_1 E) P_r}^{P_r} \frac{1}{rP(1 - P/K) - \beta PZ} dP > \frac{1}{r}\left[\ln\frac{1}{1 - c_1 E} + \ln\frac{K - (1 - c_1 E) P_r}{K - P_r}\right] > 0 \tag{6}$$

$$\int_{Z_n^+}^{Z_{n+1}} \frac{1}{Z} dZ = \int_{t_n}^{t_{n+1}}\left[(\beta_1 - \rho) P - d\right] dt \le \int_{t_n}^{t_{n+1}}\left[(\beta_1 - \rho) P_r - d\right] dt = \left[(\beta_1 - \rho) P_r - d\right](t_{n+1} - t_n) \tag{7}$$

Thus we can get

$$Z_{n+1} \le Z_n^+\left(\frac{K - (1 - c_1 E) P_r}{(1 - c_1 E)(K - P_r)}\right)^{\frac{(\beta_1 - \rho) P_r - d}{r}} = (1 - c_1 E) Z_n\left(\frac{K - (1 - c_1 E) P_r}{(1 - c_1 E)(K - P_r)}\right)^{\frac{(\beta_1 - \rho) P_r - d}{r}}$$

By $(\beta_1 - \rho) P_r - d < 0$ and $(1 - c_1 E) P_r < P_r$, it leads to $\lim\limits_{n \to \infty} Z_n = 0$. Moreover, $\lim\limits_{t \to \infty} Z(t) = 0 \cdot$

Let $Z \equiv 0$ for $t \in [0, \infty)$ in system (5), we can get the following reduce system.

$$\begin{cases} \dfrac{dP}{dt} = rP\left(1 - \dfrac{P}{K}\right), P < P_r \\ \Delta P = -c_1 EP, \qquad P = P_r \end{cases} \tag{8}$$

It can be easily calculated that system (8) with $P_0 = (1 - c_1 E) P_r$ have a

solution $P(t) = Ke^{rt}/(C + e^{rt})$ where $C = (K - (1 - c_1 E) P_r)/(1 - c_1 E) P_r$. Let

$$T = \frac{1}{r} \ln \frac{K - (1 - c_1 E) P_r}{(1 - c_1 E)(K - P_r)}.$$

Then we have $P(T) = P_r$, $P(T^+) = P(T) - c_1 EP(T) = (1 - c_1 E) P_r = P_0$. Thus

system (5) has a semi-trivial periodic solution for $t \in (nT, (n+1)T]$ $(n = 0, 1, \cdots)$:

$$\begin{cases} \phi(t) = \dfrac{Ke^{r(t-nT)} (1 - c_1 E) P_r}{K - (1 - c_1 E) P_r + P_r (1 - c_1 E) e^{r(t-nT)}} \\ \varphi(t) = 0 \end{cases} \tag{9}$$

Theorem 3 The semi-trivial periodic solution is orbital asymptotically stable for any $P_r < P^*$.

Proof. Suppose that $\left(\tilde{P}(t), \tilde{Z}(t)\right)$ is a solution of small-amplitude perturbation of periodic solution with initial value $\left(\tilde{P}(0), \tilde{Z}(0)\right) = \left((1 - c_1 E) P_r, \tilde{Z}_0\right)$, which first intersects Γ_2 at point $A_1\left(P_r, \tilde{Z}_1\right)$ and then jumps to point $A_1^+\left((1 - c_1 E) P_r, \tilde{Z}_1^+\right)$. Further, it intersects Γ_2 at $A_2\left(P_r, \tilde{Z}_2\right)$ again. Repeating the process above, we have two point sequences $\left\{\left(P_r, \tilde{Z}_n\right)\right\}$ and $\left\{\left((1 - c_1 E) P_r, \tilde{Z}_n^+\right)\right\}$ with $\tilde{Z}_n^+ = (1 - c_2 E) \tilde{Z}_n$. By Theorem 2, there is $\lim\limits_{t \to \infty} \tilde{Z}(t) = 0$. Thus the solution (9) is orbital asymptotically stable.

Case of $P^* < P_r < K$. We know that there is a point $E_0\left((1 - c_1 E) P_r, \delta\right)$ such that trajectory $O^+(E_0, t_0)$ is tangent to Γ_2 at the intersection $E_1\left(P_r, r(1 - P_r/K)/\beta\right)$ and $\dot{P} = 0$ where $\delta = \delta(P_r, c_1)$. If $(1 - c_2 E) r (1 - P_r/K)/\beta < \delta$, then trajectory $O^+(E, t_0)$ starting from

$E\left(P_r, y\right)\left(y < r\left(1 - P_r/K\right)/\beta\right)$ will intersect with Γ_2 infinitely many times due to control strategies $\Delta P = -c_1 EP$ and $\Delta Z = -c_2 EZ$.

Theorem 4. For any $c_1, c_2 \in (0,1)$ and $P^* < P_r < K$, if $(1 - c_2 E) r\left(1 - P_r/K\right)/\beta < \delta$, system (5) admits an order-1 periodic solution which is orbitally asymptotically stable.

Proof. Suppose that trajectory $O^+\left(C_0, t_0\right)$ starting from $C_0\left(P_r, Z_0\right)\left(Z_0 < r\left(1 - P_r/K\right)/\beta\right)$ jumps to $C_0^+\left(\left(1 - c_1 E\right)P_r, \left(1 - c_2 E\right)Z_0\right)$ on Γ_1 and reaches Γ_2 at $C_1\left(P_r, Z_1\right)$ due to $\left(1 - c_2 E\right)r\left(1 - P_r/K\right)/\beta < \delta$, and then jumps to $C_1^+\left(\left(1 - c_1 E\right)P_r, \left(1 - c_2 E\right)Z_1\right)$ on Γ_1. Further, it intersects Γ_2 at $C_2\left(P_r, Z_2\right)$. By the Poincaré map, $Z_1 = F\left(Z_0\right)$, $Z_2 = F^2\left(Z_0\right)$ and then $Z_n = F^n\left(Z_0\right)\left(n = 1, 2, \cdots\right)$. Particularly, system (5) has a positive order-1 periodic solution when $Z_0 = Z_1$. Now we consider the situation $Z_0 \neq Z_1$:

(i) $Z_0 < Z_1$, we can get $Z_0 < Z_1 < Z_2 < \cdots < Z_n < r\left(1 - P_r/K\right)/\beta$

(ii) $Z_0 > Z_1$, we can get $r\left(1 - P_r/K\right)/\beta > Z_0 > Z_1 > Z_2 > \cdots > Z_n$

For case (i), we have $\lim\limits_{n \to \infty} Z_n = \theta^*$, where $0 < \theta^* < r\left(1 - P_r/K\right)/\beta$. Hence $\theta^* = F\left(\theta^*\right)$. So system (5) has an orbitally asymptotically stable positive order-1 periodic solution. Similarly, for case (ii).

4. Existence of Periodic Solution with $P_r > P_C$

When $P_r > P_C$, then from system (1), if $P > P_C$, we have

$$
\left[
\begin{array}{l}
\left.\begin{array}{l}
\dfrac{dP}{dt} = rP\left(1 - \dfrac{P}{K}\right) - \beta P_C Z \\[2mm]
\dfrac{dZ}{dt} = \beta_1 P_C Z - \rho P_C Z - dZ
\end{array}\right\} P < P_r \\[6mm]
\left.\begin{array}{l}
\Delta P = -c_1 EP \\
\Delta Z = -c_2 EZ
\end{array}\right\} P = P_r
\end{array}
\right.
\tag{10}
$$

if $P < P_C$, we can get system (5).

By section 3, system (2) has an interior equilibrium $E_2(P^*, Z^*)$ when $P^* < P_C$, so supposing $P^* < P_C$.

Theorem 5. For $\forall c_1, c_2 \in (0,1)$, $\forall \hat{Z} \in \left[0, \left(1 - (1 - c_1 E) P_r / K\right) r (1 - c_1 E) P_r / \beta P_C\right]$, $P > P_C$, if $P_r < K, P^* < P_C$, then one of the following statements holds.

(a) If trajectory $O^+(E_0, t_0)$ starting from $E_0\left((1 - c_1 E) P_r, \hat{Z}\right)$ is tangent to $P = P_r$ at $E_1\left(P_r, (1 - P_r / K) r P_r / \beta P_C\right)$ and $(1 - c_2 E)(1 - P_r / K) r P_r / \beta P_C < \hat{Z}$, then system (10) admits an orbitally asymptotically stable order-1 periodic solution.

(b) Suppose that trajectory $O^+(A_0, t_0)$ starting from $A_0\left((1 - c_1 E) P_r, Z_0\right)$ intersects $P = P_r$ at $(P_r, \hat{Z}_0)$. If $(1 - c_2 E)\hat{Z}_0 \leq \left(1 - (1 - c_1 E) P_r / K\right) r (1 - c_1 E) P_r / \beta P_C$, then system (10) admits an orbitally asymptotically stable order-1 periodic solution.

(c) Suppose that trajectory $O^+(F, t_0)$ starting from $F\left((1 - c_1 E) P_r, Z\right)$ does not intersect with $P = P_r$, then system (10) has no positive order-k (k=1,2,...) periodic solution.

Proof. Suppose that there is a constant $\hat{Z} \in \left[0, \left(1 - (1 - c_1 E) P_r / K\right) r (1 - c_1 E) P_r / \beta P_C\right]$ satisfying the condition given in statement (a).

Since $(1 - c_2 E)(1 - P_r / K) r P_r / \beta P_C < \hat{Z}$, then we have $(1 - c_2 E) Z < \hat{Z}$ for $\forall E(P_r, Z) \in \Gamma_2$ and $Z < (1 - P_r / K) r P_r / \beta P_C$. Therefore, trajectory $O^+(E, t_0)$ starting from $E(P_r, Z)$ will intersect with Γ_2 infinitely many times due to state feedback control strategy $\Delta P = -c_1 E P$ and $\Delta Z = -c_2 E Z$. Considering $\forall D_i(P_r, Z_i), D_j(P_r, Z_j)$ on Γ_2, where $0 < Z_i < Z_j < (1 - P_r / K) r P_r / \beta P_C$, it is obviously that $D_i^+\left((1 - c_1 E) P_r, (1 - c_2 E) Z_i\right)$ is below $D_j^+\left((1 - c_1 E) P_r, (1 - c_2 E) Z_j\right)$ in view of control strategy $(1 - c_1 E) Z_i < (1 - c_1 E) Z_j < (1 - P_r / K) r P_r / \beta P_C$. Further, trajectories $O^+(D_i, t_0)$ and $O^+(D_j, t_0)$ intersect Γ_2 at $D_{i+1}(P_r, Z_{i+1})$ and $D_{j+1}(P_r, Z_{j+1})$, respectively.

We claim that $Z_{i+1} < Z_{j+1}$. Suppose that trajectory $O^+(C_0, t_0)$ starting from $C_0(P_r, Z_0)$ jumps to $C_0^+\left((1 - c_1 E) P_r, (1 - c_2 E) Z_0\right)$, and then arrives Γ_2 at

$C_1(P_r,Z_1)$ again due to the fact that $(1-c_1E)(1-P_r/K)rP_r/\beta P_C < \hat{Z}$ and then jumps to $C_1^+((1-c_1E)P_r,(1-c_2E)Z_1)$ on Γ_1. At state C_1^+, trajectory $O^+(C_0,t_0)$ intersects Γ_2 at $C_2(P_r,Z_2)$, where $Z_1,Z_2 \in (0,(1-P_r/K)rP_r/\beta P_C)$. By the Poincaré map, it follows that $Z_1 = F(Z_0)$, $Z_2 = F^2(Z_0)$ and $Z_n = F^n(Z_0)$ $(n = 0,1,2,\cdots)$. Particularly, system (10) has a positive order-1 periodic solution when $Z_0 = Z_1$. Now we consider the general situation for $Z_0 \neq Z_1$:

(i) $Z_0 < Z_1$, we can get $Z_0 < Z_1 < Z_2 < \cdots < Z_n < (1-P_r/K)rP_r/\beta P_C$;

(ii) $Z_0 > Z_1$, we can get $(1-P_r/K)rP_r/\beta P_C > Z_0 > Z_1 > Z_2 > \cdots > Z_n$.

In case (i), we have $\lim_{n\to\infty} Z_n = \theta^*$ where $0 < \theta^* < (1-P_r/K)rP_r/\beta P_C$. Hence $\theta^* = F(\theta^*)$. So system (10) has an orbitally asymptotically stable positive order-1 periodic solution. Similarly, for case (ii). The same, conclusion (b), (c) can be proved.

5. Conclusion

The complex dynamic behaviour of a toxin producing phytoplankton-zooplankton model with Holling I function response and state feedback control strategies is analyzed systemically in this paper. Firstly, we analyze the existence and stability without impulsive effect of system (1). The locally and global stability of the interior equilibrium is studied when it exists. Next, the existence of periodic solution with $P_r \leq P_C$ is discussed. The system (5) has a semi-trivial periodic solution which is orbital asymptotically stable for any $P_r < P^*$ and it admits an order-1 periodic solution which is orbitally asymptotically stable for any $P^* < P_r < K$. Finally, we studied conditions on the existence and orbitally asymptotical stability of periodic solution with $P_r > P_C$ by the Poincaré map.

Acknowledgement

This work was funded in part by the National Natural Science Foundation of China (No. 11401068) and the Liaoning Province Natural Science Foundation of China (No. 2014020133).

References

1. Y.F. Lv, Y.Z. Pei, S.J. Gao, C.G. Li, Harvesting of a phytoplankton-zooplankton model, Nonlinear Analysis: Real Word Applications 11,3608-3619, 2010.

2. Li L,YanY.Global stability of mathematical model of plankton. Journal of the Graduate School of the Chinese Academy of Sciences, 2008,25(3):305-312.
3. S.R.J. Janga, J. Baglama, J. Rick, Nutrient phytoplankton-zooplankton models with a toxin. Math. Comput. Modeling. 43 (2006) 105-118.
4. B. Mukhopadhyay, R. Bhattacharyya, Modeling phytoplankton allelopathy in a nutrient plankton model with spatial heterogeneity, Ecol. Model. 198(2006) 163-173.
5. S.Gao, L Chen, The effect of seasonal harvesting on a single-species discrete population model with stage structure and birth pulses, Chaos Solitons Fractals 24(2005) 1013-1023.
6. T.Saha, M.Bandyopadhyay, Dynamical analysis of toxin producing phytoplankton-zooplankton interactions, Nonlinear Anal. RWA 10(2009)314-332.
7. Y. Pei, L. Chen, Q. Zhang, C. Li, Extinction and permanence of one-prey multi-predators of Holling type II function response system with impulsive biological control, J. Theoret. Biol. 235 (2005)495-503.
8. L.S. Pontryagin, V.S. Boltyanski Gamkrelidze, E.F. Mishchenco, The Mathematical Theory of Optimal Processes, Wiley, New York, 1987.

Nonlinear Control Allocation for a Blended Wing Body Aircraft

Jian Yi[1,*], Xin-min Dong[1], Yong Chen[1], Jian-hui Zhi[1],
Bing-xu Zhou[2] and Xian-chi Huang[2]

*1College of Aeronautics and Astronautics Engineering,
AFEU, Xi'an 710038, China
2PLA Air Force the 95503 unit,
Zunyi, 563127, China*

*yijian_2015@sina.com

This paper describes a piecewise linear control allocation method based on Computational Fluid Dynamics (CFD) aero-data for a blended wing body (BWB) aircraft with redundant control surfaces. The conventional control allocation methods usually assume that linear efficiency existence between the control surfaces deflection and the virtual control torque. But from the calculated CFD aero-data of the BWB aircraft, the characteristic of the elevons and split rudders to the control torque is nonlinear, which may bring control errors by using conventional linear methods. The piecewise linear approximate method is used to estimate the nonlinear control efficiency of surface deflection. In order to obtain the surface commands, a piecewise linear programing control allocation (PLPCA) scheme is applied by considering the constraints of the deflection position and rate. The simulation result shows that the PLPCA scheme can allocate the control surfaces reasonably and generate better virtual moment command than that of the linear programing control allocation (LPCA).

Keywords: Blended wing body aircraft; Control allocation; Piecewise linearity; CFD.

1. Introduction

Modern combat and civilian aircraft employ more than three primary control surfaces, mainly to maximize mission performance and enhance survivability. The redundant control surfaces significantly make the control system design complicated [1], due to (a) the assignment of command to control surfaces and (b) the requirements of the control surface constraints on deflection position and rate.

Control allocation [1,2] is an effective method to distribute the virtual control requirements to the control surfaces in the best possible manner while accounting for their constraints, and has extend to filed such as automobile [3], marine vessel [4], spacecraft [5] and so on. To obtain a solution for control allocation problem, many practical approaches have been provided, like direct allocation [6], weighted pseudo- inverse [7], and optimize arithmetic using mathematics programming [8]. As mentioned above, the general methods are performed under the assumption that a linear relation- ship exists between the control induced moments and the control effector displacements, despite the fact that the forces and moments produced by aircraft control surfaces are almost always nonlinear functions of control surface displacement, which may be valid locally. The researches in [1, 2] point out that the yaw moment can be regard as a quadratic function of the surface displacement, and a highly nonlinearity may also appear in large deflection of control surface. The linear approximation may not be sufficiently accurate for the vehicle in this instance, and may cause improper control allocating result.

In this paper, a PLPCA scheme proposed in [9] is applied in the presence of nonlinear control efficiency surface. This paper is organized as follows. Firstly, the BWB aircraft model used in this study is described and aerodynamics of the control surface is studied via CFD. Secondly, the description of the piecewise linear method is proposed and the PLPCA scheme is introduced then. Thirdly, a numerical simulation is dedicated, and the paper is ends with a conclusion in the last section.

2. Control Efficiency Analysis of the BWB Aircraft

2.1 Description of the BWB Aircraft

The BWB aircraft layout used in the study is presented in Fig. 1. Which has a fully symmetric configuration with 3 groups of elevons and a split rudder placed on each side of wing, the control surface is placed parallel to the trailing edge. The deflection limit of position and rate for elevon is $\pm 30°$ and $150°/s$, the split rudder: $0° \sim 60°$, and $100°/s$.

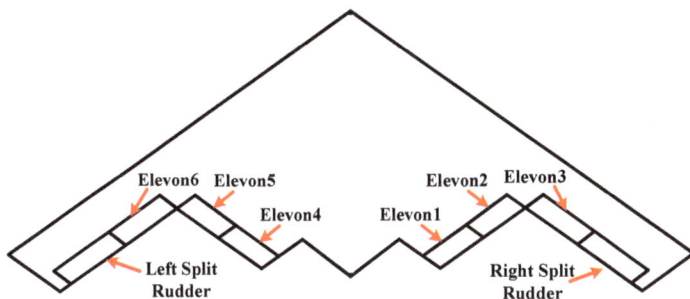

Figure 1. The blended wing body (BWB) aircraft

The downward deflection of the elevon is specified as positive, and the deflection of the split rudder in the right side is specified as negative. The deflections of each control surface is represented as: $\delta_1, \delta_2, \delta_3, \delta_{rs}, \delta_4, \delta_5, \delta_6$ and δ_{ls} respectively. Considering to the fully symmetric configuration of the BWB aircraft, 3 groups of elevons and a split rudder in the right side are studied via CFD[10]. The simulation condition is set as follows: $\alpha = 2°$, Ma = 0.75, $H = 10000m$.

3. Aerodynamics Analysis of the Surfaces

3.1 *Aerodynamics Analysis of the Elevons*

The control moment coefficients for roll, pitch and yaw and its linear approximate result are shown in Fig. 2. In Fig. 2(a) and 2(b), the moment coefficient for roll and pitch can be seen to increase fairly linearly at small deflection position. At larger deflection, the effectiveness of the control surfaces decreases. In Fig. 2(c) *the yaw moment can be seen to change almost quadratically with the elevon deflection and linear approximate gives a very poor estimation for the yaw moment curve.* In this situation, current control allocation algorithms based on linear assumption cannot account for this effect and may results in inconsequent allocating instructions.

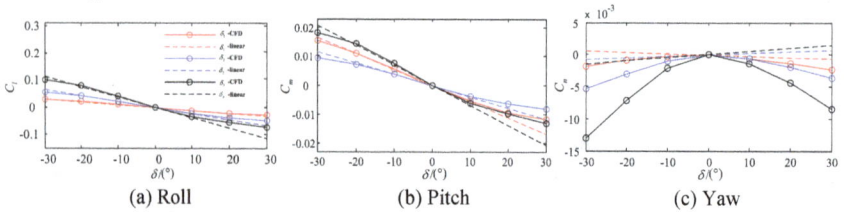

(a) Roll   (b) Pitch   (c) Yaw

Figure 2. Control moment curves due to elevons deflections and linear approximate result

Therefore, a piecewise linear method is used and the yaw control characteristics is approximated linearly respectively in both positive and negative deflection of the elevons, as shown in Fig. 3. A better result is obtained compare to the linear approximate show in Fig. 2(c).

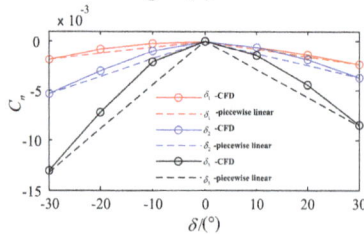

Figure 3. Piecewise linear approximate for yaw control moment

4. Aerodynamics Analysis of the Split Rudder

As the innovative drag-based control surfaces, the split rudder, mostly placed in the trailing edge of the wing, is commonly used to generate the desired yawing control moment for fly wing. The control moment response for the deflection of split rudder and its linear approximate result are shown in Fig. 4. It is shown that the roll and pitch control ability of the split rudder is less than the elevons. The yaw moment is increase fairly slow at small deflection. At larger deflection, the control efficiency increase and the linear characteristic of moment control salience. Different from the elevon, the linear approximate will gives a very poor estimation at small deflection of the split ruder. Considering this, the piecewise linear is used to approximate the deflections of $0°\sim15°$and $15°\sim60°$ respectively, shown in Fig. 4, and a better result is obtained.

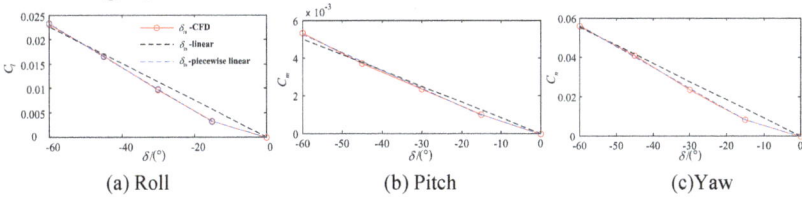

| (a) Roll | (b) Pitch | (c)Yaw |

Figure 4. Control moment curves due to split rudder deflections and linear/piecewise linear approximate

5. Control Efficiency Analysis of the Surfaces

The linear and piecewise linear approximate results of the roll, pitch and yaw moment curves are summarized as follow in tables 1~3.

Table 1. Linear control efficiency ($10\text{-}3/°$)

	C_l	C_m	C_n
δ_1	-1.2	-0.56	0.01
δ_2	-2.2	-0.39	0.025
δ_3	-3.8	-0.69	0.049
δ_{rs}	-0.38	-0.084	-0.92

Table 2. Piecewise linear yaw control efficiency of the right elevons ($10\text{-}3/°$)

	$-30°\sim0°$	$0°\sim30°$
δ_1	0.060	-0.076
δ_2	0.18	-0.12
δ_3	0.44	-0.28

Table 3. Piecewise linear control efficiency of the right split rudder (10-3/°)

	C_l	C_m	C_n
-0°~-15°	-0.22	-0.069	-0.56
-15°~-60°	-0.44	-0.095	-1.1

The linear and piecewise control efficiency of elvon4~6 and the left split rudder can be gathered symmetric as elvon1~3 and the right split rudder respectively.

6. Piecewise Linear Control Allocation for the BWB Aircraft

6.1 Description of the Piecewise Linear Method

A piecewise linear method, developed by Bolender [9], is used to deal with the non-linear control allocation problem, and is expanded to the discrete-time situation with the control surface deflection position and rate constraints enforced.

Considering an m-dimensional control space $\delta \in R^m$, by using the roll, pitch and yaw moment coefficients as the virtual control command, the discrete-time control allocation problem with nonlinearity can be written as:

$$f(\delta(k)) = v(k) \quad \text{s.t.} \quad \underline{\delta} \le \delta(k) \le \overline{\delta}, \rho \le \dot{\delta}(k) \le \overline{\rho}. \tag{1}$$

Where k means the kth time period, $\delta = [\delta_1, \cdots, \delta_m]^T$ is the deflection vector, $v = [C_l, C_m, C_n]^T$ is the virtual control command (respectively roll, pitch and yaw moment coefficients), $\overline{\delta}, \underline{\delta} \in R^{m \times 1}$ and $\underline{\rho}, \overline{\rho} \in R^{m \times 1}$ represent the upper and lower position and rate bound of the control surfaces deflection respectively.

Considering the quadratic feature of the yaw moment, linear approximate is used respectively in both positive and negative deflection of the elevons:

(1) Positive deflection of the elevons. For the kth instant, the deflection angle of elevon and the yaw moment coefficient can be given by:

$$\delta_i^-(k) = \lambda_i^{(1)}(k)\underline{\delta_i} + \lambda_i^{(2)}(k)\delta_i^d, \quad C_{n\delta_i}^-(k) = \lambda_i^{(1)}(k)C_{n\underline{\delta_i}} + \lambda_i^{(2)}(k)C_{n\delta_i^d}. \quad i = 1, \cdots, m \tag{2}$$

(2) Negative deflection of the elevons. It can be given by:

$$\delta_i^+(k) = \lambda_i^{(2)}(k)\delta_i^d + \lambda_i^{(3)}(k)\overline{\delta_i}, \quad C_{n\delta_i}^+(k) = \lambda_i^{(2)}(k)C_{n\delta_i^d} + \lambda_i^{(3)}(k)C_{n\overline{\delta_i}}. \quad i = 1, \cdots, m \tag{3}$$

Where δ_i^d is deflection angle of the breaking point (for elvon, $\delta_i^d = 0°$ and for the right split rudder, $\delta_i^d = 15°$), $C_{n\delta_i}$, $C_{n\delta_{i,d}}$ and $C_{n\overline{\delta_i}}$ are the yaw moment coefficient control surface i generated correspond to the deflection of $\underline{\delta_i}$, δ_i^d

598

and $\bar{\delta}_i$. The relaxation factors $\lambda_i^{(1)}$, $\lambda_i^{(2)}$ and $\lambda_i^{(3)}$ are subject to the following conditions:

$$0 \le \lambda_i^{(1)}, \lambda_i^{(2)}, \lambda_i^{(3)} < 1, \quad \lambda_i^{(1)} + \lambda_i^{(2)} = 1, \quad \lambda_i^{(2)} + \lambda_i^{(3)} = 1. \tag{4}$$

From Eq. 2~ Eq. 4, a unified formulation of the deflection and yaw moment coefficient is stated as follow:

$$\delta_i(k) = \lambda_i^{(1)}(k)\underline{\delta}_i + \lambda_i^{(2)}(k)\delta_i^{d} + \lambda_i^{(3)}(k)\overline{\delta}_i, \quad C_{n\delta_i}(k) = \lambda_i^{(1)}(k)C_{n\underline{\delta}_i} + \lambda_i^{(2)}(k)C_{n\delta_i^{d}} + \lambda_i^{(3)}(k)C_{n\overline{\delta}_i}. \tag{5}$$

$$0 \le \lambda_i^{(1)}, \lambda_i^{(2)}, \lambda_i^{(3)} < 1, \lambda_i^{(1)} \cdot \lambda_i^{(3)} = 0, \quad \sum_{k=1}^{3} \lambda_i^{(k)} = 1. \qquad i = 1,\cdots,m \tag{6}$$

Therefore, the roll and pitch moment coefficient can be presented as in the same way:

$$C_{l\delta_i}(k) = \lambda_i^{(1)}(k)C_{l\underline{\delta}_i} + \lambda_i^{(2)}(k)C_{l\delta_i^{d}} + \lambda_i^{(3)}(k)C_{l\overline{\delta}_i}, C_{m\delta_i}(k) = \lambda_i^{(1)}(k)C_{m\underline{\delta}_i} + \lambda_i^{(2)}(k)C_{m\delta_i^{d}} + \lambda_i^{(3)}(k)C_{m\overline{\delta}_i}. \tag{7}$$

The deflection limit of the control surface i is satisfied once the relaxation factor $0 \le \lambda_i^{(1)}, \lambda_i^{(2)}, \lambda_i^{(3)} \le 1$. $\lambda_i^{(1)} = 0$ means control surface i reach the minimal deflection $\underline{\delta}_i$, and $\lambda_i^{(3)} = 1$ means the maximum deflection $\overline{\delta}_i$. The deflection rate limit is then considered and stated as follow:

$$\underline{\rho}_i T \le \delta_i(k+1) - \delta_i(k) \le \overline{\rho}_i T \Leftrightarrow$$
$$\underline{\rho}_i T \le \{\lambda_i^{(1)}(k+1) - \lambda_i^{(1)}(k)\}\underline{\delta}_i + \{\lambda_i^{(2)}(k+1) - \lambda_i^{(2)}(k)\}\delta_i^{d} + \{\lambda_i^{(3)}(k+1) - \lambda_i^{(3)}(k)\}\overline{\delta}_i \le \overline{\rho}_i T \tag{8}$$

Where T is the update time of inner-loop flight control system.

We assign the variable $\underline{y}_i$ and $\overline{y}_i$ that corresponds to the negative and positive of piecewise linear approximation segments such that:

$$\underline{y}_i(k) = \begin{cases} 1, & \lambda_i^{(1)}(k) \cdot \lambda_i^{(2)}(k) \ne 0, \ \lambda_i^{(3)}(k)=0 \\ 0, & \text{else} \end{cases}, \overline{y}_i(k) = \begin{cases} 1, & \lambda_i^{(2)}(k) \cdot \lambda_i^{(3)}(k) \ne 0, \ \lambda_i^{(1)}(k)=0 \\ 0, & \text{else} \end{cases}. \tag{9}$$

And the constraint equation (6) can be reformed as:

$$\sum_{k=1}^{3} \lambda_i^{(k)}(k) = 1, \qquad i = 1,\cdots,m$$
$$\lambda_i^{(1)}(k), \lambda_i^{(2)}(k), \lambda_i^{(3)}(k) \ge 0, \lambda_i^{(1)}(k) \le \underline{y}_i(k), \lambda_i^{(3)}(k) \le \overline{y}_i(k), \tag{10}$$
$$\underline{y}_i(k), \overline{y}_i(k) \in \{0,1\}, \underline{y}_i(k) + \overline{y}_i(k) = 1. \quad \text{and} \quad \text{Eq.(8)}$$

Combining Eq. 5, Eq. 7 and Eq. 10, the transformed piecewise linear control allocation problem is stated as follows:

$$v(k) = \hat{B}\varLambda(k), \quad \text{s.t. Eq.10.}$$

$$\text{where: } \hat{B} = \begin{bmatrix} C_{l\underline{\delta}_1} & C_{l\delta_1^d} & C_{l\overline{\delta}_1} & \cdots & C_{l\underline{\delta}_m} & C_{l\delta_m^d} & C_{l\overline{\delta}_m} \\ C_{m\underline{\delta}_1} & C_{m\delta_1^d} & C_{m\overline{\delta}_1} & \cdots & C_{m\underline{\delta}_m} & C_{m\delta_m^d} & C_{m\overline{\delta}_m} \\ C_{n\underline{\delta}_1} & C_{n\delta_1^d} & C_{n\overline{\delta}_1} & \cdots & C_{n\underline{\delta}_m} & C_{n\delta_m^d} & C_{n\overline{\delta}_m} \end{bmatrix} \tag{11}$$

$$\varLambda = [\lambda_1^{(1)}, \lambda_1^{(2)}, \lambda_1^{(3)}, \cdots, \lambda_m^{(1)}, \lambda_m^{(2)}, \lambda_m^{(3)}]^{\mathrm{T}}$$

7. The Piecewise Linear Control Allocation Modeling

To found the proper control deflections δ that ensures the commanded virtual control v is produced jointly by the control surfaces at kth time, a mixed optimization formulation of the control allocation problem, developed by Bodson[8], can be used. The primary objective is minimization of the difference between $\hat{B}\varLambda$ and v, and the secondary objective is to drive the control effectors to some preferred position δ_p, subject to the constraints of deflection position and rate limits. Mathematically, the mixed optimization objective function is posed as:

$$J = \left\| W_v(v(k) - \hat{B}\varLambda(k)) \right\|_1 + \left\| W_\delta(\delta(k) - \delta_p) \right\|_1 \tag{12}$$

Where $W_v = \mathrm{diag}(w_{c_l}, w_{c_m}, w_{c_n})$ and $W_\delta = \mathrm{diag}(w_{\delta 1}, \cdots, w_{\delta m})$ are the diagonal weigh matrix.

Combining Eq. 10~Eq. 12, the control allocation model may be transformed into the following mixed-integer linear program (MILP) problem:

$$\min \quad J = w_v^{\mathrm{T}} \delta_{s1}(k) + w_\delta^{\mathrm{T}} \delta_{s2}(k) \tag{13}$$

$$\text{s.t. } \hat{B}\varLambda(k) + \delta_{s1}(k) \ge v(k), \quad \hat{B}\varLambda(k) - \delta_{s1}(k) \le v(k),$$

$$\lambda_i^{(1)}(k)\underline{\delta}_i + \lambda_i^{(2)}(k)\delta_i^d + \lambda_i^{(3)}(k)\overline{\delta}_i + \delta_{s2,i}(k) \ge \delta_{p,i}, \quad i = 1, \cdots, m$$

$$\lambda_i^{(1)}(k)\underline{\delta}_i + \lambda_i^{(2)}(k)\delta_i^d + \lambda_i^{(3)}(k)\overline{\delta}_i - \delta_{s2,i}(k) \le \delta_{p,i}, \quad i = 1, \cdots, m \tag{14}$$

$$\delta_{s1}(k) \ge 0, \quad \delta_{s2}(k) \ge 0, \quad \text{Eq.}^\circ(10)$$

Where $\delta_{s1} \in \mathrm{R}^{3\times1}$ and $\delta_{s2} \in \mathrm{R}^{m\times1}$ are the relaxation factors, $w_v = [w_{c_l}, w_{c_m}, w_{c_n}]^{\mathrm{T}}$ and $w_\delta = [w_1, \cdots, w_m]^{\mathrm{T}}$ are the weigh vectors correspond to δ_{s1} and δ_{s2} respectively.

From the preceding description, the deflections $\delta(k)$ for the kth instance can be obtained from Eq. 5, once the relaxation factors $\lambda_i^{(1)}(k), \lambda_i^{(2)}(k), \lambda_i^{(3)}(k)$ are solved by the MILP problems (13) ~ (14).

8. Numerical Simulation

The aforementioned PLPCA scheme is applied to the BWB aircraft model, and is compared with the LPCA method to show the effectiveness of the former. The MILP problem (13) ~ (14) is solved by the YALMIP toolbox for Matlab. The update time $T = 0.02$s and weigh matrix
$$W_v = \text{diag}(1,1,1), W_\delta = 10^{-4}\text{diag}(0.8,0.6,1.3,1,0.8,0.6,1.3,1).$$
For the desired virtual control command $v = [C_l, C_m, C_n]^T$, where C_l, C_m and C_n are set to:
$$C_l = 0.2\sin(1.26t+2.8), \quad C_m = -0.01 + 0.004t$$
and $C_n = 0.025\cos(1.26t+2.8)$, Fig. 5 shows the virtual control command requirements and the command actually attained by LPCA and the proposed PLPCA scheme.

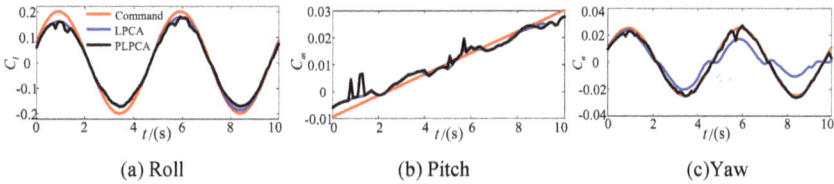

| (a) Roll | (b) Pitch | (c)Yaw |

Figure 5. Dynamic profile of the virtual control command

It is shown in Fig. 5 that if the nonlinear characteristic of the control surfaces is not taking into consideration, the LPCA scheme with linear assumption will give a really poor performance in achieving the desired moment (C_l, C_m and C_n), especially in the yawing axis. Thus the proposed PLPCA scheme is more effectively in achieving the required yaw control command. The LPCA and PLPCA scheme performance almost the same in roll and pitch control, and the control error appears at the lager control command, mostly due to the saturation of control command, which correspond to maximal deflections of the control surface, as shown in Fig. 6.

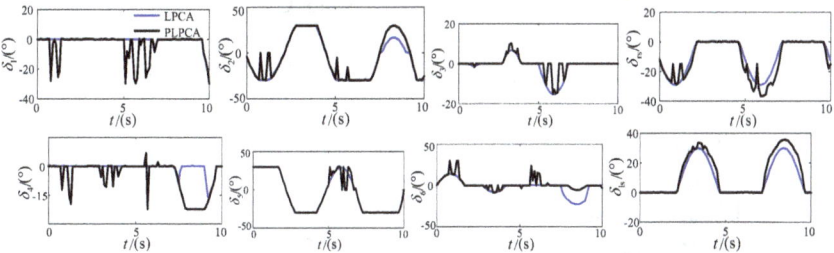

Figure 6. Deflection commands of the control surfaces

9. Conclusion

In this paper, the nonlinear aerodynamics of the elevons and the split rudder for a BWB aircraft are investigated via CFD and a PLPCA scheme is demonstrated with the nonlinearity existence between surface deflection and the virtual control torque. Compared with the LPCA method, the numerical simulations show that the PLPCA scheme is more effective in allocating the virtual control torque, and has presented less control error in the presence of the nonlinear control efficiency. In the future work, the real-time performance of the flight control system needs to be further studies.

Acknowledgement

This research was funded by the National Science Foundation of China under grants 61304120 and 61473307.

References

1. T. A. Johansen, T. I. Fossen, Control allocation - a survey, Automatic, 2013, 49(5): 1087-1103.
2. M. W. Stephen, V. Mark, L.M. Leo, et al, Control allocation performance for blended wing body aircraft and its impact on control surface design, Aerospace Science and Technology, 2013, 29(2013): 18-27.
3. J. Tjonnas, T. A Johansen, Stabilization of automotive vehicles using active steering and adaptive brake control allocation, IEEE Transactions on Control Systems Technology, 2010, 18(3): 545-558.
4. F. Scibilia, R. Skjetne, Constrained control allocation for vessels with azimuth thrusters [A]. Maneuvering and Control of Marine Craft, Arenzano, Italy: International Federation of Automatic Control, 2012: 7-12.
5. M. W. Oppenheimer, D. B Doman, A method for including control effector interactions in the control allocation problem, AIAA Guidance, Navigation and Control Conference and Exhibit [C]. South Carolina: AIAA, 2007: 1-10.
6. W. C. Durham, Constrained control allocation: three moment problem, Journal of Guidance, Control, and Dynamics, 1994, 17(2): 330-336.
7. Y. Chen, X. M. Dong, J. P. Xue, et al, Multi-objective optimization design of weight coefficients for weighted control allocation scheme, Control and Decision, 2013, 28(7): 991-1001. (In Chinese)
8. M. Bodson, Evaluation of optimization methods for control allocation, Journal of Guidance, Control and Dynamics, 2002, 25(4): 703-711.

9. M. A. Bolender, and D. B. Doman, Nonlinear control allocation using piecewise linear functions, Journal of Guidance, Control, and Dynamics, 2004, 27(6): 2017-1027.

10. A. D. Ronch, M. Ghoreyshi, D. Vallespin, K.J. Badcock, A framework for constrained control allocation using CFD-based tabular data, AIAA Aerospace Sciences Meeting, Florida: AIAA, 2011:1-18.

Fast Time-Varying Motion Parameter Estimation Based on Wavelet Transform

Lin-lin Du[1]*, Yan Ren[2] and Yu Zhang[1]

[1]*Beijing Institute of System Engineering, Beijing, P.R. China, 100101*
[2]*Engineering Headquarters of Aerospace City, Beijing, P.R. China, 100101*
Email: kddulinlin@sina.com

This paper describes a data fusion algorithm used to estimate parameter for fast time-varying motion. A method of noise variance estimation based on the characteristics of wavelet vanishing moments is analyzed, compared with the arithmetic average method, which is verified by the simulation data. The results show that the correctness and validity of the fusion algorithm.

Keywords: Motion parameters, Vanishing moment, Regularity.

1. Introduction

How to make optimal estimation of time-varying parameters and observational data that with noise in nonlinear part is important in the application of industry, traffic and military engineering. Commonly used parameter estimation method has the least square method, the maximum likelihood estimate and so on. In recent years, wavelet theory has been very rapid development, and because of its good time-frequency characteristics, wavelet theory has been widely used in the field of de-noising and parameter estimation, but most of the parameter estimation algorithms are only applicable to the static or slow time variation process, and can produce large errors in the fast time varying process. In this paper, a data fusion algorithm based on wavelet transform is used to accurately predict the characteristic parameters of the moving target.

2. Noise Model Based on Wavelet Function

The definition of Lipschitz index [1]:
If there is a coefficient K>0 and a polynomial of degree $m = \lfloor a \rfloor$, makes

$$\forall t \in R, |f(t) - p_v(t)| \le K|t - v|^a \tag{1}$$

That function f is pointwise Lipschitz a≥0 at point v.

2) If there is a coefficient K > 0 makes type (1) established for all $v \in [a,b]$, and K has nothing to do with the v, then the function f is consistent Lipschitz a in the interval [a, b].

The Lipschitz regularity of function f at point v/range [a, b] is defined as the upper bound of the parameter a that makes f Lipschitz a.

For any v, polynomial $p_v(t)$ is uniquely identified. If f is $m = \lfloor a \rfloor$ times continuously differentiable within a certain area in v, then p is the Taylor exhibition of function f at the point v. On the other hand, consistent Lipschitz index has given a more global regularity measure adapted to a given interval. If f is consistent Lipschitz a>m within a certain area of point v, then we can prove that f in this field must be m times continuously differentiable.

if $0 \le a < 1$, then $p_v(t) = f(t)$ and Lipschitz condition (1) Can be written as

$$\forall t \in R, \left| f(t) - f(v) \right| \le K \left| t - v \right|^a \tag{2}$$

The function bounded but discontinuous at point v is Lipschitz 0. If f is Lipschitz a <1 at point v, then f is non-differentiable at point v and parameter a depicts the singularity types.

The Lipschitz property (1) shows that within a certain area in v, f can be approximated by polynomial p_v as follows:

$$f(t) = p_v(t) + \varepsilon_v(t), \text{ Among them } \left| \varepsilon_v(t) \right| \le K \left| t - v \right|^a \tag{3}$$

The polynomial p_v can be eliminated when we use wavelet transform to estimate the index a.

The actual observation signals are limited bandwidth, and a wavelet function of a certain scale is equal to the band pass filter. The above properties of the wavelet function are used in the algorithm of this section, using a orthogonal wavelet function $\psi_{s,\tau}(t)$ which vanishing moment is k (k > M), this orthogonal wavelet function has the following feature [1]:

$$\int_{-\infty}^{+\infty} t^M \psi_{s,\tau}(t) dt = 0 \tag{4}$$

The M order polynomial can be eliminated using (4).

Observation sequence $O_j(t) = a_0 + a_1 t + ... + a_M t^M + n_j(t)$ convolve with wavelet function $\psi_{s,\tau}(t)$

$$W_{Y_j} = O_j * \psi_{s,\tau}(t) = W_Y + W_{n_j} \tag{5}$$

Because the wavelet function has disappeared moment with k order, it will inhibit the signal component of observation sequence and reserve the noise component, namely:

$$W_{O_j}(s, \tau) = W_{n_j}(s, \tau) \qquad (6)$$

From literature [2], the standard deviation of noise component $n_j(t)$ can be obtained by:

$$\sigma_j \approx \frac{1}{0.6745} Med(|W_{O_j}(s, t_m)|) \qquad (7)$$

In type (7), the scale factor s is defined as 1/2, time factor is the discrete express of time t, Med takes the median value of sequence $\{|W_{O_j}(s, t_m)|\}$. As can be seen from type (7), the noise standard deviation is changed over time.

3. Mathematical Model of Related Parameters

Assumption in a multi-sensor system, there are N0 sensors to observe an unknown moving object at the same time. The observations of different sensors are respectively $_{\{O_j\}}$ (j = 1, 2... N0). All the observed sequence can be decomposed into high frequency part $\{H_j(1), H_j(2),..., H_j(M)\}$ and low frequency part $\{L_j(1), L_j(2),..., L_j(M)\}$, M is the length of sequence. According to the methods of literature [3], a wavelet with long support length can be used to decompose frequency of each sensor. Compare the energy of high frequency part, if the high frequency of a sensor is seriously inconsistent with that of other sensors, then the sequence is not valid and should be excluded.

The energy of the high frequency part of each sequence is:

$$EH_j = \sum_{i=1}^{M} h_j^2(i) \qquad (j=1,2...N0), \text{ if } \frac{EH_j}{\sum_{i=1}^{N} EH_i} \times N_0 \geq E_{max}$$

, the sequence is considered invalid. E_{max} is the energy exception, that means, E_{max} times higher than the average value of high frequency energy.

It is assumed that N sequences $\{O_j\}$ (j=1,2...N) remained after screening. The observed sequences are unbiased and independent of each other, so

606

observed O can be estimated by LMS [4],

$$\hat{O} = \sum_{j=1}^{N} W_j O_j \qquad (8)$$

In (8), the weights W_j content $\sum_{j=1}^{N} W_j = 1$

The variance estimation can be achieved by the following two expressions:

$$\sigma^2 = \sum_{j=1}^{N} W_j^2 O_j^2 \qquad (9)$$

$$\sigma_{ave}^2 = \frac{1}{N^2} \sum_{j=1}^{N} \sigma_j^2 \qquad (10)$$

Equation (9) is the weighted variance estimation algorithm used in this paper, σ_j is the variance of noise parasitic in the j observation sensor; and Equation (10) made with arithmetic average to estimate the noise variance. This paper will compare the differences between the two through the experiment.

By minimizing the noise variance in (9), we can get the optimal weights:

$$W_j = \frac{1}{\sigma_j^2 \sum_{i=1}^{N} \frac{1}{\sigma_i^2}} \qquad (11)$$

By (11), the optimal weights will be adjusted with the noise standard deviation, and the noise standard deviation is change over time, so the algorithm can be applied to fast time-varying data.

At the same time, the minimum variance estimate of observation O is:

$$\sigma_{min}^2 = \frac{1}{\sum_{j=1}^{N} \frac{1}{\sigma_j^2}} \qquad (12)$$

Easy to prove that σ_{min}^2 is less than the variance of each observation sequence O_j, not only, and is less than the variance estimated by the arithmetic average method of (10). If you can get $\sigma_j^2 (j = 1, 2..., N)$, the optimization estimation of the observed sequence can be obtained by (11) and (8).

4. Simulation Results

The above algorithm is simulated and verified in this section. Assumes that the sampling frequency of the observation data is 20Hz, the recording time is 100 seconds. Suppose there are five kinds of sensors to observe the object at the same time, Gaussian white noise $\left(n_j(t)\right)$ of different variance is respectively add to the related parameters F of five kinds of sensor, to represent the different sensors system error, and the nonlinear noise error generated by clipping polynomial and disturbing force. The theory of standard deviation σ_j of $n_j(t)(j=1,2,...,5)$ is respectively 0.1, 0.3, 0.5, 0.7, and 0.9.

Measured data of the first 20 seconds is used to estimate the variance of each observation sequence, track related parameters of the 80 seconds behind is forecasted using information fusion method.

For any given order of the vanishing moment, Daubechies wavelet has the smallest set [5] relative to other orthogonal wavelet function. Choose Daubechies series of orthogonal wavelet function which vanishing moment order is five, to inhibit the signal component in the 4-order polynomial.

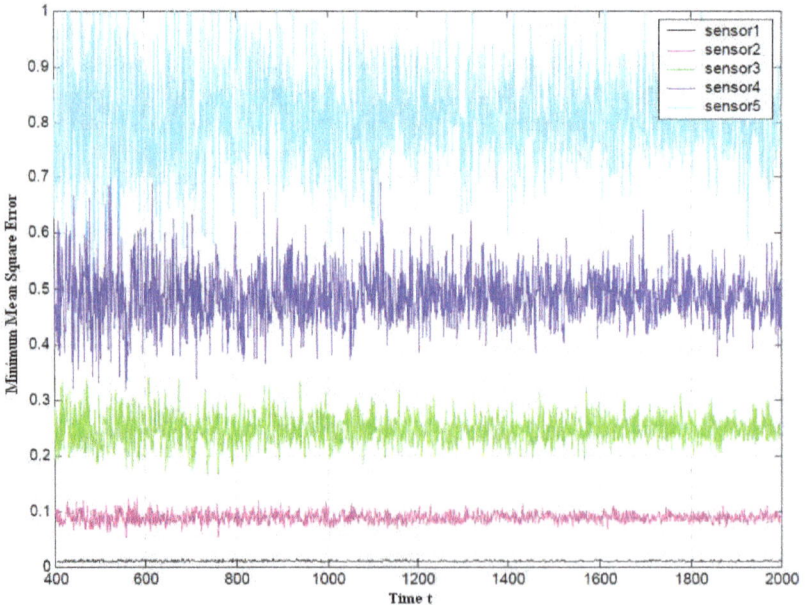

Figure 1: Minimum Mean Square Error of Different Sensors Parameters

Theoretical minimum mean square error values of various sensors are 0.01, 0.09, 0.25, 0.49, 0.91, as shown in Figure 1, the estimate value of the minimum mean square error is always fluctuating around the theoretical mean square error values, and tends to be stable, as shown in Figure 2, the variances of the estimate

608

values show a decreasing trend as time increases, namely the estimate mean square deviation value by the method is tend to be stable.

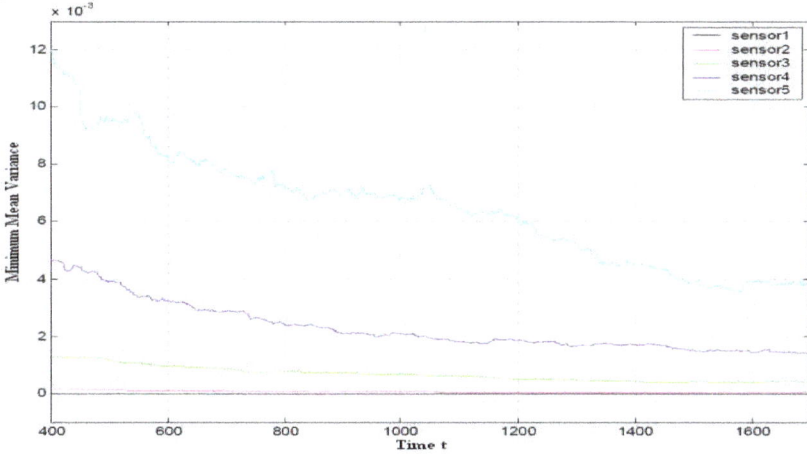

Figure 2: Minimum Mean Variance

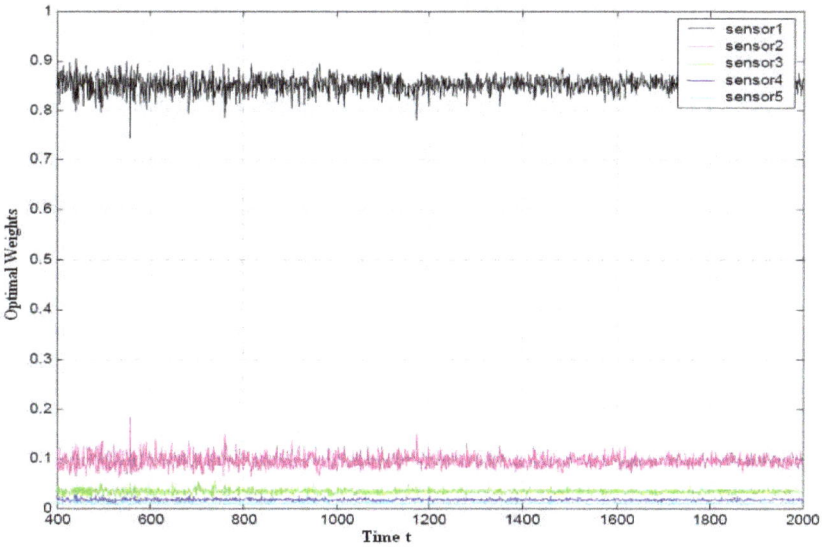

Figure 3: Optimal Weights of different sensors

As shown in Figure 3, the minimum weight is assigned to the parameter with maximum noise variance, for example, the noise variance of the fifth radar sensor parameter of is 0.81, the median value of weights assigned to it is 0.0106, its contribution to track parameter estimation is minimum; first radar sensors have the minimum noise variance 0.01, the median value of weights assigned to it is 0.8531, it will make the biggest contribution to track parameter estimation.

Figure 4 shows the original motion parameters, without adding noise.

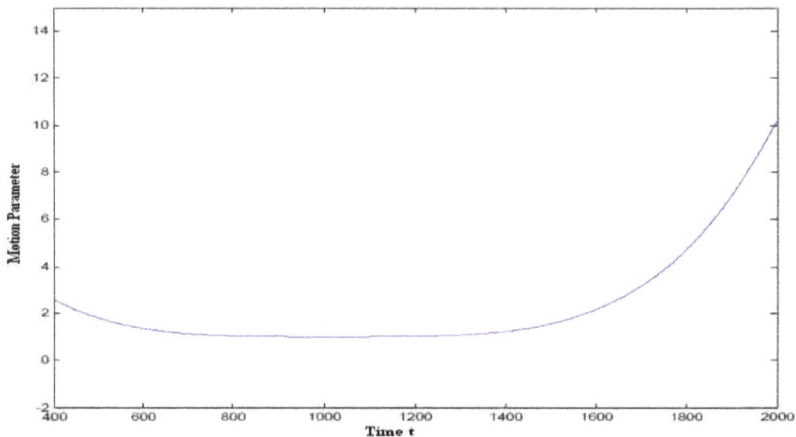

Figure 4: Original Motion Parameters

The outputs of the five sensors are shown in Figure 5 to Figure 9.

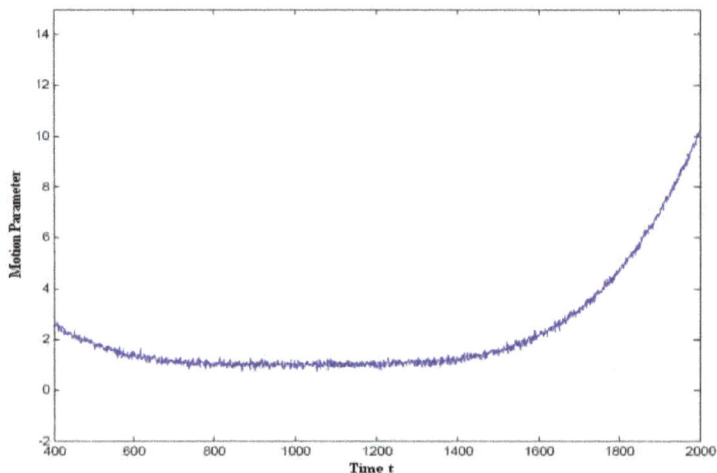

Figure 5: Output of the First Sensor

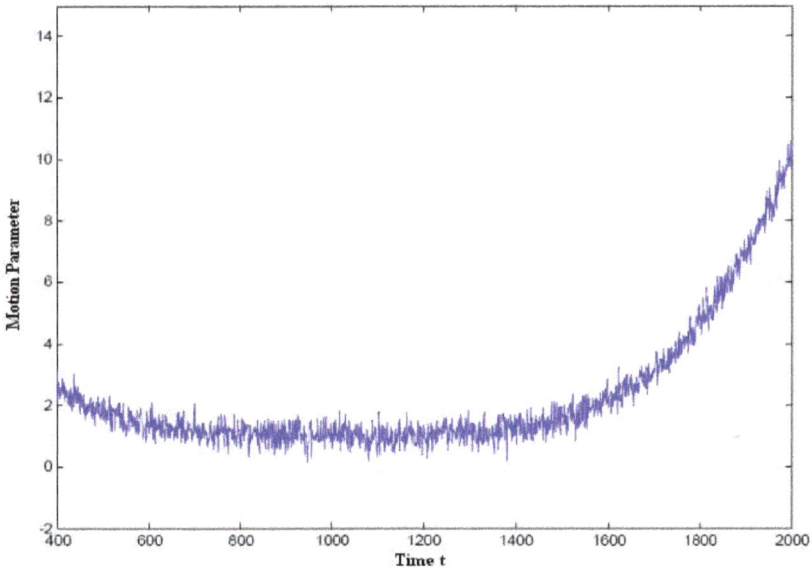

Figure 6: Output of the Second Sensor

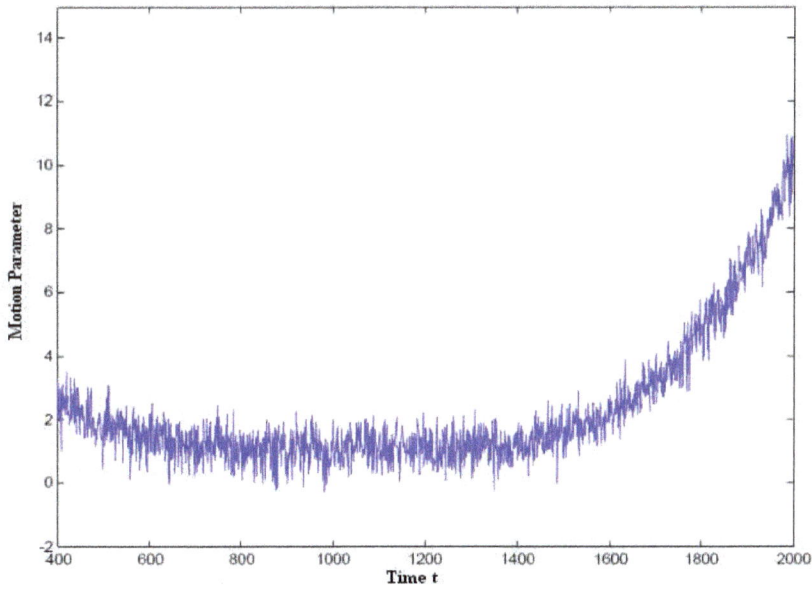

Figure 7: Output of the Third Sensor

611

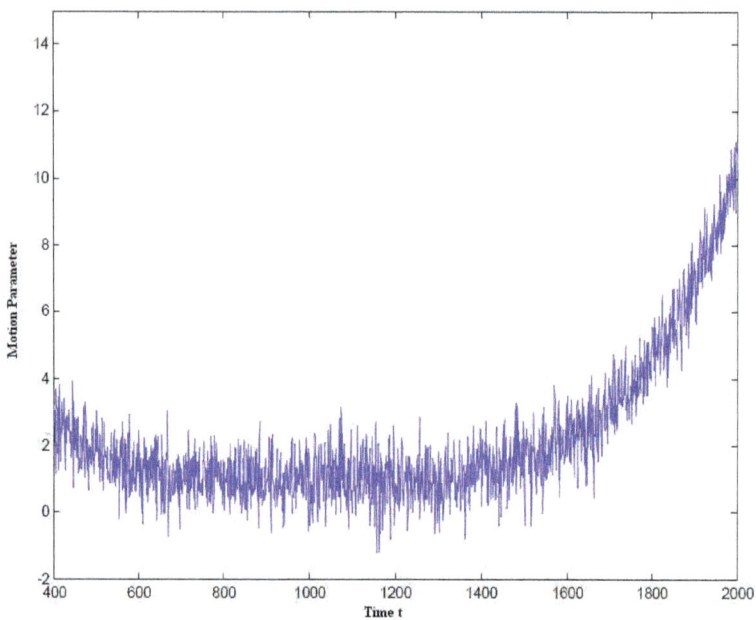

Figure 8: Output of the 4th Sensor

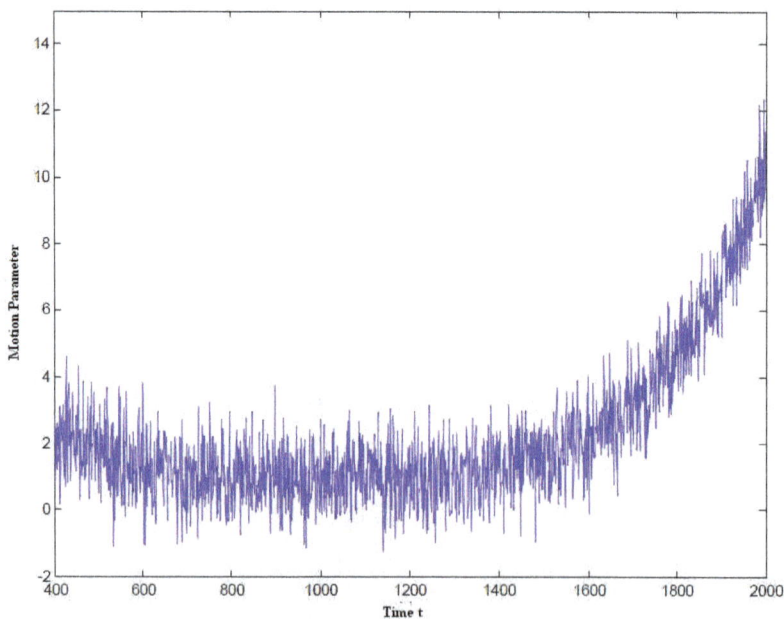

Figure 9: Output of the 5th Sensor

612

Figure 10 below shows the estimate related parameters using information fusion method. From the comparison, once could see the estimate related parameters using fusion method are closer to the original motion parameters, and its variance value is less than the variance values of any sensor output as shown in Figure 11.

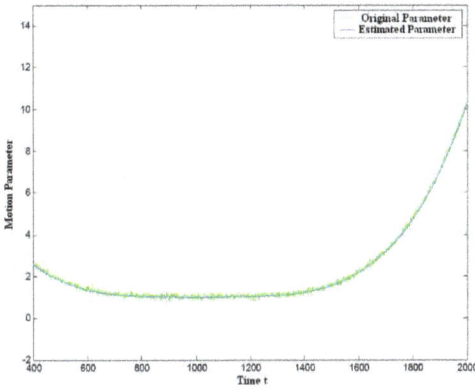

Figure 10: Compared With Original Parameters

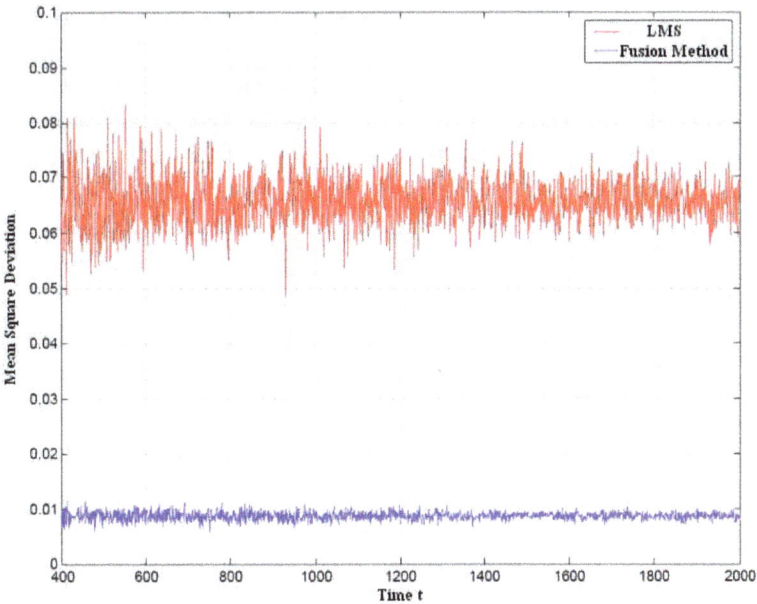

Figure 11: Results of Two Methods

As shown in Figure 11, the estimate mean square deviation using fusion method (about 0.0086) is not only less than the smallest mean square deviation (0.01) of different sensor motion parameters, that is the first radar sensor motion parameters, but also less than the mean square deviation estimation using arithmetic average method (0.0657). That fused with the proposed method track related parameter estimation to be more precise. This shows that the track parameters obtained by the fusion method proposed in this paper are more accurate.

5. Conclusion

In this paper, we mainly study the method of parameter estimation in multi-sensor detection and tracking of moving objects, to obtain a more stable and optimal estimation of motion parameters. In this paper, the multi sensor sequence is screened and the invalid sequence is removed. On the basis of this, the weights of the observation sequence of each sensor's motion parameters are calculated, and the data fusion is carried out. The results show that this algorithm is able to deal with the fast time varying data, and get the better estimation of the motion parameters.

References

1. C. L. Zhi, H. X. Wang, Y. Luo. Theory and Application of Wavelet [M]. BeiJing: Science Precess, 2004
2. D. Donoho and I. Johnstone. Ideal Spatial Adaptation via Wavelet Shrinkage. Biometrika, vol.81, 1994:425-455. Shi Wei, Peng Ju, Tan Shihai, etc. Multi-Sensor Data Fusion Method Based on Wavelets [J]. Gas Turbine Experiment and Research, 2013, 26(5):50-54.
3. Li Chao, Hu Moufa, Liu Chaojun, etc. A Wavelet-Based Multi-Sensor Space Target Data Fusion Algorithm [J]. Signal Processing, 2006, 22(2): 203-206.
4. L. H. Yang, D. Q. Dai, W. L. Huang. An Introduction of Wavelet in Signal Processing [M]. BeiJing: Metallurgical Industry Press, 2002.

Analysis of Network Dynamics for Manned/Unmanned Aerial Vehicle Networking System

Qi-feng Ren[1], Wen-qing Xi[1], Jian-dong Zhang[2,*], Zhi-yi Huang[2] and Xiao Yu[2]

[1]Science and Technology on Avionics Integration Laboratory
[2]School of Electronics and Information,
Northwestern Polytechnical University
*jdzhang@nwpu.edu.cn

This paper discusses the designing and realizing of the simulation system of UMC and FBW in an avionics integrated simulation system, primarily investigating system design and management, software development and bus information stream management. The system adopts the bus topological structure of a certain aircraft, connecting the various subsystems for the purpose of avionics integration. Under the conditions established by the ICD, this system realizes communication between the various subsystems with the use of MIL-STD-1553B Bus Data and ethernet as the transmission medium.

Keywords: UMC; FBW; MIL-STD-1553B bus data.

1. Introduction

As an emerging cross discipline, the research for complex network and its dynamics is to study the topological structure and dynamic characteristics of the complex network existing in the real world as a whole [1]. At present, the domestic empirical researches for complex network is still in the primary stage and the studies of dynamics based on Manned/Unmanned aerial vehicle networking system is even fewer.

2. Empirical Analysis

See MAV, UAV as network nodes and the communication data links as sides in a micro view so that the Manned/Unmanned aerial vehicle networking system can be abstracted into a complex network system as shown in Fig. 1, which is composed of 11 UAV nodes, 3 MAV nodes and 20 connecting edges. The reason why MAV nodes M_1, M_2, M_3 and the UAV node U_6 have more connecting edges is that the MAV nodes control the entire network system layout and commands and the UAV node U_6 acts as information relay.

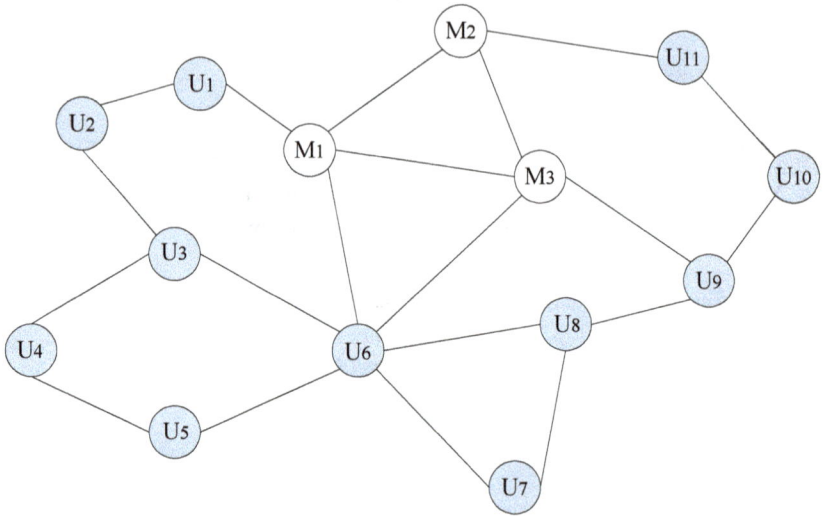

Figure 1. The network topology of Manned/Unmanned aerial vehicle networking system

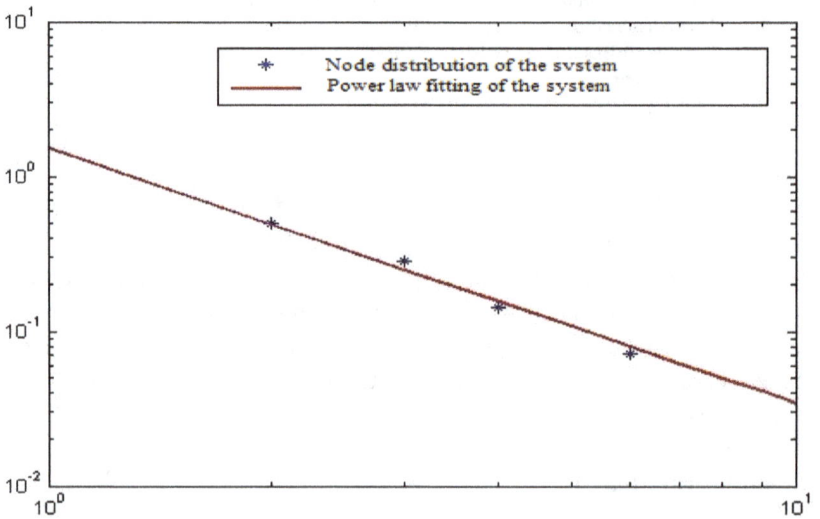

Figure 2. Node distribution probability of the Manned/Unmanned aerial vehicle networking system

3. Degree and Degree Distribution

In Manned/Unmanned aerial vehicle networking system, the degree value is the number of nodes connected to the aircraft i, which, to a certain extent, measures the importance of the node in the network. The degree of each node in the network model of the Manned/Unmanned aerial vehicle networking system is shown in Table 1.

Table 1. Degree of the network nodes in Manned/Unmanned aerial vehicle networking system

Node	Degree	Node	Degree	Node	Degree
U_1	2	U_6	6	U_{11}	2
U_2	3	U_7	2	M_1	4
U_3	2	U_8	3	M_2	3
U_4	2	U_9	3	M_3	4
U_5	2	U_{10}	2		

Degree distribution represents the probability distribution function $P(k)$ of the node degree in the subsystems of Manned/Unmanned aerial vehicle networking system. Degree distribution indicates the probability of that the node i has k sides connected to it. The degree distribution of each node in the network model of Manned/Unmanned aerial vehicle networking system is shown in Table 2.

Table 2. Degree distribution of the network nodes in Manned/Unmanned aerial vehicle networking system

Degree of Node	Number of Nodes	Node Degree Distribution
2	7	7/14
3	4	4/14
4	2	2/14
6	1	1/14

Secondly, the dot matrix diagram of the node degree distribution of Manned/Unmanned aerial vehicle networking system model can be plotted in double logarithmic coordinate system according to the calculated degree distribution shown in Table 2. The node degree in Table 2 is set to x, and the degree distribution of the node is set to y, as shown in Fig. 2.

By using the method of power law fitting in MATLAB, we can fit Fig. 2 and get the linear equation of power law linear fitting $p(k) = 1.531k^{-1.65}$.

$\gamma = 1.65$ is the degree distribution index of Manned/Unmanned aerial vehicle networking system, which indicates that the network is close to the complete network. In order to ensure the smooth communication between these nodes, there are more links between nodes, and the communication range is more extensive, so the reliability and efficiency of the system are ensured [8].

3.1. *Average Path Length*

The maximum distance between any two nodes in a network is called the diameter of the network, denoted as D. The average network distance is the average value of the distance between any two nodes in the network. Fig. 1 indicates that in the Manned/Unmanned aerial vehicle networking system model,

$$D = \max d_{ij} = 6, L_{ij} = 2.3846.$$

4. Robustness Analysis

From the angle of fault and attack strategies and starting with static robustness this paper respectively analyzes the influence of random faults and intentional attacks on Manned/Unmanned aerial vehicle networking system, and calculates the robustness of the Manned/Unmanned aerial vehicle networking system.

In order to simulate the influence of random fault and intentional attack on the network connectivity, two node removal strategies---Random failure strategy and Intentional attack strategy are considered

4.1. *Basic Parameters of Static Robustness*

4.1.1. *Maximum Connectivity*

G_{max} ($G_{max} = \dfrac{N'}{N}$) After the failure of some nodes in the Manned/Unmanned aerial vehicle networking system, the original network is decomposed into a plurality of connected small groups. Suppose the number of the nodes of the remaining functional maximal connected branch is N', and G_{max} is the ratio of the number of nodes contained in the largest group (N') and the original number of nodes (N).

4.1.2. *Connectivity Factor*

τ ($\tau = \dfrac{m}{n}$) Suppose the network is divided into m unconnected groups before an intentional attack or a random fault and n unconnected groups after an intentional attack or a random fault.

4.1.3. Failure Probability

$P_l(P_l = \dfrac{n}{N})$ Suppose the number of network nodes before an intentional attack or a random fault is N and the number of failed network nodes after an intentional attack or a random fault is n.

4.2. Modeling and Simulation of the Random Fault Model

Firstly, number the nodes of the Manned/Unmanned aerial vehicle networking system. Remove 5 of the nodes randomly, and the simulation result is shown below:

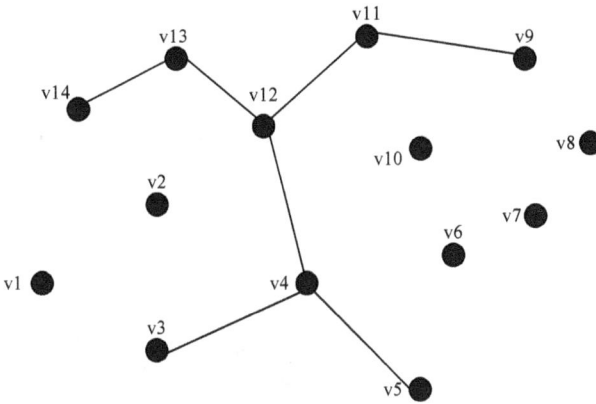

Figure 3. The network topology of Manned/Unmanned Manned/Unmanned aerial vehicle networking system after deleting 5 nodes randomly

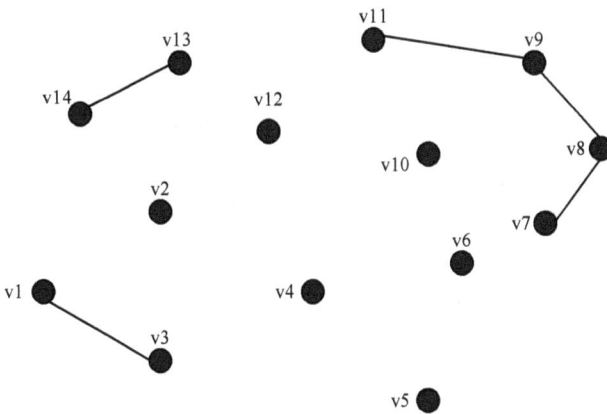

Figure 4. The network topology of aerial vehicle networking system after deleting the node v_6

619

In Fig. 3, the robustness of Manned/Unmanned aerial vehicle networking system after random failures is shown as follows:

$$G_{\max} = \frac{N'}{N} = \frac{8}{14} \approx 0.571 \; ; \; \tau = \frac{m}{n} = \frac{1}{1} = 1 \; ; \; P_l = \frac{n}{N} = \frac{6}{14} = 0.4286$$

Fig. 3 shows $G_{\max} \approx 0.5714$, $\tau \approx 0.143$, $P_l = 0.4286 < p_c$,

The whole Manned/Unmanned aerial vehicle networking system is still a connected network, and the network is still able to communicate properly

4.3. *Modeling and Simulation of the Intentional Attack Model*

From Table 1, we can see that the five nodes with the largest initial degree in the Manned/Unmanned aerial vehicle networking system are

$$v_4, v_{10}, v_{12}, v_2, v_6,$$

And therefore the sequence of the targets being intentionally attacked is

$$v_4 \rightarrow v_{10} \rightarrow v_{12} \rightarrow v_2 \rightarrow v_6.$$

After deleting the node v_6, the topology of the network is shown in Fig. 4. In Fig. 4, the vulnerability of the Manned/Unmanned aerial vehicle networking system under deliberate attacks is:

$$G_{\max} = \frac{N'}{N} = \frac{4}{14} \approx 0.286 \; ; \tau = \frac{m}{n} = \frac{1}{3} \approx 0.333 \quad P_l = \frac{n}{N} = \frac{6}{14} = 0.4286$$

$$G_{\max} \approx 0.286, \; \tau \approx 0.111 \text{ and } P_l = 0.4286 < p_c,$$

They indicate that the majority of the nodes in the network of Manned/Unmanned aerial vehicle networking system have lost their function, and the network is difficult to accomplish its combat task. It also can be seen that the node v_5 is affected by the node v_6, and become a failure node sequentially.

5. Synchronization Analysis

In the synchronous model of the Manned/Unmanned aerial vehicle networking system, assume that the whole network has 3 message transmission paths, the transmission paths are abstracted as 3 nodes in the synchronous model of the Manned/Unmanned aerial vehicle networking system, the dynamic equations of the 3 nodes are given, and then analyze the synchronization performance of the whole system.

In order to build a reasonable and accurate network model, some assumptions about the network synchronization model of the Manned/Unmanned aerial vehicle networking system are made as follows:

1. In the Manned/Unmanned aerial vehicle networking system, there are 3 message transmission paths, and each message transmission path is abstracted into a node of the Manned/Unmanned aerial vehicle networking system.

2. In order to clear the difference between the data processing speed and the transmission speed and the influence of the uncertain disturbance factors in the system, here we introduce the definition of time-varying delay to establish the network model of Manned/Unmanned aerial vehicle networking system.

3. Assume that the change of the message flow of the 3 nodes in the Manned/Unmanned aerial vehicle networking system is satisfied with the chaotic system Rossler, and then the equation for the i th node in the network system is as follows:

$$\begin{cases} \dot{x}_{i1} = -(x_{i2} + x_{i3}) \\ \dot{x}_{i2} = x_{i1} + ax_{i2} \\ \dot{x}_{i3} = x_{i3}(x_{i1} - c) + b \end{cases} \tag{1}$$

Based on the above assumptions, a dynamic equation of the continuous-time time-varying delay model composed of 3 identical nodes is given:

$$\dot{x}_i = f(x_i(t)) + \hat{\varepsilon} \sum_{j=1}^{N} \hat{a}_{ij} \hat{\Gamma} x_j(t - \tau(t)), \quad i = 1, 2, 3 \tag{2}$$

Where: vector $x_i = (x_{i1}, x_{i2}, ..., x_{in})^T \in R^n$,

x_i is the state variable of node i; $\dot{x}_i$ represents the message volume change rate of the node i; constant $\varepsilon > 0$, constant $\hat{\varepsilon} > 0$ is the internal coupling strength of the Manned/Unmanned aerial vehicle networking system;

$$\Gamma = (\zeta_{ij})_{n \times n} \in R^{n \times n}, \quad \hat{\Gamma} = (\hat{\zeta}_{ij})_{n \times n} \in R^{n \times n}$$

are the internal coupling matrices between the state vectors of each node. The definition of ζ_{ij} and $\hat{\zeta}_{ij}$ is as follows: if there is a connecting side between the node i and the node j ($i \neq j$), m is the number of the same subsystems that are included between the node i and the node j,

Then $\zeta_{ij} = m$, $\hat{\zeta}_{ij} = m$.

621

When there is no connecting side between the node i and the node j $(i \neq j)$, $\zeta_{ij} = \zeta_{ji} = 0, \hat{\zeta}_{ij} = \hat{\zeta}_{ji} = 0$.

In order to meet the dissipative coupling condition

$$\sum_j \zeta_{ij} = 0, \sum_j \hat{\zeta}_{ij} = 0,$$

it is assumed that the diagonal elements of the internal coupling matrix Γ, $\hat{\Gamma}$ are

$$\zeta_{ii} = -\sum_{\substack{j=1 \\ j \neq i}}^{N} \zeta_{ij} = -\sum_{\substack{j=1 \\ j \neq i}}^{N} \zeta_{ji}, i = 1, 2, \cdots, N, \hat{\zeta}_{ii} = -\sum_{\substack{j=1 \\ j \neq i}}^{N} \hat{\zeta}_{ij} = -\sum_{\substack{j=1 \\ j \neq i}}^{N} \hat{\zeta}_{ji}, i = 1, 2, \cdots, N;$$

$\tau(t)$, the time varying function of t, is the delay time of the system;

The internal coupling matrix of the three message transmission paths is as follows:

$$\hat{\Gamma} = \Gamma = \begin{bmatrix} \zeta_{11} & \zeta_{12} & \zeta_{13} \\ \zeta_{21} & \zeta_{22} & \zeta_{23} \\ \zeta_{31} & \zeta_{32} & \zeta_{33} \end{bmatrix} = \begin{bmatrix} -6 & 3 & 3 \\ 3 & -5 & 2 \\ 3 & 2 & -5 \end{bmatrix}$$

In order not to lose the generality, make $\tau(t) = 0.5 - e^{-t}$ here. ε is the internal coupling strength coefficient, and here make ε equal to 0.2. Here gives the waveform of $e_1(t)$ and the 2-norm waveforms of $e_1(t), e_2(t), e_3(t)$, as shown in Figs. 5 and 6. The waveforms of $e_2(t)$ and $e_3(t)$ are similar to $e_1(t)$.

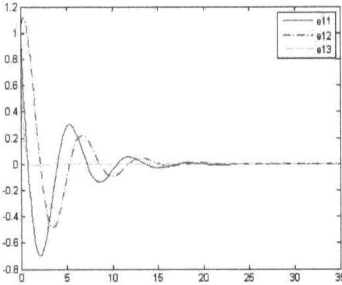

Figure 5. Waveform of $e_1(t)$

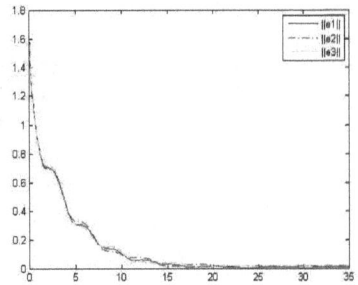

Figure 6. Waveform of $\|e_1\|, \|e_2\|, \|e_3\|$

From Fig. 5, the oscillation amplitude of e_1, e_2, e_3 gradually become smaller and tend to be stable with the increase of time. From Fig. 6, $\|e_1\|, \|e_2\|$, and $\|e_3\|$ gradually decrease with the increase of time and finally tend to zero, which indicates that the system tends to synchronize finally. At this time t is about 25.

6. Conclusion

This paper studies the dynamic characteristics of the network system, focuses on the robustness and the synchronism of the system, and establishes the random failure model, the intentional attack model, and the synchronization model respectively.

Acknowledgement

The work was funded by the Science and Technology on Avionics Integration Laboratory, and Aerospace Science Foundation of China under Grant 20135553035. Our sincere gratitudes go to all the staff of the laboratory and the Foundation.

References

1. Ding-ding Han. Topological and dynamical behavior of complex networks and its Empirical Study. Doctoral Dissertation of East China Normal University, 2008.
2. Hai-mei Qian, Ding-ding Han, Yu-gang Ma. Dynamic evolution of open complex aeronautical network system [J]. Journal of Physics, 2011, 60(9): 098901-1-098901-6.
3. Daniel Aguilar-Hidalgol, Antonio C, Rdoba Zurita, Ma Carmen Lemos Fern, Ndez. Complex networks evolutionary dynamics using genetic algorithms [J]. SCI, EI, International Journal of Bifucation and Chaos. 22(7): 1250156-1-12. 2012.

Research on Simulation Modeling and Evaluation Technology of Airborne Embedded Training System

Yong Wu[1,*], Bo Li[2], Xiang-bo Wang[2], Feng-xia Zhang[2] and Zhao-jun Yan[1]
1Northwestern Polytechnical University,
Luoyang, China
2Luoyang Institute of Electro-Optical Devices,
Luoyang, China
*jdzhang@nwpu.edu.cn

Aiming at the typical structure of military aircraft avionics systems, this paper presents a design of the airborne embedded training system. In this paper, the design requirements of embedded training system are discussed, which lay the groundwork for the improvement of the airborne embedded training system design plan. The overall architecture of airborne embedded training system is designed in details. The Arena simulation software is used to simulate the integration process between airborne embedded training system and other avionics systems. The evaluation of quota system of training is formulated. VC++6.0 is used to make simulation for performance evaluation process.

Keywords: Embedded training; Training evaluation; Simulation modeling; Avionics system.

1. Introduction

The Embedded Training (ET) of modern fighter/attacker offered training function, which could be embedded into the combat system, subsystem and equipment of a fighter. The function is used for maintaining and enhancing the pilot trainees' fighter skill under a certain given situation. Embedded training not only offers the trainees psychological and physical suitability that suits the actual combat, but enlarges training effectiveness. Embedded Training System (ETS) is the major part of the whole training system. It could be configured with weapons at the same time, and be used in routine military operations and training during peacetime [1].

2. Airborne Embedded Training System

Embedded Training System (ETS) define as an Operating System can be embedded into military system [2]. It could generate virtual combat environment by the computer on the aircraft, which makes trainer interact with pre-planned situations. Specially, for fighter, ETS generates virtual objects, which could interact with pilots. These virtual objects also make pilots immerse in vivid training system. Therefore, a pilot can utilize his own ship's all functions to participate in virtual combat including air threat and surface threat.

3. Design requirements of ETS

The total goal of ETS including achieving pilots who carry on flying cooperate with airborne avionics though a virtual external world by computer, cursing and fighting training courses. Besides, the purpose of ET contains improving fidelity, enhancing efficiency, decreasing cost of pilot training, also providing data store, download, replay during training time for estimation [3 - 5].

1) Administrators' requirements

   The main work of administrators including routine maintenance and system upgrade of Airborne Embedding Training System that makes system work safely and reliably. So, the Administrators' requirements are as follows: Install system conveniently and efficiently.

2) Users' requirements

   Pilots are the users of airborne ETS. They operate training system to complete established targets. Here are the users' requirements: The system should run stably, crash and stutter won't happen and it's easy to operate without any fuzzy operation instruction [6, 7].

3) Decision-makers' requirements

   Decision-makers are the upper-level managers of system. They are responsible to set schemes for training system and evaluate training effect after training. Decision-makers' requirements are: Have the ability to set fighter's training schemes and the setup is simple and manageable.

4. Overview of Embedded Training System

4.1. Basic Architecture of Embedded Training System [8]

The functions of ETS could be partitioned as followings:
1) Virtual target/threat simulation;
2) Virtual sensor simulation;
3) Virtual weapon simulation;
4) ET data link;
5) Ground station support.

1), 2), and 3) mean the function of making virtual model by the computer simulation, and 4) means the information sharing for ET environment by real equipment and 5) means the planning, debriefing, and reviewing for the flight training with the ETS. Above all, the fundamental of ETS is making of virtual target and threat.

4.2. *The Design of Embedded Training System Simulation*

The simulation classification of ETS is shown in Table 1.

Table 1. The function range of ETS

ET Function	Detail Function
Virtual Target Generation	The generation of virtual air targets and virtual ground targets.
Sensor Simulation	The simulation of Air-to-Air Fire Control Radar, Air-to-Ground Fire Control Radar and Identification Friend or Foe System.
Virtual Weapon Simulation	The simulation of Within Visual Range Missile, Beyond Visual Range Missile, Air-to-Air Gunnery, General Bomb, Air-to-Ground Gunnery, Electro-Optical Missile.

The ETS simulates the behaviors of virtual hostile aircrafts, virtual bomb targets, and virtual SAM and so on in a battle, which is seen as a basic element of ETS environment. Those behaviors can be pre-written or generated immediately by artificial intelligence. All information of virtual battlefield is generated by a scenario editor.

The Virtual Weapon also needs to be managed as a module. All weapons have their own features, such as mass, aerodynamic drag, fuze type, damage area, automatic flight control logic, and so on. These features are grouped by the existence of guidance logic. Non-guided weapon-general purpose bomb/ gunnery/rocket has the object parameters as mass, drag equation, fuze type or thrust profile. ETS can simulate trajectory of non-guided weapon based on the parameters and provide the reference of scoring criteria that is the miss-distance between the point of impact and the ground target. In the case of air-to-air gunnery, the scoring result shows hit or miss the target based on hit judgment circle at the center of the target image.

5.   The Simulation of ETS

5.1. *The Choice of Simulation Software*

The data delivery and process of ETS can be simplified as a process of a lot of successive modules. From the software users can see the influence by the change of supply chain, industry manufacture, processing project, logistic, distribution, storage, and so on. The result of simulation is presented in a way of dynamic animation. Finally, the software analyzes the result of simulation.

5.2. Integration Simulation of ETS

First, the simulation of the system's information processing needs the integration of ETS Fire Control Radar module, Rader Warning Receiver module, and Counter Measure Dispenser System module, and the separation of ET state and Non-ET state. Then, the model of system logic is built as shown in Figure 1.

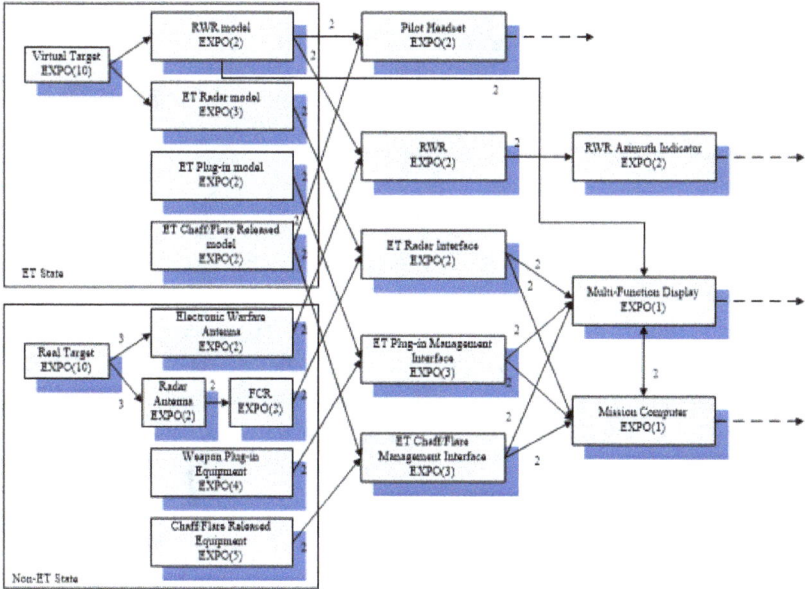

Figure 1. The logic of system simulation

In Arena, EXPO(β) is

$$f(x) = \begin{cases} \dfrac{1}{\beta} e^{-x/\beta} & x > 0 \\ 0 & others \end{cases}$$

β ---- time of information processing.

Then, with the simulation logic model and Arena software, the simulation system is built as shown in Figure 2.

627

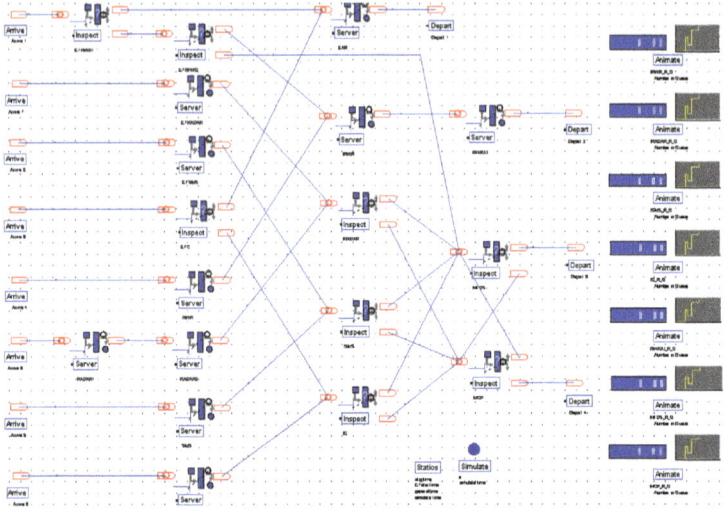

Figure 2. ET Arena Simulation System

In the first 200 seconds, it is represented that the real equipment of the aircraft is on working, and the ETS is in the state of waiting. When the Service module and the Inspect module are blue, it is represented that the related equipment is spare. When the Service module and the Inspect module are red, it is represented that the related equipment is busy. After the first 200 seconds, the system of ETS in ET state simulation starts. In this case, the real equipment doesn't produce data any more. The simulation result is shown in Figure 3.

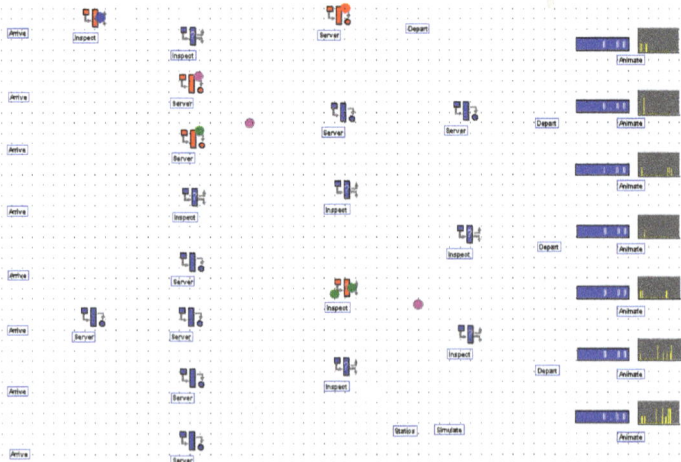

Figure 3. ETS simulation result

628

6. Effect Evaluation of the Airborne Embedded Training System

6.1. The Criteria System for Training Evaluation

The establishment of the Criteria of training evaluation is based on the subjects involved in flight operation training mainly. The whole training process could be divided into five stages. They are take-off slide, climb, flying, level flight, descent and landing. The stage of flying is divided into the embedded training stage and the non-embedded training stage mainly. The score of each item of the evaluation index system and the total score of the embedded training stage should be listed in detail in the assessment report as a part of the result of the whole training evaluation. According to the principles of making the evaluation index, and combined with the actual training process, the evaluation criteria system of the embedded training stage of fighter plane are shown in table 2.

Table 2. The evaluation index system for the embedded training stage

Sub-index	Child Pointer
Aircraft search maneuver	Height, airspeed, heading angle, pitch angle, roll angle
Aircraft combat maneuver	Height, airspeed, lifting speed, heading angle, pitch angle, roll angle
Tactical decision	Target recognition time, target recognition decision, fire distribution scheme, attack position, attack time, damage degree of the aircraft
The use of weapons	Attack mode, the set of firing elements, firing time, the number of weapons consumed, target damage degree
The use of sensors	using method, the number of lost target, the distance between target and the aircraft, the number of times to be locked

6.2. The Weight Coefficient and Scoring Method of Evaluation Criteria

Because the degree of importance of each criterion is different in the flight training, the score of each index in the proportion of the total score is not the same. This requires a reasonable distribution of the weight coefficient for each evaluation criterion to accurately estimate training results.

6.3. Simulation with Software

In VC ++ 6.0 development platform, program simulation with the aircraft stick of Saitek X52 and the emulator for throttle lever, then set up a training evaluation software to simulate the training process, and display the performance criteria of the aircraft and targets. The simulation interface is shown as Figure 4.

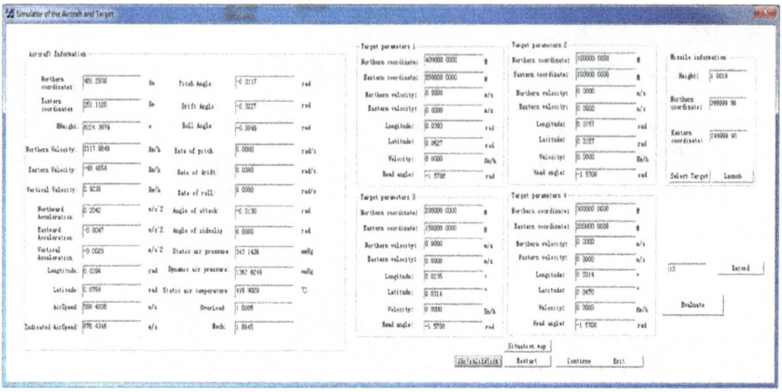

Figure 4. The simulation interface

7. Conclusion

This paper proposes the overall architecture of the airborne embedded training system [9, 10] and the design schemes of the related functions in the system. This paper also makes simulation and analysis for the integrated situations of the airborne embedded training system and the airborne equipment. In this paper, we propose to evaluate the criteria of training results, and then design a kind of simulation software based on these evaluating criteria to simulate the process of the training evaluation.

Acknowledgment

The work was funded by the Science and Technology on Electro-optic Control Laboratory, and Aerospace Science Foundation of China under Grant 20135153031. The gratitude goes to all the staff of the laboratory and the Foundation.

References

1. Kai Yuan, Rennong Yang, Jialiang Zuo and so on. Research on the embedded training system of air combat aircraft [J]. Fire Control & Command Control, 2011, 36(9): 165-167.
2. Tianqing Chang, Bo Zhang, Qinzhao Wang. Application and development of embedded training technology [J]. Journal of The Academy of Armored Forces Engineering, 2008 22(6): 1-6.
3. Liang Chen, Qin Yang, Xiao Cao. The Enlightenment of overseas simulation training to the development of our military navigation simulator [J]. Ship Electronic Engineering, 2011, 31(5): 12-14.

4. Jiye Qiu. The application of embedded training system on the Training plane [J]. Training plane, 2012 (2): 49-50.
5. Wedzinga G. E-CATS: First time demonstration of embedded training in a combat aircraft [J]. Aerospace science and technology, 2006, 10(1): 73-84.
6. Yoon K S, Yang S W, Song C I. New architecture for improving performance in embedded training system using embedded virtual avionics [C]. Digital Avionics Systems Conference, 2009. DASC'09. IEEE/AIAA 28th. IEEE, 2009: 6. C. 3-1-6. C. 3-8.
7. Yoon K S, Kim I G, Ryu K S. An efficient embedded system architecture for pilot training [C]. Digital Avionics Systems Conference, IEEE/AIAA. IEEE, 2011.
8. Joe H, Shin D, Jo J, et al. Air-to-air and air-to-ground engagement modeling for the KAI embedded training system [C]. Proc. of the AIAA Modeling and Simulation Technologies Conference and Exhibit. 2008.
9. Bremer L, Schreutelkamp E. Application of EuroSim in the F-35 Joint Strike Fighter Embedded Training solution [J].
10. Bills C G, Kern S, Flachsbart B, et al. F-35 Embedded Training [R]. Lookheed martin corp akron oh, 2009.

Simulation of TCM Pulse Wave at *Cunkou* Area Based on One-Dimensional Model of Left Upper Extremity

Jian-jun Yan[1, †], Min-sheng Liu[1], Rui Guo[2, 3, *], Hai-xia Yan[2] and Yi-qin Wang[2]

[1]Laboratory of Mechanical and Electronic Engineering,
East China University of Science and Technology,
Shanghai 200237, China
[2]Comprehensive Laboratory of Four Diagnostic,
Shanghai University of Traditional Chinese Medicine,
Shanghai 201203, China
[3]Science and Technology Information Center,
Shanghai University of Traditional Chinese Medicine,
Shanghai 201203, China
jjyan@ecust.edu.cn[†];
guoruier@sina.com[*]

In order to investigate the mechanism of variation in pulse wave during the pulse diagnosis in cunkou area, a one-dimensional model of left upper extremity is created to simulate the pulse wave under different conditions of finger distribution during pulse taking. The results show that the reflected wave caused by variation in lumen area would increase the pressure in proximal region while decline the pressure in the distal region. With the increasing force, the reflected wave makes the waveforms in three regions; with each associated with different organs in our body. Simulation results had confirmed that the pressure waveform is similar to the lumen cross-section area curves. When the blood flow is basically blocked, findings show that the sharp changes in the blood flow rate that determines the level of blood pressure, and the area curves at blocking site present a negative wave. Meanwhile, the ascending branch of the pulse wave in the distal region spends more time and the curve gets smoother, which had been called pulse-taking piezoresistive effect. This paper promotes clearing the similarities and differences between pulse wave in three regions and the engineering analysis research about traditional Chinese medicine.

Keywords: Model of left upper extremity; Pulse wave in *cunkou* area; Simulation of pulse wave; Pulse-taking piezoresistive effect.

1. Introduction

As one of the important diagnostic methods in traditional Chinese medicine (TCM), doctors analyze physiological state of the human body by the finger feelings of pulsation at the radical artery, which is defined as pulse wave in pulse

---

[*, †]Corresponding authors.

diagnosis. The *cunkou* area is located at the medial aspect of styloid process of the radial artery, and each side of which is classified into three regions: the proximal region named *chi*, the middle region named *guan*, the distal region named *cun*. The physicians are supposed to locate the *guan*-region with their middle fingers first, and then to locate the *cun*-region with their index fingers and *chi*-region with their ring fingers.

In order to obtain the optimal pulse wave for a comprehensive analysis, which means to achieve the most satisfactory finger sensation, physicians need to adjust their fingers on three regions of *cunkou* area as constantly as to analyze the physiological and pathological conditions of the human body in TCM pulse diagnosis. This method of pulse taking requires a long-term practice, and there would be great difference in the diagnosis accuracy of the same object along with different physicians. With the objective study of TCM pulse diagnosis, it is important to find the distribution law of pulse taking means and the corresponding mechanism of changes in pulse wave during the pulse diagnosis in *cunkou* area.

2. Theory and Method

2.1. Arteries of Left Upper Extremity and the Cunkou Area

From the anatomical structure consideration, the blood flow in radial artery comes from the branch of aortic arch so that the pulse wave at radial artery can sensitively reflect the changes of cardiac function [1] and other organs [2]. Arteries of left upper extremity including the left subclavian artery, axillary artery and left arm brachial artery, deep brachial artery, ulnar artery, radial artery and so on.

The *cunkou* area is the distal segment of radial artery, which is close to the surface and easy to be detected. The physicians are supposed to locate the *guan*-region first in the medial aspect of styloid process of the radial artery, the upstream is *chi*-region and the downstream is *cun*-region as showed in Fig. 1. Experiment studies have shown that the length of three regions is 2.77±0. 63(cm) and the diameter of *cunkou* area is consistent on the same side and different from the other side.

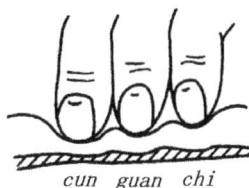

cun　guan　chi

Figure 1. Method of pulse taking in cunkou area

During the pulse diagnosis in *cunkou* area, the physicians search the most obvious characteristic of pulsation from gentle to heavy force, or from heavy to light force with individual finger palpation or simultaneous palpation with 3 fingers. With optimal pulse graph in *cunkou* area obtained by different pulse-taking methods, the physicians analyze the multidimensional information of physiological state of people. The pulse wave is affected not only by patient-specific physiological differences, but also by the distortion of local vascular as the result of the different pulse taking method, which had been called the pulse-taking piezoresistive effect [3].

2.2. Changes of Pulse Wave Caused by Vascular Distortion

Based on Riemann problem [4], the governing system of pulse wave in the vessel can be interpreted in terms of a forward, and a backward travelling waves. With the pulse diagnosis in *cunkou* area, there would be a vascular distortion and each interface separates two constant states, (A_L^u, U_L^u) and (A_R^u, U_R^u), at arterial junctions. The remaining equations follow from continuity of the total pressure at the interface,

$$P(A_L^u) + \rho(U_L^u)^2 / 2 = P(A_R^u) + \rho(U_R^u)^2 / 2. \tag{1}$$

with $P(A)$ expressed through the pressure-area relationship,

$$P = \beta(\sqrt{A} - \sqrt{A_0}) / A_0, \beta = 4\sqrt{\pi}hE / 3. \tag{2}$$

Here, we consider a simple compliant tube, h denote the vessel thickness and E is the Young modulus, A_0 denote the sectional area at the reference state (P, U)=(0,0).

As a framework using a discontinuous Galerkin projection, *PulseWaveSolver* is offered in Nektar++ to simulate the propagation and reflection of arterial pulse wave in 1D geometry. In this paper, we created a one-dimensional model of left upper extremity to get the pulse wave under different conditions of finger distribution during pulse taking and study the mechanism of changes in pulse wave during the pulse diagnosis in *cunkou* area.

There are many factors interacting during the formation of pulse wave in *cunkou* area, such as the thickness, physiological variation of the radial artery and the distance between the deep artery and bone surface. Therefore, the physicians have got a multitude of names pulse wave by the pulse diagnosis in *cunkou* area. As known that the anatomical locations of three regions is discriminate, so that the force acting on the three regions of *cunkou* area would be different for obtaining the optimal pulse-taking pressure. A corresponding change in the lumen cross-section of radial artery in different regions would occur as the result. In this paper, we assume that the pulse-taking process result in a different vessel diameter of the three regions of *cunkou* area, and observe the changes of pulse wave in the three regions with different lumen diameters and same compliance.

3. Simulation and Analysis

3.1. Simulation of the One-Dimensional Model of Left Upper Extremity

According to the anatomy structure and physiological parameters of left upper extremity arteries, which is shown in Table 1, we establish the left upper extremity arteries branching model and set geometrical parameters of each segments in the input file. The input data is provided by an equivalent electrical network model.

Table 1. Model parameters of left arm artery segments

Arterial segment	Length [cm]	Area [cm²]	Arterial segment	Length [cm]	Area [cm²]
L. Subclavian	45.6	0.510	L. Radial	23.5	0.106
L. Ulnar I	6.7	0.145	L. Interosseous	7.9	0.031
L. Ulnar II	17.1	0.133			

In this paper, we convert the wave at subclavian artery based on pressure from an equivalent electrical network model to area according to Eq. 2. Then fit the wave about area data as a function of time varying by Fourier transform as the input of one-dimensional model of the left upper extremity. Pressure and flow velocity waveform at radial artery have been obtained as shown in Fig. 2.

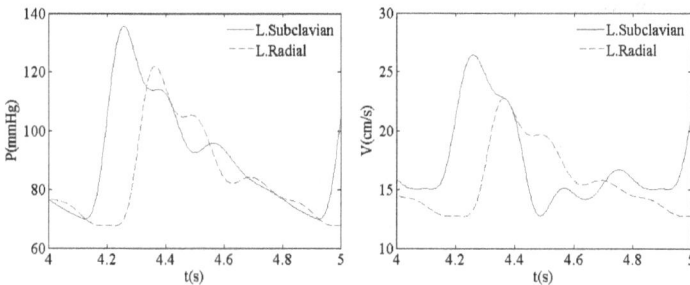

Figure 2. Pulse wave simulation of the arterial segments

Studies have shown that, compared with the direct method, cuff and other indirect measurement of arterial blood pressure would underestimate the systolic

635

pressure and overestimate diastolic pressure, besides, compared with the pressure fluctuation in aorta, the peripheral arteries perform a higher systolic and a lower diastolic blood pressure [5, 6]. Owing to the resistance of the blood vessels, the pressure gradually decreases with the blood flow along the arteries to periphery. Since the reduction in blood pressure is proportional to the resistance of the arteries to blood flow, the radial artery, as a branch of subclavian artery, perform a lower systolic blood pressure.

With a similar anatomical structure, saphenous artery is the branch of the femoral artery, it has been proved the simulation results consistent with its actual oscillation flow rate curve [2]. With increasing distance from the root of the aorta, the oscillation of flow rate becomes smaller. It is consistent with the trend that the cardiac ejection evolves into the steady flow in small vessels in the role of contraction and expansion of the elastic arteries. The reason for this phenomenon is that, with the blood flow along the arteries to periphery, resistance gradually increases and the influence of the reflected wave is more significant.

3.2. Simulation of the Pulse Wave in Cunkou Area

Force is applied when taking a pulse diagnose in *cunkou* area, so the cross-sectional area of the site being pressurized would be smaller, that is produced stenosis, and the reflected wave would superimpose on the pulse wave propagating along the arterial tree.

Since the end of radial artery in the model of left upper extremity, we consider the vessel segments length of 1 cm upward along the artery as *cun*-region, *guan*-region, and *chi*-region respectively. With the pulse-taking force change from light to heavy, the lumen cross-sectional area of each region is supposed to be 0.08, 0.06, 0.04 cm² respectively. In order to investigate the influence of variation in lumen area, we set the terminal reflection coefficient as zero. Fig. 3 compares pressure waveforms at three regions along the *cunkou* area under different force (A denotes light, B denotes gentle, C denotes heavy) with individual finger palpation.

It can be clearly seen from simulation results that the systolic pressure increases while the diastolic pressure not changing much. With the force increasing from light to heavy, the waveform features of pulse wave change significantly and the ascending steepen. In the case of light and gentle force, the waveforms keep the shape in the *chi*-region. Appling the heavy force, the dicrotic notch descended obviously and the dicrotic wave became broad in the influence of the reflected wave superposition. Closed to the end of radial artery, the effect of reflected wave in *guan*-region is more prominent. In this paper, we have considered the end of radial artery segment of 1 cm as *cun*-region, so that the reflected wave caused from variation in lumen area acts directly on the main wave and dicrotic anterior wave and performs different characters from the pulse wave in the other regions.

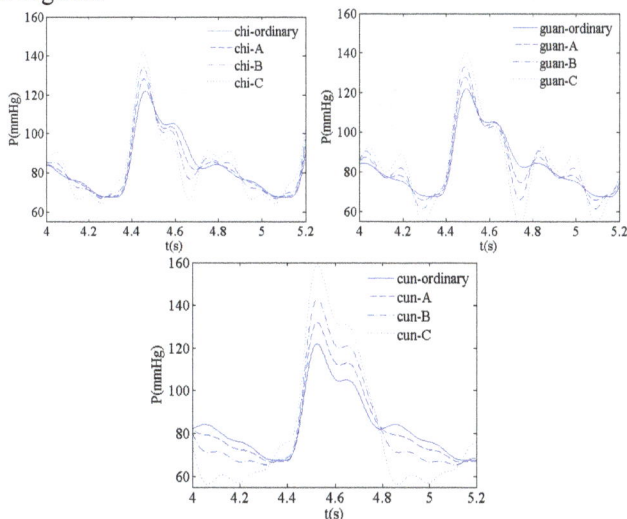

Figure 3. Pressure waveforms at three regions under different force with individual finger palpation

3.3. *Simulation of the Pulse Wave in Three Regions Under Protruding Probe*

Experiments of pulse acquisition in *cunkou* area have proposed that, when the blood flow in radial artery were blocked by the protruding probe, the other component of pulse wave, apart from the pressure wave, makes the pulse waves in three regions exhibits different features. In this paper, we consider to set the lumen area of *guan*-region as 0.01 cm^2, simulating the stenosis caused by protruding probe. Comparison in waveforms of three regions under protruding probe showed in Fig. 4.

With significantly reduce of the lumen area of the *guan*-region, the blood flow in radial artery nearly blocking, the reflected wave caused by the variation take an effect on the pulse wave. As a result, the main wave and dicrotic anterior wave fused to form a broad main wave, the dicrotic notch descended and dicrotic wave increased, which is in accordance with the change trend of the features obtained from pulse acquisition [3]. Furthermore, with the pulse wave propagating to *cun*-region, the systolic pressure decreased while the diastolic pressure increased, and the amplitude of main wave reduced significantly and the ascending branch of the pulse wave spent more time. The conclusions we get from the simulation conform to experiments of pulse acquisition, which have been described, as the pulse wave in the distal region would be smoother.

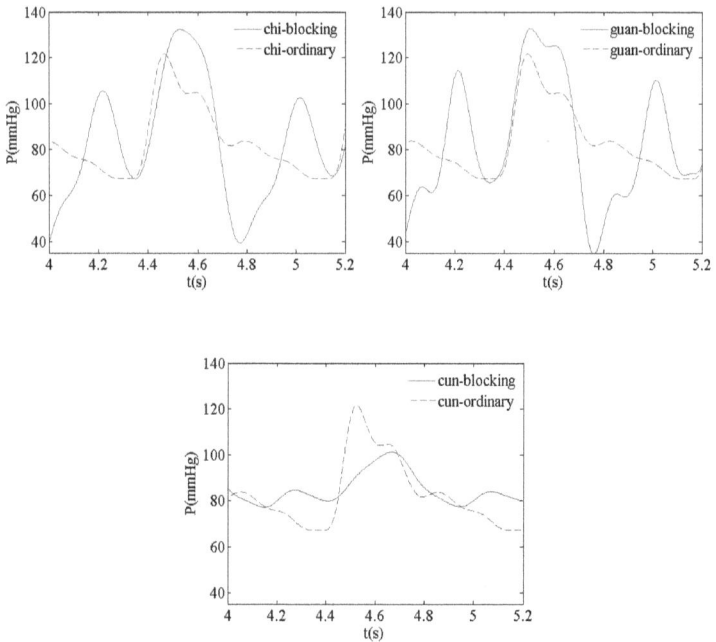

Figure 4. Waveforms in three regions under protruding probe compared with normal condition

Described as Eq. 1, the total pressure (PT) in the vessels contents the pressure component (PA) caused by changes in the lumen area and the flow component (PU) produced by the change of blood flow velocity, as illustrated in Fig. 5.

In the role of external forces, the lumen area of radial artery decreased in *guan*-region. As a result, the flow rate increased and the role of flow component enhanced. In addition, the pulse wave component caused by changes in the lumen area at blocking site present a negative wave. Compared the pulse wave in three regions, it is apparent that the pulse wave is mainly affected by the fluctuation of vessel wall, in other words, the pressure waveform is similar to the lumen cross-section area curves. It can be promoted that, when a protruding transducer probe used to detect the vessel diameter curves over time in three regions, the pulse wave detected in *guan*-region and *cun*-region would be weak or even disappear.

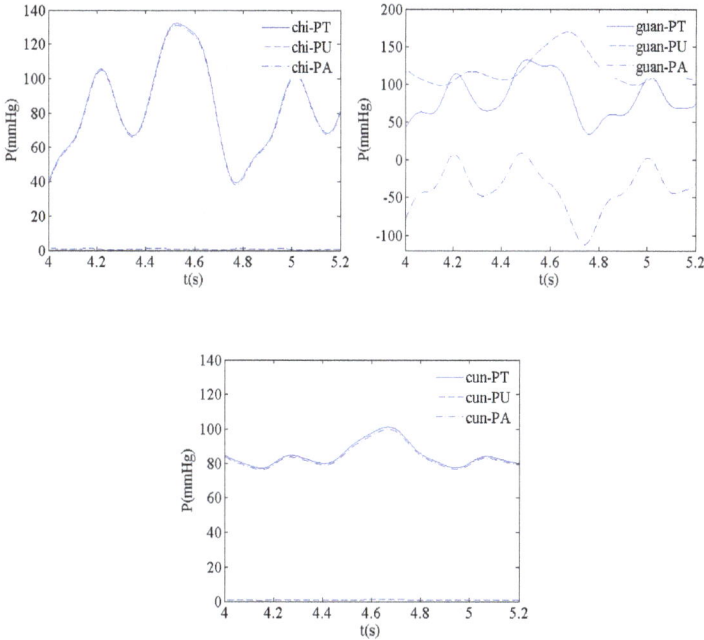

Figure 5. Decomposition of pulse wave in three regions

4. Conclusion

The *cunkou* area of the radial artery is close to the surface and easy to be detected. From the perspective of information extraction, the TCM physicians applying different forces from different directions on the radial artery, that is an effective method to take pulse waves and get detailed features. In this paper, we established a one-dimensional model of left upper extremity to simulate the pulse wave under different conditions of finger distribution during pulse taking.

The results show that the pressure increase with the increase of force in the region. The reflected wave caused by variation in lumen area would increase the pressure in proximal region while decline the pressure in the distal region. With the increasing force, the dicrotic notch descend and the dicrotic wave become broad gradually.

Furthermore, when the blood flow is basically blocked, although the pressure waveform is similar to the lumen cross-section area curves changed over time, the level of blood pressure in the region mainly depends on the dramatic changes in the flow rate and it would present the pulse-taking piezoresistive effect if a protruding transducer probe had been used to detect the variation of vessel diameter.

Acknowledgement

This research was funded by the National Science Foundation of China (No. 81302913 and No. 81473594).

References

1. Gault JH, Ross JJ, Mason DT, Patterns of brachial arterial blood flow in conscious human subjects with and without cardiac dysfunction, Circulation, 1966, 34(5): 833-848.
2. Nichols WW, O'Rouke MF, Vlachopoulos C. McDonald's blood flow in arteries: theoretical, experimental and clinical principles 6th ed., CRC Press, 2010.
3. Chen Xian-Nong, Luo Zhi-Cheng. Palmar arches by-pass effect and pulse-feeling piezoresistive effect in traditional Chinese medicine sphygmology. Chinese Journal of Biomedical Engineering, 1984, 3(2): 95-100. (In Chinese)
4. Alastruey AJ. Numerical modelling of pulse wave propagation in the cardiovascular system: development, validation and clinical applications. London: University of London, 2006.
5. Sherwin SJ, Formaggia L, Peiro J, et al. Computational Modeling of 1 D Blood Flow with Variable Mechanical Properties and Application to the Simulation of Wave Propagation in the Human Arterial System. International Journal for Numerical Methods in Fluids, 2003, 43(6-7): 673-700.
6. Wang Ya-li, Chen Xin-yi. The relation between central aortic pressure and peripheral arterial pressure and the effects of drugs on it. Advances in Cardiovascular Diseases, 2002, 23(6): 376-379. (In Chinese)

Modelling the Quasi-Zero Stiffness Property of Shape Memory Alloy Spring for Ultra-Low Frequency Vibration Isolation

Ting Wu and Lin-xiang Wang*

The Institute of Design Engineering and Automation, Zhejiang University,

310027, Hangzhou, China

*wanglx236@zju.edu.cn

In the current paper, the Quasi-zero stiffness effect caused by the pseudo elastic effect of Shape Memory Alloy (SMA) spring in vibration isolator is modeled and analyzed. The recovery force of the SMA spring is modeled as a non-convex one. A macroscopic differential model is employed to describe the hysteretic dynamics and pseudo elasticity of the SMA spring. It is illustrated that the resonant frequency of the proposed isolation system is close to zero. Numerical simulations are presented in studying the influence of the excitation frequencies on vibration isolation performance. It shows that a quasi-zero dynamic stiffness is achieved; the feasibility of the proposed system for ultra-low frequency excitation isolation is validated.

Keywords: Vibration isolation, Ultra-low frequency, Shape memory alloys spring, Pseudo elasticity.

1. Introduction

With the development of modern industry, the requirement for vibration isolation of precise instruments, and important apparatus has been more and more rigid and demanding [1]. It is clear that, to enhance the vibration isolation performance, the natural frequency of the vibration isolation system should be made much lower than the frequency of the vibration excitation. Therefore, the stiffness of the isolation system should be small. Meanwhile, the isolation system need also be capable of supporting the weight of the system to be isolated in static state, which means the spring stiffness has to be strong enough to support the mass [2]. This contradiction thus makes the design of low frequency vibration isolators very challenging. Particularly, when the excitation frequency is very low, the dynamic stiffness of the isolator spring should be as low as possible in order to reduce its natural frequency to an ultra-low region.

---

* Corresponding author.

This fact makes the use of linear springs completely unfeasible for ultra-low frequency vibration isolation, since the ultra-low stiffness of the spring demanded by low natural frequency will cause unacceptably large static deflection.

Generally, a vibration isolation system can be schematically sketch is shown Fig. 1, and the corresponding mathematical model for its dynamics can be formulated as follows:

$$m\ddot{x} + c\dot{x} + kx = F\sin(\omega \times t) \tag{1}$$

Where m, c, k, F, are mass, damping coefficient, stiffness of spring, amplitude and frequency of excitation force respectively. The resonant frequency of the system can be easily calculated as, in order to obtain a good isolation performance, one need to make much smaller than the input frequency, which is not practical when the excitation frequency itself is rather low, as mentioned above. Therefore, a new type of vibration isolation spring is demanded.

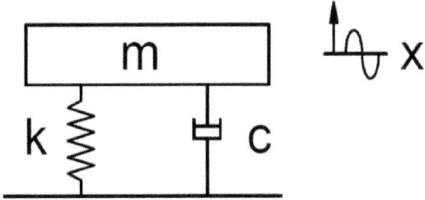

Fig. 1 Schematic sketch of conventional vibration isolation

2. The Concept of Zero-Stiffness Isolator

To fit this need, the spring used for vibration isolation should possess a High-static Stiffness whilst a Low-Dynamic Stiffness (HSLDS) so that it can support a large load statically while having low dynamic stiffness [3]. A HSLDS spring is a nonlinear spring, and it is referred as a quasi-zero-stiffness (QZS) isolator when it has zero or near zero dynamic stiffness. Many investigations have been performed on QZS spring and various innovative designs has been proposed for its implementation.

The concept of QZS isolator could be realized by including a negative stiffness mechanism to the positive stiffness structures such as common elastic element. Alabuzhev gave a detailed review of relevant design features, theory and design method about QZS [4]. Many investigations are dedicated to the inclined spring to offer the negative stiffness in order to counteract the positive stiffness. Carrella studied force-displacement characteristic and force transmissibility of a system with two oblique spring, and optimized the geometry parameter and the dynamic stiffness of such system [5-7]. Some improvements have been made to this kind of QZS structure. Yang further analyzed the nonlinearity and power flow behavior resulting from the existing mechanism [8]. Zhou designed a QSZ isolator using for horizontal spring by making a minor change in the contact type [9]. He considers a roller-cam-spring structure, which is also an alternative to the existing device with better performance than the linear counterpart.

Alternative structures have also been proposed to provide the offsetting negative stiffness to the system. Carrella designed a high-static-low-dynamic isolator comprised of two vertical mechanical springs; between which an isolated mass is mounted while outer edge of each mechanical spring is attached by a magnet [10]. A negative magnetic spring developed by Zheng is comprised of a pair of coaxial ring permanent magnets to counteract the positive stiffness caused by the normal mechanical spring [11].

Due to their internal nonlinearity and variable material properties, some smart materials are promising alternatives for constructing quasi-zero stiffness springs for ultra-low frequency vibration isolations. Among them, shape memory alloys might be the best candidate for vibration isolations and damping, due to the fact that, a QZS isolator with high-static-low-dynamic property can be easily implemented by take the advantage of the pseudo elastic effects of the SMA spring without any intricate mechanical design. The feasibility of applying shape memory alloys (SMA) isolator for reducing the on-orbit disturbance for the moment wheel assembly in space devices and has also been validated [12]. In the design of a QZS isolator, [13] a super elastic Cu-Al-Mn (SMA) bar is integrated into a tailored material design, which could convert a horizontal axial force into a vertical one.

3. Employment of SMA spring as isolating element

SMA materials has been widely employed in the application of vibration isolation, since it has a great potential in replacing the common mechanical spring as the supporting element in a vibration isolation system. SMA has two outstanding features, which are shape memory effects and pseudo elasticity. The latter is known for acquiring large recoverable strain during the process of loading-unloading. Therefore, the pseudo elasticity of SMA spring can be brilliantly utilized in the design of a quasi-zero stiffness isolator.

Since vibration isolation is an inherently dynamical process, the modeling and analysis of the isolator with SMA spring certainly should be a dynamic one. For the SMA spring, a differential model could be constructed to describe its dynamic behaviors under external excitation [14]

$$v\frac{dx}{dt} = F(\mathrm{t}) - (A_1 x + A_2 x^3 + A_3 x^5) \qquad (2)$$

where is the time constant in the current differential model, A_i are the parameters that incorporate temperature term and structural term of SMA spring. The load-deformation relation for the SMA spring at a fixed temperature can be derived as:

$$F = A_1 x + A_2 x^3 + A_3 x^5 \qquad (3)$$

It has been demonstrated that, the above differential model is capable of modeling the jump phenomenon associated with stress induced martensite transformation in SMA materials, and the martensite transformation will consequently cause hysteresis loops in the dynamics of the SMA spring. For the clarification of the discussion, a predicted jump and the resulting hystesis loop is illustrated in Fig. 2. The simulated one obtained from the proposed differential model is shown in Fig. 3.

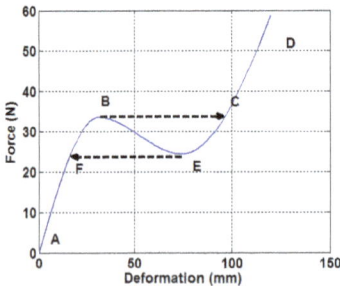

Fig. 2 Predicted jump phenomenon   Fig. 3 Simulated hysteresis of SMA spring

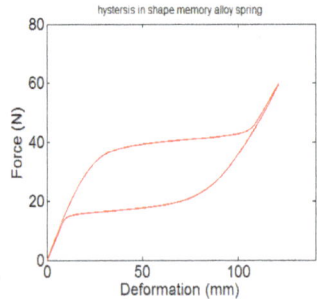

For the purpose of vibration isolation, a high static stiffness and a low dynamic stiffness is desirable, especially when both characteristics is combined together. The SMA spring can offer an optimal solution to the problem without

any intricate mechanical design. When SMA spring is settled at its initial state, it can have a considerable static stiffness to bear certain external load. When the system is subject to a constant vibration source, by suitable design, the SMA spring could be made to work in a region where martensite transformation could be induced. It is clear that when martensite transformation is induced by external stress, the deformation of the SMA would be rather large while the stress has a rather minor change, this "soft" property is exactly aquasi-zero dynamic stiffness, as sketched in Fig. 4. In other words, the desirable working area for the SMA spring is the plateau in the stress-strain relation.

4. Numerical Examples

In order to illustrate the quasi-zero stiffness feature of the SMA spring, a series numerical simulations have been performed. The one presented here is the frequency response of the isolation system under various external excitations. By scanning the excitation frequency, the natural frequency of the isolation system could be easily identified from the simulation results. Here the model parameter is chosen $m=1$kg, $c=61$N/(m/s), input frequency is in [0-50] HZ. Fig. 5 shows the result of the dynamic response corresponding to the frequency scan. It is apparent that with SMA spring used as HSLDS spring in the isolator, a near-zero resonant frequency is obtained.

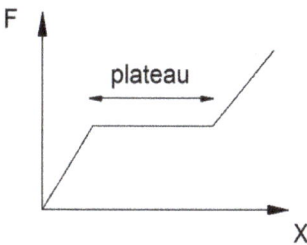

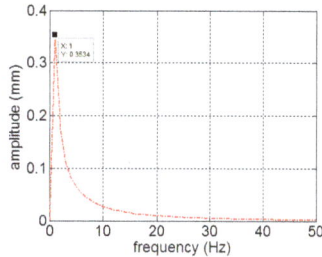

Fig. 4 Ideal working status for SMA spring   Fig. 5 Dynamic response of isolator with SMA spring

In the above simulation, the employed differential model for SMA gives a rather large static stiffness and a rather small dynamic one. It also incorporates the hysteresis loop naturally, which means it always has damping effect to vibrations.

5. Conclusion

In this paper, the SMA spring is proposed as a new implementation of quasi-zero stiffness for vibration isolation. The pseudo elasticity of the SMA spring has been successfully used to produce high static stiffness and low dynamic stiffness. Simulation result shows that near-zero stiffness is obtained in the proposed isolator.

Acknowledgement

This work was funded by the National Natural Science Foundation of China (Grant No. 51575478 and Grant No. 10872062).

References

1. Yuryev G S. Vibration Isolation of Precision Instruments [J]. Russian Academy of Sciences, Siberian Branch, Institute of Nuclear Physics Press, Novosibirsk, 1991: 89-146.
2. Harris C M, Piersol A G. Harris' shock and vibration handbook [M]. New York: McGraw-Hill, 2002.
3. Rivin E I, Rivin E I. Passive vibration isolation [M]. New York: Asme press, 2003.
4. Alabuzhev P M, Rivin E I. Vibration protection and measuring systems with quasi-zero stiffness [M]. CRC Press, 1989.
5. Carrella, A., et al., On the force transmissibility of a vibration isolator with quasi-zero-stiffness. Journal of Sound and Vibration, 2009. 322(4-5): p. 707-717.
6. Carrella, A., M.J. Brennan and T.P. Waters, Static analysis of a passive vibration isolator with quasi-zero-stiffness characteristic. Journal of Sound and Vibration, 2007. 301(3-5): p. 678-689.
7. Carrella, Optimization of a Quasi-Zero-Stiffness Isolator. 2007.
8. J. Yang, Y.P.X.N., Dynamics and power flow behavior of a nonlinear vibration isolation system with a negative stiffness mechanism. 2012.
9. Zhou, J., et al., Nonlinear dynamic characteristics of a quasi-zero stiffness vibration isolator with cam – roller – spring mechanisms. Journal of Sound and Vibration, 2015. 346: p. 53-69.
10. Carrella, A., et al., On the design of a high-static – low-dynamic stiffness isolator using linear mechanical springs and magnets. Journal of Sound and Vibration, 2008. 315(3): p. 712-720.

11. Zheng, Y., et al., Design and experiment of a high-static-low-dynamic stiffness isolator using a negative stiffness magnetic spring. Journal of Sound and Vibration, 2016. 360: p. 31-52.
12. Yiu, Y. C. Shape-memory alloy isolators for vibration suppression in space applications, 1995, Space Systems Division, 0173- 15
13. Araki, Y., et al., Integrated mechanical and material design of quasi-zero-stiffness vibration isolator with super elastic Cu‒Al‒Mn shape memory alloy bars. Journal of Sound and Vibration, 2015. 358: p. 74-83.
14. Wang L, Zhou C, Feng C. Nonlinear differential equation approach for the two-way shape memory effects of one-dimensional shape memory alloy structures[C]//Second International Conference on Smart Materials and Nanotechnology in Engineering. International Society for Optics and Photonics, 2009: 74931Z-74931Z-9.

A Class of Nonlinear Weakly Singular Integral Inequality and Its Applications

Wu-sheng Wang
School of Mathematics and Statistics, Hechi University,
Yizhou, Guangxi, 546300, P. R. China
Email: wang4896@126.com

In this paper, a class of nonlinear weakly singular integral inequality is investigated, which includes a non-constant term outside the integrals. The upper bounds of the embedded unknown functions are estimated explicitly by adopting novel analysis techniques, such as: the definitions and rules of conformable fractional differential and conformable fractional integration, the techniques of change of variable, and the method of amplification. The derived results can be applied in the study of qualitative properties of solutions of conformable fractional integral equations.

Keywords: Weakly singular integral inequalities; Conformable fractional integral; Conformable fractional differential; Analysis technique, Explicit bound.

1. Introduction

It is well known that integral inequalities are important tools to discuss the existence, uniqueness, boundedness, stability, oscillation and other qualitative properties of solution of differential equations and integral equations. One of the beat known and widely used inequalities in the study of nonlinear differential equations is Gronwall-Bellman inequality [1, 2], which can be stated as follows: If u and f are non-negative continuous function on an interval [a, b] satisfying

$$u(t) \le c + \int_a^t f(s)u(s)\,ds, \quad t \in [a,b],$$

(1)

For some constant $c \ge 0$, then

$$u(t) \le c \exp\left(\int_a^t f(s)\,ds\right).$$

(2)

In 2008, Ma and Pecaric [3] investigated weakly singular integral inequality

$$u^p(t) \le a(t) + b(t)\int_0^t \left(t^\alpha - s^\alpha\right)^{\beta-1} s^{\gamma-1} f(s) u^q(s)\,ds.$$

(3)

In 2011, Abdeldaim et al. [4] studied a new integral inequality of Gronwall-Bellman-Pachpatte type

$$u(t) \leq u_0 + \int_0^t g(s)u(s)\left[u(s) + \int_0^s h(\tau)\left[u(\tau) + \int_0^\tau \gamma(\xi)u(\xi)d\xi\right]d\tau\right]ds.$$

(4)

In this paper, on the basis of [3-8], we discuss weakly singular integral inequality

$$u(t) \leq e(t) + \int_a^t (s-a)^{\alpha-1} f(s)w_1(u(s))ds + \int_a^t (s-a)^{\alpha-1} g(s)w_1(u(s))\left[u(s) \right.$$
$$\left. + \int_a^s (\tau-a)^{\alpha-1} h(\tau)w_2(u(\tau))d\tau\right]ds.$$

(5)

In order to investigate the integral inequality (4), we shall state some basic notations and lemmas, which will be used in the proofs of our main results.

Definition 1 [6,7]. The (left) conformable fractional derivative starting from a of a function $f:[a,\infty) \to R_+$ of order $0 < \alpha \leq 1$ is defined by

$$T_\alpha^a(f)(t) = \lim_{\varepsilon \to 0} \frac{f\left(t + \varepsilon(t-a)^{1-\alpha}\right) - f(t)}{\varepsilon},$$

(6)

For *all* $t > 0$. If f is α-differentiable in some $[a,\infty)$, and $\lim_{t \to a^+} f^{(\alpha)}(t)$ exists, then define

$$f^{(\alpha)}(0) = \lim_{t \to a^+} f^{(\alpha)}(t).$$

(7)

Definition 2 [6,7]. The (left) conformable fractional integral starting from a of a function $f:[a,\infty) \to R_+$ of order $0 < \alpha \leq 1$ is defined by

$$I_\alpha^a(f)(t) = \int_a^t (s-a)^{\alpha-1} f(s)ds.$$

(8)

Lemma 1 [12, 13]. Let a, b, p, λ, α are real constants with $\alpha \in (0,1]$ and f, g be

$\alpha - differentiable$ at a point $t > 0$. Let $h(t) = f(g(t))$. Then

$$T_\alpha^a(\lambda) = 0, T_\alpha^a(t^p) = pt^{p-\alpha}, \ T_\alpha^a(fg) = fT_\alpha^a(g) + gT_\alpha^a(f),$$

(9)

$$T_\alpha^a(h)(t) = T_\alpha^a(f)(g(t))T_\alpha^a(g)(t)g^{\alpha-1}(t), T_\alpha^a(af+bg) = aT_\alpha^a(f) + bT_\alpha^a(g), \quad (10)$$

$$T_\alpha^a(af+bg) = aT_\alpha^a(f) + bT_\alpha^a(g), T_\alpha^a(I_\alpha^a(f))(t) = f(t), \quad (11)$$

$$T_\alpha^a(I_\alpha^a(f))(t) = f(t) - f(a), \quad (12)$$

If, in addition, f is differentiable, then

$$T_\alpha^a(f)(t) = (t-a)^{1-\alpha}\frac{df(t)}{dt}. \quad (13)$$

2. Main Result

Throughout this paper, let $R_+ = [0, +\infty)$. In order to facilitate the description of the theorem, we define three functions by W_1, W_2 in (5)

$$W_1(z) = \int_1^z \frac{(s-a)^{\alpha-1}\,ds}{w_1(s)s^{\alpha-1}}, z \in R_+, \quad (14)$$

$$W_2(z) = \int_1^z \frac{(s-a)^{\alpha-1}\,ds}{W_1^{-1}(s)s^{\alpha-1}}, z \in R_+, \quad (15)$$

$$W_3(z) = \int_1^z \frac{(s-a)^{\alpha-1}w_1\left(W_1^{-1}\left(W_2^{-1}(s)\right)\right)W_1^{-1}\left(W_2^{-1}(s)\right)ds}{s^{\alpha-1}w_2\left(W_1^{-1}\left(W_2^{-1}(s)\right)\right)}, z \in R_+, \quad (16)$$

Theorem 1. Suppose that $f, g, h, e \in C([a,\infty), R_+)$, e is a positive and nondecreasing function. Suppose that

$$w_1, w_2 \in C(R_+, R_+), w_1(z), w_2(z), w_2(z)/(zw_1(z))$$

Are all nondecreasing and positive functions with $w_1(c)/b \le w_1(c/b)$, $w_2(c)/b \le w_2(c/b)$ for all $c \ge 0, b \ge 0$. If $u(t)$ satisfies (5), then

$$u(t) \le e(t)W_1^{-1}\left\{W_2^{-1}\left[W_3^{-1}(\Xi(t))\right]\right\}, t \in [a, T_1], \quad (17)$$

Where

$$\Xi(t) \le W_3\left\{W_2\left[W_1(1) + \int_a^t (s-a)^{\alpha-1} f(s)\,ds\right] + \int_a^t (s-a)^{\alpha-1} g(s)e(s)\,ds\right\}$$

$$+\int_a^t (s-a)^{\alpha-1} h(s)\,ds, \tag{18}$$

And T_1 is the largest number such that

$$\Xi(T_1) \le W_3(\infty),\ W_3^{-1}(\Xi(T_1)) \le W_2(\infty),\ W_2^{-1}(W_3^{-1}(\Xi(T_1))) \le W_1(\infty). \tag{19}$$

Proof. Since $e(t)$ is a positive and nondecreasing function, from (5) we obtain

$$\frac{u(t)}{e(t)} \le 1 + \int_a^t (s-a)^{\alpha-1} f(s)\omega_1\left(\frac{u(s)}{e(s)}\right)ds + \int_a^t (s-a)^{\alpha-1} g(s)e(s)\omega_1\left(\frac{u(s)}{e(s)}\right)$$

$$\left[\frac{u(s)}{e(s)} + \int_a^s (\tau-a)^{\alpha-1} h(\tau)\omega_2\left(\frac{u(\tau)}{e(\tau)}\right)d\tau\right]ds. \tag{20}$$

Let $z_1(t) = u(t)/e(t)$. From (20), we have

$$z_1(t) \le 1 + \int_a^t (s-a)^{\alpha-1} f(s)\omega_1(z_1(s))ds + \int_a^t (s-a)^{\alpha-1} g(s)e(s)\omega_1(z_1(s))$$

$$\left[v(s) + \int_a^s (\tau-a)^{\alpha-1} h(\tau)\omega_2(z_1(\tau))d\tau\right]ds. \tag{21}$$

Define a function $z_2(t)$ by the right hand side of the inequality (21), i.e.

$$z_2(t) = 1 + \int_a^t (s-a)^{\alpha-1} f(s)\omega_1(z_1(s))ds + \int_a^t (s-a)^{\alpha-1} g(s)e(s)\omega_1(z_1(s))[z_1(s)$$

$$+\frac{1}{\Gamma(\alpha)}\int_a^s (\tau-a)^{\alpha-1} h(\tau)\omega_2(z_1(\tau))d\tau]ds. \tag{22}$$

Obviously, $z_2(t)$ is a positive and nondecreasing function on $[a,\infty)$. From (21) and (22) we see

$$z_1(t) \le z_2(t), t \in [a,\infty). \tag{23}$$

$$z_2(a) = 1. \tag{24}$$

By the definition 2 of the (left) conformable fractional integral starting from a of function, (22) can be rewritten to

$$z_2(t) = 1 + I_\alpha^a(f(t)\omega_1(z_1(t))) + I_\alpha^a\{g(t)e(t)\omega_1(z_1(t))[z_1(t)$$
$$+ \int_a^t (\tau - a)^{\alpha-1} h(\tau)\omega_2(z_1(\tau))d\tau]\}, t \in R_+.$$
(25)

Using the lemma 1 and the relation (23), making a conformable fractional derivative starting from of the function $z_2(t)$ of order α, we obtain

$$T_\alpha^a(z_2)(t) = f(t)\omega_1(z_1(t)) + g(t)e(t)\omega_1(z_1(t))[z_1(t)$$
$$+ \int_a^t (\tau - a)^{\alpha-1} h(\tau)\omega_2(z_1(\tau))d\tau]$$
$$\le f(t)\omega_1(z_2(t)) + g(t)e(t)\omega_1(z_2(t))[z_2(t) \qquad (26)$$
$$+ \int_a^t (\tau - a)^{\alpha-1} h(\tau)\omega_2(z_2(\tau))d\tau], t \in R_+.$$

Define a function $z_3(t)$ by

$$z_3(t) = z_2(t) + \int_a^t (\tau - a)^{\alpha-1} h(\tau)\omega_2(z_2(\tau))d\tau = z_2(t) + I_\alpha^a(h(t)\omega_2(z_2(t))), t \in R_+. \qquad (27)$$

Obviously, $z_3(t)$ is positive and nondecreasing on $[a, \infty)$. From (24) and (27) we have

$$z_2(t) \le z_3(t), t \in [a, \infty). \qquad (28)$$
$$z_3(a) = 1 \qquad (29)$$

Making a conformable fractional derivative of the function $z_3(t)$ of order α, using the lemma 1 and (26), we get

$$T_\alpha^a(z_3)(t) = T_\alpha^a(z_2)(t) + h(t)\omega_2(z_2(t))$$
$$\le f(t)\omega_1(z_2(t)) + g(t)e(t)\omega_1(z_2(t))z_3(t) + h(t)\omega_2(z_2(t))$$
$$\le f(t)\omega_1(z_3(t)) + g(t)e(t)\omega_1(z_3(t))z_3(t) + h(t)\omega_2(z_3(t)), t \in R_+. \qquad (30)$$

Dividing both sides of the above inequality by $\omega_1(z_3(t))$, using the lemma 1 and the definition of W_1, we obtain

$$T_\alpha^a W_1(z_3(t)) = \frac{z_3^{1-\alpha}(t)}{\omega_1(z_3(t))} T_\alpha^a(z_3)(t) z_3^{\alpha-1}(t)$$
$$\le f(t) + g(t)e(t)z_3(t) + h(t)\frac{\omega_2(z_3(t))}{\omega_1(z_3(t))}, t \in R_+. \qquad (31)$$

Firstly, substituting t with τ in (31), and then making a conformable fractional integral of order α for (31) with respect to τ from a to t, by lemma 1 we obtain

$$W_1(z_3(t)) \le W_1(1) + \int_a^t (s-a)^{\alpha-1} f(s)ds + \int_a^t (s-a)^{\alpha-1} g(s)e(s)z_3(s)ds$$

$$+ \int_a^t (s-a)^{\alpha-1} h(s)\frac{\omega_2(z_3(t))}{\omega_1(z_3(t))} ds \le W_1(1) + \int_a^T (s-a)^{\alpha-1} f(s)ds$$

$$\le W_1(1) + \int_a^T (s-a)^{\alpha-1} f(s)ds + \int_a^T (s-a)^{\alpha-1} f(s)ds$$

$$+ \int_a^t (s-a)^{\alpha-1} h(s)\frac{\omega_2(z_3(t))}{\omega_1(z_3(t))} ds, t \in [a,T].$$

(32)

Where $T \in [a,T_1]$ is chosen arbitrarily. Let z_4 denote the right hand side of the inequality (32), then z_4 is a positive and nondecreasing function on $[a,T]$, and

$$z_3(t) \le W_1^{-1}(z_4(t)), t \in [a,T],$$

(33)

$$z_4(a) = W_1(1) + \int_a^T (s-a)^{\alpha-1} f(s)ds.$$

(34)

Making a conformable fractional derivative of the function $z_4(t)$ of order α, using the relation (33) we have

$$T_\alpha^a(z_4)(t) = g(s)e(s)z_3(t) + h(t)\frac{\omega_2(z_3(t))}{\omega_1(z_3(t))}$$

$$\le g(t)e(t)W_1^{-1}(z_4(t)) + h(t)\frac{\omega_2(W_1^{-1}(z_4(t)))}{\omega_1(W_1^{-1}(z_4(t)))}, t \in [a,T].$$

(35)

Dividing both sides of the above inequality by $W_1^{-1}(z_4(t))$, using the lemma 1 and the definition of W_2, we get

$$T_\alpha^a(W_2(z_4(t))) = \frac{z_4^{1-\alpha}(t)}{W_1^{-1}(z_4(t))} T_\alpha^a(z_4)(t)z_4^{\alpha-1}(t))$$

$$\le g(t)e(t) + h(t)\frac{\omega_2(W_1^{-1}(z_4(t)))}{\omega_1(W_1^{-1}(z_4(t)))W_1^{-1}(z_4(t))}, t \in [a,T].$$

(36)

Substituting t with τ in (37), making a fractional integral of order α for (36) with respect to τ from a to t, and using the lemma 1, we obtain that

$$W_2(z_4(t)) \leq W_2(z_4(a)) + \int_a^t (s-a)^{\alpha-1} g(s)e(s)ds$$

$$+\int_a^t (s-a)^{\alpha-1} h(s) \frac{\omega_2(W_1^{-1}(z_4(s)))}{\omega_1(W_1^{-1}(z_4(s)))W_1^{-1}(z_4(s))} ds$$

$$\leq W_2\left(W_1(1) + \int_a^T (s-a)^{\alpha-1} f(s)ds\right) + \int_a^T (s-a)^{\alpha-1} g(s)e(s)ds$$

$$+\int_a^t (s-a)^{\alpha-1} h(s) \frac{\omega_2(W_1^{-1}(z_4(s)))}{\omega_1(W_1^{-1}(z_4(s)))W_1^{-1}(z_4(s))} ds, t \in [a,T]. \tag{37}$$

Let z_5 denote the right hand side of the inequality (37), then we have

$$z_4(t) \leq W_2^{-1}(z_5(t)), t \in [a,T], \tag{38}$$

$$z_5(a) = W_2\left(W_1(1) + \int_a^T (s-a)^{\alpha-1} f(s)ds\right) + \int_a^T (s-a)^{\alpha-1} g(s)e(s)ds. \tag{39}$$

Making a conformable fractional derivative of the function $z_5(t)$ of order α, using the relation (38) we have

$$T_\alpha^a(z_5)(t) = h(t) \frac{\omega_2(W_1^{-1}(z_4(t)))}{\omega_1(W_1^{-1}(z_4(t)))W_1^{-1}(z_4(t))}$$

$$\leq h(t) \frac{\omega_2(W_1^{-1}(W_2^{-1}(z_5(t))))}{\omega_1(W_1^{-1}(W_2^{-1}(z_5(t))))W_1^{-1}(W_2^{-1}(z_5(t)))}, t \in [a,T]. \tag{40}$$

From (40) we get

$$T_\alpha^a(W_3(z_5(t))) = \frac{\omega_1(W_1^{-1}(W_2^{-1}(z_5(t))))W_1^{-1}(W_2^{-1}(z_5(t)))}{z_5^{\alpha-1}(t)\omega_2(W_1^{-1}(W_2^{-1}(z_5(t))))} T_\alpha^a(z_5(t))z_5^{\alpha-1}(t)$$

$$\leq h(t), t \in [a,T]. \tag{41}$$

Substituting t with τ in (41), making a fractional integral of order α for (37) with respect to τ from 0 to t and using the lemma 1, we have

$$W_3(z_5(t)) \leq W_3(z_5(a)) + \int_a^t (s-a)^{\alpha-1} h(s)ds, t \in [a,t]. \tag{42}$$

From (23), (28), (33) and (38), we get

$$u(t) \leq e(t) z_1(t) \leq e(t) z_2(t) \leq e(t) z_3(t)$$

$$\leq e(t) w_1^{-1}(z_4(t)) \leq e(t) w_1^{-1}\left(w_2^{-1}(z_5(t))\right).$$

(43)

From (33), (39), (42) and (43), we have

$$u(t) \leq e(t) w_1^{-1}\left\{ w_2^{-1}\left[w_3^{-1}\left(w_3\left(w_2\left(w_1(1) + \int_a^T (s-a)^{a-1} f(s) ds \right)\right.\right.\right.$$

$$+ \int_a^T (s-a)^{a-1} g(s) e(s) ds \right) + \int_a^t (s-a)^{a-1} h(s) ds \right)\right]\right\}, t \in [a, T].$$

(44)

Because $T \in [a, T_1]$ is chosen arbitrarily, we obtain the required estimation (17). The proof is completed.

3. Conclusion

In this paper, we discuss a class of nonlinear weakly singular integral inequality

$$u(t) \leq e(t) + \int_a^t (s-a)^{a-1} f(s) w_1(u(s)) ds$$

$$+ \int_a^t (s-a)^{a-1} g(s) w_1(u(s)) \left[u(s) \right.$$

$$+ \int_a^s (\tau-a)^{a-1} h(\tau) w_2(u(\tau)) d\tau \right] ds.$$

By adopting novel analysis techniques, we obtain the upper bound of the embedded unknown function $u(t)$

$$u(t) \leq e(t) W_1^{-1}\left\{ W_2^{-1}\left[W_3^{-1}(\Xi(t)) \right]\right\}, t \in [a, T_1],$$

Where

$$\Xi(t) \leq W_3\left\{ W_2\left[W_1(1) + \int_a^t (s-a)^{a-1} f(s) ds \right] + \int_a^t (s-a)^{a-1} g(s) e(s) ds \right\}$$

$$+ \int_a^t (s-a)^{a-1} h(s) ds.$$

And T_1 is the largest number such that

$$\Xi(T_1) \leq W_3(\infty), W_3^{-1}(\Xi(T_1)) \leq W_2(\infty), W_2^{-1}(W_3^{-1}(\Xi(T_1))) \leq W_1(\infty).$$

The derived results can be applied in the study of qualitative properties of solutions of conformable fractional integral equations.

Acknowledgement

This research was funded by National Natural Science Foundation of China, (Project No. 11161018, 11561019).

References

1. T. H. Gronwall, Note on the derivatives with respect to a parameter of the solutions of a system of differential equations, Ann Math. 20 (1919) 292-296.
2. R. Bellman, The stability of solutions of linear differential equations, Duke Math. J. 10(1943) 643-647.
3. Q. H. Ma and J. Pecaric, Some new explicit bounds for weakly singular integral inequalities with applications to fractional differential and integral equations. J. Math. Anal. Appl. 341(2) (2008) 894-905.
4. A. Abdeldaim and M. Yakout, On some new integral inequalities of Gronwall-Bellman-Pachpatte type, Appl. Math. Comput. 217(2011) 7887-7899.
5. R.P. Agarwal, S. Deng and W. Zhang, Generalization of a retarded Gronwall-like inequality and its applications, Appl. Math. Comput. 165(2005) 599-612.
6. T. Abdeljawad, On conformable fractional calculus, J. Computational Appl. Math. 279(2015) 57-66.
7. [7] T. Abdeljawad, On conformable fractional calculus, J. Computational Appl. Math. 279 (2015) 57-66.
8. Q. H. Ma and J. Pecaric, Estimates on solutions of some new nonlinear retarded Volterra- Fredholm type integral inequalities, Nonlinear Anal. 69(2008) 393-407.

Research on Simulating, Preparing and Testing of Spherical Cap Piezoelectric Composites

Bo-wen Yu*, Li-kun Wang, Chao Zhong and Miao-jie Lv

Beijing Key Laboratory for Sensor,
Beijing Information Science & Technology University,
No. 35 North Fourth Ring Road,
Chaoyang District,
Beijing, China
*yubowen67@hotmail.com

In this article, a new type of spherical cap piezoelectric composite oscillator was designed. For the preparation of piezoelectric oscillator, a new process of cutting spherical piezoelectric ceramic-casting polymer was proposed. The modeling and simulation of piezoelectric oscillator were presented by using ANSYS 10.0, the resonant frequency and the admittance curve of the piezoelectric oscillator in the air were derived. The piezoelectric composite oscillator was prepared with the guide of simulation result. The result of the performance test of piezoelectric oscillator has shown that the resonant frequency of the piezoelectric oscillator in the air was 325.60kHz and the bandwidth can reach 14.17kHz. Comparing the measurement results and simulation, the relative error of resonant frequency is less than 1% and the test result turns out that the spherical piezoelectric composite oscillator is suitable for manufacturing broadband transducer.

Keywords: Spherical cap composite, Piezoelectric oscillator, Simulation, Preparation, Performance test.

1. Introduction

Up to day, the piezoelectric ceramics still occupied a vital position by virtue of excellent performance such as higher electromechanical coupling coefficient, lower losses, lower cost et al. in the piezoelectric material field [1]. Nevertheless, it was limited by the characteristic of large acoustic impedance, friable, poor impact resistance, and difficulty to match with water and biological tissue. Although the flexibility of the piezoelectric polymer was perfect, the scope of application was limited by its expensive cost, smaller piezoelectric constant d33. People were paid widespread attention to piezoelectric composites combined with excellent piezoelectric properties of ceramic and great flexibility of polymer. Transducer manufactured with piezoelectric composite was widely applied to ultrasonic nondestructive testing, biomedical imaging, underwater communications and other related fields.

The research on piezoelectric composites with flat linear structure at home and abroad has been relatively mature, with the help of series and parallel theory proposed by R.E. Newnham et al. and Finite Element Method (FEM) [2].

Although the operating frequency bandwidth of transducers made with piezoelectric oscillator of planar structure is wide, its beam angle is small, which limited the range of radiation and ability of receive information. The study on piezoelectric composites in the domestic and abroad was still focus on one-dimensional arc forming, there was less report which mentioned the design of two-dimensional arc forming, the manufacture of the transducer is likewise unnoticed. MSI (Materials Systems Inc.) has succeeded manufactured the 1-3 composite arc transducer, its operating bandwidth is 208khz-453khz, -3dB directional beam opening angle in the 375khz is 60.5 °, it can realize the emitting forward of the wide beam [3]. Yiming Zhou etc., the researchers of Tokai Research Station of Institute of Acoustics, Chinese Academy of Sciences, who used 1-3 composite plasticity process, succeeded in designing and producing a curved array transducer. The resonant frequency was 199 kHz, -3dB beam width was 144°, the non-uniformity in the emission surface of 140° is less than 2.5dB [4].

The curved surface transducers mentioned above were still one-dimensional (x direction) arc forming, which the beam-opening angle would be limited in the y direction. In order to expanding the beam opening angle of transducers and achieving the goal of two-dimensional wide beam emission, a piezoelectric oscillator of two-dimensional arc formed was designed in this thesis.

2. The Finite Element Analysis of Spherical Cap Piezoelectric Composites

2.1. The Extraction of Model and Parameter Settings of Simulation

The finite element analysis software (ANSYS10.0) was used in the computational designing the spherical cap composite transducer. The structure of spherical cap composite piezoelectric oscillator was shown in Figure 1, while the section of the structure was shown in Figure 2. PZT-5 and epoxy were selected as ceramic and polymer in the simulation analysis stage. The inner radius of the spherical cap (Ri) was 15mm, the outer radius (Ro) was 19mm, the height (H) was 11.5mm, the width of ceramic pillar (Pw) was 1.4mm, the gap of ceramic pillar (Gw) is 0.6mm. The typical parameters of piezoelectric ceramic PZT-5 and epoxy resin were shown in Table 1.

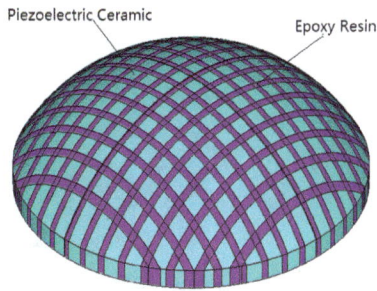

Figure 1. The structure of spherical cap oscillator

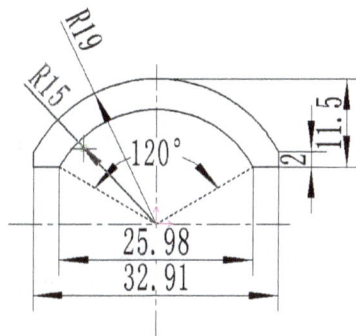

Figure 2. The section of spherical cap oscillator

Table 1. The typical parameters of piezoelectric ceramic PZT-5 and epoxy

Parameter / Material	Element type	Young's modulus (10^7N/m^2)	Poisson's ratio	Density (kg/m^3)
PZT5	Solid5	/	/	7750
Epoxy Resin	Solid45	360	0.35	1050

2.2. The Finite Element Modeling, Solving, Post-Processing and Analysis of Piezoelectric Oscillator

Considering the coordination of convenience and accuracy of modeling, we only need to select which characterize of symmetry in the center on behalf of all the piezoelectric oscillators in fact. The simplified model was shown in Figure 3.

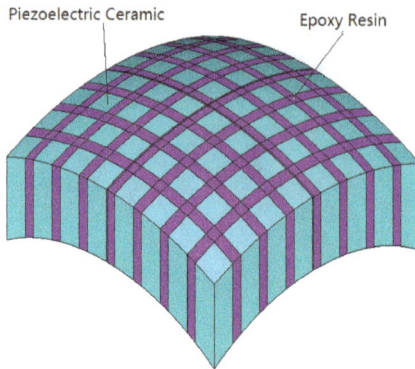

Figure 3. The simplified model of spherical cap oscillator

The generation of FEM model is actually the meshing of the solid model [5]. Normally, the sophistication of the grid and calculation speed is inversely proportional, so we need to strike a balance between the sophistication and computing speed, we selected to mesh a wavelength ten aliquots in this simulation, sweep was selected as the meshing method. The finite element model of piezoelectric oscillator generated in the air was shown in Figure 4.

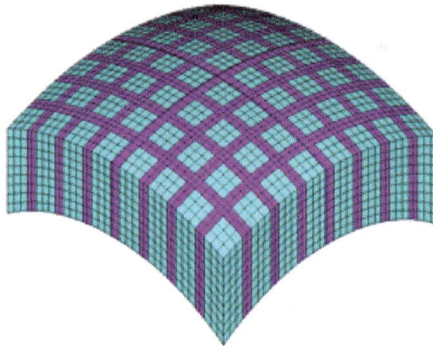

Figure 4. The finite element model of piezoelectric oscillator

Voltage was applied to the inside and outside arc surfaces of the piezoelectric oscillator, which the voltage applied to the outer surface was 1V, while the voltage applied to the inner surface was 0V, harmonic responses analysis has been applied to the finite element model. The choosing solving frequency ranges from 300 kHz to 370 kHz, set the calculation step 70, and the damping in the air was 2%. The resonance frequency of piezoelectric oscillator in the air obtained by computer solution is 326 kHz. The relationship between the admittance and the resonance frequency of piezoelectric oscillator was shown in Figure 5. The vibration mode of the piezoelectric oscillator at the resonance frequency was shown in Figure 6.

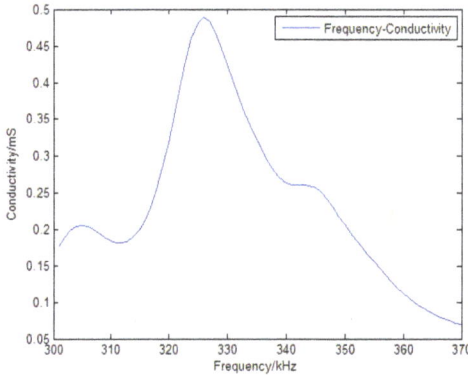

Figure 5. The relationship between the modulus of admittance and the resonance frequency

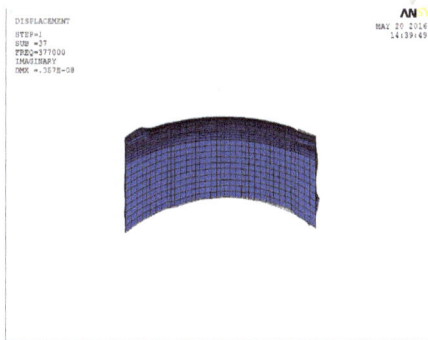

Figure 6. The vibration mode of the piezoelectric oscillator at the resonance frequency

The reason why the impure of piezoelectric ceramic occurs that situation is that the cutting direction of the piezoelectric ceramic is vertical which is different from the polar direction. It can be predictable that if the complete analysis of spherical cap was taken, the vibration will be more complex, the impurity peak will be more.

3. The Preparation of Spherical Cap Composite Piezoelectric Vibrator

The composite spherical piezoelectric oscillator was fabricated by using the new process of cutting the ceramic in the X direction-pouring the epoxy-curing-cutting the ceramic in the Y direction- secondary curing-polishing the inner and outer surfaces of spherical cap-coating electrode in the inner and outer surfaces. The specifically flowchart was shown in Figure 7.

The spherical piezoelectric ceramic was waxed on the center of 150mm × 150mm graphite plate with the help of paraffin, the cutting parameters such as cutting length, cutting depth, stepping (ceramic pillar width plus width of gap) and number of cuts was set in load pointed precision cutting machine, according to a preset program the machine will cut the ceramic into the desired structure automatically. The X-cut polyimide tape attaching to a piezoelectric ceramic was placed in pre-made Teflon mold and then casting epoxy, air bubbles was exhausted under the vacuum conditions. The excess epoxy was scooped after half an hour, then molding, curing at room temperature 12 hours. After completion of the curing piezoelectric ceramic spherical then second Y-cutting, the second casting epoxy curing to get the spherical cap composite piezoelectric vibrator, silver electrode was coated in the inner and outer surfaces of the piezoelectric oscillator. The prepared spherical cap composite piezoelectric oscillator was shown in Figure 8.

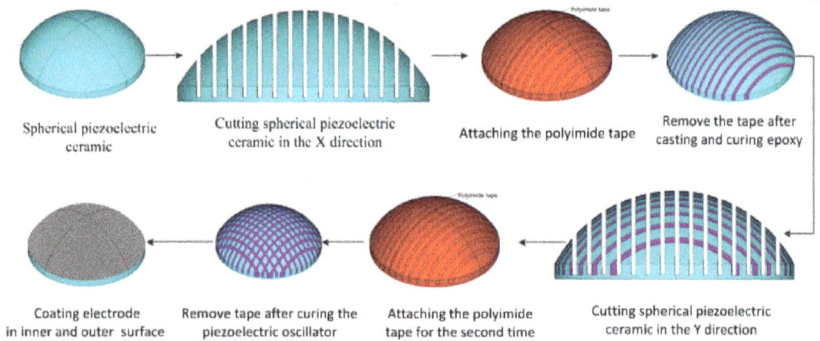

Figure 7. The preparation process of spherical cap composite oscillator

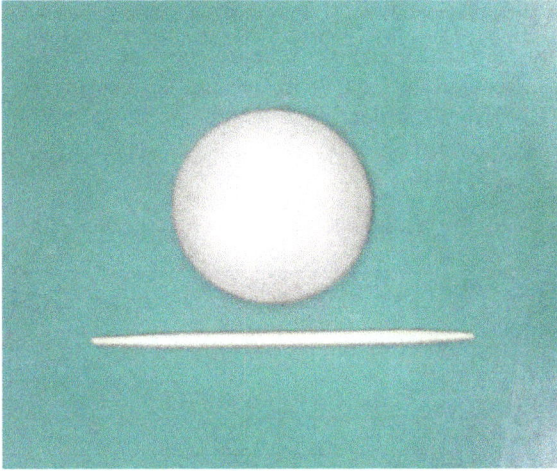

Figure 8. The prepared spherical cap composite piezoelectric oscillator

3.1. Performance Test of Spherical Cap Composite Piezoelectric Oscillator

The mass of spherical cap composite piezoelectric oscillator was measured by using a precision quality meter, and the volume could be calculated, then it is concluded that the density of piezoelectric vibrator. The piezoelectric constant of piezoelectric vibrator d33 was measured by the d33 tester produced by the Institute of Acoustics, Chinese Academy of Science. The admittance curve and the impedance curve of piezoelectric oscillator measured by the Agient4294A precision impedance analyzer was showed in Figure 9. Resonant frequency, bandwidth, Q value of the piezoelectric oscillator etc. could be derived by the test chart.

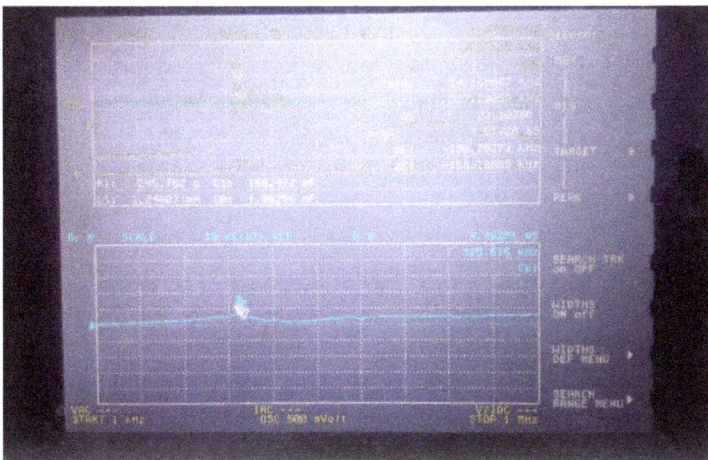

Figure 9. The test chart of spherical cap composite piezoelectric oscillator

Using the formula: $v = 2 \cdot f_p \cdot t$, $Z = \rho \cdot v$, we could derive the velocity of sound, acoustic impedance, where v represents the velocity of sound, fp represents the parallel resonant frequency, Z represents the acoustic impedance, ρ represents the density, t represents the thickness. The specific performance test of piezoelectric oscillator was summarized in Table 2:

Table 2. The specific performance test of piezoelectric oscillator

Parameters of piezoelectric oscillator	Measured value or calculated value
Density(kg/m³)	5229
Parallel resonant frequency(fp/kHz)	472.1
Series resonant frequency(fs/kHz)	325.6
Velocity of sound(m/s)	3776.8
Acoustic impedance Z(Pa*s*m⁻³)	19.75
Piezoelectric constant d_{33}/(pC/N)	349
Bandwidth(Bw/kHz)	14.17
Mechanical quality factor Q value	23.10

4. Conclusion

In this paper, the resonant frequency and vibration of spherical piezoelectric composite oscillator in the air were simulated by ANSYS 10.0, results indicated that resonant frequency in the air is 326 kHz, comparing with the test result, the relative error of resonant frequency is less than 1%. The spherical piezoelectric composite oscillator were fabricated by cutting ceramic directly, moreover, the performance of spherical piezoelectric composite were tested, test results show that the resonant frequency of piezoelectric oscillator in the air was 325.6kHz, the bandwidth was 14.17kHz, the density was 5229 kg/m³, and the acoustic velocity was 3776.8m/s, the acoustic impedance was 19.75Pa·s/m³, d_{33} constant was 349pC/N. The new technology has break through the bottleneck of that traditional piezoelectric oscillator is difficulty to two-dimensional surface forming, the piezoelectric oscillator made with this process has great advantages such as larger bandwidth, lower acoustic impedance, which is quite suitable for manufacturing broadband transducer. But considering the disadvantage of poor directivity of the transducers caused by the inconsistency of the cutting direction and polarization direction of piezoelectric oscillator, the following-up studies should improve its directivity by keeping the cutting direction and polarization direction be consistent.

Acknowledgement

This research was funded by National Natural Science Foundation of China (No. 61302015, 61471047), and the Beijing College Innovation Capability Promotion Plan of Beijing Municipal Institutions (No. 72F1610912). This research is also financially supported by the Importation and Development of High-Caliber Talents Project of Beijing Municipal Institutions (No. CIT&TCD201504053).

References

1. R. E. Newnham, D. P. Skinner, L. E. Cross, Connectivity and piezoelectric pyro-electric composites. 1978. 525-536.
2. Information on http://www.matsysinc.com.
3. Yiming Zhou, Hui Tong, Jun Dai, Xiangquan Chen, the 50th anniversary of Institute of Acoustics, Chinese Academy of Sciences of the construction and the Fifth academic communication, Beijing, 2014, In Chinese.
4. Guidong Luan, Jinduo Zhang, Renqian Wang, Piezoelectric Transducers and Transducer Array, Beijing University Press, 2005, "In Chinese".
5. ANSYS User's Manual.

A Novel Edge Detection Technique Based on Canny Algorithm and Hough Transform

Jun Wang, Gui-qin Li[†], Yang Chen, Hai Dai and Bei-bei Fan

Shanghai Key Laboratory of Intelligent Manufacturing and Robotics,
Shanghai University,
Shanghai, 200072, China
[†]E-mail: eeching@t.shu.edu.cn

Qing Guo and Jun-long Zhou

Shanghai Yanfeng Automotive Trim Systems Co., Ltd., China

Based on the strain measurement requirements of plastic materials in high-speed tensile environment, considering the difficulty of signature change in shape and the features extraction caused by the small contrast between mark features and background, a novel edge detection technique is proposed, which combines the advantages of Hough transform and canny algorithm. The designed technique can realize the feature extraction with high precision and efficiency, and meet the requirement of strain measurement in high-speed tensile test.

Keyword: Image measurement; Edge detection; Canny algorithm; Hough transform.

1. Introduction

Edge detection, as the main extraction technique of feature image analysis and recognition, has always been the research priorities and focus point in visual image processing. New edge detection methods are emerging after the edge detection operator proposed in 1965 [1]. Such as the Digital step edges from zero crossing of second directional derivatives put forward by Haralick [2]; Roberts proposed a new algorithm after the research of Machien perception of Three-Dimensional Solid Optical and Electro-Option [3]; Sobel [4] and M. K.Hu [5] both use the calculus and filtering to detect the image edge; Marr D developed LOG algorithm in Theory of Edge Detection [6]; In contrast, the Canny Operator [7] based on optimization algorithm make a good detection performance, and it has high positioning accuracy when it is used in the step-like edge detection by the white noise.

Considering the difficulty of signature change in shape and the features extraction caused by the small contrast between mark features and background, a novel edge detection technique is proposed in this paper. The first image of the collected photographs in tensile experiment will be firstly disposed by Hough transform to extract the initial position of the target feature point; then the two target feature points are dealing with the edge detection by local Canny algorithm. The designed technique can realize the feature extraction with high precision and efficiency, and meet the requirement of strain measurement in high-speed tensile test.

2. Canny Algorithm and Hough Transform

The Canny edge detection operator was developed by John F. Canny in 1986 and uses a multi-stage algorithm to detect a wide range of edges in images [8, 9]. Firstly, the images are smoothly filtering by the first derivative of a Gaussian; In order to obtain the desired final edge image, the algorithm use a technique called "non-extremum suppression "to deal with the images after filtering. The flowchart of the Canny algorithm is shown in Fig. 1.

The image is smoothed by a Gaussian function, and it can suppress image noise. After the gradient calculation completed, the data of gradient magnitude and gradient direction are obtained; "non-maximum value suppression" procedure is used to refine the gradient amplitude matrix, it can trace the possible edge points in the image; dual threshold detection is searching image edge points through double threshold recursively to complete edge detection.

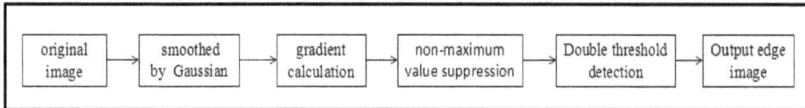

original image	smoothed by Gaussian	gradient calculation	non-maximum value suppression	Double threshold detection	Output edge image

Fig. 1 The flowchart of the Canny algorithm

Hough transform is a mapping from picture to space parameters, changing the original image into space parameters is the basic idea. The curve of the image is described by the boundary points, which meet some forms of parameters. After setting the accumulator accumulates, the required information is obtained by the peak of the corresponding point. This method is not sensitive to noise and easy to computing. It is commonly used in straight lines, circles and ellipses detection feature.

The feature extraction target is marked as the drawing on the tensile specimen. At the beginning the sharp of the marker is a regular circle. During the process of high-speed tensile plastic material, as the volume of specimen increases, the change of marker shape is growing. The contrast between mark features and background is getting smaller. It is difficult to extract the target by Canny algorithm, which is based on the global. Hough transform is no longer applicable to this case. To solve this problem, a novel edge detection technique in this paper is proposed, which combines the advantages of Hough transform and canny algorithm.

3.   The Improved Edge Detection Technique

The basic idea of the technique: Firstly, The first image of the collected photographs in tensile experiment will be disposed by Hough transform to extract the initial position of the target feature point; Based on the two target center position, two sub-regions (S1, S2) which include the first frame image feature respectively are determined to be the image processing area of the first frame image, and for the subsequent image, the previous image will be approximately marked as a target image processing region; Then the two target feature points are dealing with the edge detection by local Canny algorithm to get the center coordinates of the characteristic points. The flowchart of the improved edge detection technique is shown in Fig. 2.

The selection of the image processing area (S1, S2) actually is the key issues in the workflow. It is necessary to ensure complete information of feature points, but it cannot be too lager. Amount of computation and more noise caused by lager size are not conducive to feature extraction. After estimate of the data: the moving speed of the feature point $\leq 0.003mm/fps$, the displacement of the feature points between adjacent image is very small. Thus, the general area of the feature point in the former image can be used as the image processing area for the next image.

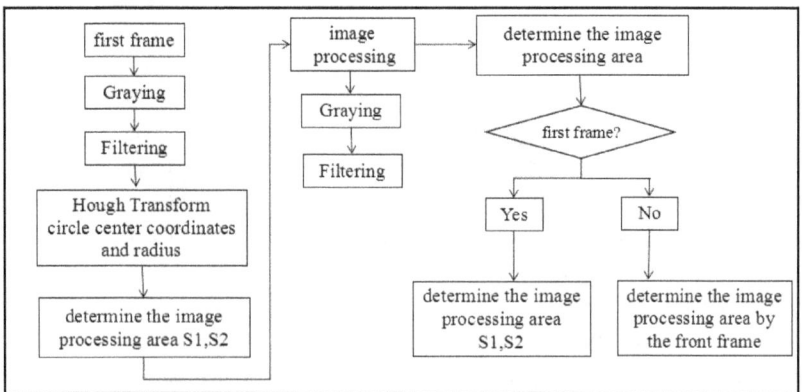

Fig. 2 The flowchart of the improved edge detection technique

For the first image, two circular target image center coordinates are extracted by the processing of Hough transform, the upper circle center coordinates are noted as $(X(1,1), Y(1,1))$, the radius is marked as $R(1,1)$; the under circle center coordinates are noted as $(X(1,2), Y(1,2))$, the radius is marked as $R(1,2)$, then:

$$S1 = rawing(h3 : h4, h1 : h2); \quad S2 = rawing(h7 : h8, h5 : h6)$$

h1=round(X(1,1)-1.6*R(1,1));

h2=round(X(1,1)+1.6*R(1,1));

h3=round(Y(1,1)-1.6*R(1,1));

h4=round(Y(1,1)+1.6*R(1,1));

h5=round(X(1,2)-1.6*R(1,2));

h6=round(X(1,2)+1.6*R(1,2));

h7=round(Y(1,2)-1.6*R(1,2));

h8=round(Y(1,2)+1.6*R(1,2)).

For the image of number i, the upper circle center coordinates are noted as $(X(i,1), Y(i,1))$, the long axis is marked as $A(i,1)$, the minor axis is marked as $B(i,1)$; the under circle center coordinates are noted as $(X(i,2), Y(i,2))$, the long axis is marked as $A(i,1)$, the minor axis is marked as $B(i,1)$, then:

$$S1 = rawing(h3 : h4, h1 : h2), \quad S2 = rawing(h7 : h8, h5 : h6)$$

h1=round(X(i-1,1)-1.6*B(i-1,1));

h2=round(X(i-1,1)+1.6*B(i-1,1));

h3=round(Y(i-1,1)-1.6*A(i-1,1));

h4=round(Y(i-1,1)+1.6*A(i-1,1));

h5=round(X(i-1,2)-1.6*B(i-1,2));

h6=round(X(i-1,2)+1.6*B(i-1,2));

h7=round(Y(i-1,2)-1.6*A(i-1,2));

h8=round(Y(i-1,2)+1.6*A(i-1,2)).

Using the second image as example, its image processing area is the determined area S1, S2 of the processed first image. Then the image area except the area S1S2 are assigned to zero and dealing with the edge detection by local Canny algorithm to get the center coordinates of the characteristic points. And so on, the target edge of all the images during the tensile processing will be extracted.

4. Experimental Results and Analysis

The experiment light source adopts the red strip LED cold light. The camera uses the phantom V5.1 high-speed digital camera, and 1394 interface was used to connect to the computer. Specimen dimensions follow German Institute for Standardization Din 53504-1994, which stipulated plastic tensile test standard. The Hough processing of the first frame image and Canny algorithm results were shown in Fig. 3.

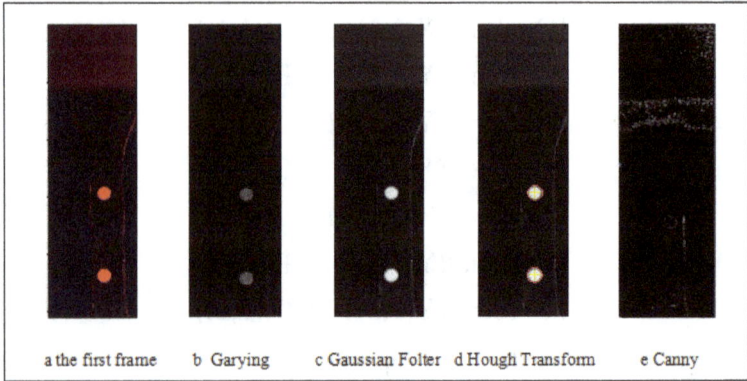

a the first frame   b Garying   c Gaussian Folter   d Hough Transform   e Canny

Fig. 3 The Hough processing of the first frame image and Canny algorithm results

Compared to Figs. 3(d) and 3(e), there were a lot of shortages about the traditional Canny algorithm to the edge of the tag feature, that is to say, the characteristic edge is not complete and irrelevant features can't be eliminated at the same time.

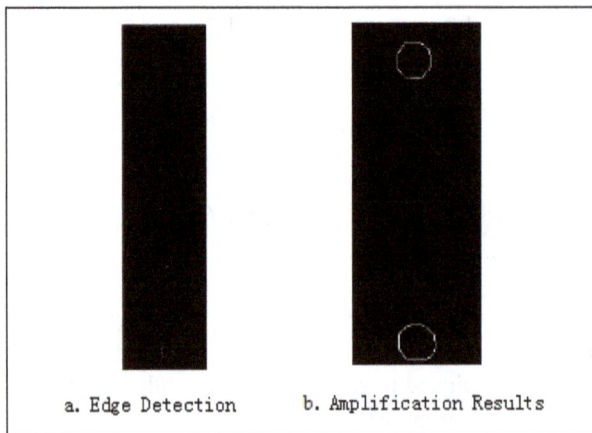

a. Edge Detection   b. Amplification Results

Fig. 4 The edge detection of the second frame

Taking the improvement of technology better achieve to detect the edge of the feature in this paper. As shown in Fig. 4, positioning feature edges accurately, continuously and completely was easy to later feature recognition.

Conclusion

In order to solve the difficulty of signature change in shape and the features extraction caused by the small contrast between mark features and background, a novel edge detection technique in this paper is proposed, which combines the advantages of Hough transform and canny algorithm. The result shows that the designed technique is stable and reliable. It can meet the requirement of strain measurement in high-speed tensile test. Due to the edge detection, only sub-areas S1, S2 in the image processing is carried out; it will reduce the range of image processing and save time.

Acknowledgement

This research was funded by Shanghai Automotive Industry Science and Technology Development Fund (Grant No. D71-0109-15-150), and Shanghai Committee of Science and Technology Project (Grant No. 14511108303).

References

1. Kwon Oh Kyu, Sim Dong Gyu, Park Rae Hong. Robust Hausdorff distance matching algorithms using pyramidal structures. Pattern Recognition, 2001, 34(7): 2005~2013.
2. Haralick R M. Digital step edges from zero crossing of second directional derivatives. IEEE Trans. OnPAMI, 1984, 6(1): 58~68.
3. Roberts. Machien perception of Three-Dimensional Solid Optical and Electro-Option.
4. Sobel, Cameral Model and Machine Perception, Ph.D dissertation, Stanford University.
5. M. K. Hu. Visual pattern recognition by moment invariants. IEEE Trans. On IT, 1962:179~187.
6. Marr D, Hildreth E. Theory of Edge Detection. London: Porc Roy Soc. 1980.
7. Canny J.A Computational approach to edge detection. IEEE Trnas on PAMI, 1986, 8 (6): 679~698.

8. Illingworth J, Kittler J. A survey of the Hough transforms. CVGIP: Image Understanding, 1988, 44(1): 87~116.
9. Zhanyi Hu, Changjiang Yang, Songde Ma. New definition of Hough transform. Chinese Journal of Computers, 1997, 20(8):744~752.

Algorithm in Searching for the Critical Point of Press-Fit Curve

Cong Tan, Guo-qing Ding* and Xin Chen

Department of Instrument Science and Engineering,
Shanghai Jiaotong University, Shanghai 200240, China
*gqding@sjtu.edu.cn

The paper designed two ways to search the critical point in the press-fit curve. The first one starts from the source data. The cubic spline interpolation method is used to deal with data, then the first derivative and second derivative are solved in turn. Finally, the maximum value of second derivative is derived, and it is the critical point. For the second way, the press-fit curve is treated as an image. Then, the Harris corner detection algorithm is applied with the image, and search the point, which has the maximum change in grayscale. Finally, the algorithm is tested in the laboratory with limited data. The algorithm is found to be useful and it can be tested with more data from factory.

Keyword: Press-fit curve; Critical point; Cubic spline interpolation; Second derivative; Harris corner detection algorithm.

1. Introduction

The bearing press-fit machine is the professional equipment in the automobile industry. The main usage of it is to press the bearing into the shaft neck. [1] At the present stage, the bearing and the shaft neck bind together with interference fit. As a result, the pressure of the process is to overcome friction of the interface. Automation of press-fit machine becomes a hot spot. The pressing process begins to pressurize while it exceeds the critical point. Therefore, judging the critical point becomes an important task.

The critical point is a point where the first derivative changes sharply. It is a critical value. From the scene of human vision, the press force changes slowly before the point, but it changes sharply after the point.

2. The Algorithm Based on Second-Order Derivative

2.1 *The Analysis of Press-Fit Curve*

With the reference of the paper, the press force of bearing, P, meet the mathematical model:

$$P = p \cdot S \cdot f = p \cdot 2\pi a \cdot f \cdot x = \frac{\pi E(b^2 - a^2)}{2b^2} \cdot \delta \cdot f \cdot x = C \cdot x \qquad (1)$$

From the formula equation above, we can conclude that, the press force and press displacement is proportional. And the press process is a uniform linear motion. Therefore, the ideal curve presents as Figure 1 [2]:

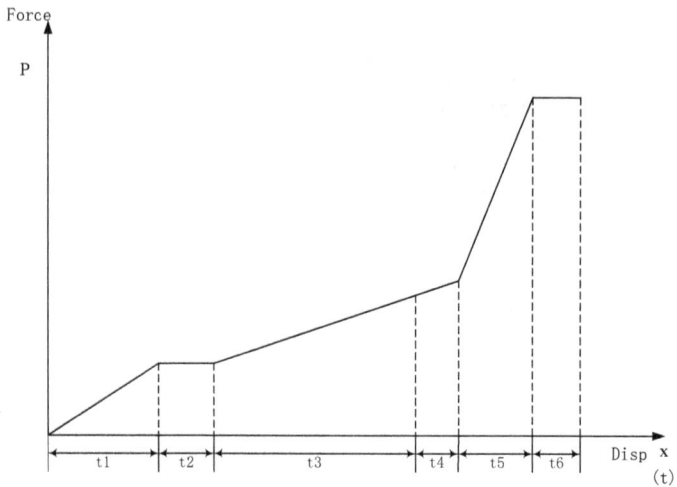

Figure 1 Force-displacement curve

3. The Pretreatment of the Data

In reality, there are some flaws in our data, and it cannot meet our need. Therefore, we need to do something to our data [3]. And it is the pretreatment. The data before interpolation is shown as Figure 2(a), and the data after the interpolation is shown as Figure 2(b).

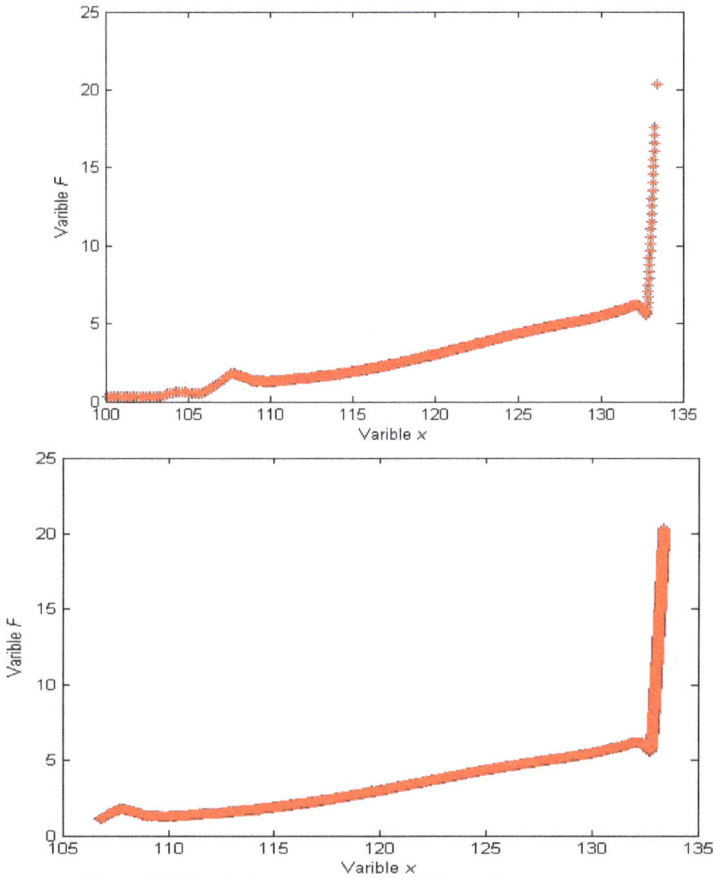

Figure 2(a) Data before pretreatment 2(b) Data after pretreatment

Concerning the smoothness of the curve, we apply the cubic spline interpolation method to the raw data. And it can do good for the continuity of the second derivative.

4. Get the First Derivative

Based on the pretreatment, we use the least square method to match the first derivative. To get rid of the influence of the glitch and disturbance, we get the first derivative in a large scale. We choose 64 points as a group to get our aim data. And the graph below is our answer:

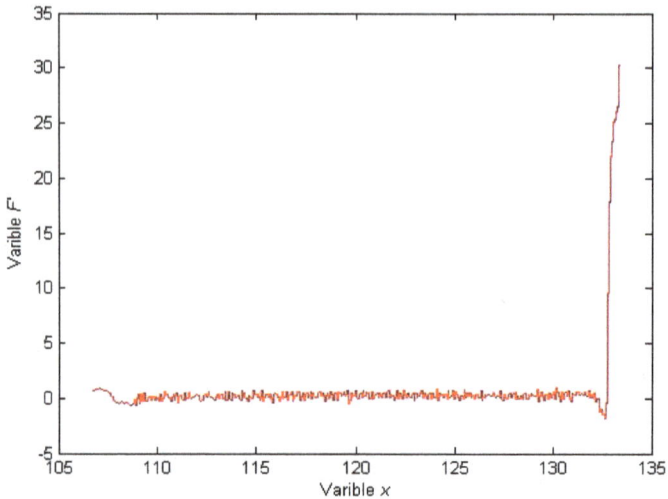

Figure 3 The diagram of first derivative

5. Get the Second Derivative

We use the definition of the second derivative to get our aim answer, like this:

$$\underset{\Delta x \to 0}{f}{}''(x) = \frac{f'(x + \Delta x) - f'(x)}{\Delta x - x} \tag{2}$$

We use the first derivative from above to get the second derivative, and the answer is shown as below:

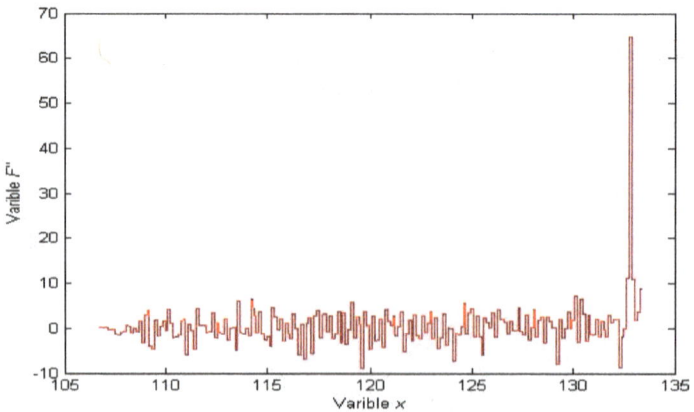

Figure 4 The diagram of second derivative

We can conclude from the data above, the maximum value of second derivative is at the end part of our data. And it just fit the physical phenomenon, which implies that the critical point is at the end part of our data. Finally, we get the point is x=133.7464, and apply it to the source data, the result is shown as Figure 5.

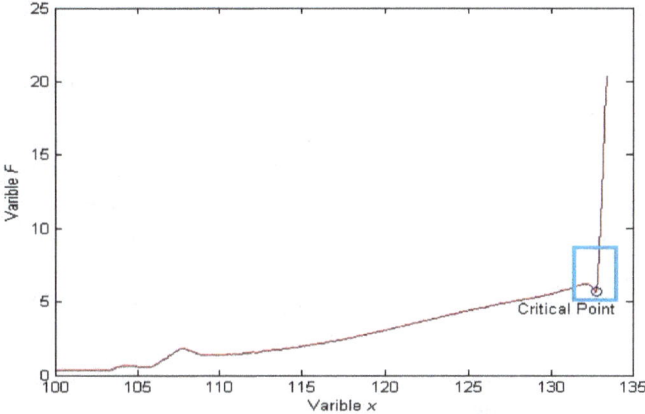

Figure 5 The location of critical point

From the graph above, we can find that the point fits the point, which is calculated by our experience algorithm. And plus our frame function (shown as the blue frame above), when our point is in the frame, we can conclude the press process is qualified and we can say that the algorithm is valid.

6. Image Detection Algorithm Based on Human Vision

6.1 *Introduction about Harris Corner Detection Algorithm*

From the graph above, we can see from the picture that the critical point is where the curve changes greatly. Therefore, from the view of human vision, we can abandon the complicated data. And we can concern all data as a whole. Then we get another method to calculate the critical point: Harris corner detection algorithm. [5]

Harris corner detection algorithm is an effective algorithm based on the change of the gray scale of the image. [4] Harris operator defines a moving window in the graph, and studies the change of the gray scale when the window moves in a certain step length. We can divide the problem into three situations. The graph below shows them [7, 8].

677

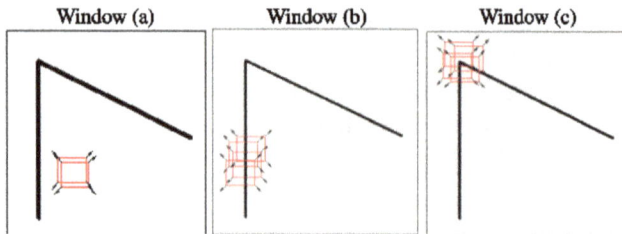

Figure 6 Changes of the grey scale of the image as the results of Harris Corner Detection Algorithm

When our window moves in the smooth area, like Figure 6(a), there is no change in the gray scale. [10, 12] When our window moves in the margin edge, like Figure 6(b), there is no change in the gray scale when the moving direction is parallel to the edge. When the window moves near the corner, like Figure 6(c), the change of the gray scale is obvious no matter our window moves in which direction. [8] Harris corner detection algorithm applies it in calculating the corner. [6,7]

7. The Process to Calculate the Corner

The process is shown as Figure 7:

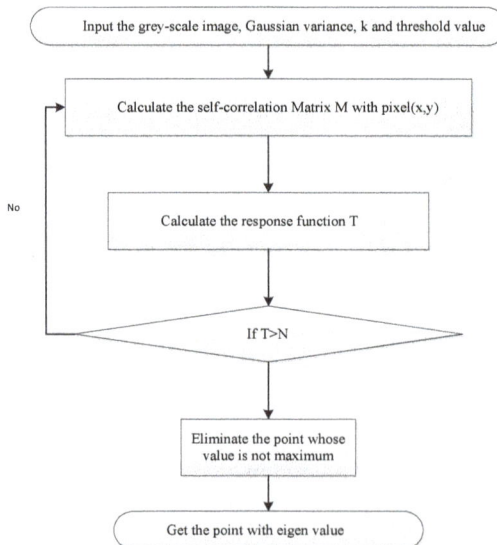

Figure 7 Block diagram of algorithm

When we apply the algorithm to the reality, we find that if we compress our image before we start and we can get some surprise unexpectedly. Firstly, we

678

can get the corner more effectively. Secondly, we can decrease our operation time, which can reduce the burden of our computer. The Figure 8(a) is the answer we get with the Harris corner detection algorithm:

Harris   Harris-Kuang

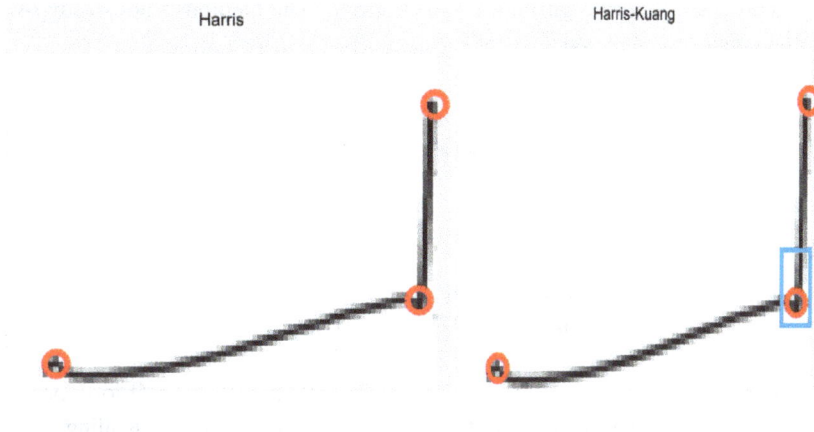

Figure 8(a) Corner detected by Harris   8(b) The selected Harris corner

We find three points in the image, three points distribute in three places in our curve. There are at the start position, the end position and the corner position. It reflects that our opinion about the change of the grey scale is correct. We can see from the image, and the three points are where the grey scale changes greatly.

Plus, our frame function, seen from Figure 8(b), if the corner is in our frame, we are sure that the process is qualified. Therefore, we can conclude that the detection algorithm based on vision is qualified in detecting the critical point of curve.

8. Comparison

Table1. Comparison between four methods

Name	Qualified data (70)		Unqualified data (30)		Success rate
	Qualified	Unqualified	Qualified	Unqualified	
Experiential	68	2	1	29	97%
Quadratic fit	50	20	12	18	68%
Second derivative	70	0	1	29	99%
Harris corner	70	0	1	29	99%

9. Conclusion

1. The Experiential algorithm is steady, but it is not smart. And it cannot satisfy the intelligent need of industrial field.
2. The Quadratic fit algorithm is Quadratic fit. The method is not steady. And it cannot fit our need.
3. The two new algorithms can fit the process. And its success rate is satisfying, which exceeds the algorithm we used before. Therefore, we can apply our algorithm to industrial field and test them in large quantities of data.

References

1. Xiao-zhu Sun. Finite element simulation and experiment research of press-mounting system of the freight train rolling bearing[D]. Harbin: Harbin Institute of Technology, 2006. In Chinese
2. Hong-wei Hu. The research about the pressing qualification of lorry bearing with the force-displacement curve [J]. Locomotive & Rolling Stock Technology, 2000(1): 15-18In Chinese
3. Chapman S J. MATLAB programming for engineers [M]. Cengage Learning, 2008.
4. Serre T, Wolf L, Poggio T. Object recognition with features inspired by visual cotex. In: Computer Vision and Pattern Recognition, 2005, vol. 2: 994~1000.
5. Harris C, Stephens M. A combined corner and edge detector [C]//Alvey vision conference. 1988, 15: 50.
6. Lu Yu, Xing-wen Hao, Yong-jun Wang. Research on Moravec and Harris Corner Algorithm [J]. Computer Technology and Development, 2011, 21(6): 95-97. In Chinese
7. Liang Yang, Research on primary corner detection algorithm. [D][D], 2012.
8. Yu-zhu Wang, Research on Image Corner Detection Algorithm [D]. Chongqin: Chongqin University, 2007. In Chinese

Building Product Simulation Models Using Simscape for Rehabilitation Training

Huan-yu Guo* and Yue-min Hou

School of Mechatronic Engineering, Beijing Information Science and Technology University, Beijing 100192, China

*Email: huanyuguo01@163.com

Simulation has been a powerful planning, strategic and visualization tool in many fields. A simulation method for multi domain physical system in MATLAB/Simscape is proposed in this paper, and the inverse kinematical simulation model of the rehabilitation training equipment is established. Multiple motional trajectories of the end effector are obtained and a detailed analysis is made on one set of the motion trajectories. The simulation results show that the rehabilitation training equipment can ensure the smooth operation of the end effector and the feasibility and reasonableness of the trajectory. Simscape, for analysis, including parametric studies and optimization, and the results can then be used to validate and make improvements to the design.

Keywords: Rehabilitation training equipment; Simscape; Simulation model; Training modes.

1. Introduction

Stroke is one of the leading causes of severe disability in the world, with up to 15 million of people suffer stroke every year, which brings a great burden to the family and society [1]. Therefore, the correct and scientific rehabilitation training on motor function recovery play a very important role. The rehabilitation training equipment, as a kind of automation equipment, can help patients with scientific and effective rehabilitation training [2]. Rehabilitation robot is a robot controlled by a computer, equipped with sensors, and corresponding security system, automatically adjust the motion parameters according to the actual situation of the patients.

Rehabilitation robot is an important branch of medical robot covering the mechanics, electronics, materials science, computer science and robotics, and many other fields; it has become one of the research hotspots in the international robot field [3]. At present, rehabilitation robot has widely applied to rehabilitation

681

nursing, prosthetics and rehabilitation therapy. This will not only promote the development of rehabilitation medicine, also contributed to the development of new technologies and theories of related fields. However, one of the main problems of rehabilitation robot is the complex structure and control, which lead to operation difficulty and high cost [4]. In order to solve the problems of complicated structure and high cost, an assisted rehabilitation training method using simple mechanical equipment is proposed, and the key method is use of the four bar linkage curve, which can satisfies the requirements of specific rehabilitation training trajectories [5]. Therefore, the simulation model is established to obtain training trajectories and the mechanism parameters. This paper established the motion analysis model in MATLAB/Simscape, and applied to the motion analysis of spatial rehabilitation equipment.

2. Modeling Methods

The system modeling of the complex products mainly include the following four parts: functional/behavior analysis, requirement analysis, physical structure analysis and system analysis and verification [6]. Figure 1 shows the system modeling.

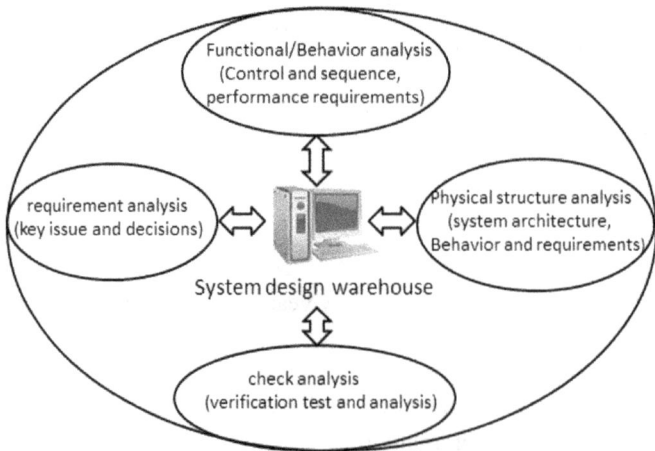

Figure 1. System modeling chart

Many engineering organizations face major challenges when designing and delivering new products that are dynamic in nature. One of the most significant of these is the discovery of design issues very late in the development cycle, introducing unbudgeted costs and delays into the project [7]. These delays often cause projects to overrun significantly, especially when the design issues are discovered during the prototyping or integration stages where real hardware is involved, so we need turn to the use of system-level modeling to develop "virtual" prototypes of systems.

Simscape is a system-level modeling and simulation tool that combines physical modeling with advanced symbolic computation techniques [8], it enables designers to evaluate mechanical, electrical, hydraulic or pneumatic sub-system designs jointly, in a virtual prototype, long before any physical prototyping or testing is possible, it also can quickly find the integration problem, and optimizes system level performance much faster and with greater rigor than was possible before [9]. Simscape toolbox has a series of module, therefore, it does not need to carry on the derivation to the traditional kinematical model by using Simscape [10]. Simscape can help product development in several ways through the use of virtual prototypes within the design process. During the initial, conceptual stages, engineers can gain early insight into how the proposed design will fulfill the requirements for the system, or the subsystem, they are working on. Figure 2 shows the Virtual Prototyping and modeling process using model-based design by means of Matlab / Simscape [11].

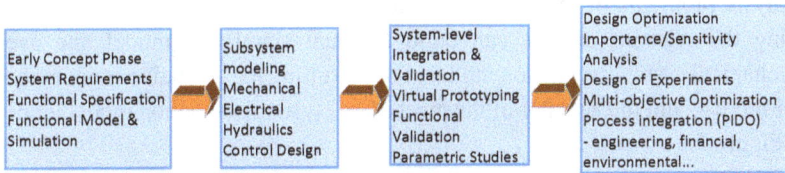

| Early Concept Phase System Requirements Functional Specification Functional Model & Simulation | Subsystem modeling Mechanical Electrical Hydraulics Control Design | System-level Integration & Validation Virtual Prototyping Functional Validation Parametric Studies | Design Optimization Importance/Sensitivity Analysis Design of Experiments Multi-objective Optimization Process integration (PIDO) - engineering, financial, environmental... |

Figure 2. Virtual Prototyping and modeling process

2.1. Simulation Model of Rehabilitation Training Equipment

2.1.1. Physical Model of Spatial Rehabilitation Training Equipment

The rehabilitation training equipment consists of vertical plane and horizontal plane. The vertical plane comprises three mechanisms, respectively a crank rocker mechanism; crank and oscillating block mechanism and crank slider mechanism. The horizontal plane includes 2-DOF five-bar mechanisms, double side five-bar mechanism and crank rocker mechanism. The rehabilitation training equipment can be configured in different forms according to the specific circumstances of the patient, a model of the rehabilitation training equipment is shown in Figure 3, the electric motor in lower case body transmits power to the drive shaft through the gear system and then drives the five bar mechanism to move.

Figure 3. Rehabilitation training equipment

3. Mechanical System Model

The aim of this section is to create the mechanical system model of the rehabilitation robot. For this purposes, the Simscape toolbox will be used. The toolbox is an object-oriented program able to perform simulations of most motion analysis problems of machines without a need of dynamic equations. Directly using the relevant module can build a visual simulation model, the overall mechanical system simulation model of the resulting rehabilitation robot is shown in Figure 4, and the actuator of the mechanical system simulation model is shown in Figure 5.

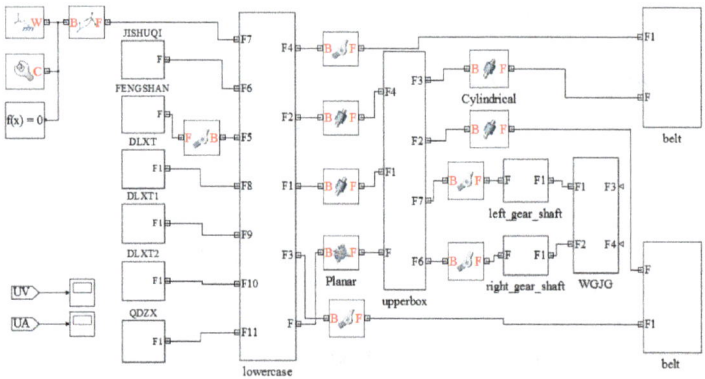

Figure 4. Mechanical system structure model of the rehabilitation training equipment

684

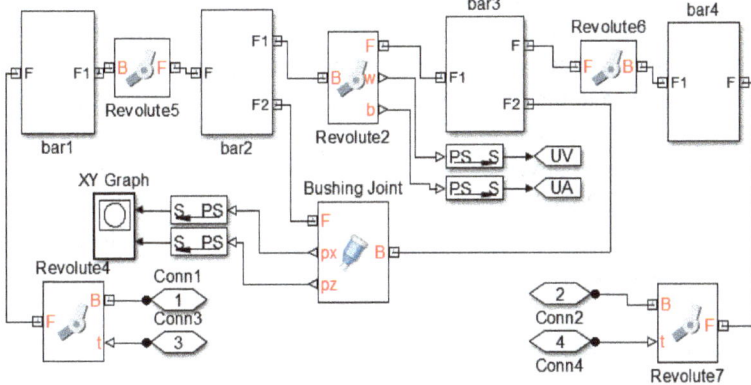

Figure 5. The actuator mechanical system structure model

4. Electromechanical Conversion Model

Driving device is controlled by the servomotor, which converts the electric signals into the output of angular displacement or angular velocity through the intermediate medium to drive the mechanical system. The control model of the schematic diagram is established by the SimElectronics toolbox, it realized the connection of electrical model and mechanical system. Figure 6 shows the electromechanical conversion model.

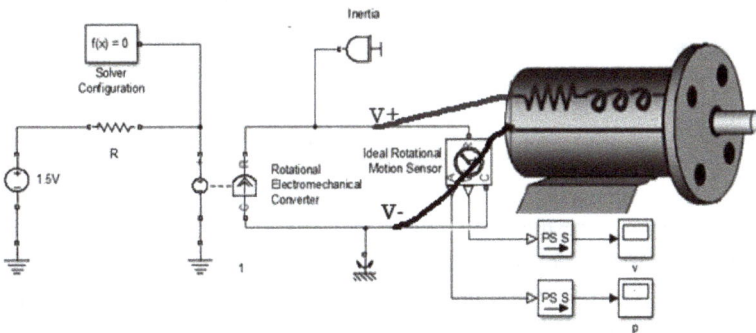

Figure 6. Electrical mechanical system conversion model

5. Simulation Results and Discussion

After setting the parameters according to the simulation model of rehabilitation training equipment and adjusting the input signal, it will automatically generate

3D View Plot when run the SimMechanics model. The simulation results can be directly output to the oscilloscope and can also be saved to the Matlab space. The simulation results can be output to the working space by calling the plot command when save the data to the Matlab space. Compared with the above two methods, by calling the plot command to output the simulation results to the working space is flexible, and it is convenient to view real-time parameters change and data analysis. One end effector's velocity, acceleration, and trajectory of the model are simulated. Figure 7 shows the velocity and acceleration curves of the end effector and Figure 8 shows the motion trajectory curve.

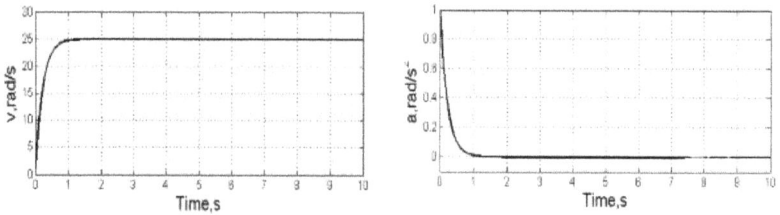

Figure 7. Velocity and Acceleration

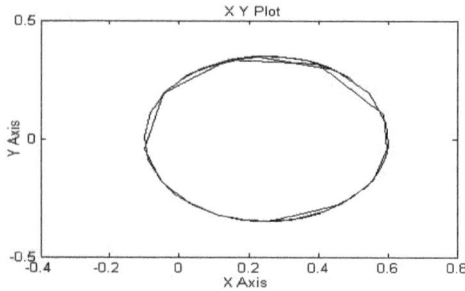

Figure 8. End Effector trajectory

The simulation results show that the velocity, with short adjusting time and fast response, is smooth without mutate site. It shows that the velocity of the end effector will soon reach the state of motion in Figure 7. The trajectory of the end effector can carry out some simple training in patients with arm and can be changed according to the corresponding parameters of SimElectronics module to do some training in different restoration stages.

Below are some kinds of training trajectories to achieve rehabilitation exercise for eating disorders and moving obstacles in Figure 9. It can be carried out some feed training according to the specific circumstances for the upper limb hemiplegia, including pick up and hold tableware (bowl, spoon, etc.) in Figures 9(a)-(d). Simple muscle training of upper and lower limbs can be executed in Figures (e) and (f), by the way, hand eye limbs dysfunction training, writing and drawing.

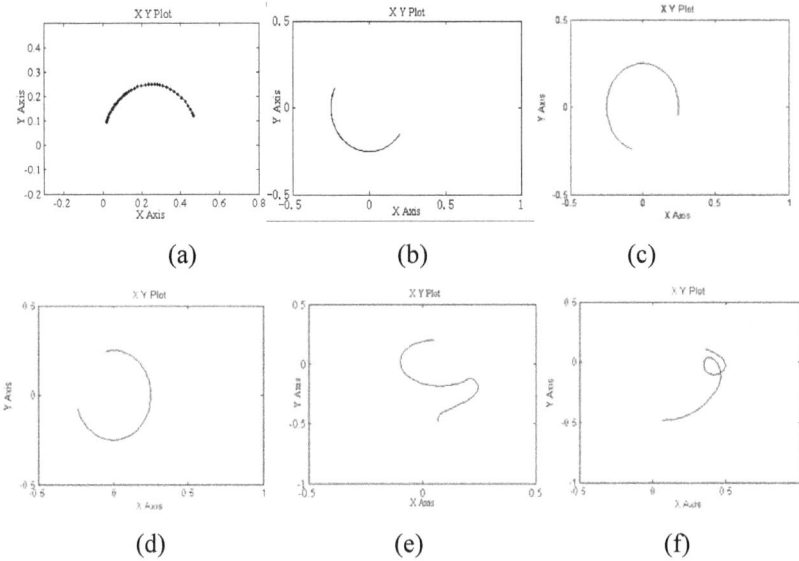

Figure 9. A set of training modes

6. Conclusion

It shows that the designer can easily complete visual simulation and analysis and does not need to carry out mathematical derivation of the kinematical model, and will obtain a couple of trajectories by analysis of different position points of the actuator. It can provide intuitive reference for fine control design of rehabilitation training equipment, and provide new concept for design direction of rehabilitation robot by analysis of the real-time data of the end actuator.

Acknowledgement

The research was funded by National Natural Science Funding of China (NSF) : (No. 51175284), and Science Research Program of Education Committee Beijing Municipality: (Grant No. SQKM201211232002), and Innovation Design Program of the Education Committee of Beijing Municipality.

References

1. Brodtmann, Amy. "IJS announces journal series on stroke, cognition and vascular dementia." International Journal of Stroke 6.5(2011): 375-375.
2. Kim, Hyunchul, et al. "Kinematic data analysis for post-stroke patients following bilateral versus unilateral rehabilitation with an upper limb wearable robotic system." IEEE Transactions on Neural Systems & Rehabilitation Engineering A Publication of the IEEE Engineering in Medicine & Biology Society 21.2(2013): 153-64.
3. Xiong, Youlun. "Control methods for exoskeleton rehabilitation robot driven with pneumatic muscles." Industrial Robot 36.3(2013): 210-220.
4. Marchal-Crespo, Laura, et al. "The effect of haptic guidance and visual feedback on learning a complex tennis task." Experimental Brain Research 231.3(2013): 277-91.
5. Hou Yuemin, Wang Xueyan, Li Hongliang. "Rehabilitation training assistant equipment of space movement for upper limbs and lower limbs", China 201410400703.3. August 14, 2014.
6. Yang, Weixia. "Modeling of Complex System of Waste Electronic Product Reclamation by Building TPL Enterprise." International Conference on Computer Science & Electronics Engineering IEEE Computer Society, 2012:219-222.
7. Repenning, N. "Understanding fire fighting in new product development", Journal of Product Innovation Management 18.5(2001): 285–300.
8. Cao, Yue, Y. Liu, and C. J. J. Paredis. "System-level model integration of design and simulation for mechatronic systems based on SysML." Mechatronics 21.6(2011): 1063-1075.
9. Cao, Yue, Y. Liu, and C. J. J. Paredis. "Integration of System-Level Design and Analysis Models of Mechatronic System Behavior Based on SysML and Simscape." ASME 2010 International Design Engineering Technical Conferences and Computers and Information in Engineering Conference 2010:1099-1108.
10. Cao, Yue, Y. Liu, and C. J. J. Paredis. "Integration of System-Level Design and Analysis Models of Mechatronic System Behavior Based on SysML and Simscape." ASME 2010 International Design Engineering Technical Conferences and Computers and Information in Engineering Conference 2010:1099-1108.
11. Arvin, R. S., et al. "Modeling and control design for rotary crane system using MATLAB Simscape Toolbox." Control and System Graduate Research Colloquium IEEE, 2014:170-175.

An Improved Incremental Conductance MPPT Algorithm for Photovoltaic System

Xue-cui Jia[1], Yue Li[2], Bo Li[2] and Xiang-jun Li[1,*]

[1]State Key Laboratory of Control and Operation of Renewable Energy and Storage Systems, China Electric Power Research Institute, Beijing, China
[2]School of Automation Engineering, Northeast Dianli University, China

*li_xiangjun@126.com

Maximum Power Point Tracking (MPPT) methods play an important role in Photovoltaic (PV) system. They maximize the output power of a PV array for a given set of conditions. Incremental Conductance (INC) MPPT is one of the most popular detection methods because of its simple implementation and high accuracy. In this paper, a new improved INC method is proposed based on the PV output power and it is suitable for practical operating conditions due to a wider operating range of irradiation changes. Simulation results confirm that the proposed algorithm reduces voltage fluctuation in comparison with conventional fixed step size INC algorithm.

Keywords: Maximum Power Point Tracking, Photovoltaic Module, Incremental Conductance Method, DC/DC Converter.

1. Introduction

The operating point of a PV generator, which is connected to a load, is determined by the intersection point of its characteristic curves. In general, this point is not the same as the generator's maximum power point. This difference means loss in the system performance. Peak energy harvesting in PV systems requires fast and effective maximum power point tracking (MPPT) [1]. Different algorithms have been proposed for MPPT. The well known perturb and observe (P&O) method [2-6]. For PV MPPT has been extensively used in practical application because of simple in idea and implementation. However, as reported by Wasynezuk [7], the P&O method is not able to track peak power conditions during periods of varying insolation, because it cannot quickly follow the change of PV characteristic and the change of disturbed PV voltage (ΔV_{pv}). An Incremental-Conductance (INC) MPPT method that is implemented by detecting the harmonic components of the PV module voltage and current was first proposed to track accurately the peak power of PV systems that are subjected to random variations in insolation. INC algorithm is based on the fact that the slope

689

of the power versus voltage curve corresponding to PV array is zero at the MPP, and it has been proposed to improve the tracking accuracy and dynamic performance under rapidly varying conditions [8-10]. Although the INC method is a little more complicated compared to the P&O strategy, it can be easily implemented due to the advancements of digital signal processors or microcontrollers. The INC MPPT algorithm usually has a fixed iteration step size [11], determined by the requirements of the accuracy at steady state and the response speed of the MPPT. Thus, the trade-off between the dynamics and steady state accuracy has to be addressed by the corresponding design. To solve these problems, some algorithms with variable step size are proposed in literatures [12-14]. The step size is automatically tuned according to the inherent PV array characteristics. If the operating point is farther from MPP, it increases the step size, which enables a fast tracking ability. On the other hand, if the operating point is close to the MPP, the step size becomes very small that the oscillation becomes well reduced contributing to a higher efficiency.

In this paper, an improved variable step size algorithm, which is based on the incremental conductance algorithm is proposed to adjust the step size using the slope of power versus voltage curves of PV array and also to adapt step size according to sun irradiation levels using PV output current. Proposed method can increase convergence speed of MPPT for a wide operation range without degrading steady state response.

2. Model of a PV Module

2.1. *Mathematical Model of a Solar Cell*

An ideal PV cell can be modeled as a current source in parallel with a diode; however, no real solar cell is ideal, so a shunt resistance R_{sh} and a series resistance R_s are always added to the model, as shown in Fig. 1.

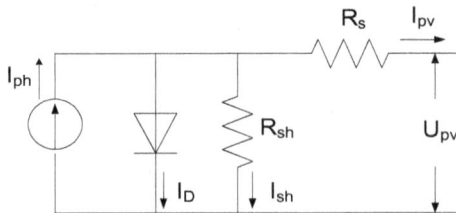

Figure 1. Typical model of a PV cell with single diode [14]

690

The typical $I_{pv} - V_{pv}$ relationship of a practical PV cell, can be described by

$$I_{pv} = I_{ph} - I_D - I_{sh} \qquad (1)$$

$$I_D = I_o \left[\exp\left(\frac{qU_D}{AkT} \right) - 1 \right] \qquad (2)$$

$$U_D = U_{pv} + I_{pv}R_s \qquad (3)$$

$$I_{pv} = I_{ph} - I_o \left[\exp\left(\frac{q(U_{pv} + I_{pv}R_s)}{AkT} \right) - 1 \right] - \frac{U_{pv} + I_{pv}R_s}{R_{sh}} \qquad (4)$$

where I_{ph} denotes light-generated current, I_o is the reverse saturation current, I_{pv} is the PV electric current, V_{pv} is the PV voltage, R_s is the series resistance, A is the ideality factor, k is Boltzmann's constant($1.3806503 \times 10^{-23} J/K$), T (in Kelvin) is the temperature of the p-n junction, and q is the electronic charge($1.60217646 \times 10^{-19} C$).

The terms related to the R_{sh} in Eq. 4 can be generally ignored, because the value of it in PV arrays is relatively high in the majority of practical cases; therefore $\frac{U_{pv} + I_{pv}R_s}{R_{sh}} \to 0$, and Eq. 4 can be simplified into:

$$I_{pv} = I_{ph} - I_o \left[\exp\left(\frac{q(U_{pv} + I_{pv}R_s)}{AkT} \right) - 1 \right] \qquad (5)$$

A single solar cell can only produce a small amount of power[15]. To increase the output power of a system, solar cells are generally connected in series or parallel to form PV modules. The main equation for the output current of a module is:

$$I_{pv} = N_p I_{ph} - N_p I_o \left[\exp\left(\frac{q(U_{pv} + I_{pv}R_s)}{N_s AkT} \right) - 1 \right] \qquad (6)$$

where N_s represents the number of PV cells connected in series, and N_p represents the number of such strings connected in parallel.

2.2. Temperature and Irradiation Effects

Eq. 4 specifies that solar irradiance and temperature are the two main factor affecting PV cell output. Therefore, it is predictable that the PV output will vary with changing atmospheric conditions. As irradiation depends on angle of incidence of the sunrays with respect to the panel. These factors directly affect the output, altering the P-V and I-V characteristics. Figs. 2(a)-(d) display the significant variation in the outputs based on a PV module with nominal parameters under Standard Testing Condition (STC).

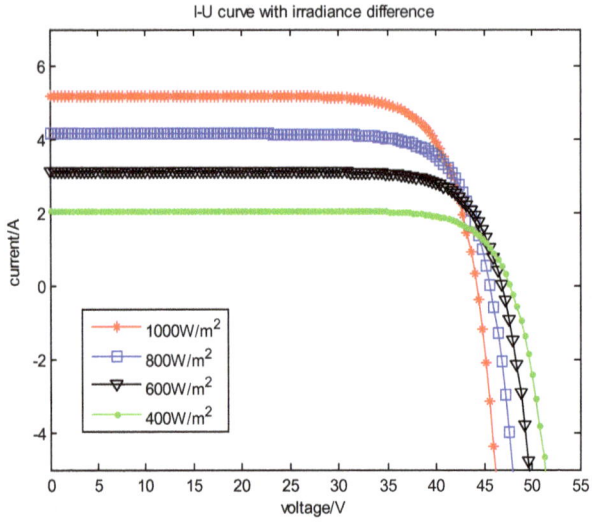

(a) I-U curves under the same ambient temperature ($25^{O}C$) and different irradiation level

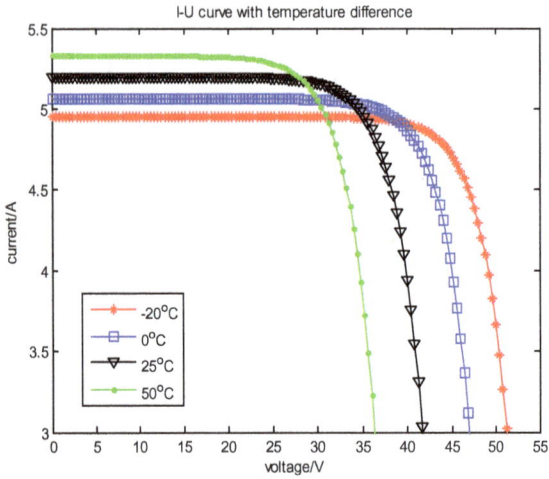

(b) I-U curves under the same irradiation level ($1000W/m^2$) and different ambient temperature

Figure 2. A set of PV module characteristics

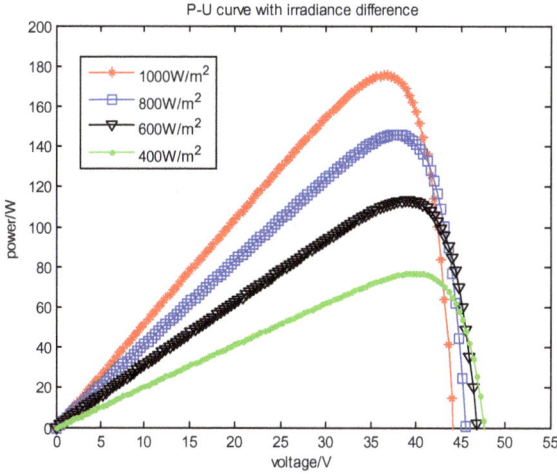

P-U curve with irradiance difference

(c) P-U curves under the same ambient temperature ($25^O C$) and different irradiation level

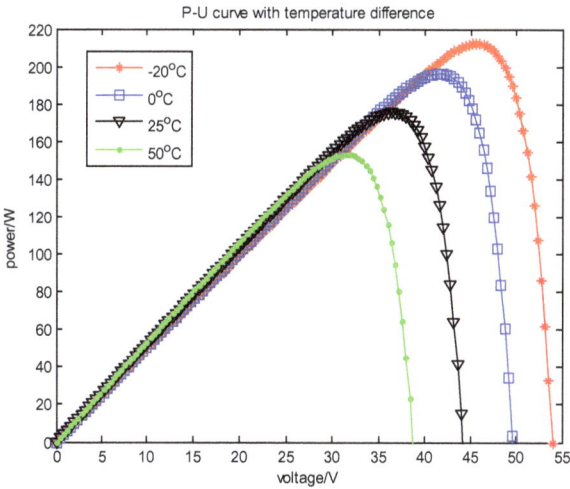

P-U curve with temperature difference

(d) P-U curves under the same irradiation level ($1000W / m^2$) and different ambient temperature

Figure 2. A set of PV module characteristics (*continued*)

The electrical characteristics of this modul for $T = 25$, $I_r = 1000W / m^2$ are listed, voltage at $P_{\max} : V_{mpp} = 35.2V$, current at $P_{\max} : I_{mpp} = 4.95A$, open-circuit voltage: $V_{oc} = 44.2V$, short-circuit current: $I_{sc} = 5.2A$.

The performance of a PV system relies on the operating conditions. The maximum power extracted from the PV generator depends strongly on three factors: insolation, load profile (load impedance) and cell temperature (ambient temperature). The variation of the output I-V, P-V characteristic of a commercial PV module as a function of temperature and irradiation is shown in Fig. 2. It can be observed that the temperature changes mainly affect the PV output voltage, while the irradiation changes mainly affect the PV output current.

As shown in Fig. 2, an operating power point exists for each given set of temperature and solar irradiation levels at which maximum output power can be harnessed from the PV module. This unique operating power point is known as the MPP and it is attained for the particular output voltage and current corresponding to the MPP.

3. Incremental Conductance MPPT Algorithm

The INC MPPT algorithm is based on the fact that the slope of the PV array power curve is zero at the MPP, positive on the left-hand-side of the MPP, and negative on the right-hand-side of it (as shown in Fig. 3). Therefore we can search for the location of MPP calculating the slope of this curve.

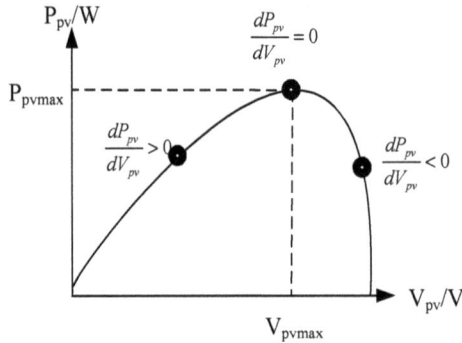

Figure 3. The relationship between the sign dP_{pv}/dV_{pv} and the operating point on P-V curve

The output power from the PV array can be given by

$$P_{pv} = V_{pv}I_{pv} \tag{7}$$

and

$$\frac{dP_{pv}}{dV_{pv}} = \frac{d(V_{pv} \cdot I_{pv})}{dV_{pv}} = I_{pv} + V_{pv} \cdot \frac{dI_{pv}}{dV_{pv}} \stackrel{\sim}{=} I_{pv} + V_{pv} \cdot \frac{\Delta I_{pv}}{\Delta V_{pv}} \tag{8}$$

The MPP in the PV module occurs when

$$\frac{dP_{pv}}{dV_{pv}} = 0 \tag{9}$$

$$\Rightarrow \frac{\Delta I_{pv}}{\Delta V_{pv}} = -\frac{I_{pv}}{V_{pv}} \tag{10}$$

According to the Eq. 10, MPP can be tracked by comparing instantaneous conductance (I_{pv}/V_{pv}) and the incremental conductance ($\Delta I_{pv}/\Delta V_{pv}$).

$$\begin{cases} \dfrac{\Delta I_{pv}}{\Delta V_{pv}} > -\dfrac{I_{pv}}{V_{pv}} & \textit{Left of MPP} \\[3mm] \dfrac{\Delta I_{pv}}{\Delta V_{pv}} = -\dfrac{I_{pv}}{V_{pv}} & \textit{at MPP} \\[3mm] \dfrac{\Delta I_{pv}}{\Delta V_{pv}} < -\dfrac{I_{pv}}{V_{pv}} & \textit{Right of MPP} \end{cases} \tag{11}$$

The flowchart for incremental conductance method is shown in Fig. 4.

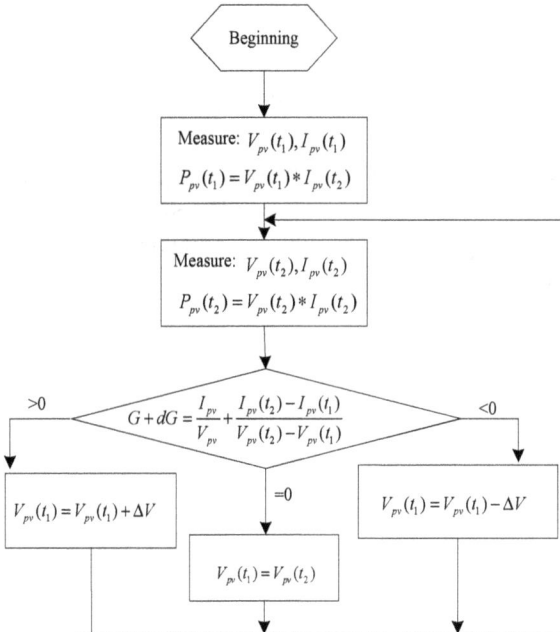

Figure 4. The flowchart for incremental conductance method

The basic method of INC MPPT usually uses a fixed iteration step size to change the duty cycle of DC/DC converter. The step size for the INC MPPT determines how fast the MPP is tracked. Fast tracking can be achieved with bigger increments, but the system might not run exactly at the MPP and oscillate around it; thus there is to be a comparatively low efficiency. This situation is inverted when the MPPT is operating with a smaller increment. Therefore, a trade-off between the dynamics and oscillations has to be made for the fixed step-size MPPT. Generally, the fixed step MPPT can only take a compromise between the dynamic response and the steady state oscillation, and cannot achieve the optimization. To solve this problem, a different INC MPPT algorithm with variable step is proposed.

4. Variable Step INC MPPT

In INC MPPT method, there is a potential relationship between the maximum power point and the sum of the conductance and its increment. The step size of MPPT algorithms is the variable increment of duty ratio ΔD.

$$\Delta D_k = \pm N \left| \frac{P_k - P_{k-1}}{V_k - V_{k-1}} \right| \tag{12}$$

where k is the step label, N is the scaling factor which is tuned at the design stage to adjust the step size.

The performance of the MPPT system is essentially decided by the scaling factor (N) for the variable step-size MPPT algorithm. To guarantee the convergence of the MPPT update rule, the variable step rule must meet the following inequality:

$$N \times \left| \frac{dP}{dV} \right| < \Delta D_{max} \tag{13}$$

where ΔD_{max} is the largest step size for fixed step-size MPPT operation and is chosen as the upper limit for the variable step size INC MPPT method. Therefore, the scaling factor can be obtained as

$$N < \Delta D_{max} / \left| \frac{dP}{dV} \right| \tag{14}$$

In the variable step size algorithm, the PV output power directly controls the power converter duty cycle. The flowchart for the proposed method is shown in Fig. 5.

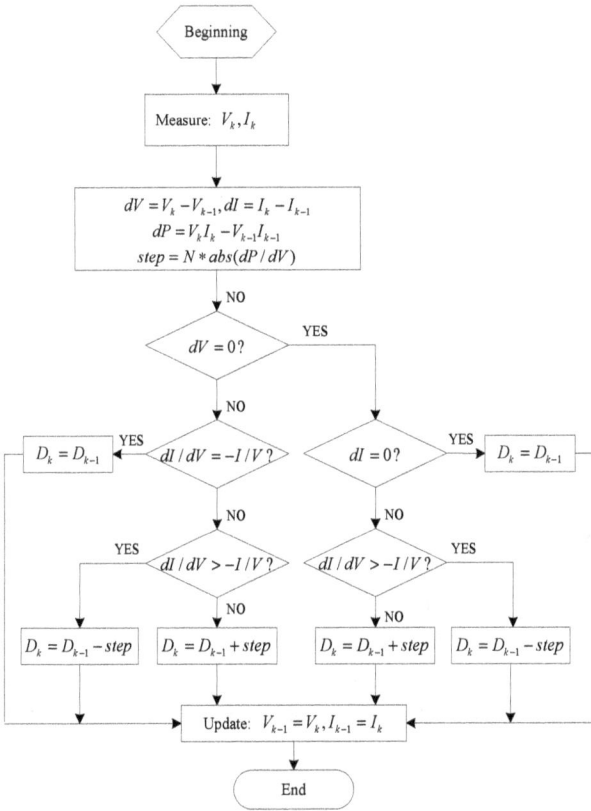

Figure 5. The flowchart for variable step size INC MPPT method

5. Simulation Results

A stand-alone PV system with MPPT has been designed and simulated in MATLAB-Simulink software for assessing the proposed MPPT method. Designed system consists of PV, MPPT algorithm (improved variable step size INC), a buck converter and also a resistive load as shown in Fig. 6.

697

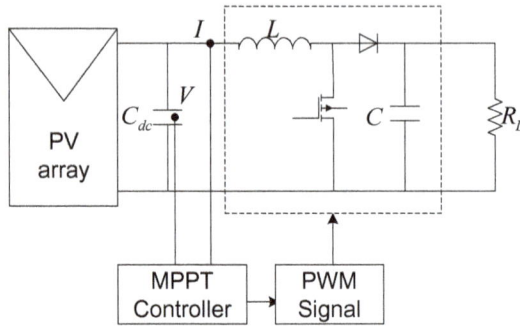

Figure 6. Stand-alone PV system with MPPT

The components used in buck converter in simulation are as follows:

1) Inductor: $L = 0.5mH$

2) Capacitors: $C_{dc} = 2\mu F$, $C = 1\mu F$

3) Resistive loud: $R_L = 2\Omega$

To test the system operation and comparing the proposed variable step size INC algorithm with fixed step size INC, we set temperature constant at $20\,^{\circ}C$ and assume that sun irradiation level varies between $200W\,/\,m^2$ and $1000W\,/\,m^2$ in order to model a wide range operation point. Fig. 7 shows the comparing result of PV array output voltage using fixed step size and variable step size INC MPPT algorithms, and Fig. 8 given some details in Fig. 7.

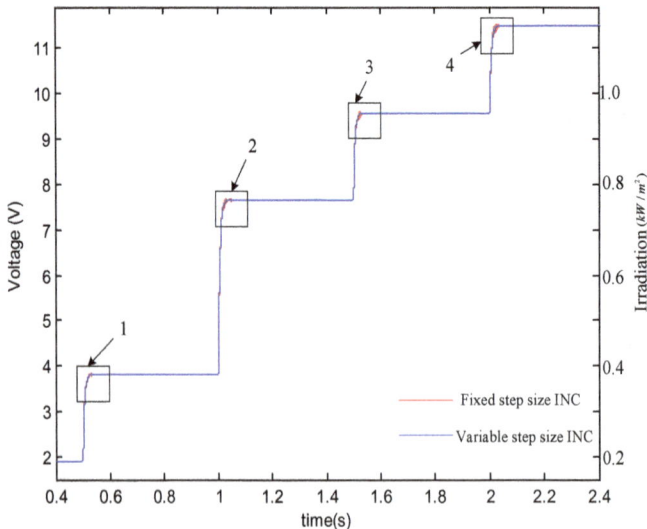

Figure 7. PV array output voltage, fixed step size and variable step size INC

698

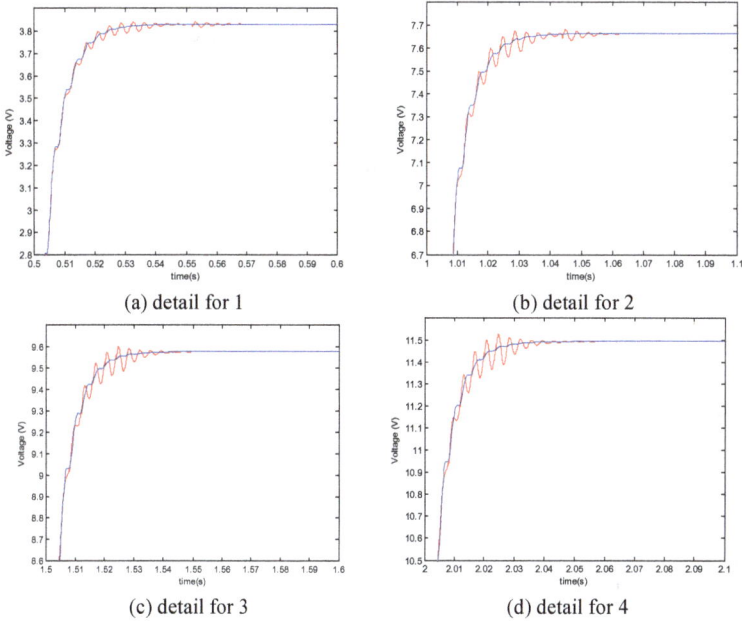

(a) detail for 1   (b) detail for 2

(c) detail for 3   (d) detail for 4

Figure 8. Detail drawing for 1 to 4 in Figure 7

According to the obtained results, two algorithms reached the same steady state response. It is clear that proposed method can reduce the voltage fluctuation in comparison with conventional fixed step size INC algorithm after irradiation changing.

6. Conclusion

In this paper an improved variable step size INC MPPT algorithm has been proposed. The proposed algorithm simulated in MATLAB Simulink software and compared with conventional fixed step size INC algorithm. Simulation results confirm that the proposed method can reduce voltage fluctuation in comparison with conventional fixed step size INC algorithm after irradiation changing.

Acknowledgment

This project was funded by Beijing Nova Program (Z141101001814094), and by Science and Technology Foundation of STATE GRID Corporation of China (DG71-15-039).

References

1. Francisco Paz, Martin Ordonez. Fast and Efficient Solar Incremental Conductance MPPT Using Lock-In Amplifier. IEEE 6th International Symposium on Power Electronics for Distributed Generation Systems, 2015:1-6.
2. R Ahmed, A Namaane, NK M'Sirdi. Improvement in Perturb and Observe Method Using State Flow Approach. Energy Procedia, 2013,42:614-623.
3. Nicola Femia, Giovanni Petrone, Giovanni Spagnuolo, et al. Optimization of Perturb and Observe Maximum Power Point Tracking Method. IEEE Transactions on Power Electronics. 2005,20(4): 963-973.
4. Sonali Surawdhaniwar, Ritesh Diwan. An Improved Approach of Perturb and Observe Method Over Other Maximum Power Point Tracking Methods. International Journal of Recent Technology and Engineering(IJRTE). 2012,1(3): 137-144.
5. KL Lian, JH Jhang, IS Tian. A Maximum Power Point Tracking Method Based on Perturb-and-Observe Combined with Particle Swarm Optimization. IEEE Journal of Photovoltaics. 2014,4(2): 626-633.
6. K Marom, V Levy, G Pillemer, et al. Maximum Power Point Tracking Method Using a Modified Perturb and Observe Algorithm for Grid Connected Wind Energy Conversion Systems. Renewable Power Generation Let.2015, 9(6): 682-689.
7. O. Wasynezuk, Dynamic behavior of a class of photovoltaic power system. IEEE Trans. Power App. Syst., Vol, PAS-102, no.9, pp.3031-3037, Sep.1983.
8. Mei, Q., Shan, M., Liu, L., et al. A novel improved variable step-size incremental-resistance MPPT method for PV systems. IEEE Transactions on Industrial Electronics, 2011,58(6): 2427-2434.
9. Tey, K.S., Mekhilef S. Modified incremental conductance MPPT algorithm to mitigate inaccurate responses under fast changing solar irradiation level. Solar Energy, 2014,101:333-342.
10. Tey K.S., Mekhilef S. Modified Incremental Conductance algorithm for photovoltaic system under partial shading conditions and load variation. IEEE Transactions on Industrial Electronics. 2014,61(10): 5384-5392.
11. Safari, A., and Mekhilef, S., Simulation and hardware implementation of incremental conductance MPPT with direct control method using cuk converter. IEEE Transactions on Industrial Electronics, 2011, 58(4): 1154-1161.

12. Liu F., Duan S. Liu B. et al. A variable step size INC MPPT method for PV systems. IEEE Transactions on Industrial Electronics, 2008, 55(7): 2622-2628.

13. Yie-Tone Chen, Yi-Cheng Jhang, Ruey-Hsun Liang. A fuzzy-logic based auto-scaling variable step-size MPPT method for PV systems. Solar Energy, 2016, 126:53-63.

14. A. Amir, J. Selvaraj, N.A. Rahim. Study of the MPP tracking algorithms: Focusing the numerical method techniques. Renewable and Sustainable Energy Reviews. 2016, 62:350-371.

15. Rauschenbach H. Solar cells array design handbook: the principles and technology of photovoltaic energy conversion. Van Nostrand Reinhold. 1980.

Sealing Ring Reliability Assessment for Hydraulic Systems Based on BP Neural Networks

Feng Zhang*, Wei-wei Deng, Qi Zhang, Lei-lei Zhang
and Kai-liang Luo
*Institute of Aircraft Reliability Engineering, School of Mechanics,
Civil Engineering and Architecture, Northwestern Polytechnical University,
Xi'an, 710129, China*
Email: 18392993424@163.com

In this paper, reliability assessment of sealing ring in hydraulic systems is carried out based on the Back Propagation (BP) neural networks. An O-ring Finite Element Model (FEM) is established based on ANSYS software, by which the calculated maximum contact stress is taken as failure criteria. The diameter, oil pressure, the amount of compression, and modulus of elasticity are treated as random variables to take into account the parameters variations for studying the effect of parameter uncertainty on the sealing performance. Simulation data that generated from the FEM model is trained by using the artificial neural network toolbox to fit the relationship between the input data and output response. The trained neural network model is finally combined with the importance sampling method, to study the effect of the parameter variations on the reliability of the seal.

Keywords: Sealing structure; Failure probability; BP neural network.

1. Introduction

Sealing rings are one of the most important types among components in aircraft hydraulic systems, and its performance directly affects the reliability of the entire hydraulic system, even endanger safety of the flight [1]. Currently, O-ring is widely used, if the structure of O-ring is damaged or failed, it may cause leakage in hydraulic piping systems and aircraft out of order, which could lead to plane crash[2]. The "Challenger" space shuttle of America wrecked, because a sealing failure on the surface of solid rocket motor deformed. Therefore, it is necessary to carry out reliability assessment of the ring's structure.

In order to assess the sealing reliability of the hydraulic systems [3-8], a lot of research has been carried out on the sealing ring performance analysis. Pressure exists on both sides of seal ring, so structural failure of the ring occurs easily, and media leaks through the gap. Chen [3] used ABAQUS software to establish O-ring

---

* Corresponding author.

finite element analysis model of hydraulic gas distribution systems, the effect on sealing O-ring in hydraulic valve system by different parameters were analyzed. Sui [5] examined the effect of temperature on the leakage rate, and used an equipment to test the rate basis on a trial, then simulated working condition of seals in the explosion container. Dong [6] calculated the model O-ring's fatigue life of nodes on the dangerous interface points, the experimental data is fitting and classifying according to the strain energy release rate of fracture mechanics. Zhai [7] studied the aging process and the failure law of nitrile rubber in seawater. Li [8] used ABAQUS software to build O-ring axisymmetric model, and studied the distribution of maximum contact pressure and the maximum Von Mises stress on the seal surface with the effect of different compression ratio and oil pressure. Currently, there is less consideration of parameter fluctuations calculation in static sealing gasket design about reliability analysis.

This paper aims at sealing problems of aircraft hydraulic systems, and select O-ring as the research object, which is currently used widely. Take the geometric shape of the ring's structure, material and boundary conditions into consideration, when analyze finite element of the sealing equipment, we use BP artificial neural network with ANSYS through co-simulation method of seal to intensity reliability analysis. Finally, we combine with the importance sampling method, and study the effect on reliability model of ring when parameters designed are changed, and verify the reliability of the seal.

2. The performance analysis of Sealing equipment

As shown in Fig. 1, sealing equipment of an aircraft hydraulic system include the O-ring, piston rod and seal groove etc. The material of O-ring is nitrile rubber. Take the four key parameters of sealing equipment into consideration, which are diameter D, elastic modulus E, the amount of compression l and working oil pressure P.

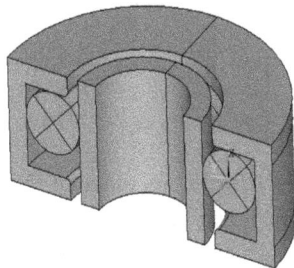

Fig. 1 Three-dimensional of O-ring model

When analyze finite element of the sealing equipment, consider the piston rod and sealing groove as a whole, build its finite element analysis model by using ANSYS software, as shown in Fig. 2. The FEM model of the O-ring parametric model consists of 1426 units, 1633 nodes.

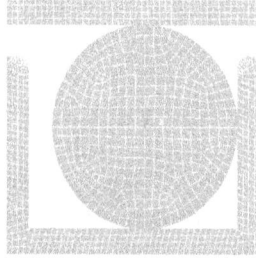

Fig. 2 O-ring parametric model

As research shows, leakage mainly occurs in the sealing interface. Oil film is formed between piston rod and seal groove in working. It is lubricating, and acts as a seal to isolate liquid of the chamber. When the maximum contact pressure σ is greater than the working oil pressure P between the ring and cylinder, it can play a role in sealing, if the maximum contact pressure is less than the working oil pressure, the leakage of hydraulic pipeline system occurs. Thus we can establish the function g about sealing failure of performance as the following Eq. (1).

$$g = \sigma - P \qquad (1)$$

Where σ is implicit function of diameter ring structure D, the elastic modulus E, the amount of construction structure l and working oil pressure P, we can get it by finite element analysis through ANSYS software. Sealing structure is failing or not will be described by the following Eq. (2):

$$\begin{cases} g < 0 & \text{failed} \\ g = 0 & \text{limited state} \\ g > 0 & \text{safe} \end{cases} \qquad (2)$$

We can get contact stress distribution cloud of ring structure by finite element analysis, as shown in Fig. 3. It can be seen that maximum stress appears in the upper right part of the sealing ring, it is 7.577 Mpa, which is greater than the oil pressure applied as 5Mpa, in this situation, the leakage problem does not occur.

704

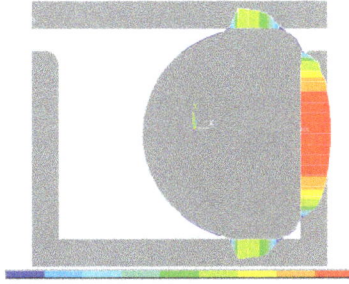

Fig. 3 O-ring stress cloud

3. Reliability analysis of the sealing structure

In actual project, it can fluctuate in the size parameters of the sealing structure [9], material parameters etc, its distribution type and parameters are as shown in Table 1. We can see that the four basic variables follow normal distribution, they are independent, which makes the function g of the sealing structure as a random variable.

Table 1. Ring's structure parameters and distribution

variables	identifiers	distribution type	means	variance
Elastic modulus E (MPa)	x_1	normal	14.4	1.189
Ring's diameter D (mm)	x_2	normal	4	0.3304
The amount of compression l (mm)	x_3	normal	0.4	0.03304
Oil pressure P (MPa)	x_4	normal	5	0.413

$$D_f = \{x : g = \sigma(E,D,l,P) - P = \sigma(x_1,x_2,x_3,x_4) - x_4 < 0\} \tag{3}$$

$$P_f = \int \cdots \int_{D_f} f_x(x)dx \tag{4}$$

Failure region of the sealing structure D_f and probability P_f can be expressed as Eqs. (3) and (4), wherein, $f_x(x)$ is the joint probability density function of the variables (x_1,x_2,x_3,x_4).

In connection with failure probability of the Eq. (4), it extracts samples through Monte Carlo method of the joint probability density function, then use the extracted sample to estimate the failure probability, and estimated result is shown as Eq. (5).

$$\hat{P}_f = \frac{1}{N} \sum_{j=1}^{N} I(x_j)$$
(5)

where, x_j is the j th sample which extracted in the joint probability density function $f_x(x)$, N is the total extraction of samples, $I(x)$ is the indicator function, if $x \in D_f$ set up, $I(x) = 1$, or $I(x) = 0$

The Monte Carlo method is based on law of large numbers, which makes an estimation to converge value of the failure probability through a large number of samples. In order to reduce the time of calculation, this paper introduces the importance sampling function $h_x(x)$, the failure probability of sealing structure can be rewritten as Eq. (6).

$$P_f = \int \cdots \int_{D_f} f_x(x)dx = \int \cdots \int_{D_f} [f_x(x)/h_x(x)]h_x(x)dx$$
(6)

Sealing structure failure probability that bases on an estimate of the importance sampling method can be written as Eq. (7).

$$\hat{P}_f = \frac{1}{N} \sum_{j=1}^{N} \left[I(x_j) \frac{f_x(x_j)}{h_x(x_j)} \right]$$
(7)

Where x_j is the jth sample to be extracted of the joint probability density function $h_x(x)$.

Traditional importance sampling method constructs an important function in the design of the sampling point of the limit state equation, which takes extra work to calculate the design point, especially implicit limit state equation. Another idea to build the construction of importance sampling function is to keep the center point of the mean unchanged, construct the importance sampling function by increasing in the standard deviation of variables, so that the standard deviation of new function is times of the original probability density function. The method does not need to find design point of the limit state equation, either the number of limit state equation. For the structure of sealing in this paper, the importance sampling function can be written as Eq. (8).

$$h_x(\lambda, x) = \frac{1}{(\sqrt{2\pi})^4 \prod_{i=1}^{4} (\lambda_i \sigma_i)} \exp \left\{ \sum_{i=1}^{4} -\frac{(x_i - \mu_i)^2}{2(\lambda_i \sigma_i)^2} \right\}$$
(8)

Where μ_i, σ_i are respectively mean and standard deviation of x_i ($i = 1, 2, 3, 4$).

During reliability assessment of sealing structure, we obtain samples directly through importance sampling method, then use ANSYS software to get maximum contact stress of the sealing structure, this requires a large amount of computation. In connection with implicit limit state equation, in this paper, BP artificial neural network fit relationship of the input parameters (x_1, x_2, x_3, x_4) and output (maximum contact stress σ), then combine with the importance sampling method and analyze reliability of sealing structure, which can greatly reduce the amount of calculation.

BP neural network topology has three layers, which are the input layer, hidden layer and output layer [12-14]. As shown in Fig. 4.

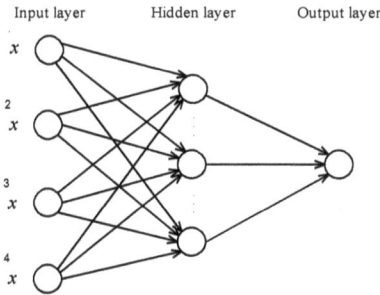

Fig. 4 BP neural network topology

In the process of building BP neural network model, we need to normalize the data firstly, it shortens the learning time of the neural network, which improves the computing efficiency. Then establish input of BP network training samples (x_1, x_2, x_3, x_4), output vector is σ, the variable input parameters are (x_1, x_2, x_3, x_4), the maximum contact stress σ is an output parameter. The choice of hidden layer transfer function is S-type logarithmic function, the output layer selection of the transfer function is purelin linear transfer function, trainlm function as training function, and then we can train the sample data. With 350 groups data as a learning sample, we can determine the fit function expression, and make prediction through trained network.

After 16-step of training, the error of neural network reaches the set value 10^{-4}, and global error is 2.5168×10^{-6}, which meets the requirement of accuracy, then end up training. Error curve is shown in Fig. 5.

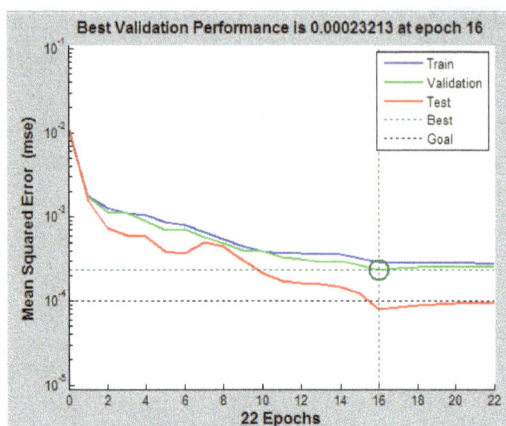

Fig. 5 Training error curve

We extract samples that combined BP neural network with the importance sampling method; the number of samples is 5×105. Fig. 6 shows the change rule of the failure probability with the number of samples. When the failure probability is 2.66×10-4, it meets the accuracy requirement.

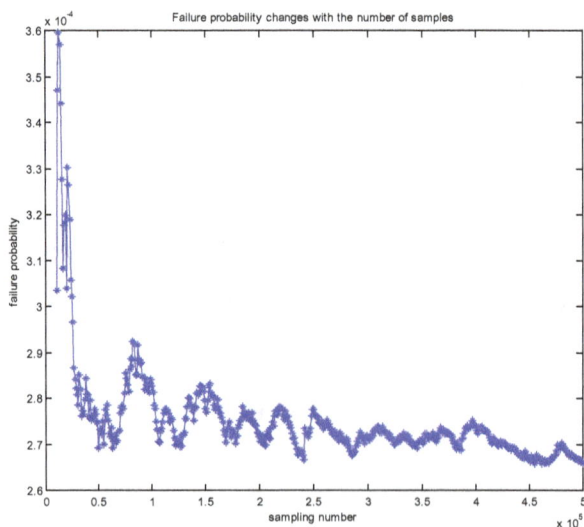

Fig. 6 Failure probability changes with the number of samples

4. Conclusion

1) Based on BP neural network, the relationship of ring's maximum contact stress and input parameters was built. After evaluation, it has good fitting accuracy.
2) In this paper, the new importance sampling method increases variance based on Monte Carlo method, which doesn't need design point and the method is simple, it can also be extended to more failure modes in reliability analysis.
3) Compared with the directly calling finite element, combining importance sampling method with BP neural networks can reduce the amount of calculation greatly.

Acknowledgements

Authors gratefully acknowledge the support of the foundation of the NPU-FFR-JC20100232, NPU-FFR-3102015BJ(II)JL04, specialized research fund for the doctoral program of higher education (20136102120032) and natural science foundation of Shaanxi Province (2016JQ5109).

References

1. Wang YX. Aircraft control system and hydraulic system design[M], Beijing: Aviation Industry Press,2003.
2. Guo JJ, Liu W. Reliability calculation method of electromechanical system based on random fault injection combined with artificial neural network [J], China Mechanical Engineering, 2015,26(9):1221-1226.
3. Chen JD, Xie QS. Analyze of dynamic sealing characteristics of O-ring in hydraulic valve system [J], Lubrication Engineering, 2014,39(11):63-65.
4. Sui YG. An equipment for leakage rate testing of owing in the flange [J], Lubrication Engineering, 2013,38(12),76-79.
5. Sui Y G, Tang SY, Ma YJ. Experimental used study on leakage rate of O-ring used for explosion container [J], Lubrication Engineering, 2012,37(10), 66-69.
6. Dong ZJ, Wu X. Research on fatigue performance of O-ring based on fracture mechanics [J], Lubrication Engineering, 2014,39(11), 59-62.
7. Zhai ZS, Zhong X. Predicting the service lifetime of O-rings in seawater [J], Synthetic Materials Aging and Application,2011,40(6),30-42.
8. Li ZT, Sun XH. Finite element numerical simulation of the sealing performance of owing seals [J], Lubrication Engineering. 2011,36(9) :102-105.

9. Lu ZZ, Li LL, Song SF. The importance of uncertainty structure analyzes, theory and Solution [M]. Beijing: Science Press, 2015.
10. Liu Z, Zhang JG. Reliability hybrid optimization algorithm based on Kriging model importance sampling method and structure [J], Acta Aeronautica et Astronautica Sinica, 2013, 34(6), 1347-1355.
11. Zhang F. Research on optimization algorithm of structural reliability [D], Xi'an: Northwestern Polytechnical University, 2007.
12. Cheng J, Li QS, Xiao R. A new artificial neural network-based response surface method for structural reliability analyze [J], Probabilistic Engineering Mechanics, 2008, 23(1): 51-63.
13. Yin G, Li MH, Li W. Model of coal gas permeability prediction based on improved BP neural network [J], Journal Of China Coal Society. 2013,38(7), 1179-1182.
14. Xu LM, Wan Q, Chen JP. Forecast for average velocity of debris flow based on BP neural network [J], 2013,43(1)186-189.

Chapter 5

Testing and Imaging

Study on Deepwater Testing Tubing Safety Near Subsea Wellhead

Bao-kui Gao*, Jing-wei Ren and Liang Gao
*Department of Petroleum Engineering,
China University of Petroleum Beijing,
Beijing, 102249, China*
*gaobaokui@126.com

With regard to wellbore structure near subsea wellhead, differential equation was established to describe the testing string deformation just above the wellhead. Effects of key parameters, such as axial force and riser flexible joint deflection, on local testing string mechanical behavior were investigated. Taking maximum bending stress and tubing-flexible joint contact as criterions, confines of key factors were expressed though a safe range chart to prevent tubing failure. According to calculation, tubing string has two potential risks: stress yielding at fixed end and collapsing from contact with flexible joint. The angular deflection of flexible joint is the dominating parameter over tubing string safety. Optimal distance from flexible joint to subsea testing tree was proposed, and strategy for promoting string safety was recommended.

Keywords: Deep water, Testing string, Wellhead, Riser, Flexible joint.

1. Introduction

Oil well testing is a important operation [1, 2]. The testing string usually consists of many tools besides tubing. In deepwater well testing, there are various controlling lines and safeguarding units as auxiliaries from below mud line to platform [3, 4]. The whole technological procedure is very complex, and any hidden trouble may lead to great loss [5]. Subsea testing tree (STT) is an important apparatus, which locates in blowout preventer (BOP) and is fixed to wellhead [6, 7]. As a connecting link, STT divides the main string into two parts: below mud line and above mud line. The diameter of STT is much larger than that of tubing, so the bottom end of upper tubing is actually a fixed end [8, 9]. From STT, the tubing extends into riser and upwards to platform. Between BOP and riser there is a flexible joint with the maximum angular deflection being 7° to 12° [10, 11]. The flexible joint can greatly

---

* Corresponding author.

reduce bending stress of riser, but at the same time it brings troubles for tubing string. In a limited range, tubing must adjust its deformation in accordance with riser displacement. The tubing usually bears tension even at its fixed bottom end, together with high internal pressure, so tubing safety in this place is noticeable. Although there have been accidents report and investigation published for drill string and casing string, local mechanical behavior of testing string has not been studied up to now [12,13].

In this paper, tubing string lateral displacement differential equation was established and solved by means of numerical method. STT, flexible joint and riser were treated as structural constraints, and axial force and operation process were also considered. By calculating tubing stress and deformation, impact of structural features on tubing safety was discussed. Finally, predominance-range chart was proposed for determining safe working environments.

2. Tubing String Mechanical Model

2.1. *Wellbore Description*

Local wellbore and tubing string structure can be simplified as that in Fig. 1.

The sinuosity in Fig. 1a) indicates flexible joint, above it is riser and below it is BOP and STT. The riser has no steady state. In order to describe its restraint to working string, riser lateral deformation is expressed as sine curve and only a small part of one period will be used.

Planar rectangular coordinates are chosen as that in Fig. 1a). Then in a limited length above origin of X coordinate wellbore lateral displacement can be expressed as

$$\begin{cases} Y_r(X) = 0, & 0 \le X \le H \\ Y_r(X) = A \, \sin\frac{2\pi(X-H)}{L}, & X \ge H \end{cases} \qquad (1)$$

Where H is distance from flexible joint to origin of coordinates, m; A and L is amplitude and wavelength of riser wave respectively, m.

3. Loads on Tubing String Section

Tubing string is a slender pipe; its lower end is treated as fixed and after a suspending section it becomes continuous contact with riser from tangent point. Intercepting tubing section from origin to tangent point, loads on the separate section are shown in Fig. 1b). Where F_i, N_i and M_i (i=0, 1) is axial load (kN), lateral load (kN) and bending moment (kN·m) respectively, subscript 0 and 1 denotes loads at fixed end and at tangent point respectively; w is distributed load from pipe gravity, kN/m. All these loads act in the X-Y plane.

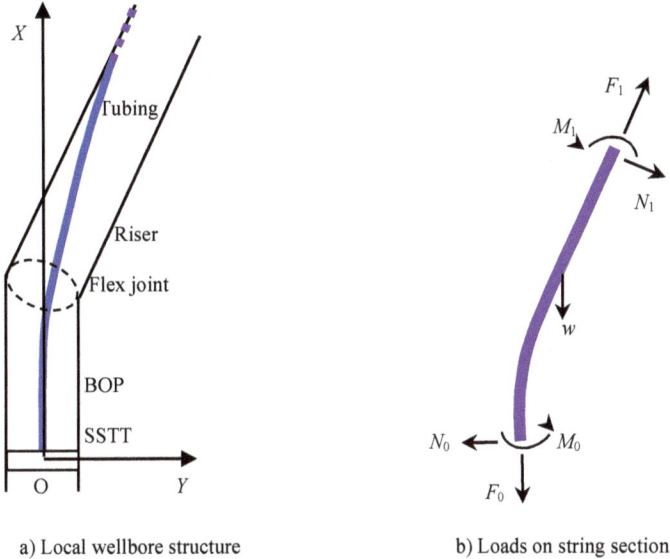

a) Local wellbore structure   b) Loads on string section

Figure 1. Local string mechanical model.

4. Tubing String Displacement Equation

Riser lateral displacement is still in small deformation category. Riser diameter is far larger than that of tubing string, therefore riser deformation does not influenced by tubing string.

Considering load conditions, tubing string lateral displacement equation can be established as

$$EIY''' - (F_0 + wX)Y' = -N_0. \tag{2}$$

Where EI is the tubular bending stiffness, kN·m².

5. Restraint Conditions

At fixed end, known parameters include: displacement $Y(0) = 0$, angular deflection $Y'(0) = 0$ and axial force $F(0) = F_0$.

Normally, tubing string axial force is tension and contacts with wellbore at upper side, as shown in Fig. 1a). The tangent point is not a determinate position, so in order to make use of it, a supposed position, X_T, is given as a variable. Then boundary conditions at tangent point can be expressed as follows:

Lateral displacement

$$Y(X_T) = b - a + A \sin\frac{2\pi(X_T-H)}{L}. \tag{3}$$

Where b is tubing string outside radius, m; a is riser internal radius, m.

Angular deflection

$$Y'(X_T) = A\frac{2\pi}{L}\cos\frac{2\pi(X_T-H)}{L}. \tag{4}$$

Bending moment

$$M(X_T) = -EI \cdot A\left(\frac{2\pi}{L}\right)^2 \sin\frac{2\pi(X_T-H)}{L}. \tag{5}$$

Where M is tubing string bending moment, kN·m.

Tubing string deformation is controlled by gravity, axial force and riser displacement. It is a static indeterminacy problem and the differential equation is transcendental.

The riser usually deviates from vertical position due to ocean current. In order to control its displacement and prevent it from buckling, tension must be imposed. Even so, riser bending and swaying cannot be eliminated completely. Main concerns of this paper include angular deflection of flexible joint and riser curvature above the joint. They are important factors to tubing deformation.

6. Tubing String Deformations and Safe Conditions

In the following discussion pipe dimensions are: Riser inside diameter 509.0 mm; tubing outside diameter 114.3mm (4.5″), inside diameter 88.9mm.

7. Tubing Displacement and Bending Stress

Let riser wavelength L=200m, half amplitude A=2.8m (corresponding to a 5° flexible joint angular deflection); distance from flexible joint to fixed end H=3m; axial force F_0=300kN.

From above data numerical solutions can be obtained. Position of tangent point takes place at X_T =26.83m with lateral displacement 1.69m and bending

716

moment 2.13kN·m. The theoretical solution of bending moment at tangent point is 2.03kN·m, so the proportional error is 4.93%. Maximum bending stress takes place at fixed end and the value is 395.5MPa.

Extensive computation produces the following features:

- Maximum bending stress varies nearly directly as angular deflection of flexible joint.
- A larger axial force leads to a higher tangent point and a greater bending stress.
- Increasing the distance from flexible joint to fixed end can reduce bending stress.

Investigation also shows that the bending string can contact with salient angle of flexible joint. This contact can be regarded as a danger because tubing may collapse at this point. Larger angular deflection, larger axial force and higher flexible joint position are all propelling factors to the contact.

So the tubing string dangers near subsea wellhead are stress yielding at fixed end and collapsing from contact with salient angle of flexible joint.

In order to prevent tubing failure, relationship among key factors should be founded. The following discussions will focus on relations among angular deflection, axial force and flexible joint position.

8.   Establishment of Safe Parameter Range Graph

8.1. *Confines of Maximum Bending Stress*

Tubing bears tri-axial stresses that are influenced by many factors. In this section the maximum bending stress is used as the criterion in order to emphasize the importance of local structure. Take 500MPa as critical value, relationship among angular deflection, axial force and flexible joint position is shown in Fig. 2. Three curves in Fig. 2 correspond to angular deflection to be 6°, 8° and 10° respectively. One curve is a boundary indicates that, under the given angular deflection, in the top-left area the bending stress is less than 500MPa.

9.   Tubing-Joint Contact Conditions

Take critical point of tubing-joint contact as the criterion, similar curves can be obtained as shown in Fig. 3. Where one curve is a boundary indicates that, under the given angular deflection, in the lower-left area the contact does not take place.

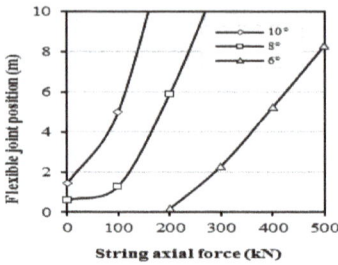

Figure 2. Max bending stress confines.

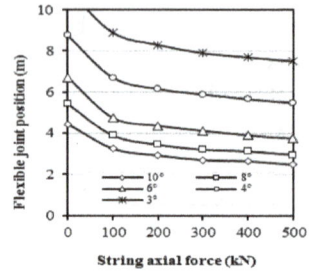

Figure 3. Tubing-joint contact conditions.

10. Safe Range Chart

Combining information in figures above safe range chart is obtained as Fig. 4a). Fig. 4a) shows:

- When angular deflection is greater than 8°, bending stress surpasses 500MPa even axial force is zero.
- When angular deflection is greater than 6°, proper selected axial force and joint position can be used safely.
- When angular deflection is less than 6°, bending stress confine fades away and contact becomes the only restraint.

Take flexible joint as delimitative parameter a similar safe range chart is obtained as Fig. 4b), lower-left area of each curve is safe region. It can be found that when flexible joint position is about 3m the safe region reaches the turning point. Before this point the confine is controlled by bending stress and after this point the confine is controlled by contact.

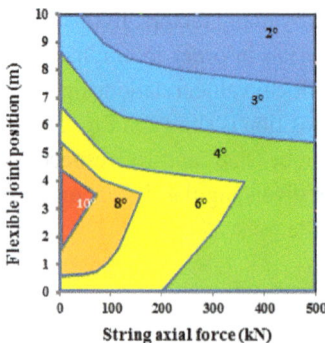

a) Delimited with angular deflection

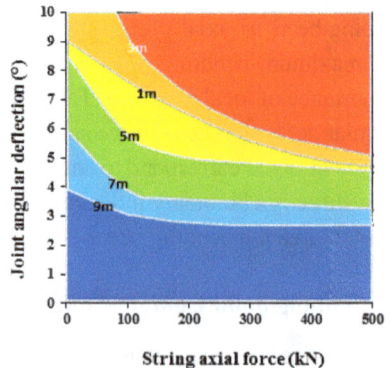

b) Delimited with joint position

Figure 4. Safe range delimited with different parameters.

11. Discussion

11.1. *Tubing String Design and Operation*

Based on data above, the following matters should be taken into account during tubing string design and operation:

- After the fluted hanger has set, a 300kN axial load should be imposed to tubing near subsea wellhead by adjusting the top tensioning system to prevent tubing string buckling.
- Optimal distance from flexible joint to subsea testing tree is 3m to 4m.
- During well testing the flexible joint angular deflection should be limited in 5°. Otherwise high strength thick wall tubing should be used.
- To use protection fittings, such as centralizer, in a range of about 30m from flexible joint is inadvisable. If necessary, the diameter and position should be optimized.

12. Management of Sensitive Parameters

With regard to deepwater well testing technological procedures, management of sensitive parameters should be considered as follows:

- Flexible joint position is determined by dimensions of BOP and STT, and can only be changed before installation, so it must be considered in design stage.
- Angular deflection of flexible joint is controlled by platform shift, top tension, wave and ocean current. It is the most sensitive parameter to tubing failure, and is the most difficult parameter to predict. So it must be determined by professional study.
- Tubing axial force can be adjusted through top tensioning system and flowing pressure, so it can be regarded as real time adjustable parameter. But when surface shutting-in, or when plugged by hydrate, inside pressure cannot be controlled. So initial axial force should be imposed based on the maximum inside pressure throughout testing procedure.

13. Safety Criterions

Tubing safety criterions used above have great limitation. They should be more concrete in practice, especially in the following aspects:

- In this paper maximum axial stress or equivalent stress is not used as criterion because they relate to many factors, such as actual force, inside and outside pressure, water depth, flow drag, tubing string deformation, etc. Equivalent stress should be used as criterion when all these factors are known.

- Tubing-joint contact is only a reference point, doesn't mean tubing failure. More work should be done to find out whether the tubing will collapse. But in view of inaccurateness of engineering data, the preceding dealing method is practical.

14. Conclusion

- Tubing string close to subsea testing tree has two dangers: excessive axial stress at fixed end and collapse due to contact with salient angle of flexible joint.
- Tubing axial force, angular deflection and position of flexible joint are key factors to tubing string safety. Angular deflection is the most sensitive parameter to tubing failure, and must be determined by professional study.
- Confines of key factors can be expressed clearly though safe range chart to prevent tubing failure.
- Optimal distance from flexible joint to subsea testing tree is from 3m to 4m.
- Protection fittings are not recommended in a range of about 30m from flexible joint.
- Equivalent testing string stress should be used as strength criterion when all important engineering parameters are known.

Acknowledgement

This work was funded by "National Basic Research Program of China" (973 Program, 2015CB251205).

References

1. C. Wendler, M. Scott, Testing and Perforating in the HPHT Deep and Ultra-Deep Water Environment, SPE 158851 (2012).
2. Beibit, Akbayev, Y. Shumakov, Efficient Deepwater Well Testing, SPE 176782 (2015).
3. M. Garrison, M. Cox, Reinventing Deepwater Exploratory Testing, OTC 21277 (2011).
4. Liu Xiaofeng, Technique of APR Testing Pressure Control, Inner Mongolia Petrochemical Industry Magazine, 24 (2010) 145-147.
5. D.L. Mason, W. Tharp, C. L. Wilie, Surface BOP: Testing and Completing Deepwater Wells Drilled with a Surface-BOP Rig, IADC/SPE 87111 (2005).
6. C.F. Etcheverry, R.F.T Maia, C. Wendler, Innovative Adaptation of Sub-Sea Test Tree Successfully Tests Ultra Deep-Water Wells in the Campos Basin, Brazil, OTC 10972 (1990).

7. Jin Liping, Lv Yin, Ren Yonghong, Liu Bo, The First National Independent Deep-Water Testing Technology, Oil Well Testing, 24(2015) 50-53.

8. Chang Yuanjiang, Duan Menglan, Design and Operation of Marine Drilling Risers: A Case History of Deepwater Drilling in the South China Sea, Natural Gas Industry. 34(2014) 106-111.

9. Qi Meisheng, Wen Jihong, Chen Guoming, Chang Yuanjiang, Stiffness Analysis of Marine Riser Flexible Joint Elastomer, China Petroleum Machinery, 42(2014) 71-74.

10. R.J. Stomp, G. J. Fraser, S.C. Actis, L. F. Eaton, K. C. Freedman, Deepwater DST Planning and Operations from a DP Vessel. SPE 90557 (2004).

11. ISO 13624-1-2009 Petroleum and natural gas industries: Drilling and production equipment (part 1): Design and operation of marine drilling riser equipment. Geneva: International Organization for Standardization, 2009.

12. R.C.S. Bueno, C.K. Morooka, Analysis method for contact forces between drillstring -well -riser, SPE 28723 (1994).

13. Xu Liangbin, Zhou Jianliang, Wang Rongyao, Analysis of deep water drilling platform evacuation from imminent typhoons with riser hang-ing-off in the South China Sea, China Offshore Oil and Gas, 27(2015) 101-107.

14. V.V. Arrieta, A.O. Torralba, P.C. Hemandez, C.T. Maia, M. Guajardo. Case History: Lessons Learned From Retrieval Coiled Tubing Stuck by Massive Hydrate Plug When Well Testing in an Ultra-Deep Water Gas Well in Mexico. SPE/IADC 140228 (2011).

Test of Localized Subway Vehicle Axle Box Bearing

Hong-yan Wang[†], Xi Li and Xiao-fei Hu

Subway Operation Technology Centre,
Mass Transit Railway Operation Corporation LTD,
Vehicle and equipment technology department,
Beijing, China
[†]Email: 15010664116@163.com

In order to reduce the purchase and maintenance cost of the subway vehicle axle box bearings and shorten the maintenance period, the localized vehicle axle box bearings applied to a line of Beijing subway were prepared. The main design technical parameters, life calculation, life test, performance test and the post-installation main track testing of the manufactured localized axle box bearings are induced in this paper. Results of all the tests show that the localized axle box bearings can satisfy the vehicle running requirements in a line of Beijing subway.

Keywords: Subway vehicle; Localized axle box bearings; Track testing.

1. Introduction

The axle box bearings are very critical basic parts in the subway vehicle equipment, whose running state affects directly the passenger comfort and the vehicle safety. At present, most of Beijing subway vehicle axle box bearings are imported products, mainly including Nippon Seiko KK (NSK), Svenska Kullager Fabriken (SKF), Schaeffler FAG (FAG) and so on. The orbit traffic bearings of these main production enterprises occupied more than 80% international market. Due to technical reasons, especially rolled steel quality, the performance of the imported bearings has certain advantages compared with domestic products.

However, the foreign bearings are expensive. Their delivery time has a long cycle and the after-sales service is not guaranteed. Once appearing early failure, the bearings need to be replaced; the long purchasing cycle of the foreign bearings can affect the normal management production. The development and cooperation of new vehicles or equipment is very difficult. Basically the developers choose bearing models from the existing bearing series, there are very few developers to research new type bearing for certain a new vehicle or equipment. Bearings developers often faced with some contradictory problems of the bearing geometry size and loading in matching.

To reduce the procurement and maintenance cost effectively and shorten the maintenance period of the axle box bearings, the Beijing subway relevant staff members researched new localized axle box bearings for a line of Beijing subway and certain trial vehicle with localized axle box bearings were tested.

2. Main Technical Parameters and Life Calculation

Based on the installation dimensions and technical requirements of the vehicle axle box bearing in a line of Beijing subway as well as the national standard [1-4], the localized axle box bearings were designed with the double row cylindrical roller bearing with seal structure in both ends. The bearing rated dynamic loading: C=843kN, the rated static loading: $C0r$ = 1370kN; the radial clearance: 0.11~0.15, the axial clearance: 0.55~1.10. The grease type is railway Ⅳ. The localized axle box bearing structure is shown in Fig. 1.

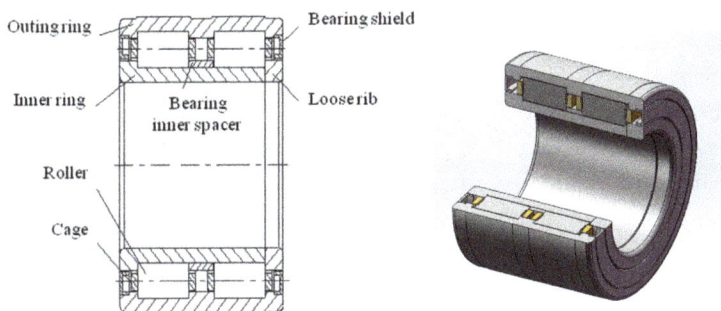

Figure 1. Structure of the localized axle box bearing

Due to the particularity of the subway bearings running environment, the vehicle axle box bearings must have very high reliability [5-6] and stable life. The subway vehicle axle box bearings have been designed at the highest reliability coefficient 99%, and the theory design life can reach more than 2 million kilometers. The reliability coefficient Re=0.21 when the highest reliability is 99%, the theory design life can be written as

$$Lna=ReL10k. \qquad (1)$$

L_{10k} is considered

$$L10k=nD\left(\frac{C}{P}\right)^{\frac{10}{3}}.$$ (2)

In which, C is the rated dynamic loading and C=843kN, D is the wheel diameter and D=840mm, P is the equivalent dynamic loading and P=7t=7×9.8=68.6kN, when the maximum speed v_{max}=80km/h.

The theoretical calculation life of the localized axle box bearings can reach 2.36 million kilometers based on the highest reliability, and the result completely can satisfy the operating requirement of the vehicle axle box bearings in a line of Beijing subway.

3. Life and Running-In Tests

After the localized subway vehicle axle box bearings were completed, to verify the reliability and the actual working life of the localized bearings, the two sets of new bearings have completed life test for 1.2 million kilometers on ZS60/120 bearing testing machine. At the same time, China Bearing Quality Supervision and Inspection Center were entrusted to conduct the same test. The whole test process did not replace or add grease. The bearing dimension basically kept the original state after the life test, and the hardness of the tested axle box bearings did not change, only appearing normal running traces on the bearings.

After the change of the localized and the imported axle box bearings, the axle completed the no-loading running-in test. All the temperature rise of the axle box bearings stabilized in 22~24 °C after the no-loading running-in test. The temperature rise can satisfy the requirement of national standard [7].

4. Running Test

The test content of the localized vehicle axle box bearings is as follows:
1) Be careful to check the bearings appearance and mainly observe appearance, grease leakage and seal status;
2) The detection of the shock, vibration and the temperature rise of the localized vehicle axle box bearings during the running process;
3) Compare with other the line vehicles installed the JK10450 systems.

The running test on trunk line [8] were divided into two stages combined with the actual application: 1000 kilometers running test without passengers(including no-loading test and over-loading test) and 100000 kilometers running test on the trunk line. The JK10450 system monitors the shock, vibration and the temperature rise of the vehicle axle box bearings during the running process. The experimental vehicle with the localized axle box bearings has completed the test mileage after 10 months.

During the test process, there were no any alarm information based on the overall shock data in the test vehicle. Analyzing the vibration and temperature data of all the positions and calculating the highest temperature and temperature rise and the maximum vibration effective value in the no-loading, over-loading and 100000 kilometers running test on the trunk line by month, we can obtain from the Fig. 2 and Fig. 3:

1) The maximum vibration effective value shows a trend of increase and the maximum value is 56.99g;

2) The maximum temperature rise (17℃ in the no-loading test, 25℃ in the over-loading test) can satisfy the requirement of national standard [8, 9]. The highest temperature rise of the localized axle box bearings is in the first month in the 100000 kilometers running test on the trunk line and the highest value is 18.3℃. The temperature rise later goes down from month to month and stabilizes in 14~16℃. The result complies with the requirement of national standard [8, 9].

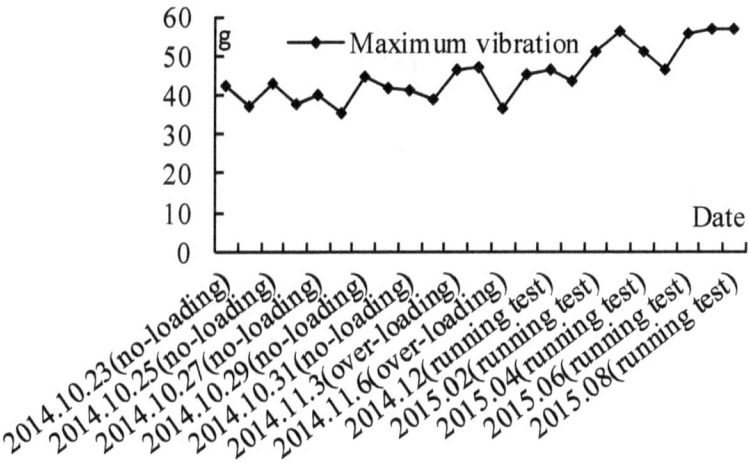

Figure 2. Vibration of the localized axle box bearings during the test process

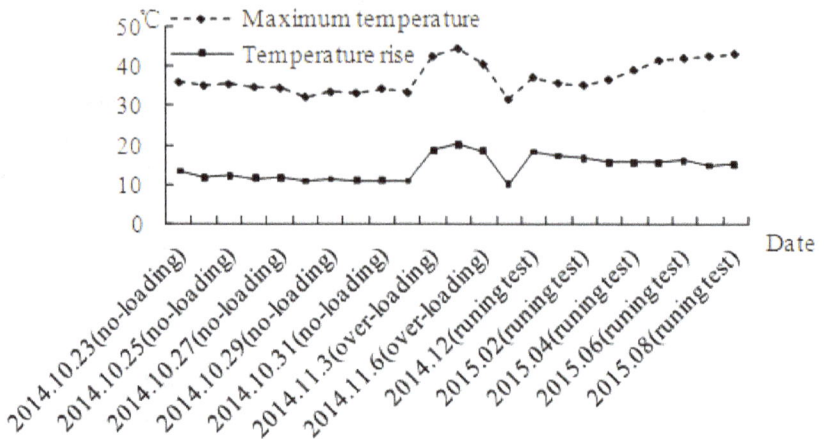

Figure 3. Temperature rise of the localized axle box bearings during the rest process

According to whether the localized axle box bearings were replaced in the test vehicle, the vibration and the temperature trend are shown as Fig. 4 and Fig. 5:

1) Vibration state: Before the localized axle box bearings were replaced (Area 1), the vibration effective value is around 8g. After the localized axle box bearings were replaced (Area 2), including the no-loading, over-loading and the 100000 kilometers running test on the trunk line, the vibration state is entirety flat.

2) Temperature rise state: Before the localized axle box bearings were replaced (Area 1), the temperature rise state trend is entirety smooth. After the localized axle box bearings were replaced (Area 2), including the no-loading and the over-loading, the temperature rise of the localized axle box bearings is higher. The temperature rise trend is steady in 14~16℃ in the 100000 kilometers running test on the trunk line(Area 3).

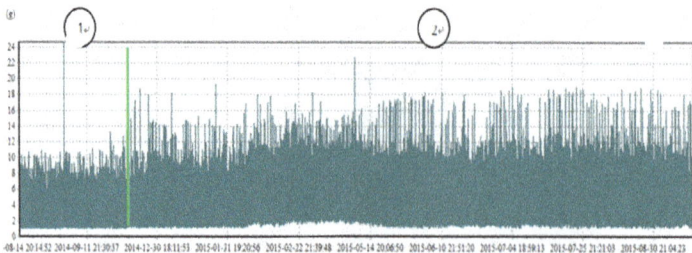

Figure 4. Vibration comparison before and after the localized axle box bearing were replaced

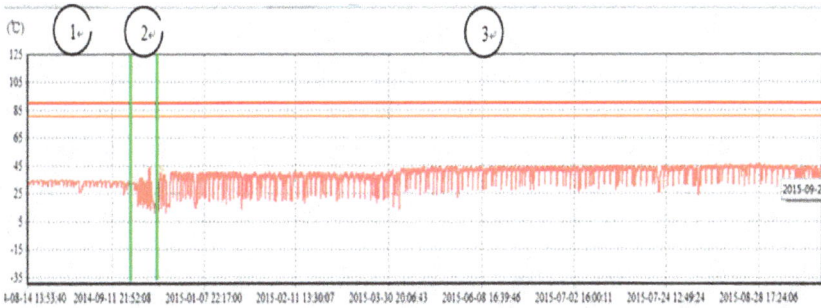

Figure 5. Temperature rise comparison before and after the localized axle box bearing were replaced

In order to further compare the localized axle box bearings with the NSK bearings, we contrasted the temperature rise and vibration data in the test vehicle during the test process(including the no-loading, over-loading and 100000 kilometers running test on the trunk line) with those data in the other two vehicles with JK10450 system in the line of Beijing subway. The following conclusions can be obtained:

1) Vibration state: The highest vibration value of the localized axle box bearings is steady in 40~60g with the other two vehicles in this line. After six months, the vibration value of the localized axle box bearings is higher than the other two vehicles.

2) Temperature rise state: The highest temperature rise value of the localized axle box bearings goes down and it is steady in 14 ~16℃. But the highest temperature rise of the localized axle box bearings is on the high side 8~10℃ than the other two vehicles in this line.

5. The Third Party Test after the Test

The average temperature rise of the 48 localized axle box bearings in the test vehicle were analyzed after the 100000 kilometers running test on the trunk line. The four axle box bearings both of which were in the same axle were teardowned, and whose temperature rise was relatively high. Then we entrusted the National Quality Supervision and Inspection Center of Luoyang Bearing Research Institute laboratory for testing the performance parameters of the four localized axle box bearings. The comprehensive judgement by inspection is that the performance parameters of the four localized axle box bearings conform to the requirement of the technical conditions and the existing standards.

727

6. Conclusion

The tests indicate that the overall performance parameters of the localized axle box bearings developed for a line of Beijing subway are quite good, good lubrication and smooth operation. The localized axle box bearings can satisfy the operating requirements of the line vehicle of Beijing subway. They can also effectively reduce the purchasing and maintenance cost, and shorten the repair time. The localized axle box bearings will be applied to Beijing subway. Furtherly, they will be popularized to other domestic urban subway at the same time. The localized axle box bearings possess a broad market prospect.

References

1. GB/T 307.3-2005, Rolling bearings–General technical regulations (In Chinese) [S].
2. GB/T 283-2007, Rolling bearings–Cylindrical roller bearings–Boundary dimensions (In Chinese) [S].
3. GB/T 273.3-1999, Rolling bearings–Radial bearings–Boundary dimensions, general plan (In Chinese) [S].
4. GB/T 274-2000, Rolling bearings–Chamfer dimension — Maximum values (In Chinese) [S].
5. X.W Yang: Submitted to Journal of Bearing (2013).
6. X.Z Zeng, L.M Sun, X.D Guo etc.: Submitted to Journal of Bearing (2014).
7. GB/T 7928-2003, General technical specification for metro vehicles (In Chinese) [S].
8. TB/T 3057-2002, Specification of alarm device for monitoring locomotive bearing temperature (In Chinese) [S].
9. H.Y Wang, X Li and L.M Zong: submitted to Journal of Railway Computer Application (2015).

Tracking Imaging Feedback Attitude Control
of Video Satellite

Xue-yang Zhang* and Jun-hua Xiang

*College of Aerospace Science and Engineering, National University of Defense
Technology, Changsha, China, 410073*

zxy1135@qq.com

Tracking imaging attitude control of video satellite for uncooperative moving object is
studied in the paper. The effect of satellite attitude adjustment during tracking imaging on
image is analyzed quantitatively, and based on it tracking imaging feedback control
strategy considering the delay of actuator is proposed in the case of having detected the
object in video image. The control law for the reaction wheels is designed, whose
asymptotic stability is proved using Lyapunov stability theory. Numerical simulation
results show that the presented approach is effective for video satellite achieving tracking
imaging of uncooperative moving object.

Keywords: Video satellite, Moving object, Tracking imaging, Attitude control.

1. Introduction

As a new observation method, video satellite has been widely concerned by
domestic and foreign researchers, and has become a hot field in micro-satellite
research. Compared with traditional satellites, video satellite can provide
real-time video images, get more information than a single image. It can provide
dynamic information of the object and detect the occurrence of dynamic events,
which is of great help to disaster relief, wartime surveillance, and planning
decision [1]. Now several video satellites have been launched into orbits, such as
Skysat-1 and 2 by Skybox Imaging, TUBSAT series satellites by Technical
University of Berlin [2], and the video satellite by Chang Guang Satellite
Technology. They can obtain video images with different performance on orbit.
In September 8, 2014, Tiantuo-2 (TT-2) designed by National University of
Defense Technology independently was successfully launched into orbit, which
is the first Chinese interactive earth observation microsatellite using video
imaging system.

Video satellite can realize continuous tracking and monitoring of moving object based on interactive control strategies with human in the loop. But human's participation increases reaction time, and can do nothing for high speed moving object. For realizing automatic tracking imaging of moving object, it is necessary to detect the object and generate control strategy in real time. Control instruction is transmitted to the controller, and the controller controls the actuator to complete attitude adjustment, keeping the object in video image.

Video satellite's attitude tracking control of staring imaging of ground fixed target has been deeply studied on the mathematical model, the controller and the simulation. The continuous pointing problem of low earth orbit satellite to the ground fixed target was transformed into attitude tracking problem by Li, et al. [3], and a control law for the reaction wheels was developed. The attitude tracking model was established by Chen et al. [4] and the control law and steering law were introduced based on Double-Gimbaled Control Moment Gyroscopes. Dynamic equation of attitude tracking was obtained based on the error Euler angle by Sun, et al. [5] and a fuzzy controller for staring-imaging satellite was given out. Dynamic equation and kinematics equation of attitude tracking were obtained based on the error quaternion and error angular velocities by Liang, et al. [6], and a robust controller based on the PD control law was proposed, which had excellent robustness with parameter uncertainty and external disturbances. The influence of the satellite's attitude point precision and stability in the mode of staring-imaging on image quality was studied based on the staring imaging model and the coordinate transformation method by Yang, and Lin [1]. An attitude tracking finite time control method driven by Double-Gimbaled Control Moment Gyroscopes (DGCMG) and reaction flywheels was proposed by Feng, et al. [7].

However, for uncooperative moving object, whose kinematic equations are completely unknown, the desired attitude can't be calculated by the position of the object in real time. Feedback control strategy based on the deviation between the object and the video center is a feasible way to achieve tracking imaging, but the delay of actuator needs to be considered. Aiming at tracking imaging attitude control of video satellite for uncooperative moving object, this paper analyses the effect of satellite attitude adjustment during tracking imaging on image, proposes feedback control strategy, and designs the control law for the reaction wheels. The approach is verified by numerical simulation.

2. Effect of Satellite Attitude Adjustment on Image

The satellite body frame is defined as $O_b\text{-}x_by_bz_b$, by the origin at the mass center of the satellite and three axes aligned with the principal axes of the satellite. Pixel frame is defined as $I\text{-}xy$, by the origin at upper left corner of the image plane and using pixel as coordinate unit. x, y represent the number of column and row of the pixel in the image respectively. Translating this frame to the image center gives O-xy, as shown in Fig. 1. For the convenience of caculaion and discussion, the pixel frame to be mentioned later refers to O-xy. Assume that the camera is fixed on the satellite and the line of sight (LOS) is coincident with the O_bZ_b axis of the satellite body frame, the O_bX_b axis aligned with $-OX$, and the O_bY_b axis aligned with $-OY$.

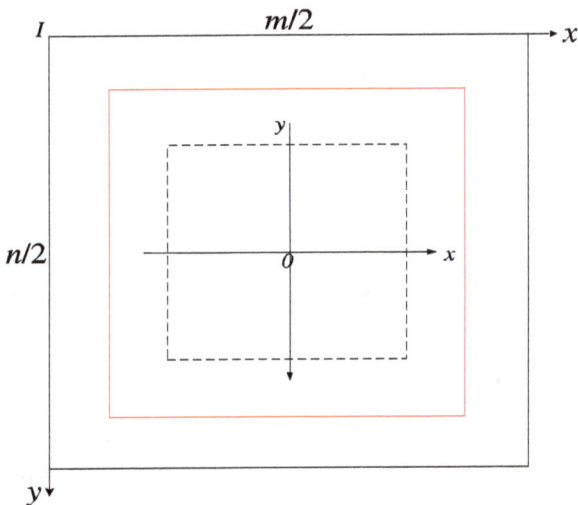

Figure 1. Pixel frame

Video satellite can achieve tracking imaging of moving object by attitude adjustment. Consider the effect of satellite attitude adjustment on image. Assume that the coordinate of the point P in the pixel frame is (u_1,v_1) before attitude adjustment and (u_2,v_2) after attitude adjustment respectively, and the 3-2-1 Euler angle of attitude adjustment is $(\Delta\psi,\Delta\theta,\Delta\varphi)$. In general, lense is not rotated. Thus $\Delta\psi=0$. $\Delta\theta$ and $\Delta\varphi$ are usually small quantity. Then coordinate transformation equation of pixel frame caused by attitude adjustment is given by

$$\begin{pmatrix} u_2 \\ v_2 \end{pmatrix} = \begin{pmatrix} u_1 \\ v_1 \end{pmatrix} + \frac{1}{d}\begin{pmatrix} -\Delta\theta \\ \Delta\varphi \end{pmatrix}. \tag{1}$$

where d is the size of pixel.

3. Tracking Imaging Feedback Control Strategy

Let the resolution of video image be $m \times n$. Without loss of generality, let $m \geq n$. Central area S is defined as the rectangle with vertices $(-m/4, n/4)$, $(m/4, n/4)$, $(m/4, -n/4)$, $(-m/4, -n/4)$. The area enclosed by dashed line in Fig. 1 is S. The cordon is connected by $(-m/2+n/8, 3n/8)$, $(m/2-n/8, 3n/8)$, $(m/2-n/8, -3n/8)$, $(-m/2+n/8, -3n/8)$, as shown the red line in Fig. 1.

It's derived from Eq. 1 that

$$\begin{pmatrix} \Delta\theta \\ \Delta\varphi \end{pmatrix} = d \begin{pmatrix} u_1 - u_2 \\ v_2 - v_1 \end{pmatrix} = d \begin{pmatrix} -\Delta u \\ \Delta v \end{pmatrix}. \tag{2}$$

Assuming that the object has been detected and its coordinate is (u, v), the 3-2-1 Euler angle of attitude adjustment is $(0, du, -dv)$ obtained from Eq. 2 in order to make the object return to the center. The attitude adjustment consists of a rotation around the O_bY_b axis with angular du and a following rotation around the O_bX_b axis with angular $-dv$. But because of the delay of actuator, the object may be away from the field of vision when rotating around O_bY_b. This paper propose a feedback control strategy considering the delay of actuator as following.

- Step 1. When the object is away from central area, i.e. $(u,v) \notin S$, compare the value of $m/2-|u|$ and $n/2-|v|$. If $m/2-|u| \geq n/2-|v|$, $i=1$, else $i=2$.
- Step 2. If $i=1$, adjust the attitude to let $|v| \rightarrow 0$, i.e. transmit the control instruction of rotation around O_bX_b with angular $-dv$ to the controller. If $i=2$, adjust the attitude to let $|u| \rightarrow 0$, i.e. transmit the control instruction of rotation around O_bY_b with angular du to the controller.
- Step 3. When the controller controls the actuator to adjust the attitude, if the object reaches the cordon, i.e. $|u|>m/2-n/8$ or $|v|>3n/8$, go to Step 1.
- Step 4. After attitude adjustment, if $(u,v) \in S$, feedback control ends and video satellite observes the object in attitude stabilization state, else go to Step 1.

Thus tracking imaging of the object is achieved by feedback control strategy. The flow chart of the feedback control strategy is shown in Fig. 2.

4. Design of Controller

4.1 Attitude Dynamics Equation

The attitude dynamics equation of the video satellite with 3 reaction wheels is given by

$$J\dot{\boldsymbol{\omega}}_b = -\boldsymbol{\omega}_b \times (J\boldsymbol{\omega}_b + \mathbf{h}) + \mathbf{u} + T_d. \tag{3}$$

732

where J is the moment of inertia of the video satellite, ω_b is angular velocity of the satellite body frame, $\mathbf{h}$ is wheel angular momentum, $\mathbf{u}$ is control torque, T_d is external disturbance torque, and

$$\mathbf{u} = -\dot{\mathbf{h}} \, . \tag{4}$$

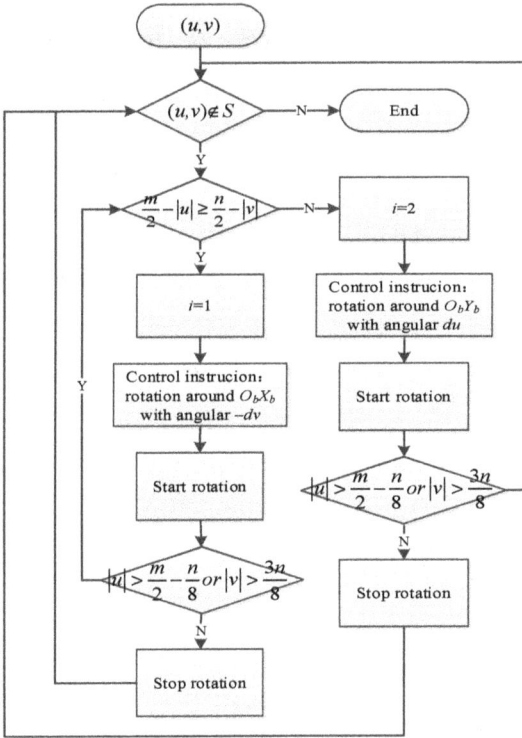

Figure 2. Flow chart of feedback control strategy

5. Controller Design

PD controllers are widely used in attitude control of operational satellites, because they are simple to implement and present low implementation risk for a space mission [8]. According to tracking imaging feedback control strategy, attitude adjustment only consists of rotation around O_bX_b (i=1) or O_bY_b (i=2) with angular θ. Ignoring external disturbance torque T_d, the PD controller is given by

$$u_i = -k_p(\theta_i - \theta_0 - \theta) - k_d\omega_i \, . \tag{5}$$

where u_i is the control torque generated by the wheel on the ith axis, ω_i is the ith component of satellite body frame's angular rate, θ_i is the ith axis Euler angle of satellite body frame respect to the inertial frame, θ_0 is the ith axis Euler angle of satellite body frame respect to the inertial frame before attitude adjustment, k_p and k_d are positive constants.

Substituting Eq. (5) into Eq. (3) gives the closed-loop system governed by

$$J_i\dot{\omega}_i = -k_p(\theta_i - \theta_0 - \theta) - k_d\omega_i. \tag{6}$$

Define the Lyapunov function:

$$V = \frac{1}{2k_p}J_i\omega_i^2 + \frac{1}{2}(\theta_i - \theta_0 - \theta)^2. \tag{7}$$

Obviously, $V \geq 0$, and the equality holds if and only if $\omega_i = 0$ and $\theta_i = \theta_0 + \theta$. Taking the time derivative of V and substituting Eq (6) gives

$$\dot{V} = \frac{1}{k_p}J_i\omega_i\dot{\omega}_i + (\theta_i - \theta_0 - \theta)\dot{\theta}_i$$

$$= \frac{\omega_i}{k_p}(-k_p(\theta_i - \theta_0 - \theta) - k_d\omega_i) + (\theta_i - \theta_0 - \theta)\omega_i. \tag{8}$$

$$= -\frac{k_d}{k_p}\omega_i^2$$

For k_p and k_d are positive constants, $\dot{V} \leq 0$. Therefore, the controller is asymptotic stable by Lyapunov stable theory.

6. Simulation Results

Fig. 3 is obtained by overlaying 125 sequent frame images from TT-2. The resolution is 960×576. The trajectory of detected object is in the white box. The object was moving towards lower left and would be away from the field of vision.

Use feedback control strategy proposed in this paper to achieve tracking imaging of the object. The moment of inertia tensor of TT-2 is $J=diag$ (2.2021,3.2358,2.1760) kg·m². The control gains are set to k_p=0.5, k_d=5. d in Eq. 2 is 8.33μm. The coordinate of the object is (-295,153), away from central area. Assume that the velocity of the object in pixel frame is (-3.42, 2.37). The velocity is only used for simulation. It's unknown to video satellite. The procedure of feedback control is as the following.

1) According to feedback control strategy, $m/2-|u|\geq n/2-|v|$, $i=1$. Control instruction was rotation around O_bX_b with angle $-dv=-4.38$'.
2) But after rotating -4.23', the object reached the cordon. At this moment the coordinate of object is (-408, 83). Then control instruction was rotation around O_bY_b with angle $du=-11.68$'.
3) After rotation the coordinate of object is (-143,178), still away from central area. Control instruction was rotation around O_bX_b with angle $-dv=-5.09$'.
4) After rotation the coordinate of object is (-280, 97), still away from central area. Control instruction was rotation around O_bY_b with angle $du=-8.02$'.
5) After rotation the coordinate of object is (-115,168), still away from central area. Control instruction was rotation around O_bX_b with angle $-dv=-4.8$'.
6) After rotation the coordinate of object is (-217, 78), in central area. Feedback control ended.

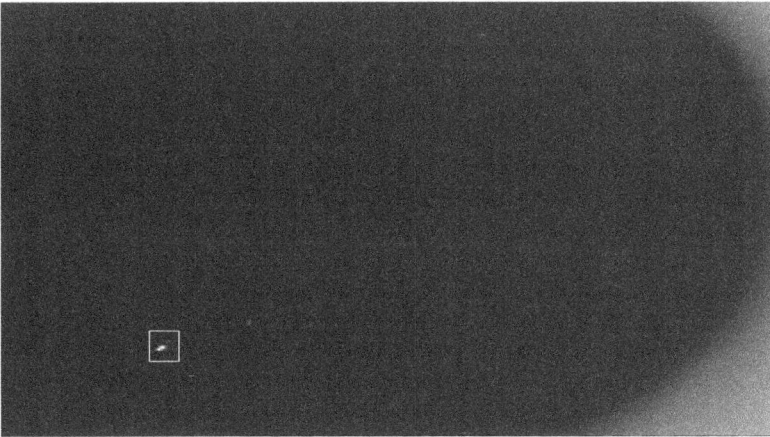

Figure 3. Overlay of 125 sequent frame images from TT-2

The trajectory of object in lower left part of pixel frame is shown in Fig. 4. Central area is enclosed by dashed line and the cordon is denoted by red.

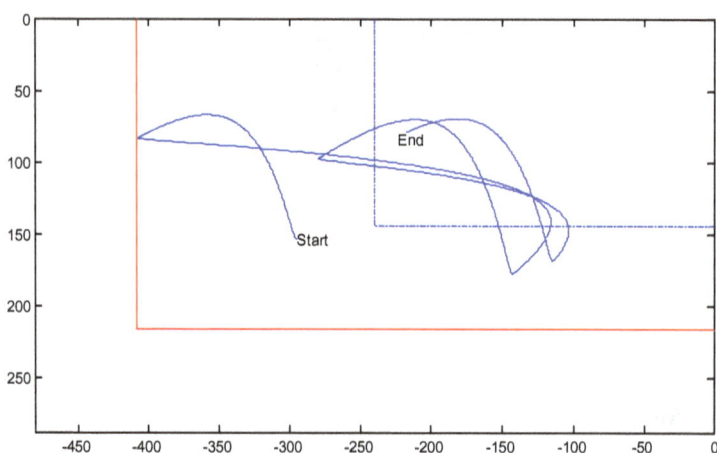

Figure 4. Trajectory of object in lower left part of pixel frame

7. Conclusion

Tracking imaging attitude control of video satellite for uncooperative moving target is studied in the paper. The effect of satellite attitude adjustment during tracking imaging on image is analyzed quantitatively, and based on it tracking imaging feedback control strategy considering the delay of actuator is proposed. The control law for the reaction wheels is designed, whose asymptotic stability is proved by Lyapunov stability theory. Numerical simulation results show that the presented approach is effective for video satellite achieving tracking imaging of uncooperative moving target.

Acknowledgement

This work was funded by the Major Innovation Project of NUDT (No. 7-03), the Young Talents Training Project of NUDT, and the High Resolution Major Project (GFZX04010801).

References

1. X. Yang, X. Lin, Analysis of influence of LEO staring satellite dynamic tracking on imaging, Infrared and Laser Engineering, S1 (2014) 203-208.

2. M. Steckling, U. Renner, H.P. Röser, DLR-TUBSAT, qualification of high precision attitude control in orbit, Acta Astronautica, 39 (1996) 951-960.

3. J. Li, M. Xu, W. Steyn, Attitude tracking maneuvers of a low Earth orbit spacecraft, Journal of Tsinghua University (Sci & Tech), 41 (2001) 102-104.

4. X.Q. Chen, Y.H. Geng, F. Wang, L.I. Dong-Bai, Staring imaging attitude tracking control of agile small satellite, Optics & Precision Engineering, 124 (2011) 143-148.

5. Z.W. Sun, H.Z. Liang, W.U. Shu-Nan, Attitude Control of a Staring-Imaging Satellite in LEO Using Fuzzy Controller, Aerospace Shanghai, (2010).

6. H.Z. Liang, J.Y. Wang, Z.W. Sun, New simple robust attitude controller of staring-imaging satellite in LEO, Journal of Harbin Institute of Technology, 43 (2011) 26-30.

7. Y. Feng, K. Liu, W. Zhang, Simulation of Staring Imaging Attitude Tracking Finite Time Control of TV Satellite, Journal of System Simulation, 28 (2016) 226-234.

8. N.M. Horri, P.L. Palmer, M. Roberts, Energy optimal spacecraft attitude control subject to convergence rate constraints, Control Engineering Practice, 19 (2011) 1297-1314.

Variable Coefficients PD Adaptive Attitude Control of Video Satellite for Ground Multi-Object Staring Imaging

Kai-kai Cui[†] and Jun-hua Xiang

School of Aerospace Science and Engineering,
National University of Defense Technology,
Changsha 410073, China
[†]Email: ckk_nudt@163.com

In this paper, the attitude control problem of the video satellite on the ground multi-object staring imaging is studied. On condition that the expected yaw angle is always zero, the expected attitude and angular velocity of the video satellite are solved. The mathematical model of attitude tracking is established based on quaternion and angular velocity. The asymptotic stability of the control model is also proved by the Lyapunov function. Then a variable coefficient PD adaptive controller is designed based on genetic algorithm. The simulation results show that under the condition of disturbances and errors, this method can realize the staring imaging of ground object with zero yaw angle. For the task of ground multi-object staring imaging, the control time can be greatly reduced under the premise of accuracy requirement by using this control method, which will improve the efficiency of the satellite to star imaging significantly.

Keywords: Video satellite; Multi-object staring; Attitude tracking; Coefficient optimization; Adaptive control.

1. Introduction

Video satellite is a new type of Earth observation satellite proposed in recent years. It can achieve continuous observation of a certain region and acquire video information. Comparing to traditional Earth observation satellites, it can get more dynamic information about the object area [1]. Staring imaging of video satellite to the ground object refers to that its observation axis of the imaging device always points to the ground object when the imaging satellite is in orbit. Because of the orbital motion of the satellite and the ground object movement caused by Earth rotation, the attitude of the satellite must be adjusted in real time to ensure that the observation axis always points to the direction of the object. For single ground object, the dynamic model, kinematics model and controller design of staring imaging have been studied deeply.

Feng, et al. [1] proposed a video satellite attitude tracking model with quaternion and a finite time controller based on homogeneous method of nonlinear system was proposed. Chen, et al., [2] used the orbit information to compute the excepted attitude angle and attitude angular velocity of staring imaging. Liang, et al [3]. proposed a new robust controller, which consists of a PD part and an additional term. And simulation results show that the controller is qualified with excellent robustness and satisfactory control effect. According to the attitude kinematic and dynamic equations derived based on Euler angle representation, Guo, and Wang [4] proposed a design of the fuzzy controller for staring imaging satellite.

Adaptive control is a control method, which can automatically maintain the optimal working state when the structure parameters, the initial condition or the extreme point of the objective function is changed [5]. For the attitude tracking control problem, the process of staring imaging to ground multi-object can be divided into two sub-processes, observation axis pointing at the object and transfer between two object points. The control objectives of the two sub-processes are not the same. The former requires high tracking accuracy and small steady state error, while the latter requires rapid control, so that the satellites can switch quickly between different object points.

For the attitude tracking control problem of video satellite staring imaging to ground multi-object, the expected attitude angles and attitude angular velocity of satellite were computed. The mathematical model of attitude tracking control is established. And the attitude control law is designed based on the reaction flywheel actuator. Then the control coefficients are optimized according to different objective functions by using genetic algorithm and a variable coefficient PD adaptive control method is designed. Finally, the control method and the model are simulated. The simulation results show that under the condition of disturbance torque and actuator error, this method can realize staring imaging to ground object point less than zero yaw angle condition. For the task of ground multi-object staring imaging, the control algorithm of this paper can take both the control time and precision into account. In this task, the control time can be greatly reduced under the premise of accuracy requirement by using this control method, which will improve the efficiency for the satellite to star imaging significantly.

2. Coordinate System Definition

2.1. Satellite Orbit Coordinate System (O)

Taking the centroid of the satellite S as the origin, ZO axis points to the center of the earth, XO axis is perpendicular to ZO and points to the direction of satellite motion in the orbit plane. And YO axis is defined by the restriction of right-handed coordinate system.

2.2. Satellite Body Coordinate System (B)

Taking the centroid of the satellite S as the origin, the three axes XB, YB and ZB are fixed on the satellite body and coincide with the inertia principal axes of the satellite.

2.3. Reference object coordinate system (T)

Taking the centroid of the satellite S as the origin, ZT axis points to the object T, YT axis is defined as the cross product of YO and ZT. And XT axis is defined by the restriction of right-handed coordinate system [6].

3. Expected Attitude Model

3.1. Expected Attitude Angle Determination

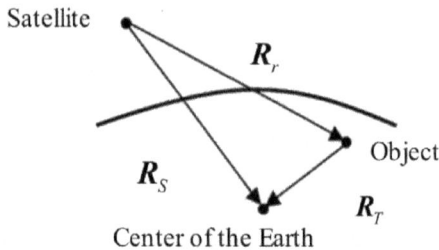

Figure 1. The schematic diagram of satellite staring

Assuming that the position of the satellite RS and the velocity VS in inertial coordinate system have been known. The position of the object point in inertial coordinate system RT is also available.

In the process of staring to the objects, to determine the excepted attitude information, it is necessary to obtain the vector from the video satellite to the object. As shown in Figure 1, we can know the vector from the satellite to the object in inertial coordinate system Rr=RS-RT.

Calculate the unit vector of YO of satellite orbit coordinate system at present time:

$$y_o = -h_I = \frac{R_S \times V_S}{\|R_S \times V_S\|}$$

(1)

Assuming that the observation axis of the camera is coincident with zb axis of the satellite body coordinate system. Calculate the unit vectors of the three axes of reference object coordinate system in the condition of zero yaw angle by Eq. 2.

$$\begin{cases} z_t = R_r / \|R_r\| \\ x_t = y_o \times z_t \\ y_t = z_b \times x_t \end{cases}$$
(2)

The transfer matrix from the orbit coordinate system to the reference object system can be calculated by Eq. 3.

$$B_{ot} = \begin{bmatrix} x_t^T \\ y_t^T \\ z_t^T \end{bmatrix} \begin{bmatrix} x_o & y_o & z_o \end{bmatrix} = \begin{bmatrix} x_t^T x_o & x_t^T y_o & x_t^T z_o \\ y_t^T x_o & y_t^T y_o & y_t^T z_o \\ z_t^T x_o & z_t^T y_o & z_t^T z_o \end{bmatrix} \equiv \begin{bmatrix} a_{11} & a_{12} & a_{13} \\ a_{21} & a_{22} & a_{23} \\ a_{31} & a_{32} & a_{33} \end{bmatrix}$$
(3)

In Eq. 3, the symbolic $\equiv$ represents "denoted as", and aij (i, j=1,2,3) are the short for elements in the transfer matrix. This is the excepted attitude matrix of staring imaging, which can be transformed into the quaternion from the reference object system to the orbit system qto.

Calculate the three excepted staring attitude angles defined by 3-2-1 rotation order:

$$\begin{cases} \psi^* = \operatorname{atan}2(a_{12}, a_{11}) = 0 \\ \theta^* = \operatorname{asin}(-a_{13}) \\ \phi^* = \operatorname{atan}2(a_{23}, a_{33}) \end{cases}$$
(4)

It is known from Eq. 4, the excepted attitude angle in yaw direction is always 0. It means that the definition of the reference object coordinate system in this paper can ensure the observation axis of the camera do not rotate in the process of staring imaging, which can improve the imaging quality.

3.2. Expected Angular Velocity Determination

By means of the difference of excepted attitude angles, the expected attitude angular change rate at current moment can be calculated as Eq. 5:

$$\begin{cases} \dot{\psi}^* = (\psi_+^* - \psi_-^*)/\delta = 0 \\ \dot{\theta}^* = (\theta_+^* - \theta_-^*)/\delta \\ \dot{\phi}^* = (\phi_+^* - \phi_-^*)/\delta \end{cases}$$
(5)

In Eq. 5, $(\cdot)+$ represents the excepted attitude angles at $t+=t+\delta/2$ moment, $(\cdot)-$ represents the excepted attitude angles at $t-=t-\delta/2$ moment, t is the current moment, δ is a short period of time, and usually ranges from 0.01s to 0.5s.

According to the Euler kinematic equation, the expected angular velocity of staring imaging $\omega_{to}^t = [\omega_x^* \quad \omega_x^* \quad \omega_x^*]^T$ can be calculated as Eq. 6:

$$\begin{cases} \omega_x^* = \dot{\phi}^* - \dot{\psi}^* \sin\theta^* - \dot{f}\sin\psi^* \cos\theta^* \\ \omega_y^* = \dot{\theta}^* \cos\phi^* + \dot{\psi}^* \sin\phi^* \cos\theta^* - \dot{f}\left(\cos\phi^* \cos\psi^* + \sin\phi^* \sin\psi^* \sin\theta^*\right) \\ \omega_z^* = -\dot{\theta}^* \sin\phi^* + \dot{\psi}^* \cos\phi^* \cos\theta^* + \dot{f}\left(\sin\phi^* \cos\psi^* - \cos\phi^* \sin\psi^* \sin\theta^*\right) \end{cases}$$

(6)

In Eq. 6, $\dot{f}$ is the instantaneous angular velocity of the satellite orbit:

$$\dot{f} = \frac{\|h_I\|}{R_s \cdot R_s}$$

(7)

4. The Mathematical Model

4.1. Satellite Attitude Kinematics and Dynamics Model

Denote q_{ob} as the quaternion from orbit coordinate system to body coordinate system, then the quaternion from reference object coordinate system to body coordinate system which we call staring error quaternion can be expressed as Eq. 8:

$$q_e = q_{tb} = q_{to} \circ q_{ob} = [q_0 \ q_1 \ q_2 \ q_3] = [q_0 \ \boldsymbol{q}_v]$$

(8)

In Eq. 8, q_{to} represents the quaternion from reference object coordinate system to orbit coordinate system, and the symbol $\circ$ represents quaternion multiplication.

And the staring angular velocity error can be expressed as Eq. 9:

$$\boldsymbol{\omega}_e = \omega_{bt}^b = \omega_{bo}^b - \boldsymbol{B}_{tb}\boldsymbol{\omega}_{to}^t$$

(9)

In Eq. 9, ω_{bo} represents the angular velocity of body coordinate system relative to orbit coordinate system, and $\boldsymbol{B}_{tb}$ represents the transfer matrix from reference object coordination system to the body coordination system.

To achieve staring imaging of video satellite to ground objects, qv and ω_e need to converge to zero. According to the kinematic equations described by attitude quaternion as Eq. 10:

$$\dot{q} = \frac{1}{2}q \circ \omega$$

(10)

In Eq. 10, q is the quaternion between two coordinates, and ω is angular velocity between the two coordinate systems expressed in body coordinate. Bring Eq. 8 and Eq. 9 into Eq. 10:

$$\dot{q}_e = \frac{1}{2} q_e \circ \boldsymbol{\omega}_e$$

(11)

In Eq. 11, $\boldsymbol{\omega}_e$ needs to be regarded as a zero-scalar quaternion. Using zero momentum reaction flywheel as attitude control actuator, the attitude dynamic formula can be denoted as Eq. 12:

$$I\dot{\boldsymbol{\omega}}_{bi} + \boldsymbol{\omega}_{bi} \times (I\boldsymbol{\omega}_{bi}) + \boldsymbol{\omega}_{bi} \times (I_w\Omega_w) = \boldsymbol{L}_e + \boldsymbol{L}_c$$

(12)

In Eq. 12, I is the total inertia tensor of the satellite, $I_w = diag[J_x \ J_y \ J_z]$ represents the inertia tensor of the flywheels, and $\Omega = [\Omega_x \ \Omega_y \ \Omega_z]^T$ is the angular velocity of the flywheel relative to the satellite body. $\boldsymbol{L}_c = -I_w\dot{\Omega}_w = -[J_x\dot{\Omega}_x \ J_y\dot{\Omega}_y \ J_y\dot{\Omega}_y]^T$ Represents the flywheel control torque, $\boldsymbol{L}_e$ represents the disturbance torque. And $\boldsymbol{\omega}_{bi}$ represents the angular velocity of satellite body coordinate system relative to inertial coordinate system, which can be expressed as Eq. 13:

$$\boldsymbol{\omega}_{bi} = \boldsymbol{\omega}_{bt} + \boldsymbol{\omega}_{ti} = \boldsymbol{\omega}_e + \boldsymbol{\omega}_{ti}$$

(13)

Bring Eq. 13 into Eq. 12, the dynamic model of staring imaging can be expressed as:

$$I\dot{\boldsymbol{\omega}}_e = -I\dot{\boldsymbol{\omega}}_{ti} - (\boldsymbol{\omega}_e + \boldsymbol{\omega}_{ti}) \times (I(\boldsymbol{\omega}_e + \boldsymbol{\omega}_{ti}) + I_w\Omega_w) + \boldsymbol{L}_e + \boldsymbol{L}_c$$

(14)

4.2. Design of Feedback Control Law for Satellite Attitude Control System

According to the kinematics and dynamics model Eq. 11 and Eq. 14, the staring imaging PD controller can be designed as Eq. 15:

$$\boldsymbol{L}_c = -K_p q_v - K_D \boldsymbol{\omega}_e + I\dot{\boldsymbol{\omega}}_{ti} + (\boldsymbol{\omega}_e + \boldsymbol{\omega}_{ti}) \times (I(\boldsymbol{\omega}_e + \boldsymbol{\omega}_{ti}) + I_w\Omega_w)$$

(15)

In Eq. 15, K_P and K_D are symmetric positive definite matrices. Bring the controller equation into the Eq. 14, ignoring outside disturbance torque, the closed-loop equation of the control system can be expressed as Eq. 16:

$$I\dot{\boldsymbol{\omega}}_e = -K_p q_v - K_d \boldsymbol{\omega}_e$$

(16)

It is known from Eq. 16 that the system is in stable state only if q_v and ω_e are both zero. Analysis the stability of the controller, taking Lyapunov function as Eq. 17:

$$
\begin{aligned}
V &= \frac{1}{2}\omega_e^T K_P^{-1} I \omega_e + q_v^2 + [1-(1-q_v^2)^{1/2}]^2 \\
&= \frac{1}{2}\omega_e^T K_P^{-1} I \omega_e + q_v^2 + (1-q_0)^2 \\
&= \frac{1}{2}\omega_e^T K_P^{-1} I \omega_e + 2(1-q_0)
\end{aligned}
\tag{17}
$$

Obviously, Eq. 17 holds if and only if $q_v=\omega_e=0$, and V is positive definite. Differentiate V with respect to time:

$$
\begin{aligned}
\dot{V} &= \omega_e^T K_P^{-1} I \dot{\omega}_e - 2\dot{q}_0 \\
&= \omega_e^T K_P^{-1}(-K_P q_v - K_d \omega_e) + \omega_e^T q_v \\
&= -\omega_e^T K_P^{-1} K_D \omega_e
\end{aligned}
\tag{18}
$$

Because $K_P^{-1} K_D$ is positive definite matrix, $\dot{V} \le 0$, the equality holds if and only if $\omega_e = 0$, which means $\dot{V}$ is negative definite. According to the Lyapunov stability theory, we know that the control law designed in this paper is asymptotically stable.

5. The Simulation Condition Setting:

5.1. *The State of the Satellite Orbit*

Time(UTCG)	26 Jun 2014 03:35:19.84
Position in Inertial System (km)	[1945.26 6468.35 1233.56]
Velocity in Inertial System (km/s)	[1.3479 1.0103 -7.4319]

5.2. *Controller Indexes*

Maximum control torque	20 mN·m
Maximum momentum	0.65 Nms
Flywheel speed range	-6000 ~ 6000 rpm
Inertia of the flywheel rotor	1.0×10^{-3} kg.m2

5.3. Error Setting

Rotation rate tracking error	< 1 rpm
Accuracy of attitude determination	0.05°(3σ, 3 axes)
Accuracy of angular rate measuring	0.005°/s(3σ)
Error of Satellite Position	30m(3σ, 3 axes)
Error of Satellite velocity	0.1m/s(3σ, 3 axes)

5.4. The Size of the Satellite

The satellite weight is about 65kg. Satellite envelope size: 515mm (x) × 524mm (y) × 685mm (z). The inertia matrix I=diag [2.2.201, 3.2358, 2.1760](unit: kg·m2).

At the initial moment, the attitude angle and attitude angular velocity of the satellite are [0.1, 20, -13] deg and [-0.1, 0.2, 0.1] deg/s respectively, and the positions of two ground object in the inertial coordinate system are [2056.31, 5941.23, 1074.71]km and [1716.35, 6089.84, 803.64]km respectively. During the first 80 seconds, the satellite needs to star at the object 1, and during 80-160 seconds it needs to star at object 2.

6. Coefficient Optimization

6.1. Considering the Rapidity Only

The values of K_P and K_D in formula (19) are not decided yet, and they play a crucial role in the performance of the control system. The process of the staring imaging to ground multi-object can be divided into two sub-processes, observation axis staring at the object and transfer between two object points. The control objectives of the two sub-processes are not the same. The former requires high tracking accuracy and small steady state error, while the latter requires rapid control. So we optimize the coefficients according to the different objective functions.

Firstly, considering the rapidity of the control system only, chose the coefficients to shorten the settling time as much as possible. Define the settling time Ts as the time it takes to make the attitude angle error less than 0.5 deg. Optimize the control coefficients to get the smallest Ts by genetic algorithms, setting the population size as 20, elite count as 2, and crossover fraction as 0.8. The bounds of the coefficients are [1, 10] and [1, 50] respectively. Then we obtain the optimization results:

$$\begin{cases} \mathbf{K}_{P1} = diag\begin{bmatrix} 13.56 & 12.64 & 7.19 \end{bmatrix} \\ \mathbf{K}_{D1} = diag\begin{bmatrix} 56.52 & 55.06 & 40.43 \end{bmatrix} \end{cases} \qquad (19)$$

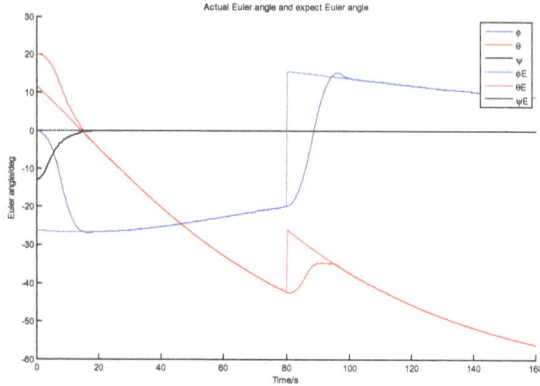

Figure 2. The performance of the control system

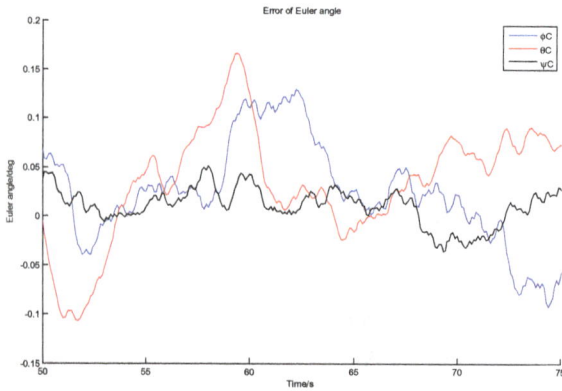

Figure 3. The errors of the attitude angles

Using the optimization results in Eq. 19, the system is simulated and the performance of the system is carried out as Fig. 2 and Fig. 3.

We can learn from Fig. 2 and Fig. 3 that the settling time can be limited in 20 seconds and the steady-state error is about ± 0.17 degree using the coefficients in Eq. 19.

6.2. Considering the Steady-state Error Only

Secondly, considering the steady-state error of the control system only, chose the coefficients to reduce the steady-state error, which means that the observation axis will star at the object points more accurately. Define the steady-state error Es as the average error in the last 10 seconds. The condition setting is the same with former case. Then we obtain the optimization results as Eq. 20:

746

$$\begin{cases} \mathbf{K}_{P2} = diag \begin{bmatrix} 13.56 & 17.70 & 10.82 \end{bmatrix} \\ \mathbf{K}_{D2} = diag \begin{bmatrix} 21.52 & 18.53 & 25.13 \end{bmatrix} \end{cases} \tag{20}$$

Using the optimization results in Eq. 20, the system is simulated and the performance of the system is carried out as Fig. 4 and Fig. 5.

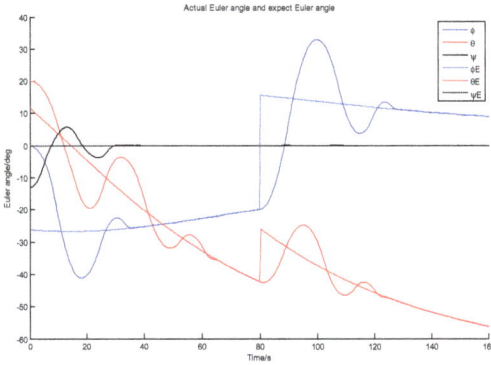

Figure 4. The performance of the control system

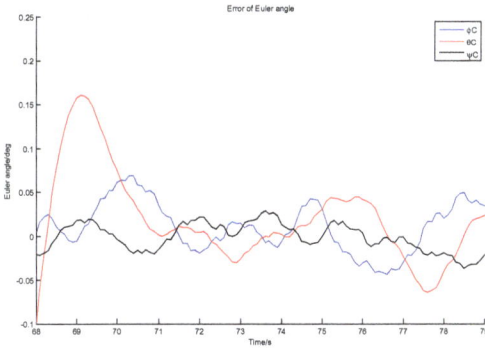

Figure 5. The errors of the attitude angles

Fig. 4 and Fig. 5 show that the steady-state error is about ±0.07 degree, which is less than half of that of the former case. While the settling time has taken too long to be accepted.

For satellite attitude PD control algorithm, usually the settling time can be reduced by increasing the proportion of control coefficient, but if the proportion coefficients are too large, the system shock will be intensified. The ratio of the proportion coefficients to the differential coefficients in Eq. 20 is much larger than that in Eq. 19, which means the proportion coefficients are too large relatively. So the system shock is intensified and the settling time becomes much longer.

6.3. A New Control Method

According to the analysis above, a new variable coefficients PD adaptive attitude control method is proposed. The new control method can be divided into two sub-processes: firstly, control the Euler angle deviation into 1 degree rapidly, and then star at the ground objects accurately. That is:

$$\text{If } \left|\varphi-\varphi^*\right|>1° \text{ and } \left|\phi-\phi^*\right|>1° \text{ and } \left|\theta-\theta^*\right|>1°$$

$$\text{Then } \begin{cases} \mathbf{K}_P = \mathbf{K}_{P1} \\ \mathbf{K}_D = \mathbf{K}_{D1} \end{cases}, \text{ else } \begin{cases} \mathbf{K}_P = \mathbf{K}_{P2} \\ \mathbf{K}_D = \mathbf{K}_{D2} \end{cases}.$$

According to the new control strategy, the system is simulated and the performance of the system is carried out as Fig. 6 and Fig. 7.

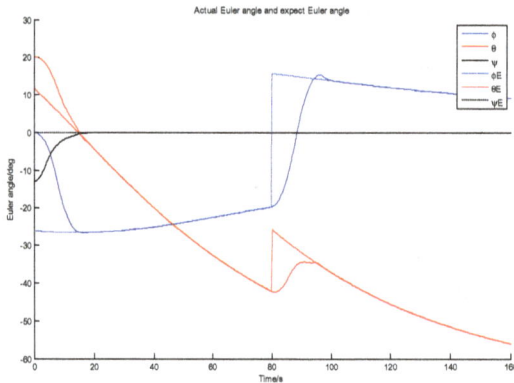

Figure 6. The performance of the control system

We can learn from Fig. 6 and Fig. 7 that the settling time can be limited in 20 seconds and the steady-state error is about ±0.07 degree by using the new control strategy. That means by this new control strategy, we can greatly shorten the control time, improve the efficiency of the task under the premise of ensuring the staring accuracy.

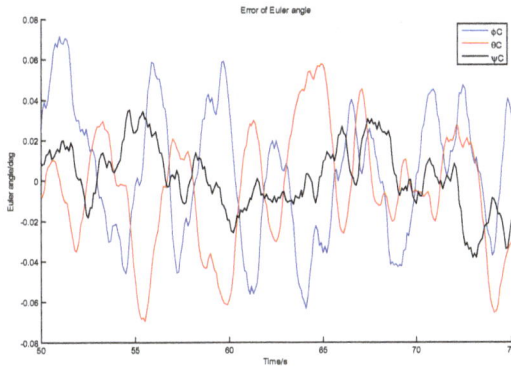

Figure 7. The errors of the attitude angles

7. Conclusion

For the attitude tracking control problem of video satellite staring imaging to ground multi-object, the expected attitude angles and attitude angular velocity of satellite are computed. The mathematical model of attitude tracking control is established. And the control coefficients are optimized according to different objective functions by using genetic algorithm and a variable coefficient PD adaptive control method is designed. The simulation results show that under the condition of disturbance torque and actuator error, this method can realize the staring imaging to ground object point under zero yaw angle condition. The control algorithm of this paper can take both the control time and precision into account. And the settling time can be greatly reduced under the premise of accuracy requirement by using this control method, which will improve the efficiency for the satellite to star imaging significantly.

References

1. Y. Feng, K. Liu, W. Zhang, Simulation of Staring Imaging Attitude Tracking Finite Time Control of TV Satellite, Journal of System Simulation, 28 (2016) 226-234.
2. X. Q. Chen, Y.H. Geng, F. Wang, L.I. Dong-Bai, Staring imaging attitude tracking control of agile small satellite, Optics & Precision Engineering, 124 (2011) 143-148.
3. H.Z. Liang, J.Y. Wang, Z.W. Sun, New simple robust attitude controller of staring-imaging satellite in LEO, Journal of Harbin Institute of Technology, 43 (2011) 26-30.
4. Z.W. Sun, H.Z. Liang, W.U. Shu-Nan, Attitude Control of a Staring-Imaging Satellite in LEO Using Fuzzy Controller, Aerospace Shanghai, (2010).

5. T. Guo, W. Wang. The Research and Development on Adaptive Control Method. Journal of Anyang Teacher College. 5 (2009): 81-84.

6. Y.J. Lian, G.Q. Zeng, J.H. Xiang, et.al. China. Patent CN201510602717.8. (2015).

Vibration Test Method Based on Synthesis of Displacement Vectors for Transport Packaging

Li-ming Shen[1], Gui-qin Li[2,†], Bei-bei Li[1], Zhen-wei Hu[1] and Chao Wang[1]
[1]School of Mechatronic Engineering and Automation,
Shanghai University,
Shanghai 200072, China
[2]Shanghai key laboratory of intelligent manufacturing and robotics,
Shanghai University,
Shanghai 200072, China
†Email: leeching@t.shu.edu.cn

Vibration test is a scientific method to measure anti-vibration performance of product. All current vibration test standards are based on three kinds of vibration test tables as crank rotary type, electromagnetic type and electro hydraulic type. Because of the different flaws of these kinds of vibration tables, the extension of vibration test in professional field is restricted. In this paper, a new vibration test method is proposed. The vibration table for the new method is composed of several excitation units, which are based on existing crank rotary test table. The compound frequency vibration is synthesized with displacement vectors of multiple excitation units by levers. The new method has both a clear output as crank rotary type and a frequency range following the real distribution as electromagnetic type and electro hydraulic type. With further research on related standard test method, the new method proposed will be a trend in professional field. The new method we proposed makes a foundation for the conclusion of vibration test from qualitative to quantitative. It will also help to promote the concept of proper packaging and green packaging.

Keywords: Anti-vibration performance of product; Vibration for transport; Vibration test method; Vibration table.

1. Introduction

Vibration test is a scientific method to measure anti-vibration performance of product in transport [1]. Although there are several standards and equipment of vibration test for package product system [2, 3], vibration test has not been applied widely [4]. Most designs of transport packaging are based on empirical data. Excessive packaging is common to ensure anti-vibration performance of package product system. Although some enterprises have vibration test equipment, they are continually using the vehicle and other practical means of transport to test the anti-vibration performance [5, 6]. Hence, the aim of this paper is to explore a vibration test method, which can be applied widely.

2. Analysis of Existing Vibration Test Methods

Transport vibration test is an equivalent reconstruction transport vibration based on vibration test technology and vibration test table. It is an effective way to explore the vibration test by the analysis of transport vibration parameters and the existing vibration table characteristics.

2.1. *Analysis of Environment Parameters of Transport Vibration*

Freight transportation can be divided into road, railway, shipping, air transportationetc,.The distribution environmental parameters (such as the pavement local defect, etc.) is a major factor that cause vibration. For example as Fig. 1, curve is one-dimensional equivalent space function r (x) for one section of the road.When a vehicle crosses this section,the vertical vibration accelerationand frequency are as a function of vehicle speed. As transport speed is influenced by many factors,simulate one actual transportation vibration parameters apparently cannotassess the anti-vibrationperformance.Hence, the transportation vibration parameters are obtained by Statistics method. The study found that underthe effect ofvehicle suspension system, the vibration acceleration power spectrum of damages mainly distributesin the limited central area of Fig. 2. It can be concluded that vibration of product package system arecomposited with multiplevibration frequency in any afinite period of time.

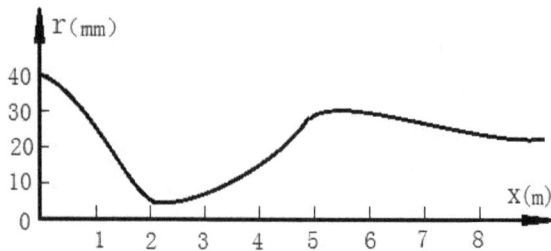

Fig. 1 The sample of equivalent fluctuation of pavement

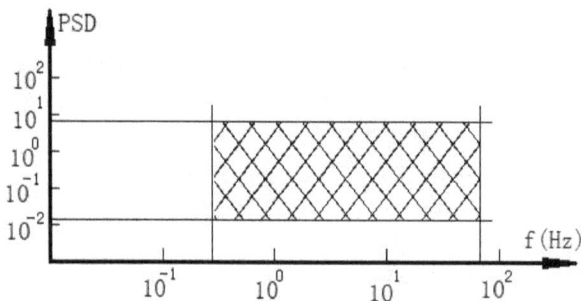

Fig. 2 Main influence area of power spectrum

2.2. Characteristics of Existing Transport Vibration Test Table

There are three kinds of vibration test table for different test standards, such as crank rotary type, electromagnetic type and electro hydraulic type. Crank rotaryvibration test table has a determined function with displacement and time, so vibrations parameters can be standardized. However, crank rotary vibration test table do an experiment in a singal frequency. The acceleration-frequency function curve is varying with the crank length, as show in Fig. 3, curve i (i=1-4). Obviously, curves 1 to 4 have different frequency ranges from $\Delta f1$ to $\Delta f4$ between the minimun and maximum values of the available acceleration. It is found that the frequency range $\Delta f1$ is about 2~5Hz when the crack length is 10mm (Curve 1 in Fig. 3), and the frequency range $\Delta f4$ can be covered 50-100Hz when the crank length is less than 0.1mm. While the effective vibration of frequency of real transport distribution is in 2~100Hz. Obviously, there will be apparent difference between test experience and real distribution environment. Both electromagnetic vibration test table and electro hydraulic vibration test table can output compound vibration frequency which cover real distribution frequency range. But, as the vibration displacement closed-loop control is restricted by dynamic characteristics, a definite vibration displacement time function is unable to be given, vibration parameters of the excitation of the same signal is discrete with different vibration tables.

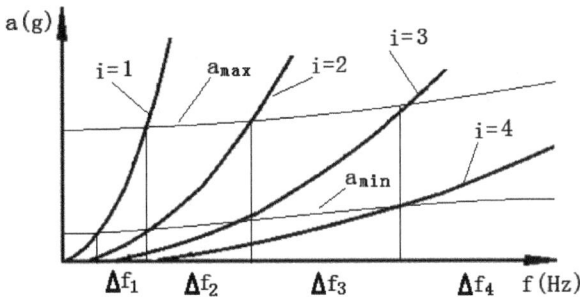

Fig. 3 The relationship between crank length and frequency range

As mentioned above, the existing three kinds of vibration table for all kinds of test standard finally gives two resultsof qualified or unqualified.Further more, with discrete of vibration parameters and difference of anti-vibration preformance based on specific products, the reference to the significance of the test results has become much more limited which eventually led to the popularity of existing vibration test standards in the field is restricted.

3. Vibration Test Method Based on Displacement Vector Synthesis

By analyzing of the characteristics of the existing vibration table, a new vibration table must be explored to make a popularization for the vibration test in professional field. The output vibration not only covers the frequency range of distribution, but also has a definite function between the vibration displacement and time for multiple compound frequencies. As a result, the grade of vibration strength can be given in the open loop, and the result of the vibration test is developed from qualitative to the quantitative.

3.1. Principle of vibration test based on displacement vector synthesis

The vibration test based on displacement vector synthesis is consist with nrotary vibration generating units. The vibration displacement vector of each vibration output unit is synthesized by level mechanism. The synthesis displacement y is:

$$y = \sum_{i=1}^{n} k_i y_i \cos(2\pi f_i t + \varphi_i)$$

(1)

Where, i is the number of vibration generating, y_i is the crank length of No i unit , k_i is the equivalent contribution coefficient of each unit vibration to total output vibration. Reasonable allocation of the k_i coefficients of each unit and the vibration speed ratio of each unit can obtain the frequency range shown in Fig. 4. By dividing rotating speed, a clear and definite output of multiple compound vibration frequency can be obtained with the vibration table.

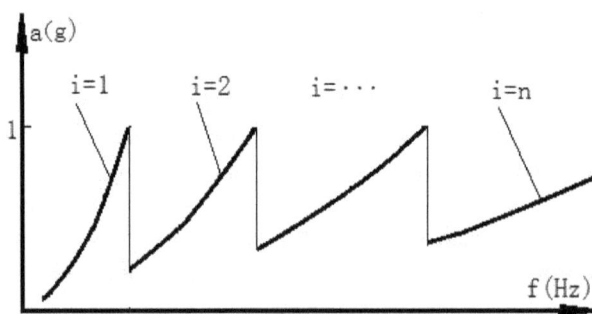

Fig. 4 The frequency range of multi units

3.2. Implementation System and Characteristics of Vibration Table Based on Displacement Vector Synthesis

Fig. 5 shows a verification system with three rotary vibration generating units to support the compound vibration of three frequencies; increase from 5-7Hz to 25Hz, on specimen weight up to 3kg, with synthesis peak acceleration increases from 1g of the existing vibration test method up to 2.3g. The system was on display at Shanghai International Industry Exibition 2014.

Fig. 5 The vertification system based on three rotary vibration generating units

Fig. 6 shows an improved design ables to handle specimenis up to a maximum weight of 30kg, and support the graded 5 strength vibration output through synchronous speed adjustment; it consits of 4 vibration units to operate up to 50Hz, to support the synthetic peak acceleration up to 4.5g.

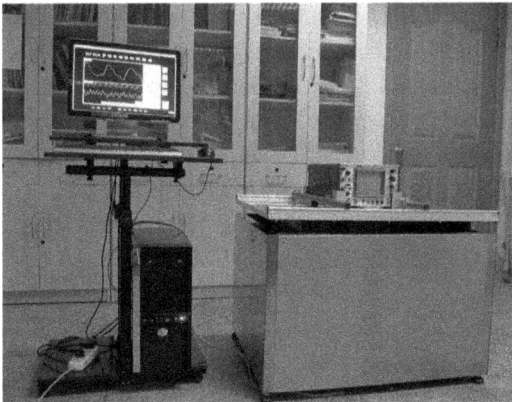

Fig. 6 The practical system based on four rotary vibration generating units

4. Conclusion

In this paper, a new vibration test method is proposed. The vibration table for the new method is composed of several excitation units, which are based on existing crank rotary test table. The compound frequency vibration is synthesized with displacement vectors of multiple excitation units by levers.

From the study, the vibration table based on displacement vector synthesis could support up to three kinds of existing vibration frequencies, as well as multi-frequency compound vibration to prove a clear and reliable vibration output to cover the real distribution environment, These information could offer insight into the classification of vibration strength, in the form of vibration displacement and time, so much so that it enable operators to optimate packaging costs, paving the way for better packaging and environmental friendly packaging.

Acknowledgement

This work was funded by Shanghai Municipal Commission of Economy and Information (Grant No. CXY-2013-37).

References

1. F. Dong, J.C. Yang, T.Z. Gao, Y. Xie, L. Guo, H.Y. Zhang, S.M. Hu, G.Q. Zhou, The Research and Design of Virtual Spindle Vibration Test System, Applied Mechanics and Materials. 556 (2014) 2903-2905.
2. Y.T. Wen, S.C. Liu, Y.Y. Zhang, Z.C. Wang, L.M. Zhao, H.B. Wang, A rail vibration test and analysis method for the electromagnetic launching process, Measurement. 85 (2016) 232-238.
3. G. Shen, Z.C. Zhu, X. Li, Y. Tang, D.D. Hou, W.X. Teng, Real-time electro-hydraulic hybrid system for structural testing subjected to vibration and force loading, Mechatronics. 33(2016) 49-70.
4. J. Lepine, V. Rouillard, M. Sek, Review Paper on Road Vehicle Vibration Simulation for Packaging Testing Purposes, Packag. Technol. Sci. 28(2015) 672-682.
5. R. Zhou, L. Yan, B. Li, J. Xie, Measurement of Truck Transport Vibration Levels in China as a Function of Road Conditions, Truck Speed and Load Level, Packag. Technol. Sci. 28(2015) 949-957.
6. G. La Scalia, G. Aiello, A. Miceli, A. Nasca, A. Alfonzo, L. Settanni, Effect of vibration on the quality of strawberry fruits caused by simulated transport, J. Food. Process. Eng.39 (2015) 140-156.

Control System for Subsurface Safety Valve Testing Based on LabVIEW

Li-xin Lu, Jun Hua and Gui-qin Li[†]

Shanghai Key Laboratory of Intelligent Manufacturing and Robotics,
Shanghai University, Shanghai, China
[†]E-mail: leeching@t.shu.edu.cn

Shu-xun Li, Ru-dong Lu and Yun-jiang Dong
Shanghai Extrong Oilfield Technology Co., Ltd.,
Shanghai, China

Based on the key technology of subsurface safety valve testing, the control system processing real-time signal sampling automatically with high precision is proposed. Furthermore, a stable and accurate signal conditioning circuit is proposed to process the sampling signal. The main part of the system is control software with function of high-speed data acquisition, real-time display and automatic memory, which are based on virtual instrument software platform.

Keywords: Subsurface safety valve; Control system; LabVIEW; Data acquisition.

1. Introduction

It's particularly important to perform test of subsurface safety valve efficiently, accurately and rigorously as it is one of the key tools in the well completion system. Are liable test system is fundamental to obtaining accurate test results. British researchers develop automatic test system to simulate extreme subsurface environment working well at pressure of 135MPa and temperature of 200 °C [1]. Wang T develops safety valve testing prototype, whose structure of PLC + virtual instrument makes this system have good stability and strong anti-jamming capacity [2].

This paper demonstrates the development of LabVIEW [3-5] used as the core of subsurface safety valve testing control system. This system can complete the function testing and performance analysis of the safety valve using the virtual instrument technology with a high degree of automation, good reliability, high precision, and high system integration.

2. The Hardware Design of Control System

Hardware design of the control system mainly divided into two parts, one is to determine the control system hardware structure scheme, while the other is to design the signal conditioning circuit for collecting physical signals in safety valve testing and converting analog signals to digital signals.

The control system based on virtual instrument + PLC, or virtual instrument + PCI slot board can be chosen to achieve control tasks of the safety valve testing. However, the PLC lacks of functions of analog input and output, and analog quantity processing needs extension module. What's more, the serial port communication rate of PLC is not high enough. On the contrary, PCI slot board can communicate with computer at high speed of data transmission by PCI bus technology. And the combination of analog and digital quantity makes PCI have many data processing functions, which leads to forming a powerful data acquisition system.

Signal conditioning circuit can transform the signals from sensors to data acquisition card and match the input parameters, realizing purposes of amplification, attenuation and incentives.

In order to prevent the impedance of the line affecting signal value and ensure the accuracy of measurement, adopting the standard transmission type with4-20 m A current output is necessary. But the data acquisition card only collects voltage signal, it's essential to transform current signal into voltage signal with signal conditioning circuit at the receiving end. The signal conditioning circuit is shown in Fig. 1.

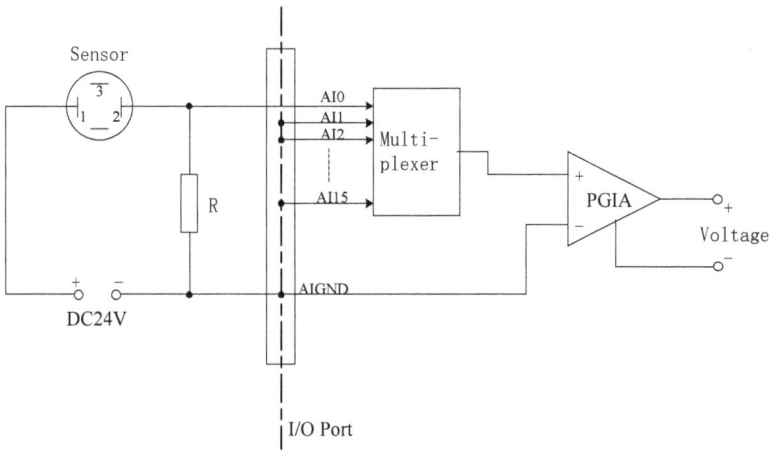

Figure 1. Signal conditioning circuit

The positive electrode accesses AI0-AI15 ports of acquisition card, and measuring signal is corresponding to each analog input port. Multiplexer is used to transmit signal from ports to the anode of PGIA chip by turns. Then PGIA performs differential operation to the input voltage in order to eliminate common-mode interference produced by excitation power supply. Finally, high-speed data acquisition of PGIA output is transmitted to the computer through PCI bus to be recorded and stored.

3. The Software Design of Control System

Connection between the hardware equipment and software is built through driver functions inside LabVIEW. On the input side, the data acquisition card reads data from sensor and transmits to the computer's memory. Then the software will process the data received, including digital filter, storage and display of the curve. At the output terminal, the system controls digital output port of the acquisition card through the switch on the software interface, so as to control the electromagnetic valve and pneumatic booster pump. The whole software is developed according to the modular approach, which consists of four function modules, including the data acquisition module, digital filtering module, data storage and curve display module, digital output module.

3.1. Data Acquisition Module Design

Rapid VI is chosen as the driver of the data acquisition card, which can be used easily and reliably. The data acquisition module bases on PCI-1716 data acquisition card, working together with the rapid VI DAQNavi Assistant, which is produced for the series of the above acquisition card. The rapid VI can drive the data acquisition card once its icon is put into the program block diagram and its parameters are configured. The process of this module's implementation is shown in Fig. 2.

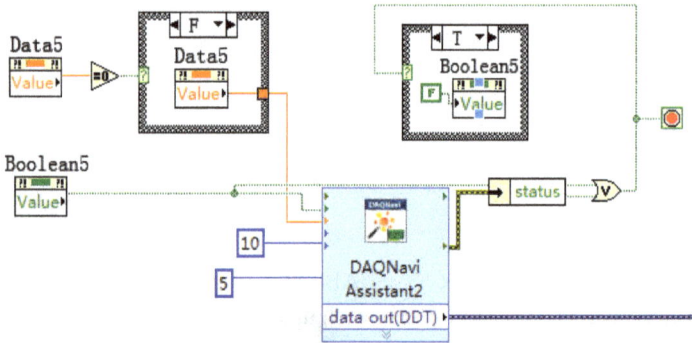

Figure 2. The program of data acquisition

3.2. Digital Filter Module Design

Due to the electromagnetic environment of System application is quite complex, the digital filtering is adopted to make the testing results become closer to the real data.

Here, the testing signal can be a periodic and infinite extended, and its main frequency band is lower than 30Hz. So, the low-pass filter internal integrated in LabVIEW can fit the need of digital filter, and the cut-off frequency is set to 40Hz. Since the system collects the real-time testing signal, which is infinite in duration, the Infinite Impulse Response filter is chosen but not Finite Impulse Response filter. The chebyshev filter is used as its stop band attenuates faster than filters with other topological structures.

3.3. Curve Display and Data Storage Module Design

The signal preprocessed can be real-time displayed in the form of curve for real-time observation by testers (done by curve display module). It can be stored in the form of value in the file as well, and can be accessed after testing by the data storage module.

The signal after preprocessing is voltage signal of 1-5V represented by u. A numeric conversion according to certain proportion is needed if the signal is to be displayed as pressure or flow rate value represented by y. Conversion formula is shown as the following.

$$y = (u-1) \times \frac{a}{u_{\max} - 1}$$

(1)

In the formula, $u_{\max} = 5$, y - actual measured values; a - the corresponding full scale value of the sensor

The data after conversion is in type of double precision which has more significant digits. The data is truncated with retaining two decimal places in precision in order to be displayed or stored conveniently. Fig. 3 shows the data storage and display program as the data from a pressure sensor.

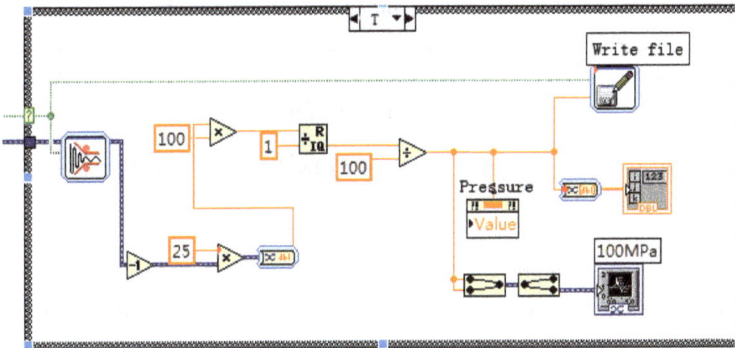

Figure 3. Storage and real-time display program of data from 100MPa sensor

Considering system's sampling rate keeps constant during a trial, the waveform chart adopted in the program, which can receive dynamic data type is fit for real-time monitoring. Waveform diagram has a good performance in drawing the curve of the single value function, when the sampling rate keeps constant, namely for the type of single value function $y = f(x)$, sampling points distributed evenly on the x-axis. Purposes here are to display and refresh the data real-time, as well as save the data to buffer. Program of viewing the curve is shown in Fig. 4.

761

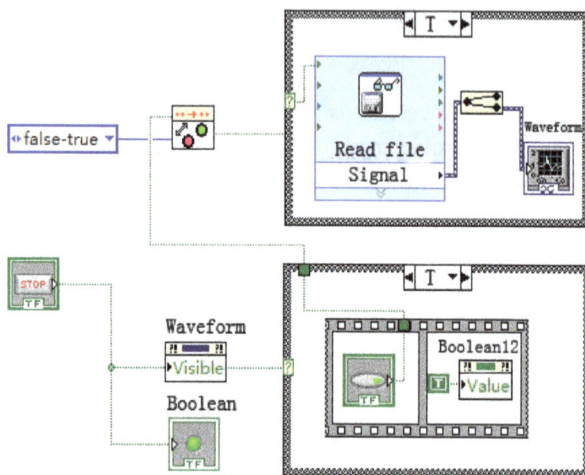

Figure 4. The program of viewing the curve

3.4. *Digital Output Module Design*

Digital output module is used to take independent control of each solenoid valve and pump on and off. As section above describes, rapid VI DAQ Navi Assistant is used to drive acquisition card to collect analog input signal. DAQ Navi Assistant is used to drive the acquisition card forgitadil output in the same way.

The control of digital quantity is taken by Boolean switch, and a Boolean switch corresponds to a digital output. DAQ Navi Assistant only accepts 8-bit unsigned single-byte integer data type; therefore a cluster is needed to convert Boolean quantity to eight bytes. Expanding cluster is useful to place more Boolean switch, so as to enhance the system's scalability. The program of digital output module is shown in Fig. 5.

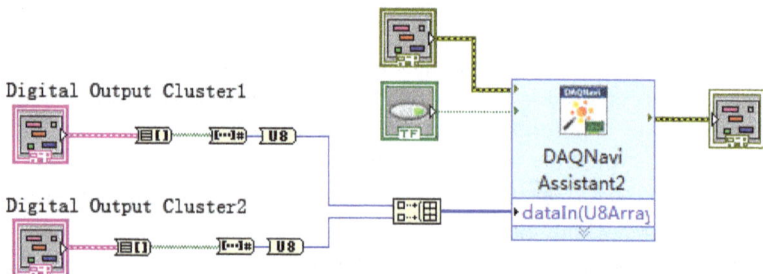

Figure 5. The program of digital output module

762

All of the descriptions above complete the development of each module of the testing system's control software.

4. Testing

The experimental test systems suitable for operation condition are constructed for testing the system's performance. The real-time monitoring of various measured data by system's display interface is performed during the test. After the test, the data during the test is stored or displayed in form of a graph. The data curve of the trial has shown in Figs. 6-7. After run-time testing, the system has been testified to be satisfying and effective in real-time processing.

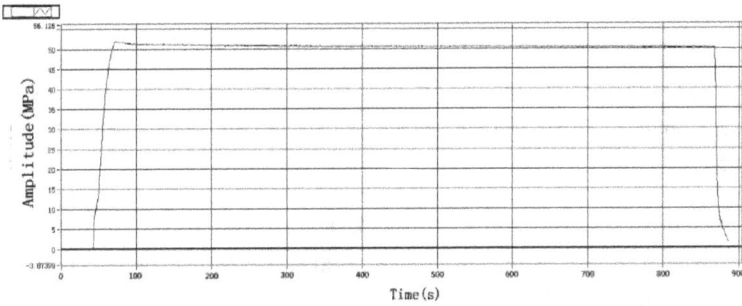

Figure 6. Curve recorded in test (1)

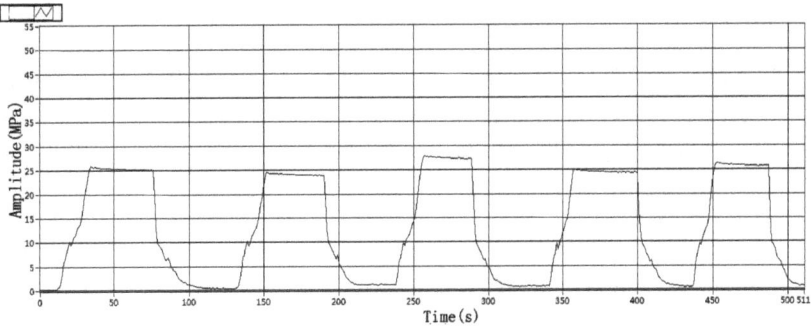

Figure 7. Curve recorded in test (2)

763

5. Conclusion

This paper aims at the subsurface safety valve testing and designs the control system's hardware structure with stable and reliable performance, together with the corresponding signal conditioning circuit for signal conversion. The software with virtual instrument is designed, including the data acquisition module, digital filtering module, and data storage and curve display module, digital output module. Lastly, the result of the trial demonstrates that both the hardware and software can work reliably and fully meet the needs of the actual production.

Acknowledgement

This research was funded by Shanghai Committee of Science and Technology Project (Grant No. 14DZ1204203).

References

1. Shang C M, Zhang D M, Zhang X M, Subsurface Safety Valve Automation Test Systems Design, J. Applied Mechanics & Materials. 543-547 (2014) 1188-1191.

2. Wang T, Zhang D M, Shang C M, The Design of the Subsurface Safety Valve Vydraulic Control Experimental System, J. Chinese Hydraulics & Pneumatics. 03 (2011) 90-92.

3. Ziemer R E, Tranter W H, Fanin D R, Signals and systems: Continuos and discrete, Macmillan, New York, 1993.

4. Chompoo-Inwai C, Mungkornassawakul J, A Smart Recording Power Analyzer Prototype Using LabVIEW and Low-Cost Data Acquisition (DAQ) in Being a Smart Renewable Monitoring System, C. IEEE. Green Technologies Conference. (2013).

5. Yang K T H F, Research on Processing Data from Dewesoft with Labview, J. Electrical Automation. (2012).

A Subsurface Safety Valve Test System Based on Hydraulic System

Li-xin Lu, De-shuai Shen and Gui-qin Li[†]

Shanghai Key Laboratory of Intelligent Manufacturing and Robotics,
Shanghai University, Shanghai, China
[†]*Email: leeching@t.shu.edu.cn*

Shu-xin Li, Ru-dong Lu and Yun-jiang Dong

Shanghai Extrong Oilfield Technology Co., Ltd.,
Shanghai, China

A new subsurface safety valve test system, including hydraulic test unit, liquid test unit, gas test unit and control unit is proposed in this paper. The control unit is realized by using virtual instruments and PCI cards. The test system can provide hydraulic oil to power the safety valve, as well as high-pressure liquid and gas to simulate the working environment. For power source, low pressure gas is pressurized by a booster pump to supply three units, thus making the system more compact with lower cost.

Keywords: Subsurface safety valve; hydraulic system; virtual instrument.

1. Introduction

Subsurface safety valve is a downhole toolto prevent the blowout and ensure safe production, and it is a special column of oil and gas wells. It will be put into the well in the process of completion,opening and colsing in the cotrol of ground hydraulic system.

Subsurface safety valve testing system needs to complete the function testing and performance analysis of safety valve to ensure the normal operation and avoid leaks. Much research have been done about testing [1] system both in China and abroad. WangTao et al control test stand by the use of hydraulic system,and the control system is realized through PLC and display instruments[2]. Liang W et al design a control platform to measure high temperature superconducting devices based on virtual instrument, and design the power conditioning module by combining programmable power with the equivalent load network[3]. These methods have both advantages and disadvantages, and can not fully meet the requirements of this test.

2. Principle of the Test System

According to the test requirements, test system needs to provide three different pressure medium: hydraulic oil, liquid (water) and air.hydraulic oil is to power safety valve to open and close, high-pressure liquid and gas test the seal of safety valve. Instead of providing separate power source for three media,our system takes advantage of low-pressure gas as the only power source, then provides high-pressure hydraulic oil, water and gas by several gas booster pumps so that the system is more compact, easier to maintain and less costly.Test system includes a hydraulic control unit, liquid and gas test unit and a control unit.

3. Hydraulic control unit

The hydraulic control unit powers valve to open and close, and complete the work of rate controlled hydraulic booster and relief. The hydraulic drawing of this unit is shown as Fig. 1. To improve the flexibility of the test system, two modes are designed: manual control mode and automatic control mode.

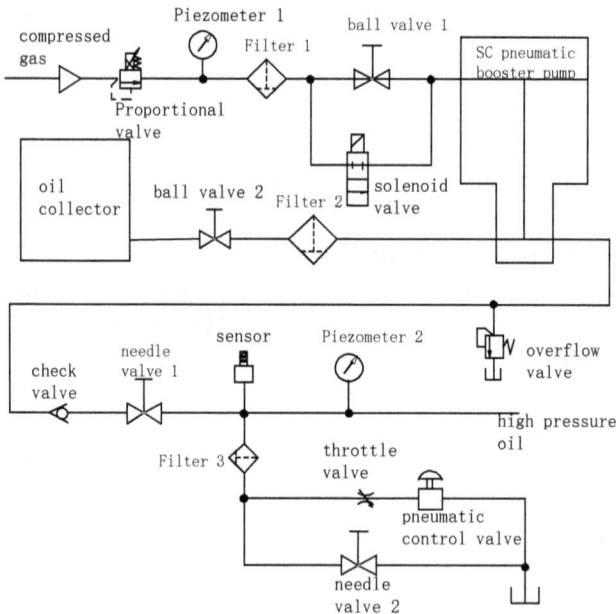

Figure 1. Hydraulic drawing of Hydraulic control unit

In manual mode, when need a boost, open ball valve 1, ball valve 2, needle valve 1, and close needle valve 2. After setting the proportional valve needed for working pressure, low-pressure compressed air enters the inlet chamber of the pneumatic booster pump,and booster will pressurize oil in a fixed pressure ratio.When the output of the oil pressure reaches the set value, the booster pump will automatically stop working and to maintain pressure balance;Conversely, when the output of oil pressure falls below the set value, the booster pump will automatically start and the pressure on both sides of the piston are balanced again.When need relief, close the needle valve 1 and open needle valve 2.

In automatic mode, close ball valve 1, needle valve 2, and open ball valve 2, needle valve 1, then set the proportional valve to operating pressure by control system, turn on solenoid valve, finally the booster pump starts to work. When the output pressure reaches set value of electrical proportional valve, it will stop increasing automatically. When need relief, close the needle valve 1 and openpneumatic control valve.

4. Liquid and Gas Test Unit

Liquid and gas test unit needs to accomplish the following tasks: working pressure test, non-equilibrium open test, liquid flow test, liquid and gas leak detection [4 - 6].

Working pressure test is to simulate the case about downhole pressure of safety valve. Firstly,inject liquid into safety valve and exert 25% of rated working pressure, then record the pilot pressure during opening or closing safety valve and check the compliment with requirement.In addition,repeat trials with 75% of rated working pressure.

Non-equilibrium open test is to detect if the pilot pressure meet compliance requirements when the safety valve opens instantly in the case that the difference of pressure between input and output is maximal.

Liquid flow test is to detect the capability of safety valve of blocking flow when the valve closes quickly after it opens for a while.

Liquid and gas leak detection is the most important test of this system.It means that record the amount of leakage of liquids or gases in five minutes after exerting 100% of rated working pressure, then calculate the leakage rate. The gase used in this test is nitrogen gas. The hydraulic drawing of this unit is shown as Fig. 2. Each state of control valve in these tests is shown in Table 1.

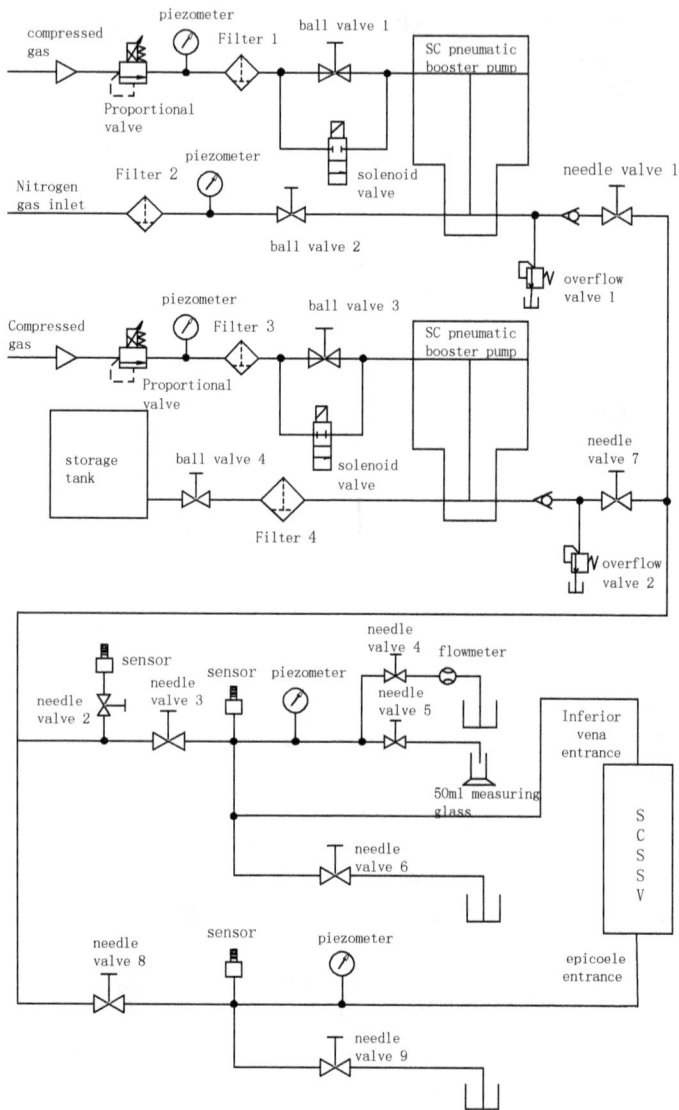

Figure 2. Hydraulic drawing of Liquid and gas test unit

Table 1. State of control valve in four tests

Hydraulic component	Working pressure test	Non-equilibrium open test	Liquid flow test	Liquid and gas leak detection	Upstreamrelief	Downstream relief
ball valve 1 (manual mode)	✓	✓	✓	✓	O	O
ball valve 2	✓	✓	✓	✓	O	O
ball valve 3 (manual mode)	O	O	O	O	O	O
ball valve 4	O	O	O	O	O	O
needle valve1	✓	✓	✓	✓	O	O
needle valve 2	O	O	O	O	O	O
needle valve 3	O	✓	×	×	O	×
needle valve 4	O	×	✓	✓	O	O
needle valve 5	O	×	×	×	O	O
needle valve 6	O	×	×	×	O	✓
needle valve 7	×	×	×	×	O	O
needle valve 8	✓	✓	✓	✓	×	O
needle valve 9	O	O	O	O	✓	O
solenoid valve 1 (auto- mode)	✓	✓	✓	✓	O	O
solenoid valve 2 (auto-mode)	O	O	O	O	O	O

✓ represents open, × represents close, O represents no effect

5. Control Unit

In order to collect and process data quickly, the control unit of our test system is based on virtual instrument and PCI bus. Virtual instrument software sends control signals to the acquisition card by PCI bus, then acquisition card control pumps and electromagnetic valves. Meanwhile, virtual instrument display output data and results in real time.

According to the conventional industry standards and subsurface safety valve test requirements, main technical parameters of test system are shown in Table 2.

Table 2. Main technical parameters

Working medium	Hydraulic oil ISO 46#, Water, Nitrogen
Work stress	Oil pressure:70MPa Hydraulic pressure:65MPa Air pressure:70MPa
Pressure measurement accuracy	≤0.5%F.S
Compressed air pressure	≤0.7MPa
Compressed air flow	≥0.6NM³/Min
Compressed air filtration precision	≤5um

1) Pneumatic booster pump

   Test system chooses US HASKEL's pneumatic booster and models in three units are AW-122, M-110, AG-122.They are both single-acting and dual gas drive. The highest output pressure is 103 MPa, 85 MPa, 110 MPa.

2) Pressure regulating valve

   In order to meet the basic technical parameters, the pressure regulating valve in our test system is ITV3051 proportional control valve of Japanese SMC Corporation. The maximum pressure of this valve is 0.9MPa, maximum input flow 3000L/min, accuracy±0.3%.

3) Pneumatic valve and relief valve

   30-11HF4-NO normally closed gas control valve of US HIP company is chosen in our system, as the maximum pressure of this valve is 70MPa. Relief valve is MV9462342 model of Singapore MACIS Company.

4) Pressure duct

For easy installation, pressure ducts in both Hydraulic control unit and Liquid and gas test unit are 30-125-316-30000PSI model steel tube of US HIP company. The maximum pressure it can withstand is 413MPa and it has high resistance to fatigue.

6. Result

Based on the hydraulic drawings, connect piping and other hydraulic components and Fig. 3 describes the connection of pneumatic booster set and tubing. The test of safety valve includes working pressure test, non-equilibrium open test, liquid and gas leak detection, etc. After multiple tests, the test system can work properly and better meet the test requirements.

Figure 3. The connection of pneumatic booster set and tubing

7. Conclusion

According to the experimental requirements, a new subsurface safety valve test system is proposed. It makes better detection of the basic function, mechanical characteristics and leak proofness of subsurface safety valve. Meanwhile, the application of virtual instruments reduces the test cost and makes it easier to control the valves.

Acknowledgement

This research was funded by Shanghai Committee of Science and Technology Project (Grant No: 14DZ1204203).

References

1. Ishida M, Ichiyama K, Watanabe D, et al. Real-time testing method for 16 Gbps 4-PAM signal interface, (2011).
2. Wang T, Zhang D M, Shang C M. The design of the subsurface safety valve hydraulic control experimental system, J. Chinese Hydraulics and Pneumatics, 03 (2011) 90-92.
3. Wen L, Jin J X, Chen X Y, et al. A Universal LabVIEW-Based HTS Device Measurement and Control Platform and Verified Through a SMES System, J. Applied Superconductivity, 24 (2014) 1-5.
4. Chou J, Lin C, Liao Y, et al. Data Fusion and Fault Diagnosis for Flexible Arrayed pH Sensor Measurement System Based on LabVIEW. J. Sensors Journal, 5 (2014) 1405-1411.
5. Shang C M, Zhang D M, Zhang X M. Subsurface Safety Valve Automation Test Systems Design. J. Applied Mechanics & Materials, 543 (2014) 1188-1191.
6. Bane D, Peoples B. Subsurface Safety Valve Control System for Ultra deep water Applications. (2009)

Design and Dynamic Analysis of a TBM Cutterhead Driving System Test-Bench

Hong-hui Ma*, Wei Sun, Xin Ding and Lin-tao Wang

School of Mechanical Engineering, Dalian University of Technology, Dalian,116024, China

*Email:ma_honghui@163.com

In Cutterhead Driving System (CDS) of Tunnel Boring Machine (TBM), failure of key component often occurs owning to the severe vibration. To investigate the dynamic characteristics, a scale CDS test-bench is designed based on similarity theory, which can control the load torque precisely and obtain dynamic response of CDS. Based on virtual shaft equivalent method, a fully coupled dynamic model of test-bench is established, which includes multi-stage gear transmission system and cutterhead. By taking account of the nonlinear factors of gear meshing and external excitation, dynamic responses of CDS test-bench are obtained, which mainly vibrate near mesh frequency and the frequencies of load torque. The external excitation and dynamic responses analyzed in this paper provide guideline in actual testing process.

Keywords: Tunnel boring machine, Cutterhead driving system, Test-bench design, Dynamic characteristics.

1. Introduction

Tunnel Boring Machine (TBM) is a large and high-tec construction equipment which unifies the functional principles of tunneling, rock cutting and segmental lining [1]. As a key subsystem of TBM, cutterhead-driving system (CDS) provides enough driving torque to make the cutterhead rotate and break rock. As shown in Fig. 1, CDS is a complex coupled system, which mainly consists of redundant driving system, multi-stage gear transmission system and cutterhead.

**Corresponding author.

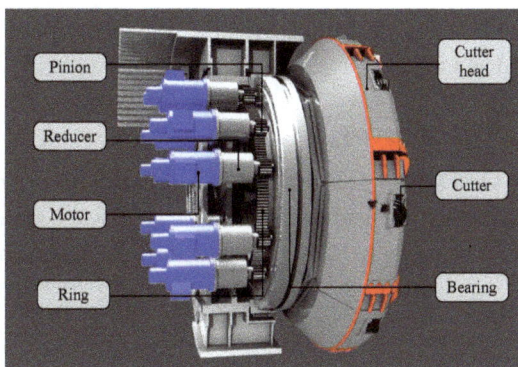

Figure 1. Cutterhead driving system in TBM

During tunneling process, cutterhead often suffers complex and volatile load torque owning to the abruptly changing geological conditions. Under such strong excitation caused by rock breaking, severe vibration often occurs in CDS and leads to failures of key components such as excessive wear of gear tooth, breakage of shaft and imbalance of redundant driving torques [2]. To avoid these problems, anti-vibration design should be conducted based on the dynamic characteristics of CDS. However, due to the confined space of TBM work environment and high operating cost, dynamic responses of key components are difficult to obtain through field test. Therefore, a CDS test-bench ought to be designed primarily and imperatively to study the dynamic characteristics.

In recent years, various TBM test-benches have been designed and built based on the research requirements of different components. Numerous researches have focused on cutting performance of TBM disc cutter which contains mechanism of rock breaking [3], prediction of cutting force [4] and optimization of cutter layout [5]. In addition, TBM driving system also attracts more and more attentions. Liu et al. designed a hybrid CDS which combined hydraulic motors and inverter motors[6]. Rao et al. studied the properties of hydraulic system in propelling system based on test-bench [7]. All these studies have made fruitful efforts on design of key components in TBM. However, the CDS test-bench, which can be applied to study the dynamic characteristic of CDS, has not been designed. Therefore, it is of great significance to design and build a test-bench for dynamic analysis of CDS.

In this paper, a scale CDS test-bench is designed based on the actual structure of TBM, which can control the load torque precisely and obtain dynamic response of CDS. A fully coupled dynamic model of test-bench is established based on virtual shaft equivalent method. By taking account of the nonlinear factors of

773

gear meshing and external excitation, dynamic responses of key components in CDS are obtained and dynamic characteristics are analyzed.

2. Design of CDS Test-bench

As shown in Fig. 2, a scale CDS test-bench is designed according to the actual structure of CDS, which is composed of redundant driving system, multi-stage gear transmission system, cutterhead, hydraulic loading system and support base.

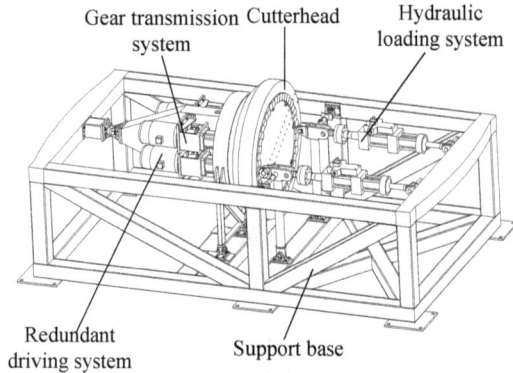

Figure 2. Structure of CDS test-bench

Cutterhead is mounted on the bearing outer race and connected with the gear transmission system through bearing inner race and flange. Based on the similarity theory, the design parameters of cutterhead can be calculated and listed in Table 1[8].

Table 1. Design parameters of cutterhead in CDS test-bench

Parameters	CDS in TBM	CDS test-bench
Diameter (mm)	8530	1200
Rated power (kW)	3440	1694
Rated speed (rpm)	5.6	10
Rated torque (kNm)	5573	5.39

According to the design parameters of cutterhead, 4 inverter motors are chosen as redundant driving system, the rated power of which is 3kW. Master-slave control structure is applied to provide driving torque synchronously. Gear transmission system is designed according to the actual gear ratio of CDS in TBM, which is composed of 2-stage planetary reducer and 1-stage ring-pinion. The parameters of gear transmission system are listed in Table 2.

Table 2. Parameters of gear transmission system

Parameters	2-stage planetary reducer						Ring-Pinion	
	1th-stage			2th-stage			3th-stage	
	Sun	Planet	Ring	Sun	Planet	Ring	Pinion	Ring
Tooth number	12	47	108	54	26	108	17	80
Module	1.25			1.25			8	
Tooth width (m)	0.06			0.06			0.08	
Pressure angle	25			25			20	

Since the existing TBM test-benches use soil-box to provide load torque for cutterhead, such load torque cannot be calculated accurately. In this test-bench, a novel hydraulic loading system is designed to control the size and frequency of load torque precisely. In hydraulic loading system, loading roller is connected with hydraulic actuator and load ring is fixedly mounted on cutterhead. As shown in Fig. 3, load torque is provided through the contact of loading roller and loading ring, which can be expressed as follows:

$$T_l = FR\sin(2\pi f_1 t + \varphi)\tan\theta \qquad (1)$$

where F is the amplitude of force provided by hydraulic actuator; R is the radius of loading ring; f_1 is the frequency of force provided by hydraulic actuator; θ is the contact angle of loading roller and loading ring, which can be expressed as follows:

$$
\begin{cases}
\theta = \arcsin\left[\pi R\left(\dfrac{4i-3}{2b} - \dfrac{n}{30}t\right)\bigg/ r_l\right] & \dfrac{60i-60}{bn} \le t < \dfrac{60i-30}{bn} \\
\theta = \arcsin\left[\pi R\left(\dfrac{n}{30}t - \dfrac{4i-1}{2b}\right)\bigg/ r_l\right] & \dfrac{60i-30}{bn} \le t < \dfrac{60i}{bn}
\end{cases}
\qquad (2)
$$

where b is the number of arc on loading ring; n is the speed of cutterhead; r_l is the radius of arc; i denotes the ith arc which contacts with loading roller at some time.

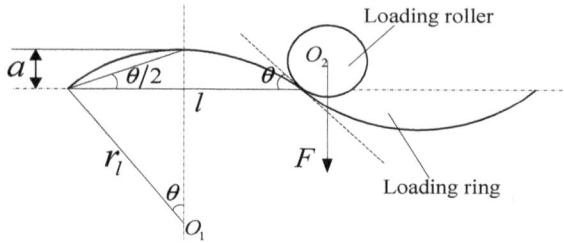

Figure 3. Contact of loading roller and loading ring

Based on Eq. 1 and Eq. 2, main design parameters of hydraulic loading system such as arc number b, arc amplitude a and arc radius r_l can be expressed as shown in Table 3.

Table 3. Design parameters of hydraulic loading system

Parameters	Calculation formula	Calculation value
Arc number b	$b = \dfrac{60(f - f_1)}{n}$	24
Arc amplitude a (m)	$a = \dfrac{\pi R\sqrt{F^2 R^2 + T_{max}} - FR}{2bT_{max}}$	0.05
Arc radius r_l (m)	$r_l = \dfrac{l}{2\sin\theta} = \dfrac{\pi R}{2b\sin\left[2\arctan\left(2ab/\pi R\right)\right]}$	0.12

In Table 3, f is the frequency of load torque; T_{max} is the maximum of load torque, which is equal to rated torque; l is the chord length of arc on loading ring. According to the requirements of load torque, main design parameters of hydraulic loading system are calculated and listed in Table 3.

3. Dynamic Model of CDS Test-bench

Due to the complex mechanical structure of CDS test-bench, it is difficult to establish and solve the fully coupled dynamic model based on traditional modeling method. In this paper, virtual shaft equivalent method is proposed to establish the dynamic model of CDS test-bench. As shown in Fig. 4, the cylindrical structures of CDS test-bench can be equivalent to Timoshenko beam element with two nodes and each node has 6 degrees of freedom, which can be defined as follows [9]:

$$X_{j,j+1} = [x_j, y_j, z_j, \theta_{xj}, \theta_{yj}, \theta_{zj}, x_{j+1}, y_{j+1}, z_{j+1}, \theta_{xj+1}, \theta_{yj+1}, \theta_{zj+1}]^T \quad (3)$$

where $X_{j,j+1}$ is displacement vector of Timoshenko beam; $x_i, y_i, z_i, \theta_{xi}, \theta_{yi}, \theta_{zi}(i = j, j+1)$ are the linear and rotational displacements along axes directions of three-phase coordinate respectively; j is the number of node.

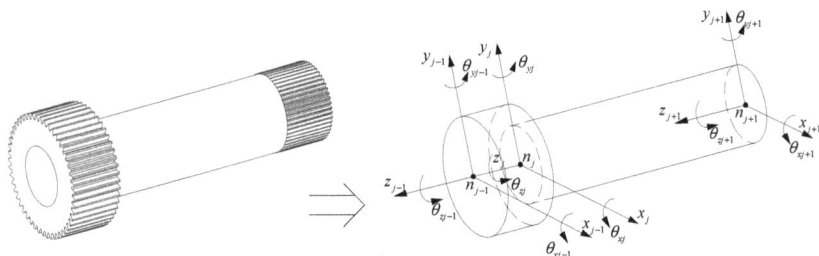

Figure 4. Virtual shaft equivalent example of pinion

The modeling process is as follows: CDS test-bench is divided into one cutterhead-ring subsystem and several identical driveshaft subsystems. The key components such as cutterhead, bearing, flange, ring and pinion in each subsystem are equivalent to Timoshenko beam elements. The dynamic model of planetary reducers is established based on lumped mass method. According to the conventional assembly principle of finite element method, the mass, stiffness and damping matrices of cutterhead-ring subsystem can be obtained by assembling the matrix of each beam element. The FE model of cuntterhead-bearing subsystem is coupled with particular number of driveshaft subsystems by gear meshing. Thus, the dynamic model of CDS test-bench can be established as shown in Fig. 5 and dynamic equation of the whole model can be expressed as follows [10]:

$$M\ddot{X} + KX + C\dot{X} = T \tag{4}$$

where M, K, C are the mass, stiffness and damping matrices respectively.

As shown in Fig. 5, nonlinear factors of gear meshing such as time-varying mesh stiffness k_m and mesh error e_m are considered in dynamic model. k_m can be expressed by means of the Fourier series expansion as follows [11]:

$$k_m(t) = \bar{k}_m + \sum_{n=1}^{N} B_n \cos i\omega_m (t + \varphi) \tag{5}$$

where $\bar{k}_m$ is average mesh stiffness which can be obtained based on gear standards such as AGMA ISO 1328-1 and DIN3990; B_n is the n-rank harmonic amplitude in Fourier series, ω_m is mesh frequency.

Transmission error e_m is approximated as superposition of harmonic function of mesh frequency and rotation frequency of shaft [12]:

$$e_m = 0.5F_p \sin\left(2\pi\omega_s t + \varphi_s\right) + 0.5f_p' \sin\left(2\pi\omega_m t + \varphi_m\right) \qquad (6)$$

where F_p is total cumulative pitch error; f_p' is tangential tolerance of single tooth; ω_s is rotation frequency; φ_s, φ_m is initial phase of shaft and mesh phase.

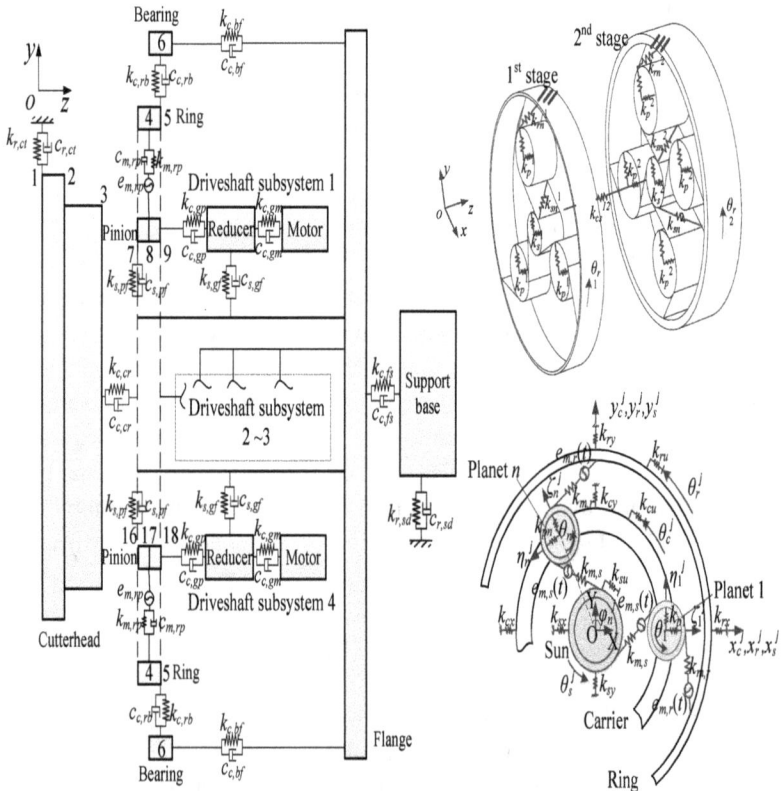

Figure 5. Equivalent dynamic model of CDS test-bench

4. Dynamic Result and Analysis

4.1 Excitations of Dynamic Model

As discussed above, since gear meshing exists in ring-pinion and planetary reducer, periodical internal excitation would be caused by the nonlinear factors of

778

gear meshing. Taking the ring-pinion as an example, internal excitation can be expressed as follows:

$$F_m = k_m(r_r\theta_{zj} - r_p\theta_{zj+3} + e_m) + c_m(r_r\dot{\theta}_{zj} - r_p\dot{\theta}_{zj+3} + \dot{e}_m) \qquad j = 4 \quad (7)$$

where r_r is radius of ring gear; r_p is radius of pinion gear; c_m is mesh damping.

The external excitation of dynamic model is provided by hydraulic loading system, which can be expressed as Eq. 1 shows. Based on the main design parameters of hydraulic loading system listed in Table 3, load torque can be obtained and shown in Fig. 6.

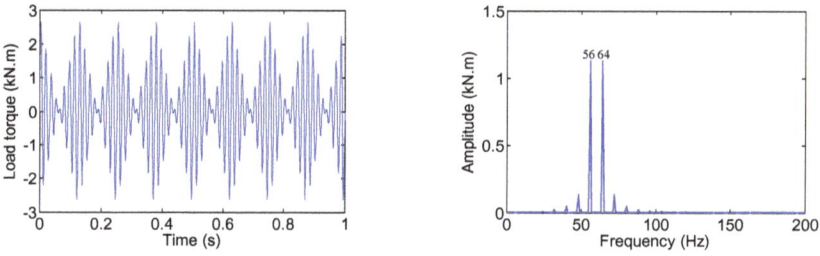

Figure 6. External excitation of dynamic model

5. Dynamic Response of Key Components

Dynamic responses under excitations can be obtained by solving Eq. 4. Due to the complexity of the fully coupled model, precise time-integration method is applied to solve the dynamic model in 1s [13]. Taking the key components such as cutterhead, ring and pinion as example, vibration displacements of these components are shown in Fig. 7 and Fig. 8.

In time domain, the dynamic responses of cutterhead and ring along torsional direction are similar owning to the bolted connection, dynamic response of pinion is smaller than the other two components. In frequency domain, the main vibration frequencies of all components appear in the low frequency domain and include mesh frequency of ring-pinion (f_m=16Hz) and frequencies of load torque (f_1=56Hz, f_2=64Hz). Compared with the internal excitation, the vibrations of these components are influenced more by external excitation.

779

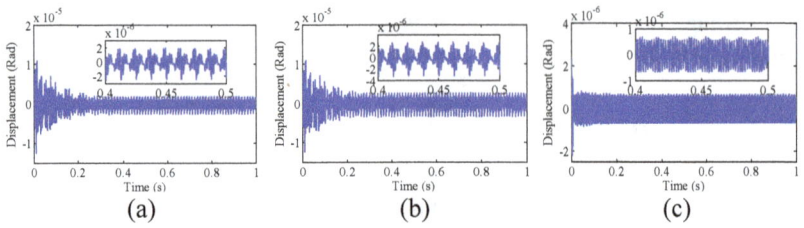

Figure 7. Torsional displacements in time domain, (a) Cutterhead, (b) Ring, (c) Pinion

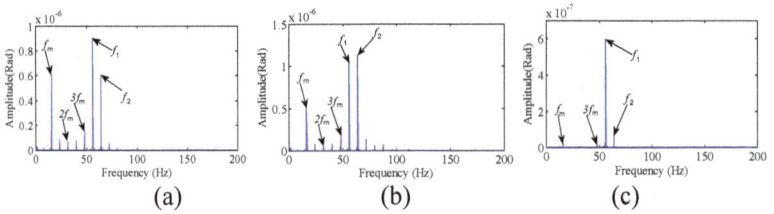

Figure 8. Torsional displacements in frequency domain, (a) Cutterhead, (b) Ring, (c) Pinion

6. Conclusion

According to the mechanical structure of actual CDS in TBM, a scale CDS test-bench is designed based on similarity theory. Hydraulic loading system is specially designed to control the external excitation precisely and the load torque is simulated based on the test requirement.

Taking the nonlinear factors of gear meshing such as time-varying mesh stiffness and transmission error into account, a fully coupled dynamic model of CDS test-bench is established. Dynamic responses of key components in time and frequency domain are calculated and analyzed. The main vibration frequencies appear in the low frequency domain, which are near mesh frequency of ring-pinion (16Hz) and frequencies of load torque (56Hz, 64Hz).

The external excitation and dynamic characteristics analyzed in this paper provide guideline in actual testing process. In future research, dynamic response will be measured through the CDS test-bench to verify the analysis results in this paper.

Acknowledgement

The National Basic Research Program (973 Program) of China (Grant No. 2013CB035400), China Postdoctoral Science Foundation (Grant No. 2015M570245), and Open Foundation of the State Key Laboratory of Fluid Power Transmission and Control of Zhejiang University of China (Grant No. GZKF201414) are acknowledged for their financial supports.

References

1. J. Wei, Q. C. Sun, W. Sun, J. Cai, and J. Zeng, "Dynamic analysis and load-sharing characteristic of multiple pinion drives in tunnel boring machine," Journal of Mechanical Science and Technology, vol. 27, no. 5, pp.1385-1392, 2013.

2. A. Delisio, J. Zhao, and H. H. Einstein, "Analysis and prediction of TBM performance in blocky rock conditions at the Lötschberg base tunnel," Tunnelling and Underground Space Technology, vol. 33, pp. 131-142, 2013.

3. Ozdemir. L and Miller. R, "Cutter performance study for deep based missile egress excavation," Golden, Colorado, Earth Mechanics Institute Colorado School of Mines, pp. 105-132, 1986.

4. Q. M. Gong, J. Zhao, and Y. S. Jiang, "In situ TBM penetration tests and rock mass boreability analysis in hard rock tunnels," Tunnelling and Underground Space Technology, vol. 22, no. 3, pp.303-316, 2007.

5. J. Z. Huo, J. Yang, and W. Sun, "Simulation and optimization design of cutter group with three-dimensional rotating cutting action under different model," Journal of Harbin Engineering University, vol. 35, no. 11, pp.1-6, 2014.

6. T. Liu, G. F. Gong, Z. Zhang, Z. Peng, and W. Q. Wu, "Design and simulation analysis of hybrid cutterhead driving system for TBM test bed," Chinese Journal of Engineering Design, vol. 22, no. 5, pp.438-444, 2015.(in Chinese)

7. Y. Y. Rao, G. F. Gong, Z. Zhang, T. Liu, and W. Q. Wu, "Design and analysis of thrust system for 2m size-reduced TBM test stand," Test and Research, vol. 46, no. 4, pp.32-38, 2015. (in Chinese)

8. J. Z. Huo, G. Q. Li, H. Y. Wu, W. Sun, and X. L. Sun, "TBM cutterhead reduced scale test bench design and static/dynamic characteristics study," Journal of Harbin Engineering University, vol. 37, no. 5, 2016. (in Chinese)

9. Y. Ren, Y. Zhang, and X. Zhang, "Vibration and stability of internally damped rotating composite Timoshenko shaft," Journal of Vibro engineering, vol. 17, no. 8, pp.4404-4420, 2015.

10. W. Sun, X. Ding, J. Wei, X. B. Wang, and A. Q. Zhang, "Hierarchical modeling method and dynamic characteristics of cutter head driving system in tunneling boring machine," Tunnelling and Underground Space Technology, vol. 52, pp. 99-110, 2016.

11. R. G. Parker, and J. Lin, "Mesh phasing relationships in planetary and epicyclic gears," Journal of Mechanical Design, vol. 126, no. 2, pp. 365-370. 2004.

12. D. T. Qin, Z. M. Xiao, and J. H. Wang, "Dynamic characteristics of multi-stage planetary gears of shield tunneling machine based on planet mesh phasing analysis," Journal of Mechanical Engineering, vol. 47, no. 23, pp. 20-29. 2011.
13. W. X. Zhong, "One precise integration method," Journal of Computational and Applied Mathematics, vol. 163, no. 1, pp. 59-78. 2004.

Chapter 6

Robotics

An Efficient Method of Recognition Human Motion in Human-Robot Interaction System

Wei Cheng and Ye Xue
Chien-Shiung Wu College,
Southeast University, Nanjing, China
Email: allyandx@hotmail.com

Chuan Liu
School of Mechanical Engineering,
Southeast University, Nanjing, China

Ying-zi Tan
School of Automation,
Southeast University, Nanjing, China

A full-body recognition method of human motions is proposed in this paper. It allows human to control the humanoid robot to imitate his motions with full-body language. In proposed Human-Robot Interaction system (HRI), real human manipulation motions are acquired by Microsoft's Kinect sensor. To improve the efficiency of the motion recognition, the particle filter algorithm is proposed. In terms of the challenge for a humanoid robot to imitate full-body human motions, two different strategies are used in upper body and lower body respectively. The upper-body system includes a series of angle calculations while a Finite-State Machine (FSM) is introduced in lower-body recognition process. The corresponding simulations and experiments are implemented with both software and a self-made humanoid robot. The results indicate the proposed method is stable and efficient.

Keywords: Human-robot interaction; Kinect; Motion recognition.

1. Introduction

Recent years have witnessed the growing concentration on vision-based human-robot interaction (HRI). Though HRI can be used for remote medical, rehabilitation engineering, remote home services, distance education, and so many other fields, with wide application prospects, it is still a relatively young field that began to emerge in the 1990s [1,2]. One vital problem in advanced HRI system is human posture recognition with which users are just required to perform in the front of the sensor devices without wearing or operating any controldevices to achieve their control purposes.

Many efforts have been made to enable human–robot interaction using posture recognition [3-5]. In particular, Microsoft's Kinect sensor, which offers new capabilities for3D image capture and human gesture recognition, has been used to localize in three-dimensional indoor environments [6] or to avoid collisions [7]. Rossmannacquired real human manipulation gestures using Kinect sensor and project them in real-time into the workspace of a robot manipulator to superimpose human motion remotely. [8] Also,the results from testing the Kinect sensor on an autonomous ground vehicle was given in [9].

In humanoid robot control field, Nguyen proposed a system to reproduce imitated motions of human during continuous and online observation with a humanoid robot [10]. In addition, Yang used Kinect to control both robot model and humanoid robot NAO based on the human motion to achieve the human-machine interaction [11].

However, the previous research experimented the HRI system with Kinect mainly on the upper body with some existed humanoid robots like Darwin-OP [10] and NAO [11]. The lack of the simulated data from the lower body makes previous methods unconvinced. Furthermore, the way to transform a captured motion data into the motion data that humanoid robot can execute remains to be another key problem.

Thus, in this paper, we proposed a full-body imitation method in HRI. In the proposed system, manipulators can control the humanoid robot by upper- body postures, lower-body postures, and full-body postures. In this way, a wider range of the applications of HRI can be achieved.

The remainder of this paper is organized as follows. The overview of the proposed system is illustrated in the first section. In the second section, the methods and the algorithms of identifying motions are presented. Simulation and analysis of experiments are given in the third section. Finally, the fourth section concludes the entire paper.

2. System Architecture

The proposed HRI system involved three main components. To begin with, the postures of themanipulator are captured by the Kinect, which is regarded as the first component (*Motion Capture*). Then, those data containing skeletal information are sent to the PC, where the filtering, calculation and identification works are done.This is the second component (*Motion Recognition*). Finally, after format transfer, the motion control information is transmitted to the robots as well as the simulator,this is the final component (*Motion Control*). The outline of the system structure is shown in Fig. 1.

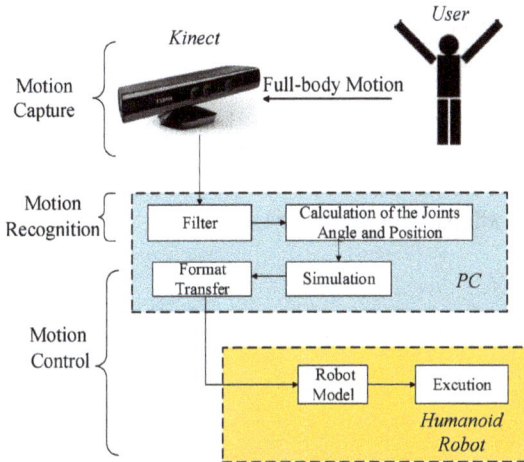

Fig. 1. System structure

In this paper, we mainly focus on the posture recognition problem; thus, the operation process in *Motion Recognition* will be illustrated in details while the other components will be discussed briefly.

3. Motion Recognition

3.1. *Filtering*

The joints andtheir trajectory are captured through SDK provided by Microsoft. However, the Kinect has default tracking mode and seated trackingmode. Becausethe full-body posture is required in theproposed system, the default tracking mode is used. In this mode, twenty skeletaljoints are tracked. However, in the experiment, the skeletal point is not always stable, so the data shows discontinuously and the sudden changes may cause the controlled process instability. Therefore, the particle filter is introduced into this system.

3.1.1. *Model Description*

To track the observation point, a set of sample particle is defined as: $S=\{s_k|s_k(x_k,y_k,z_k,v_k)\}$, where x_k,y_k,z_k describe the position of s_k with v_k as its distribution weight. The set of sample particles is updated through the *State Equation,* which can be expressed as:

$$S_i = AS_{i-1} + w_{i-1}, \qquad (1)$$

where i is the iteration index, A is the state transition matrix and w denotes the gaussian noise. Also, we define the observation point as:$P(x_p,y_p,z_p)$.

787

According to the observation points, the distribution of sample particle can be given as:

$$\Phi(s_k) = \frac{1}{\sqrt{2\pi}\sigma} e^{-\frac{|s_k - P|^2}{2\sigma^2}}, \tag{2}$$

where s_k is the kth particle sample in set S.

3.1.2. Particle Filter Algorithm

Each particle sample in set S has its probability value according to Eq. 2. Therefore, the weight of each particle sample can be expressed as:

$$v_k = \frac{1}{\sqrt{2\pi}\sigma} e^{-\frac{|s_k - P|^2}{2\sigma^2}}. \tag{3}$$

Then the iteration process can be illustrated as:

- Step1 resampling: Calculate the sum of the weight of each sample:

$$T = \sum_{k=1}^{n} v_k,$$

  compare T with the preset threshold T*. If T>T*, get into Step3; else turn to Step2.

- Step2 propagation: Update the data in set S in terms of the Eq. 1. In this paper we use the following way to simplify the construction of the state transition matrix A. To be more specifically, n knew particle samples in Gaussian distribution are generated according to

$$n_k = \frac{v_k n}{T},$$

  where n denotes the total number of the particle samples in each iteration. Then turn back to Step1.

- Step3 tracking target: Calculate the normalized probability according to

$$c_k = \frac{v_k}{T},$$

  then the target position can be worked out with the $\overline{S_i}$ denotes the expectation of S_i.

4. Joints Angle Calculation and Motion Identification

4.1. *The Upper Body*

To calculate the data which can describe the motion of user properly, the angle is animportant parameter.According to themodel of upper body (given in Fig. 2) and the basic equations in vectors:

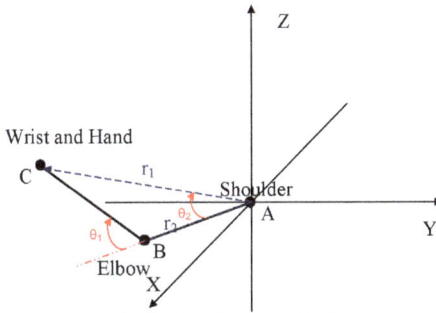

Fig. 2. Model of upper body

$$\vec{r_1} \cdot \vec{r_2} = |r_1||r_2|\cos\theta_2 \tag{4}$$

$$CB^2 = AB^2 + |r_1|^2 - AB|r_1|\cos\theta_2 \tag{5}$$

We can easily get the angle joints in the upper body so that the upper body motion can be described and recognized.

4.1.1. *The Lower Body*

Due to the difficulties to balance the robot, when it is imitating human motion with the lower body. A Finite-State Machine (FSM) is introduced to identify the motion together with a similar calculation method in the upper body. The model of the lower body is shown in Fig. 3. In the lower body, the relative position is another important parameter, which, together with the angle, can provide enough information for us to distinguish different motions.

789

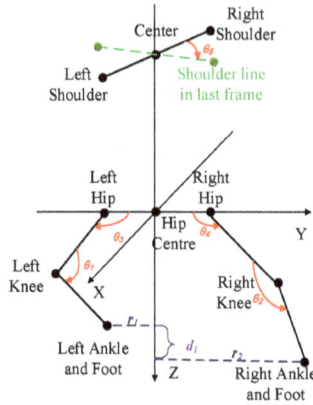

Fig. 3. Model of lower body

According to the angle and relative position, motions are identified by corresponding criterions. Fig. 4 illustrates the Finite-State Machine and the identification process.

Based on Fig. 3 and Fig. 4, the criterions in FSM can be expressed as:

- *Criterion1 or 5:*θ_1 (θ_2) $>120°$, d_1 <half length of calf and θ_5 $>30°$;
- *Criterion2 or 6:*θ_1 (θ_2) $<120°$, d_1 > calf length ;
- *Criterion3 or 7:*θ_1 (θ_2) $>120°$, d_1 < calf length and $r_1(r_2)$<shoulder width ;
- *Criterion4 or 8:* θ_1 (θ_2) $>120°$, d_1 < calf length and $r_1(r_2)$ >shoulder width ;
- *Criterion9:*θ_1 and θ_2<90°, θ_3 and θ_4<90°;
- *Criterion10:* θ_1 and θ_2<120°, θ_3 and θ_4>135°;

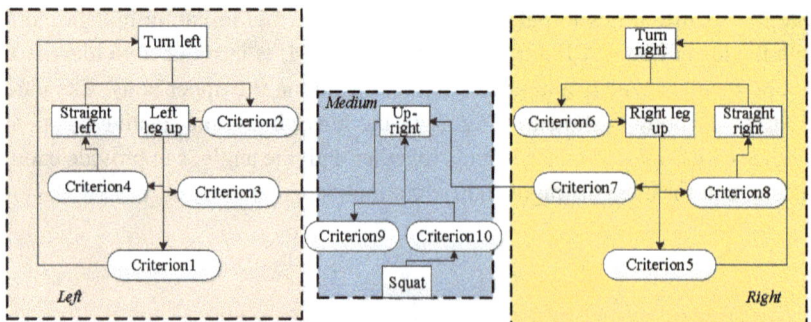

Fig. 4. Architecture of Finite-State Machine for lower-body motion recognition

790

5. Simulation and Experiments

5.1. *Simulation*

The simulation platform is written by QT with Opengl, which is regarded as the preparation of the experiments on real robots.

6. Experiments

Experiments are implemented on our self-build humanoid robot which is equipped with a GIGABYTE Mini PC as its main processor and a self-made motion control unit. The motion control unit is based on Kinetis—K60 MCU, which can drive the twenty Dynamixel actuators (eight of them are Dynamixel RX-28M, twelve are Dynamixel RX-64R) in the full body of the robot.

To improve the ability of imitation, a series of gait planning algorithm and balanced strategies are built in the processor of the robot. In the experiment, both upper-body motion and lower-body motion are involved. Fig. 5 shows the case in which user's upper-body motion is imitated. The skeleton pose of the user and the corresponding simulation movement are shown in Fig. 5(a), Fig. 5(b) respectively. The synchronous control is shown in Fig. 5(c).

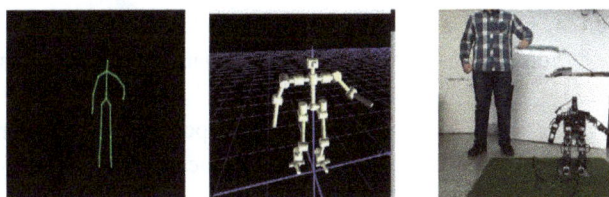

(a)The human skeleton pose (b)Motion simulation (c)The synchronous control

Fig. 5. Imitation of the upper-body motion

The result of lower-body motion imitation is given in Fig. 6.

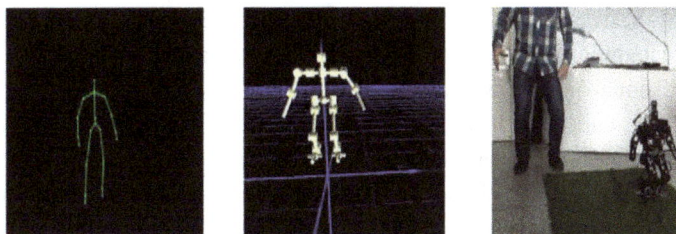

(a)The human skeleton pose   (b)Motion simulation   (c)The synchronous control

Fig. 6. Imitation of the lower-body motion

7. Conclusion

This paper proposed a full-body motion recognition method in an HRI system.By particle filter, the data used in calculation and identification is more stable. In addition, apart from straightforward geometry method in theidentification of the upper-body motion, a novel method of the recognition of the lower-body motion is introduced,which include an efficient FSM.The experimental results informed that users can control the humanoid robot through our system by full-body motion efficiently and stably.

Acknowledgements

This work was funded by the National Undergraduate Training Programs for Innovation and Entrepreneurship (No. 201510286113). Furthermore, we are grateful for the RoboCup Training Base in Southeast University.

References

1. Dautenhahn, K. Socially intelligent robots: dimensions of human–robot interaction. Philosophical Transactions of the Royal Society B: Biological Sciences, 362(1480), 2007, pp.679–704.

2. Goodrich, M. A., & Schultz, A. C. (2007). Human-robot interaction: a survey. Foundations and Trends in Human-Computer Interaction, 1(3), 203–275.

3. C. F. Juang and C. M. Chang "Human body posture classification by a neural fuzzy network and home care system application", IEEETrans. Syst. Man Cybern., A, vol. 37, 2007, pp.984.

4. L. H. W. Aloysius, G. Dong, Z. Huang and T. Tan. "Human Posture Recognition in Video Sequence using Pseudo 2-D Hidden Markov Models," Proceedings of International Conference on Control, Automation, Robotics and Vision Conference, Vol. 1, 2004, pp. 712–716.

5. M. Van den Bergh, E. Koller-Meier, and L. Van Gool. "Real-time body pose recognition using 2d or 3d haarlets". International Journal of Computer Vision, 83(1): 72-84, June 2009.

6. M. F. Fallon, H. Johannsson, and J. J. Leonard, "Efficient scene simulation for robust monte carlo localization using an RGB-D camera," in Proc. of the IEEE Int. Conf. Robotics and Automation, Saint Paul, 2012, pp. 1663-1670.

7. T. Petric and L. Zlajpah, "Smooth transition between tasks on a kinematic control level: Application to self collision avoidance for two Kuka LWR robots," in Robotics and Biomimetics (ROBIO), 2011 IEEE Int. Conf. on, Dec. 2011, pp. 162 –167.

8.   J. Rossmann, E. G. Kaigom, L. Atorf, M. Rast, and C. Schlette, "A virtual test bed for human-robot interaction," in Proc. of the IEEE Int. Conf. Computer Modelling and Simulation, Cambridge, 2013, pp.277-282.

9.   R. A. El-laithy, H. Jidong, and M. Yeh, "Study on the use of Microsoft Kinect for robotics applications," in Position Location and Navigation Symposium (PLANS), 2012 IEEE/ION, pp. 1280-1288, 2012.

10. Van Vuong Nguyen and Joo-Ho Lee, "Full-body imitation of human motions with Kinect and heterogeneous kinematic structure of humanoid robot," in Proc. of the IEEE Int. Symposium. System Integration, Fukuoka, 2012, pp. 93-98.

11. Yang, Ningjia, et al. "A study of the human-robot synchronous control system based on skeletal tracking technology." Robotics and Biomimetics (ROBIO), 2013 IEEE International Conference on IEEE, 2013:2191-2196.

A Novel Hybrid Architecture for Distributed Knowledge Discovery in Scientific Environment

Gang Chen*, Bao-ran An and Si-feng Zeng

*Institute of Computer Application, China Academy of Engineering,
Mianyang 621900, P. R. China*
Email: zjukevin@163.com

Managing data and knowledge sharing in scientific environment is a great challenge, especially in today's big data environment. Traditional monolithic Business Intelligence (BI) systems are difficult to scale up to the massive data sets distributed across a large number of sites. Targeting this challenge, the data mining research has to overcome two major obstacles; mmimg in a distributed environment, and mining on the fly in order to pursue knowledge discovery. Unfortunately, due to highly domain dependency and monolithic deployment, the integration of such large-scale mining on big data is proof to be a tall order. In this paper, we propose a flexible mining system with multi-tier architecture to alleviate these challenges. The system meets the diverse needs of scientific data mining and knowledge sharing. Each service represents the result of an orchestration of reusable building blocks as well as predefined tasks. Its main advantage is that it can deliver BI process as Software-as-a-Service and supports adaptive data mining processes.

Keywords: Distributed Data Mining, Service-Oriented, Agent, Knowledge Discovery.

1. Introduction

We are living in a heterogeneous and distributed data world. Fast-growing volumes of complex data present some huge opportunities for business intelligence (BI) systems. As well known, a BI system involves a wide range of techniques and technologies to gather, provide access to and analyze the voluminous data. It can provide accurate and comprehensive knowledge for decision makers [1]. Equally, in scientific environments, such as NASA Earth Observing System (EOS) [2], BI techniques are increasingly utilized and embraced to make better decisions for researchers, mainly due to the fact that they can improve the quantitative and qualitative values of the knowledge.

Most solutions to manage such computationally intensive knowledge discovery techniques are developed for integrating data into one centralized system [3]. However, in most cases, some issues indicate that this type of solution is gradually becoming impossible, such as complex data integration, computational resource, data privacy concerns, and data transmission bandwidth [4]. That means, the single data repository and monolithic mode are not suitable for distributed and massive data any more. Motivated by the development of Parallel Knowledge Discovery (PKD) and Distributed Knowledge Discovery (DKD) [5], some methods have been proposed to reconcile these issues. The term of Distributed Data Mining (DDM) is demonstrated as the related pattern extraction problem in BI [6-8] and expected to perform partial data analysis at individual sites and the global result is consolidated from partial analysis results. Unfortunately, these methods are still far from having resolved the mentioned issues.

In addition, the paradigm of Service Oriented Architecture (SOA) and Application Service Providers (ASP) has emerged recently to the needs of BI applications [9, 10]. That is a great inspiration to deliver data processing tasks to individual services. For example, in some scientific scenarios, since the data, which research organizations have in their transactional repositories, as well as, the knowledge they want to discovery have a lot in common, generic models for data mining and analysis can be designed to satisfy the user's requirements. One easy way to define these models is by means of various templates. The templates specify the data file locations to be processed, the required knowledge, the mining tasks and the mining algorithms to be used. And then the templates are implemented in some architecture and published as services or analytic applications as service. The templates would be defined by domain experts or technical data miners, and exploited by all the common users who access the proposed service.

This in turn leads us to construct a unified architecture for satisfying the requirements of common researchers to obtain informed knowledge from distributed scientific data sets. We utilize the concepts of agent technology in this architecture with respect to the mentioned challenges. The inherent features of intelligent agents make it quite natural to collaborate with agents in data mining systems for distributed and voluminous scientific data environments. Meanwhile, our proposed architecture follows a service-oriented model with purpose of being easily configured, deployed and published in the web as a service. With this software architecture, we can:

(1) Deploying data mining services on local sites and reducing the volume of data to be transmitted over the network.
(2) Utilizing intelligent agent to collaborate mining algorithms for autonomy and privacy protecting applications.
(3) Supporting on-demand and adaptive data mining services for common end users.

2. Related Concepts and Works

Many concepts and applications have been proposed in the literature. This section will briefly review some studies about the concept of BI and evolutions of DDM systems.

3. Concept of BI

Generally, BI is expected to reveal useful pattern from data. It involves a wide range of techniques, such as information extraction, data mining, machine learning and visual techniques [11-13]. Using BI, large volume of data can be integrated and converted into knowledge. According to U.M. Fayyad and H. Yang [14], the BI process can be summarized as follows.

(1) Fully understand the application domain and requirements of users.
(2) Construct the target data sets from complex data sources. Some data preprocessing are required in this step, such as data cleaning, transformation, formatting and semantic information adding.
(3) Predefine the goals of mining tasks. The goals reveal the inherent nature of this process, for example, descriptive or predictive.
(4) Design the algorithms and select appropriate parameters to finish these tasks.
(5) Assemble the workflow of mining tasks and execute them on the target data sets.
(6) Interpret and consolidate the outcome to provide decision-making support for users.

4. Concept of DDM and Existing Systems

Traditionally, BI systems have been constructed upon the data resources, such as database or data warehousing. The BI system copes with the discovery process in a scalable manner [15]. When the data and computing resources are distributed in diverse sites, DDM is expected to provide a framework for discovery services. DDM is available when the data scenario has the following features:

(1) The data system consists of multiple independent data sites interconnected and communicated by messaging technologies.
(2) The transmission bandwidth is limited and the mutual communication cost is high.
(3) Sites have privacy concerns.
(4) Sites have resource constraints.

Some prominent and representative DDM systems have been proposed in the literature. In this subsection, we provide a brief review of these systems according to (a) the architecture and models in due course of BI tasks to be pursued by the systems and (b) the data mining techniques, the privacy protection, the data pre-processing, the coordination and control in the systems.

PADMA copes with the DDM issues of DDM from distributed data sites [6, 17]. It is a multi-agent based framework designed for parallel data mining[16]. This framework offers parallel discovery processes to access, analyze, mine and visualize the hidden pattern in the data. The appropriate computing model is guaranteed by multi-agent and messaging technologies. It offers messaging techniques and runtime environment for agents to execute target tasks in individual local sites. The main goal of PADMA is to utilize parallel intelligent agents to realize DDM for efficiency improvement.

BODHI is a framework for specific mining tasks on distributed data sites, such as supervised inductive distributed function learning and regression [18]. It utilizes mobile agents to implement DDM. The agents have different local functionalities and retain its data, status, parameters and knowledge with themselves and work together in conjunction with each other. A central facilitator agent is maintained for initializing and coordinating discovery tasks and the individual agents perform their local functionalities to realize DDM.

JAM is a distributed, scalable and portable meta-learning DDM system based on Java platform and multi-agent systems [19]. Meta-learning is a general approach developed to address the problem of constructing a global classifier for the distributed heterogeneous data. Different learning algorithms can be employed to independent and inherently distributed databases by any JAM agent that is either executed locally on one site or is being migrated from other remote sites in the system. A set of classification models are constructed for agents and different techniques are utilized to implement classification algorithms for different agents. A special distributed mechanism is utilized to combine multiple models at different sites into one global meta-learning classifier that improves the overall predictive accuracy in many applications.

SOA4KD is a proposal for assisting users to build their own knowledge discovery applications on service oriented architecture [20]. It uses meta-learning methods and semantic service oriented architecture to select and execute the knowledge discovery algorithms. The user mining task is imported into the system and then partitioned into two separate parts, a content part and a quality part. Additionally, an Extended Knowledge Discovery Task Ontology (EKDTO) is proposed to capture the user requirements by a semantic web interface implemented based on natural language. A term of Knowledge Discovery Service (KDS) quality ontology is proposed in this study to improve the quality of KDS from the aspect of unique features of KDS as well as the general services. The most appropriate KDS is suggested by meta-learning methods according to the user requirements and a new reference model to implement service-oriented architecture for knowledge discovery.

5. Issues in DDM for Scientific Data

In scientific environment, due to some reasons, most of the daily produced data, such as diagnostic data and measured values are stored in distributed sites. In most cases, different research departments maintain different scientific data. Many of these sites are interconnected by network and provide constrained access to data for some users. This scenario provides the computing environment for DDM on scientific data. Moreover, due to the reasons of security and confidentiality, only authorized user can be allowed to access the entire database. Most users are admitted to access a subset of the scientific data. In reference to some studies for addressing the problems of DDM systems, some issues of DDM for scientific data are listed as follows for improving the DDM performance.

Heterogeneous and Data Integration. In traditional knowledge discovery systems, the target of the work is homogeneous data, which are maintained by the same database schema and management model, and are concentrated and centrally stored, such as data warehousing. If the data are distributed and hetero-geneous, the local data at different sites should be preprocessed and integrated into a global management environment before conducting the knowledge discovery processes. Otherwise, the discovery process may be struggling and contradictions may occur.

Data Dynamics. The development of data gathering and diagnostic techniques make it available that the data environment for scientific data mining is not regarded as static any more. Variety has been one of the essence features and time-related knowledge discovery becomes a great challenge for scientific data.

Knowledge Discovery Goes Service-oriented. Service-Oriented Architecture (SOA) provides seamless integration of computing services for goal-directed tasks. The disadvantages of monolithic deployment mode, as well as the variety of research requirements, roused the development of service oriented applications. With SOA, we only focus on implementing data mining services rather than dealing with the interacting details, such as messaging protocols. And SOA makes DDM applications more scalable and extensible by simply creating or constructing new discovery services. Moreover, SOA makes on-demand DDM available and emancipates users from the implementation of data mining algorithms. The users can pay more attention to their business or science problems.

Knowledge Integration. Most users demand global knowledge from distributed data. The final purpose of DDM is to fetch global results from integrating the local analysis results on individual sites. The mining of data sets on each local site makes use of traditional centralized mining methods and the integration of local results is a great issue for DDM. Just for this reason, some mining tasks cannot work on DDM architecture. For example, some local results may lose their value at the global level and the global results may have less metrics at the local level. The local mining and global requirement should be considered together for DDM.

6. Basic Requisites of DDM Architecture

The knowledge discovery from data draws upon the studies on data mining, data processing, machine learning, visualization and high-performance computing. In these studies, high performance is a key factor, especially when the target data is massive and distributed. For DDM architecture, a set of basic requirements should be considered in order to support massive and large-scale distributed data knowledge discovery.

- *Flexibility for various data mining requirements:* Different users have different mining purpose when they need to discover knowledge from data. And they might use different techniques to complete the same mining task. That requires the architecture should be flexible for various data mining techniques and demands.
- *High scalability for massive data:* Performance is an important factor for a mining system. Some specific techniques should be utilized to improve the performance for acceptable response time, such as parallel techniques and optimization strategies.
- *Full control of data mining services:* The system should provide a variety of data mining services and each one should be independent from the other. The service definition has feasible predefined rules and regulars. It is convenient for users to append one or more services to the system, without doing any substantial change.
- *Flexibility for system updates:* It is flexible and convenient for different users to configure the system factors, such as algorithm parameters, scheduling strategies and constrains on data preprocessing.

Moreover, a generic architecture for knowledge discovery from data may have the following components.

(1) An explicit data resource that maintains the data to be analyzed.
(2) A knowledge integrator and interpreter. This component is used to build a comprehensive knowledge for users and evaluate the discovered knowledge.
(3) A knowledge discovery operating system. It provides various strategies for data mining and knowledge exchange. All mining algorithms and assisted tools should be run in this system.
(4) A graphical interface should be provided for the interaction with users.

7. A Novel Hybrid DDM Architecture for Scientific BI

The term SOA describes a concept for aligning the distributed computing environment with the corresponding BI process. It benefits from the loosely monolithic and collaborative atomic services that can be flexibly combined with each other. Before starting with the architecture description, we must note that the essential features of intelligent agents, such as autonomy, proactive, reactive and sociality, inspire the design and implementation of the entire architecture.

In this section, we propose a novel Hybrid DDM Architecture for Parallel Scientific Data Analysis (HMSD). This system adopts web services and intelligent multi-agent technologies, and an optimizer for cost estimation of DDM tasks. It allows the BI process to be performed at remote sites and provides flexible access for comprehensive knowledge. That copes with the problem of heterogeneous data sources and varied user needs. In order to explain HMSD architecture with a reference framework, we adopt the proposal by Arsanjani [21]. Fig. 1 depicts the hierarchical architecture of our system. The principle of designing this hybrid model is adopting the most suitable approaches for DDM tasks with the user and resource constrains. Therefore, web Services, multi-agent services and client-service model are commonly used in this system.

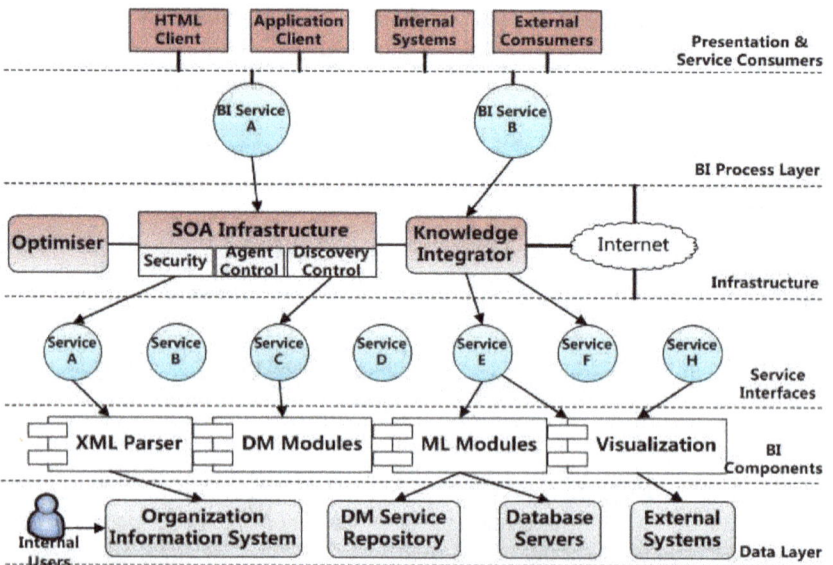

Figure 1. Hybrid DDM Service Architecture of HMSD

8. Architecture of HMSD

The service architecture of HMSD is partitioned into six layers. A brief introduction of each layer in the HMSD is presented as follows:

- **Data Resource Layer:** It is the first layer and provides the data resources required in the whole system, including data gathering and storing, data transmission and transformation and data distribution. Here, we should note that the data transmission is implemented on enterprise component, which is a wrapper mediator for transforming incoming requests into a query format understood by database services and packaging the result into a message format understood by BI components.

- **BI Component Layer:** It maintains a set of BI components that are responsible for fulfilling service needs of the functionalities and providing interfaces for the exposed services in the third layer. This layer currently involves three kinds of services: message format parser, DM-related modules, such as machine learning modules and pre-processing modules, and the last one is visualizing and rendering components for the results.

- **Service Interface:** It contains a number of entity-based services. Each service is combined with a functional boundary directly associated with a specific parent BI task or process. It can be dynamically discovered and statically bound by other services. The local result of each service can be consolidated into a composite service through knowledge integrator in the infrastructure layer.

- **Infrastructure and BI Process Layer:** The two layers together provide a mechanism for estimation, partition and composition of the BI process or tasks. The management tools for security, agent control and mining process control are integrated in the infrastructure. The functionalities of each portion are discussed in detail below.

- **Presentation and Service Consumers Layer:** This layer provides varieties of application and services, such as HTML clients and end-user application clients.

9. Key Components of HMSD

The hybrid model involves some crucial concepts and components. Generally, there are two types of components, management tools and intelligent agents respectively. We present an outline of the tasks and functionalities of the components.

- **User Manager:** The user manager receives requests from user and performs the following functions: parsing and profiling the requests in terms specifying the user requirements, connection monitoring and user authentication.

- *Algorithm Manager:* The main function of algorithm manager is to maintain the data mining algorithms utilized in the system. The algorithms are independent from the HMSD system and users can register any predefined algorithms in the system. A template profile goes with the algorithm, which records the running environment, the input parameters, the output results and other information of the algorithm. When an algorithm is registered into the system, the algorithm manager extracts all the information and delivers it to the discovery process manager.

- *Discovery Control:* It is the core part of HMSD system. It can be considered as a dynamic status register system for all discovery processes. From discovery control, the system can obtain the information about the components and the system regards all the aspects, such as running status, result production, resource consumption and communication information.

- *Knowledge Integrator:* It tries to solve the knowledge integration issues of DDM. It combines various local mining results from different data sources and provides a comprehensive result to the service interface. More details will be discussed in the following subsections.

- *Agent Control:* It is a manager for all the agents running in the system. It is responsible for creating/activating/messaging other agents required for the discovery process. Some representative types of agents are described as follows.

- *User Agent:* The primary task of user agent is to provide information of the system and tasks for user and deliver the final results back to users.

- *Resource Monitoring Agent:* A resource monitoring agent is assigned to each type of resources, such as network, data sources, data access and critical computing resources. It continuously monitors the status of the resource and provides such information to the infrastructure.

- *Data Mining Agent:* There are a large number of data mining agents in the system. Each one is an instance of a mining algorithm required by a discovery task.

- *Local Managing Agent:* A local managing agent is created and maintained at each site. It is responsible for gathering, recording and publishing the local result of data mining agents. And it is also in charge of communicating with global managing agent and corresponding data mining agents.

- *Global Managing Agent:* It is responsible for integrating and delivering global knowledge for end users. It communicates with services and all the local managing agents.

- *Assembling Agent:* It is an important agent in this system. Its function is to travel to remote sites for gathering and determining the information about data set size, computing resources, data preprocessing and report to infrastructure.

10. Knowledge Integrator and Service Composition

What the end users demand is the global and comprehensive knowledge integrated from multiple sites. Intelligent agents utilized in our system can share local knowledge with each other. As mentioned before, the local managing agents can gather local knowledge from data mining agents. If one local managing agent needs knowledge, it will first try to ask other local managing agents for it in roll polling mode. If so, the knowledge is directly transferred to the local managing agent. Otherwise, it will fetch such knowledge from its corresponding data mining agents. The knowledge will finally be transferred to the global managing agents and the global knowledge is then produced.

Basically, whether local managing agents or global managing agents, they are all connected to services provided for discovery applications. Meanwhile, a complicated mining task may need multiple kinds of service to work together. Therefore, in HMSD, the discovery processes are implemented as web services and the completion of a discovery task requires a series of subsequent steps.

- *Requirement Parsing:* When a discovery task is submitted to the system, the corresponding template is parsed and the related parameters, constrains and result requirements are extracted and delivered to the discovery control.
- *Workflow Construction:* An agent in the discovery control further examines the input requirements and constructs the workflow of functional units.
- *Scheduling Strategy:* According to the current data mining agents and the constructed workflow, discovery control estimates the cost of execution and determines the most suitable workflow.
- *Workflow Execution:* The workflow is registered in discovery control component and finally executed to produce knowledge.
- *Result Integration:* The knowledge is integrated by knowledge integrator and provided to the service interface so that it can be fetched by users.

11. Conclusion

To achieve good performance for knowledge discovery from distributed and voluminous scientific data, we propose a hybrid architecture, which is combined with hierarchical structure, multi-agent techniques and specific data mining services, which makes it easily accessible from any client application. The extension of other data mining algorithms can be effortlessly incorporated in the system since the architecture is designed following a service-oriented mode.

Currently, our focus is only on the architecture construction and wrapper various BI algorithms into the system. In the future, we will apply the architecture design to an actual use case and study the security issues in communication messages and data files when our solution is deployed in cloud environment.

Acknowledgement

This work was funded by National High Technology Research and Development Program of China (2007AA1236), and Development Foundation of China Academy of Engineering Physics (14-FZJJ-0422).

References

1. D.Veste, EXCERPT worldwide business intelligence tools 2009 vendor shares, IDC, 223725E1, 2010.
2. Information on http://eospso.gsfc.nasa.gov/
3. O.Rud, Business intelligence success factors: tools for aligning your business in the global economy, Wiley&Sons, Hoboken, 2009.
4. H. Dam, H. Abbass, C. Lokan, DXCS: an XCS system for distributed data mining, in Proceedings of the 2005 Conference on Generic and Evolutinary Computation, Washington DC, USA, 2005: 1883-1890.
5. H.Kargupta, B. Park, D. Hershberger, E. Johnson, Collective data mining: a new perspective towards distributed data mining, in Advances in Distributed and Parallel Knowledge Discovery. AAAI/MIT Press, 2000: 131-178.
6. D. Cheng, On logic-based intelligent systems, in Proceedings of 5th International Conference on Control and Automation, 2005: 71–75.
7. M. Pipattanasomporn, H. Feroze, S. Rahman, Multi-agent systems in a distributed smart grid: design and implementation, in Power Systems Conference and Exposition, Washington, USA, 2009: 1-8.
8. A. Prodromidis, P. Chan, S. Stolfo, Meta-learning in distributed data mining systems: issues and approaches, in Advances in Distributed and Parallel Knowledge Discovery, MIT Press, 2000: 81-114.
9. V. Rao, Multi-agent based distributed data mining: an overview, in International Journal of Computing, 2009: 83-92.
10. P. Clark Dickson, Flag fall for application retal, in System, 1999: 23-31.
11. J. Morency, Application service providers and E-business, in Network World Fusion Newsletter, 1999.
12. Z. Gong, M. Muyeba, J. Guo, Business information query expansion through semantic network, in Enterprise Information System, 2010: 1-22.
13. M. Petrini, M. Pozzebon, Managing sustainability with the support of business intelligence: integrating socio-environmental indicators and organizational context, in Journal of Information Systems, 2009: 178-191.

14. P. Trkman, K. McCormack, M. Olivera, M. Ladeira, The impact of business analytics on supply chain performance, in Decision Support Systems, 2010: 318-327.

15. U. M. Fayyad, G. Piatetsky Shapiro, P. Smyth, Advances in knowledge discovery and data mining, Boston, USA, AAAI/MIT Press, 1996.

16. J. Han, M. Kamber, Data mining: concepts and techniques, San Francisco: Morgan Kaufman Publishers, 2001.

17. A. Brintrup, Behaviour adaptation in the multi-agent, multi-objective and multi-role supply chain, in Computing Industry, 2010: 636-645.

18. H. Kargupta, B. Park, D. Hershberger, E. Johnson, Collective data mining: a new perspective toward distributed data mining, in Advances in Distributed and Parallel Knowledge Discovery, AAAI/MIT Press 2000: 131-178.

19. S. Stolfo, L. Prodromidis, S. Tselepis, W. Lee, D. Fan, P. Chan, JAM: Java agents for meta-learning over distributed databases, in Thrid International Conference on Knowledge Discovery and Data Mining (KDD), Newport Beach, California, USA, 1997: 74-81.

20. L. Yang, C. Zuo, Y. Wang, Research and implementation of service oriented architecture for knowledge discovery, in Chinese Journal of Computers, 2005: 445-457.

21. A. Arsanjani, Service-oriented modeling and architecture:how to identify, specify, and realize services for your SOA retrieved, 2011.4.

A PD Control Strategy Applying to Omnidirectional Lower Limbs Rehabilitation Robot

Ying Jiang*, Wen-dan Zhao and Shuai Li

*College of Information Engineering, Shenyang University
of Chemical Technology, Shenyang 110142, China*

Email: hao770203@163.com

The global population aging is increasingly serious and patients with lower-limb dysfunction caused by various diseases, injuries and traffic accidents have increased significantly. Motion control of lower limbs rehabilitation robot plays an important role. This paper deals with the tracking control of an omnidirectional lower limbs rehabilitation robot. The main research contents of this paper consist of two parts: 1) the kinematic and dynamic model is integrated within the control framework. 2) A PD control scheme is used to achieve the desired trajectories. Simulation tests are performed and demonstrate the feasibility and efficacy of the proposed method.

Keywords: Omnidirectional lower limbs rehabilitation robot, Tracking control, PD control.

1. Introduction

Aging of societies is becoming a major concern. Among different impairments being reported in elderly population, acute blood vessel of brain disease, Parkinson sickness and coronary disease are counted as ones of the most serious[1]. These patients often develop disabling movement disorders, especially walking ability. However, the ability to walk is the most basic and fundamental function, which enables a human being to enjoy life. It was shown [2] that the patient can expect a better recovery if rehabilitation training starts early after injury and involves highly repetitive walking training. In medical institutes or rehabilitation centers, a walking rehabilitation program often requires many physiotherapists and nurses, the training process may be delayed and disabled elder people may lose their best chance to recover. So, robots can assist in walking rehabilitation training have played an important role in recent years. Some examples on the development of robotic walker could be obtained [3-7].

Robotic walkers may be classified into types according to the technology applied to exert the relief force [8]. Most lower limbs rehabilitation systems proposed so far are ones that patients are supported in a harness over a treadmill. These systems focus on relieve the patients of their partial weight by hanging mechanism [9, 10]. However, the normal person can move within any radius toward any direction. In [11], it is suggested that the mode and quality of movement is crucial for recovery. It is also suggested there, that for lower limbs rehabilitation, omni- directional walking training is more efficient to recovery of every parts of body.

One of the major problems in realizing rehabilitation by walker is the tracking control. This paper is organized as follows. Section 2 describes robotic structure design and kinematics, dynamics analysis. Section 3 presents a PD control scheme to adapt to the robot and simulation tests are performed show the feasibility of the proposed approach for tracking control of robot in Section 4. A brief conclusion is given in Section 5.

2. Robotic Structure Design

As shown in Fig. 1, the robot has omnidirectional mobile structure. This structure enables the robot to move in any direction without swerving on the level ground. The developed system consists of a mobile base with four-driving wheels and a support plane, which can be adjusted to the patient's height. The robot doesn't involve suspending the patient in a harness.

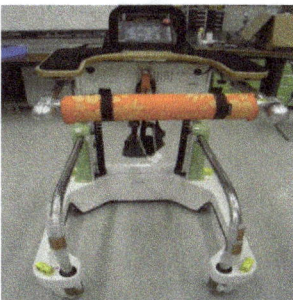

Fig. 1 Omnidirectional lower limbs
rehabilitation robot

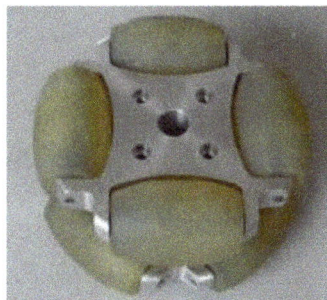

Fig. 2 Omnidirectional wheel

As shown in Fig. 2, this kind of omnidirectional wheel is composed of two parts, namely the driving wheel hub and the passive roller, which distributes along the wheel hub outflow boundary. The spool thread direction of passive roller and the wheel hub end surface is parallel. When the wheel revolves, the wheel central speed is the synthesis of the wheel hub speed and the passive roller spin rate.

3. Kinematics and Dynamics Analysis

According to robot's wheeled structure modality, we can carry on the kinematics analysis to the walker as follows: In Fig. 3, v and θ present the central point speed and the advance angle of walker respectively, α_i is the angle between the ith arm and the x axis positive direction, l is the distance between the arm and the central point.

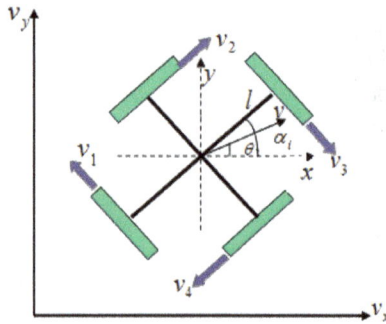

Fig. 3 Robot coordination

The velocities of the wheels are:

$$\begin{cases} v_1 = -v_x \cos(\frac{\pi}{2} - \alpha_3) + v_y \cos\alpha_3 - \dot{\theta}l \\ v_2 = v_x \cos\alpha_3 + v_y \cos(\frac{\pi}{2} - \alpha_3) - \dot{\theta}l \\ v_3 = v_x \cos(\frac{\pi}{2} - \alpha_3) - v_y \cos\alpha_3 - \dot{\theta}l \\ v_4 = -v_x \cos\alpha_3 - v_y \cos(\frac{\pi}{2} - \alpha_3) - \dot{\theta}l \end{cases}$$

(1)

With $\quad v_x = v\cos\theta; \quad v_y = v\sin\theta$ (2)

The restraint equation of robot velocity:

$$v_1 + v_3 = v_2 + v_4$$

(3)

Despite of the friction, the robot kinetic energy mainly is:

$$T = \frac{1}{2}M(v_x^2 + v_y^2) + \frac{1}{2}I\dot{\theta}^2$$

(4)

The robot dynamics has the form:

$$
\begin{cases}
\dot{v}_x = \dfrac{1}{M}[-f_1\cos(\dfrac{\pi}{2}-\alpha_3)-f_2\cos\alpha_3+f_3\cos(\dfrac{\pi}{2}-\alpha_3)-f_4\cos\alpha_3] \\[2mm]
\dot{v}_y = \dfrac{1}{M}[f_1\sin(\dfrac{\pi}{2}-\alpha_3)+f_2\sin\alpha_3-f_3\sin(\dfrac{\pi}{2}-\alpha_3)+f_4\sin\alpha_3] \\[2mm]
I\ddot{\theta} = -lf_1-lf_2-lf_3-lf_4
\end{cases}
\tag{5}
$$

Where, v_x and v_y are the velocities with respect to x-axis and y-axis direction respectively; M is the mass of walker and I is the inertia of mass. The subscripts of lf1, lf2, lf3 and lf4 represent the torque acted on the wheels.

4. A PD Controller

In this section, we present a PD control scheme applying to omnidirectional lower limbs rehabilitation robot for tracking control.

The continuous form of a PD controller, with input $e(\cdot)$ and output $u(\cdot)$, is generally given as:

$$
U = K_{pi}E + K_{di}\frac{dE}{dt}
\tag{6}
$$

where $E = X_d - X$, $\dot{E} = \dot{X}_d - \dot{X}$, $X_d = \begin{bmatrix} x_d & y_d & \theta_d \end{bmatrix}^T$, $X = \begin{bmatrix} x & y & \theta \end{bmatrix}^T$, k_{pi} is the proportional gain and k_{di} is the derivative gain, $U = [u_1 \quad u_2 \quad u_3]$ represents the output of the PD controller, $F = [f_1 \quad f_2 \quad f_3 \quad f_4]$ represents the torque acted on the wheels.

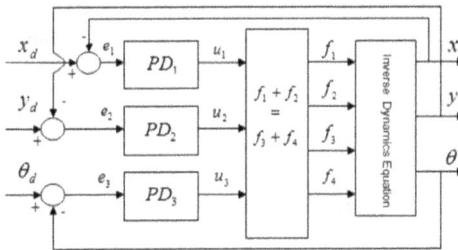

Fig. 4 The PD control strategy

Fig. 4 illustrates the closed loop structure of the system. Three PD controllers are used to control four wheels of the robot to achieve the necessary motion.

5. Simulation

5.1. Simulation Trajectory Setting

We simulate the robot's movement along a linear The reference trajectory is uniformly continuous to satisfy the stability of PD controller. The trajectory is described by (7), as follows:

$$
\begin{cases}
x_d(t) = 10 \times (1 - e^{\alpha t}) \\
y_d(t) = 10 \times (1 - e^{\alpha t}) \\
\theta_d(t) = \dfrac{\pi}{2}
\end{cases}
\tag{7}
$$

where $\alpha = -0.02s^{-1}$.

5.2. Simulation Results

The PD controller is applied to the tracking of a linear path with position $x_{cd}(0) = 0m$, $y_{cd}(0) = 0m$, $\theta_{cd}(0) = \pi/2\,rad$, $x_c(0) = 1m$, $y_c(0) = 1m$, $\theta_c(0) = \pi/3\,rad$, as shown in Fig. 5 and Fig. 6. The optomal value of the PD design parameters is obtained by iterative adjustment for the case $m = 0kg, r_0 = 0.00m, and\ \beta = 0rad$ in the following simulation. Fig. 5 shows the simulation results without the centre-of-gravity shift and load change. In contrast, Fig. 6 shows the simulation results with the centre-of-gravity shift $r_0 = 0.2m, and\ \beta = 2.36rad$ and load change $m = 45kg$.

In Fig. 5 (a), the horizontal axes indicate time and the vertical axes indicate the x position. In Fig. 5 (b), the horizontal axes indicate time and the vertical axes indicate the y position. In Fig. 5 (c), the horizontal axes indicate time and the vertical axes indicate the tracking performance of the orientation angle. Fig. 5 (d) shows the tracking and orientation of the robot. The dotted line represents the reference response and the solid line represents the PD control response. The reference track and PD control response are shown for 300s.

As shown in Fig. 5, the PD control method can successfully track the reference target for the x position, y position, and orientation angle under the condition $m = 0kg, r_0 = 0.00m, and\ \beta = 0rad$.

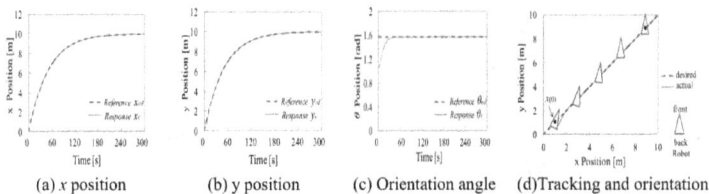

(a) x position   (b) y position   (c) Orientation angle   (d)Tracking and orientation

Fig. 5 Simulation results of PD control with $m = 0kg, r_0 = 0.00m, and\ \beta = 0rad$

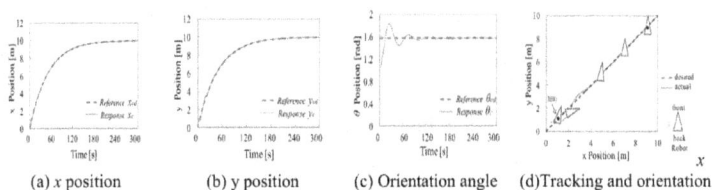

(a) x position   (b) y position   (c) Orientation angle   (d)Tracking and orientation

Fig. 6 Simulation results of PD control with $m = 45kg, r_0 = 0.2m, and\ \beta = 2.36rad$

In Figs. 6(a)-(c), the path tracking error is lager in the first 120 s than that in Figs. 5(a)-(c). But only after approximately 150 s the robot can track the reference path. Fig. 5(d) and Fig. 6(d) show the path and orientation of the omnidirectional robot. Both the path and orientation can be tracked after 150 s.

6. Conclusion

Tracking control of lower limbs rehabilitation robot is an actual problem with robot. In this paper, we used a PD controller design approach for the tracking control of an omnidirectional lower limbs rehabilitation robot. The PD controller can improve the system performance obviously even if in the presence of the uncertainty of the model parameters.

Acknowledgement

This research was funded by the Doctoral Foundation of Liaoning Provincial (201501073).

References

1. J.P.A. Dewald, P. S. Pope, J. D., "Abnormal muscle coactivation patterns during isometric torque generation at the elbow and shoulder in hemiparetic subjects", Brain, vol. 118, pp. 495–510, 1995.

2. M. E. Smith, W. M. Garraway, D. L. Smith, and A. J. Akhtar, "Therapy impact on functional outcome in a controlled trial of stroke rehabilitation", Arch. Phys. Med. Rehabil, vol. 63, no. 1, pp. 21–4, 1982.

3. Nemoto Y, Egawa S, Koseki A, Hattori S, Ishii T, Fujie M, "Power-assisted walking support system for elderly", 2@ Ann. Inl. Con of the IEEE Eng. in Med and Biol. Soc, vol. IO, no.5, pp. 2693-2695.

4. R.D. Schraft, C. Schaeffer, and T. May, "The Concept of a System for Assisting Elderly or Disabled Persons in Home Environments", IECON: Proc.of the IEEE 24th Annual Conf, vol. 4. pp. 2476-2481, 1998.

5. S. Dubowsky, F. Genot, S. Godding, Kozono H, A. Skwersky, H. Yu, and L. S. Yu, "PAMM-A Robotic Aid to the Elderly for Mobility Assistance and Monitoring: A "Helping-Hand" for the Elderly", Proceedings of IEEE International Conference on Robotics and Automation, pp. 570–576,2000.

6. Chuy, Y. Hirata, and K. Kosuge, "Control of a Walking Support System Based on Variable Center of Rotation", Proceedings of the IEEE/RSJ Int'l Conference on Intelligent Robots and Systems, Japan, September 2004.

7. Shuoyu WANG, Koichi KAWATA, Yoshio INOUE, Kenji ISHIDA and Tetsuhiko KIMURA, "A proposal of Omni-directional Mobile walker", Proceedings of annual Conference of Technology, pp 48(2001).

8. Bani F, Fadda A, Torre M, Macellari V (2000) WARD, "A pneumatic system for body weight relief in gait rehabilitation", IEEE Tr. on Rehabilitation Engineering, vol. 8, no. 4, pp. 506-513.

9. Hesse S, BeRelt C, Schafrin A, Maleic M, Mauritz KH (1994), "Restoration of gait in non ambulatory hemiparetic patients by treadmill training with partial body-weight support", Arch. Phys. Med Rehobil, vol. 75, pp. 1087-1093.

10. Norman K. E, Pepin A, Ladouceur M, Barbeau H (1995), "A treadmill apparalus and harness support for evaluation and rehabilitation of gait", Arch. Phys. Med. Rehab, vol. 76, pp. 772-778.

11. V. Dietz and S. J. Herkema, "Locomotor activity in spinal cord-injured persons", J. Appl. Physiol, vol. 96, pp. 1954–60, 2004.

Path Planning for Robot Based on Modified Ant Colony Algorithm

Cheng Chen, Yuan-liang Zhang[†], Jun Gu and Yong-qiang Liu

Huaihai Institute of Technology,
Jiangsu Institute of Marine Resources,
Lianyungang, Jiangsu, 222005, China

[†]*E-mail: 407973063@qq.com*

For the path planning application of the basic ant colony algorithm, it is easy to fall into the local optimal solution. And the long searching time is also a problem. In order to solve these problems, this paper presents a modified ant colony algorithm. At the beginning, in order to make the ants broaden the search and avoid falling into local optimum, a piecewise function using the state transition probabilities and the priority to balance the pheromone of all paths is introduced. And after searching a certain region, in order to speed up the convergence, an acceleration function is used. The simulation results show that the algorithm can quickly find the optimal path in the model grid map.

Keywords: Path planning; Ant colony optimization; Piecewise function; Acceleration function.

1. Introduction

Robot technology is an important research field, and the path planning of the mobile robot is its basic research content. The task of path planning is to find a collision free path from the starting point to the end point, according to the certain requirements in the environment with obstacles. Current research methods of path planning include artificial potential field method [1], visual graph method [2], grid method, genetic algorithm [3], A*(A-star) algorithm [4], and ant colony algorithm[5].

Ant colony algorithm was first applied to solve the traveling salesman problem (TSP), which showed a good ability to solve the problem. So that the ant colony algorithm has been widely developed and used in the path planning of the mobile robot, which has the advantages of good searching ability, positive feedback, distributed computing and etc. [6]. But the traditional ant colony

---

[†]Corresponding author.

algorithm also has some shortcomings, such as slow convergence, easy to fall into the local optimal solution and the phenomenon of stagnation [7]. In response to these shortcomings, based on modified ant colony algorithm for mobile robot path planning, a piecewise function is introduced in this paper. When the robot selects a node, the piecewise function and probabilistic combinatorial optimization of node selection are used to recalculate the probability of the candidate nodes, so that the problem of trapping into local optimal can be avoided. When the search area reaches a certain range, an acceleration function is used to speed up the convergence rate. Simulation results show that the modified ant colony algorithm can make the robot to choose an optimal path quickly, and it shows a good performance in the convergence speed.

2. The Basic Principle of Ant Colony Algorithm

In reality, the ants search the path through the pheromone on the path. The ants will release pheromone about the length of the path. The following ants will choose the path with a large amount of pheromone, which will form a positive feedback mechanism. Ants choose the path basing on the amount of pheromone on the path, which is the basic principle of ant colony algorithm.

The basic mathematical model of ant colony algorithm is as follows. m is the number of ants, $b_i(t)$ is the number of ants in the element i at t moment, and $\tau_{ij}(t)$ is the amount of pheromone from the node i to node j at t moment. So the Eq.(1) is obtained.

$$m = \sum_{i=1}^{n} b_i(t) \ . \tag{1}$$

In order to satisfy the constraint that each ant must go through n different nodes, a data structure for each ant is produced. A data structure called $\text{tabu}_k k = (1,2,\cdots m)$ which records the nodes have been gone through at t moment is built. It can forbid the ant to go through these nodes in this data structure again. The translation probability from node i to node j at t moment for the k ant is defined as:

$$P_{ij}^k(t) = \begin{cases} \dfrac{\tau_{ij}^{\alpha}(t)\eta_{ij}^{\beta}(t)}{\sum \tau_{is}^{\alpha}(t)\eta_{is}^{\beta}(t)}, j \in allowed_k \\ 0 \qquad\qquad otherwise \end{cases} \tag{2}$$

Where $allowed_k$ is the next step for ant k to choose the node, α is the factor of heuristic pheromone, β is the expected factor of heuristic pheromone and η_{ij} is a reflection of the heuristic function, which can be expressed as follows.

$$\eta_{ij} = 1/d_{ij} \tag{3}$$

Where d_{ij} is the distance between node i to node j.

When an iteration cycle is completed, the pheromone update is performed on each path. Update formula is as follows.

$$\tau_{ij}(t+n) = (1-\rho)\tau_{ij}(t) + \Delta\tau_{ij}(t,t+n), \rho \in (0,1) \tag{4}$$

$$\Delta\tau_{ij}(t,t+n) = \sum_{k=1}^{m} \Delta\tau_{ij}^{k}(t,t+n) \tag{5}$$

Where ρ is the volatile coefficient of pheromone, $\rho \in (0,1)$. $\Delta\tau_{ij}(t,t+n)$ is the total amount of increased pheromone in the path (i,j), and $\Delta\tau_{ij}^{k}(t,t+n)$ is the amount of increased pheromone on edge (i,j) by the kth ant between time t and $t+n$. There are three-updated model about the pheromone, which are the ant-cycle system, the ant-quantity system and the ant-density system. For path planning application, the ant-cycle system has a better performance. Its formula is as follows.

$$\Delta\tau_{ij}^{k} = \begin{cases} \dfrac{Q}{L_K}, & \textit{If kth ant uses path in tour} \\ 0, & \textit{otherwise} \end{cases} \tag{6}$$

3. Algorithm Design

3.1. Environment Modeling

Before the path planning of the mobile robot, it is required to observe the working environment, and establish the environment model of the robot. There are many environmental modeling methods, such as raster method, topology map, visual map, geometric information method and so on. Grid method is the most widely used environment modeling method [8]. The grid represents a two-dimensional environment of the robot's motion, and the robot can reach the neighboring grid from the current grid along eight different directions.

A robot's activity area is a finite two-dimensional plane. If there is no obstacle in a grid, it is called a free grid. If there is an obstacle in a grid, it is called the obstacle grid. Coordinate axis is used to express the partition of each grid, assuming that the upper left corner is the origin of coordinates, horizontal to the right is the positive X-axis direction and vertically downward is the Y-axis direction. Define the upper left corner of the first grid number is 1, as shown in Figure 1.

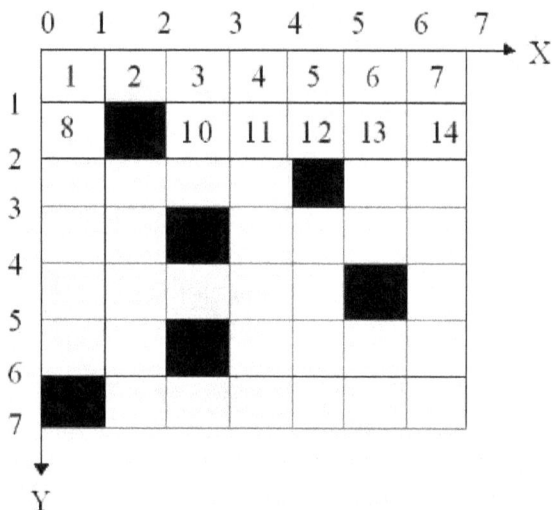

Figure 1. Grid coordinates and the corresponding relationship between the serial number

3.2. Modified Ant Colony Algorithm

In order to make the ant search more paths at the beginning, a piecewise function is introduced, so that the robot can select the effective path at the initial time to find the global optimal solution.

3.3. Piecewise Function

At the beginning, the amount of pheromone on the path is zero. With the increase of time, the amount of pheromone will be more and more which can bring the local optimal problem. In order to avoid this situation, the parameter λ $(0 < \lambda < 1)$ is introduced to balance the pheromone on the path. The parameter λ can be divided into three grades $\lambda_1, \lambda_2, \lambda_3$, so that ants find the best path, The formula is as follows.

$$\lambda_\chi = \begin{cases} 0.25, & \chi = 1 \\ 0.5, & \chi = 2 \\ 0.75, & \chi = 3 \end{cases}$$

(7)

Function x can make the ants search range as large as possible and find as much as possible the effective path. The ants from the current node i to node j, do not choose the maximum probability of node. They select the node with the combination optimization of probability and piecewise based on the combinatorial optimization of the segmentation and state transition probabilities as shown in the following formula.

$$P_{ij}^k(t) = \begin{cases} \dfrac{\tau_{ij}^\alpha(t)\eta_{ij}^\beta(t)}{\sum \tau_{is}^\alpha(t)\eta_{is}^\beta(t)} \cdot \lambda_\chi, & j \in allowed_k \\ 0 & otherwise \end{cases} \tag{8}$$

Where $\chi = 1$ is the highest of the alternative nodes, and when $\chi = 3$ is the minimum of the alternative nodes.

3.4. Acceleration Function

In order to search a large enough range, at the same time to accelerate the convergence, an acceleration function H is used which is shown as follows.

$$H = \frac{d_{i,start}}{d_{start,end}} \tag{9}$$

Where $d_{i,start}$ represents the distance from the current node i to the starting point, and $d_{start,end}$ is the distance from the starting point to the point of termination.

For a path-planning environment, $d_{start,end}$ is a determined value, and $d_{i,start}$ is a variable value. When the $d_{i,start}$ is larger, the current point is farther to the starting point. The closer to the target point, the greater the p_{ij}, so the probability of selecting the node j is greater. By using the H function, the probability formula is as follows.

$$P_{ij}^k(t) = \begin{cases} \dfrac{\tau_{ij}^\alpha(t)\eta_{ij}^\beta(t) \cdot H}{\sum \tau_{is}^\alpha(t)\eta_{is}^\beta(t)}, & j \in allowed_k \\ 0 & otherwise \end{cases} \tag{10}$$

3.5. Simulation

In this simulation the 20*20 grid map is used, and the basic ant colony algorithm and the modified ant colony algorithm are applied, respectively. The simulation results are shown in figures 2 and 3. And the corresponding obtained path data by using these two methods are compared in Table 1.

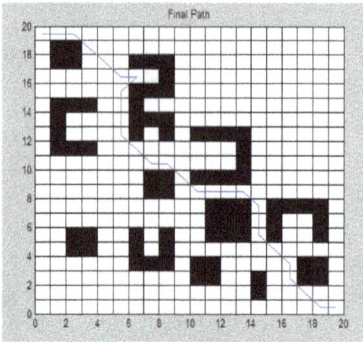 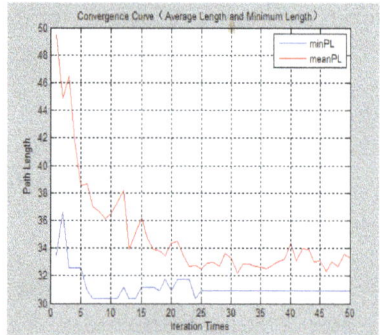

(a)   (b)

Figure 2 Searching result of the traditional ant algorithm
(a)Shows the final path and (b)Shows the average and the best path length

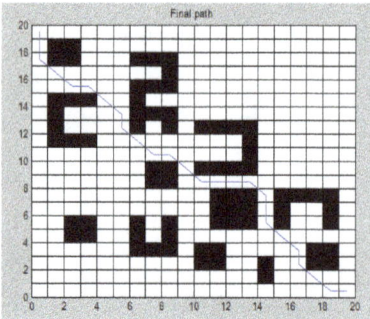 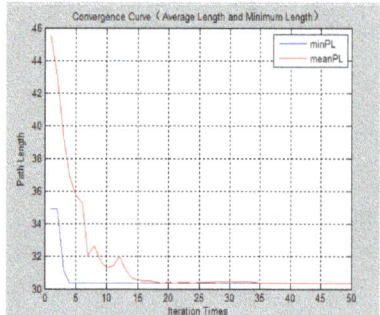

(a)   (b)

Figure 3 Searching result of the modified ant algorithm
(a)Shows the final path and(b)Shows the average and the best path length

Table 1 Data for ant colony algorithm and the modified ant colony algorithm

	Shortest path length	Average path length	Iteration number
Basic ant colony algorithm	31.2	33.1	25
modified algorithm	30.2	30.2	5

From figure 2 and figure 3, we can see that the basic ant colony algorithm is easy to fall into local optimum. At the same time the modified ant colony algorithm can quickly find the optimal path. The simulation results show that the improvement is effective.

4. Conclusion

In this paper, the grid method is used to establish the environment of the robot. Firstly, the basic ant colony algorithm is used for path planning. And then the basic ant colony algorithm is modified to improve the path planning performance. In order to expand the scope of the search, the piecewise function λ_χ is introduced. In order to speed up the search speed, acceleration function H is used. Finally, the modified ant colony algorithm can improve the path planning performance. Simulation results show that the effectiveness of the modified algorithm.

Acknowledgement

This paper is supported by Qing Lan Project, and Jiangsu Institute of Marine Resources Open Fund Project (JSIMR201407).

References

1. C. W. Warren, Global path planning using artificial potential fields, IEEE International Conference on Robotics and Automation, 1989, pp.316-321.
2. Maron, O and Lozano-Pwrez. T, Visible Decomposition: Real-Time Path Planning in Large Planar Environments, AI Memo 1683, January, 1996.
3. Oscar. C, Leonardo. T, Patricia. M, Multiple objective genetic algorithms for path-planning optimization in autonomous mobile robots, Soft computer, 2007, 11,11:259-279.
4. C. W. Warren, Fast path planning using modified A* method. IEEE. Robotics and Automation, Atlanta, USA.1993.
5. Zhao Juan-ping, Gao Xian-wen, Liu Jing-gang, Chen Ying-qiao, Research of path planning for mobile robot based on improved ant colony optimization algorithm, In The 2nd IEEE International Conference on Advanced Computer Control, 2010, pp.241-245.
6. Duan Hai-bin, Wang Dao-bo, Zhu Jia-qiang, Development on Ant Colony Algorithm Theory and Its Application, Control and Decision, Vol.9, No.12, 2004, pp.1321-1326.
7. Zhao Ji-Dong, Ant colony algorithm improvement strategies. Computer knowledge and technology in Chinese, 2014, 10 (28): 6674-6676.
8. Boschian. V and Pruski. A, Grid modeling of robot cells: A memory-efficient approach, Journal of Intelligent and Robotic Systems, vol.8, No.2, 1993, pp.201-223.

Decision-Making Model for Preventive Periodic Maintenance Cycle of Complex System

Jia Liu[1,2], Qu-li Ma[1] and Zong-ren Xie[1,†]

[1]Dept. of Management Science,
Naval Univ. of Engineering,
Wuhan 430033, China
[2]Institute of Navy Equipment and Technology,
Beijing 102442, China,
†email: wx18392015@163.com

For optimization of preventive periodic maintenance cycle of complex system, two patterns are developed for the preventive periodic maintenance of complex equipment system. With the purpose of minimizing the safety, task and economic risks of system, a decision-making model is built for the preventive periodic maintenance cycle of complex equipment system. As decision-making model is based on the calculation of the optimal equilibrium solution in Pareto optimal solution set. For the reliability of the system to be used in solving the decision-making model, GO method is combined with Extend model to simulate the reliability of complex equipment system. The determined preventive periodic maintenance cycle of complex equipment system can lower the safety and task risks of system to the maximum, while considerably reducing the maintenance cost, so it plays an important part in the decision-making of maintenance.

Keywords: Complex system; Preventive periodic maintenance; Maintenance cycle; Decision-making.

1. Introduction

Since a complex equipment system consists of multiple components, and preventive maintenance must be implemented according to the same maintenance cycle for all these components, two patterns are developed for preventive periodic maintenance of complex equipment system, i.e. the same contents are implemented in each maintenance cycle and the contents in the previous cycle are contained in the current cycle. On this basis, a decision-making model for preventive periodic maintenance cycle of complex equipment system is developed with an aim to minimize the safety, task and economic risks of

---

†Corresponding author.

system [1-3]. As the decision-making model for system maintenance cycle and its sub-objectives feature multi-objective decision-making, the method for calculating the optimal equilibrium solution of Pareto optimal solution set is employed to calculate the optimal equilibrium solution of each sub-objective and then general objective, so as to solve the decision-making model for preventive periodic maintenance cycle of complex equipment system. Meanwhile, reliability simulation can be performed to guarantee the reliability of the system to be used in the solution of decision-making model for system cycle [4, 5]. Nevertheless, Monte Carlo method and other simulation methods are normally unable to simulate a complex system like complex equipment system. For this reason, GO method and Extend model are combined to simulate the reliability of complex equipment system.

2. Decision-making Model for Preventive Periodic Maintenance Cycle of Complex System

Complex equipment system contains a large number of components, so preventive periodic maintenance should be arranged for these components in the same period [6]. In other words, preventive maintenance should be performed according to the same maintenance cycle.

The preventive maintenance of system involves two cases. In the first case, the preventive maintenance has the same contents in each cycle, i.e. the same components involved in each preventive maintenance, $S(T^S, A^S)$. In the second case, each preventive maintenance has different contents. In practice, the contents of maintenance are often difference between two cycles, i.e. $S((T^S, A_1^S), (2T^S, (A_1^S, A_2^S)))$, but the contents in the second preventive maintenance cover the contents in the first preventive maintenance. In other words, the components involved in the first maintenance receive the preventive periodic maintenance in the cycle T^S, while the components involved in the second maintenance have preventive periodic maintenance in the cycle $2T^S$. The cases are often identified based on the distribution of optimal maintenance cycle of components.

Within the task cycle T_m, the optimal maintenance cycle T^S of the system S minimizes the failure risk $R^S(T^S, T_m)$. In other words, it leads to the minimum failure safety risk $S^S(T^S, T_m)$, down time $D^S(T^S, T_m)$ and maintenance cost $C^S(T^S, T_m)$.

$$\min R^S(T^S, T_m) = \min(S^S(T^S, T_m), D^S(T^S, T_m), C^S(T^S, T_m)) \tag{1}$$

2.1. Decision-making Model for Preventive Periodic Maintenance Cycle of Complex System

2.1.1. Solution of the Model

The above model is subject to multi-objective decision-making, so it can be solved by calculating the Pareto optimal solution set and then obtain the optimal equilibrium solution among the Pareto optimal solution set.

Assuming that the system S contains N components

$$S = (P_1, P_2, \cdots, P_N,),$$

to minimize the safety risk $S(T^S, T_m)$ of the system in the task cycle T_m, it is necessary to minimize the cumulative probability $F_n(T^S, T_m)$, $n = 1, 2, \cdots, N$, cumulative down time $D_n(T^S, T_m), n = 1, 2, \cdots, N$ and cumulative maintenance cost $C(T^S, T_m)$ of these components $P_n, n = 1, 2, \cdots, N$.

$$\min S(T^S, T_m) = \min(F_1(T^S, T_m), F_2(T^S, T_m), \cdots, F_N(T^S, T_m)) \tag{2}$$

$$\min D(T^S, T_m) = \min(D_1(T^S, T_m), D_2(T^S, T_m), \cdots, D_N(T^S, T_m)) \tag{3}$$

$$\min C(T^S, T_m) = \min(C_1(T^S, T_m), C_2(T^S, T_m), \cdots, C_N(T^S, T_m)) \tag{4}$$

The model given in the formula 2-4 is also subject to multi-objective decision-making, so it can be solved in the same manner.

Based on the above analysis, the algorithm of decision-making model for maintenance cycle of complex system is as follows:

Step 1: Calculate the maintenance cycle T^F to minimize the safety risk of the system $S(T^F, T_m)$

Within the task cycle T_m, the safety risk of the system S is the sum of safety risks of all components. The safety risk of each component is determined by the failure probability and the severity of safety consequence. Since the severity of safety consequence cannot be changed through preventive periodic maintenance, it can be represented by the total failure probability of components, and the failure risk of component is taken as the weight coefficient $W = (\omega_1, \omega_2, \cdots, \omega_N,)$. Therefore, the formula 6.5 is employed to represent the safety risk of system:

$$S(T^F, T_m) = \sum_{n=1}^{N} \omega_n F_n(T^F, T_m) \tag{5}$$

In the Pareto optimal solution set, we obtain the

$$T^F \in \left[\min(T_1^F, T_2^F, \cdots, T_N^F), \max(T_1^F, T_2^F, \cdots, T_N^F) \right]$$

To minimize

$$S(T^F, T_m) = \sum_{n=1}^{N} \omega_n F_n(T^F, T_m),$$

So T^F is the optimal equilibrium solution to minimize safety risk of the system $S(T^F, T_m)$.

Step 2: Calculate the maintenance cycle T^S to minimize the cumulative down time of the system $D(T^D, T_m)$

Within the task cycle T_m, the cumulative down time $D(T^D, T_m)$ of the system S is determined by the total down time of components. As preventive periodic maintenance is employed, the maintenance of components starts at the same time. Hence, the down time of the system during maintenance depends on the component that takes the longest time to maintain. The time of post-event maintenance is ignored, and the failure risk of component is taken as the weight coefficient $W = (\omega_1, \omega_2, \cdots, \omega_N,)$, so the formula 6.6 is employed to simplify the calculation of the cumulative down time of the system.

$$D(T^D, T_m) = \max(\omega_1 D_1(T^D, T_m), \omega_2 D_2(T^D, T_m), \cdots, \omega_N D_N(T^D, T_m)) \quad (6)$$

In the Pareto optimal solution set,

$$T^D \in \left[\min(T_1^D, T_2^D, \cdots, T_N^D), \max(T_1^D, T_2^D, \cdots, T_N^D) \right]$$

Is obtained to minimize

$$D(T^D, T_m) = \max(\omega_1 D_1(T^D, T_m), \omega_2 D_2(T^D, T_m), \cdots, \omega_N D_N(T^D, T_m))$$

So T^D is the optimal equilibrium solution to minimize the cumulative down time of the system $D(T^D, T_m)$.

Step 3: Calculate the maintenance cycle T^C to minimize the cumulative maintenance cost of the system $C(T^C, T_m)$.

Within the task cycle T_m, the cumulative maintenance cost $C(T^C, T_m)$ of the system S is the sum of maintenance costs of all components. The failure risk of component is taken as the weight coefficient $W = (\omega_1, \omega_2, \cdots, \omega_N,)$, so the formula 6.7 can be used to calculate the cumulative maintenance cost of the system.

$$C(T^C, T_m) = \sum_{n=1}^{N} \omega_n C_n(T^C, T_m)$$

$$(7)$$

In the Pareto optimal solution set,

$$T^C \in \left[\min(T_1^C, T_2^C, \cdots, T_N^C), \max(T_1^C, T_2^C, \cdots, T_N^C) \right]$$

is obtained to minimize

$$C(T^S, T_m) = \sum_{n=1}^{N} \omega_n C_n(T^S, T_m),$$

so T^C is the optimal equilibrium solution to minimize the cumulative maintenance cost of the system $C(T^C, T_m)$.

Step 4: Calculate the maintenance cycle T^S to minimize the failure risk of the system $R^S(T^S, T_m)$.

Due to different measuring units for failure probability, down time, breakdown loss and maintenance cost, it is necessary to process the data and obtain the non-dimensionalization expressions for failure probability, down time, breakdown loss and maintenance cost respectively,

i.e. $F_S^{'}(T^S, T_m)$, $D_S^{'}(T^S, T_m)$ and $C_S^{'}(T^S, T_m)$.

Based on the analysis in Section 4.3, it is calculated to obtain the weight coefficients

$$W^S = (\omega_s, \omega_m, \omega_c)$$

of safety risk, task risk and economic risk. Then, the failure risk of equipment system can be presented in the following way:

$$R^S(T^S, T_m) = \omega_s F_S^{'}(T, T_m) + \omega_m D_S^{'}(T, T_m) + \omega_c C_S^{'}(T, T_m). \quad (8)$$

In the Pareto optimal solution set,

$$T \in \left[\min(T_S^S, T_D^S, T_C^S), \max(T_S^S, T_D^S, T_C^S) \right]$$

is obtained to minimize

$$T \in \left[\min(T_S^S, T_D^S, T_C^S), \max(T_S^S, T_D^S, T_C^S) \right],$$

so T^S is the optimal equilibrium solution of the decision-making model for the system, and also the preventive periodic maintenance cycle to minimize the failure risk of the system.

2.2. Modeling for Task Reliability Simulation of Complex Equipment

While solving the above model for the preventive maintenance cycle of complex equipment system, it is necessary to calculate the failure probability or task reliability of complex equipment system. The failure probability of components in complex equipment subject to preventive periodic maintenance is normally believed to follow the normal distribution or Weibull distribution, which are both difficult to calculate the system reliability. Hence, simulation must be introduced.

For simple system, reliability simulation can be carried out with Monte Carlo method. For complex equipment system consisting of numerous subsystems and components, it is also necessary to consider multiple possible states of equipment during the task period, which is impossible to calculate the reliability of task simply with Monte Carlo method. For this reason, a system simulation model must be developed to reflect the overall reliability of equipment.

Extend Sim software features simple modeling and strong simulation capability, and can be combined with GO method to complement each other in the modeling and simulation for reliability of complex equipment during task period. Therefore, the GO method and Extend model are combined in this paper to build a failure simulation model for complex equipment system. The specific steps are as follows:

(1) Step 1: Build a GO model for complex system.

Analyze the structure of complex equipment system, the functional relations among all components, create the schematic diagram of complex equipment system functions, and convert it into the GO diagram on this basis. The types, symbols and roles of operator symbols in GO diagram are presented in Table 1.

Table 1 Operator Symbols of GO Diagram

Operator Symbol	Sign	Description
Signal Generator (cateogry 5)	R—Output signal	There is no input, but output. It is used to indicate the external event outside system or the signal sent by another system, which is normally used as the input of the system
Two-state Unit (category 1)	S-Input signal R—Output signal	It indicates the unit with only two states. If sigal can pass, it is the success state. If signal cannot pass, it is the fail state.
AND Gate (category 10)	$S_1, S_2, \cdots, S_N$—Input signal, R—Output signal	It indicates multiple inputs and one output, and their logical relations.
Route Separator (category 12)	S-Input signal R—Output signal	It indicates one input and mutliple outputs, and their logical relations.

To be specific, use the operator symbols of GO diagram to represent the functions of components in the system and the logical relations between input and output; use the signal flow in GO diagram to represent the functional relations among the components in the system, which are the link

825

operator symbols. In this way, the schematic diagram of system functions is converted into GO diagram. In a complex system, signal generator is normally employed to represent the input of system. Two-state unit (category 1) is used to represent the normal state and fail state of system. AND gate indicates the logical relations of the system with multiple inputs and one output. Path separator (category 12) indicates the logical relations of the system with an input and multiple outputs.

(2) Step 2: Build an Extend simulation model for complex system.
The GO model is converted into Extend model. Based on the analysis of the GO method and Extend model, the GO model can be converted into Extend model by changing the operator symbols in GO model into the modules in Extend model. The specific conversion is presented in Table 2.

Table 2 Corresponding Relations between GO Operator Symbols and Extend Modules

No.	GO Operator Symbol	Extend Module	Extend Module Symbol	Extend Module Function
1	Signal Generator	Create		Create an object
2	Two-state Unit	Activity		Activity of object processing
3	AND Gate	Math		Mathematic calculation module
4	Route Separator	Activity		Activity of object processing

In the process, Create module is used to indicate the signal sent by an independent system, which is identical to the function of signal generator (category 5), so they can be exchanged. As two-state unit (category 1) is used to indicate two states, i.e. normal and fail, Activity module can be employed by setting the input "1" for blockage. Hence, Activity module can be also exchanged with the two-state unit (category 1). Similarly, the function of route separator (category 12) for one input and multiple outputs can be achieved by setting different inputs for the corresponding outputs in the Activity module, so Activity module can replace the route separator (category 12). AND gate (category 10) indicates multiple inputs and one output, which can be realized using the Math module through the Boolean algebra computation for setting logical AND. Then, Math module and AND gate (category 10) can be exchanged with each other.

(3) Step 3: Perform reliability simulation of complex system during task period

The reliability simulation of complex equipment is performed by setting the task cycle T_m, failure law of each component, e.g. parameter of Weibull distribution, and simulation times as the inputs of simulation model.

3. Case Analysis

Under the background of task execution by complex equipment, the preventive periodic maintenance cycle of fuel system in diesel engine is analyzed. The task cycle is 1 year, i.e. $T_m = 8760h$. The optimal maintenance cycles of all components in the fuel system show two distributions, so the second pattern is employed. In other words, the second maintenance cycle covers the contents of maintenance in the first maintenance cycle. The contents of maintenance in the first cycle include self-cleaning fuel filter, magnetic safety filter, fuel coarse filter, pump pre-filter and fuel injector. Apart from the contents in the first cycle, the second cycle also contains pressure regulating valve, one-way valve marked 0.5-1bar, overspeed safety valve and parking control valve. In other words, the components in the first maintenance cycle receive preventive periodic maintenance according to the cycle T^S, while the components in the second maintenance cycle have preventive periodic maintenance according to the cycle $2T^S$.

Based on the above analysis, the decision-making model for the preventive periodic maintenance cycle of fuel system is obtained as follows:

$$\min R^S(T^S, 8760) = \min(S^S(T^S, 8760), D^S(T^S, 8760), C^S(T^S, 8760)) \qquad (9)$$

Before solving the model, it is necessary to employ the reliability simulation model for complex equipment to perform the reliability simulation of fuel system. According to the operating principles of diesel engine, batteries provide the power supply for startup of diesel engine, air needed in combustion of diesel engine is taken from external environment, regulating device control the fuel injection pump, and fuel tank supplies the fuel. All these are the inputs provided by external environment or other components to diesel engine, so they can be represented by signal generator. Fuel injection pump can be started up only after transmission screw functions and igniter starts ignition, so an "AND gate" is provided. Fresh water pump and filter must function simultaneously to cool down the combustion chamber, so an "And gate" is provided. Fuel injector and combustion chamber must operate simultaneously to guarantee normal combustion and push the piston, so an "AND gate" is provided. Crankcase is responsible for output power and waste gas. Based on the failure mode and influence analysis of diesel engine, if cooling system fails, the diesel engine can output 70% of rated power at most. Therefore, there are two possible cases for power output of crankcase, so "route separator" is obviously suitable. All other components have simple input-output relations; so two-state unit operator symbol can be used. In the GO model for diesel engine, the corresponding relations between GO operator symbols and Extend simulation modules as presented herein above are used to obtain the Extend model for the reliability simulation of diesel engine.

The reliability simulation of fuel system carried out by setting the failure law of components, e.g. parameters of Weibull distribution, and simulation times as the input of simulation model.

According to the results of simulation, it is calculated to obtain the maintenance cycle

$$T^F = 2000h$$

to minimize the safety risk of the system

$$S(T^F, 8760),$$

the maintenance cycle

$$T^S = 1750h$$

to minimize the cumulative down time of the system

$$D(T^D, 8760),$$

and the maintenance cycle

$$T^S = 3000h$$

to minimize the failure risk of the system

$$R^S(T^S, 8760).$$

Also, the cost within the cycle is 50,000RMB, the reliability is 0.920, and the decision-making indicator is 0.184.

4. Conclusion

This paper analyzes the features of preventive periodic maintenance of complex equipment system, and develops two patterns for preventive periodic maintenance of complex equipment system, i.e. the same contents are implemented in each maintenance cycle and the contents in the previous cycle are contained in the current cycle. With the purpose of minimizing the safety, task and economic risks of system, a decision-making model is built for the preventive periodic maintenance cycle of complex equipment system on this basis. As the decision-making model for system maintenance cycle and its sub-objectives feature multi-objective decision-making, the method for calculating the optimal equilibrium solution of Pareto optimal solution set is employed to calculate the optimal equilibrium solution of each sub-objective and then general objective, so as to solve the decision-making model for preventive periodic maintenance cycle of complex equipment system. With regard to the reliability of system to be used in the solution of decision-making model for system cycle, Monte Carlo method cannot realize the simulation for the complex equipment system, so the GO method is combined with Extend model to simulate the reliability of complex equipment system. The determined preventive periodic maintenance cycle of complex equipment system can reduce the safety risk and task risk of the system to the maximum, and considerably lower the maintenance cost, so it plays an important role in the decision-making on maintenance.

Acknowledgements

This work was funded by National Natural Science Foundation of China (71401171), PLA General Armament Department Pre-research Fund (9140A19030214JB11273), and Military Universities 2110 Projects Phase III (4142D4557).

References

1. Jovanovic A. Risk-based inspection and maintenance in power and process plants in Europe [J]. Nuclear Engineering and Design, 2003, 226:165-182.
2. Badescu A, Drekic S, Landriault D. On the analysis of a Multi-Threshold Marko-vian risk model [J]. Scandinavian Actuarial Journal, 2007, 248-260.
3. Haijun H, Guangxu C, Yun L, et al. Risk-based maintenance strategy and its applications in a petrochemical reforming reaction system [J]. J. Loss Prevent. Process Ind. 2009, 22:392-397.
4. Reza Seyedshohadaie S, Damnjanovic I, Butenko S. Risk-based maintenance and rehabilitation decisions for transportation infrastructure networks [J]. Transportation Research Part A. 2010, 44:236-248.
5. Wu X, Li S. On the discounted penalty function in a discrete time renewal risk model with general interclaim times [J]. Scandinavian Actuarial Journal, 2009, 281-294.
6. Chen Y, Yuen K, Ng K, Asymptotics for ruin probabilities of a two-dimensional renewal risk model with heavy-tailed claims [J]. Applied Stochastic Models in Business and Industry, 2011, 27, 290-300.

Study on Fuzzy-PID Control System of T-S Fuzzy Model

Shu-bin Wang*, Xin Zuo, Yan-yun Wang and Lu Zheng

*College of Geophysics and Information Engineering, China University
of Petroleum, Beijing, 102249*

*Email: wsbwyy@126.com

A fuzzy-PID control strategy based on T-S fuzzy model is proposed by combining fuzzy control and Conventional PID control. PID controller is developed for each subsystem of T-S fuzzy model. In terms of fuzzy-control theory, each subsystem's control law and membership function can be used to synthetically calculate the whole control system's control law, which control the whole system. The fuzzy-PID control strategy is also applied to three-tank water level nonlinear system. The control results show the effectiveness and feasibility of the proposed method.

Keywords: Fuzzy control, PID, T-S model.

1. Introduction

Since Zadeh put forward fuzzy set theory, two kinds of fuzzy models have been studied widely.

- T-S fuzzy model [1] was put forward in 1985 by Takagi and Sugeno.
- Fuzzy-relation model was put forward in 1984 by Pedrycz.

The T-S fuzzy model is used as a powerful structure for representing nonlinear dynamic systems. The main essence of T-S fuzzy model is that the input space is divided into a lot of fuzzy subspaces. And a simple linear model of input-output is founded in each fuzzy subspace. T-S fuzzy model, which can easily denote the dynamic characteristics of complicated system, is essentially a sort of nonlinear model. The front part of fuzzy rule denotes fuzzy subspace, while the rear part denotes the relation of input-output in the fuzzy subspace. Since the rear part is described by linear equation, a good bridge is founded to combine the fuzzy-control and the tradition-control.

PID control is by far the most common method. Because of its high reliability, simplicity and algorithm robustness, it can be widely used in process control, especially for chemical and metallurgical industry. PID in industrial process control applications nearly has a hundred years of history, and during this period, despite the advent of many control algorithms, thanks to the long-term use of the PID algorithm and its own characteristics, coupled with people accumulating a wealth of experience it is widely used in industrial control.

But with the continuous development of science and technology, people have increasingly high requirement for industrial process control system. Such as the common PID control can't get satisfied effect when the operating point of the controlled process has been changed. So a lot of people have studied on PID control with other advanced control technology and received great theory fruits [2-6] in recent years. Most of them focus on the setting and optimization of the three parameters; of the P, I and D [7, 8]. But in this paper by taking PID control as the bottom control of fuzzy control, a new control algorithm-fuzzy-PID control -which can successfully control nonlinear process control system can be invented.

2. Fuzzy-PID Control Based on T-S Fuzzy Model

2.1. Discrete T-S Fuzzy Model

The general form of a discrete T-S fuzzy model is described as follows:

$$R_p^i : \ if \ \ x_1(k) \ \ is \ \ M_1^i \ \ and \cdots and \ \ x_n(k) \ \ is \ \ M_n^i$$
$$then \ \ x(k+1) = A_i x(k) + B_i u(k) \qquad i = 1, 2 \cdots l \tag{1}$$

where R_p^i denotes the i-th rule of T-S fuzzy model, $M_j^i (j = 1, 2 \cdots n)$ denotes fuzzy set defined by its membership function, l denotes the number of rules, $A_i \in R^{n \times n}$ $\quad B_i \in R^{n \times m}$ $x(k) = [x_1(k), x_2(k), \cdots x_n(k)]^T \in R^n$.

Let $\mu_j^i(x_j(k))$ denote the membership function of $x_j(k)$ belonging to M_j^i and $M^i = M_1^i \times \cdots M_n^i$ adopt multiplication.

$$\text{Then} \qquad \mu^i(x(k)) = \prod_{j=1}^{n} \mu_j^i(x_j(k)) \tag{2}$$

where $\mu^i(x(k))$ denotes the grade of membership of $x(k)$ in M^i.

The whole system state equation can be expressed as:

$$x(k + 1) = \sum_{i=1}^{l} h_i(k)[A_i x(k) + B_i u(k)] \qquad (3)$$

where $h_i(k) = \dfrac{\mu^i(x(k))}{\displaystyle\sum_{j=1}^{l} \mu^j(x(k))}$

Considering $\mu^i(x(k)) \geq 0, \displaystyle\sum_{j=1}^{l} \mu^j(x(k)) > 0$, then $0 \leq h_i(k) \leq 1$,

and $\displaystyle\sum_{i=1}^{l} h_i(k) = 1$.

The characteristic of this fuzzy modeling is that a nonlinear dynamic model can be taken as the approximate model of a lot of local linear model. The fuzzy model can reach high accuracy if we take enough model rules.

3. PID Control

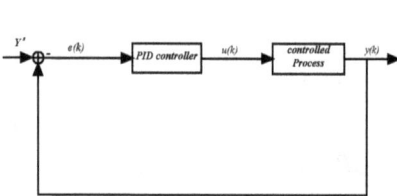

Figure 1. The block scheme of PID control

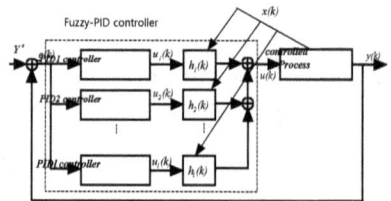

Figure 2. The block scheme of fuzzy-PID control based on T-S model

Fig. 1 shows the block scheme of PID control. PID control algorithm discrete cycle uses the following formula:

$$u(k) = K_p e(k) + K_I \sum_{i=0}^{k} e(i) + K_D [e(k) - e(k - 1)] \qquad (4)$$

4. The Fuzzy-PID Control System Based on T-S Model

Fig. 2 shows the theory of the fuzzy-PID control system.

We apply the PID control to the subsystems of the T-S model:

$$u_i(k) = K_{pi} e(k) + K_{Ii} \sum_{i=0}^{k} e(i) + K_{Di} [e(k) - e(k - 1)] \ (i = 1,2, \cdots l) \qquad (5)$$

Take the total control law as

$$u(k) = \sum_{i=1}^{l} h_i(k)u_i(k) \qquad (6)$$

where $h_i(k) = \dfrac{\mu^i(x(k))}{\displaystyle\sum_{j=1}^{l} \mu^j(x(k))}$, $\mu^i(x(k))$ denotes the membership of the

i-th rule.

Eq. 6 shows that the total control law of the system is the weighted linear combination of the PID control laws. Where $h_i(k)$ varies with $x(k)$ adopting the actual dynamic characters of non-linear process. Thus this algorithm is also a self-adoption PID control by self-tuning the weighted parameter $h_i(k)$.

The steps of this control algorithm:

a. To build the discrete T-S fuzzy model.

b. To design the PID controller by choosing the parameter K_{pi}, K_{Ii}, K_{Di} for each subsystems of the T-S model.

c. To read the history input data, the present and history state and output data.

d. To calculate $h_i(k)$, $(i = 1,2,\cdots l)$;

e. To calculate $u(k)$ using Eq. 6 and apply it to the controlled process.

f. Repeat step d and e.

5. Example

5.1. The Introduction of the Three-Tank Water Level Control System

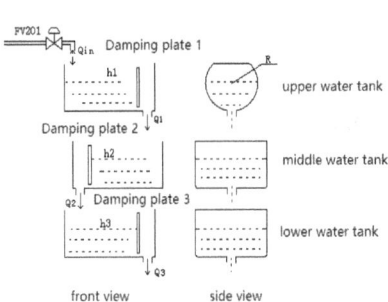

Figure 3. Process flow diagram of three-tank water level system

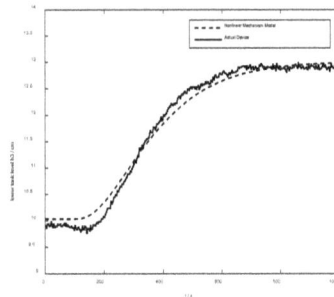

Figure 4. Step response comparison of the mechanism model and the actual device

Fig. 3 shows the process flow diagram of three-tank water level system. The three water tanks are vertically series connected. Upper water tank is a horizontal cylindrical tank and the others are rectangular tank. Manipulated variable and Controlled variable of the whole system are entrance the flow Q_{in} and lower tank level h_3 respectively. The lower tank level can be controlled by tuning regulating valve (FV201) opening.

6. Building the Mechanism Model of the Three-Tank Water Level Control System

According to the dynamic material balance and formula deduction, the following nonlinear mechanism model:

$$\left\{ \begin{array}{l} \dfrac{dh_1}{dt} = \dfrac{Q_{in} - K_1\sqrt{h_1}}{2l\sqrt{R^2 - (R - h_1)^2}} \\[3mm] \dfrac{dh_2}{dt} = \dfrac{K_1\sqrt{h_1} - K_2\sqrt{h_2}}{S_2} \\[3mm] \dfrac{dh_3}{dt} = \dfrac{K_2\sqrt{h_2} - K_3\sqrt{h_3}}{S_3} \end{array} \right.$$

(7)

Where h_i -the level height of i -th tank

K_i -the flow characteristic coefficient of the i -th damping plate,

$K_1 = 3.474 \times 10^5 cm^{5/2} / h$, $K_2 = 4.383 \times 10^5 cm^{5/2} / h$,

$K_3 = 3.224 \times 10^5 cm^{5/2} / h$

$l = 45cm$ -the length of the upper tank

$R = 13.9cm$ -the radius of the upper tank

S_i -the cross-sectional area according to the i -th level height

$$S_1 = 2l\sqrt{R^2 - (R - h_1)^2}, \quad S_2 = 1912.9cm^2, \quad S_3 = 1880cm^2$$

In order to verify the validity of the established mechanism model, the simulation was carried out by MATLAB and compared with the actual device.

Fig. 4 shows that the dynamic response of the continuous nonlinear mechanism model and the actual three tank liquid level system is basically the same.

7. Building the T-S Fuzzy Model of the Three-Tank Water Level Control System

According to the actual device in the low, medium and high level of the water tank to get the three balance point:

$$u_0 = 0.85, x_0 = \begin{bmatrix} 6.0036 & 3.8794 & 7.1215 \end{bmatrix}^T, y_0 = 7.1215$$

$$u_0 = 1.00, x_0 = \begin{bmatrix} 8.2551 & 5.2805 & 9.6628 \end{bmatrix}^T, y_0 = 9.6628$$

$$u_0 = 1.20, x_0 = \begin{bmatrix} 11.8136 & 7.4839 & 13.6531 \end{bmatrix}^T, y_0 = 13.6531$$

By using Eq. 4 the above balance point linearization, and selecting the membership function, three T-S fuzzy model of liquid level system as follows:

$$R_p^i : if \quad x_3(k) \quad is \quad M_i$$

$$then \; x(k+1) = A_i x(k) + B_i u(k)$$

$$y(k+1) = C \sum_{i=1}^{l} w_i [A_i x(k) + B_i u(k)]$$

Where $l = 3$

$$A_1 = \begin{bmatrix} 0.9070 & 0 & 0 \\ 0.0480 & 0.9192 & 0 \\ 0.0021 & 0.0802 & 0.9539 \end{bmatrix} \quad B_1 = \begin{bmatrix} 1.2853 \\ 0.0334 \\ 0.0010 \end{bmatrix} \quad C_1 = \begin{bmatrix} 001 \end{bmatrix} \quad D_1 = 0$$

$$A_2 = \begin{bmatrix} 0.9278 & 0 & 0 \\ 0.0416 & 0.9304 & 0 \\ 0.0015 & 0.0694 & 0.9603 \end{bmatrix} \quad B_2 = \begin{bmatrix} 1.1705 \\ 0.0259 \\ 0.0006 \end{bmatrix} \quad C_2 = \begin{bmatrix} 001 \end{bmatrix} \quad D_2 = 0$$

$$A_3 = \begin{bmatrix} 0.9437 & 0 & 0 \\ 0.0353 & 0.9412 & 0 \\ 0.0011 & 0.0588 & 0.9665 \end{bmatrix} \quad B_3 = \begin{bmatrix} 1.0910 \\ 0.0202 \\ 0.0004 \end{bmatrix} \quad C_3 = \begin{bmatrix} 001 \end{bmatrix} \quad D_3 = 0$$

$$w_i = \frac{\mu_i(x(k))}{\sum_{j=1}^{l} \mu_j(x(k))}$$

The membership function:

$$\mu_1(x_3) = \exp(-\frac{(x_3 - 6.12)^2}{2.828^2}), \quad \mu_2(x_3) = \exp(-\frac{(x_3 - 9.66)^2}{4.2426^2})$$

$$\mu_3(x_3) = \exp(-\frac{(x_3 - 14.65)^2}{2.828^2})$$

In order to verify the validity of the established T-S fuzzy model, the simulation was carried out by MATLAB and compared with the actual device.

Fig. 5 shows that the dynamic response of the T-S fuzzy model and the actual three tank liquid level system is basically the same.

8. Fuzzy-PID Control Simulation of the Three-Tank Water Level Control System

Designing the PID controller by choosing the parameter K_{pi}, K_{Ii}, K_{Di} for each subsystems of the T-S model.

$$K_{p1} = 0.1135, K_{I1} = 0.0008, K_{D1} = 0,$$

$$K_{p2} = 0.0946, K_{I2} = 0.0004, K_{D2} = 0$$

$$K_{p3} = 0.0843, K_{I3} = 0.0003, K_{D3} = 0$$

The Fuzzy-PID control law as Eq. 6: $\quad u(k) = \sum_{i=1}^{l} h_i(k)u_i(k)$

where $h_i(k) = \dfrac{\mu^i(x(k))}{\sum\limits_{j=1}^{l} \mu^j(x(k))}$, $\mu^i(x(k))$ is the membership of the i-th rule

which is the same as T-S fuzzy model's membership.

For comparing the control effect with common PID control, the Fuzzy-PID control and PID control are also applied to the mechanism model, Eq. 7 of the three-tank water level control system at the same condition. The output under the different set-point $\begin{cases} 5 & 0 \le t \le 3000 \\ 11 & 3000 < t \le 6000 \\ 9 & t > 6000 \end{cases}$ are shown in Fig. 6.

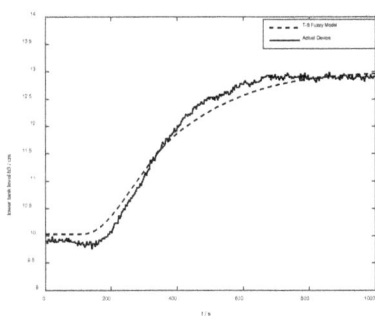

Figure 5. Step response comparison of the T-S model and the actual device

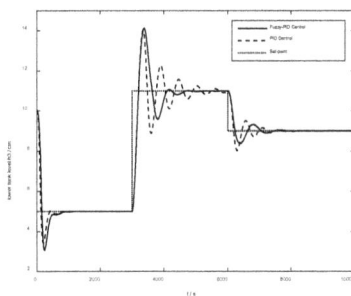

Figure 6. The output of Fuzzy-PID control and PID control

It is evident that when the set-point is changed, fuzzy-PID control can also control the three water tank system well without adjusting the parameters. The common PID can get well control effect at set-point 5. But when the set-point is changed 11 or 9, the control effect become worse.

9. Conclusion

Based on T-S fuzzy model, a new fuzzy-PID control algorithm is proposed. It gives a new method to combine fuzzy control with PID control. Since T-S models can perfectly describe the dynamic characteristic of non-linear systems, the combination of T-S models and PID can realize the optimal control of non-linear systems. The simulating results of the three tank water system show that this algorithm has a good control effect.

Acknowledgement

This research was funded by the Science Foundation of China University of Petroleum, Beijing (No. 2462015YQ0501).

References

1. Pedrycz W, An identification algorithm in fuzzy relational systems. Fuzzy Sets and Systems, 1984, 13(2): 153-167.
2. Jia-jia He, Zai-en Hou, A algorithms for Parameters Optimization of PID Controller. Control and Instruments in Chemical Industry, 2010, 37(11): 1-4. (In Chinese)

3. Jian-dong Chang, En-dian Hu, Wen-xian Zhao, Design and implementation of liquid-level control system based on PID parameters self-tuning. Modern Electronics Technique, 2016, 39(5): 152-154. (In Chinese)

4. Guo-zhen Bai, Jie-hao Yu, Self-tuning of PID parameters based on modified fuzzy neural network. Application Research of Computers, 2016, 33: 66-71. (In Chinese)

5. You-hong Wan, Xin-hua Li, PID Tuning Based On Genetic Algorithms. Control Theory and Applications, 2004, 23(7): 7-8. (In Chinese)

6. Yuanmei Li, Hong-li Zhang, Optimization of PID Controller Parameters Based on Improved Glowworm Swarm Algorithm. Computer Simulation, 2015, 32(9): 356-359. (In Chinese)

7. Chao-Da Chen, Zhi-Sheng Lv, Temperature Control In Agricultural Preservation Application Research. International Journal of Advancements in Computing Technology, 2012, 4(1), 470-476.

8. Jun Bi, Dongfusheng Liu, Kexin Zhan, PID Parameters Optimization for Liquid Level Control System Based on Genetic Algorithm. JDCTA, 2012, 6(1), 361-368.

Robust Stability Analysis of Uncertain Time-Delay System

Shu-bin Wang*, Hai-yan Wan and Yan-yun Wang
*College of Geophysics and Information Engineering,
China University of Petroleum, Beijing, 102249*
*Email: wsbwyy@126.com

In the industry process, the delayed time is possible to change, and the uncertainty of delayed time can sharply increase the difficulty of the control. Based on Time-Delay Dependent State Feedback Predictive Control (DDSFPC) algorithm and Lyapunov functional method, this thesis deals with the system with uncertain state time-delay. The delayed time is led into the matrix, a sufficient condition for the range of the uncertain delayed time is provided via LMI. It also provides the range of the delayed time, which can make sure of the stability of the system. The validity of the above condition is proved by representative example.

Keywords: Predictive control, Lyapunov function, LMI, Robust stability.

1. Introduction

Analysis and synthesis of time-delay system has been a hot issue in the field of control theory and control engineering, because the existence of time-delay especially the uncertain time-delay makes the analysis and synthesis of the system become more complex and difficult [1-3]. Many researchers study on robust stability analysis and robust H_∞ control for discrete stochastic systems with time-varying delays and time-varying norm bounded parameter instability [4, 5]. The problem of robust stability is mainly focused on the design of a state feedback controller, which guarantees the robust stability of the closed loop system for all admissible uncertainties. The sufficient conditions of delay dependence are obtained by using LMI to solve the above problems. A new design method of controller for uncertain input delay system is proposed by Dong [6]. This method does not need to know the exact value of time-delay, and it is less conservative, and the LMI stability criterion can be solved by convex optimization algorithm. But at present the research results about the control of time-delay system and time-delay system focus on model parameters due to the mismatch of the robust stability analysis and robust control for uncertain time delay, there is little research on the uncertain robust stability analysis.

2. Problem Description

Consider the following discrete time-delay system:

$$\begin{cases} X(k+1) = AX(k) + \sum_{i=1}^{L_a} A_i X(k - \tau_i) - BU(k) \\ Y(k) = CX(k) \end{cases} \qquad (1)$$

Where $X \in R^n$ is the state variable, $Y \in R^r$ is the output variable, $U \in R^m$ is the manipulated variable, $A \in R^{n \times n}$, $A_i \in R^{n \times n}$, $i = 1, 2, \cdots L_a$, L_a is the maximum number of time-delay, $B \in R^{n \times m}$, $C \in R^{r \times n}$ is the model parameter matrix, $\tau_1 < \tau_2 < \cdots < \tau_{L_a}$ is the time-delay step of the state variable and $\tau_i > 0$.

By using the delayed state feedback predictive control algorithm [7], we get the following form of closed loop system:

$$X(k+1) = \overline{A}X(k) + \sum_{i=1}^{N} \overline{A}_i X(k - d_i) \qquad (2)$$

Where $\overline{A}$ and $\overline{A}_i$ ($i = 1, 2, 3, \cdots N$) is the parameters of the closed loop system.

$$d_1 < d_2 < \cdots < d_i < \cdots < d_N$$

3. Time-delay Uncertain System Stability Condition

Lemma [5]. *For any appropriate dimension vector* a, b *and matrix* $X, \Upsilon, Z,$ *where* $X,$ *and* Z *are symmetric matrices, if*

$$\begin{pmatrix} X & \Upsilon \\ \Upsilon^T & Z \end{pmatrix} \geq 0$$

then $\quad -2a^T N b \leq \inf_{X, \Upsilon, X} \begin{pmatrix} a \\ b \end{pmatrix}^T \begin{pmatrix} X & \Upsilon - N \\ \Upsilon^T - N^T & Z \end{pmatrix} \begin{pmatrix} a \\ b \end{pmatrix} \geq 0 \qquad (3)$

Theorem 1 *Considering the closed-loop system (2), if there exists a scalar* $d_m > 0$ *(d_m is the maximum time-delay), matrix* $P > 0$, $Q_i > 0$, $Q_i = Q_i^T$, *($i = 1, 2, 3, \cdots N$),, $X = X^T$, $Z = Z^T$, symmetric matrix* Υ *make*

$$\begin{pmatrix} X & \Upsilon \\ \Upsilon^T & Z \end{pmatrix} \geq 0 \tag{4}$$

$$\begin{pmatrix} \Pi_{11} & \Pi_{12} & \Pi_{13} & \Pi_{14} & \cdots & \Pi_{1N+1} \\ & \Pi_{22} & \Pi_{23} & \Pi_{24} & \cdots & \Pi_{2N+1} \\ & & \ddots & & & \vdots \\ & & & \Pi_{ii} & \cdots & \Pi_{iN+1} \\ & & & & \ddots & \\ & & & & & \Pi_{N+1N+1} \end{pmatrix} < 0 \tag{5}$$

Then the range of time delay d_m *can be obtained, so that the discrete system (2) is asymptotically stable.*
Where

$$\Pi_{11} = \bar{A}^T P \bar{A} - \bar{A} P \sum_{i=1}^{N} \bar{A}_i + \sum_{i=1}^{N} \bar{A}_i^T P \bar{A} + N d_m X + 2N\Upsilon$$

$$+ N d_m (\bar{A}^T Z \bar{A} - \bar{A}^T Z - Z \bar{A} + Z) + Q_1 + Q_2 + \cdots + Q_N - P$$

$$\Pi_{12} = \bar{A}^T P \bar{A}_1 - \Upsilon + N d_m (\bar{A}^T Z \bar{A}_1 - Z \bar{A}_1)$$

$$\Pi_{13} = \bar{A}^T P \bar{A}_2 - \Upsilon + N d_m (\bar{A}^T Z \bar{A}_2 - Z \bar{A}_2)$$

$$\Pi_{14} = \bar{A}^T P \bar{A}_3 - \Upsilon + N d_m (\bar{A}^T Z \bar{A}_3 - Z \bar{A}_3)$$

$$\vdots$$

$$\Pi_{1N+1} = \bar{A}^T P \bar{A}_N - \Upsilon + N d_m (\bar{A}^T Z \bar{A}_N - Z \bar{A}_N)$$

$$\vdots$$

$$\Pi_{ii} = \bar{A}_{i-1}^T P \bar{A}_{i-1} + N d_m \bar{A}_{i-1}^T Z \bar{A}_{i-1} - Q_i$$

$$\Pi_{ii+1} = \bar{A}_{i-1}^T P \bar{A}_i + N d_m \bar{A}_{i-1}^T Z \bar{A}_i$$

$$\vdots$$

$$\Pi_{N+1N+1} = \bar{A}_N^T P \bar{A}_N + N d_m \bar{A}_N^T Z \bar{A}_N - Q_N$$

Where $i = 2, 3, \cdots N$

Proof:

Define

$$X(k-d_i+1)-X(k-d_i)=\overline{A}X(k-d_i)+\sum_{i=1}^{N}\overline{A}_iX(k-2d_i)-X(k-d_i)=\Delta X(k-d_i)$$

Then

$$X(k)-X(k-d_i)=\sum_{\theta_i=-d_i}^{-1}\Delta X(k+\theta_i)\Rightarrow X(k-d_i)=X(k)-\sum_{\theta_i=-d_i}^{-1}\Delta X(k+\theta_i)\qquad(6)$$

By formula (2) and (6)

$$X(k+1)=(\overline{A}+\sum_{i=1}^{N}\overline{A}_i)X(k)-\sum_{i=1}^{N}\sum_{\theta_i=-d_i}^{-1}\overline{A}_i\cdot\Delta X(k+\theta_i)\qquad(7)$$

Definition discrete Lyapunov function

$$V(X(k))=V_1(X(k))+V_2(X(k))+V_3(X(k))$$

$$\text{Where}\quad V_1(X(k))=X^T(k)PX(k)$$

$$V_2(X(k))=\sum_{l=1}^{N}\sum_{i=-d_l}^{-1}\sum_{j=k+i}^{k-1}[\Delta X^T(j)E\Delta X(j)]$$

$$V_3(X(k))=\sum_{i=1}^{N}\sum_{\theta=k-d_i}^{k-1}X^T(\theta)Q_iX(\theta)$$

P, E and Q_i are the Lyapunov parameters.

Forward difference for $V_1(X(k))$:

$$\Delta V_1(X(k))=V_1(X(k+1))-V_1(X(k))$$
$$=X^T(k+1)PX(k+1)-X^T(k)PX(k)$$

By (7)

$$\Delta V_1(X(k))=X^T(k)(\overline{A}+\sum_{i=1}^{N}\overline{A}_i)^T P(\overline{A}+\sum_{i=1}^{N}\overline{A}_i)X(k)$$

$$-2X^T(k)(\overline{A}+\sum_{i=1}^{N}\overline{A}_i)^T P\sum_{i=1}^{N}\sum_{\theta_i=-d_i}^{-1}\overline{A}_i\Delta X(k+\theta_i)+\sum_{i=1}^{N}\sum_{j=1}^{N}X^T(k)\overline{A}_i^T P\overline{A}_jX(k)$$

$$-X^T(k)PX(k)+\sum_{i=1}^{N}\sum_{j=1}^{N}X^T(k-d_i)\overline{A}_i^T P\overline{A}_jX(k-d_j)-2\sum_{i=1}^{N}\sum_{j=1}^{N}X^T(k)\overline{A}_i^T P\overline{A}_jX(k-d_j)$$

Application lemma 1:

$$-2X^T(k)(\overline{A} + \sum_{i=1}^{N}\overline{A}_i)^T P \sum_{i=1}^{N}\sum_{\theta_i=-d_i}^{-1}\overline{A}_i \Delta X(k+\theta_i)$$

$$\leq (d_1 + d_2 + \cdots d_N)X^T(k)XX(k) + 2NX^T(k)\Upsilon X(k) - 2X^T(k)\Upsilon\sum_{i=1}^{N}X(k-d_i)$$

$$+2X^T(k)(\overline{A} + \sum_{i=1}^{N}\overline{A}_i)^T P \sum_{i=1}^{N}\overline{A}_j X(k-d_j) - 2X^T(k)(\overline{A} + \sum_{i=1}^{N}\overline{A}_i)^T P \sum_{i=1}^{N}\overline{A}_j X(k)$$

$$+\sum_{i=1}^{N}\sum_{\theta_i=-d_i}^{-1}\Delta X^T(k+\theta_i)Z\Delta X(k+\theta_i)$$

Forward difference for $V_2(X(k))$:

$$\Delta V_2(X(k)) = V_2(X(k+1)) - V_2(X(k))$$

$$=(d_1 + d_2 + \cdots d_N)[2X^T(k)\overline{A}^T E \sum_{i=1}^{N}\overline{A}_i X(k-d_i) + X^T(k)(\overline{A}^T E\overline{A} + \overline{A}^T E + E\overline{A} + E)X(k)$$

$$+\sum_{i=1}^{N}\sum_{j=1}^{N}X^T(k-d_i)\overline{A}_i^T E\overline{A}_j X(k-d_j) - \sum_{l=1}^{N}\sum_{i=-d_l}^{-1}[\Delta X^T(k+i)E\Delta X(k+i)]$$

$$-2X^T(k)E\sum_{i=1}^{N}\overline{A}_i X(k-d_i)]$$

Forward difference for $V_3(X(k))$:

$$\Delta V_3(X(k)) = V_3(X(k+1)) - V_3(X(k))$$

$$=\sum_{i=1}^{N}X^T(k)Q_i X(k) - \sum_{i=1}^{N}X^T(k-d_i)Q_i X(k-d_i)$$

Let $E = Z$. d_m *is the time-delay boundary which is wanted to get.*

Then $(d_1 + d_2 + \cdots d_N) \leq Nd_m$

Therefore, the above results are used. Get

$$\Delta V(X(k)) = \Delta V_1(X(k)) + \Delta V_2(X(k)) + \Delta V_3(X(k))$$

$$\leq \begin{pmatrix} X(k) \\ X(k-d_1) \\ \vdots \\ X(k-d_i) \\ \vdots \\ X(k-d_N) \end{pmatrix}^T \begin{pmatrix} \Pi_{11} & \Pi_{12} & \Pi_{13} & \Pi_{14} & \cdots & \Pi_{1N+1} \\ & \Pi_{22} & \Pi_{23} & \Pi_{24} & \cdots & \Pi_{2N+1} \\ & & \ddots & & \vdots & \\ & & & \Pi_{ii} & \cdots & \Pi_{iN+1} \\ & & & & \ddots & \\ & & & & & \Pi_{N+1N+1} \end{pmatrix} \begin{pmatrix} X(k) \\ X(k-d_1) \\ \vdots \\ X(k-d_i) \\ \vdots \\ X(k-d_N) \end{pmatrix}$$

If

$$\begin{pmatrix} \Pi_{11} & \Pi_{12} & \Pi_{13} & \Pi_{14} & \cdots & \Pi_{1N+1} \\ & \Pi_{22} & \Pi_{23} & \Pi_{24} & \cdots & \Pi_{2N+1} \\ & & \ddots & & & \vdots \\ & & & \Pi_{ii} & \cdots & \Pi_{iN+1} \\ & & & & \ddots & \\ & & & & & \Pi_{N+1N+1} \end{pmatrix} < 0$$

Then $\Delta V(X(k)) < 0$, so by the Lyapunov stability theorem shows system (2) is asymptotically stable.

4. Example

Consider the following open-loop unstable double input double output system.

$$\begin{cases} X(k+1) = AX(k) + A_1 X(k-3) + A_2 X(k-9) + BU(k) \\ Y(k) = CX(k) \end{cases}$$

Where

$$A = \begin{bmatrix} 0.2 & 0.2 & 0.3 \\ 0.2 & 0.1 & 0.8 \\ 0.1 & 0.2 & 0.4 \end{bmatrix}, \ A_1 = \begin{bmatrix} 0 & 0.1 & 0 \\ 0 & 0.1 & 0.1 \\ 0 & 0 & 0 \end{bmatrix}, \ A_2 = \begin{bmatrix} 0 & 0 & 0 \\ 0.2 & 0.1 & 0 \\ 0 & 0 & 0 \end{bmatrix},$$

$$B = \begin{bmatrix} 1 & 0 \\ 0 & 0.5 \\ -1 & 1 \end{bmatrix}$$

$$C = \begin{bmatrix} 0 & 0.1 & 1 \\ 0.2 & 0.1 & 0 \end{bmatrix}$$

The open step response curve of the controlled process is shown in Figure 1. If the set-point $Y_S = \begin{bmatrix} 1 & -1 \end{bmatrix}^T$, the delayed state feedback predictive control algorithm is applied to the controlled process. Then apply theorem 1 to the closed system. We get

$$P = \begin{bmatrix} 467.4523 & -80.9011 & 320.7153 \\ -80.9011 & 495.5138 & 209.3125 \\ 320.7153 & 209.3125 & 1553.7702 \end{bmatrix},$$

$$Q_1 = \begin{bmatrix} 70.8932 & -6.7937 & -3.6836 \\ -6.7937 & 91.5760 & 22.0493 \\ -3.6836 & 22.0493 & 77.8472 \end{bmatrix}$$

$$Q_2 = \begin{bmatrix} 98.0831 & 14.3999 & -1.9080 \\ 14.3999 & 52.2022 & -0.4872 \\ -1.9080 & -0.4872 & 61.6416 \end{bmatrix},$$

$$X = \begin{bmatrix} 18.4442 & -13.2000 & 8.9383 \\ -13.2000 & 18.0208 & -0.6725 \\ 8.9383 & -0.6725 & 62.4353 \end{bmatrix}$$

$$Y = \begin{bmatrix} -1.9312 & 1.4724 & 2.1822 \\ 1.4724 & 0.2842 & 3.9124 \\ 2.1822 & 3.9124 & 13.1167 \end{bmatrix},$$

$$Z = \begin{bmatrix} 18.2769 & -0.4109 & 20.3902 \\ -0.4109 & 13.0667 & 14.7001 \\ 20.3902 & 14.7001 & 59.4301 \end{bmatrix}, \ d_m = 26$$

So maximum uncertain time-delay $\Delta d = 26\text{-}9 = 17$. Figure 2 shows the closed response curve at different uncertain time-delay. The following conclusions are obtained from it. (1) If the uncertain time-delay $\Delta d \leq 17$, the closed system is asymptotically stable. If the uncertain time-delay $\Delta d > 17$, the asymptotically stable of the closed system cannot be guaranteed. (2) The response time and the setting time of the system are both increased and the system performance becomes worse when the system's time-delay is uncertain.

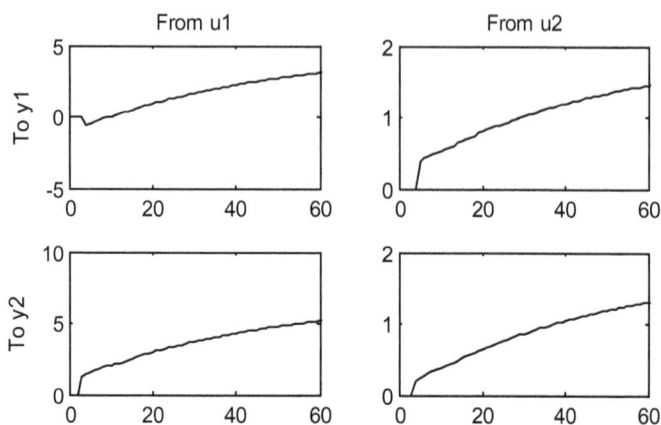

Figure 1. The step response curve of the controlled process

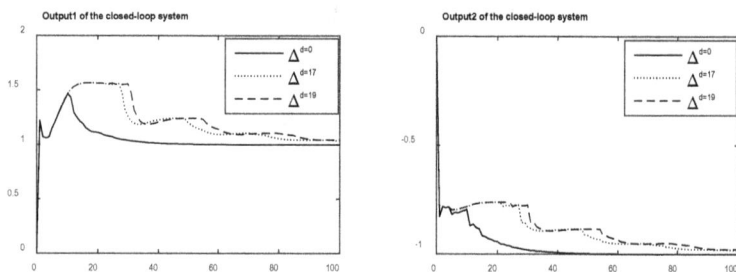

Figure 2. The closed response curve at different uncertain time-delay

5. Conclusion

For uncertain state time-delay system, a sufficient condition for the time range of uncertain time-delay is given by using the Lyapunov function method and linear matrix inequality method. And the maximum delay time is also given to ensure the stability of the closed system. The correctness of this sufficient condition is verified by the simulation example.

Acknowledgement

This research was funded by the Science Foundation of China University of Petroleum, Beijing (No. 2462015YQ0501).

References

1. Dong-Gang Cui, Yan-Bin Zhang, Yan-Min Su. On Control Algorithm for Processwith Pure Time Delay. Control Engineering of China, 2005, 12(4), 368- 369. (In Chinese)
2. He Zhao, Yi Liang. Lag synchronization of delayed neural networks via intermittent control with two pe-riods. Computer Engineering and Applications, 2016, 1-6. (In Chinese)
3. Peng-song Wu, Chao-he Wu, Dong-hua Zhou. Research on Signal-decoupling Internal Mode Control System with Big Time Lag. Chemical Automation and Instrumentation, 2012, 9, 1115-1117. (In Chinese)
4. Shengyuan Xu, James Lam, Tongwen Chen. Robust H_∞ control for uncertain discrete stochastic time-delay systems. Systems & Control Letters, 2004, 51(3), 203- 215.
5. S. J Dan, S. X. Yang, F. Wei. Lag synchronization of coupled delayed chaotic neural networks by periodically intermittent control [J]. Abstract and Applied Analysis, 2013, 5: 045011. (In Chinese)
6. Yue Dong. Robust stabilization of uncertain systems with unknown input delay. Automatica, 2004, 40(2), 331- 336. (In Chinese)
7. Pin-Hui Hu, Feng Yan. Predictive control to multi-time delayed systems. Control and Decision, 2004, 19(11), 1294~1296. (In Chinese)

Robust Square-Root Cubature Kalman Filtering for Relative Orbit Determination of Non-Cooperative Maneuvering Target

Guang-de Xu[†], Fan-da Meng, Zhong-qiu Gou and Bai-nan Zhang
China Academy of Space Technology,
Beijing, 100194, China
[†]Email: xuguangde.2007@163.com

In this paper, a novel-filtering algorithm for the relative orbit determination of non-cooperative maneuvering target is proposed. The algorithm is based on the Strong Tracking Filter (STF) and the Square-root Cubature Kalman Filter (SCKF), called Robust Square-root Cubature Kalman Filter (RSCKF). With the nonlinear orbital dynamics equation as the state model, radar located on the chaser spacecraft as the measurement equipment, the RSCKF is applied for the relative orbit estimation. Simulation results show the superiority of RSCKF over SCKF.

Keywords: Robust square-root cubature Kalman filter; Relative orbit determination; Non-cooperative target; Orbit maneuver.

1. Introduction

Relative orbit determination between spacecraft has been more and more desirable in the past decades for many spacecraft missions, such as space rendezvous [1], formation flying [2] and so on. In the past decades, many scholars have investigated the issue. But few literatures discussed the relative orbit determination problem of a non-cooperative target, which may undergo orbit maneuver during the tracking process.

The relative orbit determination is a typical nonlinear estimation problem. The most classical nonlinear estimation method is the extended Kalman filter [3] (EKF), but high running complexity induced by the computation of the Jacobian matrix limit its application. Julier and Uhlmann proposed a new algorithm named unscented Kalman filter (UKF) [4], which does not need to calculate the Jacobian matrix. When the system state dimensionality is relatively high, the performance of UKF will obviously decrease. A kind of Monte-Carlo method, Particle filter (PF) greatly improves the estimation precision of the nonlinear filter, but leads to tremendous computational complexity. Recently, the square-root cubature Kalman filter (SCKF) was proposed by Arasaratnam and Haykin [5], which uses the spherical cubature rule and radial rule to optimize the sigma points and weights and clearly improves the estimation precision and stability.

By analysis, all of above algorithms lack the ability to track sudden state change. Strong Tracking Filter (STF) is first proposed in 1990 by Zhou [6], which has better robustness and tracking ability. In the paper, I try to combine the STF with the SCKF to solve the relative navigation problem of non-cooperative maneuvering target.

2. Robust Square-root Cubature Kalman Filtering

2.1 Square-root Cubature Kalman Filtering

Consider a nonlinear continuous-discrete system with additive process and measurement noises:

$$x_k = f(x_{k-1}, u_{k-1}) + w_{k-1}.$$ (1)

$$z_k = h(x_k, u_k) + v_k.$$ (2)

where x_k and z_k are the n-dimensional state vector and l-dimensional measurement vector. w_k and v_k are the n-dimensional process noise and l-dimensional measurement noise respectively. $f(\cdot)$ and $h(\cdot)$ are the state function and measurement function of the nonlinear system. w_k and v_k are uncorrelated white Gaussian noise with the constant statistical properties, w_k satisfies $N(0, Q_k)$, v_k satisfies $N(0, R_k)$.

The specific SCKF algorithm process is given as follows:

(a) Initialization:
The initial value $S_{k|k}$ is the square-root factor of the error covariance matrix

$$S_{k|k} = \left[chol(P_{k|k}) \right]^T.$$ (3)

(b) Time update:
Evaluate the cubature points

$$X_{i,k|k} = S_{k|k}\xi_i + \hat{x}_{k|k}.$$ (4)

Evaluate the propagated cubature points

$$X^*_{i,k+1|k} = f_k(X_{i,k|k}), i = 1, 2, \cdots m.$$ (5)

849

where $X_{i,k|k}$, $X^{*}_{i,k+1|k}$ are the cubature points, m=2n.

Evaluate the predicted cubature points

$$\hat{x}_{k+1|k} = \frac{1}{m}\sum_{i=1}^{m} X^{*}_{i,k+1|k} .$$ (6)

Calculate the triangular square-root matrix of the predicted error covariance matrix $P_{k|k}$.

$$S_{k+1|k} = Tria([X^{*}_{i,k+1|k}, S_{Q,k}]) .$$ (7)

where $Q_k = S_{Q,k}S^{T}_{Q,k}$, $S_{Q,k}$ is the square root factor of Q_k , and the centered matrix $X^{*}_{k+1|k}$

$$X^{*}_{k+1|k} = \frac{1}{\sqrt{m}}[X^{*}_{1,k+1|k} -\hat{x}_{k+1|k}, X^{*}_{2,k+1|k} -\hat{x}_{k+1|k}, \cdots X^{*}_{m,k+1|k} -\hat{x}_{k+1|k}].$$ (8)

(c) Measurement update:
Recalculate the cubature points

$$X_{i,k+1|k} = S_{k+1|k}\xi_i + \hat{x}_{k+1|k}.$$ (9)

Evaluate the measured cubature points

$$Y_{i,k+1|k} = h_{k+1|k}(X_{i,k+1|k}).$$ (10)

Evaluate the predicted measurement vector

$$\hat{z}_{k+1|k} = \frac{1}{m}\sum_{i=1}^{m} Y_{i,k+1|k} .$$ (11)

Estimate the square-root of the innovation covariance matrix

$$S_{zz,k+1|k} = Tria([Y_{k+1|k}, S_{R,k+1}]).$$ (12)

where $S_{R,k+1}$ denotes a square-root factor of R_{k+1} and the weighted, centered matrix

$$Y_{k+1|k} = \frac{1}{\sqrt{m}}[Y_{1,k+1|k} -\hat{z}_{k+1|k}, Y_{2,k+1|k} -\hat{z}_{k+1|k}, \cdots Y_{m,k+1|k} -\hat{z}_{k+1|k}].$$ (13)

Estimate the innovation covariance matrix

$$P_{zz,k+1|k} = S_{zz,k+1|k} S_{zz,k+1|k}^{\mathrm{T}} . \tag{14}$$

$$P_{z,k+1|k} = P_{zz,k+1|k} - R_{k+1} . \tag{15}$$

Estimate the cross-covariance matrix

$$P_{xz,k+1|k} = X_{k+1|k} Y_{k+1|k}^{\mathrm{T}} , \tag{16}$$

where the weighted, centered matrix

$$X_{k+1|k} = \frac{1}{\sqrt{m}} [X_{1,k+1|k} - \hat{x}_{k+1|k}, X_{2,k+1|k} - \hat{x}_{k+1|k}, \cdots X_{m,k+1|k} - \hat{x}_{k+1|k}]. \tag{17}$$

Estimate the Kalman gain

$$K_{k+1} = (P_{xz,k+1|k} / S_{zz,k+1|k}^{\mathrm{T}}) / S_{zz,k+1|k} . \tag{18}$$

Estimate the updated state as

$$\hat{x}_{k+1|k+1} = \hat{x}_{k+1|k} + K_{k+1}(z_{k+1} - \frac{1}{m}\sum_{i=1}^{m} h_{k+1|k}(X_{i,k+1|k})). \tag{19}$$

Estimate the square root factor of the corresponding error covariance

$$S_{k+1|k+1} = \mathrm{Tria}([X_{k+1|k} - K_{k+1}Y_{k+1|k}, K_{k+1}S_{R,k+1}]). \tag{20}$$

3. Robust SCKF Based on Strong Tracker Filter

A robust SCKF based on strong tracking filter can be got by introducing a fading factor λ_{k+1} in the estimation of the predicted error covariance matrix. λ_{k+1} can be calculated as follows

$$\lambda_{k+1} = \begin{cases} C_{k+1} & C_{k+1} > 1 \\ 1 & C_{k+1} \leq 1 \end{cases} . \tag{21}$$

$$C_{k+1} = tr\left(N_{k+1}\right) / tr\left(M_{k+1}\right) \tag{22}$$

$$N_{k+1} = V_{k+1} - R_{k+1} - \left[P_{xz,k+1|k}^{(l)}\right]^{\mathrm{T}} \left[(P_{k+1|k}^{(l)})^{-1}\right]^{\mathrm{T}} Q_k \left[P_{k+1|k}^{(l)}\right]^{-1} P_{xz,k+1|k}^{(l)} . \tag{23}$$

$$M_{k+1} = P^{(l)}_{zz,k+1|k} - V_{k+1} + N_{k+1}.$$

(24)

$$V_{k+1} = \begin{cases} \varepsilon_1 \varepsilon_1^{\mathrm{T}} & k = 0 \\ \dfrac{pV_k + \varepsilon_{k+1}\varepsilon_{k+1}^{\mathrm{T}}}{1+p} & k \geq 1 \end{cases}$$

(25)

where p is the forgetting factor, generally $p = 0.95$; V_k is the covariance matrix of actual measurement innovation; $P^{(l)}_{k+1|k}$ is the predicted error covariance matrix; $P^{(l)}_{zz,k+1|k}$ is the innovation covariance matrix; $P^{(l)}_{xz,k+1|k}$ the cross-covariance matrix; $tr(\cdot)$ is the trace operation; ε_{k+1} is the measurement innovation, which is defined as

$$\varepsilon_{k+1} = z_{k+1} - \frac{1}{m}\sum_{i=1}^{m} h_{k+1|k}(X_{i,k+1|k}).$$

(26)

Through calculating fading factor λ_k and estimation of the predicted error covariance matrix, the total robust square-root cubature kalman filter process is given as follows:

(a) **Time update:**

The state estimation $\hat{x}_{k|k}$ and $S_{k|k}$ at t_k are known. Use Eq. (6) and Eq. (7) to get the state prediction $\hat{x}_{k+1|k}$ and the square root factor of the error covariance matrix prediction $S_{k+1|k}$. Calculate the predicted error covariance matrix $P^{(l)}_{k+1|k}$

$$P^{(l)}_{k+1|k} = S_{k+1|k}S^{\mathrm{T}}_{k+1|k}.$$

(27)

(b) **Calculation of Fading Factor:**

Recalculate the cubature points

$$X_{i,k+1|k} = S_{k+1|k}\xi_i + \hat{x}_{k+1|k}.$$

(28)

Evaluate the propagated cubature points

$$X^{*}_{i,k+1|k} = f_k(X_{i,k|k}), i = 1, 2, \cdots m.$$

(29)

852

Evaluate the predicted measurement vector

$$\hat{z}_{k+1|k} = \frac{1}{m}\sum_{i=1}^{m} Y_{i,k+1|k}.$$

(30)

Estimate the square-root factor of the innovation covariance matrix

$$S_{zz,k+1|k} = Tria([Y_{k+1|k}, S_{R,k+1}]).$$

(31)

Estimate the innovation covariance matrix

$$P_{zz,k+1|k} = S_{zz,k+1|k} S_{zz,k+1|k}^{T}.$$

(32)

Estimate the cross-covariance matrix

$$P_{xz,k+1|k} = X_{k+1|k} Y_{k+1|k}^{T}.$$

(33)

Calculate fading factor λ_{k+1} using Eqs. (21) - (25) and Eq. (32) and Eq. (33). Use fading factor λ_{k+1} to modify the error covariance matrix, we get

$$P_{k+1|k} = \lambda_{k+1}(P_{k+1|k}^{(l)} - Q_k) + Q_k.$$

(34)

(c) Measurement Update:
Decompose the error covariance matrix $P_{k+1|k}$

$$S_{k+1|k} = \left[chol(P_{k+1|k}) \right]^{T}$$

(35)

Use Eq. (9) - Eq. (18) to calculate the square-root factor of the innovation covariance matrix $S_{zz,k+1|k}$ and the cross-covariance matrix $P_{xz,k+1|k}$.

After getting the new measurement z_{k+1}, use Eq. (17) - Eq. (21) to realize measurement update.

4. Relative Orbit Estimation

4.1 Nonlinear Equations of Motion

Nonlinear equations of motion for the relative motion of the target with respect to the chaser are derived as follows using Lagrangian mechanics [7].

$$\ddot{x} = 2n\dot{y} + n^2(x+r_c) - \frac{\mu(x+r_c)}{((x+r_c)^2 + y^2 + z^2)^{3/2}}$$

$$\ddot{y} = 2n\dot{x} + n^2 y - \frac{\mu y}{((x+r_c)^2 + y^2 + z^2)^{3/2}}$$

$$\ddot{z} = -\frac{\mu z}{((x+r_c)^2 + y^2 + z^2)^{3/2}}. \tag{36}$$

5. Measurement Model

The radar system located on the chaser satellite provides the measurements including the relative distance ρ between the chaser and the target, the azimuth angle α, and the elevation angle β. The measurement equation is given as follows:

$$\begin{cases} \rho = \sqrt{x^2 + y^2 + z^2} + v_\rho \\ \alpha = \arcsin\dfrac{z}{\sqrt{x^2 + y^2 + z^2}} + v_\alpha \\ \beta = \arctan\dfrac{y}{x} + v_\rho + v_\beta. \end{cases} \tag{37}$$

where $\rho = [x, y, z]^T$ is the relative position vector from the chaser to the target.

In order to reduce calculation, the relative position vector ρ is used to establish the measurement model. We can get ρ from the vector $[\rho \ \alpha \ \beta]^T$ as the following equation

$$\rho = \begin{bmatrix} x \\ y \\ z \end{bmatrix} = C_{ob} \begin{bmatrix} \rho\cos\beta\cos\alpha \\ \rho\cos\beta\sin\alpha \\ -\rho\sin\beta \end{bmatrix}. \tag{38}$$

854

where C_{ob} is the transfer matrix from the body frame of the chaser to the LVLH frame.

By defining the measurement vector as

$$z(t) = [x, y, z]^T.$$ (39)

We get $z(t) = Cx(t)$.

The above equation can be written in discrete form as

$$z_k = Cx_k + v_k.$$ (40)

where v_k is the measurement noise vector, $C = [I_{3\times3} \quad 0_{3\times3}]$.

6. Simulation Results

The relative motion scenario is natural approach of two satellites in different orbit, the initial orbit elements are listed as follows

Table 1. Initial orbit elements

Satellite	Semi-major axis/km	Eccentricity	Inclination (deg)	Right ascension of ascending node (deg)	Argument of perigee (deg)	True anomaly (deg)
Chaser	7178	1.0e-5	98	0	0	0
Target	7228	1.0e-5	98	90	0	2

The simulation time for relative motion of the two satellites is 4000 seconds. Supposing that the target starts orbit maneuver at 1600 seconds and lasts for 200 seconds, the acceleration is 1.5m/s² at the x-axis direction.

The initial relative position and velocity error is [100 100 100] km, [0.1 0.1 0.1] km/s, $P_{0|0} = diag[100\ 100\ 100\ 1\ 1\ 1]$, Q_k is $10^{-4} I_3$, R_k is $diag([\sigma_\alpha^2\ \sigma_\beta^2\ \sigma_\rho^2])$, $\sigma_\alpha = \sigma_\beta = 1mrad$, σ_ρ is 0.5m, p is set as 0.95.

The relative navigation performance using different filter algorithms is given in Figs. 1-2. Fig. 1 shows the relative position and velocity estimation error of SCKF. It can be seen that the error of the SCKF increases markedly in the interval 1600–1800 s after the orbit maneuver of the target occurs and diverges subsequently. In Fig. 2, we can see that although the error of the RSCKF also increases after orbit maneuver occurs, the relative position error and the relative velocity error converge rapidly to the error before orbit maneuver.

The estimated state of the robust SCKF can track the true relative position and velocity of the target very well, but the estimated state of the standard SCKF loses the trace ability. It is proved that the robust SCKF can handle the unexpected situations which are not included in the nonlinear system model by the regulation of the fading factor in the strong tracking filter algorithm.

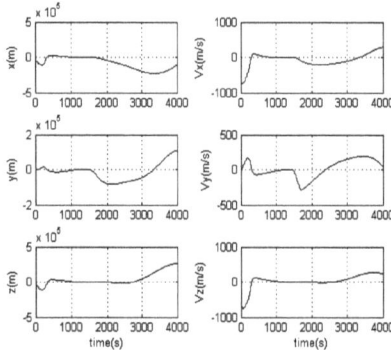

Figure 1. Estimation errors of SCKF   Figure 2. Estimation errors of RSCKF

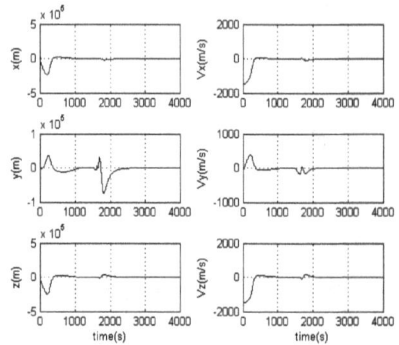

7. Conclusion

To overcome the flaw of traditional SCKF, lack of the ability to track sudden state change, the paper proposed a robust SCKF algorithm based on the strong tracking filter algorithm and applied it in the relative navigation problem of non-cooperative maneuvering target satellite. Simulation results validate the ability to track the sudden change caused by the orbit maneuver of the target satellite.

References

1. Sabatini M, Palmerini G B, Gasbarri P. A testbed for visual based navigation and control during space rendezvous operations [J]. Acta Astronautica, 2015, 117:184-196.
2. Park H E, Kim Y R. Relative navigation for autonomous formation flying satellites using the state-dependent Riccati equation filter [J]. Advances in Space Research, 2015, 57.
3. Daum, F., 2005. Nonlinear filters: beyond the Kalman filter. IEEE Aerosp. Electron. Syst. Mag., 20(8): 57-69.
4. Julier, S.J., Uhlmann, J.K., 2004. Unscented filtering and nonlinear estimation. Proc. IEEE, 92(3): 401-422.
5. Arasaratnam, I., Haykin, S., 2009. Cubature Kalman filters. IEEE Trans. Autom. Control, 54(6): 1254-1269.

6. D.H. Zhou, Y.G. Xi, Z.J. Zhang, A suboptimal multiple fading extended Kalman filter, Acta Autom. Sin. 17 (6) (1991) 689 – 695.

7. Okasha M, Newman B. Modeling, Dynamics and Control of Spacecraft Relative Motion in a Perturbed Keplerian Orbit [J]. International Journal of Aeronautical & Space Sciences, 2015, 16(1): 77-88.

Product Family Planning Game Based on Quality Preference

Zhi-liang Wang
School of Mechanical Engineering,
Nanjing Institute of Technology, Nanjing, 211167, China
Email: wwangzzll@njit.edu.cn

As to product family customization planning, enterprises not only need to determine the performance locations of the customized products, but also face the market price game between the customized products. Considering the consumers' preference for high quality products, firstly, the planning game model is built. Secondly, based on the models (simulation) calculation and analysis, price equilibrium and performance location equilibrium can be gained in turn by backward induction method. At last, it is proved that the quality factor can drive the equilibrium location to the high end, and then through the price game, companies are impelled to produce the higher quality customized products. Game analysis will help enterprises employ the appropriate strategies of product family customization planning.

Keywords: Mass customization; Hotelling's model; Price equilibrium; Location equilibrium; Quality preference.

1. Introduction

Effective product family planning can appropriately balance the product family universality degree and the customer personalized preference [1], so the customized product family can gain more customer satisfaction degree [2, 3]. When the supply chain is incorporated into the product family planning [4], in order to gain more competition advantage, customized products supply chain not only need to consider both the supply chain profits and the customer interests, also deal with the game (among themselves) of the product family performance configuration and pricing [5].

Since Hotelling creatively put forward the product differentiation competition model [6], many scholars have expanded it from various perspectives. For example, considering the market competitors' diversification, Economides creativity constructed a game model containing multi vendors based on linear space and profit function, and then got the equilibrium price, but did not achieve the game equilibrium location [7].

Based on the differences in product quality, brand preferences and other factors [8], some scholars analyzed these factors role in the space location game by constructing different game models [9], but the conclusions is not completely consistent [10]. So, the role of the quality factors in product differentiation competition remains the further study.

Under supply chain competition, considering consumer preference for high quality products, this paper explores the product family planning strategies (including performance location and pricing) and influence of quality difference on the planning, strives to enhance the competitiveness of customized products.

2. Customized Planning Game Model Building and Solution

2.1. The Model of Customized Planning Game Based on Quality Preference

Assuming that there is a certain customization attribute for X, the competing manufacturers draw up n customization points, and product n corresponding customization modules $P_i, i = 1,2,...,n$. Through the conversion, this attribute's customizable interval is mapped to the interval $[0,1]$ and n customization points respectively to $x_1, x_2,..., x_n$. Let $0 \le x_1 < x_2 < ... < x_n \le 1$ and the larger the variable subscript is, the better the corresponding module performance is, as shown in Fig. 1.

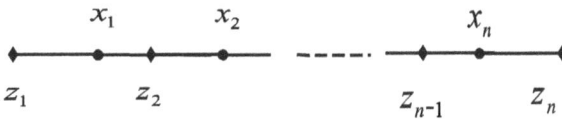

$$x_1 \qquad x_2 \qquad\qquad x_n$$

$$z_1 \qquad z_2 \qquad\qquad z_{n-1} \qquad z_n$$

Figure 1. Location point and marginal point layout

Assuming that the customers' preference points uniformly distribute in customizable interval; when customization point x_j deviates from a customer's preference point, this deviation will bring utility cost, this paper adopts linear utility cost function, denote the unit preference difference utility cost by α.

Defining the i-th customization module price as $p_i, i = 1,2,...,n$, there is a relationship between customization points x_{i-1}, x_i and marginal point z_i:

$$p_{i-1} + \alpha(z_i - x_{i-1}) = p_i + \alpha(x_i - z_i)$$

Furthermore, if $p_{i-1} = p_i$, then $z_i = (x_{i-1} + x_i)/2$, z_i is in the middle of x_{i-1} and x_i. But in fact, as the performance of customization module P_i is better than P_{i-1}, the price p_i at which the customers on the middle point $z_i = (x_{i-1} + x_i)/2$ are willing to buy P_i is higher than p_{i-1}; so, the different attribute values will

859

bring differences in the customized products' performance (quality), customers are willing to pay more additional money for high-quality products and thus this preference brings quality difference utility cost, marginal points' determination must consider the quality difference utility cost between different customized products. When the linear cost function is used and unit quality difference utility cost is denoted by β, the relationship between customization point x_{i-1}, x_i and marginal point z_i can be rewritten as:

$$p_{i-1} + \alpha(z_i - x_{i-1}) = p_i + \alpha(x_i - z_i) - \beta(x_i - x_{i-1}) \tag{1}$$

$$z_i = \frac{\alpha(x_i + x_{i-1}) - \beta(x_i - x_{i-1}) + (p_i - p_{i-1})}{2\alpha} \tag{2}$$

Let the two ends of customizable mapping interval are z_1, z_n respectively and

$$z_1 = 0, z_n = 1.$$

Therefore, the total social surplus of customization module i is:

$$\pi_i = (p_i - C_i) \int_{z_i}^{z_{i+1}} dz - \int_{z_i}^{x_i} \alpha(x_i - z)dz - \int_{x_i}^{z_{i+1}} \alpha(z - x_i)dz \tag{3}$$

In Eq. 3, C_i is the cost of i-th customization module, the second and the third are respectively the total utility cost of two customer groups which locate in the left or the right of the customization point x_i.

2.2. Customization Planning Price Game

Multiple manufacturers bear the games of the product family customization planning in two stages. In the first stage, each manufacturer confirms the attribute values of the own product family; in the second stage, each manufacturer determines the customized product family competitive price. We will use the backward induction to analyze and solve this dynamic game.

After multiple manufacturers confirm their respective attribute values of product family, product family price competition should maximize the social surplus. So, taking Eq. 2 into Eq. 3, and then making the partial derivative of the total social surplus with respect to the customization price and the derivative equal to zero, we can get:

$$P_1 - \frac{3}{5}P_2 = \frac{3(\alpha-\beta)x_2 + (\alpha+3\beta)x_1}{5} + \frac{2}{5}C_1, i = 1,$$

$$-\frac{3}{10}P_{i+1} + P_i - \frac{3}{10}P_{i-1} = \frac{3(\alpha-\beta)x_{i+1} + 6\beta * x_i - 3(\alpha+\beta)x_{i-1}}{10} + \frac{2}{5}C_i, i = 2,3,...,n-1$$

$$P_n - \frac{3}{5}P_{n-1} = \frac{4\alpha - (\alpha-3\beta)x_n - 3(\alpha+\beta)x_{n-1}}{5} + \frac{2}{5}C_n, i = n$$

Rewriting the above n equations in matrix form:

$$AP = D \qquad (4)$$

In Eq. 4,

$$A = \begin{pmatrix} 1 & -\dfrac{3}{5} & 0 & 0 & \cdots & 0 & 0 & 0 \\ -\dfrac{3}{10} & 1 & -\dfrac{3}{10} & 0 & \cdots & 0 & 0 & 0 \\ & -\dfrac{3}{10} & 1 & -\dfrac{3}{10} & \cdots & 0 & 0 & 0 \\ \vdots & \vdots & \vdots & \vdots & \vdots & \vdots & \vdots & \vdots \\ 0 & 0 & 0 & 0 & \cdots & -\dfrac{3}{10} & 1 & -\dfrac{3}{10} \\ 0 & 0 & 0 & 0 & \cdots & 0 & -\dfrac{3}{5} & 1 \end{pmatrix}_{n \times n}$$

$$P = (P_1, P_2, ..., P_{n-1}, P_n)'$$

$$D = \begin{pmatrix} \dfrac{3(\alpha-\beta)x_2 + (\alpha+3\beta)x_1}{5} + \dfrac{2}{5}C_1 \\ \vdots \\ \dfrac{3(\alpha-\beta)x_{i+1} + 6\beta * x_i - 3(\alpha+\beta)x_{i-1}}{10} + \dfrac{2}{5}C_i \\ \vdots \\ \dfrac{4\alpha - (\alpha-3\beta)x_n - 3(\alpha+\beta)x_{n-1}}{5} + \dfrac{2}{5}C_n \end{pmatrix}$$

In order to solve the linear Eq. 4, denote the inverse matrix of the matrix A as $B = A^{-1}$, $B = (b_{ij})_{n \times n}$. Considering $AB = E$, as to j-th column of matrix B, multiplying i-th ($i = 1,2,...,n$) line of matrix A in sequence, we can get two order linear difference equations with constant coefficients, e.g., $j = 1$, $-\frac{3}{10}b_{i,1} + b_{i+1,1} - \frac{3}{10}b_{i+2,1} = 0$.

By the general solution, we can get each column of matrix B in sequence:

$$b(i,1) = 15*(3^i + 9^n 3^{-i})/(4*(-9+9^n))$$

$$b(i,n) = 5*(9*3^{n-i} + 3^{n+i})/(4*(-9+9^n))$$

$$b(i,j) = 5*(3^{n+j} + 9^n 3^{n-j})*(3^{n-i} + (1/3)^{n-i})/(4*(-9+9^n))$$
$$i \geq j, j = 2,3,...,n-1$$,

$$b(i,j) = 15*(3^j + 9^n 3^{-j})*(3^{i-1} + (1/3)^{i-1})/(4*(-9+9^n))$$
$$i \leq j, j = 2,3,...,n-1$$,

Therefore, one-dimensional customization planning game has equilibrium price:

$$p^* = BD$$

Summary 1:   (5)

For product family planning game, while the product family performance location is given, the manufacturers can achieve price equilibrium based on the social surplus value.

The example of equilibrium performance locations is given at the next section.

Taking p^* into Eq. 2 and Eq. 3, we can get the maximum profit in the stage of price game:

$$\pi_i^P = (p_i^* - C_i)(z_{i+1}^* - z_i^*) - \frac{1}{2}\alpha(x_i - z_i^*)^2 - \frac{1}{2}\alpha(z_{i+1}^* - x_i)^2$$ (6)

2.3. Customization Planning Performance Location Game

Planning product family performance location is to determine the optimal attribute value of each customized product under the maximum social surplus. So, taking partial derivative of the maximum profit (Eq. 6) with respect to the corresponding customization point x_i and the partial derivative equal to zero,

$$\frac{\partial \pi_i^P}{\partial x_i} = 0, i = 1, 2, ..., n$$

As this equation set complexity, it is difficult to directly deduce the solution expression by manual calculation; here, the machine proof method is employed to demonstrate the existence of solution with symbolic solving function "solve()" in Matlab while the number of customized products was respectively 2, 3, 5, 8, 20, 50,

$$X^* = \{x_i^*, i = 1,2,...,n\}$$

$$\exists X = X^*, \frac{\partial \pi_i^P}{\partial x_i} = 0, i = 1, 2, ..., n$$ (7)

Summary 2:

For product family planning game, the manufacturers can achieve the performance location equilibrium based on the social surplus value.
The example of equilibrium customization prices is given at the next section.

3. The Driving Role of Quality Difference on Customization Planning

In order to explore the impact of quality difference on customization planning game, making the partial derivatives of the location equilibrium X^* with respect to the unit quality difference utility cost β, (in order to analyze conveniently, let $Ci=0$),

$$f_i(\alpha, \beta) = \frac{\partial x_i^*}{\partial \beta}, i = 1, 2, ..., n \tag{8}$$

Here, the numerators and denominators of each expression are multi-order polynomial of α and β, they are too long and omitted.

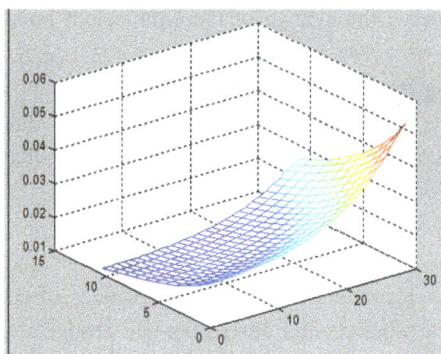

Figure 2. The partial derivatives of equilibrium performance location with respect to β

In order to judge the value of $f_i(\alpha, \beta)$ is positive or negative, considering that the values of α, β and p^* can change as different measurement unit, according to the corresponding relationship among them, this paper assumes, $\alpha \in [10, 20]$, $\beta \in [0, 30]$ ($\beta = 0$ means no quality difference). Calculating the values of $f_i(\alpha, \beta)$ according to Eq. 8 and drawing 3-D surface charts; hereinto, Fig. 2 is the 3-D surface chart of $f_3(\alpha, \beta)$ which is the third customized product while $n = 5$ (the others is similar).

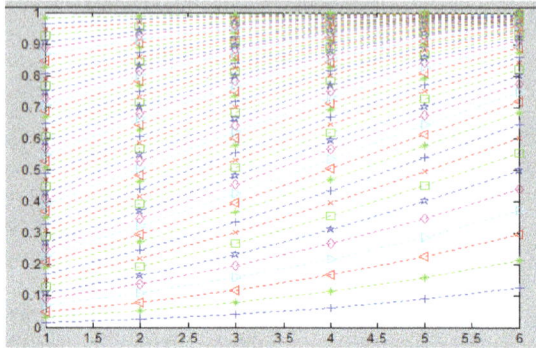

Figure 3. Equilibrium performance locations with 50 products

From the simulation data and graphics, when the quality difference utility cost does not exceed a certain multiple of the preference difference utility cost (if the value of β is too big, marginal point z_i will be not between x_{i-1} and x_i, smaller than x_{i-1}, our discussion do not consider such a situation), $f_i(\alpha, \beta) \geq 0$. It means that the value of customization point will get larger with the improvement of the product quality.

In order to observe the change of the equilibrium point x_i with β more specifically, taking $\alpha = 20$, when $\beta = 3*i, i = 0, 1, 2, 3, 4, 5, 6$, according to Eq. 7, calculating respectively the values of each x_i and drawing, taking 50 customized products as a example, as shown in Fig. 3. Data and graphics show that there is the continuous driving role of product quality to equilibrium performance location X^*.

Meanwhile, in order to explore the impact of the quality difference on the equilibrium price p^*, also taking $\alpha = 20$, when $\beta = 3*i, i = 0, 1, 2, 3, 4, 5, 6$, according to Eq. 5, calculating respectively the values of each p_i^* and drawing, taking 5 customized products and 50 customized products as two examples, as shown in Fig. 4 and Fig. 5.

In Fig. 4 and Fig. 5, the dash dot blue line (vertex mark '+'), the dashed black line (vertex mark '*') and the dotted red line (vertex mark '✿') respectively corresponds to the distribution of the equilibrium price while $\beta = 0, \beta = 6$, $\beta = 15$. It shows that the equilibrium price of low-end products will rise and high-end products will decline with the increasing of β. But Eq. 1 shows that the product price will rise with the improvement of product quality.

The source that appears this seemingly contradictory phenomenon is, we limit the customizable area $x \in [0,1]$ and the customers are evenly distributed in this area. When the customization point X^* increases with the increasing of

864

β and tends to 1, the number of customized products in high-end will increase, it leads to intense competition, so the product price will decline and the profit will decrease.

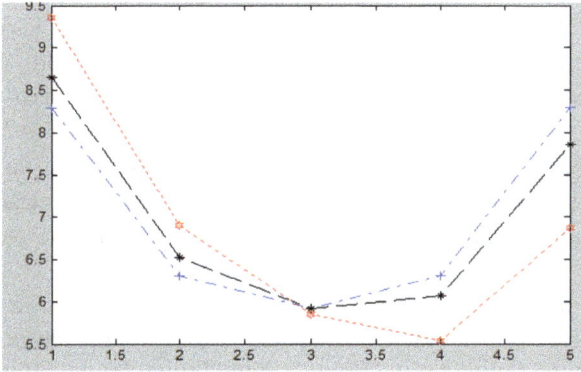

Figure 4. Equilibrium customization prices with 5 products

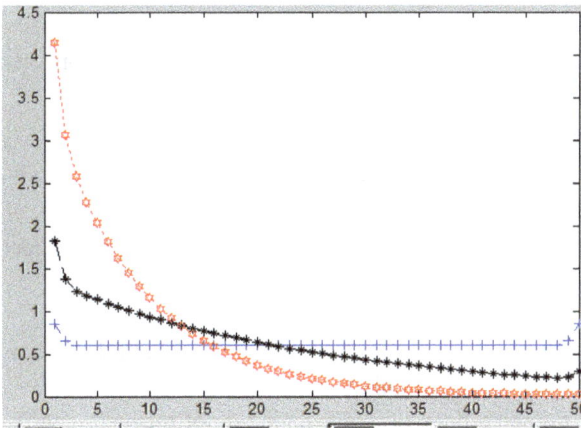

Figure 5. Equilibrium customization price with 50 products

But in fact, as the high quality products can obtain additional high profits, this drives the middle-end product manufacturers to produce current high quality products in the future, high-end product manufacturers will produce higher quality products, low-end product manufacturers will lose market share as the change of consumer preference; therefore, market competition drives customized area to move to the higher end and still make high quality products to obtain extra high profits. Therefore, we can get:

Summary 3:

The quality factor drives the equilibrium performance location of the customization planning game to move to high-end, and then impels manufacturers to produce the higher quality customized products through the price game.

Just as there is such a cycle drive, the market evolves to the higher end constantly. Therefore, this conclusion explains the internal mechanism of the market's continuous evolution moving to higher end from the perspective of quality.

In addition, contrasting Fig. 4 and Fig. 5, the equilibrium price is uniformly distributed when $\beta = 0$; when the number of customized products is relatively small, the closer to two end points the equilibrium performance location is, the higher the equilibrium price is; and the equilibrium price at the middle-end is the lowest; the reason is that the competition strength at both ends is weaker. When the number of customized products is large, the edge effect is weakened or even disappeared; the equilibrium price at the middle tends to be consistent.

4. Conclusion

Market preference for high quality drives enterprises to consider the impacts of quality on product family planning.

1) Based on the market preference for high quality products, under the objective of social total surplus, the extended Hotelling spatial model is built.

2) Through solving the game model, it is proved that the manufacturers can gain the price equilibrium.

3) At the same time, by the machine proof method, with the help of the symbolic solving function in Matlab, it is proved that the manufacturers can gain performance location equilibrium.

4) Considering that customers are willing to pay more additional money for high-quality products, it is proved that the quality difference plays the continuous driving role to the equilibrium performance location and the conclusion is gained: the quality factor drives customization planning equilibrium location move to high-end and then drives the manufacturers to produce higher quality products by the price game. So, this paper interprets the internal mechanism that the market evolves constantly to high-end from the quality.

The customization product family planning based on quality preference will enhance the market competitiveness of manufacturers greatly.

Acknowledgement

This research was funded by the Humanities and Social Sciences Planning Foundation of the Ministry of Education, China (Grant No. 12YJAZH151, 12YJCZH209) and Major Project of Innovation Foundation of Nanjing Institute of Technology, China (Grant No. CKJA201208).

References

1. Liu Z, Wong YS, Lee KS. A manufacturing-oriented approach for multi-platforming product family design with modified genetic algorithm [J]. Journal of Intelligent Manufacturing, 2011(22): 891-907.
2. Huang J, Qi ES, Liu L. Product scale and scheduling optimization based on customer preference and complexity evaluation manufacturing [J]. Computer Integrated Manufacturing Systems, 2015, 21 (4): 992-1001(In Chinese).
3. Ma JM, Kim HM. Product family architecture designs with predictive, data-driven product family design method [J]. Research in Engineering Design, Published online: 09 September 2015.
4. Fujita K, Amaya H, Akai R. Mathematical model for simultaneous design of module commonalization and supply chain configuration toward global product family [J]. Journal of Intelligent Manufacturing, 2013(24): 991-1004.
5. Wang ZL, Yang ZY, Bai SB, Wang YX. Joint economic lot sizing model of delayed customization product' supply chain network, solution and supply chain collaboration [J]. Systems Engineering, 2016, 32(5), accepted (In Chinese).
6. Hotelling H. Stability in competition [J]. Economic Journal, 1929, 39(10): 41-57
7. Economides, N. Hotelling's main street with more 2 competitors [J]. Journal of Regional Science, 1993, .33(3): 303-319.
8. Peng J, Ma JH. Hotelling Game Model Based on Consumers' Brand Preference [J]. Journal of Xidian University (Social Sci. Edition), 2010, 20(1): 30-34(In Chinese).
9. Leng MM, Parlar M, Zhang DF. The Retail Space-Exchang Problem with Pricing and Space Allocation Decisions [J]. Production and Operations Management, 2013, 22(1): 189-202.
10. Xu B. An Extended Hotelling Model with Different Product Quality [J]. Mathematics in Practice and Theory, 2007, 37(22): 26-31(In Chinese).

Singular Configurations of a Novel 9-DOF Robotic Arm

Di Bao[1,*], Xue-qian Wang[2] and Bin Liang[3]

[1]*Graduate School at Shenzhen, Tsinghua University, Shenzhen, China*

[2]*Shenzhen Laboratory of Space Robotics & Telescience, Shenzhen, China*

[3]*Tsinghua National Laboratory for Information Science and Technology,*

Beijing, China

[*]Email: baodi23@163.com

All the singular or velocity-degenerate configurations of a novel 9-DOF (Degree of Freedom) robotic arm are presented in this paper. The 9-DOF robotic arm consists of four uniform decentered joints and a single joint. A reciprocity-based methodology for finding the 1-DOF loss singular configurations of joint-redundant robotic arm is used for the 9-DOF robotic arm. Fifteen conditions of singular configurations are listed.

Keywords: Singular configurations, Reciprocity-based, Joint screws.

1. Introduction

In this paper, a novel 9-DOF robotic arm will be introduced. This robotic arm is consist of four uniform decentered joints and a single joint. This design makes the robotic arm flexible and possess the ability of reconfiguration because the position of any uniform joint can be changed easily. The schematic diagram of the robotic arm is Figure 1.

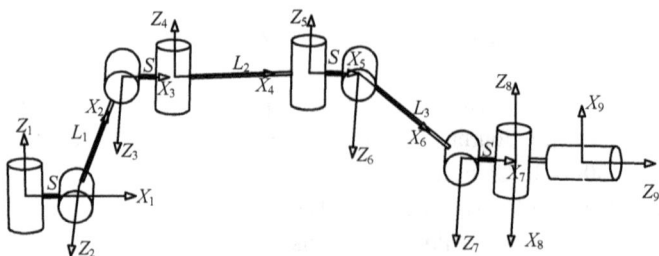

Figure 1 Schematic diagram of the 9-dof robotic arm

---

[*]Corresponding author.

The decentered joint has two DOF which is perpendicular to each other. The distance between two axes is four millimeters. The joint makes the robotic arm's structure very simple, and also make the robotic arm very flexible.

The research of singularity of robotic arm has last for a very long time. The singular configurations is also called special configurations [1,2]. The most common method is to make the determinant of the Jacobian matrix of the robotic arm equal to zero [3-7]. But this method is just fit for non-redundant robotic arms, because the Jacobian matrix of a redundant robotic arm is not a square matrix. The robotic arm we introduce in this paper is a 9-DOF robotic arm, and it is a redundant robotic arm. When the redundant robotic arm is in a singular configuration, the Jacobian matrix is not a column line nonsingular matrix.

When a robotic arm is in a singular configuration, a very small displacement of the end of the robotic arm will need a huge joint velocity which may damage the robotic arm [8]. And if the actuator cannot provide the joint velocity, the end of the robotic arm will deviate the anticipatory trajectory. So the singularity of the robotic arm will influence the control strategy. Singularity analysis of a robotic arm is to find the conditions in which the robotic arm is in singular configuration. And it is also very important for the kinematics of the robotic arm.

2. Methodology

In this paper, reciprocity-based methodology which was presented by Nokleby et al. will be used to find all the 1-DOF loss singular configurations of the robotic arm [9-11]. The screw can be presented as (1).

$$\$ = \begin{bmatrix} \mathbf{s} \\ \mathbf{s}_0 \end{bmatrix} = \begin{bmatrix} \mathbf{s} \\ \mathbf{r} \times \mathbf{s} + h\mathbf{s} \end{bmatrix}. \tag{1}$$

where $\mathbf{s}$ and $\mathbf{s}_0$ are the unit screw coordinates; $\mathbf{r}$ is an arbitrary point's position vector; h is the pitch of the screw. The joints of the robotic arm in this paper can be regarded as a screws with zero pitch.

The Jacobian matrix can be expressed as $[J] = [\$_1, \$_2, \ldots, \$_k]$, where k is the total number of joints and $\$_i$ is the i th joint of the robotic arm. When the robotic arm is in a singular configuration, $[J]$ is not a column nonsingular matrix.

As mentioned earlier, reciprocity-based methodology will be used to get all the singular configurations of the robotic arm. At first, six joint screws are chosen to form a six-joint sub-group Jacobian matrix, $\mathbf{J}_{sub}$. Note that the joints can be any six joints of the robotic arm there are $k-6$ joints left. These joints can be taken as redundant screws $\$_{r1}, \$_{r2}, \ldots, \$_{rk-6}$. The determinant of $\mathbf{J}_{sub}$ have to be zero ($|\mathbf{J}_{sub}| = 0$) to make the six-joint sub-group to be in a singular configuration.

In a singular configuration there exists a screw ($\mathbf{W}_r$) which is reciprocal to all the joint screws of the robotic arm [12], ie.

$$\mathbf{W}_r \otimes \$_i = 0, i = 1, 2, \ldots, k \ . \tag{2}$$

In this way, $\mathbf{W}_{r1}, \mathbf{W}_{r2}, \ldots, \mathbf{W}_{rm}$, can be found, when there are n singular configurations. The $\mathbf{W}_{ri}$ is reciprocal to the six joints which comprise $\mathbf{J}_{sub}$ when i th $\mathbf{J}_{sub}$ degeneracy condition is true. The robotic arm may not in singular configuration because $\mathbf{W}_{ri}$ may not reciprocal to the redundant screws. Therefore additional conditions must be added to make $\mathbf{W}_{ri}$ reciprocal to the redundant screws, ie.

$$\begin{aligned} \mathbf{W}_{ri} \otimes \$_{r1} &= 0 \\ \mathbf{W}_{ri} \otimes \$_{r2} &= 0 \\ &\ldots \\ \mathbf{W}_{ri} \otimes \$_{rk-6} &= 0 \end{aligned} \tag{3}$$

These is the additional conditions which can make sure that the robotic arm is in a singular configuration.

3. Modeling

The robotic arm is consist of revolute joints and links. So the D-H parameters is in Table 1 [13].

Table 1 D-H parameters of the 9-DOF robotic arm

i	a_{i-1}	α_{i-1}	d_i	θ_i
1	0	0	0	θ_1
2	S	90°	0	θ_2
3	L_1	0	0	θ_3
4	S	-90°	0	θ_4
5	L_2	0	0	θ_5
6	S	90°	0	θ_6
7	L_3	0	0	θ_7
8	S	-90°	0	θ_8
9	0	-90°	0	θ_9

S is the distance between the axis of a joint, and L_i is the length of the ith link.

The 5th joint frame will be the reference frame to make calculation simple. And the joint screws are

$$\$_1 = \begin{bmatrix} s_{23}c_{45} \\ -s_{23}s_{45} \\ c_{23} \\ L_1c_2s_{45} + L_2s_5c_{23} + Ss_{45} + Ss_{45}c_{23} \\ L_1c_2c_{45} + L_2c_5c_{23} + Sc_{45} + Sc_{45}c_{23} \\ L_2s_4s_{23} \end{bmatrix}$$

$$\$_2 = \begin{bmatrix} -s_{45} & -c_{45} & 0 & L_1 s_3 c_{45} & -L_1 s_3 s_{45} & L_1 c_3 + L_2 c_5 + S \end{bmatrix}^T$$

$$\$_3 = \begin{bmatrix} -s_{45} & -c_{45} & 0 & 0 & 0 & L_2 c_4 + S \end{bmatrix}^T$$

$$\$_4 = \begin{bmatrix} 0 & 0 & 1 & L_2 s_5 & L_2 c_5 & 0 \end{bmatrix}^T$$

$$\$_5 = \begin{bmatrix} 0 & 0 & 1 & 0 & 0 & 0 \end{bmatrix}^T$$

$$\$_6 = \begin{bmatrix} 0 & -1 & 0 & 0 & 0 & -S \end{bmatrix}^T$$

$$\$_7 = \begin{bmatrix} 0 & -1 & 0 & L_3 s_6 & 0 & -L_3 c_6 - S \end{bmatrix}^T$$

$$\$_8 = \begin{bmatrix} -s_{67} & 0 & c_{67} & 0 & -L_3 c_7 - S c_{67} - S & 0 \end{bmatrix}^T$$

$$\$_9 = \begin{bmatrix} -c_{67} s_8 \\ c_8 \\ -s_{67} s_8 \\ -L_3 s_6 c_8 - S s_{67} c_8 \\ L_3 s_7 s_8 + S s_{67} s_8 \\ L_3 c_6 c_8 + S c_{67} c_8 + S c_8 \end{bmatrix} \qquad . \quad (4)$$

where $s_i = \sin(\theta_i), c_i = \cos(\theta_i), s_{ij} = \sin(\theta_i + \theta_j), c_{ij} = \cos(\theta_i + \theta_j)$.

4. Singular Analysis

$\$_3, \$_4, \$_5, \$_6, \$_7, \$_8$ are chosen to be the sub-group, so the $\mathbf{J}_{sub}$ is

$$\mathbf{J}_{sub} = \begin{bmatrix} -s_{45} & 0 & 0 & 0 & 0 & -s_{67} \\ -c_{45} & 0 & 0 & -1 & -1 & 0 \\ 0 & 1 & 1 & 0 & 0 & c_{67} \\ 0 & L_2 s_5 & 0 & 0 & L_3 s_6 & 0 \\ 0 & L_2 c_5 & 0 & 0 & 0 & -L_3 c_7 - S c_{67} - S \\ L_2 c_4 + S & 0 & 0 & -S & -L_3 c_6 - S & 0 \end{bmatrix} . \quad (5)$$

The determinant of $\mathbf{J}_{sub}$ is

$$|\mathbf{J}_{sub}| = L_2 L_3 \left[(L_3 c_7 + S c_{67} + S) s_{45} s_5 c_6 - (L_2 c_4 + S c_{45} + S) c_5 s_6 s_{67} \right]. \quad (6)$$

Then 15 conditions can be found to make the $\mathbf{J}_{sub}$ degenerated.

1. $L_3 c_7 + S c_{67} + S = 0, L_2 c_4 + S c_{45} + S = 0$

In this condition the $\mathbf{W}_{r1}$ is

$$\mathbf{W}_{r1} = \begin{bmatrix} c_5 c_6 & -s_5 c_6 & c_5 s_6 & 0 & -S c_5 s_6 & 0 \end{bmatrix}^T. \quad (7)$$

The additional conditions are

$$\begin{cases} \mathbf{W}_{r1} \otimes \$_1 = L_1 c_2 s_4 c_6 + L_2 s_{23} s_4 c_5 s_6 + S s_4 c_6 + S s_4 c_{23} c_6 + S s_{23} s_{45} c_5 s_6 = 0 \\ \mathbf{W}_{r1} \otimes \$_2 = L_1 s_3 c_4 c_6 + \left(L_1 c_3 + L_2 c_5 + S \right) c_5 s_6 + S c_{45} c_5 s_6 = 0 \\ \mathbf{W}_{r1} \otimes \$_9 = -L_3 s_7 s_8 s_5 s_6 - S c_5 s_7 c_8 - S s_{67} s_8 s_5 c_6 = 0 \end{cases} \qquad (8)$$

2. $L_3 c_7 + S c_{67} + S = 0, c_5 = 0$

In this condition the $\mathbf{W}_{r2}$ is

$$\mathbf{W}_{r2} = \begin{bmatrix} 0 & 1 & 0 & 0 & 0 & 0 \end{bmatrix}^T . \qquad (9)$$

The additional conditions are

$$\begin{cases} \mathbf{W}_{r2} \otimes \$_1 = L_1 c_2 c_{45} + S c_{45} + S c_{45} c_{23} = 0 \\ \mathbf{W}_{r2} \otimes \$_2 = -L_1 s_3 s_{45} = 0 \\ \mathbf{W}_{r2} \otimes \$_9 = L_3 s_7 s_8 + S s_{67} s_8 = 0 \end{cases} \qquad (10)$$

3. $L_3 c_7 + S c_{67} + S = 0, s_6 = 0$

In this condition the $\mathbf{W}_{r3}$ is

$$\mathbf{W}_{r3} = \begin{bmatrix} c_5 & -s_5 & 0 & 0 & 0 & 0 \end{bmatrix}^T \qquad (11)$$

The additional conditions are

$$\begin{cases} \mathbf{W}_{r3} \otimes \$_1 = L_1 c_2 s_4 + S s_4 + S c_{23} s_4 = 0 \\ \mathbf{W}_{r3} \otimes \$_2 = L_1 s_3 c_4 = 0 \\ \mathbf{W}_{r3} \otimes \$_9 = -L_3 s_7 s_8 s_5 - S s_{67} c_8 c_5 - S s_{67} s_8 s_5 = 0 \end{cases} \qquad (12)$$

4. $L_3 c_7 + S c_{67} + S = 0, s_{67} = 0$

In this condition the $\mathbf{W}_{r4}$ is

$$\mathbf{W}_{r4} = \begin{bmatrix} s_{45} c_5 c_6 \\ -s_{45} s_5 c_6 \\ s_{45} c_5 s_6 \\ \left(L_2 c_4 + S c_{45} + S \right) c_5 s_6 \\ -S s_{45} c_5 s_6 \\ 0 \end{bmatrix} . \qquad (13)$$

The additional conditions are

$$
\begin{cases}
\mathbf{W}_{r4} \otimes \$_1 = L_1 c_2 s_4 s_{45} c_6 + L_2 s_{23} c_5^2 s_6 + \left(s_4 s_{45} c_6 + c_{23} s_4 s_{45} c_6 + c_{45}^2 s_{23} c_5 s_6 + s_{45}^2 s_{23} s_5 c_6 + s_{23} c_{45} c_5 s_6 \right) S = 0 \\
\mathbf{W}_{r4} \otimes \$_2 = L_2 s_{45} c_5 s_6 \left(c_5 - c_4 \right) + L_1 s_{45} \left(s_3 c_4 c_6 + c_3 c_5 s_6 \right) = 0 \\
\mathbf{W}_{r4} \otimes \$_9 = -L_3 s_7 s_8 s_{45} s_5 c_6 - L_2 c_4 c_5 s_6 c_{67} s_8 + \left(c_{67} c_8 s_{45} c_5 s_6 - c_{45} c_5 s_6 c_{67} s_8 - c_5 s_6 c_{67} s_8 \right) S = 0
\end{cases}
\qquad (14)
$$

5. $s_{45} = 0, L_2 c_4 + S c_{45} + S = 0$

In this condition the $\mathbf{W}_{r5}$ is

$$
\mathbf{W}_{r5} =
\begin{bmatrix}
c_5 c_6 s_{67} \\
-s_5 c_6 s_{67} \\
c_5 s_6 s_{67} \\
-\left(L_3 c_7 + S c_{67} + S \right) s_5 c_6 \\
-S c_5 s_6 s_{67} \\
0
\end{bmatrix}
\qquad (15)
$$

The additional conditions are

$$
\begin{cases}
\mathbf{W}_{r5} \otimes \$_1 = \left[\left(L_1 c_2 + S c_{23} + S \right) s_{67} - \left(L_3 c_7 + S c_{67} + S \right) s_{23} \right] c_{45} s_5 c_6 + \left(L_2 s_4 + S s_{45} \right) s_{23} c_5 s_6 s_{67} = 0 \\
\mathbf{W}_{r5} \otimes \$_2 = \left[L_1 s_3 c_{45} c_6 + \left(L_1 c_3 + L_2 c_5 + S + S c_{45} \right) s_6 \right] c_5 s_{67} = 0 \\
\mathbf{W}_{r5} \otimes \$_9 = \left[\left(L_3 c_7 + S c_{67} + S \right) c_{67} - \left(L_3 s_7 + S s_{67} \right) s_{67} \right] s_5 c_6 s_8 + S c_5 s_6 c_{67} s_{67} c_8 - S c_5 c_6 s_{67}^2 c_8 = 0
\end{cases}
\qquad (16)
$$

6. $s_{45} = 0, c_5 = 0$

In this condition the $\mathbf{W}_{r6}$ is

$$
\mathbf{W}_{r6} = \begin{bmatrix} 0 & -s_{67} & 0 & L_3 c_7 + S c_{67} + S & 0 & 0 \end{bmatrix}^T.
\qquad (17)
$$

The additional conditions are

$$
\begin{cases}
\mathbf{W}_{r6} \otimes \$_1 = \left(L_3 c_7 + S c_{67} + S \right) s_{23} c_{45} - \left(L_1 c_2 + S c_{23} + S \right) s_{67} c_{45} \\
\mathbf{W}_{r6} \otimes \$_2 = 0 \\
\mathbf{W}_{r6} \otimes \$_9 = -\left(L_3 c_6 + S c_{67} + S \right) s_8
\end{cases}
\qquad (18)
$$

7. $s_{45} = 0, s_6 = 0$

In this condition the $\mathbf{W}_{r7}$ is

$$\mathbf{W}_{r7} = \begin{bmatrix} c_5 s_{67} \\ -s_5 s_{67} \\ 0 \\ (L_3 c_7 + S c_{67} + S) s_5 \\ 0 \\ 0 \end{bmatrix}. \tag{19}$$

The additional conditions are

$$\begin{cases} \mathbf{W}_{r7} \otimes \$_1 = -(L_1 c_2 + S c_{23} + S) s_5 s_{67} c_{45} + (L_3 c_7 + S c_{67} + S) s_{23} c_{45} s_5 = 0 \\ \mathbf{W}_{r7} \otimes \$_2 = L_1 s_3 c_{45} c_5 s_{67} = 0 \\ \mathbf{W}_{r7} \otimes \$_9 = -S c_5 s_{67}^2 c_8 - (L_3 c_6 + S c_{67} + S) s_5 s_8 = 0 \end{cases}. \tag{20}$$

8. $s_{45} = 0, s_{67} = 0$.

In this condition the $\mathbf{W}_{r8}$ is

$$\mathbf{W}_{r8} = \begin{bmatrix} 0 & 0 & 0 & 1 & 0 & 0 \end{bmatrix}^T. \tag{21}$$

The additional conditions are

$$\begin{cases} \mathbf{W}_{r8} \otimes \$_1 = s_{23} c_{45} = 0 \\ \mathbf{W}_{r8} \otimes \$_2 = 0 \\ \mathbf{W}_{r8} \otimes \$_9 = -c_{67} s_8 = 0 \end{cases}. \tag{22}$$

9. $s_5 = 0, L_2 c_4 + S c_{45} + S = 0$

In this condition the $\mathbf{W}_{r9}$ is

$$\mathbf{W}_{r9} = \begin{bmatrix} c_6 & 0 & s_6 & 0 & -S s_6 & 0 \end{bmatrix}^T. \tag{23}$$

The additional conditions are

$$\begin{cases} \mathbf{W}_{r9} \otimes \$_1 = (L_1 c_2 + S c_{23} + S) s_{45} c_6 + (L_2 s_4 + S s_{45}) s_{23} s_6 = 0 \\ \mathbf{W}_{r9} \otimes \$_2 = L_1 s_3 c_{45} c_6 + (L_1 c_3 + L_2 c_5 + S c_{45} + S) s_6 = 0 \\ \mathbf{W}_{r9} \otimes \$_9 = -S s_7 c_8 = 0 \end{cases}. \tag{24}$$

10. $s_5 = 0, s_6 = 0$

In this condition the $\mathbf{W}_{r10}$ is

$$\mathbf{W}_{r10} = \begin{bmatrix} 1 & 0 & 0 & 0 & 0 & 0 \end{bmatrix}^T. \tag{25}$$

The additional conditions are

$$\begin{cases} \mathbf{W}_{r10} \otimes \$_1 = L_1 c_2 s_{45} + S s_{45} + S s_{45} c_{23} = 0 \\ \mathbf{W}_{r10} \otimes \$_2 = L_1 s_3 c_{45} = 0 \\ \mathbf{W}_{r10} \otimes \$_9 = -L_3 s_6 c_8 - S s_{67} c_8 = 0 \end{cases} \tag{26}$$

11. $s_5 = 0, s_{67} = 0$

In this condition the $\mathbf{W}_{r11}$ is

$$\mathbf{W}_{r11} = \begin{bmatrix} s_{45} c_6 \\ 0 \\ s_{45} s_6 \\ \left(L_2 c_4 + S c_{45} + S \right) s_6 \\ -S s_{45} s_6 \\ 0 \end{bmatrix}. \tag{27}$$

The additional conditions are

$$\begin{cases} \mathbf{W}_{r11} \otimes \$_1 = \left(L_1 c_2 + S c_{23} + S \right) s_{45}^2 c_6 + \left(L_2 c_5 + 2S \right) s_{23} s_6 = 0 \\ \mathbf{W}_{r11} \otimes \$_2 = \left(L_1 s_3 c_{45} c_6 + L_1 c_3 s_6 + L_2 c_5 s_6 - L_2 c_4 s_6 \right) s_{45} = 0 \\ \mathbf{W}_{r11} \otimes \$_9 = \left(S s_{45} c_8 - L_2 c_4 s_8 - S c_{45} s_8 - S s_8 \right) c_{67} s_6 = 0 \end{cases} \tag{28}$$

12. $c_6 = 0, L_2 c_4 + S c_{45} + S = 0$.

In this condition the $\mathbf{W}_{r12}$ is

$$\mathbf{W}_{r12} = \begin{bmatrix} 0 & 0 & 1 & 0 & -S & 0 \end{bmatrix}^T. \tag{29}$$

The additional conditions are

$$\begin{cases} \mathbf{W}_{r12} \otimes \$_1 = L_2 s_{23} s_4 + S s_{23} s_{45} = 0 \\ \mathbf{W}_{r12} \otimes \$_2 = L_1 c_3 + L c_5 + S c_{45} + S = 0 \\ \mathbf{W}_{r12} \otimes \$_9 = S c_{67} c_8 = 0 \end{cases} \tag{30}$$

13. $c_6 = 0, c_5 = 0$

In this condition the $\mathbf{W}_{r13}$ is

$$\mathbf{W}_{r13} = \begin{bmatrix} 0 \\ -\left(L_2c_4 + Sc_{45} + S\right)s_{67} \\ \left(L_3c_7 + Sc_{67} + S\right)s_{45} \\ \left(L_2c_4 + Sc_{45} + S\right)\left(L_3c_7 + Sc_{67} + S\right) \\ -S\left(L_3c_7 + Sc_{67} + S\right)s_{45} \\ 0 \end{bmatrix} \tag{31}$$

The additional conditions are

$$\begin{cases} \mathbf{W}_{r13} \otimes \$_1 = \left(L_2c_5 + Sc_{45} + S\right)\left(L_3c_7 + Sc_{67} + S\right)s_{23} - \left(L_1c_2 + Sc_{23} + S\right)\left(L_2c_4 + Sc_{45} + S\right)s_{67}c_{45} = 0 \\ \mathbf{W}_{r13} \otimes \$_2 = \left(L_2c_4 + Sc_{45} + S\right)L_3s_3s_{45}s_{67} + \left(L_3c_7 + Sc_{67} + S\right)\left(L_1c_3 - L_2c_4\right)s_{45} = 0 \\ \mathbf{W}_{r13} \otimes \$_9 = -\left(L_2c_4 + Sc_{45} + S\right)\left(L_3s_7 + Ss_{67}\right)s_{67}s_8 + \left(L_3c_7 + Sc_{67} + S\right)\left[Sc_{67}s_{45}c_8 - \left(L_2c_4 + Sc_{45} + S\right)c_{67}s_8\right] = 0 \end{cases} \tag{32}$$

14. $c_6 = 0, s_{67} = 0$

In this condition the $\mathbf{W}_{r14}$ is

$$\mathbf{W}_{r14} = \begin{bmatrix} 0 & 0 & s_{45} & L_2c_4 + Sc_{45} + S & -Ss_{45} & 0 \end{bmatrix}^T. \tag{33}$$

The additional conditions are

$$\begin{cases} \mathbf{W}_{r14} \otimes \$_1 = L_2c_5s_{23} + Ss_{23}c_{45} + Ss_{23} = 0 \\ \mathbf{W}_{r14} \otimes \$_2 = \left(L_1c_3 + L_2c_5 - L_2c_4\right)s_{45} = 0 \\ \mathbf{W}_{r14} \otimes \$_9 = Sc_{67}c_8s_{45} - \left(L_2c_4 + Sc_{45} + S\right)c_{67}s_8 = 0 \end{cases} \tag{34}$$

15.

$$\left(L_3c_7 + Sc_{67} + S\right)s_{45}s_5c_6 - \left(L_2c_4 + Sc_{45} + S\right)c_5s_6s_{67} = 0,$$
$$\left(L_3c_7 + Sc_{67} + S\right)s_{45}s_5c_6 \neq 0, \left(L_2c_4 + Sc_{45} + S\right)c_5s_6s_{67} \neq 0$$

In this condition the $\mathbf{W}_{r15}$ is

$$\mathbf{W}_{r15} = \begin{bmatrix} s_{45}c_5c_6 \\ -s_{45}s_5c_6 \\ s_{45}c_5s_6 \\ \left(L_2c_4 + Sc_{45} + S\right)c_5s_6 \\ -Ss_{45}c_5s_6 \\ 0 \end{bmatrix}. \tag{35}$$

The additional conditions are

$$
\begin{cases}
\mathbf{W}_{r15} \otimes \$_1 = L_1 c_2 s_4 s_{45} c_6 + L_2 s_{23} c_5^2 s_6 \\
\quad + \left(s_4 s_{45} c_6 + c_{23} s_4 s_{45} c_6 + c_{45}^2 s_{23} c_5 s_6 + s_{45}^2 s_{23} s_5 c_6 + s_{23} c_{45} c_5 s_6 \right) S = 0 \\
\mathbf{W}_{r15} \otimes \$_2 = L_2 s_{45} c_5 s_6 \left(c_5 - c_4 \right) + L_1 s_{45} \left(s_3 c_4 c_6 + c_3 c_5 s_6 \right) = 0 \\
\mathbf{W}_{r15} \otimes \$_9 = -L_2 c_4 c_5 s_6 c_{67} s_8 - L_3 s_{45} s_5 c_6 s_7 s_8 \\
\quad - \left(s_{45} c_5 s_7 c_8 + c_{45} c_5 s_6 c_{67} s_8 + c_5 s_6 c_{67} s_8 \right) S = 0
\end{cases}
$$

(36)

5. Conclusion

Joint screws and reciprocity-based methodology are introduced in this paper. All the 1-DOF loss singular configurations for the novel 9-DOF robotic arm are presented in this paper by using reciprocity-based methodology. Conditions for each singular configuration have been listed.

Acknowledgement

In this paper, the research was funfed by the Nature Science Foundation of Guangdong Province (Project No. 2015A030313881).

References

1. K. H. Hunt, "Special configurations of robot-arms via screw theory," Robotica, vol. 4, pp. 171-179, 1986.
2. K. H. Hunt, "Special configurations of robot-arms via screw theory," Robotica, vol. 5, pp. 17-22, 1987.
3. B. Gorla, "Influence of the control on the structure of a manipulator from a kinematic point of view," in Proc. 4th Symp. Theory and Practice of Rob. Manipulators, 1981, pp. 30-46.
4. H. Lipkin and J. Duffy, "Analysis of industrial robots via the theory of screws," in Proceedings of the 12th International Symposium on Industrial Robots, 1982, pp. 359-370.
5. R. P. Paul and C. N. Stevenson, "Kinematics of robot wrists," The International journal of robotics research, vol. 2, pp. 31-38, 1983.
6. K. Waldron, S.-L. Wang, and S. Bolin, "A study of the Jacobian matrix of serial manipulators," Journal of mechanisms, transmissions, and automation in design, vol. 107, pp. 230-237, 1985.

7. K. Hunt, "Robot kinematics—a compact analytic inverse solution for velocities," Journal of mechanisms, transmissions, and automation in design, vol. 109, pp. 42-49, 1987.

8. S.-L. Wang and K. J. Waldron, "A study of the singular configurations of serial manipulators," Journal of mechanisms, transmissions, and automation in design, vol. 109, pp. 14-20, 1987.

9. S. Nokleby and R. Podhorodeski, "Methods for resolving velocity degeneracies of joint-redundant manipulators," in Advances in Robot Kinematics, ed: Springer, 2000, pp. 217-226.

10. S. B. Nokleby and R. P. Podhorodeski, "Reciprocity-based resolution of velocity degeneracies (singularities) for redundant manipulators," Mechanism and machine theory, vol. 36, pp. 397-409, 2001.

11. S. B. Nokleby and R. P. Podhorodeski, "Velocity degeneracy determination for the kinematically redundant CSA/ISE STEAR testbed manipulator," Journal of Robotic Systems, vol. 17, pp. 633-642, 2000.

12. K. Sugimoto, J. Duffy, and K. Hunt, "Special configurations of spatial mechanisms and robot arms," Mechanism and Machine Theory, vol. 17, pp. 119-132, 1982.

13. J. Denavit, "A kinematic notation for lower-pair mechanisms based on matrices," Trans. of the ASME. Journal of Applied Mechanics, vol. 22, pp. 215-221, 1955.

Review on Machine Learning Approaches in Maritime Anomaly Detection Based on AIS Data

Zheng-yang Zhang[†] and Wen-zhong Tang
School of Computer Science and Engineering,
Beihang University, Beijing, China
[†]zhangzhengyang@buaa.edu.cn

Yan-yang Wang
School of Aeronautic Science and Engineering,
Beihang University, Beijing, China

Zhao Chen
Institute of Computing Technology,
Chinese Academy of Sciences,
Beijing, China

With the development of maritime trade, more and more people draw attention to the safety of maritime transportation. Anomaly detection of vessels is important for maritime safety. Most vessels are equipped with Automatic Identification System (AIS), which provides the kinetic information of vessels. It is possible to detect anomaly by mining the massive AIS data with various machine-learning algorithms. In this paper, the background and the conceptual model of anomaly detection are introduced. Then, the machine learning algorithms that are used for vessel motion analysis and anomaly detection are reviewed. Meanwhile, the features and problems of each approach are discussed. At last, the future research direction of maritime anomaly detection is summarized and discussed.

Keywords: AIS; Anomaly Detection; Motion Analysis; Machine Learning.

1. Introduction

Anomaly detection or outlier detection is an important research field in data mining. Abnormal behavior can indicate events of interest in a wide variety of domains, such as credit card, tax fraud detection and many other areas [1]. In the maritime domain, vessel anomaly has a great impact on the safety of maritime and ground facilities. Vessel anomaly contains out of speed anomaly, deviations from standard routes, and so on.

AIS is a technology that installed on the vessels, which sends messages to ground-based stations or satellites. The sending messages include static information and dynamic information about the vessels [2]. With the widely use of AIS technology, people is able to get lots of AIS messages. Using these data, researchers can model vessel's behavior and detect anomaly [3-6], which is helpful to counter piracy and organized crime, prevent terrorism [7-11].

The remainder of the paper is organized as follows. Concept of anomaly introduces the definition of anomaly, the classification of vessel anomaly and the conceptual model of anomaly detection. Motion analysis of vessel presents and

880

discusses the approaches of machine learning to model the motion pattern of vessels. Subsequently, we reveal the approaches to detect anomaly. In the last section, we make a prospect of the future of maritime anomaly detection.

2. Concept of Anomaly

Based on Chandola [12], "anomalies are patterns in data that do not conform to a well-defined notion of normal behavior". Fig. 1 illustrates anomalies in a 2-dimensional dataset. N1 and N2 are the normal regions, while A1 and A2 are anomaly instances, which are not lie close to the normal regions.

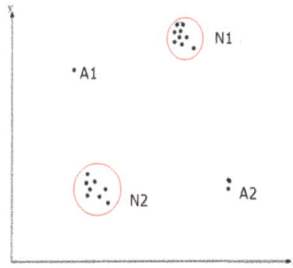

Figure 1. An example of anomaly

For vessel anomaly, if the trajectory of the ship does not match the historic trajectory, then we announce that it is an anomaly. Davenport [13] gets two main types of anomalies that are motion anomaly and location anomaly. Specifically, motion anomaly contains 7 types of anomaly and location anomaly contains 9 types of anomaly.

The purposes of maritime anomaly detection are manpower optimization and Maritime Situation Awareness (MSA). Manpower optimization means that let the machine do the filtering work and people do the reasoning work. Automatic surveillance is needed in both specific requirements. While the remaining requirements are decision support and prediction. The analysis model and the detection method will be used in the next part to detect anomaly. Next we will describe these two parts in detail. The conceptual model of anomaly detection is displayed in Fig. 2.

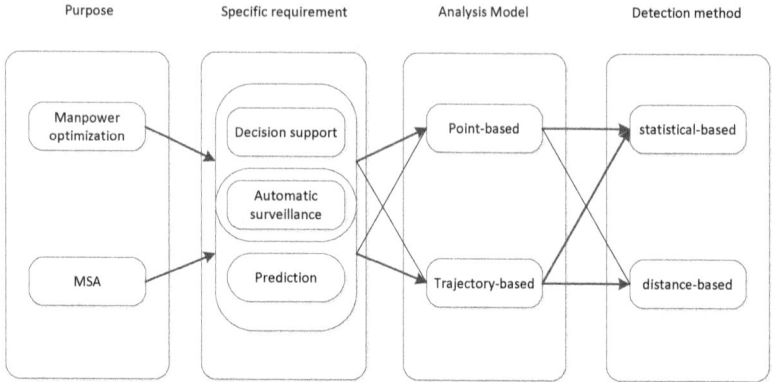

Figure 2. Conceptual model of anomaly detection

3. Motion Analysis of Vessel

More and more researchers apply machine learning [14,15] approaches to vessel motion analysis and anomaly detection. These approaches could be divided into two categories: point-based approaches and trajectory-based approaches.

3.1. Point-based approaches

When we use point-based machine learning approaches, it usually means that we regard each AIS dynamic message or each geographical cell as an independent point [16,17]. In most cases, the cell is a square grid [18] with related AIS static information, such as the number of ships and average speed.

Zhu [19] divides the area into grids by the dimensions

$$0.02° \times 0.02°$$

and trajectory can be represented by

$$\{(lat_1, lon_1), (lat_2, lon_2), ..., (lat_i, lon_i), ...\},$$

then he uses association rule algorithm [20] to discover ship spatial trajectory patterns. Deng [21] makes this further. He takes the time scale of tracks into consideration by applying Fp-Growth algorithm in the mining of AIS data and uses Markov model to supplement for the results of association rules sufficiently. Shortly afterwards, he uses prefix-span algorithm [22] to guarantee the time scale of the vessel trajectory [23]. The use of prefix can reduce the space and time scale, which can also significantly support the anomaly detection for vessel motion. Fabbri [24] combines terrestrial and satellite AIS data. The paper divides the Indian

Ocean with $0.25° \times 0.25°$. When predicting which area is high risk, they take wind speed, wave height and wave peak period into account.

3.2. Trajectory-based approaches

Compared to point-based approaches, trajectory-based approaches need more complex mathematical computing. Trajectory-based approaches pay more attention to estimating trajectories rather than each independent AIS message. Lee [25] proposes a novel framework, which is consisted of two phases: the partitioning phase and the detection phase. This framework can effectively detect abnormal sub-trajectory from trajectory.

Vries [26] uses similarity-based and kernel-based machine learning methods to cluster vessel traffic and detect abnormal vessel behavior. Soleimani [27] takes use of the entire trajectory to analysis. They compare the trajectory of the ship with the near-optimal path which get by A* algorithm and get the anomaly score.

Point-based approaches ignore the connection between grids, while trajectory-based approaches need more complex computing. Pallotta has made great contributions to the ship trajectory analysis. One contribution is TREAD [28], which is short for Traffic Route Extraction for Anomaly Detection. TREAD is a hybrid approach, which combines point-based approaches and trajectory-based approaches. It means that this framework does not only rely on the entire trajectories, but also take the individual AIS message into consideration. Later, She [29] proposes a data-driven approach to associate the track to the model. Based on TREAD, Liu [30] proposes a new clustering algorithm that can take speed and direction into consideration.

4. Anomaly Detection

Based on the results of vessel motion analysis mentioned above, we can judge whether a sequence of AIS messages are abnormal or not. Generally, there are two major methods to detect anomaly: statistical-based approaches and distance-based approaches.

4.1. Statistical-based approaches

Statistical-based approaches [31,32] assume the distribution for the given dataset. Anomalies are these points that do not fit the distribution. Laxhammar [32] gets abnormal vessels if the number of vessels in the grid exceeds the given threshold. Ristic [17] divides the maritime area into grids and treats each grid, c_{ij}, $i = 1$, ..., M, $j = 1$, ..., N. They believe the AIS data is Poisson distribution with Eq. 1. If the AIS message is beyond or below the Poisson distribution that they got, then it is considered an abnormal message.

$$\rho(n;\lambda T) = \frac{(\lambda T)^n}{n!} e^{-\lambda T}, \quad n=0,1,2,\dots.$$

(1)

4.2. Distance-based approaches

Distance-based approaches [26,33-35] are the most commonly used method. The distance is computed with Euclidean distance or Manhattan distance. If the distance is beyond the threshold formed by the normal data, the vessel is considered to be an abnormal vessel. The notion of distance-based anomaly does not assume any underlying data distributions. What is more, distance-based methods are more efficient than the statistical-based methods.

Soleimani [23] uses the length of the trajectory (L), the area under the curve (A) and the gradient of trajectory(G), to calculate scores to sort the trajectories of vessels with Eq. 2. Op and re in the upper right corner of the equation respectively represent the optimized path and the real path. If the score is negative, it means that the ship did not choose the near-optimal path and its trajectory may be abnormal.

$$Score = \frac{L^{op} - L^{re}}{L^{op}} + \eta \left(\frac{A^{op} - A^{re}}{L^{op}} + \frac{G_\theta^{op} - G_\theta^{re}}{L^{op}} + \frac{G_\phi^{op} - G_\phi^{re}}{L^{op}} \right).$$

(2)

5. Conclusion

In this paper, we introduce the background and conceptual model of anomaly detection, besides we attempted to provide an overview of the existing methods of vessel motion analysis and anomaly detection. For each category of motion analysis and anomaly detection, we discussed the features and the latest researches about them.

There are several promising directions for further research in maritime anomaly detection. Contextual anomaly detection techniques are quickly developing. The context influences the vessel motion greatly, so it will be helpful to understand the vessel motion by adding more factors like time, weather and tide to anomaly detection, and will improve the accuracy of maritime anomaly detection.

Acknowledgement

The authors are grateful to the supervisor and the colleagues. The work was funded by National Natural Science Foundation of China (No. 51475025).

References

1. Kalinichenko L, Shanin I, Taraban I. Methods for Anomaly Detection: a Survey [J].

2. International Telecommunications Union, "Technical characteristics for an automatic identification system using time division multiple access in the vhf maritime mobile band (Recommendation ITU-R M.1371-4),"2012.

3. Mazzarella F, Vespe M, Damalas D, et al. Discovering vessel activities at sea using AIS data: Mapping of fishing footprints [C]//Information Fusion (FUSION), 2014 17th International Conference on. IEEE, 2014: 1-7.

4. Le Guillarme N, Lerouvreur X. Unsupervised extraction of knowledge from S-AIS data for maritime situational awareness [C]//Information Fusion (FUSION), 2013 16th International Conference on. IEEE, 2013: 2025-2032.

5. Vandecasteele A, Napoli A. Spatial ontologies for detecting abnormal maritime behaviour [C]//OCEANS, 2012-Yeosu. IEEE, 2012: 1-7.

6. Mascaro S, Nicholso A E, Korb K B. Anomaly detection in vessel tracks using Bayesian networks [J]. International Journal of Approximate Reasoning, 2014, 55(1): 84-98.

7. R. Lane and K. Copsey, "Track anomaly detection with rhythm of life and bulk activity modeling," in Information Fusion (FUSION), 2012 15th International Conference on, July 2012, pp. 24–31.

8. Kazemi S, Abghari S, Lavesson N, et al. Open data for anomaly detection in maritime surveillance [J]. Expert Systems with Applications, 2013, 40(14): 5719-5729.

9. Shahir H Y, Glasser U, Shahir A Y, et al. Maritime situation analysis framework: Vessel interaction classification and anomaly detection [C]//Big Data (Big Data), 2015 IEEE International Conference on. IEEE, 2015: 1279-1289.

10. Patroumpas K, Alevizos E, Artikis A, et al. Online Event Recognition from Moving Vessel Trajectories [J]. arXiv preprint arXiv:1601.06041, 2016.

11. Cazzanti L, Millefiori L M, Arcieri G. A document-based data model for large scale computational maritime situational awareness [C]//Big Data (Big Data), 2015 IEEE International Conference on. IEEE, 2015: 1350-1356.

12. Chandola V, Banerjee A, Kumar V. Anomaly detection: A survey [J]. ACM computing surveys (CSUR), 2009, 41(3): 15.

13. Davenport M. Kinematic Behaviour Anomaly Detection (KBAD)-Final Report [J]. DRDC CORA report KBAD-RP-52-6615, 2008.

14. Obradović I, Miličević M, Žubrinić K. Machine Learning Approaches to Maritime Anomaly Detection [J]. Naše more, Znanstveno-stručni časopis za more i pomorstvo, 2014, 61(5-6): 96-101.

15. Cazzanti L, Pallotta G. Mining maritime vessel traffic: Promises, challenges, techniques [C]//OCEANS 2015-Genova. IEEE, 2015: 1-6.

16. Papa G, Horn S, Braca P, et al. Estimating sensor performance and target population size with multiple sensors[C]//Information Fusion (FUSION), 2012 15th International Conference on. IEEE, 2012: 2102-2109.

17. Ristic B. Detecting anomalies from a multitarget tracking output [J]. Aerospace and Electronic Systems, IEEE Transactions on, 2014, 50(1): 798-803.

18. Pham D S, Venkatesh S, Lazarescu M, et al. Anomaly detection in large-scale data stream networks [J]. Data Mining and Knowledge Discovery, 2014, 28(1): 145-189.

19. Feixiang Z. Mining ship spatial trajectory patterns from AIS database for maritime surveillance[C]//Emergency Management and Management Sciences (ICEMMS), 2011 2nd IEEE International Conference on. IEEE, 2011: 772-775.

20. Rajaraman A, Ullman J D. Mining of massive datasets [M]. Cambridge: Cambridge University Press, 2012.

21. Deng F, Guo S, Deng Y, et al. Vessel track information mining using AIS data [C]//Multisensor Fusion and Information Integration for Intelligent Systems (MFI), 2014 International Conference on. IEEE, 2014: 1-6.

22. Pei J, Han J, Mortazavi-Asl B, et al. Mining sequential patterns by pattern-growth: The prefixspan approach [J]. Knowledge and Data Engineering, IEEE Transactions on, 2004, 16 (11): 1424-1440.

23. Sun F, Deng Y, Deng F. Mining Spatio-temporal AIS Data Using Prefix-span//Proceedings of The Fourth International Conference on Information Science and Cloud Computing (ISCC2015). 18-19 December 2015. Guangzhou, China. Online at http://pos. sissa. it/cgi-bin/reader/conf. cgi? confid= 264, id. 5. 2015.

24. Fabbri T, Vicen-Bueno R, Grasso R, et al. Optimization of surveillance vessel network planning in maritime command and control systems by

fusing METOC & AIS vessel traffic information [C]//OCEANS 2015-Genova. IEEE, 2015: 1-7.

25. Lee J G, Han J, Li X. Trajectory outlier detection: A partition-and-detect framework [C]//Data Engineering, 2008. ICDE 2008. IEEE 24th International Conference on. IEEE, 2008: 140-149.

26. [26] De Vries G K D, Van Someren M. Machine learning for vessel trajectories using compression, alignments and domain knowledge [J]. Expert Systems with Applications, 2012, 39(18): 13426-13439.

27. Soleimani B H, De Souza E N, Hilliard C, et al. Anomaly detection in maritime data based on geometrical analysis of trajectories [C]//Information Fusion (Fusion), 2015 18th International Conference on. IEEE, 2015: 1100-1105.

28. Pallotta G, Vespe M, Bryan K. Vessel pattern knowledge discovery from ais data: A framework for anomaly detection and route prediction [J]. Entropy, 2013, 15(6): 2218-2245.

29. Pallotta G, Jousselme A L. Data-driven detection and context-based classification of maritime anomalies[C]//Information Fusion (Fusion), 2015 18th International Conference on. IEEE, 2015: 1152-1159.

30. Liu B, De Souza E N, Matwin S, et al. Knowledge-based clustering of ship trajectories using density-based approach [C]//Big Data (Big Data), 2014 IEEE International Conference on. IEEE, 2014: 603-608.

31. Bamnett V, Lewis T. Outliers in statistical data [J]. 1994.

32. Laxhammar R. Anomaly detection for sea surveillance [C]//Information Fusion, 2008 11th International Conference on. IEEE, 2008: 1-8.

33. Vespe M, Visentini I, Bryan K, et al. Unsupervised learning of maritime traffic patterns for anomaly detection [C]//Data Fusion & Target Tracking Conference (DF&TT 2012): Algorithms & Applications, 9th IET. IET, 2012: 1-5.

34. Sun F, Deng Y, Deng F, et al. Unsupervised maritime traffic pattern extraction from spatio-temporal data [C]//Natural Computation (ICNC), 2015 11th International Conference on. IEEE, 2015: 1218-1223.

35. Laxhammar R, Falkman G. Sequential conformal anomaly detection in trajectories based on hausdorff distance [C]//Information Fusion (FUSION), 2011 Proceedings of the 14th International Conference on. IEEE, 2011: 1-8.

Environment Perception and Target Recognition Algorithm for Desktop Grade Manipulator Based on Machine Learning

Bo-wen Shang[1,†], Pei-dong Zhu[1] and Luo-lan Yang[2]

[1] College of Computer Science and Technology,
National University of Defense Technology,
Changsha, 410072, China
[2] Beijing Research Institute of Telemetry,
Beijing 100094, China
†Email: 819089115@qq.com

The desktop grade manipulator refers to the manipulator with smaller volume, lower cost and multi-function relative to the industrial manipulator, used in smart home and office. It can help users with daily work like taking and moving objects, etc. Currently, the desktop grade manipulator can only be controlled directly by program or manual and not based on the data from environmental awareness and target recognition, through the way of autonomous learning to control actions and services for the user. The industrial manipulator control system, while the precision is high, but its ability to adapt to the different environment and the autonomous learning ability is insufficient. This article aims at the environmental awareness and target recognition problem in the desktop grade manipulator control field, designs and implements environment perception and 3D modeling algorithm based on distance sensor. It is also by 3D modeling design on the target search algorithm based on camera and distance sensor. Finally, using the search results to improve the target recognition algorithm based on machine learning. This paper uses low cost open source software and hardware in making the prototype system. It realizes the surrounding environment perception and automatic search to identify and pick the unknown location object.

Keywords: Machine learning; Desktop grade manipulator; Environment perception and recognition.

1. Introduction

With the development of computer technology and automatic control technology, the robot and manipulator used in the field of industry, has been more and more popular in people's lives. Desktop grade manipulator, which can be placed at home, office and other places, to help people deal with daily affairs. Its

---

†Corresponding author.

representatives are uarm, 7bot, Dobot, etc., 7bot using six servo motor driver, can be programmed in advance to control. Dobot uses three stepper motor with planetary deceleration drive, its repeated positioning accuracy is better than 7bot, but the degree of freedom is lower than 7bot. The current desktop grade manipulator only can be controlled by program, or controlled by people directly, so it can not automatically complete the task like "Pick the apple." if it does not know the position of the target object and the environmental. The reason is that the control system of the existing desktop grade manipulator cannot feel the environment automatic and know the location of the target object. And a practical desktop grade manipulator should be able to assist the user to work, such as implement the user's voice commands, to pick something on the table [1]. Compared with the industrial manipulator, the desktop grade manipulator does not require high positioning precision and large load capacity. The desktop grade manipulator should have the ability to adapt to different environments as the environment change such as the user put different things on the table. The desktop grade manipulator should have the ability to adapt different environments, because user put different things on the table can change the environment. it also needs the ability that search and distinguish different target object as there are many different objects on the table [2].

This paper uses machine learning and spatial modeling methods, designed and implemented the desktop grade manipulator environment perception and target recognition algorithm based on the distance sensor [3] and the camera and realize the system prototype by open source software and hardware. This has a very important significance for the realization of the automatic control of the manipulator.

This paper uses machine learning and spatial modeling methods, designed and implemented the desktop grade manipulator environment perception and target recognition algorithm which is based on the distance sensor, realizes the system that prototype by open source software and hardware. This has a very important significance for the realization of the automatic control of the manipulator [4].

2. Environment Aware Algorithm for Desktop Grade Manipulator

Desktop level manipulator's environment aware is to perceive the surrounding environment and establish the 3D model. Due to the environment around the manipulator changing every time, so we need to sense the environment in real time. However, if the environment around the manipulator need to be modeled, the cost of the time is so much. To solve this problem, this article use method of Step-perception. As the target object can only appear on some particular plane in the environment and cannot fly in the air, we call the particular plane to load plane. First, use the environment aware algorithm to find the fixed obstacles and loading plane around the manipulator, and then use the loading plane search algorithm to search the target object on the loading plane.

Though the environment aware algorithm, we can fix the obstacle and measure the distance from the cradle of the manipulator. After the aware of the environment, we need to establish the spatial rectangular coordinate system to build the 3D model like Figure 1. The manipulator is placed at the origin of the coordinates.

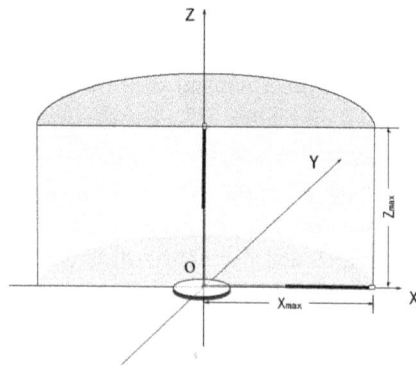

Figure 1. The spatial rectangular coordinate system

Suppose manipulator placed on the desktop, and its range of motion does not exceed the range of the desktop, so that its scope of activities contained by the first and second quadrants of the coordinate system. Assuming the maximum height of the manipulator is Z_{max}, the maximum lateral extension distance is X_{max}, so that the scope of activities included in the half cylinder that with the origin of coordinates as the center, with X_{max} as the radius, with Z max as the high. The coordinates expressed as shown:

$$\begin{cases} x^2 + y^2 < X_{max}^2 \\ 0 \le z < Z_{max}^2 \end{cases}, (y > 0)$$

(1)

We need to aware the fixed obstacles and loading plane, then we can find the target object on the loading plane. We use the distance sensor to aware the environment. As there may be a blind zone behind the fixed obstacles if we use the distance sensor aware the environment in a single direction, so we should aware the environment in different. In this way, we can reduce the blind zone.

However, there are two problems: How can we get the minimal blind zone with the least aware times and how can we build the 3D model with the different aware results?

In this chapter, we use the horizontal direction aware and the vertical direction aware to minimal the blind zone and propose a 3D modeling algorithm.

2.1. Environmental Perception in the Horizontal Direction

(1) Lets the distance sensor locate at the origin of coordinates O(0,0,0) .

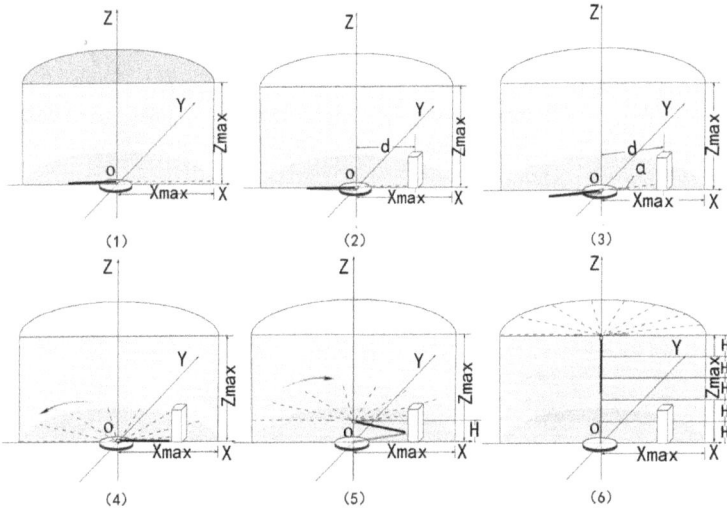

Figure 2. The horizontal direction perception

Is shown in Figure 2, thick real line is the manipulator, and the distance sensor is on the head of the manipulator. We can freely control the position and the detection direction of the distance sensor.

(2) Let the detection direction of the distance sensor towards the x-axis positive direction, measuring the distance between the O and the obstacle as measurement d, if $d < X_{max}$, then save d, else if $d > X_{max}$, then return X_{max}.

(3) Let the distance sensor rotate a small angle α, again for ranging.

(4) Repeating process (3), until the direction of the detection direction is negative direction of the x-axis.

(5) Select fixed height H, then let the distance sensor rises along the z-axis to stop at(0,0, H), repeat process (2), (3), (4).

(6) Repeat the process (5) until the height of the distance sensor is Z_{max}.

2.2. Environmental Perception in the Vertical Direction

As shown in Figure 3:

   (1) Let the distance sensor to locate on the position that directly above of the origin (0,0, Zmax).

   (2) Let the detected direction of the distance sensor is the negative direction of the z-axis, this time we get the measuring distance d. Assuming that current distance sensor height is H, the height of obstruction h = H-d.

   (3) Keep the length of the manipulator, rotate to the positive x-axis direction around the y-axis, the rotation angle is α, measuring the height h of the obstruction.

   (4) Keep the length of the manipulator and the angle α, rotate to the x-axis positive direction around z-axis.the rotation angle is β, measuring the height h of the obstruction and calculates the spatial coordinates of the measurement point.

   (5) Repeat the process (4) until the projection of manipulator falls on the negative direction of x-axle.

   (6) Add the angle α and repeat, until α = 90 .

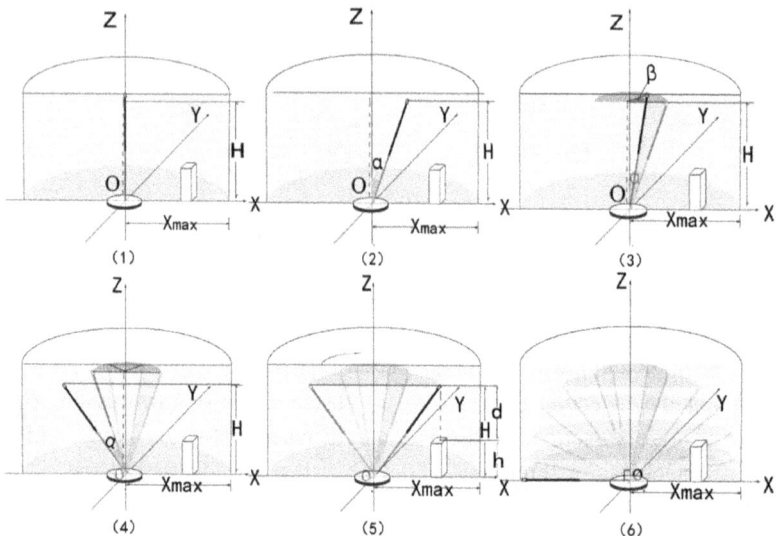

Figure 3. The vertical direction perception

2.3. Identify the Loading Plane and Calculate the Volume of Obstacles

When there is an obstacle, we need to measure the volume of the obstacle, the position of the obstacle, and the loading plane above the obstacle.

We get a set of point from the environmental perception in the vertical direction, each point's spatial coordinates is known. We can find the loading plane by the points and calculate the volume of the obstacles. We define the loading plane is a point set that the Z coordinates of the points is the same and the points connect to each other. To find every loading plane, we should find the point sets that meet the above conditions. We use the loading plane identify algorithm to classify all the points is in the set S into different set P that meet the definition of the loading plane.

(1) Get a point a in the set S, remove a in the set S.

(2) Find all the points the Z coordinates is the same as a and the points set is D, remove all the points that in set D from set S.

(3) Get a point b from the set D, remove b in the set D and put b in the set L.

(4) Find all the point adjacent to point b in the set D, if we can't find four point adjacent to point b, then add b to the set E. Remove all the adjacent point in set D and add them to set K and L.

(5) For each point c in the set K, find all the point adjacent to point c in the set D, if we can't find four point adjacent to point c, then add c to the set E. Remove all the adjacent point in set D and add them to set K. Remove c in the set K and L.

(6) Repeat (5) until the set K is empty, then return the set L and set E. Remove all the point in set L and E.

(7) Goto (3), if the set D is empty, goto (1).

(8) If the set S is empty, end.

The set L is the loading plane and the set E is the edge of the loading plane.

In this way, we also can find the vertical plane with the points from the horizontal direction.

After we get the loading plane and the edge of the plane, we can calculate the volume of obstacles V. As we get the loading plane of the obstacle and the height of the obstacle. Suppose the length of the distance between two adjacent points is l. If a point is in the loading plane and not in the edge. This point must belong to the collection L and not belong to the collection E, we call this kind of point inside point and we call the point belong to the collection E edge point.

To calculation the volume of the obstacle, we use the point number in the point set L and E. Suppose the number of the point in the set L is x and the number of the point in the set E is y. The height of the point in the set is h. Then the volume of the obstacle below the loading plane L is:

$$V = (x - y) \times l^2 \times h + 0.5 \times y \times l^2 \times h \tag{2}$$

If the upper surface of the obstacle is curved surface. It can be closely approximated by a number of planes. Suppose the volume of the part of the obstacle below the plane is v_i. Then the volume of the obstacle is:

$$V = \sum_{i=1}^{n} v_i$$

(3)

There may be blind area blow the loading plane that can't be detected by the vertical direction perception. So we need to detect the blind area by horizontal direction perception or other direction perception. Through the horizontal direction perception, we can reduce the blind area but cannot remove them.

Though the horizontal direction perception, we get a point set H and though the vertical direction perception we get a loading plane P, the point set L_p and E_p. Suppose that we do the horizontal direction perception n times and the height difference between two adjacent order detection is Δh. We define a set function $L_p(h)$ that change every point's Z coordinates in the set L_p to h and a set function $E_p(h)$ that change every point's Z coordinates in the set E_p to h.

For any height h less than the loading plane P, we can get a plane P' that E'= $E_p(h)$ and L'= $L_p(h)$. Then we find all the point in the set H that the Z coordinates is h and put them into a set N.

Then we find all the point in the set N and in the set L', put them into the set C. We connect all points in the set C to the point (0,0,h) and put all the points on the connect line and in the set L' into the point set Q. Suppose the area of the plane Q is S_i, then $\Delta h \times S_i$ is the volume that should be reduce. Then all the volume should be reduced is:

$$V_r = \sum_{i=0}^{H} \Delta h \times S_i$$

(4)

Then the volume of the obstacle V_o is:

$$V_o = V - V_r$$

(5)

2.4. Target Discovery and Recognition Algorithms

In the last chapter, we find the loading plane. When we want to discover the target object, we can only detect the loading plane and not need to detect all the space.

When we discover the target object, we can use the same method in the above chapter to get the surface feature information of its upper surface, side and the volume. We build the 3D model for the target object. We build the feature vector use the 3D model and use the svm machine learn methods to recognize the target object.

(1) Put the target object A and B at different place and detect the target object.
(2) Use the detect result build the 3D model for them.
(3) Train the svm use the 3D model.
(4) Use some of the 3D model to test the method.

Though the 3D model, we can get a better result than the image recognition [5] methods. And will not be affected by the position of the object and the light.

3. Case Study

We use a six degree of freedom manipulator (6DOF), a rudder control panel to control the 6DOF manipulator and a Raspberry with the Linux operating system to run the control program in Python.

When we give a coordinate to the control system, it can control the 6DOF manipulator to the point. We also can control the direction of the manipulator head. When we vertical scan the movement space, we divided it into 100 layers; the thickness of each layer is 0.2 cm. And when we horizontal scan the movement space, we divided it into five layers in horizon, the thickness of each layer is 0.2 cm. In every layer, we use the distance sensor measure 180 times.

(1) We put a cube that the length is 5cm in the activity range of the manipulator and detect it in the horizontal direction and vertical direction, use the detected results calculate the volume 100 times, and average volume is 127.3cm^3.
(2) We use a cylinder that the height is 3cm and the diameter is 5cm, and the cube, put them at different place and detected them, use the data build 3D model for them and use the data train and test the svm. The confusion matrix is in Table 1.

Table 1. Confusion matrix

	Forecast cube	Forecast cylinder
Actual cube	20	1
Actual cylinder	2	19

4. Conclusion

In this article, we propose an environment perception and target recognition algorithm for desktop grade manipulator based on machine learning. Using this method, we can dynamic perception and modeling the environmental around the desktop grade manipulator and find the target object fast.

We also use the machine learn method train and recognize the target object and get a better result than the image recognition which will be affected by the position of the object and the light.

Acknowledgement

This project was funded by the Nature Science Foundation of China (project number 61402511, project number 61572514).

References

1. Cai C, Somani N, Knoll A. Orthogonal Image Features for Visual Servoing of a 6-DOF Manipulator with Uncalibrated Stereo Cameras [J]. IEEE Transactions on Robotics, 2016, 32(2): 452-461.
2. Chang T K, Spowage A, Chan K Y. Review of Control and Sensor System of Flexible Manipulator [J]. Journal of Intelligent & Robotic Systems, 2015, 77(1):187-213.
3. Chang C C, Song K T. Ultrasonic sensor data integration and its application to environment perception [J]. Journal of Robotic Systems, 1996, 13(10): 663–677.
4. Zhao Q, Principe J C. Support vector machines for SAR automatic target recognition [J]. IEEE Transactions on Aerospace Electronic Systems, 2001, 37(2): 643-654.
5. Boots B, Byravan A, Fox D. Learning predictive models of a depth camera & manipulator from raw execution traces [C]// 2014:4021-4028.

Blind Source Separation Using Canonical Decomposition of Higher Order Tensor

Ming-jian Zhang[t] and Jun Peng

School of Information Science and Engineering,
Central South University,
Changsha 410083, China
Email: mingjianzhang@163.com[t], pengj@csu.edu.cn

In this paper, the problem of blind source separation in the over determined mixtures case is considered. A two-step-type blind source separation method is proposed, which first identifies the mixing matrix and then performs source estimation with the identified mixing matrix. In the step of identifying the mixing matrix, a new blind identification algorithm, which is based on the canonical decomposition of a higher order tensor, is proposed. Gradient descent approach and generalized eigenvalue decomposition technique are used to perform the canonical decomposition. By using the identification detecting device, blind identification is reformulated as a constrained optimization problem. Unlike the traditional blind identification based on simultaneous matrix diagonalization, the proposed algorithm identifies the columns of the mixing matrix in a sequential fashion. Simulation results demonstrate that the proposed algorithm can achieve good performance.

Keywords: Blind source separation; Canonical Decomposition; Constrained Optimization.

1. Introduction

The problem of Blind Source Separation (BSS) involves recovery of a number of unknown source signals from their observed mixtures without prior knowledge of the mixing matrix. If the number of sensors is greater than or equal to the number of source signals, BSS is said to be Overdetermined Blind Source Separation (OBSS) [1,2]. On the contrary, if the number of sensors is less than the number of source signals, BSS is said to be Underdetermined Blind Source Separation (UBSS) [3,4].

A large number of OBSS and UBSS algorithms consist of two steps: an identification step for estimating the mixing matrix and a separation step for recovering the source signals. Therefore, the identification of the mixing matrix plays an important role in OBSS and UBSS algorithms [1,4]. Taking into account this fact, this paper aims at developing an algorithm for sequential blind

897

identification of overdetermined mixtures. Unlike the Simultaneous Matrix Diagonalization (SMD)-based blind identification algorithm [5], the proposed algorithm estimates the columns of the mixing matrix in a sequential manner. The sequential-type algorithms rely on the use of deflation technique [6]. However, as pointed out in [7,8], the conventional deflation method suffers from error accumulation, because the estimation errors generated in the previous estimates are propagated to the following estimates, and hence give rise to growing estimation errors during the successive deflation stages. To overcome this drawback, in this paper, we use a deflation scheme proposed in [8], which has the ability to avoid error accumulation during the deflation procedure.

The algebraic blind identification of mixing matrix algorithms implicitly or explicitly exploit the canonical decomposition (CANDECOMP) [5,9], also known as parallel factor (PARAFAC) decomposition, of a higher order tensor containing either second-order or higher order cumulants of the data. CANDECOMP can be performed by using the SMD technique [5] and the generalized eigenvalue decomposition (GEVD) technique [8]. In this paper, the blind identification is based on the CANDECOMP of a higher order tensor, and the CANDECOMP is performed by using the gradient and GEVD techniques.

2. Mixing Model

Consider the following mixing model

$$\mathbf{x}(t) = \mathbf{As}(t) \tag{1}$$

Where $\mathbf{x}(t) = [x_1(t), \cdots, x_J(t)]^T$ is a vector of observations, $\mathbf{s}(t) = [s_1(t), \cdots, s_J(t)]^T$ is a vector of sources, $\mathbf{A} = [\mathbf{a}_1, \cdots, \mathbf{a}_J]$ is an unknown full column rank mixing matrix. The task of blind source separation is to estimate the source signals up to amplitude scaling and permutation. For the separability of the mixing model, we assume that the source signals are zero-mean, spatially uncorrelated but temporally correlated.

The covariance matrix of $\mathbf{x}(t)$ is defined as
$$\mathbf{R}_i = E\{\mathbf{x}(t)\mathbf{x}^T(t+\tau_i)\} = \mathbf{AD}_i\mathbf{A}^T,$$
Where $\mathbf{D}_i = E\{\mathbf{s}(t)\mathbf{s}^T(t+\tau_i)\} = diag\{d_{i1}, \cdots, d_{iJ}\}$ is a diagonal matrix.

3. Identification of Mixing Matrix Based on Canonical Decomposition

Stack the set of covariance matrices $\{\mathbf{R}_i, i = 1, \cdots, K\}$ in a third-order tensor C as follows $(C)_{ijk} = (\mathbf{R}_k)_{ij}$. Construct a matrix $\bar{\mathbf{D}}$, whose elements are defined as $(\bar{\mathbf{D}})_{kr} = (\mathbf{D}_k)_{rr}$. We further stack the entries of the tensor C in the matrix $\overline{\mathbf{C}}$ as follows $(\overline{\mathbf{C}})_{(k-1)J+i,j} = (C)_{ijk}$. According to [5], the matrix $\overline{\mathbf{C}}$ can be written in the form of

$$\overline{\mathbf{C}} = \left[\overline{\mathbf{d}}_1 \otimes \mathbf{a}_1, \cdots, \overline{\mathbf{d}}_J \otimes \mathbf{a}_J \right] \mathbf{A}^T \tag{2}$$

Where $\otimes$ denotes Kronecker product, and $\overline{\mathbf{d}}_k$ is the kth column of $\overline{\mathbf{D}}$. By performing the reduced-size singular value decomposition of $\overline{\mathbf{C}}$ [5], we have

$$\overline{\mathbf{C}} = \mathbf{USV}^T \tag{3}$$

Where $\mathbf{U}$ is a columnwise or thonormal matrix, $\mathbf{S}$ is a diagonal matrix of singular values, and $\mathbf{V}$ is an orthogonal matrix. According to Eq. 2 and Eq. 3, we get

$$\left[\overline{\mathbf{d}}_1 \otimes \mathbf{a}_1, \cdots, \overline{\mathbf{d}}_J \otimes \mathbf{a}_J \right] = \mathbf{GF} \tag{4}$$

Where $\mathbf{G} = \mathbf{US}$ and $\mathbf{F} = \mathbf{V}^T (\mathbf{A}^T)^{-1}$.
Define an operator $\text{unvec}(\cdot)$ by
$$\mathbf{M} = \text{unvec}(\mathbf{m}) \Leftrightarrow (\mathbf{M})_{ij} = (\mathbf{m})_{(i-1)J+j},$$
Which stacks an IJ-dimensional vector $\mathbf{m}$ in an $I \times J$ matrix $\mathbf{M}$ [5]. Then, we have

$$\text{unvec}(\mathbf{Gf}_k) = \text{unvec}(\overline{\mathbf{d}}_k \otimes \mathbf{a}_k) = \overline{\mathbf{d}}_k \mathbf{a}_k^T \tag{5}$$

Where $\mathbf{a}_k$ and $\overline{\mathbf{d}}_k$ denote the kth columns of the matrices $\mathbf{A}$ and $\overline{\mathbf{D}}$, respectively. It is obvious to see that the columns of $\text{unvec}(\mathbf{Gf}_k)$ are all proportional.

Define $\mathbf{G}_k = \text{unvec}(\mathbf{g}_k), k = 1, \cdots, J$, where $\mathbf{g}_k$ denotes the kth column of $\mathbf{G}$. From the matrices $\{\mathbf{G}_k\}$, we construct matrices

$$\mathbf{W}_j = \left[(\mathbf{g}_j^1)^T, \cdots, (\mathbf{g}_j^J)^T \right], \quad j = 1, \cdots, K,$$

Where $(\mathbf{g}_j^k)^T$ denotes the transpose of the jth row of $\mathbf{G}_k$. Then, we obtain

$$(\mathbf{w}_s^i \mathbf{f}_k \mathbf{w}_t^j - \mathbf{w}_t^i \mathbf{f}_k \mathbf{w}_s^j + \mathbf{w}_t^j \mathbf{f}_k \mathbf{w}_s^i - \mathbf{w}_s^j \mathbf{f}_k \mathbf{w}_t^i) \mathbf{f}_k = 0 \tag{6}$$

Or

$$\mathbf{f}_k^T ((\mathbf{w}_s^i)^T \mathbf{w}_t^j - (\mathbf{w}_t^i)^T \mathbf{w}_s^j + (\mathbf{w}_t^j)^T \mathbf{w}_s^i - (\mathbf{w}_s^j)^T \mathbf{w}_t^i) \mathbf{f}_k = 0 \tag{7}$$

Where $\mathbf{w}_q^p$ denotes the pth row of $\mathbf{W}_p$.

4. Identification of the First Column of Mixing Matrix

The estimation of $\mathbf{f}_1$ can be reformulated as a constrained optimization problem:

$$\min \sum_{1 \le i < j \le J} \sum_{1 \le s < t \le J} (\mathbf{f}_1^T \mathbf{W}_{st}^{ij} \mathbf{f}_1)^2$$

(8)

s.t. $\|\mathbf{f}_1\| = 1$

Where $\mathbf{W}_{st}^{ij} = (\mathbf{w}_s^i)^T \mathbf{w}_t^j - (\mathbf{w}_t^i)^T \mathbf{w}_s^j + (\mathbf{w}_t^j)^T \mathbf{w}_s^i - (\mathbf{w}_s^j)^T \mathbf{w}_t^i$.

Applying the gradient descent approach, we obtain the update rule for $\mathbf{f}_1$ as follows

$$\nabla \mathbf{f}_1 = -\eta \left(\sum_{1 \le i < j \le J} \sum_{1 \le s < t \le J} (\mathbf{f}_1^T \mathbf{W}_{st}^{ij} \mathbf{f}_1) \mathbf{W}_{st}^{ij} \mathbf{f}_1 \right) - \mu(\mathbf{f}_1^T \mathbf{f}_1 - 1)\mathbf{f}_1$$

(9)

Where η and μ denote learning rates. The first estimated column $\hat{\mathbf{f}}_1$ of $\mathbf{F}$ is achieved when the update rule above reaches a stationary point.

An estimated column $\hat{\mathbf{a}}_1$ of $\mathbf{A}$ can be computed as the dominant left singular vector, which is associated with the largest singular value of the matrix $\mathbf{H}^T$, where $\mathbf{H} = \mathrm{unvec}(\mathbf{G}\hat{\mathbf{f}}_1)$.

5. Identification of the Other Columns of Mixing Matrix

Next, we estimate the other columns of $\mathbf{A}$ in a sequential fashion. Let us denote by $\hat{\mathbf{f}}_l$ the lth estimated column of the matrix $\mathbf{F}$ and $\hat{\mathbf{a}}_l$ the lth identified column of the mixing matrix $\mathbf{A}$. The sequential identification process can be formulated as:

Given $\hat{\mathbf{f}}_1, \cdots, \hat{\mathbf{f}}_l, \mathbf{G}$, it is desired to find $\hat{\mathbf{f}}_{l+1}$ and then identify $\hat{\mathbf{a}}_{l+1}$ [8].

Consider a variable λ to be minimized as a function of a column vector $\mathbf{z}$

$$\lambda = \frac{\mathbf{z}^T (\mathbf{P}^T \mathbf{P}) \mathbf{z}}{\mathbf{z}^T (\mathbf{u}\mathbf{u}^T) \mathbf{z}}$$

(10)

Where $\mathbf{u}$ is orthogonal to $\hat{\mathbf{f}}_1, \cdots, \hat{\mathbf{f}}_l$. If $\mathbf{z}$ satisfies the two conditions $\mathbf{z}^T (\mathbf{P}^T \mathbf{P}) \mathbf{z} = 0$ and $\mathbf{z}^T (\mathbf{u}\mathbf{u}^T) \mathbf{z} \ne 0$ at the same time, then λ takes the minimum value of 0, in which the rows of $\mathbf{P}$ are

$$\mathbf{w}_s^i \mathbf{z} \mathbf{w}_t^j - \mathbf{w}_t^i \mathbf{z} \mathbf{w}_s^j + \mathbf{w}_t^j \mathbf{z} \mathbf{w}_s^i - \mathbf{w}_s^j \mathbf{z} \mathbf{w}_t^i, 1 \le i < j \le J, 1 \le s < t \le J.$$

We iteratively search for the optimal solution of $\mathbf{z}$ by fixing the matrix $\mathbf{P}^T \mathbf{P}$ to the value given by $\mathbf{z}$ at the previous iteration and then updating $\mathbf{z}$ according to the following iterative update rule [8].

$$\mathbf{z} \leftarrow \mathbf{z} + \mathbf{v}, \quad \mathbf{z} \leftarrow \frac{\mathbf{z}}{\|\mathbf{z}\|} \tag{11}$$

From Eq. 10, it is obvious that λ is the generalized eigenvalue of the matrix pencil $(\mathbf{P}^T\mathbf{P}, \mathbf{uu}^T)$. Therefore, the vector $\mathbf{v}$ in Eq. 11 is chosen as the generalized eigenvector of $(\mathbf{P}^T\mathbf{P}, \mathbf{uu}^T)$ associated with the generalized eigenvalue with the smallest magnitude.

In order to find the vector $\mathbf{u}$ orthogonal to $\hat{\mathbf{f}}_1, \cdots, \hat{\mathbf{f}}_l$, we first stack $\hat{\mathbf{f}}_1, \cdots, \hat{\mathbf{f}}_l$ in a matrix $\hat{\mathbf{F}}^{(l)}$ As follows

$$\hat{\mathbf{F}}^{(l)} = \hat{\mathbf{f}}_l, l = 1, \hat{\mathbf{F}}^{(l)} = [\hat{\mathbf{F}}^{(l-1)}, \hat{\mathbf{f}}_l,], l = 2, \cdots, J-1.$$

Compute the following eigenvalue decomposition (EVD) [8]

$$\hat{\mathbf{F}}^{(l)}\left(\hat{\mathbf{F}}^{(l)}\right)^T = [\mathbf{U}_e, \mathbf{U}^{(l)}]\begin{bmatrix} \mathbf{\Sigma}_e & \mathbf{0} \\ \mathbf{0} & \mathbf{0} \end{bmatrix}[\mathbf{U}_e, \mathbf{U}^{(l)}]^T \tag{12}$$

Then, we can choose the vector $\mathbf{u}$ from the columns of $\mathbf{U}^{(l)}$.

After convergence of the iterative update rule in Eq. 11, $\mathbf{z}$ is $(l+1)$th estimated column $\hat{\mathbf{f}}_{l+1}$ of the matrix $\mathbf{F}$. $\hat{\mathbf{a}}_{l+1}$

Can be computed as the dominant left singular vector of $\mathbf{H}^T$,

Where $\mathbf{H} = \mathrm{unvec}(\mathbf{G}\hat{\mathbf{f}}_{l+1})$.

The sequential identification is continued until all the columns of $\mathbf{A}$ are estimated.

6. Source Estimation

In the two-step-type blind source separation, the aim of the second step is to recover source signals. By using the estimated matrix $\hat{\mathbf{A}}$, the source signals can be recovered as $\hat{\mathbf{s}}(t) = \hat{\mathbf{A}}^{-1}\mathbf{x}(t)$, where $\hat{\mathbf{A}}^{-1}$ is the inverse of $\hat{\mathbf{A}}$. It is worth noting that $\hat{\mathbf{s}}(t) = [\hat{s}_1(t), \cdots, \hat{s}_J(t)]^T$ is the estimated version of the original source signals $\mathbf{s}(t) = [s_1(t), \cdots, s_J(t)]^T$ up to amplitude scaling and permutation.

7. Computer Simulations

In this section, we illustrate the performance of our proposed algorithm. We consider 3 source signals (see Fig. 1 (a)). The source signals are mixed with a randomly generated mixing matrix

$$A = \begin{bmatrix} -1.0636 & 0.6427 & 0.2452 \\ 0.0737 & 0.3710 & 1.8902 \\ 1.9208 & -0.1454 & -0.4610 \end{bmatrix}.$$

The mixtures of source signals are shown in Fig. 1 (b). According to the definition of covariance matrix, we have

$$R_1 R_2^{-1} = A(D_1 D_2^{-1})A^{-1},$$

Where $D_1 D_2^{-1} = diag\{d_{11}/d_{21}, \cdots, d_{1J}/d_{2J}\}$

Is a diagonal matrix. It is obvious that the column vectors of A are the eigenvectors of $R_1 R_2^{-1}$. Hence, F can be roughly estimated as

$$\breve{F} = V^T (\breve{A}^T)^{-1},$$

where $\breve{A}$ is a matrix whose columns are eigenvectors of $R_1 R_2^{-1}$. In computer simulations, the initial value of $\hat{f}_1$ in the update rule of Eq. 9 and the initial value of z in the update rule of Eq. 11 are chosen from the columns of $\breve{F}$. The recovered source signals are shown in Fig. 1 (c).

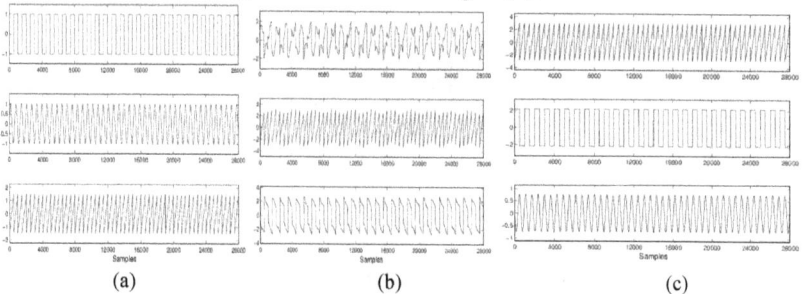

(a)	(b)	(c)

Figure 1. (a) Original source signals, (b) Mixtures of source signals, (c) Recovered source signals.

8. Conclusion

We have proposed a new two-step-type blind source separation algorithm for the overdetermined mixtures case, which consists of blind identification step and source estimation step. A novel sequential blind identification procedure, which is based on the gradient descent approach and GEVD method, is presented. Simulation results show that the proposed algorithm yields good separation performance. It is worth mentioning that the identification of the mixing matrix is important and indispensable in a large number of UBSS algorithms. With this regard, future investigation may concern the application of the proposed algorithm to the problem of UBSS.

Acknowledgement

This work was funded in part by the Humanities and Social Science Project of Ministry of Education of China under Grant 13YJCZH250, the National Natural Science Foundation of China under Grant 61471169, the Planned Science and Technology Project of Hunan Province of China under Grant 2013GK2013, the Scientific Research Project of Hunan Provincial Department of Education under Grant 13C278, the Social Science Project of Hunan Province of China under Grant 14YBA147, and the Open Research Foundation of Hunan Provincial Key Laboratory of Network Investigational Technology under Grant 2016WLZC008.

References

1. S. Saito, K. Oishi, and T. Furukawa, "Convolutive blind source separation using an iterative least-squares algorithm for non-orthogonal approximate joint diagonalization," IEEE/ACM Transactions on Audio, Speech, and Language Processing, vol. 23, no. 12, pp. 2434–2448, Dec. 2015.

2. M. J. Zhang, "Blind separation of mixed-kurtosis signals using soft switch activation function," Electronics Letters, vol. 43, no. 17, pp. 954–955, August 2007.

3. J. Cho and C. D. Yoo, "Underdetermined convolutive BSS: Bayes risk minimization based on a mixture of super-gaussian posterior approximation," IEEE/ACM Transactions on Audio, Speech, and Language Processing, vol. 23, no. 5, pp. 828–839, May 2015.

4. S. Cruces, "Bounded component analysis of noisy underdetermined and overdetermined mixtures," IEEE Transactions on Signal Processing, vol. 63, no. 9, pp. 2279–2294, May 2015.

5. L. D. Lathauwer and J. Castaing, "Blind identification of underdetermined mixtures by simultaneous matrix diagonalization," IEEE Transactions on Signal Processing, vol. 56, no. 3, pp. 1096–1105, March 2008.

6. V. Zarzoso and P. Comon, "Robust independent component analysis by iterative maximization of the kurtosis contrast with algebraic optimal step size," IEEE Transactions on Neural Networks, vol. 21, no. 2, pp. 248–261, Feb. 2010.

7. E. Ollila, "The deflation-based FastICA estimator: Statistical analysis revisited," IEEE Transactions on Signal Processing, vol. 58, no. 3, pp. 1527–1541, March 2010.

8. M. Zhang, S. Yu, and G. Wei, "Sequential blind identification of underdetermined mixtures using a novel deflation scheme," IEEE Transactions on Neural Networks and Learning Systems, vol. 24, no. 9, pp. 1503–1509, Sep. 2013.

9. A. Karfoul, L. Albera, and G. Birot, "Blind underdetermined mixture identification by joint canonical decomposition of HO cumulants," IEEE Transactions on Signal Processing, vol. 58, no. 2, pp. 638–649, Feb. 2010.

Chapter 7

Actuating and Sensing

Cutting Force Prediction Based on the J-C Constitutive Equation in High Speed Cutting

H. J. Pei*, L. F. Chen, G. A. Li, K. P. Fu, S. F. Chen and G. C. Wang
Institute of Precision Engineering,
Jiangsu University, Zhenjiang, P. R. China
*Email: hjpei@ujs.edu.cn

The steady strain rate, velocity, strain and stress fields in the first deformation zone of orthogonal cutting are established based on the J-C constitutive equation and cutting models of Tounsi and Oxley, while the steady temperature field in the first zone is described using the temperature field model of Komanduri. Utilizing the interdependent relationship between the force and temperature, the predicted value of cutting force can be gained by calculating circularly and controlling convergence errors. Finally, comparing with the experimental results, it is found that the model is applicable to high speed and little cutting thickness conditions, and the shear zone thickness is approximately half of the cutting thickness and decreases with the increment of cutting speed.

Keywords: J-C constitutive equation, Orthogonal cutting, Cutting model, Temperature field model.

1. Introduction

Cutting force, cutting temperature and other parameters in a cutting process are extremely important for the design and selection of machine tool and fixture, the prediction of cutting tool life and the selection of cutting thickness. The cutting force and cutting temperature have been deeply studied by lots of domestic and foreign scholars. For example, cutting equation was deduced by M. E. Merchant using minimum energy principle; slip line field theory was introduced by Lee and Shaffer and M. C. Shaw. However, the workpiece was ideally dealt with as rigid or plastic. Oxley [1] considered the shear zone as a parallel and narrow area, took

---

* Corresponding author.

the strain rate as well as the work hardening into account, which made the model more fit the reality. However, material constitutive equation wasn't led in to describe the effects of strain, strain rate and temperature on flow stress, which was considered as a linear function of strain by Oxley. The results show that flow stress is profound functions of strain in a certain condition of strain rate and temperature [2], and the strain and the temperature have a greater influence on the flow stress than does the strain rate which can be neglected [3]. The shear plane is not located in the middle of the shear zone but closer to the upper boundary of the shear zone at the same time [4]. The paper predicts the cutting force based on the Oxley model by introducing the material constitutive equation to establish cutting model, combining with the temperature model of Komanduri and writing computer programs by MATLAB.

2. The Building of Cutting Model and Temperature Field in the First Deformation Zone

Experience shows that deformation in continuous cutting occurs in some narrow area, which can be represented as a localized shear line and as first approximation, this line can be considered straight. The workpiece material passes through the initial shear slip line CD to enter the shear zone and is removed in the form of chips when it goes across the last shear slip line EF, as shown in Fig. 1. Take any mass point M on CD line, then take an infinitesimal element around M and study the time variation of strain rate, velocity, strain and stress in the shear zone. They are only a function of y coordinate (y coordinate corresponds to time). Strain rate field, velocity field, strain and stress fields in the first deformation zone can be obtained by solving these parameters, which can obtain the cutting force.

3. The Building of Shear Strain Field Along Slip Line Direction in the First Deformation Zone

Shear strain field along slip line direction in the first deformation zone can be expressed as [1]:

$$\dot{\gamma} = \frac{V_c \cos \gamma_o}{h \cos(\phi - \gamma_o)} \qquad 0 \le y \le h \ . \tag{1}$$

908

Where Vc is cutting speed, γ0 is rake angle, $\dot{\gamma}$ is the shear strain rate along the x direction, h is the shear zone of infinitesimal thickness and φ is shear angle.

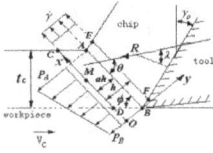

Figure 1. Orthogonal cutting model

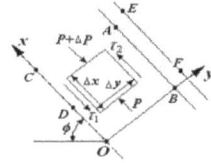

Figure 2. Stress state of infinitesimal element

4. The Building of Speed Field in the First Deformation Zone

Speed u and v can be expressed as [5, 6]:

$$u = \frac{V_c \cos \gamma_0}{h \cos(\phi - \gamma_0)} y - V_c \cos \phi \qquad 0 \leq y \leq h$$

$$v = V_c \sin \phi \qquad 0 \leq y \leq h$$

(2)

Where u is the velocity component for M along x direction and v is the velocity component for M along y direction.

5. The Building of Shear Strain Field in the First Deformation Zone

$$\gamma = \frac{\cos \gamma_0}{h \sin \phi \cos(\phi - \gamma_0)} y \qquad 0 \leq y \leq h \qquad (3)$$

Where γ is the shear strain.

6. The Building of Stress Field in the First Deformation Zone

The J-C model can be expressed as [7]:

$$\tau = [A + B(\frac{\gamma}{\sqrt{3}})^N][1 + C \ln(\frac{\dot{\gamma}}{\sqrt{3}})][D - E(\frac{T - T_{room}}{T_{melt} - T_{room}})^M] / \sqrt{3} \qquad (4)$$

Where A, B, C, D, E, N and M are constants that reflect the material properties, τ is shearing stress along x direction, and Tmelt, Troom and T are melting temperature, room temperature and cutting temperature.

909

Meanwhile, the normal stress of the top point A on the shear plane AB is [1]:

$$P_A = \tau_{AB}[1 + \pi/2 - 2\phi], \quad P_B = P_A - \frac{t_c}{\sin\phi}\frac{\Delta P}{\Delta x}\Big|_{y=ah} \quad (5)$$

Where τAB is shear stress on AB plane, PB is the normal stress of the point B, tc is cutting depth, a is the ratio of the distance between AB and CD to the thickness of the shear zone and $\frac{\Delta P}{\Delta x}$ is the normal stress change rate along the x direction on the slip plane.

$$F_c = \frac{t_c w \tau_{AB}\cos(\lambda - \gamma_0)}{\sin\phi\cos\theta}, \quad F_t = \frac{t_c w \tau_{AB}\sin(\lambda - \gamma_0)}{\sin\phi\cos\theta} \quad (6)$$

Where Fc is main cutting force, Ft is the resistance of cutting depth, w is cutting width, θ is angle between cutting force R and AB and λ is the average friction angle on the contact surface of tool and chip.

7. The Building of Temperature Field in the First Deformation Zone

The temperature of any point on shear surface AB can be obtained using the temperature field model by Komanduri:

$$T_{AB}(x) = \frac{q_S}{2\pi\lambda_W}\int_0^S K_0\left(\frac{V_c R}{2a_w}\right) + K_0\left(\frac{V_c R'}{2a_w}\right)]\exp\left[\frac{-(x + l_i\cos\phi)V_c}{2a_w}\right]dl_i + T_{room} \quad (7)$$

Where qs is the average heat flux on the shear plane AB, s is the length of the shear plane AB, λw and aw are the thermal conductivity and thermal diffusivity of workpiece material, R is the distance between the unsolved point x and shear plane, R' is the distance between the unsolved point x and the micro-segment on the mirror plane, and K0 is the zero-order modified Bessel function.

8. The Solution of the Cutting Model and the Temperature Field Model

The cutting model requires the average temperature on the shear plane and the temperature field model requires cutting force. As a result, an average surface temperature is assumed, according to which the calculated value of the average surface temperature is obtained by substituting the cutting force into the temperature field model. The solution finishes when the difference between the two temperatures is less than the set value, otherwise substitute the average value of the two temperatures into cutting model, substitute the results into the temperature model, compare and calculate until the difference is less than the set value.

It was found that the value of tc/ (h sinφ) varies from 6 to 12[1]. Based on this, the range of h under different cutting conditions is determined depending on cutting parameters, tool geometry, the basic properties of the workpiece material, chip thickness and shear angle measurements. Fc/Ft is obtained using the previous cutting model and the temperature model. Change h until the theoretical value of Fc/Ft is equal to the experimental one. Then, the prediction of cutting force can be obtained.

9. Identification and Analysis of Experiments

The SB-CNC ultra-precision machining centre, manufactured by Spinner, was used. The workpiece material was AISI P20 tool steel and tool was the uncoated carbide inserts (WC). Table 1 shows the cutting parameters and cutting force measurements (tch in Table 1 means the chip thickness); Table 2 shows the predictive values of cutting force, its error and h/tc.

Adjusting the value of h, the theoretical value of Fc/Ft can be made equal to the experimental value. However, the value is in the range only under the experimental conditions of four groups (1,5,8,9).

Table 1. Cutting variables and the experimental results of cutting forces [w =3mm]

Number	1	2	3	4	5	6	7	8	9
V_c [m/min]	200	200	200	200	300	300	300	550	550
t_c [mm]	0.025	0.051	0.075	0.1	0.025	0.051	0.075	0.025	0.051
t_{ch} [mm]	0.127	0.155	0.168	0.203	0.106	0.121	0.162	0.104	0.146
F_c [N]	389	518	658	812	330	488	597	288	488
F_t [N]	418	465	513	565	333	393	417	252	369

Table 2 shows, the higher cutting speed is, the smaller the shear zone thickness gets when other conditions unchanged. While, h/tc value is maintained at about 0.5, which is consistent with the conclusion of the literature [4].

Table 2. Predicted values and errors of cutting forces

Number	1	5	8	9
F_c [N]	356	309	286	446
F_t [N]	383	311	251	337
F_c error[%]	-9.27	-6.80	-0.70	-9.42
F_t error[%]	-9.14	-7.07	-0.40	-9.50
H / t_c	0.59	0.59	0.41	0.48

Table 2 shows that the model is suitable for high-speed, small chip thickness of the cutting conditions. Because of small chip thickness, the chip becomes very soft. Although the chip thickness increases in condition 9, the cutting speed also increases, which is bound to improve cutting temperature and thus play a role in softening chips and weakening its impact.

It is found from the calculation error of cutting force in Table 2 that the theoretical value of cutting force is always smaller than the experimental value. There may be two reasons:

1) When calculating the temperature field, ignoring the heat exchange between the workpiece and the environment, so that the average temperature on shear surface is higher and the theoretical cutting force is smaller.

2) The tool flank will have a "plow effect" on the machined surface in the actual cutting and this model does not consider this effect making the theoretical value smaller.

10. Conclusion

The paper based on the Oxley model introduced the material constitutive equations and other assumptions, established a cutting model and predicted cutting force with the temperature field model of Komanduri. The model is suitable for high-speed and small chip thickness of the cutting conditions. The error of cutting force between estimated value and the experimental value is within 10%. The calculations also confirm that the higher cutting speed is, the smaller the thickness of the shear zone gets when other conditions unchanged, and the ratio of shear zone thickness to cutting thickness maintains about 0.5.

Acknowledgment

This study was funded by the National Science and Technology Major Project of China (No. 2013ZX04009031), Natural Science Foundation of the P. R. C (No. 51075192), and Science and Technology Innovation Fund of Jiangsu Province (No. BC2014202).

References

1. Oxley P L B, Introducing strain rate dependent work material properties into the analysis of orthogonal cutting, Annals of the CIRP. 13 (1966) 127-138.
2. Tuğrul Özel and Erol Zeren, Determination of work material flow stress and friction for FEA of machining using orthogonal cutting tests, Journal of Materials Processing Technology. 153-154 (2004) 1019-1025.
3. Imed Zaghbani and Victor Songmene, A force-temperature model including a constitutive law for Dry High Speed Milling of aluminium alloys, Journal of Materials Processing Technology. 209 (2009) 2532-2544.
4. Tounsi N, Vincenti J, and Otho A, from the basic mechanics of orthogonal metal cutting toward the identification of the constitutive equation, International Journal of Machine Tools and Manufacture. 42 (2002) 1373-1383.
5. PJ. Arrazola, T.Özel, D. Umbrello, M. Davies, and IS. Jawahir1, Recent advances in modelling of metal machining processes, CIRP Annals - Manufacturing Technology. 62(2013) 695-718.
6. B. Li, X. Wang, Y. Hu, and C. Li, "A new analytical method based on unequal division shear zone model in orthogonal cutting," International Journal of Mechanical Sciences. 54(2011) 431-443.
7. B. Wang, and Z. Liu, Investigations on the chip formation mechanism and shear localization sensitivity of high-speed machining Ti6Al4V," International Journal of Advanced Manufacturing Technology. 75(2014) 1065–1076.

Study of Nuclear Explosion Infrasound Detection Based on Spherical Geometry

Hui-xing Chen[1,2,*], Yan Ren[3], Xin-liang Pang[1] and Yi Zheng[1]
[1]State Key Laboratory of NBC Protection, Beijing, China
[2]Xichang Satellite Launch Center, Sichuan Province, China
[3]Spaceflight Project Command, Beijing, China
*Email: chx0808@sina.com

Traditional location technology research on nuclear explosion infrasound detection is based on plane geometry. Taking into account the earth's sphere and ellipsoid characteristics, the traditional method inevitably exists the disadvantages such as low location accuracy, strong disturbance and limited location distance. Under the prerequisite that earth is a sphere, this paper focuses on the calculation of the distance between the infrasound source and the measurement array using spherical geometry theorem, and solves the infrasound source location (latitude and longitude) problem with the known relative azimuth and distance. Simulation comparison results show that, compared with the traditional method, the proposed infrasound detection and location method based on the spherical geometry has advantages in location accuracy and range.

Keywords: Nuclear explosion infrasound, Spherical geometry, Plane geometry, Location accuracy.

1. Introduction

Due to the absorption effect caused by atmosphere, the energy of the acoustic wave in a higher frequency decays faster, and its propagation distance is shorter than that of the acoustic wave in a lower frequency. Such as, for atmospheric absorption, 1000 Hz sound wave in the propagation of 7 kilometers will lose 90% of the energy. However, 20 Hz sound waves in the spread of 100 kilometers will also lose about 90% of the energy. Furthermore, atmospheric absorption is relatively slow to the decay of the sound waves with the frequency below 20Hz (commonly known as infrasonic waves). Such as, 1 Hz infrasound in the propagation of 3000 kilometers loses 90% of the energy, and 0.01 Hz infrasound will lose 90% of the energy after its rotation around the earth in a week [1-3]. Therefore, the detection and recognition for long-range nuclear event can be achieved by associated infrasound analysis.

According to the formation mechanism and propagation characteristics of infrasound, the infrasound detection technology may be the only monitoring mean, which can be used for the nuclear explosion detection with underground, the lower atmosphere and high altitude. In other words, the infrasound detection of nuclear explosion can be regarded as all-rounder, and the only one can be compared with the space-based nuclear explosion detection. Therefore, the study about infrasound detection and location of nuclear explosion has very important military value and significance.

In the current research of infrasound positioning technology about nuclear explosion, the most results are carried out from plane geometry. This method has the benefits with small calculating quantity and simple structure, and its detection performance is better in short distance. But, for the long-range nuclear explosion detection, commonly used formula and theorem in plane geometry bring no fruits because the earth is an ellipsoid, and the theorem in spherical geometry must be discussed.

2. Location Technology

The infrasound location technology of nuclear explosion detection resolve the position problem of explosion center. The technology can be disassembled into three sub problems, direction determination, distance measurement and location determination. Direction determination refers to the problem where the infrasound signal is from, and this problem can be regarded as azimuthal angle (True north point) problem in spherical geometry. Distance measurement is determining the distance between infrasound source (nuclear explosion center) and measurement station, and this problem can be presented to solve the length of the shortest arc between center and station in spherical geometry. Location determination is determining the position of explosion center, and it can be presented to solve the longitude and latitude coordinates of center position in spherical geometry. The subsequent chapters will discuss the above three issues.

3. Direction Determination Analysis

Single infrasound detector can only measure the pressure changes of the ambient gas, and the result cannot determine which direction the signal is from. Furthermore, the data recorded by a single detector can't be used to judge whether pressure changes are from the ambient gas disturbance or from a long-range signal. Therefore, single infrasound detector obviously can't satisfy the need of orientation function. Only through the establishment of infrasound measurement array, the azimuthal angle of nuclear explosion can be detected and determined. In addition, the measurement technology based on infrasound measurement array can improve the signal-to-noise ratio and the localization ability.

In order to simplify the calculation, this paper selects a ternary array as the base of the measuring table, as shown in Fig. 1. Meanwhile, it is supposed that the distance between nuclear explosion center and measurement array is very far, that is, the distance is more than 10 times that of array maximum baseline length. Under this assumption, infrasound through every array element can be considered as a plane wave, and the phase velocity is coincident.

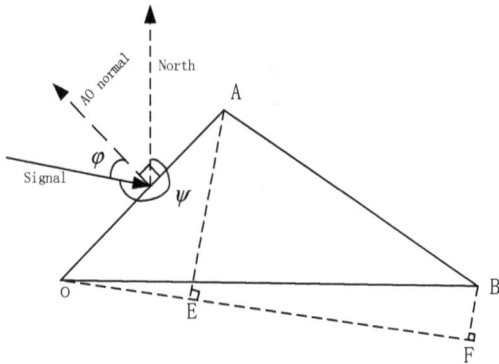

Figure 1. Schematic diagram of the 3-element array

In Fig. 1, point A, O, B, are three measurement elements of an arbitrary ternary array. It is assumed that infrasound signals are from the top left, the angle between the infrasound direction and the normal line of AO linked line is φ, OE and OF are the distances by which the infrasound signal is delivered from O to A and B, the spent time is recorded as τ_{OA} and τ_{OB}. If the baseline length of AOB array is not very long, the phase velocity v can be considered coincident. Then, according to the sine and cosine formula, we have Eq. 1 and Eq. 2.

$$\sin \varphi = \frac{OE}{OA} = \frac{v\tau_{OA}}{OA} \tag{1}$$

$$\cos \angle BOF = \sin(\varphi + \angle AOB) = \frac{v\tau_{OB}}{OB} \tag{2}$$

Combined with Eq. 1 and Eq. 2, the azimuth equation can be deduced by Eq. 3.

$$\varphi = \cot^{-1}\left(\frac{\tau_{OB}}{\tau_{OA}} \cdot \frac{OA}{OB} \cdot (\sin \angle AOB)^{-1} - \cot \angle AOB\right) \tag{3}$$

If the azimuth of the AO normal line is θ', the azimuth ψ of the measured signal source can be calculated by Eq. 4.

$$\psi = \theta' - \varphi - \gamma \tag{4}$$

In Eq. 4, γ is the angle between north direction of array's rectangular coordinate system and meridian direction. Specially, if the north direction of array's rectangular coordinate system coincides with meridian direction, the value of γ is zero and Eq. 4 can be simplified.

Based on the above analysis, if the coordinates and the distance of the ternary array elements are known, combined with time difference measurement of each measurement point, it will be able to determine the azimuth angle of the explosion center position related to the ternary array. That is, the problem of direction determination is solved.

4. Distance Measurement Analysis

According to the analysis of above section, a single ternary array can determine the direction of explosion center position, but it is difficult to realize the distance calculation. Due to the infrasound detection without prior knowledge, the position estimation of explosion center is calculated only by passive received infrasound signals.

Currently, array measurement methods are used widely for direction finding of nuclear explosion infrasound at home and abroad. Typically, the structure of dual ternary array is shown in Fig. 2. In order to improve the measurement accuracy, the array dimension of A1B1C1 and A2B2C2, is not greater than one wavelength, and the distance between two arrays is far enough to ensure that angle difference between two angles measured by each array is large. Meanwhile, infrasound signal can be approximately regarded as plane wave for each array.

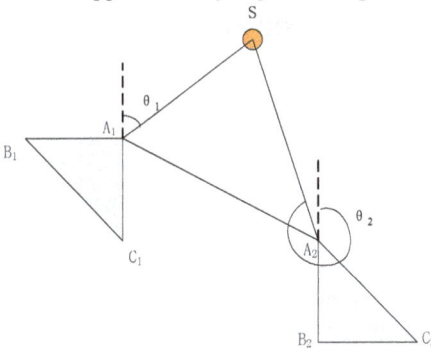

Figure 2. Schematic diagram of dual ternary array

In Fig. 2, it is assumed that azimuth obtained respectively by array A1B1C1 and A2B2C2 are recorded as θ_1 and θ_2. Combined with the position information (longitude and latitude) of point A1 and A2, it is easy to solve angle $\angle SA1A2$ and $\angle SA2A1$. For the convenience of description, we redefine the labels, earth's

917

center recorded for point O, the explosion center for A, original definition A1 for B, A2 for C, angle ∠SA1A2 and ∠SA2A1 for ∠B and ∠C respectively, then the spherical structure based point OABC is shown in Fig. 3.

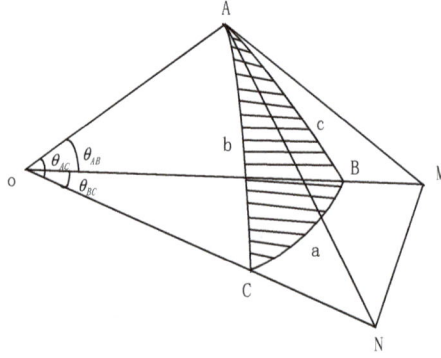

Figure 3. Spherical structure diagram of infrasound location

In Fig. 3, the distance measurement problem of nuclear explosion center is simplified as the calculation problem of the length b or c on the condition that arc a and angles (∠B and ∠C) are known. Due to the nonplanar triangle ABC, original sine and cosine theorems are not applicable and spherical geometry theorem must be adapted to solve this problem.

Based on the two five element formula of spherical geometry, the Eq. 5 and Eq. 6 can be obtained as follows.

$$\sin \angle A \cos \theta_{AC} = \cos \angle B \sin \angle C + \sin \angle B \cos \angle C \cos \theta_{BC} \quad (5)$$

$$\sin \angle A \cos \theta_{AB} = \cos \angle C \sin \angle B + \sin \angle C \cos \angle B \cos \theta_{BC} \quad (6)$$

Taking into account that arc length a is known, the center angle θ_{BC} is known, combined with sine formula of spherical triangle, the Eq. 7 can be promoted.

$$\frac{\sin \theta_{BC}}{\sin \angle A} = \frac{\sin \theta_{AC}}{\sin \angle B} \quad (7)$$

With the Eq. 5 and Eq. 7, Eq. 8 can be obtained.

$$\cot \theta_{AC} = \frac{\cos \angle B \sin \angle C + \sin \angle B \cos \angle C \cos \theta_{BC}}{\sin \angle B \sin \theta_{BC}} \quad (8)$$

According to the relationship between arc length and central angle, the radius of the earth is set to R (Known), the arc length is denoted as b. Then, the Eq. 9 can be obtained as follows.

$$b = R\theta_{AC} = R \cdot \cot^{-1}\left(\frac{\cos\angle B\sin\angle C + \sin\angle B\cos\angle C\cos\theta_{BC}}{\sin\angle B\sin\theta_{BC}}\right) \quad (9)$$

Based on the above analysis, the position of explosion center can be determined by dual ternary array, and the distance from center to array can be calculated according to Eq. 9.

5. Location Analysis

According to the analysis of above two sections, the structure of dual ternary arrays has been able to determine the location of nuclear explosion center [4-6]. But in practical applications, it is often required to be described in the form of latitude and longitude. Therefore, the location problem can be described as a mathematic problem. With known longitude latitude coordinates of the measurement point, and relative distance and azimuth angle between the infrasound source location and the measurement point, how to solve the latitude and longitude of the infrasound source location?

As shown in Fig. 3, X is for longitude coordinate, Y is for latitude coordinate. It is assumed that the coordinates of measurement point C is (XC, YC), explosion center is for B, arc length of BC is a and the azimuth angle of arc BC are known, then the latitude and longitude of the infrasound source location B (XB, YB) can be obtained by introduction of special point A (XC, 90°) [7].

Because point A and point C are at the same meridian, angle $\angle C$ is the direction angle of arc BC. Taking into Eq. 10, the value of central angle θ_{AB} can be calculated by Eq. 11 based on the cosine formula of spherical triangle. Furthermore, angle $\angle A$ can be calculated, and this value is the longitude difference between point A and B [8].

$$\theta_{AC} = \frac{\pi(90 - Y_C)}{180}, \theta_{AB} = \frac{\pi(90 - Y_B)}{180}, \theta_{BC} = \frac{a}{R} \quad (10)$$

$$\cos\theta_{AB} = \cos\theta_{AC}\cos\theta_{BC} + \sin\theta_{BC}\sin\theta_{AC}\cos C \quad (11)$$

The latitude coordinate of the point B is calculated by Eq. 12.

$$Y_B = 90 - \frac{180}{\pi}\cos^{-1}(\cos\theta_{AC}\cos\theta_{BC} + \sin\theta_{BC}\sin\theta_{AC}\cos C) \quad (12)$$

919

The longitude of the point B is calculated by Eq. 13.

$$X_B = X_C + \frac{180}{\pi}\cos^{-1}\left(\frac{cos\theta_{BC}-cos\theta_{AC}cos\theta_{AB}}{sin\theta_{AC}sin\theta_{AB}}\right)$$

(13)

Eq. 12 and Eq. 13 are the solution formula for the latitude and longitude coordinates of explosion center B.

6. Verification and analysis

In order to verify the method provided above, three measurement points are selected on the Google Earth Map, A (116°7'4.77", 40°7' 57.34"), B (116°7'35.53", 40°8'32.36") and C (116°7'45.84", 40°7'55.72"). Respectively, one point is selected as the infrasound source from point C, Guangzhou, Chengdu and Pyongyang. As known points, point A is the measurement point as well as point B. The location data is shown in Table 1.

Together the presentation from location technology, the accuracy of the algorithm is in the order of 10-3, and it is at a competitive advantage in long-distance event location demand. Furthermore, the calculation accuracy is limited to the azimuth, distance calculation accuracy and timing precision, and it is also affected by infrasound propagation characteristics, monsoon, topography and other factors.

Table 1. Calculation results

Items	Google Azimuth	Google Distance	Latitude and Longitude from Calculation	Latitude and longitude from Google Map	Deviation Accuracy
C measured by A	92.58°	971.6m	L:116°7' 45.83" W:40°7' 55.92"	L:116°7' 45.84" W:40°7' 55.72"	6.3m 6.5‰
C measured by B	167.84°	1155.3m	L:116°7' 45.82" W:40°7' 55.84"	L:116°7' 45.84" W:40°7' 55.72"	3.6m 3.1‰
Guangzhou measured by A	188.10°	1844342m	L:113°36' 11.11" W:23°42' 2.21"	L:113°35' 47.62" W:23°38' 25.10"	6746m 3.7‰
Chengdu measured by A	228.63°	1846995m	L:102°02' 14.64" W:28°15' 25.71"	L:102°01' 36.14" W:28°14' 36.26"	1854m 1‰
Pyongyang measured by A	95.22°	837836m	L:125°47' 10.94" W:39°02' 32.41"	L:125°46' 11.71" W:39°02' 59.78"	1655m 2‰

7. Conclusion

In order to solve the location problem of nuclear explosion infrasound detection, this paper puts forward a complete algorithm from the angle of spherical geometry, and the algorithm consists of direction determination, distance measurement and latitude longitude location composed of three parts. Combined with existing infrasound measurement array, test process of the infrasound location algorithm is analyzed by MATLAB simulation. The test results show that the accuracy of the proposed algorithm can be in the order of 10-3, and it has certain advantages in the long-distance event location. However, location accuracy of the algorithm is limited to azimuth, distance calculation accuracy and timing precision. On the other hand, from the view of infrasound propagation characteristics, the accuracy is susceptible to the influence of monsoon, topography and other environmental factors. The follow-up studies will focus on considering these limiting factors.

References

1. Zhong-shan Zhang et al, Nuclear Explosion Detection, third ed., National Defense Industry Press, Beijing, 2006 (In Chinese).
2. Hu-hu Chen, Earthquake, Atmosphere and Surface of the Explosion at the same time to Stimulate the Seismic Waves and Infrasound, J. Geophysical Research and Practice in Western China, (2010) 5-8 (In Chinese).
3. Fang Su, Wei Tian, Measurements and Study of the Precursory Infrasonic Waves of Strong Earthquakes from July to September 2000, J. Seismological Research, (2002)11-19 (In Chinese).
4. Ya-qin Xia, Xiao-yan Cui, Characteristic Analysis of Abnormal Infrasonic Signals, J. Chinese Catastrophe Comprehensive Prediction and Environmental Variation, (2010) 34-36 (In Chinese).
5. Xiao-yan Cui, Relationship between the Abnormal Infrasonic Signals and Earthquakes, Master's Thesis of Beijing Technology University, Beijing, 2011, pp.20-24 (In Chinese).
6. Wei Wang, Direction and Recognition Study of Infrasonic Signal in Earthquake Prediction, Master's Thesis of Beijing Technology University, Beijing, 2009, pp.15-17 (In Chinese).

7. Lin Lin, Atmospheric Low Frequency Infrasonic Observation Based on Wide Area Network Measurement Sensor Array, Master's degree thesis of Shandong Science and Technology University, Shangdong, 2010, pp.15-23 (In Chinese).

8. A. G. Milton, Infrasonic Signals Generated by Volcanic Eruptions, IEEE Internatiolal Geoscience & Remote Sensing Symposium, 2000, pp.1189-1191.

Components Reconstruction of Fluid Catalytic Cracking Products Based on Maximum Information Entropy Method

Yong Li and Ji-zheng Chu*

No. 15, Beisanhuan East Road, Chaoyang District, Beijing 100029, China
Email: chujz@mail.buct.edu.cn

A new method of Fluid Catalytic Cracking (FCC) products expression based on special pseudo-components (SPCs) is proposed in this article. The Maximum Information Entropy Method (MIEM) combined with Gaussian distribution is taken into account to solve the problem of the overlaps among components when describing products in previous FCC models based on SPCs. The successful application on the commercial FCC model constructed by Zhang et al. proves the validity of the new method.

Keywords: Fluid catalytic cracking, Special pseudo-components, Products expression, Maximum information entropy method, Gaussian distribution.

1. Introduction

FCC is the most important technology to produce liquefied petroleum gas, gasoline, diesel and propylene from heavy hydrocarbon fractions. Modeling and simulation of FCC processes have been being an important topic of research since about 1940's [1].

FCC riser models are mainly divided into three categories: lumped model [2–5], carbon centers and single-event kinetics model [6–8] and pseudo-components or SPCs model [9–11]. It is noted that the characterization methods of products in all three categories, no matter in accordance with boiling points or another common characteristic variables, lead to clear distinct boundaries among them. Thus the products obtained from these models fail to meet the qualities of the corresponding industrial output.

---

*Corresponding author.

The most challenging problem in characterizing products is the existence of the mix of several components. In other words, there are no clear boundaries among products in fact. MIEM is based on readily available information of the mixture and has got some success on reconstruction of a detailed molecular composition of relatively simple mixtures [12, 13]. According to the principle of MIEM, if only partial information concerning the possible outcomes is available, the most likely composition for reconstruction is the one to locate the maximization of the information entropy [14–16]. Given the above, using MIEM to characterize the FCC products in order to optimize SPCs modeling thoughts for FCC has been a good choice.

The rest of this paper is organized as follows: in section 2, the characterization procedure based on SPCs is established. In section 3, the reconstruction method of MIEM combined with Gaussian distribution is introduced in detail. Finally, the method is applied to Zhang's model in section 4 and the comparison results prove the validity of it. Section 6 is the conclusion.

2. Division of SPCs and the Analysis of Previous Characterization of Products

SPCs proposed by the previous study of our research group [10] appear in pairs. The two SPCs in each pair have the same normal boiling temperature or normal boiling temperature range, but different Watson characterization factors (Kw, L and Kw, H). With the normal boiling temperature and Watson characterization factor known, the properties of a SPC such as density, molecular weight, combustion heat, carbon-to-hydrogen weight ratio and liquid and vapor heat capacities can be fully determined by well-established correlations.

SPCs are distributed by their normal boiling temperatures in a range covering the true boiling point (TBP) distillation temperature ranges of all possible stock oils, and the normal boiling temperature interval between neighboring pairs of SPCs is suggested to be 20~30 °C.

With the boiling temperature ranges of the defined SPCs, the TBP distillation recovery curves of all the stock oils are cut. Distillate cut or pseudo-component i of stock oil j contribute to the contents of the i-th pair of SPCs of the same boiling temperature range in the following

$$\Delta w_{L,ij} + \Delta w_{H,ij} = w_{ij}. \tag{1}$$

$$\frac{\Delta w_{L,ij}}{\rho_{L,i}} + \frac{\Delta w_{H,ij}}{\rho_{H,i}} = \frac{w_{ij}}{\rho_{ij}} \tag{2}$$

where w is the concentration of a PC or SPC among all the PCs or SPCs in weight fraction, Δw the increment of concentration, and ρ the density in kg/m3, respectively. The concentrations of SPCs in the feed oil as a whole are calculated as

$$w_{L,i} = \sum_j \Delta w_{L,ij} \tag{3}$$

$$w_{H,i} = \sum_j \Delta w_{H,ij} \tag{4}$$

The above characterization procedure as detailed by [10] also includes light definite components hydrogen, methane, ethylene, ethane, propylene, propane, butene, butane and pentane as cracking products. Fig. 1 shows the narrow cuts of feedstock oil TBP distillation curve used to get SPCs.

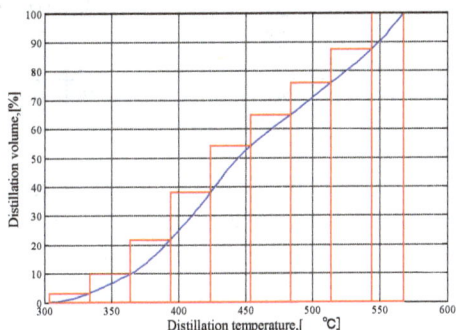

Figure 1. The narrow cuts of a feedstock oil TBP distillation curve

Table 1. Basic properties of the 30 °C spaced special pseudo-components with Kw,
L=12.6 Kw, H=10

i	$T_{B,j}$	$T_{B,j+1}$	$T_{b,j}$	$\rho_{L,j}$	$\rho_{H,j}$	$M_{L,j}$	$M_{H,j}$	$C/H_{L,j}$	$C/H_{H,j}$
10	305.22	333.15	319.185	659.7	831.3	91.361	83.318	4.839	7.488
11	333.15	363.15	348.15	679.1	855.7	102.889	92.391	4.936	7.723
12	363.15	393.15	378.15	698.1	879.6	116.127	102.711	5.037	7.967
13	393.15	423.15	408.15	716.1	902.3	130.845	114.078	5.138	8.211
14	423.15	453.15	438.15	733.2	923.9	147.224	126.614	5.239	8.455
15	453.15	483.15	468.15	749.6	944.5	165.462	140.453	5.34	8.699
16	483.15	513.15	498.15	765.3	964.2	185.781	155.739	5.441	8.943
17	513.15	543.15	528.15	780.3	983.2	208.427	172.635	5.542	9.186
18	543.15	573.15	558.15	794.8	1001.5	233.673	191.317	5.643	9.43
19	573.15	603.15	588.15	808.8	1019.1	261.823	211.984	5.744	9.674
20	603.15	633.15	618.15	822.4	1036.2	293.216	234.852	5.845	9.918
21	633.15	663.15	648.15	835.4	1052.7	328.232	260.165	5.946	10.162
22	663.15	693.15	678.15	848.1	1068.7	367.291	288.189	6.047	10.406
23	693.15	723.15	708.15	860.5	1084.2	410.865	319.221	6.148	10.65
24	723.15	753.15	738.15	872.5	1099.3	459.478	353.593	6.249	10.894
25	753.15	783.15	768.15	884.1	1114.0	513.716	391.668	6.35	11.138
26	783.15	813.15	798.15	895.5	1128.3	574.232	433.854	6.451	11.382
27	813.15	843.15	828.15	906.6	1142.3	641.755	480.600	6.552	11.625
28	843.15	873.15	858.15	917.4	1155.9	717.0992	532.407	6.653	11.869
29	873.15	903.15	888.15	928.0	1169.2	801.173	589.830	6.754	12.113
30	903.15	933.15	918.15	938.3	1182.3	894.989	653.486	6.856	12.357

Table 1 lists SPCs acquired by the cutting temperatures set to be $30\,^\circ C$ as a characterization example while Fig. 2 shows the products' characterization using boiling temperatures and the actual division for the products of a commercial fluid catalytic cracking unit in [10]. It can be seen that the products' ranges are roughly similar in 2008 and 2009 while there's a big difference between model characterization and the actual division.

Componets division in Zhang et al.(2007)	Product range in 2008	Product range in 2009		GSL

Figure top legend (right column): GSL, Mixtures of GSL & LCO, LCO, Mixtures of LCO & RO, RO, Mixtures of LCO & RO & RES, Mixtures of RO & RES, RES

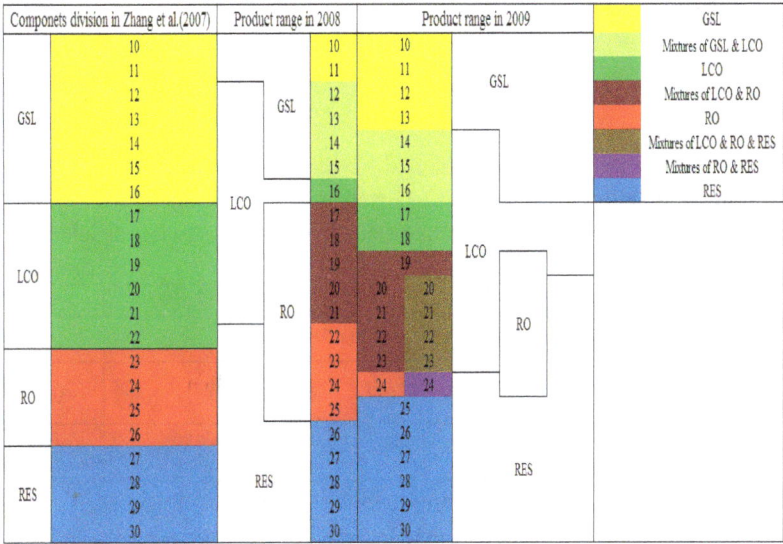

Figure 2. Products' characterization and the actual division in [10]

3. MIEM combined with Gaussian distribution

The core of MIEM is the Shannon's information entropy. According to MIEM, the way to obtain the most likely composition is to maximize the uncertainty on the missing information, which is represented by the maximum Shannon's information entropy defined in Eq.5.

$$S(\pi_i) = -\sum_{i=1}^{N} \pi_i \ln \pi_i \quad Subject\ to: \sum_{i=1}^{N} \pi_i = 1 \ . \tag{5}$$

in which S represents Shannon's entropy and π_i is the probability of a certain state.

Eq.6 gives a modified maximizing equation where the probabilities π_i are replaced by mole fractions x_i of the SPCs in the mixture.

$$MAX \qquad S(x_i) = -\sum_{i=1}^{N} x_i \ln x_i \ , \tag{6}$$

$$\text{Subject to}: f_j - \sum_{i=1}^{N} f_{i,j} x_i = 0, \quad j = 1, ..., J$$

$$0 \le x_i \le 1 \text{ and } \sum_{i=1}^{N} x_i = 1$$

where f_j is the value of constraint j, $f_{i,j}$ is the coefficient of component i for constraint j, N is the number of SPCs, J is the number of constraints. Fig. 3 gives several commonly used boundary conditions as constraints in this article.

Variable	$f_{i,j}$	f_j
Molecular weight	$M^{\exp} - M_i$	0
C/H	$\left((C/H)^{\exp} - (C/H)_i \right) M_i$	0
Specific density	$\left(1/\rho_i - 1/\rho^{\exp} \right) M_i$	0
ASTM boiling point	If boiling point $i <$ boiling point k $1 - \left(\%G_k^{\exp}/100 \right)(M_i/\rho_i)$	0
	If boiling point $i >$ boiling point k $-\left(\%G_k^{\exp}/100 \right)(M_i/\rho_i)$	0

Figure 3. Overview of several commonly constraints used in our study

It is known that Shannon's entropy is a measure of the homogeneity of a probability distribution. Besides, experimental data show that most compositions approach a Gaussian distribution of the mole fractions [13]. Hence we introduce Gaussian distribution as a constraint

$$\sigma_m^2 - \sum_{i=1}^{N} x_i \left(M_i - M^{\exp} \right)^2 = 0 \ . \tag{7}$$

Considering the global optimum of a nonconvex problem as Eq.6, [12,13,16] gave a simple and effective method using Lagrange multipliers to solve the problem. Eq.6 is rewritten as

$$\xi(x_i) = -\sum_{i=1}^{N} x_i \ln x_i + \mu \left(1 - \sum_{i=1}^{N} x_i \right) + \sum_{j=1}^{J} \lambda_j \left(f_j - \sum_{i=1}^{N} x_i f_{i,j} \right) \ .$$
$$+ v \left[\sigma_m^2 - \sum_{i=1}^{N} \left(x_i \left(M_i - M^{\exp} \right)^2 \right) \right] \tag{8}$$

928

The optimum of the function above can be found by setting the derivatives of this equation with respect to the mole fractions xi equal to 0:

$$\frac{\partial \xi}{\partial x_i} = -1 - \ln x_i - \mu - \sum_{j=1}^{J} \lambda_j f_{i,j} - v \left(M_i - M^{\exp} \right)^2 . \tag{9}$$

Re-ordering the terms in Eq.9 leads to the following expression:

$$e^{1+\mu} x_i = \exp \left[-\sum_{j=1}^{J} \lambda_j f_{i,j} - v \left(M_i - M^{\exp} \right)^2 \right] . \tag{10}$$

Sum up all the factors about structural increments and eliminate μ, resulting in:

$$x_i = \frac{\exp \left[-\sum_{j=1}^{J} \lambda_j f_{i,j} - v \left(M_i - M^{\exp} \right)^2 \right]}{\sum_{i=1}^{N} \exp \left[-\sum_{j=1}^{J} \lambda_j f_{i,j} - v \left(M_i - M^{\exp} \right)^2 \right]} . \tag{11}$$

Substitution of Eq.11 in Eq.8 leads to the following expression

$$E(\lambda, v) = \ln \left\{ \sum_{i=1}^{N} \exp \left[-\sum_{j=1}^{J} \lambda_j f_{i,j} - v \left(M_i - M^{\exp} \right)^2 \right] \right\} + \sum_{j=1}^{J} \lambda_j f_j + v \sigma_m^2 . \tag{12}$$

According to Eq.12, the most likely composition can be obtained based on the optimum values of λ_j and v.

4. Reconstruction of FCC products

In this section, production data from a commercial fluid catalytic cracking unit of SINOPEC of China is used to check the actual usability of our method. The steady state model of a FCC riser and the corresponding model solution method as detailed in [10] are also adopted in this paper.

Fig. 4 shows the optimized reconstruction of FCC products while Table 2 shows the simulation and prediction effects of the improved model compared with the previous model.

929

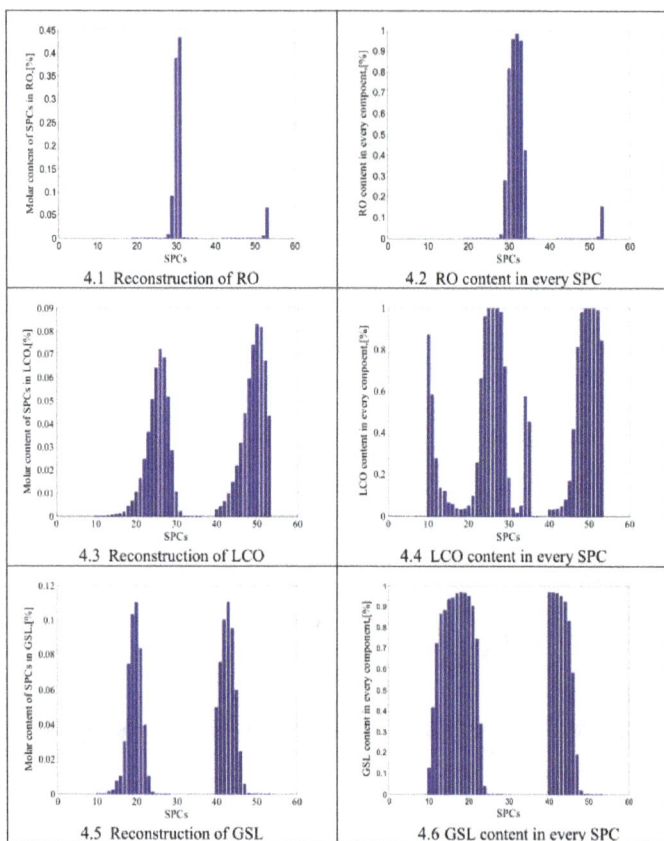

Figure 4. The optimized reconstruction of FCC products

It can be seen from Fig. 4 that the method of MIEM combined with Gaussian distribution has effectively solve the problem of the overlaps among components. Although the distribution of LCO in lighter components in Fig. 4.4 has certain differences with the industrial data, possible physical property calculation errors and the lack of average properties of the mixtures for building restrictions and the comparison results in Table 2 prove the validity of the method.

Table 2. Simulation and prediction using our model compared with Zhang's previous model

Case 1 Production data of 2008, model simulation			
	Industrial data[%]	Previous model[%]	Modified model[%]
RO	2.783	3.358	2.631
LCO	24.518	26.412	23.680
GSL	44.372	43.377	43.524
Case 2 Production data of 2009, model prediction			
	Industrial data[%]	Previous model[%]	Modified model[%]
RO	10.128	8.334	9.803
LCO	22.147	21.556	21.927
GSL	40.851	43.366	41.397

Only RO, LCO and GSL are listed in Table 2, because LPG and GAS are expressed by true components while parameters and structure of the whole model remain the same, which means LPG, GAS and the outlet temperature of the model are fixed. It is obvious that after our method applied to express products, the precision of Zhang's model has been greatly improved.

5. Conclusion

In this paper, the maximum information entropy method combined with Gaussian distribution is successfully used to solve the problem of the overlaps among components when describing products in previous FCC models based on SPCs. The reconstruction of FCC products using properties of the components and their mixtures provides a possible way to study the physicochemical properties of raw oil and the products based on SPCs. Besides, it also becomes feasible to connect a FCC unit model based on SPCs to the main fractionator thus to build a complete model of the FCC system.

Nomenclature

C/H	Carbon-to-hydrogen weight ratio
$f_{i,j}$	Stoichiometric coefficient of molecule i for constraint j variable
f_j	Value of constraint j variable
%G_k	Volume percent of the boiling point k
GSL	Gasoline
Kw	Watson characterization factor

LCO	Light cycle oil/Light diesel oil
M	Molecular weight, kg/kmol
N	Total number of components
RO	Recycle oil
RES	Residue
Tb,i	Boiling temperature of special pseudo-component i, [K]
w	Mass fraction

Subscripts and Superscripts

C	Carbon
H	Hydrogen
H	Heavy end SPC
L	Light end SPC
exp	Experimental /Industrial

Greek letters

μ	Langrange multiplier
λ	Langrange multiplier
ν	Langrange multiplier
ρ	Density, kg/m3
σ_m	Standard deviation of the distribution

References

1. A. Voorhies, Carbon formation in catalytic cracking, Ind. Eng. Chem., 4(1945) 318-322.
2. V.W., Jr. Weekman, D.M. Nace, Kinetics of catalytic cracking selectivity in fixed, moving, and fluid bed reactors, A. I. Ch. E. Journal, 3(1970) 397-404.
3. S.M. Jacob, B. Gross, S.E. Voltz, et al., A lumping and reaction scheme for catalytic cracking, A. I. Ch. E. Journal, 4(1976) 701-713.
4. I.-S. Han, C.-B. Chung, Dynamic modeling and simulation of a fluidized catalytic cracking process. Part I: Process modeling, Chemical Engineering Science, 5(2001a) 1951-1971.
5. I.-S. Han, C.-B. Chung, Dynamic modeling and simulation of a fluidized catalytic cracking process. Part II: Property estimation and simulation, Chemical Engineering Science, 5(2001b) 1973-1990.
6. D.K. Liguras, D.T. Allen, Structural models for catalytic cracking. 1. Model compound reactions, Industrial and Engineering Chemistry Research, 6(1989a) 665-673.

7. D.K. Liguras, D.T. Allen, Structural models for catalytic cracking. 2. Reactions of simulated oil mixtures, Industrial and Engineering Chemistry Research, 6(1989b) 674-683.

8. W. Feng, E. Vynckier, G.F. Froment, Single event kinetics of catalytic cracking, Industrial & Engineering Chemistry Research, 12(1993) 2997-3005.

9. R.K. Gupta, V. Kumar, V.K. Srivastava, A new generic approach for the modeling of fluid catalytic cracking (FCC) riser reactor, Chemical Engineering Science, 62(2005) 4510-4528.

10. J.R. Zhang, Z.Q. Wang, J. Hao, et al., Modeling fluid catalytic cracking risers with special pseudo-components, Chemical Engineering Science, 102(2013) 87-98.

11. H. Sildir, Y. Arkun, U. Canan, et al., Dynamic modeling and optimization of an industrial fluid catalytic cracker, Journal of Process Control, 31(2015) 30-44.

12. D. Hudebine, J.J. Verstraete, Molecular reconstruction of LCO gasoils from overall petroleum analyses, Chemical Engineering Science, 59(2004) 4755-4763.

13. K.M. Geem, D. Hudebine, M.F. Reyniers, et al., Molecular reconstruction of naphtha steam cracking feedstocks based on commercial indices, Computers and Chemical Engineering, 31(2007) 1020-1034.

14. N. Virgo, From maximum entropy to maximum entropy production: a new approach, Entropy, 12(2010) 107-126.

15. L.P. de Oliveira, J.J. Verstraete, M. Kolb, Simulating vacuum residue hydroconversion by means of Monte-Carlo techniques, Catal. Today, 220(2014) 208-220.

16. Y.B. Pan, B.L. Yang, X.W. Zhou, Feedstock molecular reconstruction for secondary reactions of fluid catalytic cracking gasoline by maximum information entropy method, Chemical Engineering Journal, 281(2015) 945-952.

Effect of Two Diffuser Types of Volute on Pressure Fluctuation in Centrifugal Pump Under Part-Load Condition

Fan Meng, Ji Pei[†] and Jia Chen

Nation Research Centre of Pumps, Jiangsu University
Jiangsu, Zhenjiang, 212013, China
[†]Email: jpei@ujs.edu.cn

In this research, two diffuser types of volute were designed, the volute with tangential diffuser and the volute with radial diffuser. Normally, different diffuser types of volute will affect the performance of pump, while in this paper, little effect can be found. Two diffuser types of volute were designed to study the effect on pressure fluctuation features of centrifugal pump under part-load condition, with the same volute design parameters and impeller parameters. The unsteady, three dimensional turbulent flow in the pump was simulated. It shows that the periodic features in pressure fluctuation near its tongue are the same for two diffuser types of volute, however in the diffuser of the volute, the values of the pressure fluctuation in radial diffuser is greater than that in tangential diffuser. The results can provide a useful reference for designing the diffuser of volute in centrifugal pump.

Keywords: Part-load condition; Tangential diffuser; Radial diffuser; Pressure fluctuation.

1. Introduction

Centrifugal pump has two important flow components that are impeller and volute casing basically which together determine the centrifugal pump performance. As an important flow component within a centrifugal pump, volute casing plays an important role of diversion and rise pressure. In order to adapt various installation positions of pipeline, the volute of centrifugal pump needs to be designed into different types.

Nowadays, CFD is rapidly used for studying the pressure fluctuation in pumps. Many researchers have studied the pressure fluctuation in centrifugal pump. Glc M et al. [1] and Kergourlay et al. [2] studied the effect of blade on pump performance and flow field. Wang Wenquan et al. [3] and Germano M [4] investigated the Large-eddy simulation of turbulent flow numerical computation. Longatte F et al research the rotor-stator-circuit interactions in a centrifugal pump [5]. Other researchers reported the numerical simulation on pressure fluctuations such as the work by Kitano Majidi et al. [6], Guo Pengcheng et al. [7] and Zhu Lei et al. [8]. However, few researchers study the effect of different diffuser types of volute on pressure fluctuation.

In this study, two diffuser types of volute were designed to study the effect on pressure fluctuation features of centrifugal pump under part-load condition, with the same volute design parameters and impeller parameters.

2. Numerical Simulation

2.1. Geometry of the Pump

The model pump consists of the impeller, suction and volute. Two diffuser types of volute are designed. The first one is the tangential diffuser type, the second one is the radial diffuser type, and the parameters of volute and impeller are the same. The models of two diffuser types of volute are shown in Figure 1. The impeller is shrouded with 6 twist and backswept blades, with a specific speed of n_s=162.3. The main design parameters of the pump are shown: n=2900rpm, Q_{DES}=330m³/h, and H_{DES}=48m.

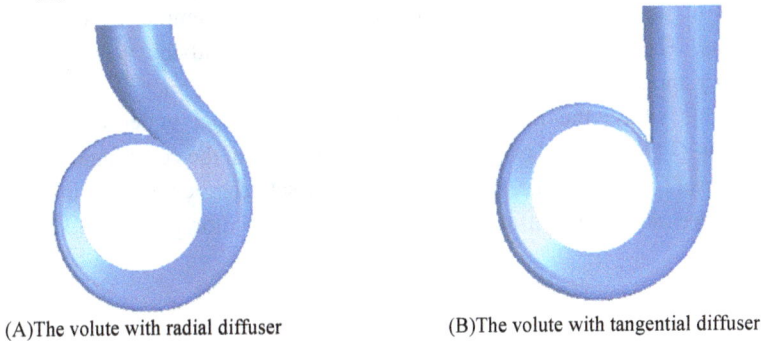

(A)The volute with radial diffuser   (B)The volute with tangential diffuser

Fig. 1 The model of volute with two diffuser types

3. Mesh Setup and Turbulence Model

The whole computational domain is made up of the suction, impeller and volute. Computational domains are created using the grid generation tool ICEM, and the mesh is refined in the walls of the channels. The number of mesh is 2630149 and the details of the mesh are shown in Figure 2. In this study, three-dimensional unsteady simulations were conducted by solving the RANS equations in the design procedure with shear stress transport (SST) k-ω turbulence model.

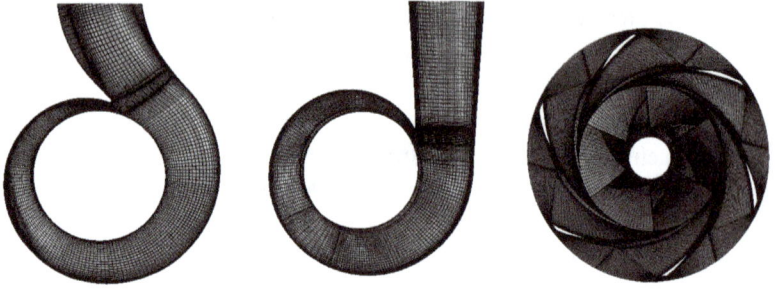

(A)The volute with radial diffuser   (B)The volute with tangential diffuser   (C)Impeller

Fig. 2 Computational mesh for two diffuser types of volute

4. Boundary Conditions

The interfaces between rotor and stator are set as 'Transient Rotor Stator'. The simulation cycle is set as 6 with 2 degrees per time step. Monitoring data of the sixth cycle are analyzed. The total pressure is used in the inlet and the mass flow rate is specified in the outlet. The wall is set as smooth wall in the physical surfaces of flow field. The high-resolution algorithm is employed to solve the equations. The maximum residuals are set to 10-5, and the values of head, efficiency and power are monitored.

In this study, the numerical analysis will be done under the part-load condition, Q/Q_{DES} =0.8. To study pressure fluctuation in the centrifugal pump for two diffuser types of volute, the point c1, c11 are set near the tongue and the point c2, c22 are set in the diffuser. The positions of monitor points are shown in Figure 3.

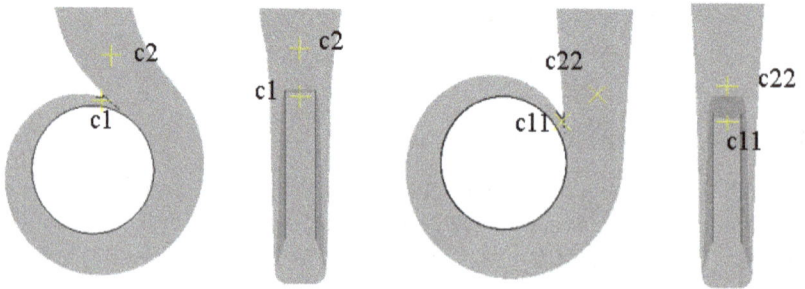

(A)The volute with radial diffuser   (B)The volute with tangential diffuser

Fig. 3 The location of monitor point

936

5. Results and Discussion

5.1. *Comparison of Performance*

By using the numerical simulations under 3 different flow rates, the predicted pump performance was obtained for two diffuser types of volute, the volute with radial diffuser and the volute with tangential diffuser. As shown in Figure 4. The calculation of each operating point value is the average head and efficiency of a period under unsteady numerical simulation.

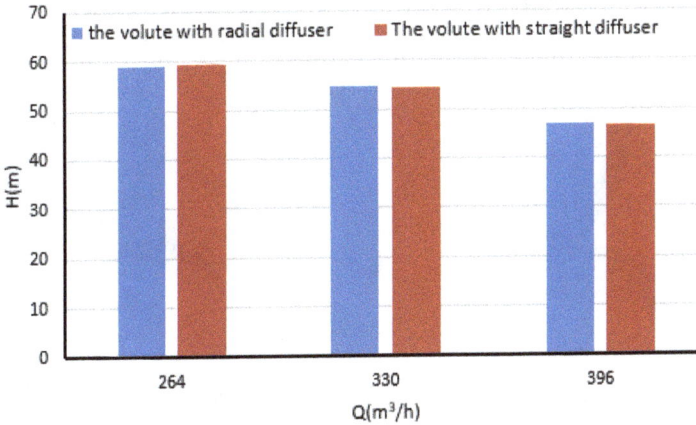

Fig. 4 The head for two diffuser types of volute

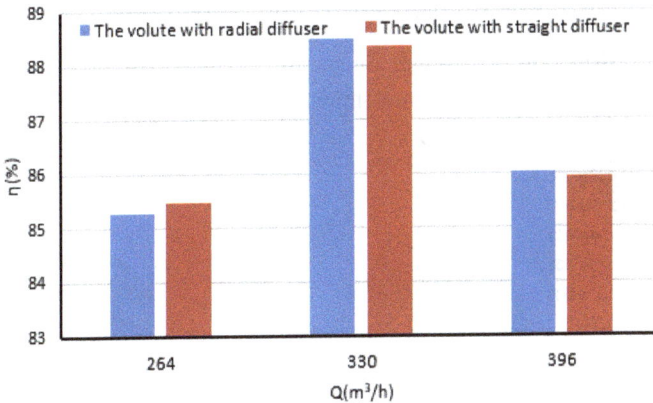

Fig. 5 The efficiency for two diffuser types of volute

As shown in Figure 4, the values of head in the volute with radial diffuser and tangential diffuser decrease gradually with the increase of flow rate. In the volute with radial diffuser, the maximum head value is 59.04m, and the minimum head value is 46.57m. In the volute with tangential diffuser, the maximum head value is 59.18m, and the minimum head value is 46.52m.

In the two diffuser types of volute, the efficiency value under part-load condition (Q/Q_{DES}=0.8) has the biggest difference and that are almost same under other condition. Therefore, it is necessary and significant to study the pressure fluctuation characteristics under part-load condition.

6. Pressure Fluctuation Intensity Distribution

Under part-load conditions, two cross section of fluid channel were created respectively in the diffuser and near the tongue of volute with two diffuser types, as shown in Figure 6.

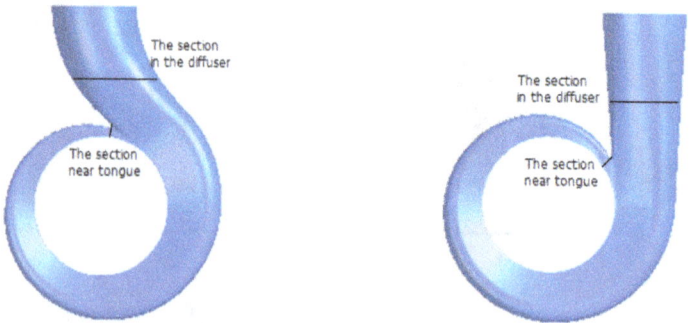

(A) The volute with radial diffuser	(B) The volute with tangential diffuser

Fig. 6 The cross section of two diffuser types of volute

The distribution of pressure fluctuation intensity in two diffuser types of volute under part-load condition is shown in Figure 7 and Figure 8.

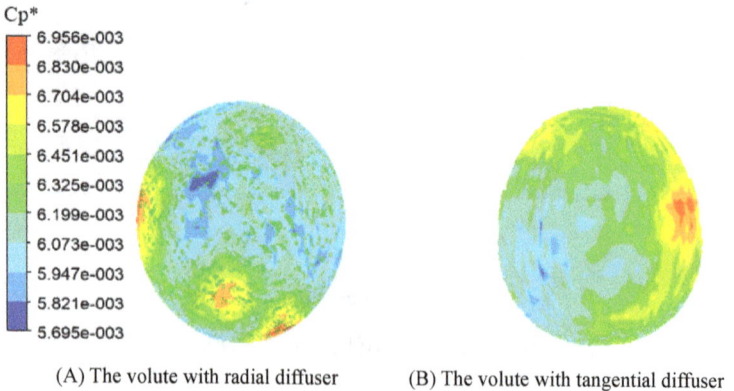

(A) The volute with radial diffuser	(B) The volute with tangential diffuser

Fig. 7 The pressure fluctuation intensity in the diffuser of volute

As shown in Figure 6, in the diffuser of volute, whether in radial diffuser or tangential diffuser, under part-load condition, the pressure fluctuation intensity of outer side is higher than that of inner side. The large gradient distribution of pressure fluctuation intensity was shown in the position close to the wall. Compare the pressure fluctuation intensity in radial diffuser with that in tangential diffuser, the distribution of pressure fluctuation intensity is more average and the gradient of pressure fluctuation is bigger.

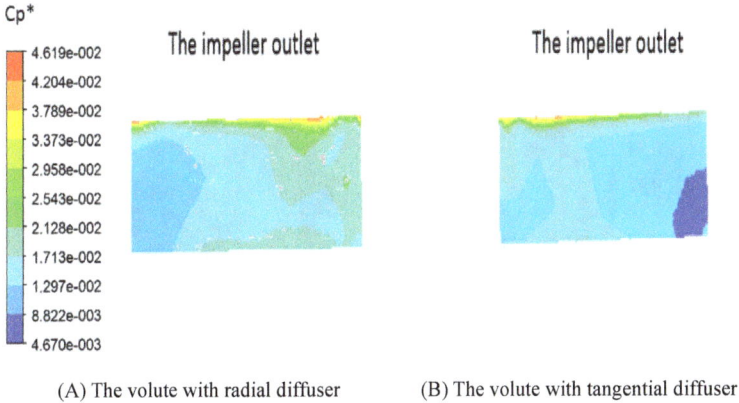

(A) The volute with radial diffuser   (B) The volute with tangential diffuser

Fig. 8 The pressure fluctuation intensity near the tongue of volute

As shown in Figure 8, near the tongue of volute, whether in radial diffuser or in tangential diffuser, under part-load condition, the large pressure fluctuation intensity was shown at the junction of the impeller outlet and the tongue inlet, which present that the rotor-stator interaction of impeller and tongue will cause large pressure fluctuation near the tongue. The pressure fluctuation intensity is smaller far away from the outlet of impeller. Compare the pressure fluctuation intensity in radial diffuser of volute with that in tangential diffuser of volute, the distribution of pressure fluctuation intensity is more average but the gradient of pressure fluctuation is smaller. The minimum value of pressure fluctuation intensity in tangential diffuser of volute is smaller than that in radial diffuser of volute.

7. Time Domain and Frequency Domain Analysis

Different diffuser types of volute have little effect on performance of pump from the Figure 4, but can exert an influence on flow characteristic in pump and piping system. Especially under part-load condition, pressure fluctuation will be affected more obviously.

When the impeller blade passes by the volute tongue, a pressure fluctuation is initiated by the rotor-stator interaction of impeller and tongue. In the case of strong impeller–volute interaction, the pressure fluctuation become high enough to induce high vibration to pump components. So under part-load condition, by using numerical analysis, in the two types of volute, the point c1 near the tongue and the point c2 in the outlet are set to obtain the time domain characteristics on pressure fluctuation, as shown in Figure 9.

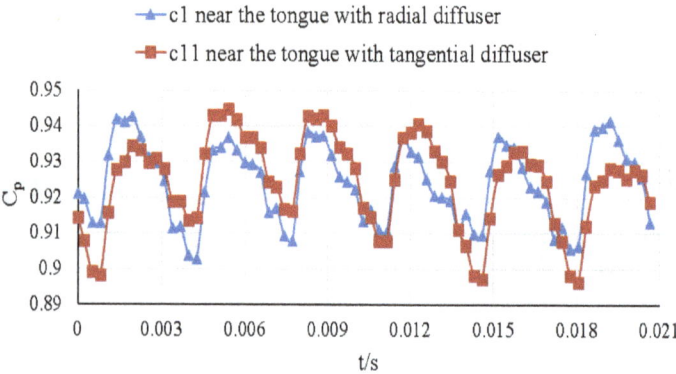

Fig. 9 The pressure fluctuation near the tongue

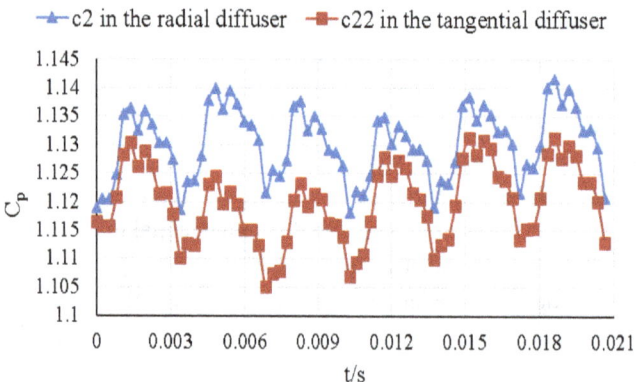

Fig. 10 The pressure fluctuation in the diffuser

As can be seen from Figure 9, under part-load condition the change trend of pressure fluctuation near the tongue of volute with radial diffuser and the tongue of volute with tangential diffuser are basically same, and exhibit an evident periodicity property. The peak value of pressure fluctuation is same near tongue of volute with two diffuser types.

As can be seen from Figure 10, under part-load condition the change trend of pressure fluctuation in the diffuser of volute with radial diffuser and the diffuser of volute with tangential diffuser are basically same, but the peak value of pressure fluctuation in the diffuser of volute with radial diffuser is greater than that of pressure fluctuation in the diffuser of volute with tangential diffuser. It also exhibits an evident periodicity property. The pressure fluctuation amplitude different is small in two diffuser types of volute.

By comparison, the conclusion can be drawn, the periodic features in pressure fluctuation near its tongue are the same for two diffuser types of volute, but in the diffuser of the volute, the peak value of the pressure fluctuation in radial diffuser is greater than that in tangential diffuser.

By using Fast Fourier Transform (FFT), under part-load condition in the volute with two diffuser types, the pressure fluctuation frequency domain graph of the point $c1$ near the tongue and the point $c2$ in the diffuser are shown in Figure 11 and Figure 12. Where the blade passing frequency is 290Hz.

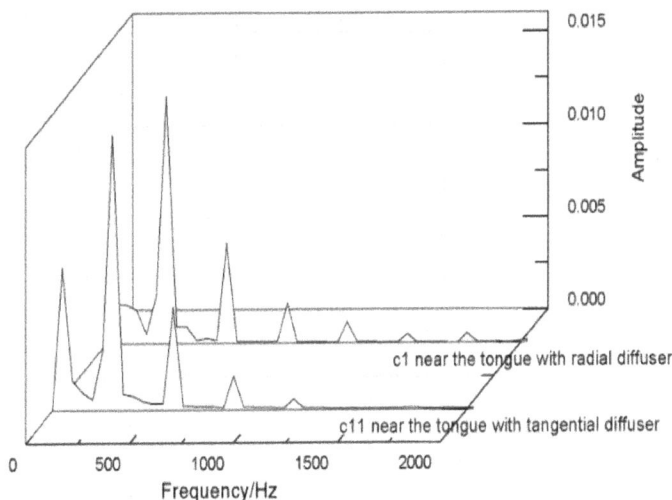

Fig. 11 The frequency domain graph near the tongue

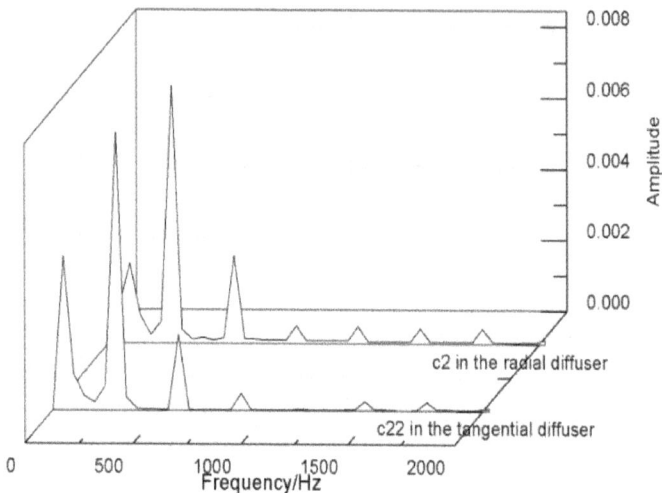

Fig. 12 The frequency domain graph in the diffuser

As shown in Figure 11, near the tongue, under part-load condition, the pressure fluctuation frequency based on blade pass frequency in two diffuser types of volute, and the pressure fluctuation amplitude in radial diffuser of volute is greater than that in tangential diffuser of volute.

As shown in Figure 12, in the diffuser of volute, under part-load condition, the pressure fluctuation frequency also is based on blade pass frequency in two diffuser types, and the pressure fluctuation amplitude in tangential diffuser of volute is greater than that in radial diffuser of volute.

8. Conclusion

In this paper, by using the unsteady numerical simulation analysis of whole flow field in two diffuser types of volute, the volute with tangential diffuser and the volute with radial diffuser were discussed and some conclusions are shown:

● The different diffuser types of volute have little effect on performance of pump. However, under part-load condition, the pressure fluctuation will be affected more obviously.

● In the diffuser, the pressure fluctuation intensity of outer side is higher than that of inner side. In radial diffuser, the distribution of pressure fluctuation intensity is more average and the gradient of pressure fluctuation is bigger.

- Near the tongue, under part-load condition the large pressure fluctuation intensity was shown at the junction of the impeller outlet and the tongue inlet. In radial diffuser of volute, the distribution of pressure fluctuation intensity is more average but the gradient of pressure fluctuation is smaller.

- Under part-load condition, whether near the tongue or in the diffuser, the pressure fluctuation frequency is based on blade pass frequency for two diffuser types of volute.

Acknowledgments

This study was funded by Natural Science Foundation of Jiangsu Province Youth Fund (Grant No. BK20140554), National Natural Science Foundation of China (Grant No. 51409123), China Postdoctoral Science Foundation (Grant No. 2014M560402), Postdoctoral Science Foundation of Jiangsu Province (Grant No. 1401069B), and the Priority Academic Program Development of Jiangsu Higher Education Institutions (PAPD).

References

1. Glc M, Usta N, Pancar Y. Effects of splitter blades on deep well pump performance [J]. Journal of Energy Resources Technology, 2007, 129: 169-176.
2. Kergourlay G, Younsi M, Bakir F, et al. Influence of splitter blades on the flow field of a centrifugal pump: Test-analysis comparison [J]. International Journal of Rotating Machinery, 2007: 85024.
3. Wang Wenquan, Zhang Lixiang, Yan, et al. Large-eddy simulation of turbulent flow considering in flow wakes in a franc is turbine blade passage [J]. Journal of Hydrodynamics, Ser. B, 2007, 19(2): 201-209.
4. Germano M, Piomelll U, Mion P, et al. A dynamic sub grid - scale eddy viscosity model [J]. Physics of Fluid A, 1991, 3(7): 1760-1765.
5. Longatte F, Kueny J L et. al. Analysis of rotor-stator-circuit interactions in a centrifugal pump [C]. Proceedings of the 3rd ASME/JSME Joint Fluids Engineering Conference, 2009: 1039-1045.
6. Kitano Majidi et al. Numerical study of unsteady flow in a centrifugal pump [J]. Journal of Turbomachinery, 2005, 127: 363-371
7. Guo Pengcheng et al. Numerical investigation and performance prediction on 3d complex viscous flows in hydromachinery [D]. Xian: Xian University of Technology, 2009. (In Chinese with English abstract)

8. Zhu Lei, et al. Numerical simulation on pressure fluctuations and radial hydraulic forces in a centrifugal pump with step-tongue [J]. Transactions of the Chinese Society for Agricultural Machinery, 2010, 41: 21-26. (in Chinese with English abstract).

Study on CSNS AC Dipole Magnet Coil Assembly Mechanics

Lei Liu[1, 2*], Yong-ji Yu[1, 2], Chang-jun Ning[1, 2] and Ling Kang[1, 2]
Institute of High Energy Physics,
Chinese Academy of Sciences (CAS), Beijing, China
[2]Dongguan Neutron Science Center, Dongguan 523803, China
[*]Email: lliu@ihep.ac.cn

China Spallation Neutron Source (CSNS) AC dipole magnet is the largest AC vibration magnet in China. The magnet is composed of an upper and a lower half iron core combined, which is an H type magnet. Two excitation coils are respectively arranged in the grooves of the upper and the lower iron cores, and are fixed by eight supports. When the ceramic vacuum box is installed, the two iron cores are required to be opened, after the vacuum box is installed, the iron core is needed to be assembled again, but due to the limited air gap of the magnet, the eight coil supports cannot be adjusted at this time. In the assembly, to ensure the coil is pushed tightly, this paper, through ANSYS analysis and experimental method to analyze the force deformation of the epoxy coil and support. Firstly, it sets the interference fit of the coil and the support, and then the upper half magnet coil directly presses on the coil support, through the coil and the support interference fit to fix coils, completing the assembly magnet and vacuum box.

Keywords: AC dipole magnet, Epoxy coil, Assembling, Deformation.

1. Introduction

CSNS AC dipole magnet is a total of 24 units, located in the accelerator ring four corner, to control beam bending, belonging to the high voltage and high current AC vibrating magnet, amplitude of 25Hz. H type structure is adopted in the dipole magnets, with a total weight of about 23 tons, consists of two half iron core, two iron cores are connected by both sides of the screw rods. Each iron core is made of 4000 pieces of 0.5mm steel, with a total length of 2060mm. Each coil weighs about 1 ton, which is located in the groove of the upper and the lower iron core; there are eight supports between coils, detailed structure as shown in Figure 1.

Because the ceramic vacuum box flange is larger than the dimension of the magnetic gap, we must first lift the upper magnet, and then install the vacuum box. When the upper magnet reset, because the vacuum box takes up space regulating coil support, coil support can no longer be any adjustment. Therefore, before installing the vacuum box, the interference of the coil support must be set in advance, and in the final assembly, the fastening of the coil can be realized by adopting the interference fit between the coil support and the coil [1].

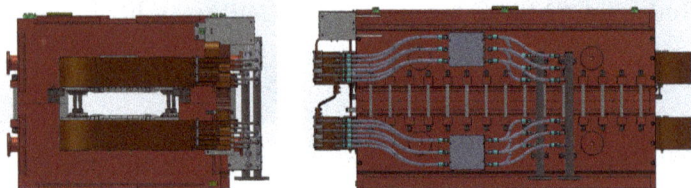

Figure 1. Dipole magnet structure

Coil support should have the regulating function of the height direction, so that in the vibration test of a magnet for height adjustment, as well as to prevent the screw loose, supports should be able to lock. Finally, after setting the height of the support, lock compact support and the backing plate. Based on the above requirements, coil support design is a claw disc structure, grooved disc in order to cut off the vortex and reduce stiffness. Support material is 304 stainless steel; the plate adopts epoxy plate. The structure is as shown in Figure 2.

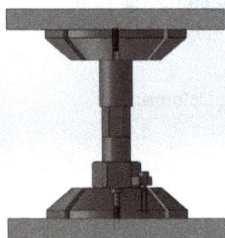

Figure 2. Coil support   Figure 3. Test piece compression test

2.   Experimental Study on the Interference Quantity

2.1. *Magnet Coil Compression Test*

Dipole magnet coil wire is a kind of special aluminum wire, inside it is a stainless steel cooling pipe, and the external of pipe winding aluminum wire, the wire is roundness inside but squareness outside structure [2]. Single coil is divided into three layers of total 30 turns, external pouring epoxy curing, because the epoxy glue is self-modulation, so it is necessary to use experimental method to obtain the coil of deformation and stress of the relationship. We use coil test piece and pressure tool to complete the experiment of pressure deformation, as shown in Figure 3. By repeated experiments, when the coil shape variable is 0.3mm, there is no crack in the coil and the epoxy layer. In order to ensure the internal coil aluminum stranded wire will not shift deformation, so we set dipole magnet coil can bear the deformation can't more than 0.3mm. The experiments also prove that deformation and pressure into a linear relationship.

3. Force Deformation Analysis of Coil Support

In the assembly of the magnet, when the upper and lower core assembly, in order to ensure that the coil support can be in a state of compression, we must analyze the coil support pressure deformation relationship. In the analysis, through the optimization of the supporting structure, we ignore the effect of epoxy plate and lock nut. Coil supports height can be adjusted by stud, in the magnetic test, locking the support nuts to prevent loosening. The height adjustment selects the M24 thread, the thread pitch 2mm, every lower plate is evenly distributed with 8 slots and 8 holes.

In the force analysis, the lower jaw plate has 8 holes and 8 grooves on the circumference of the 360 degree, thus, the range of the shape variables set by the coil support from the preload to the final pressing is fixed, and the minimum compression deformation is 1/16 of the thread pitch, so the range of deformation is (0, 0.125mm). When the claw disc is rotated from 1/16 to 1/8 ring, the corresponding compression deformation range is (0.125mm, 0.25mm). When the claw disc is rotated from 1/8 to 1/4 ring, the corresponding compression deformation range is (0.25mm, 0.375mm), and so on.

When set compression deformation amount is in the range (0, 0.125mm), due to symmetrical structure, so unilateral maximum deformation is 0.0625mm, the amount of deformation is by a coil and a coil support under the same pressure generated by the sum of the deformation. Assuming that the pressure is F, the deformation of the coil is X, the deformation of the coil support is Y, and the relationship between the deformation and the pressure of the coil is known by experiment:

$$F = 15000 * X \qquad (1)$$

By using ANSYS to analyze the mechanics of the coil support, we know that the relationship between the deformation and the pressure of the coil support is:

$$F = 46040 * Y \tag{2}$$

According to the relationship between the coil and the coil support deformation can get the following relationship:

$$X + Y = 0.0625 \tag{3}$$

Calculated by the formula (1), (2), (3):

$$F = 720N; X = 0.048 \text{ mm}; Y = 0.015 \text{mm}$$

Under the action of the 720N pressure, the deformation of the coil is 0.048mm, less than the safety value of 0.3mm, at the same time, under this pressure, we use the software to carry on the force analysis of the coil support, the results shown in Figure 4, through the analysis results show that the stress is 53MPa, less than the material allowable stress.

When setting the amount of compression deformation range is (0.125mm, 0.25mm), the results are calculated as:

$$F = 1410N; X = 0.094 \text{mm}; Y = 0.0311 \text{mm}$$

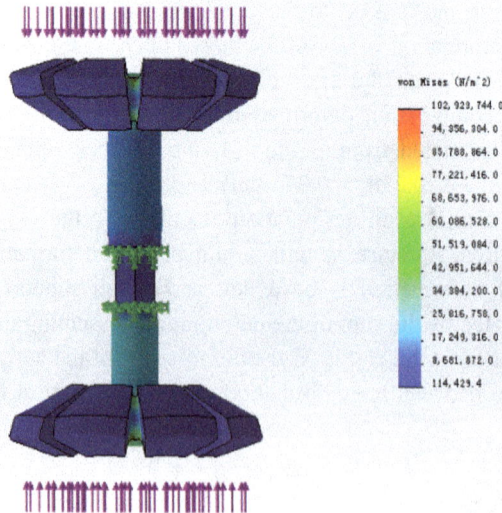

Figure 4. Coil support force analysis-720N

Shown in Figure 5, through the analysis results show that the stress is 103Mpa, under the pressure of the 1410N; the deformation of the coil is 0.094mm, less than the safety value of 0.3mm.

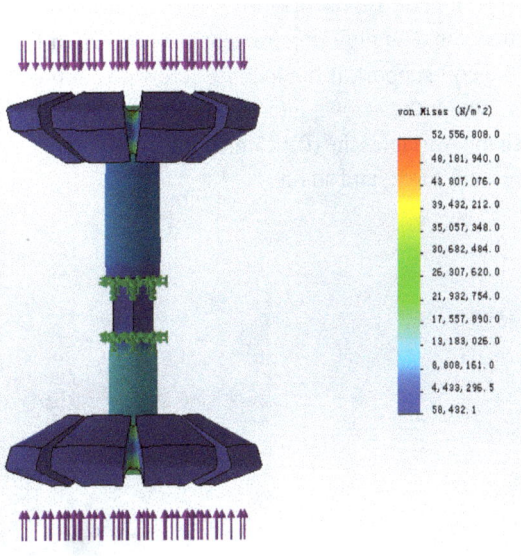

Figure 5. Coil support force analysis-1410N

Support deformation and stress is a linear relationship, the amount of compression increased by 0.125mm, the stress increased by about 52MPa, if we continue to increase the compression to 0.375mm, then the stress will reach 155MPa, more than the safety stress of the material.

4. Conclusion

By analyzing the force of the coil and the coil support, it is known that when the range of the compression deformation is set up (0, 0.125mm), the coil support pressure range is 0~720N. Consider to coil processing error and magnet vibration, the maximum amplitude will reach 0.03mm, so if we set the compression deformation is (0, 0.125mm), then in 24 magnet for a total of 192 support, will support force is too small, even without force, this will cause the uneven force of the coil, leading to increase in the magnet amplitude and the serious consequences of the coil cracking. When the compression deformation range is (0.125mm, 0.25mm), the force range of the coil support is

720N~1410N, which can ensure that the coil support is effective, and can meet the requirements of the material allowable stress. When the compression deformation increased to 0.375mm, the long-term use will make the coil support failure.

In summary, the coil support compression deformation range is (0.125mm, 0.25mm), to meet the assembly requirements. As shown in figure 6, after the pre tightening of the coil support, if the locking screw position is in the AB, then the final assembly, screw the screws into the C slot, ensure that the coil supporting and sum of shape variables in (0.125mm, 0.25 mm) range to ensure the coil support the pressing force, and so on.

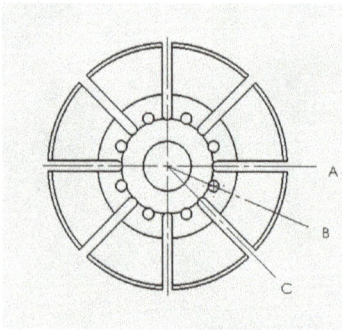

Figure 6. Claw plate diagram   Figure 7. Magnets successfully installed in the tunnel

At present, we have used the above method to complete the installation work of 24 pole magnets. After the installation of supports, we conducted 3 days of vibration test, there is no support loose phenomenon, and the support effect is good. At present, all magnets have successfully completed the tunnel installation work. Figure 7 shows the installation of the magnet in the tunnel.

Acknowledgement

This research was funded by National Natural Science Foundation of China (Project 11375217).

References

1. Chang-dong Deng, Fu-san Chen, Xian-jing Sun. Chinese Physics C, Vol. 32, Mar., 2008.

2. Chang-dong Deng. Magnet Design of Chinese Spallation Neutron Source Rapid Cycling Synchrotron, Huazhong University of Science and Technology, Wuhan, Hubei P. R. China 430074, April, 2004.

Design of RCS Magnet Girder Electromotion Adjustment Device

Guang-yuan Wang*, Jun-song Zhang and Ling Kang

Institute of High Energy Physics,
Chinese Academy of Sciences (CAS), Beijing, China
Dongguan Neutron Science Center,
Dongguan 523803, China
Email: gywang@ihep.ac.cn

CSNS-RCS magnet girder is used for supporting, adjusting and locating magnet. The precision of the adjustment will influence the magnet collimation precision. In order to optimize on manpower and time, the magnet girder electromotion adjustment device must be designed with capability to adjust girder to change the operation to electromotion. Using a mitsubishi FX system, PLC and Wein View touch screen host computer, the electromotion adjustment was designed. This system made all of the magnets in RCS be adjusted, which used with laser tracker.

Keywords: CSNS; Electromotion adjustment; Magnet girder; Control system.

1. Introduction

The CSNS is accelerators consist of an 80-MeV H-linac and a rapid cycling synchrotron of 1.6 GeV. The RCS ring is a four-folded symmetrical topological structure, which consists of four arc zones and four line segments [1, 2]. RCS magnet girder is used for supporting, adjusting and locating magnet, The precision of the adjustment will influence the magnet collimation precision. The maximal mass of magnet in RCS is 25 tons, and the minimum mass is 2.5 tons. For economizing on manpower and time, the magnet girder electromotion adjustment device must be designed, for changing the manually operation to electromotion adjusting girder.

2. Structure Design of Electromotion Adjustment Device of RCS Magnet Girder

2.1. *The Structure of Electromotion Adjustment Girder*

The weight of dipole magnet and Lambertson magnet is above 25 tons, it goes beyond manually operation force. Therefore electromotion adjustment was used in these magnet girders. For the vertical direction adjustment, there are four stepper motors driving the turbo worm lifter. And for the horizontal direction adjustment, there are three stepper motors driving ballscrew, as illustrated in Fig. 1

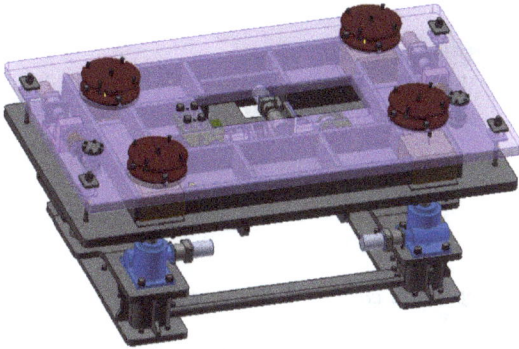

Fig. 1. The structure of Lambertson magnet grider

3. Structure of Electromotion Adjustment Base on Manually Adjustment

The quadrupole magnet mass is less than 10 tons, the primal design of the girder adopted the manually adjustment. But the vertical adjustment needs a large force to lift the magnet. The elctromotion adjustment device was designed, the structure is shown in Fig. 2. All of the magnet in RCS can be adjusted by it. It is made up by four adjustment assemblies, which include stepper motor and retarder. The adjustment assembly is shown in Fig. 3. It can drive the turbo worm lifter for vertical adjustment.

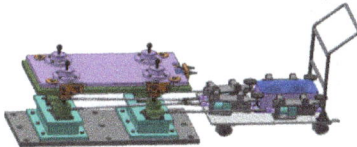

Fig. 2. Structure form of the magnet adjusting device in quadrupole QA206

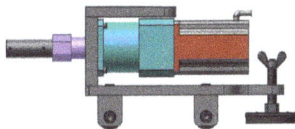

Fig. 3. Mainly components of the magnet adjusting device

The maximum load of the quadrupole magnet QB272 girder is up to 10 tons. In order to avoid the motor turned over, 4 adjustment componets shown in Fig. 4 are fixed to each other by long steel.

953

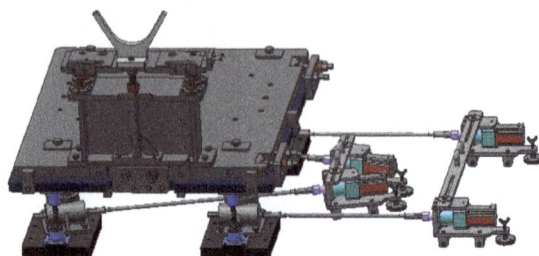

Fig. 4. Structure form of the magnet adjusting device in quadrupole QB272

4. The Design of Magnet Girder Electromotion Adjustment Device Control System

4.1. Control principle and control flow chart of the PLC control system.

Used with laser tracker and electromotion adjustment device, all magnet girder are adjusted. The control flow chart of the grider adjustment process is shown in Fig. 5.

The laser tracker will measure position bias value between the current position and the theoretical position of the magnet, and then the value will be input to the touch screed. After calculating the number of pulses and pulse frequency, PLC drives all the stepping motor and begin to adjust the magnet girder. After adjustment, the laser tracker will measure Position bias value again. The adjustment process will not stop until the bias value is within the allowable range.

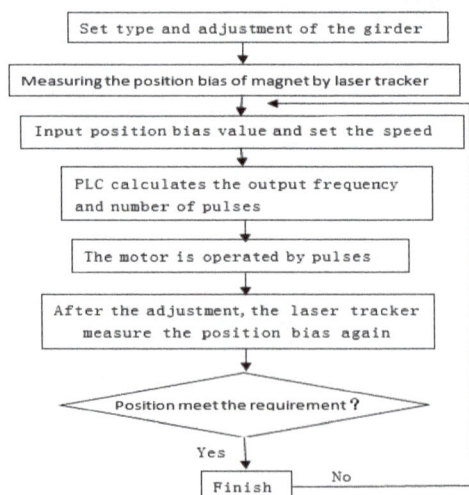

Fig. 5. Flow chart of the control system

954

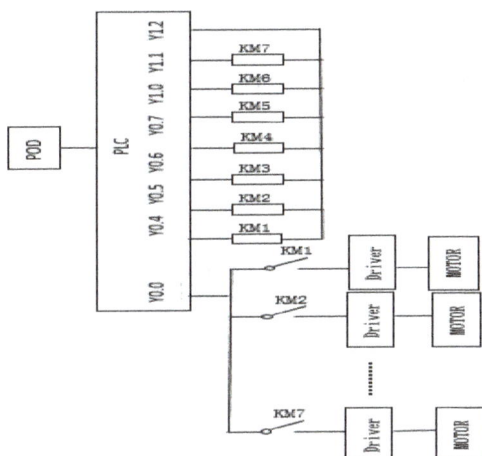

Fig. 6. Schematic of PLC control system

5. Hardware Wiring Diagram Based on the PLC Control System

RCS magnet girder electromotion adjustment device uses PC POD (touch screen) combined with programmable controller PLC. The structure is shown in Fig. 6. Each stepper motor is independently driven by a driver. The PLC only has two high frequency pulse output, seven relays were used to connect the PLC and drivers, in order to realize the synchronous work among each stepper motor.

The Y0.4 to Y1.2 of PLC output connect the coil terminal of the relays, when there is signal output of the PLC, the corresponding relay switch is closed, and PLC Y0.0 high-frequency output terminal connects to the corresponding stepper motor driver, so as to realize the purpose of each control. The Schematic of PLC control system is shown in Fig. 7.

Fig. 7. Schematic of PLC control system

955

6. I/O of the PLC Control System

The I/O settings of PLC is an important part of PLC control system, which is shown in Table 1.

Table 1. I/O of the PLC control system

	Input
M0	Set the motor A of the elevator
M1	Set the motor B of the elevator
M2	Set the motor C of the elevator
M3	Set the motor D of the elevator
M4	Set the motor E of the horizontal direction
M5	Set the motor F of the horizontal direction
M6	Set the motor G of the vertical direction
	Output
Y0.0	High frequency pulse output
Y0.4	Enable the the motor controller A
Y0.5	Enable the the motor controller B
Y0.6	Enable the the motor controller C
Y0.7	Enable the the motor controller D
Y1.0	Enable the the motor controller E
Y1.1	Enable the the motor controller F
Y1.2	Enable the the motor controller G

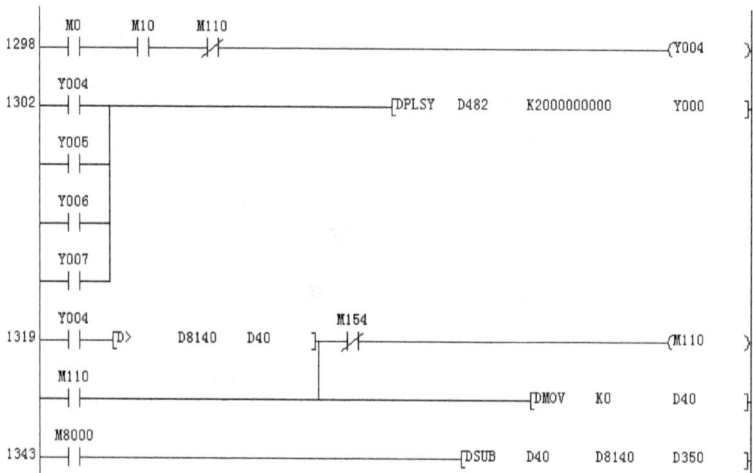

Fig. 8. PLC ladder diagram

7. PLC Program Design

Mitsubishi PLC uses ladder diagram language, which can be compiled by the GX Developer software, the ladder diagram design must combine with the design of the touch screen interface. PLC, being the control unit of the whole system, is primarily responsible for sending pulse information to the Stepper Motor Controllers so as to get the motor working. The following figure is partly about the pulse output program ladder diagram of a stepper motor in position A of an elevator.

PLC uses the DLPSY pulse output instruction to output pulse to the stepper motor, such instruction is a 32 bit high-speed pulse output instruction, the pulse output port is Y0.0.Take the 1302nd line statement as an example, the frequency value is D482, the pulse is in the form of square wave, the duty ratio is 50%. When any one of the elevator enable output takes effect, (for instance, when Y0004 takes effect), the pulse output do not stop until its amount reaches 2000000000. In order to take control of the pulse output amount of different positions, Comparative language [D>] is used, D8140 is the statistic of the output pulse amount, when D8140 is greater than the enter value D40, M110 takes effect, and the driver' s enable pin lose efficacy, thus, position A of the elevator do not make any lifting operation. As Y005, Y006, Y007 and Y004 are of coordinative relation, lose efficacy of Y004 does not have any influence. Pulse output keep sending among stepper motors of other positions, which reaches a state of parallel working. The PLC ladder diagram is shown in Fig. 8.

8. Control Interface System Design of the Touch Screen

Electric adjustment system uses WeinviewKT6070i touch screen as its upper machine, which is more humanization than tradition mechanical push button control interface that can only use PLC's24 control inputs. Using touch screen, it can call thousands of auxiliary relay as input of PLC, which greatly improve the maneuverability of the PLC controller. The interface of the touch screen is compiled by Weinview software Easy builder 8000, we can design the interface as demanded through tool invocation, set of related items as well as macro definition within the software toolbar. The touch screen communicates with the PLC through the communication cable line RS-232, at the same time, it accomplishes modification work of touch screen operation system [3].

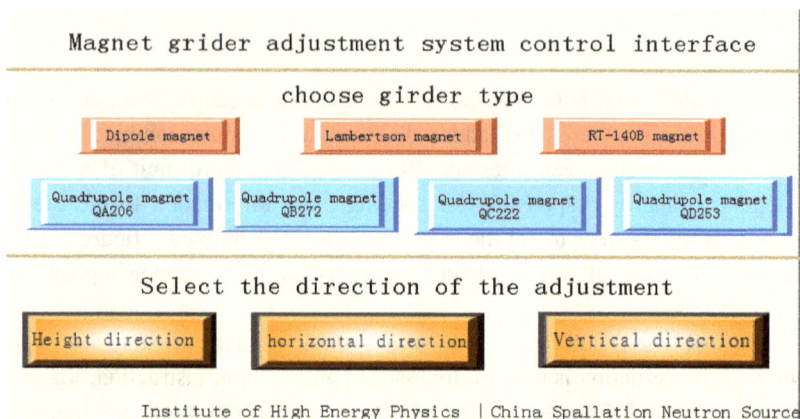

Fig. 9. Magnet girder adjustment system control interface

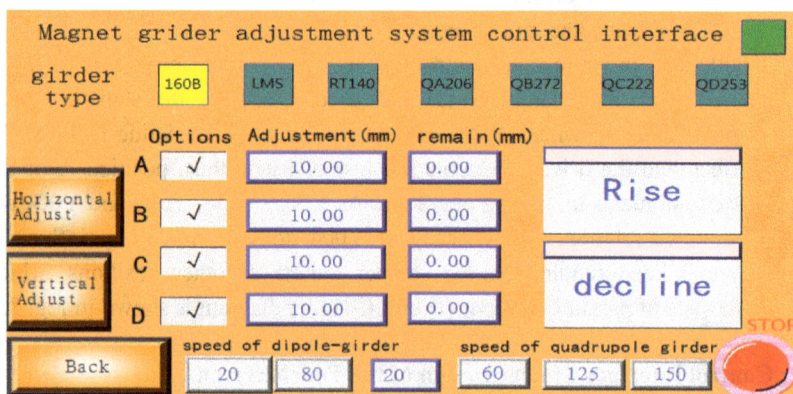

Fig. 10. Magnet girder adjustment system control interface of High direction

According to the operation process of the control system, the interface is shown in the figure below. Fig. 9 shows the girder selection interface of the magnet girder system, as different magnet girder uses different worm gear and worm, different speed reducers and different stepper motors, the PLC output frequency and pulse number also differ. So the interface is used to comfirm which girder will be adjusted.

The adjustment interface of the magnet girder in the vertical direction and horizontal direction is shown in Fig. 10. Take the vertical direction as an example, Laser tracker measures the magnets position bias value, and input the value to the control interface, the magnet position can be automatically adjusted. To adapt different adjustment condition, manual operation mode was established, which can manually input adjust value and speed.

958

9. Testing of the Magnet Girder Electromotion Adjustment Device

Taking the quadrupole magnet girder QC222 as a test object. The adjustment test process is shown in Fig. 11; the magnet position was measured in real time by laser tracker, the measuremants were then fed back to the magnet electric adjusting device to controll accordingly, so as to keep the error of the magnet alignment in within 0.05mm. The results were summarized in Table 2.

Table 2. The result of 5mm and 10mm adjustment test

Times	Target value /mm	Measure value /mm	Error /mm
1	5	4.96	0.04
2	5	5.01	0.01
3	5	5.00	0.00
4	10	9.98	0.02
5	10	9.97	0.03
6	10	9.98	0.02
7	-5	-4.98	-0.02
8	-5	-5.00	0.00
9	-5	-5.03	0.03
10	-10	-10.03	0.03
11	-10	-10.01	0.01
12	-10	-10.03	0.03

Fig. 11. Testing of the magnet girder electromotion adjustment device

10. Conclusion

In order to optimize on manpower and time, the magnet girder electromotion adjustment device should equipped with capability to control the electromotion by menas of girder. Using a mitsubishi FX system, PLC and Wein View touch screen host computer, the electromotion adjustment, the authors put forwarded a girder electromotion adjustment design with the following features to realize this goals:

(1) Design of Electromotion Adjustment Device of RCS Magnet Girder included stepper motor and retarder, was used to drive the turbo worm lifter for vertical adjustment was presented.

(2) The operation of Schematic of PLC control system in the form control flow chart was put forwared to explain the process in details.

(3) Using a touch screen interface, the operation of Stepper Motor Controllers, could be minotored to display pulse information sent from ladder diagram. The ladder diagram language was compiled using the GX Developer software to be interfaced with Weinview software Easybuilder 8000 to interface with the touch screen.

(4) The proposed magnet girder electromotion adjustment device was tested to achieve precision error of the magnet alignment within 0.05mm.

References

1. Wei Jie, Fu Shi-Nian, Fang Shou-Xian. Proceedings ofEPAC. 2006
2. Wei Jie, Chen He-Sheng, Chen Yan-Wei et al. Nuclear Instruments and Methods in Physics Research Section A, 2009,600: 10-13
3. Huang Jian-Xin, Liu Jian-Qun, Kuang Hui, Wu Shi-lin.Servo Motor Control System Based on Programmable Operation Display and PLC. Instrument Technique And Sensor, 2005, Issue 2, 44-45

Mechanical Design of Multi-Wire Profile Monitor for CSNS Injection

An-xin Wang[†], Tao Yang, Xiao-jun Nie, Jie-bing Yu,
Jia-xin Chen, Tao-guang Xu and Ling Kang

Institute of High Energy Physics,
Chinese Academy of Sciences (CAS),
Beijing 100049, China
Dongguan Neutron Science Center,
Dongguan 523803, China
[†]*Email: wanganxin@ihep.ac.cn*

A set of 3 Multi-wire Profile Monitors (MWPMs) has been installed in the injection line of Chinese Spallation Neutron Source (CSNS). In the early stage of commissioning, for establishing an optimum injection orbit, MWPMs are the most important beam monitors to measure positions and profile distribution of injection beam from LINAC. This paper demonstrates the mechanical design of MWPMs for CSNS injection. According to the physical parameters, the wire-distribution structure consist of several horizontal and vertical wires with a 30° angle is fabricated and choose the optimal wire by thermal analysis and mechanical analysis with the finite element software ANSYS. Furthermore, the most appropriate wire-clamping assembly to hold the 0.03mm tungsten wire under a 0.2~0.3N pretension force is determined. The novel MWPMs can quickly and precisely measure the beam profile density and gain the micrometre-sized minimum resolution by motor moving.

Keywords: CSNS; Multi-wire profile monitor; Beam profile; Wire-clamping assembly.

1. Introduction

There are several ways of beam profile measurement, such as fluorescent monitor, wire scanner, beam chopper, laser interferometer, and so on. Multi-wire profile monitor is adopted in this design, implemented to measure the beam profile density quickly, by means of capturing the voltage signal output of low energy electron, which comes from the bombardment of accelerator beam onto the target. [1]

There are 3 MWPMS installed in the CSNS injection region as shown in Fig. 1, while the physical parameters are shown in Table 1.

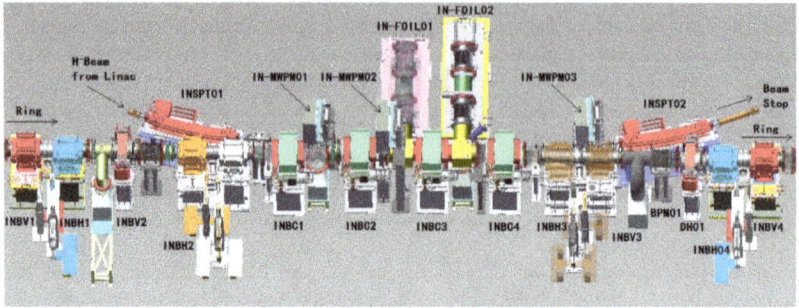

Fig. 1 The layout of MWPMS in RCS injection region.

Table 1 The physical parameters of MWPMs in injection region

MWPM	β_x/β_y (m)	D_x (m)	σ_x/σ_y (mm)	Beam-pipe size(mm)
INMWPM01	2.4/2.5	0/0	1.2/1.3	20
INMWPM02	1.7/2.0	0/0	1.0/1.1	22
INMWPM03	5.2/5.1	0/0	1.8/1.8	35

2. Selection of Wire Type

Due to the MWPMs are working in the beam existing area, wire of MWPMs shall be interacted with beam. Accompanied with ionization and radiation, energy of beam reduces. Part of decrement energy shall be deposited in the wire; as a result, the wire will be heated, so thermal analysis is essential to the optional wire.

3. Wire Materials

The main factors of wire material selection, thermo-mechanical properties, material of high melting-point, high tensile strength and low coefficient of thermal expansion are preferred.

Silicon Carbide (SiC), tungsten and titanium are considered as the candidate materials in this design. The main material properties are listed in Table 2.

Table 2. Physical properties of the three materials

Material properties	SiC	W	Ti
Density (Kg/m³)	3210	19350	4500
Melting point (°C)	3460	3410	1660
Thermal conductivity (W/cm/K)	1.1	1.3	0.22
Radiant emissivity	0.9	0.13	0.2
Heat capacity (J/g/K)	0.82	0.150	0.511
Tensile strength (N/mm²)	100~450	1500~3500	108~250

Thermal analysis is performed under 80MeV beam energy, 50μ pulse, 1Hz repetition frequency, and 30μm diameter wire in the central position of beam.

Under the same boundary condition, the temperature evolution of materials are different, as shown in Fig. 2. The maximum equilibrium temperature is about 630 K for SiC wire, 740 K for tungsten wire, and 1020 K for titanium wire, respectively. In consideration of the melting point and the respective mechanical properties, the tungsten wire is selected as the wire material.

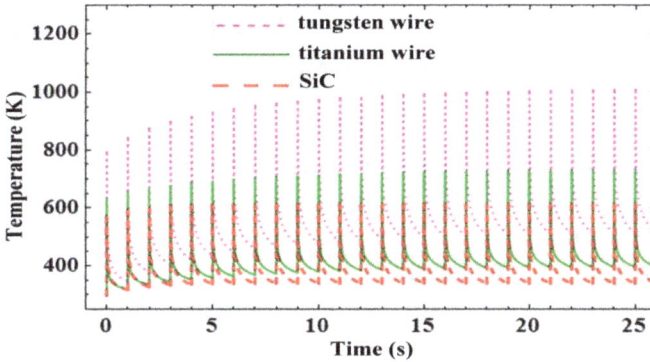

Fig. 2 Temperature evolution with time at the beam center in case of 80 MeV H- beam injection.

4. Determination of Wire Diameter

The wire diameter selection takes many factors into consideration, i.e. particle range, tensile strength, measurement requirements, etc.

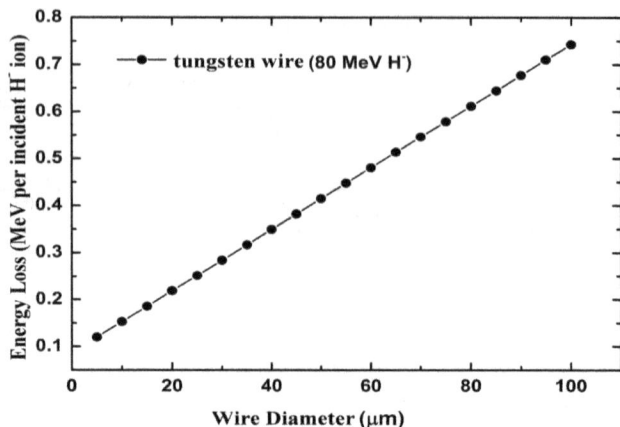

Fig. 3 Variation of energy loss and wire diameter.

Fig. 3 shows the variation of energy loss and wire diameter which presents a proportional relation when the diameter is smaller than 100 µm. The heat generation rate is proportional to the ratio of energy loss and diameter, and this ratio has a small variation in the diameter range shown in Fig. 3.

Yang had evaluated the influence of wire diameter to the maximum temperature in his research [2]. With the increasing of wire diameter, which is more than 30µm, the average stopping power has a small variation, but the surface-to-volume ratio declines continuously which brings a worse effect in the cooling performance. The maximum temperature reaches a minimum value when the wire diameter is about 30 µm. [2]

However, the wire diameter also cannot be too thin. With the diameter decreasing and temperature increasing during the measurement, the tensile strength will decline simultaneously, therefore it may cause the wire fracture.

As mentioned above, we choose the 30-µm tungsten wire as the wire material finally.

5. Mechanical Design of Wire-Clamping

Due to the extremely tiny diameter of tungsten wire, which is smaller than human hair, high performance clamping structure is demanded to correspond with the tiny shearing force.

6. Pretension force

On account of dead weight itself, as well as thermal expansion and contraction, the tungsten wire shall be loose. The loosen wire shall shake during measuring, cause the lower positional accuracy and measuring error. A structure with

964

pretension force adjusting is designed for the long term serviced wire to prevent the loosen condition.

In additional, ignore the tiny deformation caused by thermal expansion and contraction, which is about μm grade.

Table 3 shows the stress analysis of tungsten wire, take the maximum deformation as example: 204mm length, 30μm diameter and 30° angle to the horizontal plane.

Table 3. Mechanical analysis of tungsten wire

Pretension force(N)	0.2	0.25	0.3
Section deformation in vertical direction(mm)	4.72E-2	4.67E-2	4.64E-2
Section deformation in horizontal direction(m)	4.85E-3	4.59E-3	4.08E-3
Deformation in axial direction(mm)	0.289	0.361	0.433
Maximum principal stress(Mpa)	317.94	392	467.91

The Table 3 shows that, with the pretension force increasing, deformation in section direction decreasing, but the decrease range is no more than μm. In order to provide higher measurement accuracy, the section deformation (especially in vertical direction) shall be as small as possible. As the value in Table 3, the maximum deformation in vertical direction is 0.0472mm under the 0.2-0.3N pretension force, satisfied with the requirements of actual measurement accuracy.

Accompanied with the pretension force increasing, deformation in axial direction and principal stress shall increase rapidly. A reasonable value shall be introduced to prevent wire fracture: the maximum principal stresses shown in the table are all less than the tensile strength of tungsten wire (1500~3500N/mm²), satisfied with operational requirement.

In conclusion, the pretension force of spring be set as 0.2~0.3N finally.

7. Wire-clamping Structure

According to the related articles, four types of wire clamping are shown in Fig. 4:

As Fig. 4 shows four sturctures:
a) structure itself is heavy, wire may droop easily, causing low positioning accuracy;
b) structure is difficult to achieve fixing the wire and providing pretension force in parallel, dismounting with inconvenience as well;
c) plenty of soldering shall be avoided in high vacuum;
d) cracking of wire occur easily during hooking process.

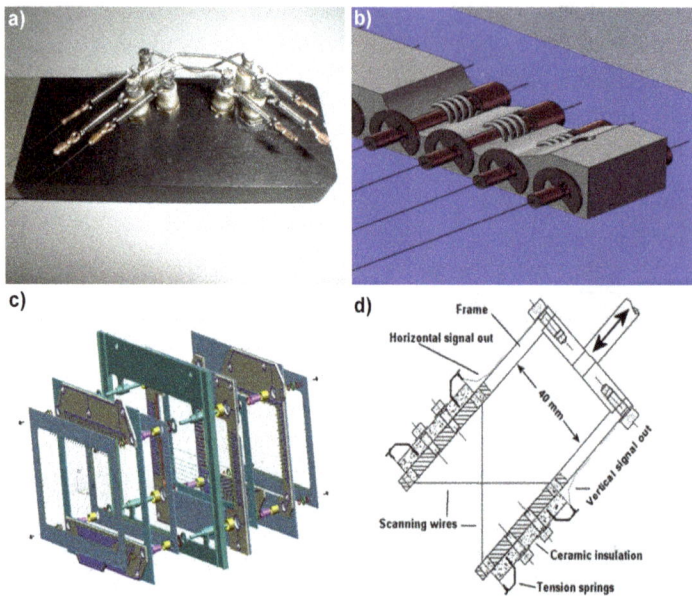

Fig. 4 Type of wire clamping: a) Clamped type [3], b) Compressed type [4], c) Welded type [5], d) Hooked type [6].

According to the four types of wire clamping above, based on b) structure, improve and redesign a new type of wire clamping structure, as show as the following Fig. 5, the main components are: ceramic bar for wire distribution, clamping spring, clamping clip and standard parts such as screw and washer. Ceramic bars are used for supporting and insulation; springs provide the pretension force; clips ensure the positioning accuracy of wire, bolts are used for wire-fixing and retaining. Presentation of the assembly process: signal wires cross through the location holes, which holes in the middle of clamping clip; reach out from screw mounted holes; then twine a couple of turns around the oxygen free copper washer, be compressed and fixed by screw at last.

The novel type of clamping structure divisive the fixing and tensioning function, and wire can be replaced quickly due to the easily assembling structure.

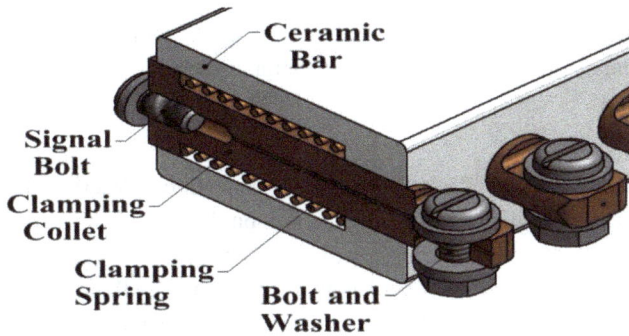

Fig. 5 Wire-clamping structure of MWPM.

8. Wire-distribution Structure

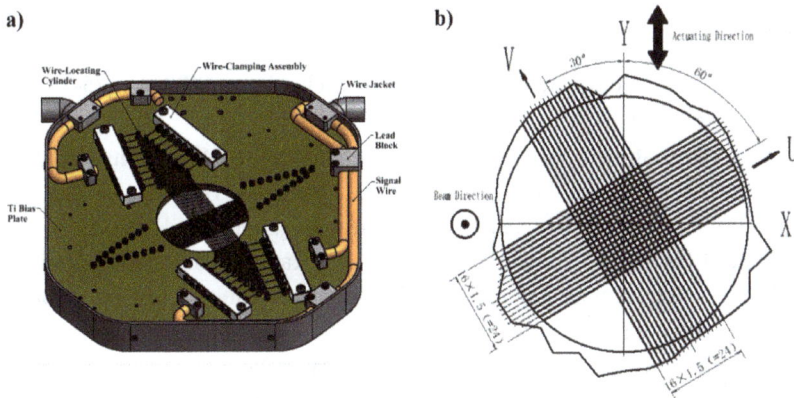

Fig. 6 (a) The wire-distribution structure of MWPM2, (b) The wire configuration of MWPM2.

The integrated layout wire-distribution structure is similar to the shape of lute, as shown in Fig. 6(a). The MWPM2 consists of a U- and a V- wire planes which both consist of 17 wires, while 5 in the 6σ region, with a tilt angle (60° and 30°) to the horizontal and vertical direction on each side of Ti bias plate, as shown in Fig. 6(b). The wire is positioned by the groove of cylinder and fixed by the clamp assembled with a 0.2~0.3N pretension force.

Due to the space constraint of injection region and mechanical structure of clamp assembly, the wire pitch on the same side is determined to 1.5mm and the wire pitch of the two layers is 14mm.

Actually, we can drive the motor moving to gain much higher precise measure data by the wire scanning. The minimum resolution can reach microns.

9. Conclusion

The mechanical design of MWPMs for CSNS injection is described in this essay. All 34 wires in two directions are distribution on the senor, which configured the actual wire pith of 1.5mm. According to the results of wire heating and pretension force evaluating, both sides of the 30μm diameter wire are fixed by new designed clamp assembly which provide about 0.2~0.3N pretension force.

Acknowledgment

The MWPMs are designed, and fabricated by Hefei Keye Electro Physical Equipment Manufacturing Co., Ltd., and expected to install into the RCS injection area in April of 2016.

References

1. Baojie Liu, Information Security and Technology. 9(2014): 75-78. (in Chinese).
2. Tao Yang, Shinian Fu, Taoguang Xu, Zhihong Xu, Ming Meng, Ruiyang Qiu, Jianmin Tian, Lei Zeng, Peng Li, Fang Li, Biao Wang, Nuclear Instruments and Methods in Physics Research A. 760 (2014) 10-18.
3. J. F. O'Hara, J. D. Gilpatrick, LA Day, JH Kamperschroer, DW Madsen, Slow wire scanner beam profile measurement for LEDA, Beam instrumentation workshop. May 8-11, (2000) 5-6.
4. J. Douglas Gilpatrick, Phillip Chacon, Derwin Martinez, John F. Power, Brian G. Smith, Mark A. Taylor, M. Gruchalla. Aug 26, (2014) 1-5.
5. MA. Plum, M. Holding, T. Mcmanamy, Beam parameter measurement and control at the SNS target, Particle Accelerator Conference. (2005) 3913-3915.
6. Strehl, Beam Instrumentation and Diagnostic, Springer Berlin Heidelberg. (2005) 116-117.

Optimization of the Vibration Isolation of AC Dipole Magnet Girder

Jun-song Zhang[†] , Guang-yuan Wang, Ling Kang and Lei Liu
Institute of High Energy Physics,
Chinese Academy of Sciences(CAS), Beijing, China
Dongguan Neutron Science Center, China
[†]Email: zhangjs@ihep.ac.cn

The dipole magnet of the China Spallation Neutron Source (CSNS) Rapid-cycling Synchrotron (RCS) will be operated at a 25 Hz sinusoidal alternating current which causes severe vibration. The vibration will influence the long-term safety and reliable operation of the dipole magnet. For the AC magnet, the girder must have vibration isolator function. It will establish the dynamic modal of the AC diploe magnet by the means of scattered system, and analyze and calculate optimal isolator parameter. Then, according to testing modal analysis and vibration response test, the optimal isolator parameter will be validated. This paper establishes one way of designing dynamic characteristic parameter and it can be used in lots of AC magnet in particle accelerator.

Keywords: CSNS, Accelerator, Diploe girder, Vibration isolator, Dynamic characteristic.

1. Introduction

The China Spallation Neutron Source (CSNS) accelerators consist of an 80 MeV H linac and a rapid cycling synchrotron of 1.6 GeV [1, 2]. The rapid cycling synchrotron (RCS) ring is a four-folded symmetrical topological structure that consists of four arc zones and four line segments. There are 24 sets of dipole mag-nets uniformly distributed in the whole RCS ring, and the magnets will be operated at a 25 Hz rate sinusoidal alternating current. The magnetic core and coils can cause severe vibration, especially at a frequency of 25 Hz, such as in the J-PARC AC dipole magnets. The vibration may influence other equipment through the magnet girder system.

The major function of the magnet girder in particle accelerator is locating and fixing the magnet. The girder has adjustment function; it can make the girder move in six freedom of motion. For the AC magnet, the girder must have vibration isolator function. There are two target of the isolator, reducing the vibration force to the outside environment, reducing the vibration force in magnet self. This paper establishes the dynamic modal of the AC diploe magnet by the means of scattered system, analyzing and calculating optimal isolator parameter.

2. The Vibrational Excitation of RCS Dipole Magnet

The CSNS/RCS dipole magnet is operated at a 25 Hz sinusoidal alternating current of 1100 DC with 816 AC, which causes severe vibration. The dipole magnet girder without vibration isolator was used for the magnetic-field measurement. The major structure of the magnet girder is shown in Fig. 1.

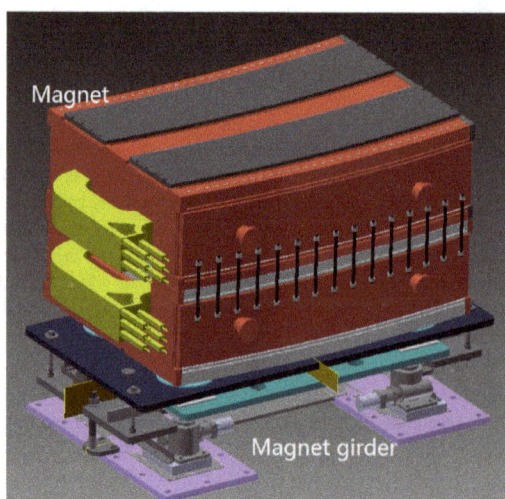

Figure 1. Diploe magnet girder

Acceleration sensors will be used to measure the vibration of the dipole. The testing project and results are shown in Fig. 2 and Fig. 3. The maximum vibration amplitude is from 4.1ηm to 32.6ηm at the vertical direction (y). The main amplitude frequency of all diploe magnet girder is 25 Hz. The amplitude frequency doubling of 25 Hz is much lower than the exciting frequency. At the same time, the vibration influences the other equipment through the magnetic measurement girder. The iron core and coil of the old CSNS/RCS dipole cracked in the magnetic measurement phase. The sixth natural frequency of the dipole-magnetic measurement girder system is 24.75 Hz, which is quite close to the exciting frequency of 25 Hz [6]. The sixth natural frequency may amplify the vibration amplitude of the dipole. So, the girder of the AC dipole must avoid resonance phenomenon through structural design or the use of a vibration isolator.

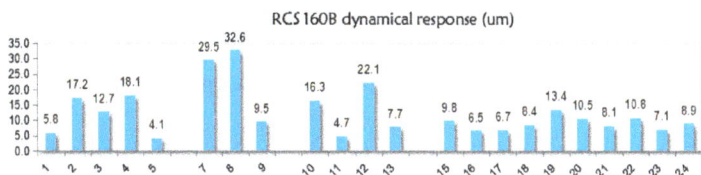

Figure 2. The maximum vibration amplitude of 22 diploes

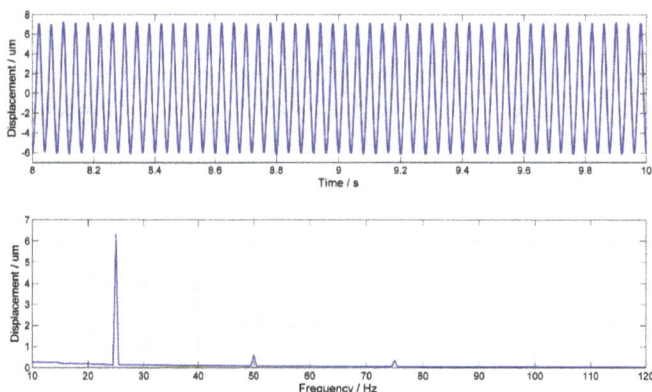

Figure 3. The main amplitude frequency of dipoles

3. Dynamical Analysis of Dipole Girder System

In dipole girder system, four or six vibration isolators are used to connecting the magnet and girder. The Fig. 4 shows the dynamic model of the dipole girder system.

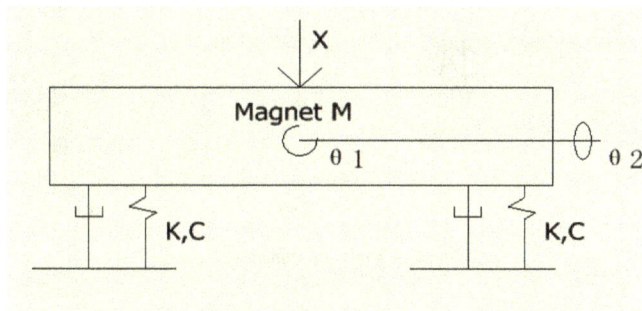

Figure 4. Simplified dynamic model of diploe girder system

971

Under the vertical vibration force, the girder has three freedoms of motion, vertical motion X, rotary motion θ1, θ2 round the normal axis in horizontal direction.

During the magnet working in AC magnetic field, the magnet generates the vibration force F, the main vibration frequency is 25Hz. Dynamic differential equation of the magnet girder system is on following equation.

$$
\begin{bmatrix} m & 0 & 0 \\ 0 & j_1 & 0 \\ 0 & 0 & j_2 \end{bmatrix} \begin{bmatrix} \bar{x} \\ \bar{\theta}_1 \\ \ddot{\theta}_2 \end{bmatrix} + \begin{bmatrix} 2C & 0 & 0 \\ 0 & 2Cl^2 & 0 \\ 0 & 0 & 2Cl^2 \end{bmatrix} \begin{bmatrix} \dot{x} \\ \dot{\theta}_1 \\ \dot{\theta}_2 \end{bmatrix} + \begin{bmatrix} 2K & 0 & 0 \\ 0 & 2Kl^2 & 0 \\ 0 & 0 & 2Kl^2 \end{bmatrix} \begin{bmatrix} x \\ \theta_1 \\ \theta_2 \end{bmatrix} = F(t)
$$

(1)

There are two assumed condition in this equation: All of the vibration isolators in one girder have same rigidity and damping [3]. The mass center of the magnet is on the center of isolator installation site.

When the vibration force transfers from magnet to the girder, the vibration acceleration and motion will be changed. It will be described by vibration transmission factor (TR). It also can assess the vibration isolation efficiency. It can be the vibration force transmissibility or the vibration motion transmissibility.

Vibration transmission factor is equal to vibration force in the girder divided by vibration force on the dipole magnet. By means of Laplace transform, transmission curve can be gotten.

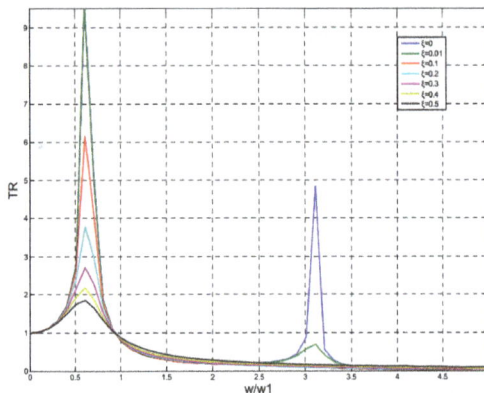

Figure 5. Vibration transmission curve

972

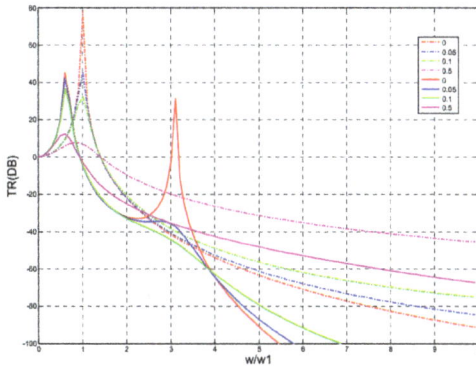

Figure 6. Rate of vertical and whirl transmission decay

In Fig. 6, the solid line means t rate of whirl transmission decay, and the imaginal line means the vertical vibration rate of decay of transmission. According to the Fig. 6, the whirl transmission decay faster than the vertical.

According to the vibration transmission of the system, it can be made a choice for right frequency ratio, and the right springiness and the damping parameter can be designed.

4. Testing and the Experiment Analysis

The vibration responses of two kinds of AC dipole girder system, girder with the vibration isolator and without the vibration isolator, were measured. Meanwhile, the modal analysis of the two kinds girder system [4, 5] was done. As illustrated in Fig. 7 and Fig. 8.

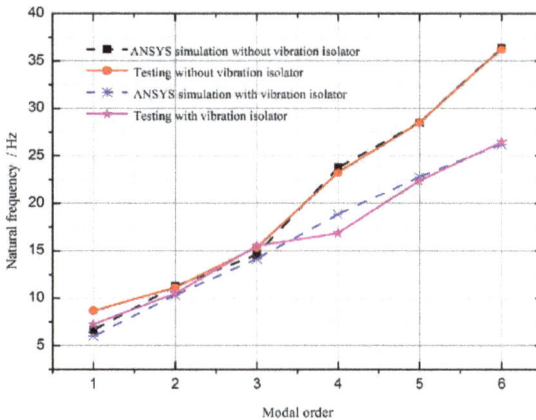

Figure 7. Modal test and analysis

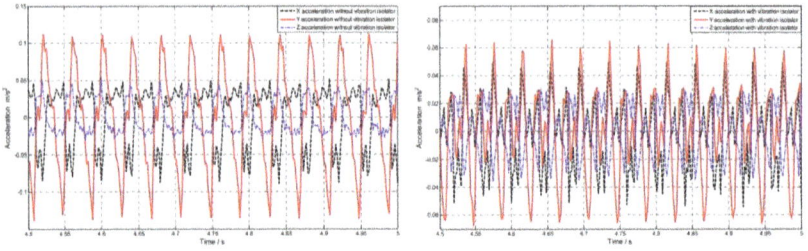

Figure 8. The vibration response test

In the modal test, the dipole girder system with the vibration isolator didn't couple the 25Hz magnet vibration frequency. The sixth frequency of the girder close the 25Hz, but the shape of this step is horizontal movement. It will not enlarge the vertical vibration. By means of modal test and vibration response test, the design parameter was validated that the isolator reduces the vibration force and movement of the AC dipole magnet. The vibration response of girder with isolator reduce about 50% to the without isolator.

5. Conclusion

The dipole magnet and girder play a very important role in accelerator of CSNS/RCS. This paper established the suitable dynamic characteristics modal of the dipole magnet, and analyzed the girder vibration isolator parameter design method. The testing and simulation results show that the dynamic modal of the system is proper. The theoretical calculation natural frequency and ANSYS modal shape are almost identical with the testing results, the simulation method can be used to estimate the dynamic characteristic of the system before manufacture. This paper provides a theoretical basis for the CSNS/RCS formal dipole magnet girder dynamic characteristics design, manufacture and reconstruction. The dynamic characteristic of the system can be veraciously obtained through scattered dynamic differential equation; the resonance phenomenon can be avoided in the design phase.

Acknowledgement

This research was financially supported by project 11375217 of National Natural Science Foundation of China.

References

1. Wei J, Fu S N, Fang S X. China spallation neutron source accelerators: design, research, and development. Proceedings of EPAC, 2006.
2. Yu C H. The amplitude-measurement of the dipole magnet report, Internal-report, 2008, IHEP.
3. Doose C, Sharma S. Investigation of passive vibration damping methods for the advanced photon source storage ring girders, MEDSI02, APS, Chicago 2002, 133–139.
4. Liu Ren-Hong, Qu Hua-Ming, Kang Ling et al. Chinese Physics C (HEP & NP), 2013, 37(8): 087002.
5. Liu Ren-Hong, Zhang Jun-Song, Qu Hua-Min, Wang Hai-Jing. Nuclear Science and Techniques, 2012, 23: 328-331.
6. Ward H, Stefan L, Paul S. Modal Analysis Theory and Testing. Beijing: Beijing Institute of Technology Press, 2001. 1-70.

Precision Control for Sextupole Magnet Yoke in Designing and Manufacturing

Chang-jun Ning[*], Chang-dong Deng, Lei Liu, Yong-ji Yu,
Hua-yan He and Ling Kang

Institute of High Energy Physics, Chinese Academy of Sciences, BeiJin, China
Dongguan Neutron Science Center, Dongguan, China
[*]Email: ningcj@ihep.ac.cn

This paper introduces the production process of sextupole magnet for China Spallation Neutron Source. The yoke of sextupole magnet is made of non-oriented silicon steel sheet, at the both side of the yoke, electrical iron has been used as end plate. The large transverse dimension and thin magnet yoke with large beam aperture is easy to cause small deformation. Optimization of the yoke structure has been made to decrease micro elastic deformation. Heat treatment and machining method of the end plate is proposed. Using the finite element method, analyze the influence of machining error and structure style on air gap precision. By using the heel block in the outside of the magnet yoke, the actual measurement of the prototype shows that the polar air gap precision is in the acceptable range.

Keywords: Sextupole magnet, Micro deformation, Mechanical manufacturing.

1. Introduction

Magnet is the key equipment for particle accelerator, the manufacturing precision of the magnet device directly determines the magnetic field properties, thereby affecting the dynamic characteristics of the particle beam. Sextupole is adopted for chromaticity correction and harmonics correction in the beam storage ring, adjusting the momentum dispersion of particles back to the ideal track [1]. There are sixteen sextupole magnet in Rapid Cycling Synchrotron for China Spallation Neutron Source, using DC power excitation supply.

Figure 1. The yoke of sextupole magnet

The yoke of the magnet is assembled by three one-third iron core, with non-oriented silicon steel thin sheet stacking, and using electrical pure iron as end plate. As shown in Fig. 1, there is clamping screw in iron core to avoid punching sheet loose, and outside the pull plate welding. The magnet yoke has a transverse size of 960×960mm and only 170mm of the thickness. For the convenience of the alignment operation, a fixed target base is arranged on the top of the iron core.

2. Iron Core Structure Characteristics

Sextupole magnet consists of three independent magnetic yoke assembly; the angle between the surfaces of the assembly is 120 degrees. The iron core is provided with a large, thin structure characteristic, each independent magnetic yoke includes two pole head, the pole head cantilever longer, the tiny displacement in the polar root will enlarge the displacement in the pole head [2]. Compared with the transverse dimension of the core structure, pole position accuracy design requirements are controlled within 0.05mm, which can be classified as structure micro elastic deformation. To limit the structure small elastic deformation, but also to avoid a significant increase in the overall weight of the core, while meeting the requirements of the magnetic circuit, a reasonable structure is essential. After the coil is installed, the iron core also generates a certain amount of micro elastic deformation under the action of gravity. The machining accuracy of the yoke pole head and assembly-binding surface requires a strict control. Silicon steel sheet is manufactured by a metal-stamping die production, and in punching a certain number of product sampling, then measuring the profile of sample using a three coordinate measuring instrument, ensuring the mold profile size within the allowable error range. And the magnet end plate is manufactured individually by using wire-electrode cutting. Thus there are a little error difference between punched steel sheet and the end plate. After the completion of the actual stacking and there is a certain deviation between end plate and punched steel sheet in the pole head surface.

3. The Influence of the Structure of the Core and the Angle Precision of the Joint Surface on the Air Gap of the Polar Head

Sextupole magnet structure is shown in Figure 2, the angle between the core joint surfaces is 120 degrees, and the outer side of the iron core is connected with a bolt and a cushion block. The contour curve of the pole head surface is given by equation [3]:

$$3x^2 y - y^3 = r^3$$

(1)

In which $r = 150$ mm. Due to the existence of machining errors in the actual production, such as the angle error of the joint surface, the contraction caused by the welding block, etc., will affect the precision of the air gap. The following through the finite element analysis method, analyzes the influence of the three factors on the air gap precision: the combination of surface angle error, have or don't have heal block between the welding block outside the iron core.

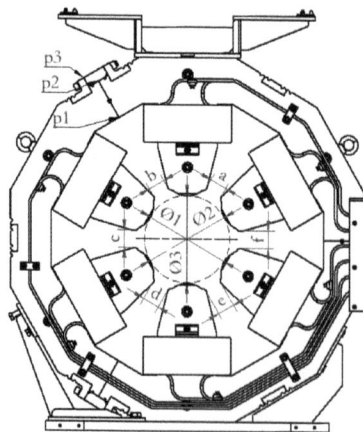

Figure 2. The structure of sextupole magnet

Because the iron core adopts the lamination process, and use the side plate welding and bolt fixing. In the analysis of the above three factors, first assume that in one-third of the inner core the punched steel sheet and the end plate is integrated as one unit, that means there is no relative sliding between the

978

punched steel sheets and the end plate, namely position of the different thickness of the iron core in the form of variables is consistent. The angle error of the joint surface is estimated according to 0.03mm of the lateral surface of the joint, and the influence of gravity is taken into account. Simulation analysis of three kinds of conditions has been executed. The size of the gap in the position of a~f (refer with Fig. 2) is shown in Table 1, in which the negative value indicates that the adjacent poles close to each other, and the positive value indicates departure.

Table 1. Effect of three factors on air gap precision

Factor	Δa (mm)	Δb (mm)	Δc (mm)	Δd (mm)	Δe (mm)	Δf (mm)
No heal block	-0.057	0.078	-0.083	0.066	-0.062	0.046
Block	-0.002	0.004	-0.006	0.006	-0.006	0.003
Angle error	-0.087	0.157	-0.119	0.100	-0.118	0.090

Analyzing the influence of various factors from the results can be qualitatively obtain the conclusion: in the case of no heal block, the yoke has been micro deformed under the action of bolt pre tightening force, and the fulcrum of the lever is at p2 point, thus larger displacement in the polar head have been amplified due to longer cantilever. Compared to non-heal block, the fulcrum of the lever is at p3 point when there have block at the outer side, the polar air gap displacement will not be amplified by cantilever. In the case of angle error analysis, the fulcrum of the lever is at p1 point, the leverage effect is more obvious comparing with the first case, so the air gap deviation greater than all else. Actually, if the outer side of the punching sheet and the end plate has an ear like protrude reserved, there will be more convinced.

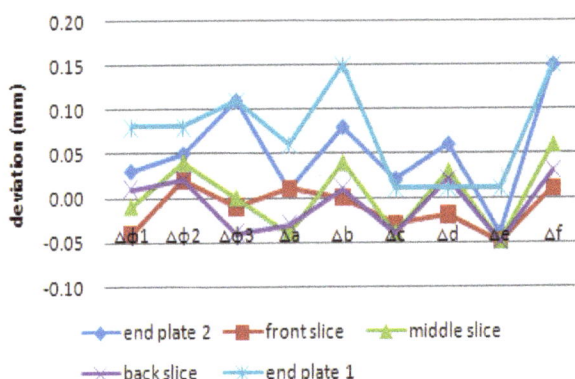

Figure 3. Air gap deviation of magnet 1

After the completion of the manufacturing equipment, precision of yoke air gap was measured using standard block with a feeler gauge measurement. In each position measured (refer with Fig. 2), there are five measurement points selected respectively in the direction of the thickness of the core, two of the five in the position of the end plate, three other arrangements in different positions on the punched steel sheet. Two magnet air gap accuracy measurement results are shown as in Figs. 3 and 4.

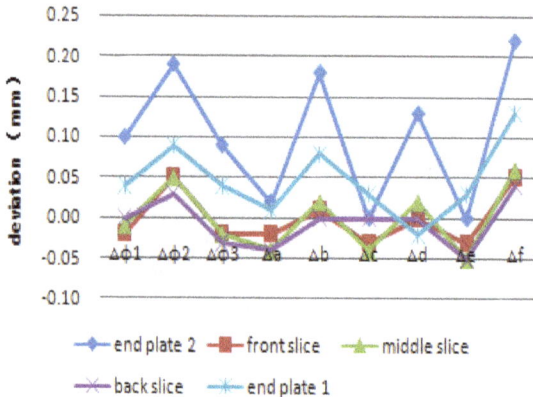

Figure 4. Air gap deviation of magnet 2

It is obviously that the air gap deviation of the steel sheet meets the accuracy requirements of the design basically, but the air gap deviation of the end plate is larger. Mainly due to the processing of the end plate are manufactured separately by wire cutting, the deviation will slightly larger; and silicon steel sheet are manufactured by punching mold processing, the precision of pieces is consistent with each other.

4. Processing Method of End Plate

In the prototype trial production process, flame cutting is used for cutting the material, large amount of material is reserved at the contour edge of the end plate, and the cutting profile is shown in Fig. 5. Then the end plate is manufactured by wire cutting after temper treatment. First we need to select a benchmark on the end plate, then the exact contour wire cutting. In order to avoid the precision deviation of the yoke pole air gap generated by the deformation of stress release, there need to 3mm material reserved in the pole face for second precision wire cutting. Actually after our first prototype assembled, measurement result shows that there have serious precision deviation in the end plate pole face. The process of manufacturing is examined, we think the reason of end plate large deviation mainly lies in the fact that the material internal stress is released in the manufacturing process and caused micro deformation.

Figure 5. End plate blanking by flame cutting

So we have improved manufacturing process, first using water cutting the contour of the end plate after which basic profile has been obtained. Tempering elevating temperature curve need to be strictly controlled, then two times wire cutting process, end plate finally meets the accuracy requirements.

5. Conclusion

The six pole magnet transverse size is large, but with thin length. In order to meet the requirements of magnetic circuit and don't increase the mass and size of the magnet device obviously. In the design and manufacturing process, the facts, which have influence on the yoke air gap precision have been considered strictly.

a) The sextupole magnet yoke have a large transverse size and thin length structure which prone to cause micro deformation in elastic range, thus there need to strictly control the processing accuracy in order to ensure the quality of the magnetic field. In the design process, the inflexibility of the yoke needs to be strictly verified. Due to the polar root is long, the micro deformation caused by the bolt pre-tightening force in the polar root tend to produce large displacement in the polar face.

b) Due to the angle error of the yoke joint face caused by welding, under the action of the bolt pre-tightening force, the air gap deviation is obviously increased.

c)   Iron core connecting bolt should be within the joint face, otherwise it is easy to cause the lever effect by pre-tightening force.

d)   The silicon steel sheet punched by mold have good consistent with each other, but the end plate is inferior a little, and there is a certain deviation between the end plate and the punching sheet.

e)   For the large and thin end plate, there will lead to larger internal stress reserved and caused micro deformation when manufacturing. Tempering temperature should be controlled strictly, the water cutting can significantly reduce the internal stress of residual. Secondary wire cutting in the polar face can better guarantee the accuracy of polar.

Acknowledgement

This research was funded by the National Natural Science Foundation of China (Project 11375217).

References

1.   S. Y. Lee, Jack, Accelerator Physics (2nd edition), Indiana University, USA 2006.

2.   Shi Cai-Tu, Ni Gan-Lin, Sun Xian-Jing, etc. Design and fabrication of the magnets for BEPC II project, J, Chines Physics C, 32(2008) 7-10.

3.   Jack Tanabe. Iron Dominated Electromagnets Design, Fabrication, Assembly and Measurement, January 2005.

Research on Incremental Support Vector Machine and Its Application in Sintering Process Modeling

Jian-xin Wang[1,*] and Yong-gang Wu[1,2]
1School of Mechanical Engineering,
Inner Mongolia University of Science and Technology, Baotou 014010
2Ironmaking Plant,
Inner Mongolia Baogang Group, Baotou 014010
Email: zhwx_335100@163.com

Traditional Support Vector Machines (SVMs) in solving large-scale sample training, with the solution to occupy a quadratic programming problem slows convergence speed. When a new sample joins together, all samples need to train together is time consuming and limit the scope of use. In view of the above defects, this paper proposed a method based on generalized KKT constraint and boundary vector incremental Support Vector Regression (SVR). The sample of a little effect is deleted timely, while retaining important information. Using the database of sintering process of this method was verified. Theoretical analysis and experimental results show that this method is a more traditional support vector regression algorithm. The training speed is faster and higher accuracy with stability.

Keywords: Incremental learning; Support vector machine; Sintering process; General KKT conditions.

1. Introduction

Support vector machines are a statistical learning method with the foundation of Structure Risk Minimum (SRM), which have solved the problem of disaster dimension, and generate an optimal separating hyper plane by minimizing the generalization error [1]. SVMs are widely applied in face recognition [2], fault diagnosis [3], water quality prediction [4], etc. However, the traditional support vector machine for solving all the training samples are required in solving the convex quadratic programming, which has involved m order matrix calculation (where m is the number of learning samples), when m is a great number, the storage and calculation of the matrix will spend a lot of machine memory and counting-time training, the training speed is slow, even which cannot calculate because of memory leak [5]. And the traditional support vector machines learning based on the historical data of the off-line, so those cannot use the real-time data for online learning. In practice, the data of the production process will always change, such as production equipment wear, aging phenomenon will make data change greatly, if the model cannot be accessed for learning online, and has not grasp these changes, the accuracy of the model will be lower. If all old and new samples have been together to retrain, which need a long time and even can't be calculated.

Many scholars aiming at solving those problems, this paper puts forward different SVM incremental learning algorithm. Syed and others proposed the Batch-SVM incremental learning algorithm [6], abandoned the non-support vector of the historical sample set, put support vector into the new sample for training, so as to achieve the purpose of incremental learning. This algorithm has decreased the number of incremental sample, which will participate in the incremental training, but hasn't selected for the new sample set. Regardless of whether it contains new information, all join incremental learning, this waste is obvious. Zeng and others proposed the mutual KKT conditions incremental learning algorithm [7] (K - ISVM), the first respectively on the sample set of initial and incremental training, gain their own classifier and support vector sets, then find out the conditions of violation of the other by each other, joining the conditions of samples of their respective support vector sets training together for the ultimate classifier. Classification accuracy has been improved, but those most likely to become the final SV haven't been considered to the violate KKT conditions samples. In order to solve these problems and apply into sintering process modeling, on the basis of above studies, I proposed a new incremental regression support vector machine (New-ISVM), which is based on the KKT conditions and boundary vector, and validated using the standard database.

2. Support Vector Machine Regression Theory

Define the search-evade antagonistic not sensitive loss function for:

$$\left| y - f(x) \right| = \begin{cases} 0; & \left| y - f(x) \right| \le \varepsilon \\ \left| y - f(x) \right| - \varepsilon; & \left| y - f(x) \right| > \varepsilon \end{cases}$$

(1)

The search-evade antagonistic not sensitive loss function is the meaning of: when x point target y and forecast f (x) of the poor shall not exceed prior given search-evade antagonistic, thinks that the point prediction of f (x) is without a loss. That is a without a loss of search-evade antagonistic-belt, its main losses from the search-evade antagonistic-with outside of the samples.

Given the training sample sets $\{(x_1, y_1), \ldots, (x_l, y_l)\} \in (x * y)^l$, such that $x_i \in R^n$ is an input, $y_i \in R$ for target, l as sample number. In order to allow certain error exists, improve the generalization ability of the SVM model introducing flabby variable factor ξ_i, ξ_i^*; deal with nonlinear problem, introducing the kernel functions to input the sample space from the low-dimensional nonlinear transform into a higher-dimensional space solving linear characteristics, including kernel function $K(x_i, x_j) = \varphi(x_i) \cdot \varphi(x_j)$, optimization problem as follows:

$$\min \quad \frac{1}{2}\|w\|^2 + C\sum_{i=1}^{l}(\xi_i + \xi_i^*)$$

$$s.t \begin{cases} y_i - w \cdot \varphi(x_i) - b \le \varepsilon + \xi_i \\ w \cdot \varphi(x_i) + b - y_i \le \varepsilon + \xi_i^* \\ \xi_i, \xi_i^* \ge 0, i = 1, 2, \ldots, l \end{cases} \tag{2}$$

The dual is:

$$\min \quad \frac{1}{2}\sum_{i,j=1}^{l}(a_i - a_i^*)(a_j - a_j^*)K\langle x_i, x_j \rangle + \varepsilon\sum_{i=1}^{l}(a_i + a_i^*)$$

$$- \sum_{j=1}^{l} y_i(a_i - a_i^*)$$

$$s.t \begin{cases} \sum_{i=1}^{l}(a_i - a_i^*) = 0 \\ a_i, a_i^* \in [0, C] \end{cases} \tag{3}$$

The regression estimation function is:

$$f(x) = \sum_{x_i \in sv}(a_i - a_i^*)K(x_i, x) + b \tag{4}$$

Example with non-zero lagrangian multiplier are called Support Vectors (SV), only a small fraction of the coefficients $a_i \neq 0$ or $a_i^* \neq 0$ will affect decision-making function, so solution is sparseness; this has guiding significance to incremental learning algorithm constructing [8].

3. Incremental Learning Algorithm

Incremental learning algorithm is described below [9]:

Prerequisite: existence initial samples is A, incremental samples B, and assume that guarantee two samples meet $A \cap B = \emptyset$, SVM_A and SV_A respectively on A support vector machine and the corresponding support vector sets,

Goal: seek in $A \cup B$ on support vector machine (SVM) and support vector sets. In the support vector regression machine, the KKT conditions for:

$$\begin{cases} a_i^{(*)} = 0 & \Rightarrow \left| y_i - f(x_i) \right| \leq \varepsilon \\ 0 < a_i^{(*)} < c \Rightarrow \left| y_i - f(x_i) \right| = \varepsilon \\ a_i^{(*)} = c & \Rightarrow \left| y_i - f(x_i) \right| \geq \varepsilon \end{cases}$$

(5)

where $a_i^{(*)}$ Including a_i and a_i^* are lagrange multipliers.

Circular points for the initial sample In figure 1, thick lines for its training get support vector machine, fell on fine solid online dot for support vector, thin lines for the area between the search-evade antagonistic - belt. $a_i^{(*)}=0$ corresponding sample distribution in the search-evade antagonistic - in the band, $0<a_i^{(*)}<C$ samples in the search-evade antagonistic - take, $a_i^{(*)}=C$ samples which are located in the search-evade antagonistic - take away. When the diamond of new sample to join, because the new sample didn't get original support vector machine, make all the training sample of new Lagrange multipliers $a_i^{(*)}$ to 0.

Violation of the support vector machine KKT conditions of new sample located in outside the search-evade antagonistic. New samples existence violate KKT conditions of samples, contains the original samples with new information and support vector machine (SVM) will change, these new information there would have been a new SV. But not all the samples are KKT conditions violation can become the new SV, only those boundary conditions of samples and violates KKT most likely to become the new SV. When new sample is original "SVM SV, become the new SV very likely, would break the KKT condition defined relaxing: | y - f (x) | p search-evade antagonistic, called against generalized regression KKT conditions. Boundary vector is not necessarily the support vector, but support vector must be boundary vector. Therefore this paper selects those already KKT conditions, and violation of boundary sample joining incremental learning vector.

Boundary vector extraction process:
1) Setting threshold $D>0(D> \varepsilon)$, define ε-deft:

$$y= (w. \ x)+b+ \varepsilon, \ y= (w. \ x)+b-\varepsilon$$

(6)

2) Define sample forgetting factor: $d \ (x_i)=y- f(x_i)$, it is the distance to $f(x)=(w. \ x)+b$,

$$\overline{d} = \frac{1}{n} \sum_{i=1}^{k} \left| d(x_i) \right|, i = 1, \cdots, k$$

(7)

3) Loot with search-evade antagonistic - with lower threshold D for new samples of upper, $B'=\{x_k / \varepsilon \leq d(x_i) \leq D\}$, Boundary vector with threshold D to change, This paper threshold D take $\overline{d}$, because mean represents violates the KKT sample concentrated trend.

Algorithm process:

1) The initial samples with X_0 training get support vector machine SVM_0 and support vector sets SV_0;

2) Test new sample X1 whether exist sample in violation of KKT conditions of SVM0 (for the return of the problem whether there is speaking, the inspection sample beyond the search-evade antagonistic - belt), if does not exist, SVM0 namely for studying the result, the training to cease, If present, find new samples of boundary vector sets B" ;

3) SV_0 and B'' constitute a new training corpus clamps $X1''$, get the new support vector machine SVM1 and support vector sets SV1;

4) When new samples joining again, repeat steps (2), (3), through multiple learning, for the ultimate support vector machine.

4. Application in Sintering Process Modeling

Take the real date of sintering process of BAOGANG Group limited company as the object. The dataset contains 532 samples, which statistic the alkalinity of sintered ore and its 17 influencing factors, there are material weight, high return weight, cold return weight, charred coal weight, dry distillation coal weight, Mongolia mine matching, mixing mine matching, quicklime weight dolomite weight, FeO, CaO, SiO2, Fe, units speed, ignition temperature, main pipe negative pressure, main pipe temperature.

In order to validate the proposed method (New-ISVM), take the RBF function as the New-ISVM kernel for its good nonlinear processing power and stable performance, using cross validation methods select support vector parameter and nuclear parameters. Model evaluation criteria are the model training speed and prediction sample mean square error:

$$mse = \frac{1}{n}\sum_{i=1}^{n}\left(y_i - y_i^*\right)^2$$

(8)

where y_i is target, y_i^* is prediction, n is the number for the prediction sample.

First take132 Initial training samples, incremental learning points on three times, and each increment join 100 samples, the remaining samples as prediction samples. Each incremental learning as the result of the next incremental learning of the initial situation, write down every training needed time and use every training model of the testing samples obtained forecast and calculation measurement errors. This paper also compared with the traditional SVM, Batch-SVM and K- ISVM method, the experimental results are shown below:

Table 1. Performance comparison of different models

The learning process	SVM		Batch-SVM		K-ISVM		New-ISVM	
	t	mse	t	mse	t	mse	t	mse
Initial training	7.9	0.035	7.9	0.035	7.9	0.035	7.9	0.035
First incremental 100	75.3	0.031	10.3	0.027	9.8	0.029	9.2	0.024
Second incremental 100	460.5	0.032	20.9	0.041	15.3	0.025	14.3	0.022
Third incremental 100	1027.6	0.029	27.4	0.037	19.5	0.024	15.9	0.022

The experiment result shows that the traditional support vector machine because is for small sample questions, when the sample is much increase, its training speed slow. Batch-SVM, K-ISVM than traditional SVM speeds although increased, but the prediction precision not much increased. It shows that New-ISVM is not only faster, but also higher accuracy. It is effective for sintering process modeling.

5. Conclusion

This paper puts forward a kind algorithm of generalized KKT constraint and boundary vector selection for incremental SVR. Considering the increment algorithm in learning new sample after accession samples, which can determine the incremental learning new SVR KKT conditions, both violation of new samples will most likely be support vector boundary vector join many study, promptly eliminate little impact on subsequent regression forecast of sample, while retaining the original samples of SV set to join incremental after new SVR training. Based on what the predictive regression Machine data and experiment show that this method can effectively improve support vector Machine learning speed and has high precision of prediction. This can be considered for support vector machine processing large sample problem and online study.

References

1. X.G. Zhang, Introduction to statistical learning theory and support vector machines [J]. Acta Automatica Sinica, 2000, 26(1): 32-42 (in Chinese).
2. E. Osuna, R Freund, Training Support Vector Machines: An Application to Face Detection [A] Proc of Computer Vision and pattern recongnition [C]. San Juan, Puerto Rico, IEEE Computer Soc, 1997: 130-136.

3. S. Yuan, F. L Chu, Support Vector Machines and its application in machine fault diagnosis [J]. Journal of bartion shock, 2007, 26(11): 29-35 (in Chinese).

4. L. Shen. Application of particle swarm optimization and support vector machine in simulating and predicting stream water quality [D]. Zhejiang Normal University, 2008 (in Chinese).

5. W.Y. Zha, Based on the incremental learning support vector machine classification algorithm research and application [D]. Kunming university of science and technology, 2008 (in Chinese).

6. N. Syed, H. Liu, K. Sung, Incremental learning with support vector machines [A]. IJCAI[C]. Morgan kaufmann publishers, 352-356, 1999.

7. W.H. Zeng, J. Ma, A novel approach to incremental SVM learning algorithm[J]. Journal of Xiamen University, 2002, 41(6): 687-691 (in Chinese).

8. N.Y. Deng, a new method of data mining: support vector machine (SVM) [M]. Beijing: science press, 2004 (in Chinese).

9. K. Li, H.K. Huang, Research on incremental learning algorithm of Support vector machine [J]. Journal of Northern Jiaotong University, 27(5): 34-37 (in Chinese).

The Research of Multi-Station High-Precision Rotary Motion of CSNS Stripper Foil System

Jia-xin Chen[1,2,*], Jie-bing Yu[1,2], Zhe-xi He[1], Guang-lei Xu[1]
and Ling Kang[1,2]
*Institute of High Energy Physics,
Chinese Academy of Sciences (CAS), Beijing, China*
²Dongguan Neutron Science Center, Dongguan 523803, China
*Email: chenjx@ihep.ac.cn

Multi-station rotary structure is used in the China Spallation Neutron Source (CSNS) stripper foil. The repeat positioning accuracy of per foil frame is more than 0.5mm, which do not meet the technical requirements, due to the rotary membrane structure and the frame accuracy. This paper researches for the multi-axis synchronous system to control the rotary membrane and achieve the repeat positioning accuracy to be better than ±0.05mm. This paper also researches on the remote control system based on the EPICS and realize the remote control of stripper foil system.

Keywords: Multi-station, High-precision rotary motion, EPICS, Stripper foil.

1. Introduction

Remote online foil changing and foil position online adjustment are the two main functions of the China Spallation Neutron Source (CSNS) stripper foil [1]. The structure of the stripper foil [2] is shown as Fig. 1. The stripper foil consists of support frame, vacuum box, foil exchange system with 22 foils, foil position adjustment system, foil storehouse and the control system. The foil exchange system is including the stainless steel, magnetic fluid seal, cam splitter, limit switch and motor. The foil positon adjustment system is including the horizontal slide table, vertical lifts, front and rear bellows. The foil number 0 is defined as the reference foil.

2. Realize of Multi-Station of the Rotary Movement in Vacuum

There are there motors to realize the foil exchange and position adjustment online. The rotary membrane rotated by the foil exchange motor, and the foil

---

*Corresponding author.

exchanged. The other two motors realize the horizontal and vertical position adjustment. The vacuum box moves overall, with the foil storehouse and stainless steel, and the foil position changed at the same time, which realized the foil positon adjustment. When the foil position adjustment, the bellows, which can achieve tension, compression and dislocation, ensure the vacuum is not destroyed. Exchange membrane motion sequence is as follows. The foil exchange motor drives the cam splitter with the accurately 180 degrees output, then the magnetic fluid seal rotary the motion to the stainless steel which stated into the vacuum inside system, and the foil number changed. The horizontal slide table and vertical lifts move at the same time to compensate the positon deviation relative to the reference.

The stripper foil uses a magnetic fluid seal to connect the movement inside and outside the vacuum. The magnetic fluid dynamic seal leakage rate can reach 1.0e-11Pa·m3/sec. The input shaft and output shaft are the same shaft, without transmission errors. To meet the requirements for resistance to radiation, the magnetic fluid has a special custom formulation, which can withstand radiation doses of more than 5.0e8Rad.

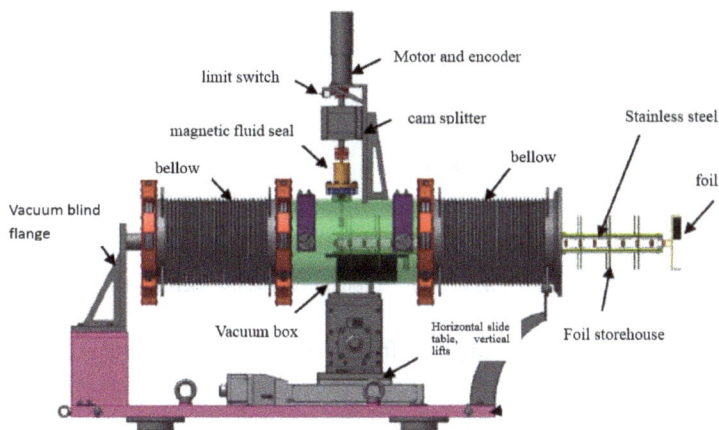

Fig. 1 The structure of the CSNS stripper foil [2]

The stainless steel is a cantilever structure part, assumed that a total length L, the angle θ, the X direction offset dimension diaphragm δx, and Z direction offset dimensions δz.

$$\delta x = L(1-cos\theta) \tag{1}$$

$$\delta z = L sin\theta \tag{2}$$

$$\theta = min(arcsin(\delta z/ \tag{3}$$

$$arccos(1-\delta x/L) \tag{4}$$

The stainless steel driving shaft and the driven shaft with a wheelbase of 836mm, take $\delta x = \delta z = 0.5mm$, θ was 0.034 °, it is necessary to strictly control the strip in order to ensure repeat positioning accuracy to make sure that the system overall unchanged.

The single foil repeat removable positioning accuracy affects the foil position compensation. The repeat positioning accuracy of each foil frame is 0.011mm at direction X and 0.013mm at direction Z, and the accuracy make little effect on the foil position.

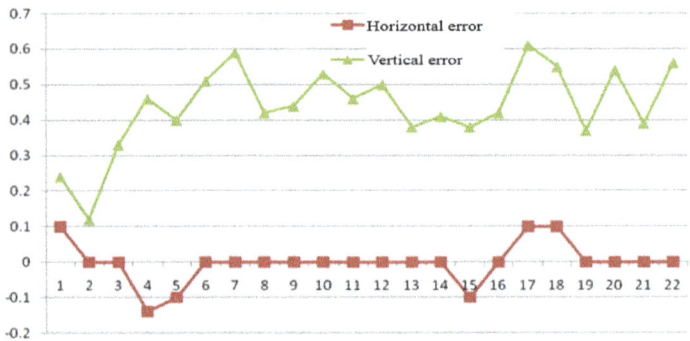

Fig. 2 Errors between 22 foil frames and the reference frame

There are 22 preinstalled foil frames and one reference frame on the stainless steel, and each foil frame is required better coincide the reference frame with the position accuracy ±0.1mm. The repeat positioning accuracy of per foil frame is more than 0.5mm, which don't meet the technical requirements, due to the rotary membrane structure and the frame accuracy. The errors between 22 foil frames and the three coordinate measuring machines had measured the reference frame, which will be the compensation values of the control system. The errors are showed in Fig. 2. As the figure shown, the biggest error is 0.6mm.

3. The Rotary Motion Control System Research

Local control system

The automated foil exchanging and position adjustment of the CSNS stripper foil [3] are realized by motor drive. The local control system is composed of Yokogawa PLC, decoder, motor driver and etc. The local control system develop the three-axis foil position compensate, the function of interlocking between the axis and the homing function. The PLC is composed of circuit board module, power supply module, CPU module, input and output of the diaphragm, the Ethernet module and the position module.

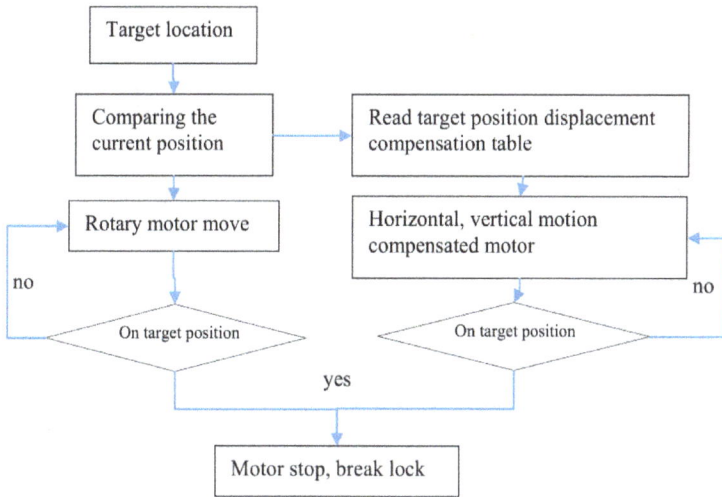

Fig. 3 Local motion control schematic

Local control program using the ladder Widefield2 compiling, and realize realize motor motion control, position compensation data storage, data input and output, online film and change position adjustment of the foil. In order to further improve the repeatability of each foil positon, the local control system increase the function of three-axis position compensation, that 22 foil position errors are storing in a register data table, using the pointer to read the different foil data, then the compensation value is written to motion control.

Remote Control System

The stripper foil uses a remote control system based on EPICS SoftIOC and Yokogawa PLC realized data connection, and via Ethernet realized hardware connection. The remote control system is shown in Fig. 4. The central control room remotely using OPI interface can directly control the movement of the [4] foil, the parameters can be read on the local controller via remote operation, or modify, set variables, modify the position compensation table data or motor brakes are closed and opened.

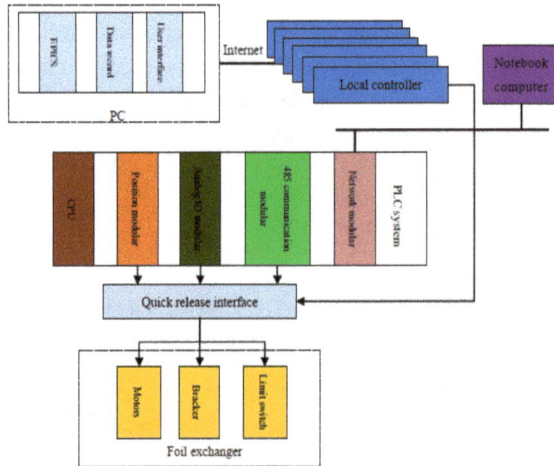

Fig. 4 Remote control schematic

4. Calibration of Precision Rotary Motion

The errors between 22 foil frames and the reference frame had been measured as shown in Fig. 2, and the errors had been written into the table. The stripper foil now is moving with the compensation and the errors between 22 foil frames and the reference frame had been measured, as shown in Fig. 5. The horizontal repeat position accuracy is 0.05mm, and the statistical accuracy is 0.02mm. The vertical repeat position accuracy is 0.03mm, and the statistical accuracy is 0.01mm.

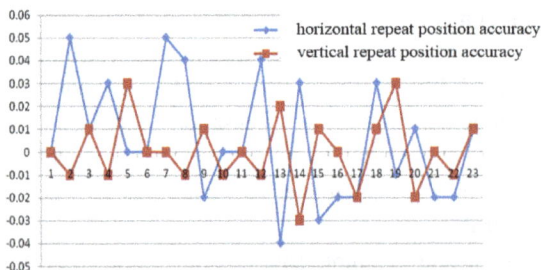

Fig. 5 The measurement result after the position compensation

5. Conclusion

This paper describes the structure of the CSNS stripper foil system and multi-station high-precision rotary motion, local control and remote control system based on the EPICS. The three-axis control system is used to realize the accuracy from ±0.5mm up to ±0.05mm, which has an important reference value to the movement inside and outside vacuum and multi-stations high-precision motion.

Acknowledgement

This research was funded by the National Natural Science Foundation of China No. 11375217.

References

1. Xu Guanglei, He Zhexi, Chen Jiaxin, Li Chunhua. Study of Control System of CSNS Main Strip Foil Base on EPICS [J]. Nuclear Electronics & Detection Technology, 2014, 02:160-163+169.
2. He Zhexi, Li Chunhua, Qu Huamin, Xu Guanglei, Zou Yiqing. Design of primary stripper foil changer for CSNS/RCS [J]. High Power Laser and Particle Beams, 2012,12: 2885-2888.
3. He Zhe-Xi, Qiu Jing, Li Chun-Hua, Chen Jia-Xin, Kang Ling, Qu Hua-Min, Zou Yi-Qin. Wrinkle analysis and mounting optimization of the primary stripper foil for CSNS [J]. Chinese Physics C, 2013,10:85-90.
4. He Zhe-xi, Qiu Jing, Li Chun-hua, Qu Hua-min, Zou Yi-qing. Analysis of Thermal Effect for China Spallation Neutron Source Primary Stripper Foil [J]. Atomic Energy Science and Technology, 2013, 10:1867-1871.

Buttons on Demand Using Sliding Mechanism

Sivaperuman Kalairaj Manivannan[1], Wei-min Huang[2,*]
and Felix Klanner[3]
*[1]School of Mechanical and Aerospace Engineering,
Nanyang Technological University, Singapore*
*[2]School of Mechanical and Aerospace Engineering,
Nanyang Technological University, Singapore*
*[3]Energy Research Institute,
Nanyang Technological University, Singapore*

Almost all the electronics and mechanical equipment have controllers to control the entire equipment or some specific component in the equipment to do a certain task. These controllers responds to various types of inputs and performs the function accordingly, one type of input given to the controller is the touch. These controllers, which use human touch by means of either touch screen or buttons. The touch screen will not give haptic feedback to the users whereas the buttons will take up more unwanted space and also causing disturbance and confusion in recognizing due to ambiguity, either buttons are is active, or inactive. In this paper we present a simple way to achieve the buttons on demand, by which the buttons remain flat at the surface when inactive and protrudes out of the surface and appears as a button for operation when active. The design and the functionaility of the controller are presented here. The design and simulation are carried out on Solid Works.

Keywords: Buttons on Demand; Surface Patterns; Sliding Mechanism.

1. Introduction

Now-a-days automobiles are used vastly in various applications from personal to commercial and the automobiles have evolved from manual to semi or fully automated and one of the most important components that aided the evolution is the controller. The controllers, which are used in automobiles are used to perform several functions ranging between turning-on to self-driving and also controllers could be either handy and portable or also fixed. The controllers are used in automobiles to control almost all the components. It could be that one controller to control all the functions in an automobile or a controller to control a specific function or a specific component in an automobile. Most of the controllers, which are used in automobiles receives signal by human touch, and they could be either touch screen or having buttons. The controllers, which have the buttons were the first of this kind and were bulky and also have several disadvantages, like ambiguity in recognizing the functions of every button and also in recognizing which is active and which is not also in recognizing which are the set of buttons that are suitable for this operation. The advanced and latest controllers are touch screen where the humans can touch and interact with the screen to perform the functions, but this kind of interaction does not provide haptic feedback to the humans, which is necessary for effective communication between the human and the machine. The buttons on demand are necessary as it increases the human-machine interaction and also not all the buttons are required during operation of vehicle. In this paper we present simple way to achieve buttons on demand using a sliding mechanism, which can be conveniently fitted into current systems.

2. Literature Survey

Patent No. US 20070247420 A1 [1] explains about a reconfigurable, shape-changing buttons on a vehicle control panel. It uses shape memory alloy and/or shape memory polymer to control the movement of the buttons during active and inactive states. During the inactive state, the button remains on the surface and when it is changed to the active state, there is a protrusion in a hemispherical or dome shaped. Whereas Patent No. US20140257567 A1 [2] explains to provide concealable physical buttons while the display associated with the buttons and surroundings remain stationary. In such a manner the buttons or hard keys are concealed when not interacting and rise above the surface when needed. This motion occurs due to the presence of motor, a threaded shaft coupled to the motor, a movable platform coupled to the threaded shaft and a hard key movably supported on the movable platform. Due to this arrangement and the movement of the motor, the hard key rises from the surface when required and this also provides haptic feedback to the end-user. This method requires more space and uses more moving parts, which require large actuation power. In another method, the buttons could rise over the surface on demand by the help of Tactus Technology [3]. In this way, the button is formed with pressure in small channels inside of the substrate and the small holes let the fluid in the area in-between the elastomer and the substrate. The rise occurs due to the expansion of the elastomer surface when the compressed fluid is allowed to pass through the small channels. The compressed air and the micro-valves are switched using energy. In this technology, the response time might be slow and liquid leakage could be a potential problem.

3. Design of the Mechanism for Buttons on Demand

3.1. *Design of the Sliding Mechanism Actuated by Gear*

In the controller, which is actuated by the sliding mechanism, there are six components and they all play a major role in the mechanism and the interaction. The six components are Top Cover, Buttons, Slider, Pinion Gear, Bottom Cover and the Mechanochromic material that is placed on the top cover over the buttons. In this mechanism, the finger could be used to rotate the pinion gear to move the slider from one end to the other as there is a rack either attached to the slider for the pinion gear and the base of the buttons are designed in such a way that the rise and fall of the buttons form a pattern. The design was carried out in such a way that the entire controller fits in the hand of the humans and the total thickness is limited to 5mm and the rise of the buttons from the surface is 0.5mm.

3.1.1. *Top Cover*

This component acts as a cover for the buttons and the slider, so the working mechanism is not visible. It has nine holes of the diameter of the buttons so that nine buttons could be used in the controller. The top cover should be flat, so that when the buttons are inactive and concealed to the surface, the controller looks as if it has no buttons. The top cover should have a provision for bolt or screws, so that the top cover could be fixed to the bottom cover such that all the rest of the components could be in-between the top and the bottom cover. The inner surface of the top cover should possess certain designs of slots, so that the buttons does not rotate about its axis during the sliding mechanism and this slot further facilitates rise and fall of the buttons. The design of both sides of top cover of the controller is shown in Figure 1.

3.1.2. *Bottom Cover*

The function of this component is to cover the controller and to act as a base for holding in the hand. The most important function of this component is to aid the slider to slide. This acts as a base for sliding the slider, and the material of the bottom cover should be selected, so that the friction between the slider and the bottom cover should be minimum. The bottom cover has the provision for attaching to the top cover at the edges or in the corners by having holes to enable bolts so that it could be tightened with nuts or by any other screws.

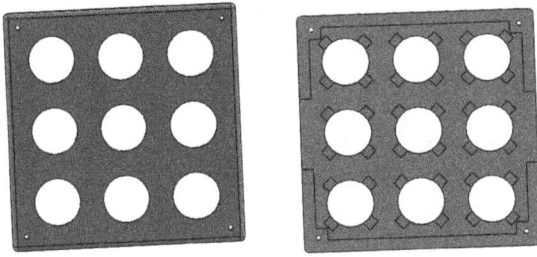

Figure 1. Design of top cover of controller (views from both sides).

3.1.3. *Buttons*

The controller has nine buttons in it and all the buttons are fitted in the holes present on the top cover and these buttons are designed such that it is balanced in the slider. The buttons are designed with smooth top surface, while the bottom surface with profile, which determines the pattern of active and inactive state of the button (three button-patterns). The profile design of the buttons with different profiles for each button is shown in Figure 2.

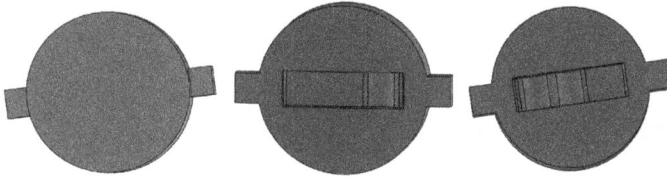

Figure 2. Design of buttons with different profiles for each buttons (top view and bottom views of two typical profiles).

Prototypes were made according to the design and some problems were encountered with the working of the prototype, such as, the buttons undergoing warp and twist about its axis rather than rising over the surface. Some buttons got eccentric and got stuck within the top cover and the slider when actuated. The reason for the instability was identified as the result of bad design in the slider and the button. The slider had protrusions in the form of a cylinder and a spherical top that would contact the profile of the button and each button had one protrusion from the slider such that they are in contact all the time. But the contact between the protrusion in the slider and the button is a point contact and it is difficult to maintain the stability when the slider is moving and this causes all the problems. Subsequently, the next design was altered but the patterns of the active and inactive states of the buttons were not changed. In the new design all the buttons have the same profile on the bottom whereas the protrusions on the slider determine the patterns of the active and inactive states of buttons. It was designed such that a line contact prevails between the protrusions in slider and the profile on the buttons. This helps in increasing the stability while slider is moving and also some more provisions are made to further increase the stability. The other important advantage of this design of profile in the button is that all the buttons have the same profile and hence the fabrication could be more ease. The button with the same profile is presented in Figure 3.

3.1.4.   *Slider*

The slider and the buttons are the two parts that plays the most important role in this mechanism. The slider is a movable part which could in-turn helps in the activation and inactivation of the buttons based on the patterns. The mechanism is designed in such a way that there are three different positions to which the slider could be moved and each position produces different patterns of buttons. The slider was designed as a flat plate that had nine cylindrical protrusions (one for each button – all same) on the top side and the top of the cylinders had a spherical top. The diameter of the cylindrical protrusion and the diameter of the sphere on top of it was designed based on buttons bottom profile since the protrusions should fit and always have a contact during working of the controller. This led to instability when the slider was moving which was due to the design of the buttons, so the design of buttons were changed to facilitate the working of the mechanism, the design of the slider was also changed. The design of the slider with similar protrusions is shown in Figure 4.

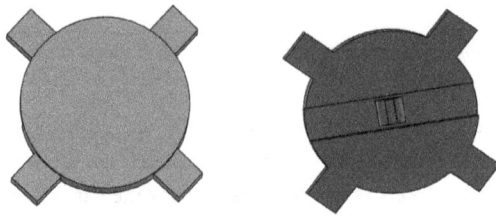

Figure 3. Design of buttons with the same profile for all buttons.

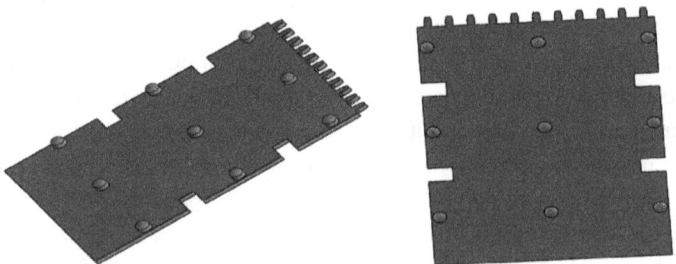

Figure 4. Design of slider with protrusions of same profile for all buttons.

The design of the protrusions were changed such that the nine protrusions were for each of the buttons were not the same and the profile of the protrusions caused the patterns in the buttons whereas the bottom profile of the buttons was changed and it was the same for all the nine buttons. To make the buttons move gradually, the protrusions in the slider should have a suitable profile. Hence, the design was carried out,so that the motion between the buttons and the protrusion in the slider undergoes a simple harmonic motion (SHM). It was designed in such a way that the protrusion acts as a wedge cam and buttons acts as a follower. SHM is preferred as it is simple to design than the other cam motion. The design of the slider with protrusions of different profiles for each buttons is shown in Fig. 5.

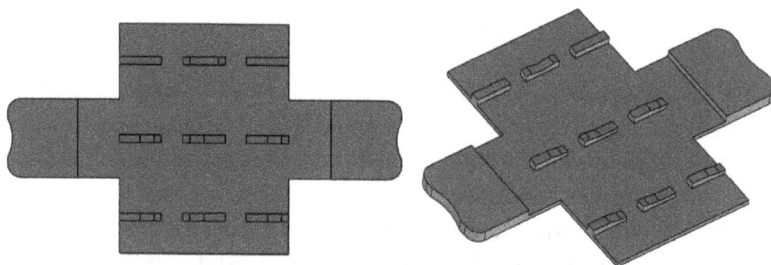

Figure 5. Design of slider with protrusions of different profiles for each buttons.

The slider is the component that actuates the buttons in the controller due to the mechanism, but the actuation of the slider to actuate the buttons could be done by various methods. The easiest method is to actuate manually through sliding motion of the slider using hands. Using this method, the part of the slider has to come out of the controller such that hands could be used to actuate the slider. To reduce the parts that come out of the controller, pinion gear could be used to actuate.

When using pinion gear, only half of the pinion gear comes out of the controller and is more stable. In this actuation, the slider should be designed in such a way that it acts as a rack for the pinion gear and moves linearly when the pinion gear is rotated and also the slider rack should be designed such that it meshes properly with the pinion gear. The three positions of activation of controller, which creates the surface patterns, are shown in Figure 6.

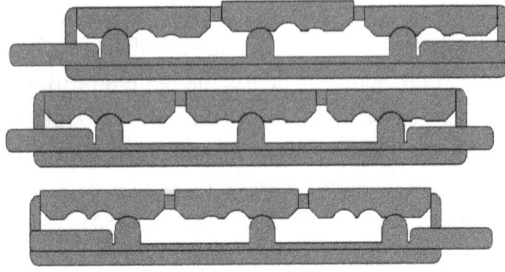
Figure 6. Various positions of button patterns.

3.1.5. *Pinion Gear*

The pinion gear is used to actuate the slider which in-turn actuates the buttons that perform the functions required in the controller. The main challenge in the design of the pinion gear is that it should be small and should act as a pinion and should help in sliding the rack slider. The pinion gear would be actuated with hand and it should be compatible with human operation. The pinion gear would be placed on the top of the slider such that it is in contact with the slider. It has a hole in the center which fits into the shaft that extends from the top cover to the bottom cover.

3.1.6. *Mechanochromic Material*

The top surface is enclosed in a thin surface of material called mechanochromic material. Mechanochromic materials are the materials that change their color of emission when an appropriate external mechanical force stimulus like stress or pressure is applied to it. [4, 5] Due to this, when the buttons rise from the surface, there is a change in color in the top surface of the button where the part of the thin film is stretched. Thus, the buttons that are above the surface and can be used is shown clearly and separately from the buttons that are concealed to the surface and are inactive.

4. Prototyping

After the design of the controller was finalized then the material and the fabricated method was determined. To get the prototype within a short period of time, 3D printing was selected for fabrication. The material used for 3D printing was photopolymer resin and the method of 3D printing that was used was Polyjet Direct 3D Printing. The photopolymer resin is a polymer that changes its properties when exposed to light such as ultra-violet or visible light. When the polymer is exposed to these lights the polymer gets cross-linked and the polymer gets hardened to a certain degree based on the characteristics of the light and the time for which the polymer is exposed. The Polyjet Direct 3D Printing process features direct inkjet printing of photopolymer resin and the successive curing using the ultra-violet lamps [7]. The machine has eight heads in which four are used for build materials and the other four are used for support materials. The 3D design could be sent to the CAM interface and then used for 3D printing. Pictures of prototype made in 3D printing are shown in Figure 7.

5. Conclusion

Advantages of sliding mechanism based for buttons on demand are:
1. It can be actuated manually or by motor(s) whenever required.
2. Nine or more button-patterns may be achieved.
3. The number of the degree of freedom of the slider is only one or two.
4. The actuation and the motion of the slider and the buttons are highly reliable.
5. The buttons provide haptic feedback to the users.
6. Touch sensing layer may be integrated atop to provide signal feedback.

A layer of mechanochromic material may be placed atop for colour change, which will aid the embedded word or image to appear and increase the human interaction.

Figure 7. Prototype fabricated using 3D printing.

In terms of applications, the levels of driving automation of on-road vehicles are of six levels according to Society of Automotive Engineers (SAE). The level varies from Level 0 (No Automation) to Level 5 (Full Automation). The buttons in the controllers are more suitable for Level 2 (Partial Automation) and Level 3 (Conditional Automation), where the operation of the buttons is automatic but the control is in the hands of the driver. More suitable operations could be like lane keeping and adaptive cruise control, [6] where the automated operations could be taken over by the driver at anytime during the operation.

Acknowledgement

This project is funded by BMW-NTU Joint R&D program.

References

1. Patent No. US 20070247420 A1, "Reconfigurable, shape-changing button for automotive use and vehicle control panel having reconfigurable, shape-changing buttons", Volkswagen Of America, Inc., Filing date. 24 Apr 2006, Publication date. 25 Oct 2007.

2. Patent No. US 20140257567 A1, Sliding hard keys, Audi Ag, Volkswagen Ag, Filing date. 28 Mar 2013, Publication date. 11 Sep 2014.

3. The Tactus tactile layer, Tactus Technology, USA, 2014.

   http://tactustechnology.com/

4. Weder C. Mechanochromic polymers, in Encyclopedia of Polymeric Nanomaterials (ed: Shiro Kobayashi and Klaus Müllen), 1-11, Springer Berlin Heidelberg

5. Peterson GI et al. 3D-Printed Mechanochromic Materials, ACS Appl. Mater. Interfaces, 2015, 7 (1), pp 577–583.

6. Marinik, A., Bishop, R., Fitchett, V., Morgan, J. F., Trimble, T. E., & Blanco, M. (2014, July), Human factors evaluation of level 2 and level 3 automated driving concepts: Concepts of operation. (Report No. DOT HS 812 044). Washington, DC.

7. Michael W. Barclift, Christopher B. Williams, Examining Variability in the Mechanical Properties of Parts Manufactured Via Polyjet Direct 3D Printing, Department of Mechanical Engineering, Virginia Tech, Aug 2012, pp 876–890.

Research of the Sterilization Temperature Detection Technology for Tunnel Drying Oven

Huai-yuan Sun[1,2] , Bo Sun[2,*], Lai-quan Song[3] and Li-ying Yang[1]

[1]Shanghai University of Medicine & Health Sciences, Shanghai 200093

[2]University of Shanghai for Science and Technology, Shanghai 200093

[3]Jining No.1 People's Hospital, Shandong Jining 272011
Email: wo4sunbo@163.com

In allusion to the verification of temperature testing and sterilization effects in the tunnel drying oven, a sterilization temperature detection system was designed, including test module, validation module and analysis module, etc. Through the no-load heat distribution test, the full load heat penetration test, the correlation data analysis and the sterilization check to verify the feasibility and rationality of the system to test the temperature of the tunnel drying oven, in accordance with the requirements of the actual temperature measurement and verification of tunnel oven.

Keywords: Tunnel drying oven, Sterilization temperature, Detection technology, Research.

1. Introduction

The tunnel drying oven is one of the main equipment for drying and sterilizing medicine packaging containers, and it is one of the important equipment to ensure the safety of drugs. The sterilization process consists of three stages: preheating, sterilization and cooling. In the process of equipment operation, the temperature uniformity in the tunnel-drying oven is very important to the sterilization process, because of its direct impact on the tunnel oven sterilization effect [1-2]. The detection of temperature is an important basis to reflect the effect of the tunnel-drying oven. According to the characteristics of internal temperature change and the application environment of the tunnel in the tunnel, we studied the sterilization temperature detection technology, selection of reliable and convenient method of temperature measurement, design of tunnel oven sterilization temperature detection system, for detection of temperature in the tunnel drying oven and validate the sterilization effect of tunnel oven whether meet the requirements of GMP (drug production quality management standards).

2. Temperature Measurement System

Temperature measurement system is based on the multi-channel temperature detector as the core, with the temperature sensor to achieve on-site temperature acquisition. The temperature sensor calibration was completed by the E450 Heimsen dry type temperature calibration instrument. The operation, analysis and operation of the system software were completed by the PC. The temperature detecting instrument was composed of the signal conditioning part and the data acquisition card, and realization of data transmission with PC machine by USB. Dry type temperature calibration instrument and the PC was connected by a serial port to achieve sensor calibration temperature data transmission. Temperature detection system structure diagram as shown in figure 1.

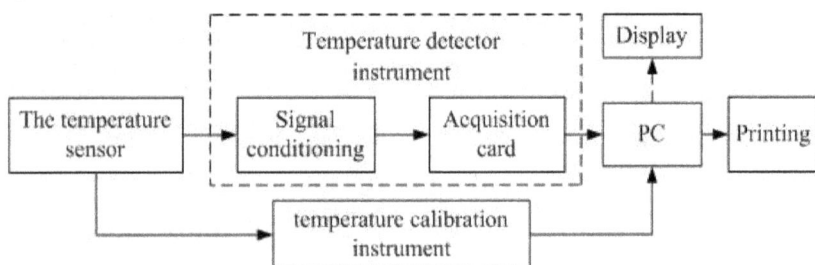

Figure 1. The block diagram of temperature detection system overall structure

3. Thermocouple Temperature Sensor

Thermocouple is one of the most common sensors used in temperature measuring system. The front end signal conditioning circuit of the thermocouple comprised a thermocouple cold end compensation circuit and a thermocouple voltage amplifying and filtering circuit. The A/D conversion of the thermocouple voltage signal was realized by the data acquisition card.

Thermocouple temperature measurement is based on the principle of thermoelectric effect. Thermoelectric effect would generate thermoelectric potential. Type T thermocouple thermoelectric potential efficiency was higher than other types of thermocouples and its thermoelectric potential around -6.25~20.87 mV.

The temperature in the tunnel oven was below 400°C while the effective measurement range of the type T thermocouple temperature sensor was about -200~400℃. Type T thermocouple can completely adapt to the internal environment of high temperature with temperature fast response and high measurement accuracy. So the 16 type T thermocouples were used as the temperature acquisition elements, and the different position of the medicine bottle in the tunnel drying oven was detected simultaneously.

4. Thermocouple Cold End Compensation

When measuring the temperature, the temperature of the cold end of the thermocouple (reference point) must be guaranteed to be 0°C or maintaining 0°C after compensation, otherwise, it can't be used directly. Therefore, in order to ensure the accuracy of temperature measurement, must take certain measures of compensation on the cold end (Reference) temperature [3-5]. In this design, the automatic compensation of thermocouple cold end temperature was realized by the method of current compensation, which was based on the analog type thermocouple cold end compensation, as shown in figure 2.

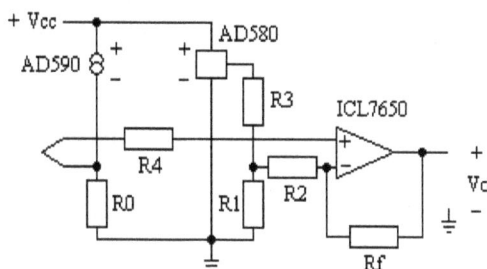

Figure 2. The principle diagram of current type thermocouple cold junction compensation circuit

5. Thermocouple Signal Amplifying Circuit

This design adopted the multistage amplifier form, and finally the thermocouple voltage signal was amplified to 0~5V. The front stage was the output stage of the thermocouple cold end compensation circuit, and the back stage used two reverse amplification circuits. In the end, the signal was enlarged to the size of the input required to meet the A/D conversion.

In figure 2, the ICL7650 was a high precision integrated operational amplifier developed by the American Intersil Company. It used the dynamic zero technique to reduce its own drift and offset, which had the characteristics of high common mode rejection ratio, high gain, low zero drift and small offset.

The low input offset voltage and open-loop voltage gain characteristics have an advantage in the amplification of weak signals generated by various sensors. Two-stage amplifier circuit was composed of ICL7650 as figure 3 shows.

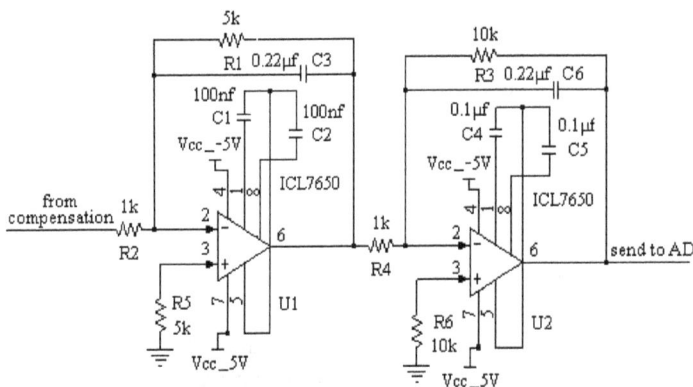

Figure 3. The amplifying circuit for thermocouple temperature signal

Front end input is the temperature signal compensated by the cold end of the thermocouple in figure 3. According to the magnification of the integrated operational amplifier circuit calculation shows that the inverse proportion of operation of U1 and U2 were 5 and 10. Since the post-stage magnification cold junction compensation is 5. So after a thermocouple amplifier, weak voltage signal generated by a thermocouple $0 \sim 20mv$ is magnified 250 times, exactly in line with the back-end data acquisition card input range of $0 \sim 5V$.

6. Data Acquisition Card

This design used Zhongtaiyanchuang USB7660 series data acquisition card. It can achieve 16-thermocouple temperature data acquisition, and control the various thermocouple voltage signal for A/D conversion. And through the USB port, the converted temperature signal was transmitted to the PC. USB7660 was a 12 bit data acquisition card. The temperature measurement range of the T thermocouple was divided into 4096 parts, about $0.1°C$. So the accuracy of the 12 data acquisition card could meet the requirements of the temperature measurement accuracy of the system.

7. System Software Design

This system was based on National Instruments (NI) Company LabVIEW software design [6]. The program was generally composed of a front panel window and program flow chart of the window. Data acquisition card USB7660 had a dynamic link library LabVIEW that can be used to call the usb7660.dll. The most important thing was to use LabVIEW call dynamic link library to achieve the related functions of the data acquisition card to call for the control of data acquisition card [7]. In order to be able to use LabVIEW software to achieve the general data acquisition card programming, LabVIEW provided users with a call library function node (CLFN) and code interface node (CIN) functions.

8. The Tunnel Drying Oven Sterilization Temperature Detection

Taking a pharmaceutical laminar hot air tunnel oven as the object, we adopted the above design of temperature detection system to detect the sterilization temperature of capacity 5ml ampoules.

9. Temperature Testing

The main purpose of the no-load test was to verify whether the temperature distribution in the horizontal position of tunnel oven meets the requirements. According to the verification requirements, the temperature difference of the same horizontal position should be less than 10°C. Otherwise, the no-load test of heat distribution failed. The 16 type T thermocouple temperature sensors were fixed on the stainless steel bracket by 25mm lateral spacing, followed by preheating, sterilization and cooling zones.

Heat penetration was a kind of effect produced by the transfer of heat through conduction convection and radiation [8]. Tunnel oven heat penetration test was also known as the distribution of heat load test, which was to measure the ampoule temperature on the belt. The purpose was to verify the uniformity of temperature distribution of glass bottles in the same conveyor belt with the maximum loading capacity of the tunnel oven. And it was verified that the time of the actual sterilization temperature can be achieved in the case of the maximum loading capacity, so that the Fh value of an important sterilization parameter could be calculated. Take the temperature measuring end of the 16 type T thermocouple temperature sensors fixed on the stainless steel bracket in tunnel oven evenly into the ampoule, and try to make the thermocouple didn't contact the bottle wall and the bottom of the bottle. In the case of full load operation, the temperature of the heating section was set to 330°C and the interval time of the recording temperature was 30s.

Data from the temperature testing can be used to draw the conclusions:
(1) The temperature difference of each point of the tunnel oven at the same time (the same cross section) was less than 10°C.

(2) The time of Tunnel oven temperature above 300 has reached 5 minutes, in accordance with the requirements of the tunnel oven sterilization.

(3) Compared with the no-load heat distribution test, the temperature in the corresponding position of the tunnel during the heat penetration test was decreased. There were reasons such as surface residual moisture when the ampoules went through the washing machine.

Through the system software, the temperature data of 16 thermocouples of the tunnel oven heat distribution test and heat penetration test were processed separately, and the data of the two groups were compared, as shown in figure 4. Obviously, the hot and cold conditions and the sterilizing condition of the same cross section in the tunnel oven were suitable for the working state of the tunnel oven.

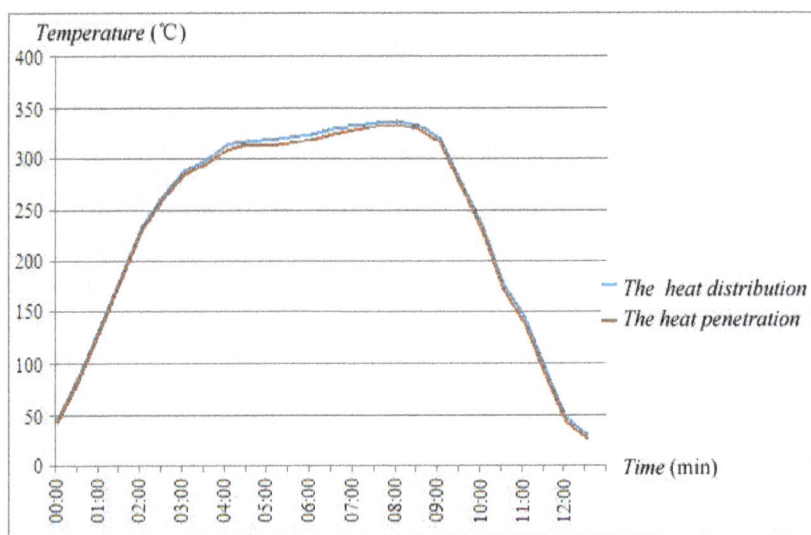

Figure 4. The data contrast diagram of heat distribution test and heat penetration test

10. Checking Sterilization

According to the Pharmacopoeia of the People's Republic of China, the dry heat sterilization of 250°C ×45min can remove drug container source matter, which reached the sterilization effect. Fh value can be used to evaluate the sterilization ability of the tunnel oven during normal operation. Fh value is calculated as:

$$Fh = \int_{t1}^{t2} 10^{(T-T0)/Z} dt \tag{1}$$

Where:
T: The actual temperature of the dry heat sterilization (Including preheating, heating, cooling temperature of the three sections);
T0: Standard temperature, 170°C;
Z: Dry heat sterilization constant 54;
t1, t2: Preheat and cooling sections from heating to 170°C.

According to the parameters and Fh value calculation formula, the Fh value is 1365. As long as the Fh value is greater than 1365, the tunnel oven is qualified.

The data from the full load heat penetration test showed that the temperature measured by the thermocouple was above 170°C. Put the data into the Fh calculation formula and get Fh value as shown in Table 1.

Table 1. The Fh value in the heat penetration stage

thermocouple	T1	T2	T3	T4	T5	T6	T7	T8
Fh	6357	8537	6337	7633	7274	7770	7121	7291

Value Fh was far greater than 1365, indicating that its sterilization effect was completely in line with the requirements.

11. Conclusion

In this paper, we designed a special system for the detection and verification of the temperature in the tunnel oven, including the temperature sensor, the temperature detector, the temperature calibration instrument and the computer software. Temperature sensor of type T thermocouple was used to realize the field temperature collection. Cold end compensation was performed using AD590 measurement at room temperature. Heimsen E450 dry type temperature calibration instrument used to calibrate the thermocouple temperature sensor. Choose high precision integrated operational amplifier ICL7650. USB7660 series of 12-bit data acquisition card was used to achieve 16-thermocouple temperature data acquisition and signal A/D conversion. Use graphical design language Labview to prepare the temperature acquisition and display process.

According to the data and analysis report of the no-load heat distribution and the full load heat penetration test of the tunnel drying oven, the temperature detection system realized the real-time collection, storage, analysis and display of the temperature in the tunnel, which verified the feasibility and rationality of the system. The sterilization temperature detection system improved the intelligence of temperature detection, greatly reduced the intervention of artificial factors, and would be widely used in actual production.

References

1. Mingwei Wan, Design and Research of the Laminar Flow Sterilizing Tunnel [D]. Shanxi University of Science & Technology, 2012.
2. Yongjie Lu, Yue Yang. Freeze-drying process and equipment validation in pharmaceutical companies [J]. Chinese Journal of Pharmaceuticals.2011, 42(4): 319-321
3. Wang Yao-guo, Wu Xiao-nan, Yu Jin-yong. Error analysis and correction method of the aeroplane engine whiff temperature measurement by thermocouple [J]. Instrumentation Technology. 2011, (1): 25-27.
4. Zhaoting Tang. Design of thermocouple temperature measuring instrument for spacecraft [D]. Shaanxi University of Science & Technology.2011.
5. Zhang Hai-tao, Luo Shan, Guo Tao. Improvement of cold-conjunction compensation of thermocouple [J]. Instrument Technique and Sensor. 2011, (7):11-14.
6. Haidong Chen. Design of automatic measurement system for thermal error of NC machine tool based on LabVIEW [D]. Hefei University of Technology.2014.
7. Li Rui, Zhou Bing, Hu Renxi. Labview 2009 Chinese version of the virtual instrument from entry to the master [M]. Machinery Industry Press.2010.
8. Huang Xiaoyong, Han Yonghong. A test method for thermal distribution and thermal penetration testing [J]. Pharmacy Today, 2012, 22(2):97-99.

Study on the Steel Latticed Arch Affected by Impact with the Multi-Scale Model

Meng-sha Liu[†], Jing-yuan Li and Jin-san Ju

Department of Civil Engineering,

China Agricultural University, Beijing, China

[†]*Email:* liumengsha@cau.edu.cn

In this study, ABAQUS was applied to carry out the numerical simulation of the steel latticed arch affected by impact. The multi-scale model suitable for the steel latticed arch was determined through the comparison of the numerical simulation results between two multi-scale models. The multi-scale model was adopted to analyze the behavior of the steel latticed arch under impact load and the results were analyzed comparing with the element beam model. From the data comparison in the action time of the impact block, the plastic deformation and the internal energy the Introduction, the multi-scale model is proven to be suitable for dynamic impact analysis.

Keywords: The multi-scale model, Steel latticed arch, Impact load.

1. Introduction

The stability of the steel structure has been the extensive research topic of many scholars and the stability of the steel arch should be studied to provide the design basis for the widely used arch structure in various kinds of buildings. There have been many research results about the static stability study of the steel arch, but the researches about the steel arch mechanics characteristics under dynamic load are relatively rare. The impact load produced by explosion shock load and impact load often bring disastrous consequences, so the researches on the dynamic stability of the steel latticed arch have the actual engineering significance and wide research prospects [1 - 3].

The finite element numerical analysis [4] is widely used in engineering design, but for some structure in huge form, the macroscopic model can't accurately response to the actual stress distribution especially the local buckling failure. The microcosmic failure mechanics can be reflected by microcosmic model set up by the solid elements, but the Computer computing ability and the complexity of modeling limit its actual use. The emergence of the multi-scale model [5 - 7] can effectively solve this problem. In the multi-scale model, the detailed model was set up in the local parts that need a detailed analysis and the other parts use macroscopic model to simulate. It can be better grasp that the whole stress characteristics of the structure and the micro failure process.

In this study, the multi-scale model was used to research dynamic response of the steel latticed arch under impact load.

2. The selection of the multi-scale model

2.1. Model geometry information

This study adopted the steel latticed arch as the research object. The geometry information is shown in Fig. 1. The arch is 10m in span length and 1.5m in height.

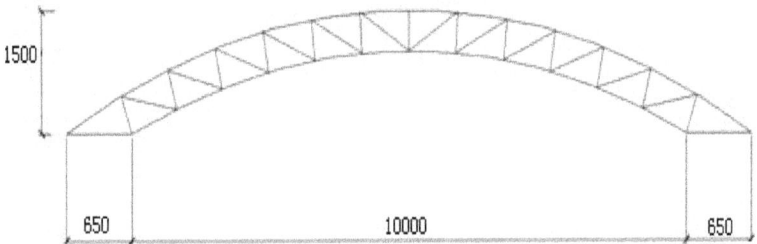

Fig. 1 The geometry information of the steel latticed arch

The cross section of the arch is a triangle whose height is 0.5m and base is 0.6m. Circular tubes, which are 0.05m in diameter and 0.003m in thickness, were used in the chord members and 0.02m in diameter and 0.001m in thickness were used in web members. The elastic-plastic constitutive model was adopted in the numerical simulations. The elasticity modulus=2.06×1011 and the Poisson ration=0.3.

2.2. The impact model of the steel latticed arch

The ABAQUS/EXPLICIT was used to carry out the stress calculation of the steel latticed arch affected by impact. The beam element model or the multi-scale model was used to set up the part of the arch. The impact block was simulated by the rigid element. In the numerical simulation [2 - 3], the arch was clamped in the arch feet and the vertical displacement limit was released for the impact block. The impact load was applied on the arch through the gravity rush from the rigid element with a certain impact velocity.

2.3. The multi-scale model of the steel latticed arch

The impact made the local buckling and plastic failure easier to occur in the arch crown. So it is necessary to set up the detailed model in the impact point for accurate force analysis. The detailed part is simulated by shell elements and the other parts are simulated by the beam elements. Coupling connects the shell elements and the beam elements. But the range of the detailed model is not apparent. In the study, two multi-scale models [6 -7] shown in Fig. 2 were established and through the dynamic response analysis, the multi-scale model suitable for this impact model was selected.

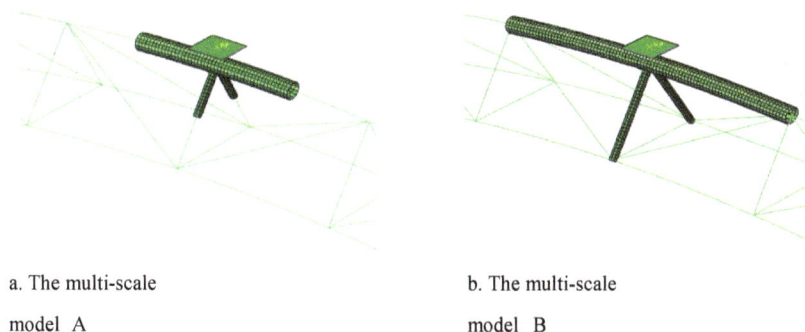

a. The multi-scale
model A

b. The multi-scale
model B

Fig. 2 The multi-scale model

In the multi-scale model A, the multi-scale model A, only half of the tubes connected to the arch crown are set up by the shell elements. The 4 tubes connected to the arch crown are simulated with the shell elements in the multi-scale model B. The impact block of 300Kg in mass and 5m/s in velocity shocked the multi-scale models.

2.4. The selection of the multi-scale model in the study

The impact load-time curves of the two models are shown in Fig. 3. It can be observed from the curves that the maximum of the impact load of the multi-scale model B is higher than it in the multi-scale model A. The reducing to zero of the impact load shows that the impact block has separated from the arch. The contact time of the model A is 0.06s and the model B is 0.045s. The multi-scale A has a more stable impact load because of its smaller impact load and longer contact time.

The time-history curves of the artifical strain energy and internal energy carried out by the two multi-scale models are shown in Fig. 4. From the Fig. 4, it can be observed that the artifical strain energy of the multi-scale model A is stable relatively during the impact process and the artifical strain energy of the multi-scale model B increased gradually. The ratio of the artificial strain energy to the internal energy with the calculation of the multi-scale model A is 1.55% and the ratio of the multi-scale model B is 14.7%. The higher artificial strain energy illustrates that the energy used to control the hourglass phenomenon is higher. In the numerical simulation, too much artificial strain energy can cause the results in a bigger error. The reasonable artificial strain energy is less than 5%, so the multi-scale B is not the ideal model and the model A can meet the requirements for computing.

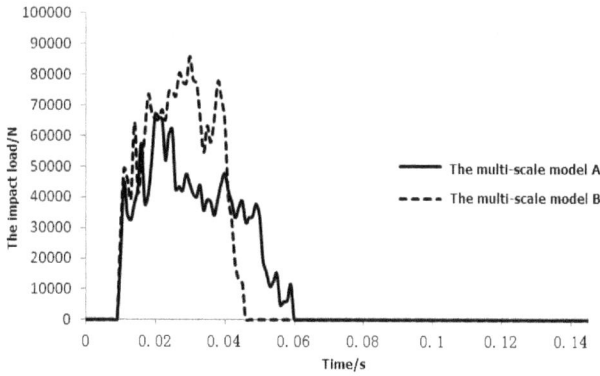

Fig. 3 The impact load-time curves of the two models

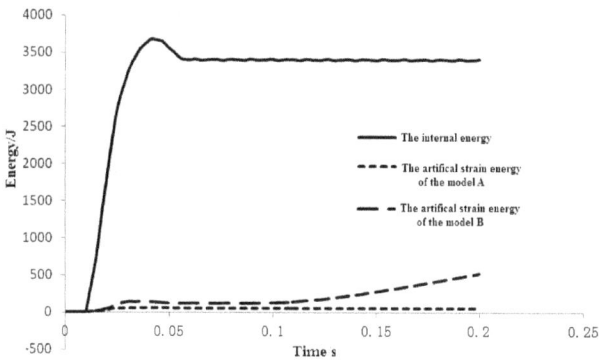

Fig. 4 The time-history curves of the artifical strain energy and internal energy

Table 1 shows the comparison of the computation time and the memory usage between the two multi-scale models. It can be concluded that the computation time and the memory usage of the model B is almost twice as the model A from the table.

Table 1. The computation time and the memory usage of the two multi-scale models

	The computation time	The memory usage
The multi-scale model A	12mins	2G
The multi-scale model B	21mins	4.1G

The local buckling calculation results of the two multi-scale models are shown in Fig. 5. The deformation and the stress results of the two models are almost the same. From Fig. 5, the local buckling of the steel arch can be observed apparently in both of the two models. So the multi-scale model A can carry out a good numerical simulation for the steel latticed arch under impact load.

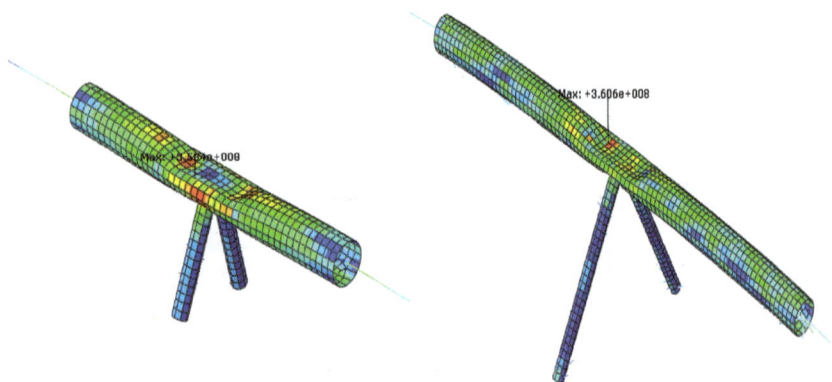

Fig. 5 The local buckling calculation results of the two multi-scale models

The multi-scale model A can meet the requirements of the computational accuracy and the computational efficiency for FEM calculation. The multi-scale model A is used to simulate the steel latticed arch in this study.

3. The dynamic response analysis

In this part, the multi-scale model and the beam element model were applied to carry out the study on the steel latticed arch affected by impact. The suitability of the multi-scale model in impact problems is analyzed with the analytic results comparison of the numerical simulation between the multi-scale model and the beam element model.

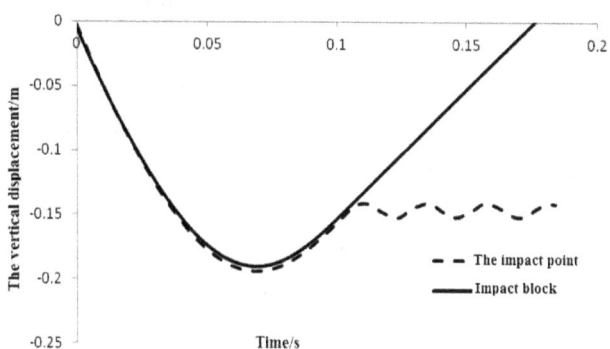

Fig. 6 The displacement and time curves of the beam element model when the velocity of the impact block is 5m/s

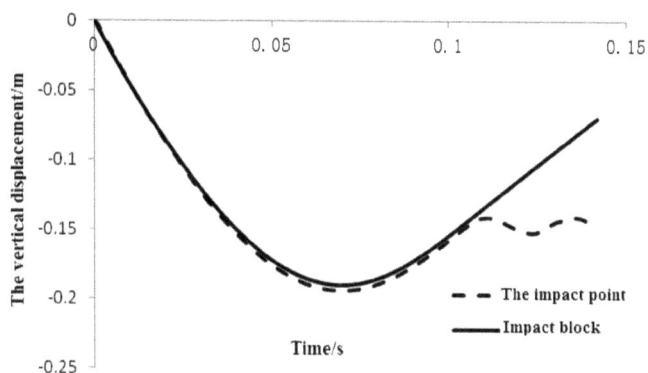

Fig. 7 The displacement and time curves of the multi-scale model when the velocity of the impact block is 5m/s

3.1. The action time analysis of the impact load

Fig. 6 and Fig. 7 show the displacement and time curves of the beam element model and the multi-scale model respectively when the velocity of the impact block is 5m/s. It can be observed from Fig. 6 and Fig. 7 that the impact point on the arch had the same displacement with the impact block before the impact block bounced off. After the displacement met the extremum, the impact block moved upwards under the reaction effect of arch. But the separation between the block and arch occurs due to the plastic deformation of the arch, which limited the displacement restoring of the arch. So from Fig. 6 and Fig. 7, the action time of the impact load in the beam element model is 0.125s and in the multi-scale model is 0.105s.

Fig. 8 and Fig. 9 show the displacement and time curves of the beam element model and the multi-scale model respectively when the velocity of the impact block is 7m/s. The same analysis as above can be carried out and it can be concluded that the action time of the impact load in the beam element model is 0.145s and in the multi-scale model is 0.125s.

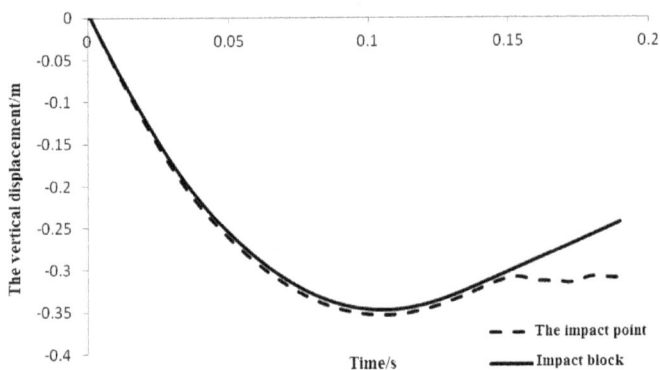

Fig. 8 The displacement and time curves of the beam element model when the velocity of the impact block is 7m/s

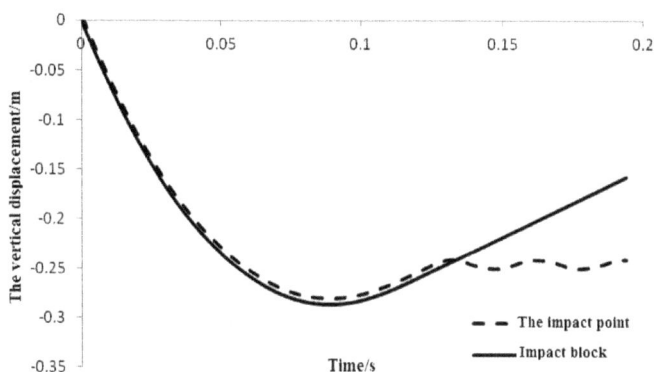

Fig. 9 The displacement and time curves of the multi-scale model when the velocity of the impact block is 7m/s

From the action time of the impact load analysis above, it can be concluded that the action time of the beam element model is longer than the multi-scale model and this means the Impact load to reach the stable time is relatively longer.

It also can be observed from Fig. 8 and Fig. 9 that the final arch crown vertical displacement of the beam element model is 0.354m and the multi-scale model is 0.245m. The final arch crown vertical displacement of the beam element is greater than the multi-scale model and that is to say the energy dissipation in the plastic deformation is greater in the beam element model.

3.2. The internal energy analysis

Fig. 10 shows the time-history curve of the internal energy when the impact block is 300Kg in mess and 5m/s, 10m/s and 20m/s in impact velocity.

It can be known from the analysis above that the energy dissipation in the plastic deformation is greater in the beam element model than multi-scale model and the impact energy is the same in the beam element model and the multi-scale model. Therefore, the internal energy of the beam element model is greater than the multi-scale model and this conclusion can be gotten from the observation of Fig. 10.

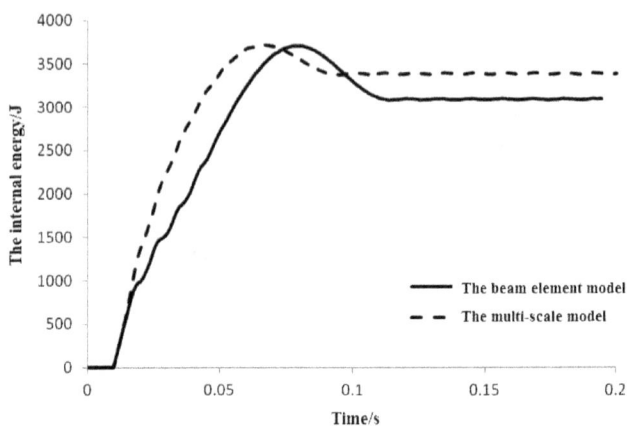

a. The time-history curve of the internal energy as the impact velocity=5m/s

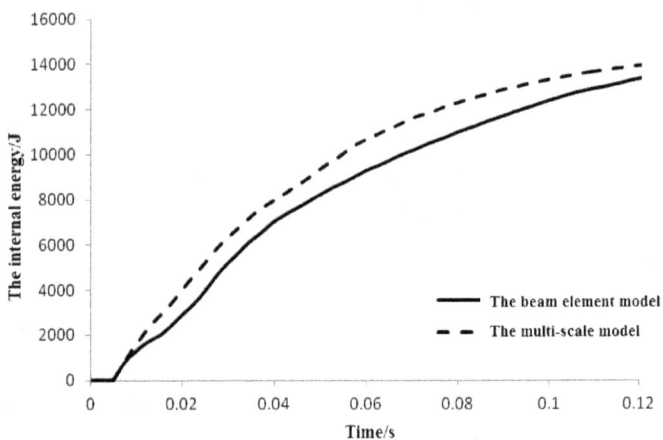

b. The time-history curve of the internal energy as the impact velocity=10m/s

Fig. 10 The time-history curve of the internal energy

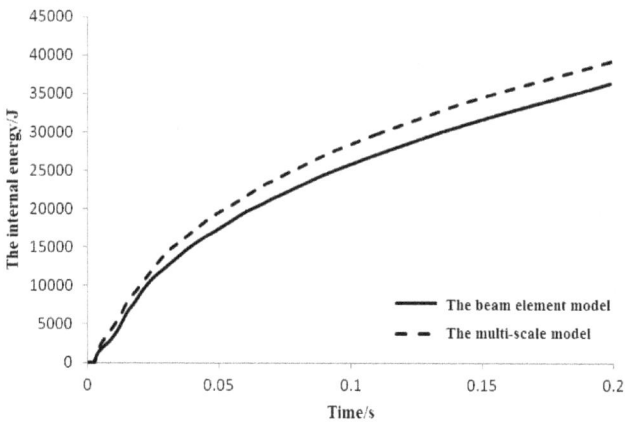

c. The time-history curve of the internal energy as the impact velocity=20m/s

Fig. 10 The time-history curve of the internal energy (*continued*)

4. Conclusion

In this study, ABAQUS was applied to carry out the numerical simulation of the steel latticed arch affected by impact. The multi-scale model was adopted to analyze the behavior of the steel latticed arch and the results were analyzed comparing with the element beam model.

At the start of the study, two multi-scale models were established for the steel latticed arch and through the dynamic response analysis, the multi-scale model suitable for this impact model was selected. The analytical results can be obtained from finite element numerical simulation as follows:

1) When the calculations are conducted under the same impact conditions, the action time of the impact block in the beam element model is longer than the multi-scale model through the FEA analysis. So it can be concluded that the time to reach the stable of the impact load is relatively longer in beam element model than multi-scale model.

2) The plastic deformation of the multi-scale model is smaller than beam element model in the same impact conditions. The local plastic deformation of the steel arch can limit the vertical deformation of the

arch crown and the analysis results in this study agreed with this mechanics characteristic.

3) The internal energy of the beam element model is greater than the multi-scale model in the same impact conditions.

With the change of the impact condition, the numerical simulation results of the multi-scale model are more reasonable than the beam element model. So the multi-scale model is suitable for dynamic impact analysis.

Acknowledgement

This work was funded by National Science Foundation of China (51279206).

References

1. Y. J. Shi, M. Wang, Y. Q. Wang. The seismic performance analysis of steel framework based on the multi-scale model [J]. Engineering Mechanics, 2011, 28(12): 20-26.
2. Y. L. Zhang, X. M. Guo. The research progresses of the multi-scale simulation and calculation [J]. Chinese Journal of Computational Mechanics, 2011, 28(B04): 1-5.
3. Khandelwal, K. Multi-scale computational simulation of progressive collapse of steel frames. Dissertations & Theses - Gradworks. 2008.
4. Z. H. Sun, Z. X. Li, H. T. Chen. The experimental study on the structure multi-scale finite element model [J]. Special Structure, 2008, 25(4).
5. Rudd, R. E., & Broughton, J. Q. Concurrent coupling of length scales in solid-state systems. physica status solidi (b), 2005, 217(1), 251-291.
6. C. Y. Wang. The analytical method of the multi-scale model [J]. Complex Systems and Complexity Science, 2004, 1(1):9-19.
7. Ladeveze, P., Nouy, A., & Loiseau, O. A multiscale computational approach for contact problems. Computer Methods in Applied Mechanics & Engineering, 2002, 191(43), 4869 - 4891.

Synchronization of Chaotic Gyros via a Novel Adaptive Wrinkling-Type Terminal Sliding Mode Control

Shih-hung Kao, Chi-ching Yang[†] , Cheng-da Shei and Geeng-jen Sheu

Department of Mechatronics Engineering,
National Changhua University of Education, Taiwan
Department and Graduate School of Electrical Engineering,
Hsiuping University of Science and Technology, Taiwan
[†]*Email: ccyang@hust.edu.tw*

The main goal of the study is to provide a novel adaptive Terminal Sliding Mode (TSM) control scheme to achieve the complete synchronization between master and slave chaotic gyros, in the effectiveness of presenting system's uncertainties and external disturbances. The proposed control scheme is based on a novel time-varying TSM, named as wrinkling-type TSM. The stability of the adaptive controller is satisfied in the sense of the Lyapunov stability theorem. Numerical simulations are given to verify the effectiveness of proposed scheme.

Keywords: Terminal Sliding Mode (TSM), Wrinkling-type TSM, Chaotic gyro.

1. Introduction

Complete synchronization between two systems is one of the important processes in the control of complex phenomena for chemical, physical, and biological systems. The gyro system [1] is one of the most attractive dynamic systems due to the attributes of great utility to navigational, aeronautical and space engineering. The non-singular TSM [2] was developed by introducing a fractional powers item into the sliding mode, which provides many good performances such as fast, finite time convergence and superior tracking precision.

In this study, a novel time-varying TSM, named wrinkling-type TSM is introduced according to the defined non-singular TSM in [2]. It can be utilized to develop the novel adaptive TSM control scheme to accomplish the complete synchronization between two chaotic gyro systems.

The rest of the study is organized as follows. Dynamics of a chaotic gyro system is introduced firstly. Secondly, the novel wrinkling-type TSM is defined and the adaptive TSM controller for achieving synchronization of two gyros is proposed. Then, the numerical simulations are performed to verify the effectiveness of the presented scheme. In the final section, the conclusions are addressed.

2. Dynamics of Chaotic Gyro

The system equation of the chaotic gyro in terms of the angular variable θ with linear-plus-cubic damping under the assumption that the gyro is in the sleep position is described by [1]

$$\ddot{\theta} + c_1\dot{\theta} + c_2\dot{\theta}^3 + \alpha^2 \frac{(1-\cos\theta)^2}{\sin^3\theta} - \beta\sin\theta = -f\sin(\omega t)\sin\theta \qquad (1)$$

where $f\sin(\omega t)$ is a harmonic parametric excitation, $c_1\dot{\theta}$ and $c_2\dot{\theta}^3$ are linear and nonlinear damping terms, respectively, and $\alpha^2(1-\cos\theta)^2/\sin^3\theta - \beta\sin\theta$ is a nonlinear resilience term. The dynamics of this system has been extensively studied [1] for a space range of the amplitude of the term f. In particular, the nonlinear gyro system eq. (1) exhibits chaotic behavior for the specific parameter values of $\alpha^2 = 100$, $\beta = 1$, $c_1 = 0.5$, $c_2 = 0.05$, $\omega = 2$, $f = 35.5$ [1].

Define the state variables $x_1 = \theta$, $x_2 = \dot{\theta}$, the normalized state equations of (1) in convenient first-order form are

$$\begin{cases} \dot{x}_1 = x_2 \\ \dot{x}_2 = \Psi(x_1, x_2, t) \end{cases} \qquad (2)$$

where

$$\Psi(x_1, x_2, t) = -c_1 x_2 - c_2 x_2^3 - \alpha^2 \frac{(1-\cos x_1)^2}{\sin^3 x_1} + (\beta - f\sin\omega t)\sin x_1.$$

In the sequel, the complete synchronization between two identical chaotic gyros is considered, even when the slave system is undergoing with system uncertainties and external disturbances.

3. Problem Formulations for Complete Synchronization

At the complete synchronization between the master and slave systems, the master system is shown in (2) and the slave system is described as follows,

$$\begin{cases} \dot{y}_1 = y_2 \\ \dot{y}_2 = \Psi(y_1, y_2, t) + \Delta h(y_1, y_2) + \delta(t) + U(t). \end{cases} \qquad (3)$$

where y_1, y_2 are the state variables and $\Psi(\bullet)$ is the nonlinear function defined in eq. (2). Considering the existence of system uncertainties and external disturbances in the practical application, the slave gyro system in eq. (3) is assumed to be subject to the system uncertainty $\Delta h(y_1, y_2) \in R$ and the external disturbance $\delta(t) \in R$. Without loss of generality, the system uncertainty and the external disturbance are bounded and satisfy

$$|\Delta h(y_1, y_2)| \leq H \in R^+ \quad , \quad |\delta(t)| \leq D \in R^+ \tag{4}$$

It is also reasonable to assume that $\Delta h(y_1, y_2)$ satisfies the conditions required to ensure system (3) has a unique solution in the time interval $[t_0, \infty)$, $t_0 > 0$ for any given initial condition. The goal of the control problem is to design an appropriate controller $U(t)$, such that for any given initial conditions, the states of slave system (3) converge to the states of master system (2), that is, $\lim_{t \to \infty}(y_i(t) - x_i(t)) \to 0$, $i = 1, 2$.

The synchronous error states between systems (2) and (3) are defined as

$$e_i(t) = y_i(t) - x_i(t), \quad i = 1, 2 \tag{5}$$

By taking the time derivatives along with (2) and (3), one can obtain the synchronous error system

$$\begin{cases} \dot{e}_1 = e_2 \\ \dot{e}_2 = \Psi(y_1, y_2, t) - \Psi(x_1, x_2, t) + \Delta h(y_1, y_2) + \delta(t) + u(t) \end{cases} \tag{6}$$

where $|\Psi(y_1, y_2, t) - \Psi(x_1, x_2, t)| \leq \lambda_0 + \lambda_1|e_1(t)| + \lambda_2|e_2(t)|$, $\lambda_0 > 0, \lambda_1 > 0, \lambda_2 > 0$ is assumed. To this end, it is clear that the problem of complete synchronization is replaced by the equivalence of stabilizing the synchronous error system (6) by applying a suitable design of the controller $U(t)$. The control goal for the current problem is to determine the appropriate controller $U(t)$ such that for any initial states of the synchronous error system (6), states of the synchronous system error converges to zeros, that is, $\lim_{t \to \infty} e_i(t) \to 0$, $i = 1, 2$.

1027

4. Wrinkling-type Terminal Sliding Mode

First of all, the nonsingular TSM $\sigma(t)$ is defined in [2] and given as the following equation:

$$\sigma(t) = e_1(t) + (1/k)|e_2(t)|^n \operatorname{sgn}(e_2(t)) \tag{7}$$

where $k > 0$, $1 < n < 2$ are positive constants and $\operatorname{sgn}(\bullet)$ denotes the signum function. Basically, TSM was developed by introduced the fractional powers item into the sliding mode, which provides many good performances such as fast, finite time convergence and superior tracking precision. The parameters, $k > 0$, $1 < n < 2$, in (7) can be chosen to achieve the system performance.

In this study, a novel TSM is defined by replacing the fractional powers item $1 < n < 2$ in (7) instead of the tuning function of time $\gamma(t)$, as follows

$$\sigma(t) = e_1(t) + (1/k)|e_2(t)|^{\gamma(t)} \operatorname{sgn}(e_2(t)) \tag{8}$$

where $1 < \gamma(t) < 2$. By setting $\sigma(t) = 0$ with $k = 10$ and different values of $\gamma(t)$, the curves of the novel TSM are depict in Fig. 1. When $\gamma(t)$ is increasing, it is shown that the curves demonstrate the behavior to be like as 'wrinkling' string. Thus, the novel TSM is named as wrinkling type TSM, which can be treated as a kind of time-varying TSM.

5. Novel Adaptive Wrinkling-Type Terminal Sliding Mode Control

In this section, the rule of thumb for the novel adaptive wrinkling-type TSM control design is introduced. Two basic steps are involved in order to accomplish the synchronization. Firstly, the appropriate wrinkling type terminal-sliding surface (8) for desired sliding motion is chosen. Then, a robust controller is designed for that any trajectory in phase space of the synchronous error system can be conducted onto and stayed in the sliding surface $\sigma(t) = 0$ even in the presence of system uncertainty $\Delta h(y_1, y_2)$ and external disturbance $\delta(t)$. It follows that states of the synchronous system error approach to zeros in the sliding surface $\sigma(t) = 0$ and the slave system is synchronous with the master system.

In the sequel, the novel adaptive TSM controller of system (6) for achieving complete synchronization is proposed in the following Theorem.

Theorem. If the controller $U(t)$ in system (6) is applied as follows:

$$
\begin{aligned}
U(t) = & -\frac{k}{\gamma(t)}|e_2|^{1-\gamma(t)} e_2 - \frac{\dot{\gamma}(t)}{\gamma(t)} e_2 \cdot \ln|e_2| \\
& -\left[k_0(t) + k_1(t)|e_1(t)| + k_2(t)|e_2(t)| \right] \cdot \operatorname{sgn}(\sigma(t))
\end{aligned}
\tag{9}
$$

where the TSM $\sigma(t)$ is defined in (8) with $1 < \gamma(t) < 2$. The positive and adaptive feedback gains, $k_0(t)$, $k_1(t)$ and $k_2(t)$ are updated according to the following adaptation algorithms [3], respectively,

$$
\dot{k}_0(t) = \left(\varepsilon_0 \gamma(t)/k \right) |\sigma(t)| |e_2(t)|^{\gamma(t)-1} \geq 0, \quad k_0(0) = 0, \quad \varepsilon_0 > 0 \tag{10}
$$

$$
\dot{k}_1(t) = \left(\varepsilon_1 \gamma(t)/k \right) |\sigma(t)| |e_1(t)| |e_2(t)|^{\gamma(t)-1} \geq 0, \quad k_1(0) = 0, \quad \varepsilon_1 > 0 \tag{11}
$$

$$
\dot{k}_2(t) = \left(\varepsilon_2 \gamma(t)/k \right) |\sigma(t)| |e_2(t)|^{\gamma(t)} \geq 0, \quad k_2(0) = 0, \quad \varepsilon_2 > 0 \tag{12}
$$

where $\varepsilon_0, \varepsilon_1, \varepsilon_2$ are the positive constants determining the adaptation process.

The time-varying function $\gamma(t)$ is tuning by fuzzy rule-based method [4, 5].
The membership functions (MF), μ_{EL}, μ_{ES}, for synchronous error states $e_1(t), e_2(t)$ and membership functions, $\mu_{\gamma L}, \mu_{\gamma S}$, for $\gamma(t)$ are defined as follows, respectively,

$$
\mu_{EL} = 1 - \operatorname{sech}(e_i), \ \mu_{ES} = \operatorname{sech}(e_i), \ i = 1, 2. \tag{13}
$$

$$
\mu_{\gamma L} = \begin{cases} 1, \ \gamma(t) = \gamma_{\max} \\ 0, \ \gamma(t) \neq \gamma_{\max} \end{cases} \quad ; \quad \mu_{\gamma S} = \begin{cases} 1, \ \gamma(t) = \gamma_{\min} \\ 0, \ \gamma(t) \neq \gamma_{\min} \end{cases} \tag{14}
$$

The fuzzy rules are defined
If error $e_1(t)$ is Large (μ_{EL}), then the $\gamma(t)$ is Large ($\mu_{\gamma L}$).
If error $e_1(t)$ is Small (μ_{ES}), then the $\gamma(t)$ is Small ($\mu_{\gamma S}$).

The $\gamma(t)$ can be derived by the max-min inference for defuzzification.

$$\gamma(t) = \gamma_{\max}\mu_{EL}(e_1) + \gamma_{\min}\mu_{ES}(e_1) = \gamma_{\max} - \Delta\gamma\mathrm{sech}(e_1), \quad \Delta\gamma = \gamma_{\max} - \gamma_{\min} \tag{15}$$

Moreover,

$$\dot{\gamma}(t) = \frac{-\Delta\gamma \cdot \sinh(e_1)}{\cosh^2(e_1)}(e_2) \tag{16}$$

Then, states of the synchronous system error (6) will asymptotically approach to and stay in the terminal sliding surface $\sigma(t) = 0$ with $\gamma(t)$ approaching to constant. It follows that both states of the synchronous error system converge to zeros in finite time and the synchronization between systems (2) and (3) can be achieved.

Proof. By selecting the Lyapunov candidate function to be

$$V(t) = \frac{1}{2}\sigma^2(t) + \sum_{i=0}^{2}\frac{1}{2\rho_i}(k_i(t) - k_i^*)^2 \geq 0 \tag{17}$$

where $\rho_i > 0, i = 0,1,2$ and $k_i^*, i = 0,1,2$ are sufficient large positive constants and satisfy the following inequalities

$$k_0^* > \lambda_0 + H + D > 0, \quad k_1^* > \lambda_1 > 0, \quad k_2^* > \lambda_2 > 0. \tag{18}$$

Taking the time derivative of (17) along with the solutions of the synchronous error system (6), the selection of the wrinkling-type TSM (8), the time derivative of (8), the time-varying function $\gamma(t)$ (15) and $\dot{\gamma}(t)$ (16), and the controller (9), it obtains the stable condition $\dot{V}(t) < 0$ in the sense of Lyapunov stability. Due to the limitation of length of manuscript, the detail procedures of proof are omitted here.

6. Numerical Simulations

In the section, numerical simulations are performed to verify and demonstrate the effectiveness of the proposed novel adaptive wrinkling-type TSM controller. For convenience in numerical studies, the system uncertainty and the external disturbance are assumed to be $\Delta h(y_1, y_2) = \sin(y_1)\sin(y_2)$ and $\delta(t) = \sin(2\pi t)$, respectively. Using the fourth-order Runge-Kutta method with initial conditions $(x_1, x_2) = (0.5, 2.0)$, $(y_1, y_2) = (1.0, 0.5)$, a time step size of 0.0001, and system parameters as in (1) to ensure chaotic dynamics of state variables, system (6) with the controller (9) are numerically solved. For the adaptive controller described in (9) associated with (10) to (16), the positive constants are chosen as $k = 15$, $\gamma_{min} = 1.20$, $\gamma_{max} = 1.95$, $\varepsilon_0 = 1.2$, $\varepsilon_1 = 1.5$, $\varepsilon_2 = 0.95$. The time responses of synchronous error states are depicted in Fig. 2. It is shown that the synchronous error states oscillate irregularly when the controller is switched off, and when the controller is in actuated at $t = 5$, the synchronous error states converge to zero and complete synchronization is accomplished. The time responses of time-varying function $\gamma(t)$, sliding mode, and control input are shown in Figs. 3 and 4, respectively. It is clear that the control signal is continuous and the chattering phenomenon is free.

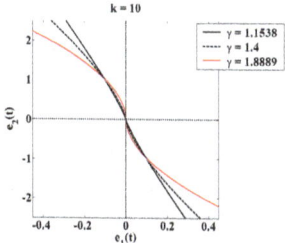

Figure 1. Wrinkling type TSM for $\gamma(t)$

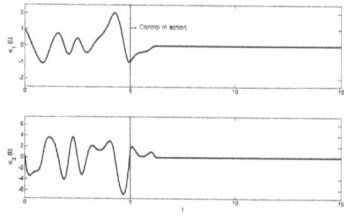

Figure 2. Time responses of $e_1(t)$, $e_2(t)$

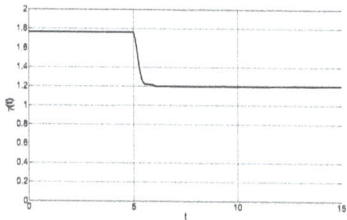

Figure 3. Time response of $\gamma(t)$

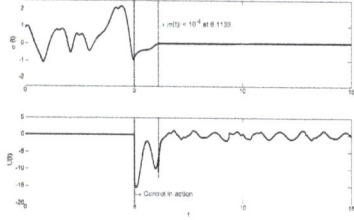

Figure 4. Time responses of $\sigma(t)$, $U(t)$

7. Conclusion

In this study, the novel time-varying TSM is firstly introduced, named wrinkling-type TSM, and an adaptive wrinkling-type TSM controller has been addressed. Techniques for achieving complete synchronization between two identical chaotic gyros with system uncertainties and external disturbances are demonstrated. Based on the Lyapunov stability theorem, the stable synchronization is provided. In addition, the performed numerical simulations demonstrate the chattering free control and verify the effectiveness and the robustness of the presented schemes.

Acknowledgements

This work was funded in part by the Ministry of Science and Technology, Taiwan, ROC., under Grant MOST 104-2221-E-164-005 -.

References

1. H.K. Chen, Chaos and chaos synchronization of a symmetric gyro with linear-plus-cubic damping, Journal of Sound and Vibration 255 (2002) 719–740.
2. Y. Feng, X. Yu, Z. Man, Non-singular terminal sliding control of rigid manipulators, Automatica 38 (2002) 2159–2167.
3. C.C. Yang, Robust Adaptive Terminal Sliding Mode Synchronized Control for a Class of Non-autonomous Chaotic System, Asian Journal of Control 15 (2013) 1-9.
4. Q.P. Ha, D.C. Rye, H.F. Durrant-Whyte, Fuzzy moving sliding mode control with application to robotic manipulators, Automatica 35 (1999) 607-616.
5. Z. Jinggang, K. Jiang, Z. Chen, Z. Zhao, Global robust fuzzy sliding mode control for a class of non-linear system, Transactions of the Institute of Measurement and Control 28 (2006) 219–227.

Research on Time Delay Control for the Vertical Vibration Characteristics of Roll System

Bin Liu[1], Peng Li[1,†], Kun Wang[2], Jia-hao Jiang[1] and Gui-xiang Pan[1]

[1]School of Electrical Engineering, Yanshan University, Qinhuangdao
[2]School of Science, Yanshan University, Qinhuangdao
†Email: 993777238@qq.com

For the vibration problem of rolling mill under the influence of dynamic characteristics of hydraulic system, time delay state feedback is introduced, and a vibration control model of roll system is established based on the nonlinear stiffness constraint of hydraulic cylinder. The amplitude frequency response of the system is obtained by the method of multiple scales. The actual parameters of rolling mill are used to analyze the impact law of time delay parameters on the mill roll system amplitude frequency characteristics and time domain features. Founding that with the increase of the control gain coefficient, the stability of the roll system is improved. The change of time delay makes the system show complex nonlinear dynamic behavior. Therefore, it is very important to choose the appropriate time delay parameter to suppress the rolling mill vibration, which provides a theoretical reference to control mill stability movement.

Keywords: Rolling mill vibration; Time delay state feedback; Control model.

1. Introduction

The rolling mill vibration has always been one of the technical problems in the steel industry. In recent years, with the improvement of the rolling speed and product quality requirements, the vibration problem of strip rolling mill has become more and more significant [1,2]. Major steel mills through research and analysis to try kinds of means to make the speed of the rolling mill and rolling precision is increased, but the roll system instability phenomenon is more and more frequent, equipment damage and other accidents occur frequently [3]. Therefore, the major steel mills from the optimization of the structure and improve the design accuracy and other ways to solve this problem, but didn't obtain good effect. With the mature of active control theory and technology, using active control to suppress the occurrence of the vertical vibration of the rolling mill has become a hot research topic.

Active control is a method to control the vibration system by means of the external energy [4]. Many mathematical models of active control are generally ordinary differential equations or partial differential equations, the dynamical system can be described by the differential equation of state variable change with time and space. These models consider that the dynamic change of the system is only related to the current state. There is also a part of the dynamical system show time delay phenomenon, its dynamic change is related to the state of the current system and the state of the past. In the control of rolling mill vibration, in recent years, some advanced control algorithms in the application of torsional vibration [5, 6], vertical vibration control is relatively less.

The control of vertical vibration system of rolling mill is studied in this paper. For the vibration of rolling mill under the influence of dynamic characteristics of hydraulic system, time-delay state feedback is introduced, and the amplitude frequency response of the system is obtained by the method of multiple scales. Finally, the appropriate time delay parameter combinations are selected, in order to improve the stability of the system.

2. Rolling Mill Vibration Control Model

In this paper, the mill vibration characteristics under the constraints of the double-acting single piston cylinders are mainly considered. This hydraulic cylinder is equipped with a piston rod in the side of the piston, thus the active area in two cavities is different, and the round-trip velocity and force are not equal. The movement of hydraulic cylinder piston changed the effective length of two-cavity liquid, the change in the length leads to the change of hydraulic oil rigidity. The equivalent stiffness of the hydraulic cylinder can be expressed as

$$k(x) = \frac{\beta_e A_1^2}{A_1(L_1 + x) + V_{1l}} + \frac{\beta_e A_2^2}{A_2(L - L_1 - x) + V_{2l}} \tag{1}$$

Where

β_e is the bulk modulus of hydraulic oil, A_i is the effective area on either side of the hydraulic cylinder piston, $i = 1, 2$, L_1 is the initial effective length of the rodless cavity, x is the roll system vibration displacement, V_{il} is the hydraulic oil volume of the hydraulic line between valve and one side of the hydraulic cylinder, $i = 1, 2$;

The Taylor series expansion of the equation (1) at the origin

$$k_i(x) = \beta_e A_i \frac{1}{L_i + x} = \beta_e A_i \left(\frac{1}{L_i} - \frac{x}{L_i^2} + \frac{x^2}{L_i^3} - \frac{x^3}{L_i^4} + O(x^4) \right) \tag{2}$$

Because the elastic potential energy of spring is symmetrical, it can be expressed as

$$U = \frac{K_1}{2} x^2 + \frac{K_2}{4} x^4 \tag{3}$$

The spring force can be expressed as

$$F_1^*(x) = \frac{dU}{dx} = \kappa_1 x + \kappa_2 x^3$$

(4)

Where $\kappa_1 = \dfrac{\beta_e A_1}{mL_1} + \dfrac{\beta_e A_2}{m(L-L_1)}$, $\kappa_2 = \dfrac{\beta_e A_1}{mL_1^3} + \dfrac{\beta_e A_2}{m(L-L_1)^3}$

According to structure's symmetry of the mill roll system, considering the nonlinear stiffness constraint of hydraulic cylinder. As shown in Fig. 1, the upper roll system dynamics control model is established. Among them, m is the equivalent mass of backup roll and work roll, c is the equivalent damping of rolled piece, k is the equivalent stiffness of the rolled piece, $k(x)$ is the equivalent stiffness of hydraulic cylinder, F is external excitation, $\lambda x(t-\tau)$ is delay state feedback of systems, λ is feedback gain factor, τ is time delay.

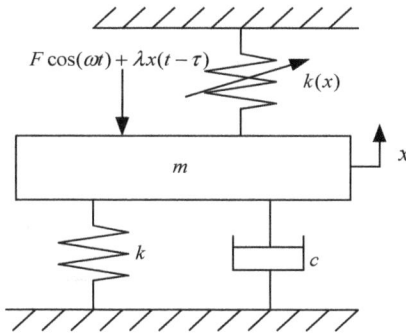

Fig. 1. Dynamic control model of the system

According to the generalized Lagrange principle, the dynamic equilibrium equation of the controlled system can be listed.

$$m\ddot{x} + c\dot{x} + kx + F_1(x) = F\cos(\omega t) + \lambda(-x + x(t-\tau))$$

(5)

Eq. (5) also can write as follow

$$\ddot{x} + \omega_0^2 x = -\delta\dot{x} - \kappa_2 x^3 + F^*\cos(\omega t) + \rho x(t-\tau)$$

(6)

Where $\omega_0 = \sqrt{\dfrac{k+\kappa_1+\lambda}{m}}$, $\delta = \dfrac{c}{m}$, $F_1^*(x) = \dfrac{F_1(x)}{m}$, $F^* = \dfrac{F}{m}$, $\rho = \dfrac{\lambda}{m}$

The amplitude frequency response of the system is obtained by the method of multiple scales

$$[a\omega_0\sigma - \frac{3}{8}\kappa_2 a^3 + \frac{1}{2}\rho a\cos(\omega_0\tau)]^2 + [-\frac{1}{2}\delta\omega_0 a - \frac{1}{2}\rho a\sin(\omega_0\tau)]^2 = \frac{1}{4}(F^*)^2 \tag{7}$$

3. Delay Feedback Control Analysis

Considering the nonlinear stiffness constraint of hydraulic cylinder, a single degree of freedom vibration model of roll system was established. The actual parameters of four roll mill are: $m = 1.44 \times 10^5$ kg, $c = 2.04 \times 10^6$ N·s/m, $k = 2.35 \times 10^{10}$ N/m, $L = 0.11$ m, $A_1 = 0.6361$ m², $A_2 = 0.3243$ m², $\beta_e = 1.6 \times 10^9$ Pa, $\varepsilon = 0.01$. Based on the numerical results, the numerical simulation of roll system vibration system is carried out. Analysis of amplitude frequency characteristics and time domain characteristics of rolling mill system with time delay state feedback.

4. Amplitude Frequency Characteristics

Fig. 2 and Fig. 3 are the amplitude frequency curves of the roll system vibration system under different parameters.

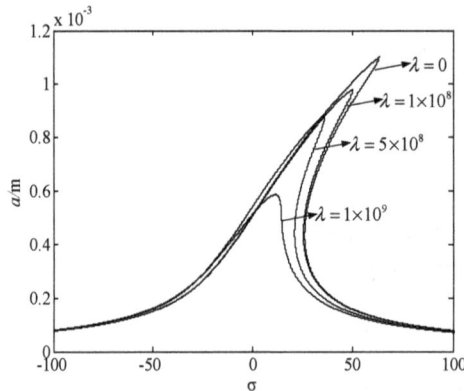

Fig. 2. Amplitude frequency curve of the system when time delay feedback gain coefficient changed

As can be seen from Fig. 2, can be found, when the parameter $\lambda = 0$, the amplitude frequency characteristic curve of the original system is not controlled. The vibration amplitude of the system decreases with time delay feedback gain coefficient λ increasing, at the same time, the common jump phenomena in nonlinear systems is obviously weakened. That is to say, the vibration of roll system can be effectively suppressed by increasing the time delay feedback gain coefficient.

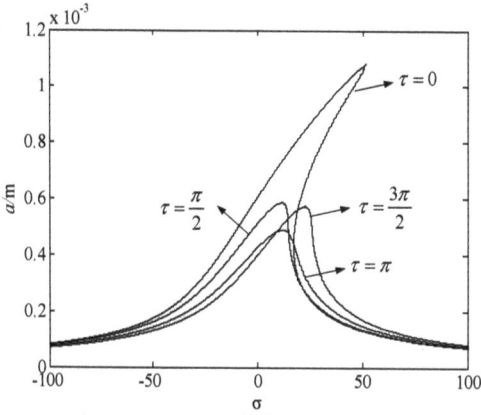

Fig. 3. Amplitude frequency curves when the time delay changed

As can be seen from Fig. 3, when the parameter $\tau = 0$, the amplitude frequency characteristic curve of the original system is not controlled. The change of time delay parameter can also control the magnitude of the vibration amplitude of the system, and affect the migration of the system vibration curve. Therefore, it is very important to choose the appropriate time delay parameter to suppress the roll system vibration.

5. Time Domain Characteristics

Fig. 5 and Fig. 4 are the time domain response of the system with different parameters.

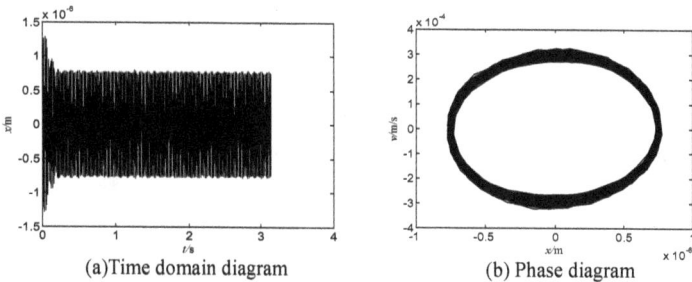

(a)Time domain diagram   (b) Phase diagram

Fig. 4. The time domain response of the system when $\lambda = 0, \tau = 0$

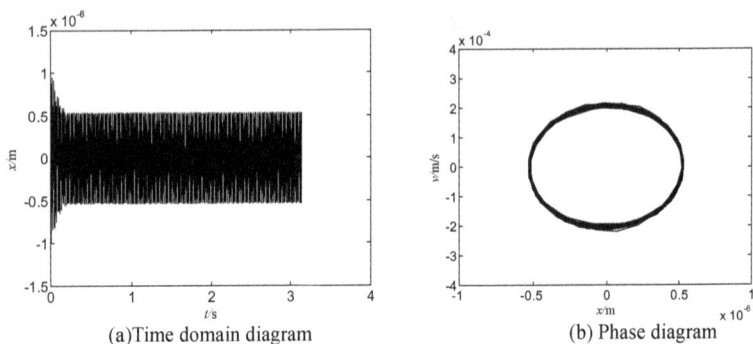

(a)Time domain diagram (b) Phase diagram

Fig. 5. The time domain response of the system when $\lambda = 9 \times 10^{10}, \tau = \pi$

Fig. 4 is the phase diagram and time domain diagram when time delay parameter $\lambda = 0, \tau = 0$, which is not controlled system. Fig. 5 is the controlled system. Compared with Fig. 4 and Fig. 5, we can find that the amplitude of the system is decreased and the stability of the system is improved after adding the time delay control. The time delay parameter λ and τ can adjust the size of the amplitude of the system, which is to select the appropriate parameters can effectively inhibit vibration of roll system and make roll system in a more stable state, improving the system stability.

6. Conclusion

This paper considered the nonlinear constraint effect of hydraulic cylinder, time delay state feedback is introduced, based on vibration control model of roll system, analyze the time domain and amplitude frequency characteristics of the system. Founding that it is very important to choose the appropriate time delay parameter to suppress the rolling mill vibration, which provides a theoretical reference to control mill stabilize motion.

Acknowledgement

The work was funded by Natural Science Foundation of Hebei Province of China through Project number E2015203349, which is gratefully acknowledged.

References

1. Bar A, Świątoniowski A. Interdependence between the rolling speed and nonlinear vibration of the mill system [J]. Journal of Materials Processing Technology, 2004, 155: 2116-2121.
2. Fujita K, Saito T. Unstable vibration of roller mills [J]. Journal of Sound and Vibration, 2006, 297(1): 329-350.

3. Hou Fuxiang, Zhang Jie, Cao Guojian and Shi Xiaolu. Review of Chatter Studies in Cold Rolling [J]. Journal of Iron and Steel Research, 2007, 19(10): 6-10.
4. Chen Liping. Research status and developing tend of active vibration control [J]. Modern Machinery, 2005, (2): 52-55. (In Chinese)
5. Yamapi R, Aziz-Alaoui M.A. Vibration analysis and bifurcations in the self-sustained electromechanical system with multiple functions [J]. Communications in Nonlinear Science and Numerical Simulation, 2007, 12(8): 1534-1549.
6. Liu Le, Fang Yiming, Li Xiaogang and Li Jianxiong. Tensiometer-free Control for a Speed and Tension System of Reversible Cold Strip Mill Based on Hamilton Theory [J]. Acta Automatica Sinica, 2015, 41(1): 165-175. (In Chinese)

The Effects of Compressed Packers Rubber Structural Parameters on Sealing Performance

Hai Dai[1], Gui-qin Li[1,†], Li-xin Lu[1], Ru-dong Lu[2],
Yun-jiang Dong[2] and Shu-xun Li[2]

*[1]Shanghai Key Laboratory of Intelligent Manufacturing and Robotics,
Shanghai University, Shanghai, 200072, China
[2]Shanghai Extrong Oilfield Technology Co., Ltd., Shanghai, China
†Email: leeching@t.shu.edu.cn*

Packer is one of the indispensable tools in oilfield operations and its sealing performance is closely related to rubber structural parameters. The rubber structural model is established by ABAQUS software. Different height, camber angle and the thickness of the rubbers are optimized in finite element analysis. By comparing the maximum contact stress between rubber and casing wall, the optimal size is selected. From analysis results, the existence of optimal structure parameters on packer rubber is proved. The maximum contact stress can be reduced by unsuitable value of height, camber angle and thickness, which is not conducive for sealing.

Keywords: Rubber; Structure size; Sealing; The maximum contact stress; Finite element analysis.

1. Introduction

Packer is one of the indispensable tools with many types in oilfield operations, and the compressed packer is most widely used in oil production process like well completion, water injection, fracturing, sand control, mechanical oil recovery and gas lift. Sealing rubber is the core component of compressed packer [1], whose performance is closely related to the contact stress between rubber and casing wall. Generally, the greater the maximum contact stress, the higher the capacity to bear working pressure [2]. The sealing performance of rubber is affected by many factors. In terms of structure parameters, height, camber angle and thickness are three key factors. In this paper, different structural parameters models are simulated by ABAQUS software and the optimal size is obtained by comparing the maximum contact stress between rubber and casing wall.

2. Theoretical Basic of Packer and Establishment of FEA Model

2.1. *Theoretical Basic*

In compressed packers, rubber is compressed by external axial pressure, so that the axial compression and radial expansion is produced to fill the annular space between the rubber and casing, which belongs to contact sealing. Relying on contact stress between the rubber and casing wall, the partition of well fluid, separation of production layer and prevention of mutual interference between laminar flow are realized. The sealing quality is mainly depended on the value of the maximum contact pressure between rubber and the casing wall, which means the bearing capacity of the rubber is directly affected by contact stress.

Packer rubber is made of rubber material that is highly nonlinear composite materials with a series of characters like geometric nonlinearity, material nonlinearity, large elongation and volume incompressible [3]. In continuum mechanics, the rubber material is called hyper-elastic material. After a long time study, many classic constitutive models are put forward like Mooney-Rivlin, Yeoh, Ogden, etc. But each model can only be in a specific environment to accurately represent the mechanical properties of rubber, otherwise, the error of the calculation is relatively large. In finite element analysis of packer rubber, material nonlinearity, geometric nonlinearity and contact nonlinearity are involved. ABAQUS is CAE software with a strong nonlinear analysis capabilities and the ability to solve complex problems, which is the most advanced large-scale general nonlinear finite element analysis software in the world. ABAQUS/Standard has a strong nonlinear solver, adjusting increment size automatically so as to achieve the better convergence [4].

2.2. *Establishment of FEA model*

Packer rubber, piston briquetting, center pipe and cashing are all axisymmetric geometries. Under ideal conditions, constraints and loads are also axisymmetric, so axisymmetric two-dimensional model is utilized in analysis [5]. Hydrogenated nitrile rubber with Shore A hardness of 84 is adopted with Yeoh constitutive model [6]. The material constant C_{10}=1.134MPa, C_{20}=-0.526MPa, C_{30}=0.528MPa, D_1=D_2=D_3=0 [7]. The other parts are structural steel, Young's modulus $E = 2.06e5$, Poisson's ratio $\mu = 0.3$. Single rubber is analyzed, and the shape and dimensions of rubber before deformation is shown in Fig. 1. Hybrid property unit CAX4RH and quadrilateral mesh are adopted for finite element analysis of rubber [8]. The mesh element size is 1mm and others are 2mm with CAX4R unit and quadrilateral mesh. The structured grid model is shown in Fig. 2.

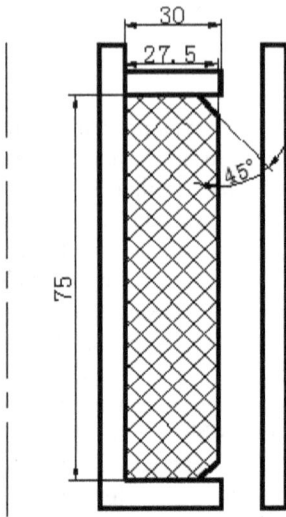

Figure 1. The structure size before deformation   Figure 2. The structured grid model

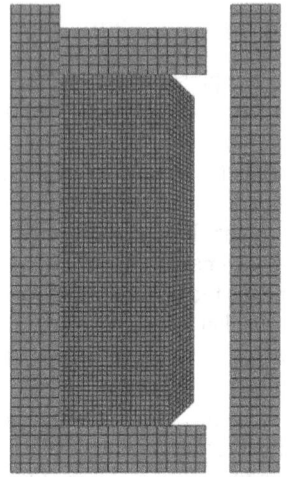

3. Analysis of Rubber Structural Parameters

3.1. *The Effect of Rubber Height on Sealing Performance*

There is a great effect of rubber height on sealing performance. To analyze the effect of height on the maximum contact stress, the rubber height of 65mm, 75mm, 85mm, 95mm are respectively selected with same thickness and camber angle. In the case of applying a 10-20MPa packer increasing force, analyzing the relationship between loads and the maximum contact stress, the maximum contact stress contrast curve is obtained as shown in Fig. 3.

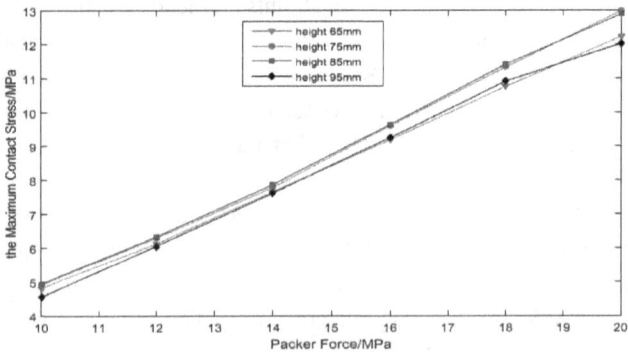

Figure 3. The maximum contact stress contrast curve under different heights

It can be seen from the figure that the maximum contact stress of 85mm and 75mm is basically the same and larger than that of 95mm and 65mm. We can draw conclusions that the maximum contact stress between rubber and casing wall is decreased no matter whether the height is too big or too small. The packer can easily lose the whole stability caused by too high height. But the force that the rubber can bear will become small if the height of rubber is too low. Therefore, 75mm or 85mm can be the best height.

4. The Effect of Rubber Camber Angle on Sealing Performance

To analyze the effect of camber angle on the maximum contact stress, the rubber camber angle of 25°, 35°, 45°, 55° are respectively selected with same thickness and height. The maximum contact stress between n rubber and casing wall is shown in Fig. 4.

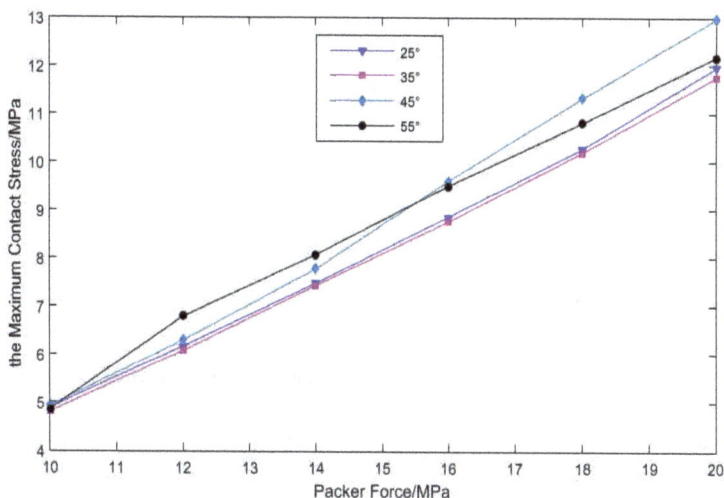

Figure 4. The maximum contact stress contrast curve under different camber angle

It can be seen from the figure that the maximum contact stress between the rubber and the casing wall is directly related to the camber angle. The larger camber angle is good for the sealing performance. The rubber sealing performance is relatively poor when the camber angle is 25° or 35°, so it is not recommended in actual application. When the force is less than 15.5MPa, the maximum contact stress of 55° is greater than 45°, but less than 45° when the sealing force is greater than 15.5MPa. Larger camber angle is easier to seal, but it is also easy to produce the stress concentration area and larger deformation.

"Convex shoulder" has been appeared when the force is not large. Therefore, 45° rubber should be chosen preferentially in the engineering application.

5. The Effect of Rubber Thickness on Sealing Performance

To analyze the effect of thickness on the maximum contact stress, the rubber thickness of 26.5mm, 27.5mm, 28.5mm, 29.5mm are respectively selected with same angle and height. The maximum contact stress between rubber and casing wall is shown in Fig. 5.

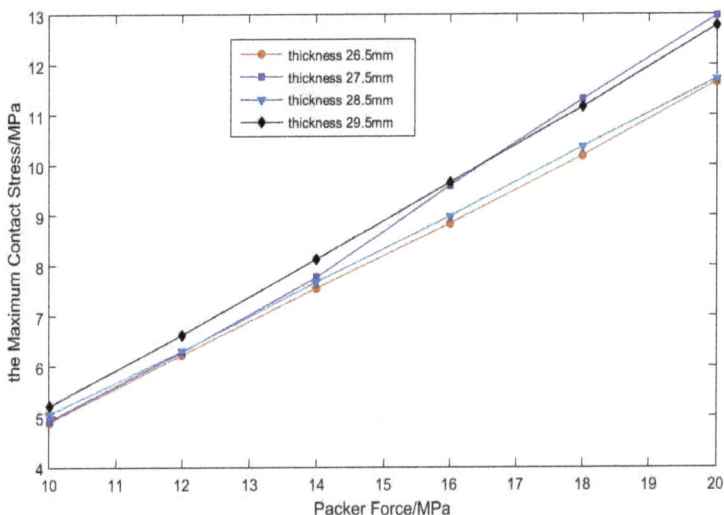

Figure 5. The maximum contact stress contrast curve under different thickness

The contrast curve shows that the maximum contact stress is nearly the same when the rubber thicknesses are 26.5mm, 27.5mm or 28.5mm but less than that of thickness 29.5mm. When packer force exceeds 14MPa, the maximum contact stress of 27.5mm and 29.5mm are significantly greater than that of 28.5mm and 26.5mm.

After the packer force is more than 16MPa, the maximum contact pressure of 27.5mm gradually exceeds that of 29.5mm. And the rubber thickness deformation of 29.5mm and 27.5mm under 20MPa is shown in Fig. 6, and Fig 7. It is clearly showed that rubber compressed distance of 27.5mm is 5mm smaller than that of 29.5mm. Smaller compressed distance is bad for sealing, so 27.5mm can be the most appropriate thickness.

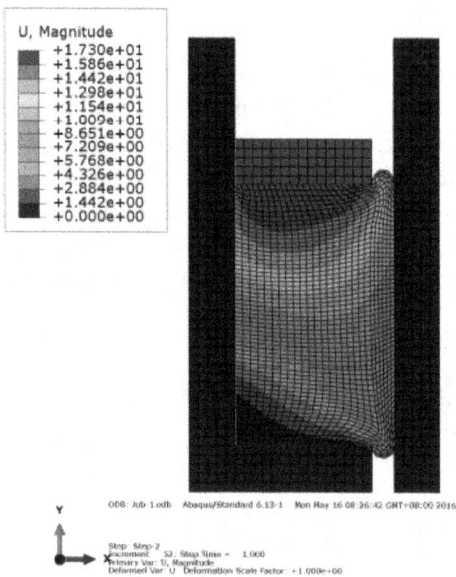

Figure 6. The rubber of thickness29.5mm deformation under 20Mpa

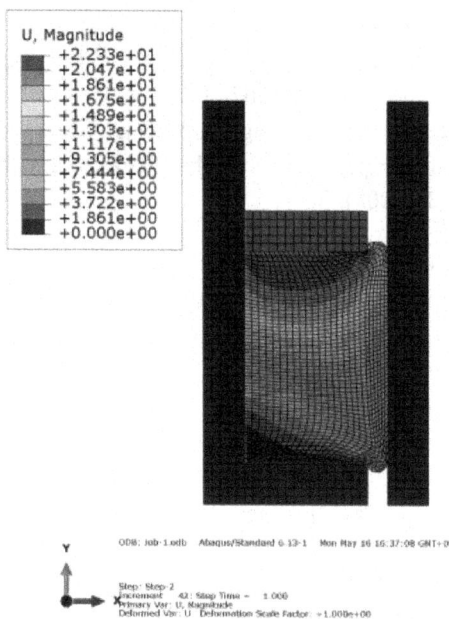

Figure 7. The rubber of thickness 27.5mm deformation under 20Mpa

6. Conclusion

The maximum contact stress between the rubber and casing wall is directly related to the height, camber angle and thickness of the rubber. Therefore, the optimal selection of the structural parameters has important significance for the sealing performance of the packer. The maximum contact stress can be reduced by too big or small value of height, camber angle and thickness. The maximum contact stress with best sealing performance can be obtained with height of 75 or 85mm, camber angle of 45°and thickness of 27.5mm.

Acknowledgement

This research was funded by Shanghai Committee of Science and Technology Project (Grant No. 14DZ1204203).

References

1. L.X. Zhang, Z.J. Sheng, Y.L. Li, G.W. Zhang, X.Q. Gao. Packer technology
2. Development and application in our country [J], Petroleum Machinery. 8 (2007)10-12.
3. Liu Song, Wu Jing, X.M. Dong. Sealing performance analysis and structure improvement on compressed packer rubber [J]. Petroleum Mining Machinery, 42(2013) 67-70.
4. Wu Jian, X.P. Xu, L.P. Wang, Q.M. Zhang. Mechanical analysis of the conventional high-pressure rubber barrel of packers [J]. Petroleum Mining Machinery, 37(2008): 39-41.
5. L.T. Zhao. ABAQUS6.6 application in mechanical engineering [M]. Beijing: China Water Power Press, (2007) 282-283.
6. Z.L. Wang, B.Q. Sun, F.T. Zhang. Optimization of packer rubber's structural parameters [J]. Inner Mongolia Petrochemical Industry, 13 (2015) 47-49.
7. X.F. Li, X.X. Yang. A review of elastic constitutive model for rubber material [J]. Elastomer, 15(2015) 50-58.
8. Chen Xin. Design and research on structure of the compressed sealing rubber of the packer [M]. China University of Geosciences, (2015).
9. Y.H. Liu, J.H. Fu, Y.H. Lin, L.Q. Bai, J.Z. Yang, Z.X. Wei, H.T. Xu, H.Y. Ma   Finite element analysis of packer's rubber sealing properties [j]. Petroleum Mining Machinery, 36 (2007) 38-41.

A PID Based Control System for Multi-Wire Saw Machine

Li-fei Ma, Yan Wang[†], Shi-bo Wei and Pan-pan Tian

School of Mechanical Engineering,
Beijing Institute of Graphic Communication,
Beijing, 102600, China
[†]*Email: wangyanzi@bigc.edu.cn*

In order to solve the problem that the steel wire gets loose or broken in Multi-Wire-Saw when it works in a high-speed, this paper introduces a control method based on PLC and servo motor, uses the separated integral PID to control the speed of the steel wire when it's taken on and off accurately. This method improves the stability of the system, even operating at high-speed multi-wire saw. The tension control system can also be used in papermaking, printing and packaging.

Keywords: Multi-Wire-Saw; Tension control; Servomotor; PID.

1. Introduction

The Multi-wire cutting technology is currently a type of relatively advanced processing technology in the world, and the principle lies in bringing the abrasive materials in the processing region to grind the cut body via the high-speed reciprocating motion of the steel wire. Therefore, as a kind of new-type cutting and processing method [1], the foregoing technology can cut the body into hundreds even thousands of slices simultaneously. During the cutting process the steel wire can form a wire mesh on the godet via the guidance and commutation of the guide pulley, and in such case the cut body can be carved into slice pieces after passing through the wire mesh. Due to the high efficiency, high capacity, high precision and other advantages, the Multi-wire cutting technology, when compared to traditional processing technology, has been widely applied to cut and process [2] the single (multi) crystal silicon, quartz, crystal, artificial gem, magnetic materials, chemical compound and other crisp and hard materials.

In the course of the cutting process, the tension wave of the cutter steel wire is likely to result in the inconsistency [3] between the cutting lines which are mutually parallel with each other on the processing roll, and as a result the cutting slice thickness is extremely different from each other. The central thickness difference, warping degree and other single-slice performance indexes between the adjacent slices will be accordingly affected when the certain line tension in the cutting groove fluctuates over time. Breaking the line is easy as long as the tension fails to be properly controlled due to the high-speed reciprocating and circle cutting in the working process. In such case will not only a large number of working hours be wasted, but also the expensive semiconductor materials will be scrapped and result in economic losses. The tension control thus appears to be of essence when the user wants to guarantee the production efficiency and product precision of the cutter. Currently the tension control is considered to be one of the key technologies for the Multi-wire saw.

2. System Hardware Design

2.1. *Overall Structure Analysis of the Multi-Wire Saw*

The Multi-wire saw is generally composed of a pay-off roller, a take-up roller (driven by the motor), and also utilizes a guide pulley, a processing roller, and workbench andother. The overall structure diagram regarding the Multi-wire saw used in the test is shown in Figure 1.

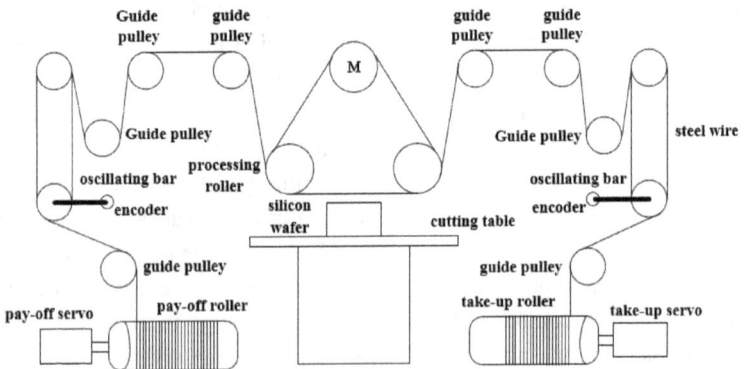

Figure 1. Structure diagram regardingthe Multi-wire saw

The steel wire, which is coated with cutting materials, shall be twined on the pay-off roller and the take-up roller [4], while the foregoing rollers are driven by two servomotors. The steel wire will be twined on the external grooves of the three processing rollers after passing through multiple guide pulleys, and as a result the steel wire will wind into a wire mesh. The foregoing three processing rollers will be handled by one servomotor and transmission mechanism. The cutting table is lifted via the servomotor and roller screw mechanism, under condition that the cut body (silicon wafer) is closely stuck by glue. While being driven by the pay-off servo, take-up servo and spindle servo, the steel wire can make reciprocating motion back and forth to cut the silicon wafer. The oscillating bar and encoder in the system are always used for inspecting the tension In addition, the foregoing encoder can also measure the displacement amount of oscillating which is caused by the steel wire tension. The electrical part of the tension control system is composed of major controller PLC, servo actuator and human-computer interface. The touch screen, operating as the system's upper computer on the human-computer interface, can not only be used for setting the system parameter, but also for monitoring the system operation. In addition, the PLC, serving as the major controller, can control various drivers in real-time way. The servomotor can serve as the power producer to drive the machine. The system applies three axes to control the overall tension and realize the simple and high-efficient processing.

2.2. Structure Analysis of the Tension Part Regarding the Multi-wire Saw

1) Tension control mechanism analysis: The tension control mechanism of the oscillating bar, which is used in the tension control system, is shown in Figure 2. The appliance of a constant-torque tension motor (servomotor) can control a light oscillating bar and change the tension on the cutting line.

If we assume that the tension oscillating bar quality is classified as m, length as L, the angular velocity of oscillating bar as ω, the equivalent rotational inertia of the tension rocker motor rotor as I, the distance from the axis center of the tension motor to the oscillating bar center as L/2, and the constant torque which is exerted by tension motor as M, the kinetic equation regarding the tension rocker shall be:

$$2FL - mgL/2 - M = I\frac{d\omega}{dt} \tag{1}$$

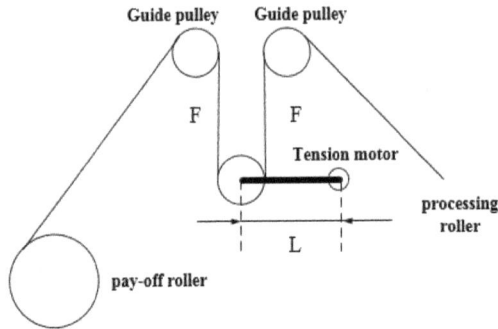

Figure 2. Tension control mechanism

Namely, the obtained tension value is calculated by using the following formula:

$$F = \frac{mg}{4} + \frac{M}{2L} + \frac{I}{2L}\frac{d\omega}{dt} \qquad (2)$$

When the tension is changed, the angular displacement will occur on the oscillating bar, while beingreciprocalto the tension variation quantity. The tension sensors are installed on the side of the pay-off roller and take-up roller in order to inspect the tension value in real-time. The encoder is also mounted on the right-hand oscillating bar axis to measure the angular displacement of oscillating bar and feed the changing information back to the PLC. The PLC will then make the calculation and adjust the linear velocity of the pay-off roller and take-up roller for the sake of preventing the tension fluctuation.

2) The Tension control system framework is mainly composed of tension inspection module, tension control module and velocity control module (as shown in Figure 3). In that matter the tension inspection module, which is composed of an encoder and oscillating bar, will inspect the tension and the inspection result will be input in the major controller PLC of the tension control module for calculation [5] after being compared with the set value. The obtained result will consequently control the servo driver and actuate the servo motor operation (velocity control module).

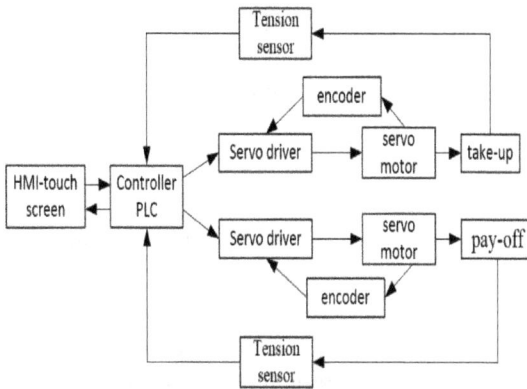

Figure 3. Constant tension control system model

3. System Software Design

3.1. *Traditional PID Control Method*

The tension control system draws lessons from the traditional PID control, one of the most commonly used control mode [6] in the current industrial control field. According to the PID parameters and tension set value, the corresponding control requirements will be generated by utilizing the PID algorithm. The system will make subtraction for the tension value which is collected from tension controller and tension set value and then input the obtained differential value as the PID control algorithm The output value of control system will be obtained, and the output value is the servo motor velocity which controls the pay-off roller and the take-up roller. Bear in mind to adjust the take-up and the pay-off servo motor velocity to regulate the tension fluctuation of steel wire.

The control rules of the PID are:

$$u(t) = k_p \left(error(t) + \frac{1}{T_1} \int_0^t error(t)dt + \frac{T_D derror(t)}{dt} \right)$$

(3)

Formula information: kp refers to the proportion coefficient; T1 refers to the integral time constant; TD refers to the calculus time constant; error(t) refers to the differential value between output value and set value (the differential value between the inspected tension value and set value).The PID control parameter can be set in the formula (2) via MATLAB. After that please select the most suitable PID parameters via the simulation data oscillogram.

In general, the PLC is equipped with self-tuning function for PID parameter, therefore it can be directly invoked [1].

3.2. *PID Integral Separation Control Method in the Machine*

The high-speed to and fro movement is realized by applying the steel wire in the system, and as a result the tension fluctuation will be accordingly large.The PID controller integral will be accumulated [7] when the traditional PID control algorithm is adopted, and the control quantity will exceed the limit control volume resulting in a large overstriking in the system.The system is therefore inclined to apply the integral separation PID regulating value, which is based on the PLC, to control the tension control system.In other words, the integral function will be cancelled as long as the difference between the measured tension value and the set value is relatively large, so that the system's stability doesn't get reduced and the overstriking quantity doesn't get increased. Additionally, it is convenient to eliminate the static error and enhance the control precision by importing the integral control when the measured value approaches the set value.

The threshold value e is assumed to be larger than 0;

When |error(k)|>e, please apply PD control to avoid the ultra-large regulation and make the system respond rapidly;

When |error(k)|≤e, please apply PID control to guarantee the system control precision.

The integral separation algorithm can be expressed as:

$$u(k) = k_p error(k) + \beta k_i \sum_{j=0}^{k} error(j)T + k_d(error(k) - error(k-1))/T \tag{4}$$

Formula information: T refers to the sampling time; β refers to the switching coefficient of integral item, and

$$\beta = \begin{cases} 1, |error(k)| \le e| \\ 0, |error(k)| > e| \end{cases} \tag{5}$$

The flow chart, which is obtained accordingly to the integral separation-type PID control algorithm is shown in Figure 4.

Processing roller motor ω_1 → $F_1(S)$ → R_1 → V_1 → ⊗ → $2/L_0$ → ω_4 → [$\Delta\omega > \omega_0$? $\Delta\theta > \theta_0$?] Yes → PD → pay-off roller motor

R_2 → V_2 ; $1/s$ → θ_4 ; no → PID

$F_2(s)$ → ω_2

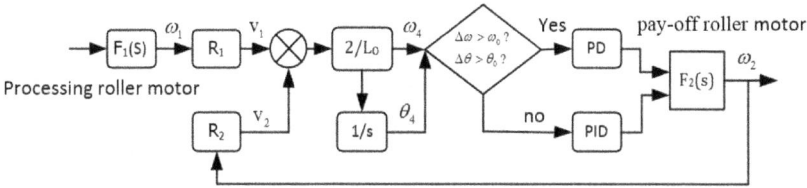

Figure 4. Integral Separation-type PID Algorithm System Control Flow Chart

4. On-site Testing

Based on the design of the system, we have undertaken an on-site testing to compare the performance of the Multi-wire cutting machine, either with or without the integral PID control system. The results are shown in Table 1.

Table 1. Before and after the improvement of the machine performance index comparison

performance index	Before improvement	After improvement	Contrast conclusions
The highest line speed (m / min)	460	500	better
Tension control accuracy(N)	within 2.5	within 1.5	better
Qualified slice rate	90%	95%	better
Sheet thickness error (mm)	0.020	0.015	better
Machining surface parallelism (mm)	0.013	0.008	better

The result shows that the control system makes the machine run smoothly, the tension fluctuation is small, the line speed can be up to 500m / min, and the cable is even, the tension control precision is within 1.5N, Cable uniformity of the outer diameter of less than 0.3mm, Minimum thickness of single slice is 0.8mm,slice qualified rate of 95%, slice thickness error of less than 0.015mm, fully meet the expected performance targets. As shown in Table 1, compared to the machine after the improvement of performance indicators before the comparison.

5. Conclusion

The system can utilize the tension sensor to monitor the tension changes in real-time, and form a constant tension controlled full-closed loop system for the output data via the algorithm treatment of the PLC controller and servo controller, while the applied integral separation PID algorithm solves the tension oscillation, which is caused by the high-speed variable of the machine. Currently the control method has been applied in the high-speed multi-wire saw, and as shown in the operation results, the control system has reached the prospective process control requirements and created service value in the multi-wire cutting industry, as well as in packaging, printing and many other fields.

Acknowledgement

This work was funded by the key project (project no. TJSHG201510015011) of enhancing the innovation ability of Beijing municipal colleges and universities, the project (project no. 04190116008/001) of Municipal scientific research projects and the project (project no. 06170115003/013) of Beijing College of young talents.

References

1. Zhang Yibing, Dai Yuxing, Yuan Julong, etc. Design and Implement of Tension Control System for Multi-wire Saw [J]. Journal of Mechanical Engineering, 2009, 45(5): 295-300.
2. Jiang Jin, Dai Yuxing, Peng Siqi. Development of Control System for Multi-wire Saw [J]. China Mechanical Engineering, 2010, 21(15): 1780-1784
3. He Jingliang, Du Kaixun, Wang Xuejun. High-Speed Wire Cutting System Tension Control Research [J]. Manufacturing Automation, 2011, 33(9): 6-8
4. Wu Yi, Xu Zhenyue, Li Junqiang. Design of Pay-off and Take-up Machine of Constant Tension Control System Based on PLC [J]. Industrial Control and Computer, 2006, 19(12): 70-71.
5. Dorf, Bi Xiaopu. Modern Control Systems [M]. Version 11. Beijing: Publishing House of Electronics Industry, 2011:490-510.
6. Fu Ji, Lin Lizhi, Zhou Xilong, et al. A macroscopic nondestructive testing system based on the cantilever-sample contactresonance [J]. Review of Scientific Instruments, 2012,83,123707.

Research on Design Methods of Drive Cam Contour of Scrape-Clearer of Die-Cutting Machine

Ning-ning Guo†, Ju-ying Qian, Ying-min Han and Shi-bo Wei

School of Mechanical Engineering,
Beijing Institute of Graphic Communication,
Beijing, 102600, China
†*Email:791504485@qq.com*

In order to improve the scrap-cleaning precision of die-cutting machine, five-polynomial laws are used to fit framework motion curve based on the moving principle of drive cam in scrape-clearer. According to two limiting positions at the start and end of motion, complex number vector method is used to integrate cam link mechanism and calculate cam contour, which has theoretical guidance significance to improving the performance of die-cutting machine contour and scrap-cleaning precision.

Keywords: Die-cutting Scrap-cleaning; Cam Link Mechanism; Complex Number Vector Method.

1. Introduction

Die-cutting machine is a kind of surface decoration and processing equipment [1], which is used in die cutting and creasing in printing and packaging industry. A complete die-cutting machine generally consists of three parts, namely paper-feeding system, host and scrap-cleaning part [2]. Die-cutting machine may be divided into manual, semi-automatic and full-automatic die-cutting machine according to the functions of scrape-clearer. As one part of die-cutting machine, scrape-clearer is of great importance to die cutting. Scrape-clearer of die-cutting machine is shown in Figure 1, which consists of three frames in the upper, middle and lower and its driving mechanism - cam link mechanism. Where, the upper and lower framework moves advance and back at high speed in vertical direction and the middle frame is fixed.

The main principle of scrap-cleaning of die-cutting machine is to complete with the upper and lower thimble of upper and lower framework- scrap-cleaning thimble. The scrap-cleaning thimble of upper framework is fixed and inflexible. The scrap-cleaning thimble of lower framework is flexible, as shown in Figure 2. The fixed scrap-cleaning thimble in upper wise and the flexible scrap-cleaning thimble installed on spring will clean [2] the scraps via the hole on scrap-cleaning formwork of middle framework with two actions of "clamp" and "punch".

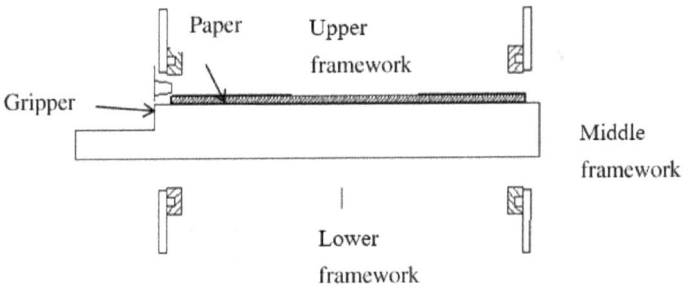

Figure 1. Schematic diagram of three frameworks of scrape-clearer

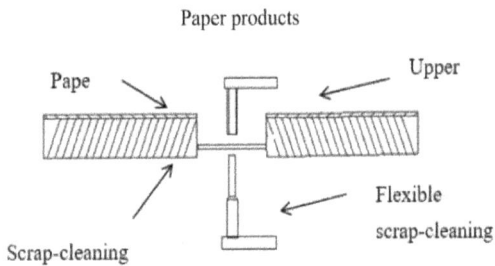

Figure 2. Schematic diagram of working of scrap-cleaning thimble

Figure 3 shows the schematic diagram of upper framework cam link mechanism. The cam rotates at even angular velocity ω and drives the oscillating bar swing. Passing 1 and 2 four-link mechanism and 3 rocker mechanism, driving framework structure G may achieve to-and-fro movement ups and downs, and framework movement, and complete scrap-cleaning work of die-cutting machine.

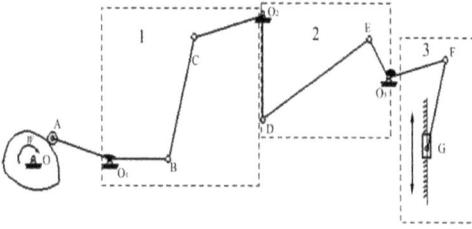

Figure 3. Diagram of upper framework cam link mechanism

The paper analyzes the working principle of scrape-clearer and uses displacement vector method to conduct theoretical calculation on cam of scrape-clearer according to the requirements of die-cutting process. Under the condition of meeting die-cutting process, the paper gets the theoretical contour and actual contour curve.

2. Rule Analysis of Rocker Mechanism of Driven Member

According to die-cutting process and scrap-cleaning principle, the framework moves cyclically in "stop-rise-stop". The motion belongs to high speed and middle load. As for the motion, select five-item motion rule [3], namely

$$s = c_0 + c_1\varphi + c_2\varphi^2 + \cdots + c_5\varphi^5$$

Oscillating bar is connected to upper framework with link mechanism. With the same motion laws, from the start and end position of framework (boundary conditions), use vector triangle method [4] to conduct displacement analysis and calculate the boundary conditions of oscillating bar (cam driven member).

Establish vector triangle for four-link mechanism and oscillating bar sliding mechanism in figure 3, conduct displacement analysis. The vector triangle is shown in Figures 4 ~ 6.

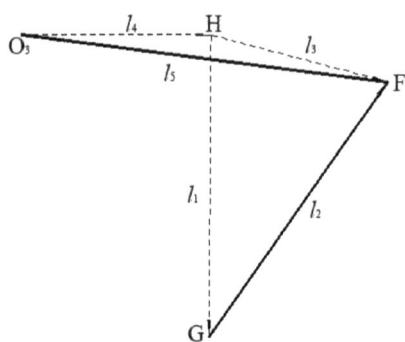

Figure 4. Vector diagram of sliding mechanism

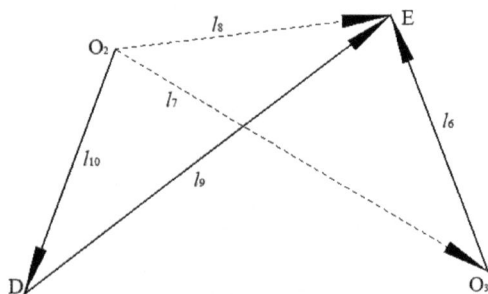

Figure 5. Vector diagram of connecting part

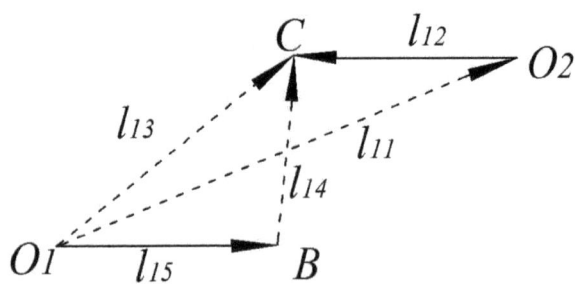

Figure 6. Vector diagram of drive cam

$HF = HG + GF, l_3 = l_1 + l_2,$

$l_1(\cos\alpha_1 + i\sin\alpha_1) + l_2(\cos\alpha_2 + i\sin\alpha_2) = l_3(\cos\alpha_3 + i\sin\alpha_3).$ (1-1)

$O_3F = O_3H + HF, l_5 = l_4 + l_3,$

$l_4(\cos\alpha_4 + i\sin\alpha_4) + l_3(\cos\alpha_3 + i\sin\alpha_3) = l_5(\cos\alpha_5 + i\sin\alpha_5).$ (1-2)

$O_2E = O_2O_3 + O_3E, l_8 = l_7 + l_6,$

$l_7(\cos\alpha_7 + i\sin\alpha_7) + l_6(\cos\alpha_6 + i\sin\alpha_6) = l_8(\cos\alpha_8 + i\sin\alpha_8).$ (1-3)

$O_2D = O_2E + ED, l_{10} = l_8 + l_9,$

$l_8(\cos\alpha_8 + i\sin\alpha_8) + l_9(\cos\alpha_9 + i\sin\alpha_9) = l_{10}(\cos\alpha_{10} + i\sin\alpha_{10})$ (1-4)

$O_1O_2 + O_2C = O_1C, l_{13} = l_{11} + l_{12},$

$l_{11}(\cos\alpha_{11} + i\sin\alpha_{11}) + l_{12}(\cos\alpha_{12} + i\sin\alpha_{12}) = l_{13}(\cos\alpha_{13} + i\sin\alpha_{13})$ (1-5)

$O_1C + CB = O_1B, l_{15} = l_{13} + l_{14},$

$l_{13}(\cos\alpha_{13} + i\sin\alpha_{13}) + l_{14}(\cos\alpha_{14} + i\sin\alpha_{14}) = l_{15}(\cos\alpha_{15} + i\sin\alpha_{15})$ (1-6)

As shown in Figure 3, AO_1B, CO_2D and EO_3F are leverage components, namely

$$\alpha_6 = \alpha_5 + \angle EO_3F$$ (1-7)

$$\alpha_{12} = \alpha_{10} - \angle CO_2D$$ (1-8)

Substitute l_1 (vertical distance of upper framework deviating from center of rotation axis O_3) into (2-1). According to the vector triangle principle, solve the above equations and get scope of pivot angle of oscillating bar is $\psi1 - \psi2$. According to the boundary conditions [3]

When $\phi = 0, \psi = 0, \dfrac{d\psi}{d\phi} = 0, \dfrac{d^2\psi}{d\phi^2} = 0$

When $\phi = \varphi, \psi = \psi_m, \dfrac{d\psi}{d\phi} = 0, \dfrac{d^2\psi}{d\phi^2} = 0; (\psi_m = \psi_2 - \psi_1)$

The motion equation of pull process is:

$$\psi = \psi_m[15(\frac{\phi}{\varphi})^2 - 54(\frac{\phi}{\varphi})^3 + 30(\frac{\phi}{\varphi})^4 - 10(\frac{\phi}{\varphi})^5], (0 < \phi < \varphi)$$

The boundary conditions of return period are:

$$\phi = \varphi, \psi = \psi_m, \frac{d\psi}{d\phi} = 0, \frac{d^2\psi}{d\phi^2} = 0; (\psi_m = \psi_2 - \psi_1)$$

When $\phi = 2\pi, \psi = 0, \dfrac{d\psi}{d\phi} = 0, \dfrac{d^2\psi}{d\phi^2} = 0;$

The motion equation of return period is:

$$\psi = \psi_m[1 - 15(\frac{\phi}{\varphi})^2 + 54(\frac{\phi}{\varphi})^3 - 30(\frac{\phi}{\varphi})^4 + 10(\frac{\phi}{\varphi})^5], \varphi < \phi < 2\pi$$

Upon derivation, get the oscillating bar speed corresponding to cam angle

Pull period

$$\frac{d\psi}{d\phi} = \frac{2\psi_m}{\varphi}[15(\frac{\phi}{\varphi}) - 81(\frac{\phi}{\varphi})^2 + 60(\frac{\phi}{\varphi})^3 - 25(\frac{\phi}{\varphi})^4], (0 < \phi < \varphi)$$

Return period

$$\frac{d\psi}{d\phi} = \frac{2\psi_m}{\varphi}[15(\frac{\phi}{\varphi}) - 81(\frac{\phi}{\varphi})^2 + 60(\frac{\phi}{\varphi})^3 - 25(\frac{\phi}{\varphi})^4], (\varphi < \phi < 2\pi)$$

3. Calculation of Drive Cam Contour

3.1. *Theoretical Contour*

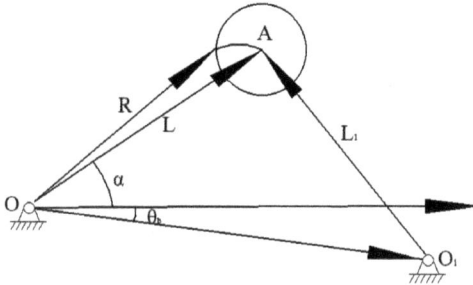

Figure 7. Cam rocker mechanisms

Figure 7 shows one moment of cam oscillating bar, where O, O1 are respectively rotating center of cam and swinging center of oscillating bar. Establish vector triangle OFO1, L=L0+L1, namely

$$l(\cos\alpha + i\sin\alpha) = l_0(\cos\theta_0 + i\sin\theta_0) + l_1(\cos\beta + i\sin\beta)$$

(2-1)

Including: θ_0 is the vector direction angle of rack; β is the vector direction angle of driven member, namely pivot angle of oscillating bar.

3.2. *Tangent line and common normal line of theoretical profile and actual profile*

Theoretical profile refers to the track of roller center if adding ω on the whole mechanism; namely the track S formed in roller center in "inversion" process, as shown in Figure 8.

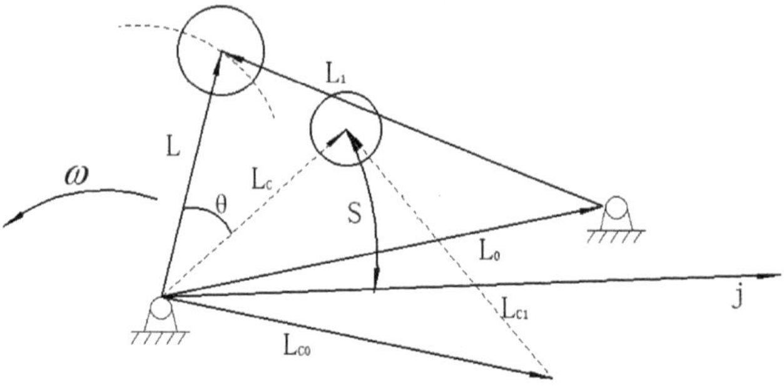

Figure 8. Reversal method

L_C, L_{C0} and L_{C1} indicate the theoretical contour radius vector of cam, rack and driven oscillating bar vector when forming roller center track (theoretical profile) in "reverse" process.

$$L_C = L_{C0} + L_{C1}$$

Namely
$$l = l_0 e^{i(\theta_0 - \theta)} + l_1 e^{i(\beta - \theta)} \qquad (2\text{-}2)$$

Where: θ is reverse angle

Upon derivation

$$\dot{L}_C = s e^{i\varphi_C} = -l_0 \omega e^{i(\theta_0 - \theta - \frac{\pi}{2})} + l_1(\dot{\beta} - \omega) e i^{(\beta - \theta - \frac{\pi}{2})} \qquad (2\text{-}3)$$

Where: ϕ_c is the direction angle in tangent line of theoretical contour in "reverse" process; ω is actual angular velocity of cam.

As shown in the equation

$$tg\phi_c = \frac{-l_0 \omega \cos(\theta_0 - \theta) + l_1(\dot{\beta} - \omega) \cos(\beta - \theta)}{l_0 \omega \sin(\theta_0 - \theta) - l_1(\dot{\beta} - \omega) \sin(\beta - \theta)} \qquad (2\text{-}4)$$

Direction angle of normal of cam is:

$$\eta = \varphi_c + \theta + SGN(\omega) \cdot \frac{\pi}{2}$$

(2-5)

3.3. Calculation of Actual Contour

As shown in Figure 8, R=L+F, namely

$$re^{ir} = le^{i\alpha} - fe^{i\eta} = le^{i\alpha} - fe^{i[\varphi + SNG(\omega) \cdot \frac{\pi}{2}]}$$

(2-6)

Get r and τ is in term of the fixed coordinate system. The radius vector of actual contour on cam is indicated with τ_{cj}

$$\tau_{cj} = \tau_j - SNG(\omega) \cdot \frac{\pi}{2} \cdot j$$

(2-7)

Get cam contour with r and τ_{cj} corresponding angle.

4. Analytic Computation

As shown in Figure 9, substitute the parameters in the mechanism into (2-7) and calculate cam contour profile.

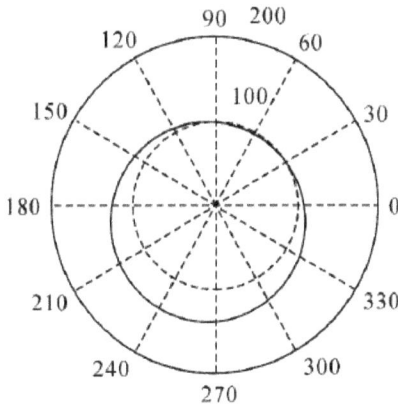

Figure 9. Upper framework cam contours

Figure 10 shows the curve of displacement speed of cam upper oscillating bar and upper framework when the die-cutting speed is 4,000 pieces/hour according to five-polynomial motion law. The pressure angle of cam rocker mechanism is shown in Figure 10.

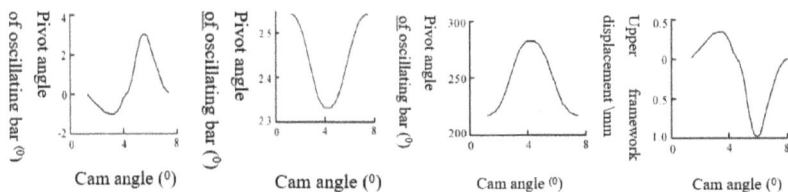

Figure 10. Displacement speed curve of oscillating bar and upper framework

As for the driven member cam mechanism of roller, the pressure angle is

$$\alpha = \arctan[\frac{L_{OA}\cos(\psi_b + \psi) - L_{AB}(1 - \eta\frac{d\psi}{d\phi})}{L_{OA}\sin(\psi_b + \psi)}]$$

Where: $\psi_b = (\frac{L^2_{OA} + L^2_{AB} - R^2_b}{2L_{OA}L_{OB}})$, $\psi_b \in [0, \pi]$,

R_b is radius of base circle, L_{AB} is length of oscillating bar, L_{OA} is center distance, ψ and ψ_b are respectively driven member angular velocity and similar angular velocity.

5. Conclusion

The paper analyzes the working principle of scrape-clearer and uses complex number vector method to conduct theoretical calculation on cam of scrape-clearer according to the requirements of die-cutting process. Under the condition of meeting die-cutting process, the paper gets the theoretical contour and actual contour curve. The number of discrete points may increase or decrease as required to obtain the motion curve. Fit the upper framework motion curve according to seven-polynomial motion law to meet the motion requirements of upper framework mechanism and complete scrap-cleaning action. The method may prevent the impact caused by velocity jump to some extent.

Acknowledgement

This work was funded by the key project (project no. TJSHG201510015011) of enhancing the innovation ability of Beijing municipal colleges and universities, the project (project no. 04190116008/001) of Municipal scientific research projects and the project (project no. KM201510015005) of Beijing City Board of Education Science and Technology and the strategic project.

References

1. Chen Kai. Analysis research on cam driving mechanism of moving platform conjugation of flat die-cutting machine [D] Xi'an: Xi'an University of Technology, 2009
2. Zhu Jun. Coordinated motion control of high-speed automatic die-cutting machine scrape-clearer [D] Shanghai: Tongji University, 2007
3. Shi Yonggang, Wu Weifang, Design and application innovation of cam mechanism [M] Beijing: China Machine Press, 2007
4. Wei Tingbin, Zhao Qinghai, Zhang Haiyan, Characteristic analysis of toggle link mechanism of die-cutting machine [J] Machine Design and Research, 2005 (1): 76.
5. Zhang Tianxuan, Innovation design and innovation of automatic flat die-cutting machine main structure [D] Taiyuan: North University of China, 2014
6. Yuan Yongbao, Shen Jingfeng and Huang Yiqing: Cam improvement design of flat die-cutting machine scrape-clearer based on ADAMS [J]. Shanghai Mechanical drive, 2014

Modal Analysis and Optimization for the Feed Roller of the Inspection Machine

Ju-ying Qian[†], Ying-min Han, Ning-ning Guo and Li-fei Ma

School of Mechanical Engineering,
Beijing Institute of Graphic Communication,
Beijing, 102600, China
[†]Email:860786281@qq.com

The rotation of the feed roller on the inspection machine is achieved by the friction between the tensioning belt and the shaft-end on one side. But in the process of transmission, the uneven force on the shaft-end of both sides causes the deformation of the roller that increases the circular runout of the roller, which results in the uneven force on the printed products and affects the quality of the printed products. This article uses Ansys Workbench to conduct the modal analysis on the roller for the determination of the natural frequency and vibration mode of the roller. It is in accordance with the modal vibration of the roller, conducting the structural optimization of the position with larger amplitude of vibration.

Keywords: Carrier roller; Modal analysis; Structural optimization; Natural frequency.

1. Introduction

For the force on one end of the roller, the modal analysis of the roller belongs to the modal analysis of prestressing force [1-2]. The material of the roller uses No. 45 steel, which is treated with chrome on the surface, take the modal vibration mode of the previous 4 stages of the roller of which the vibration mode is shown as Figures 1~4.

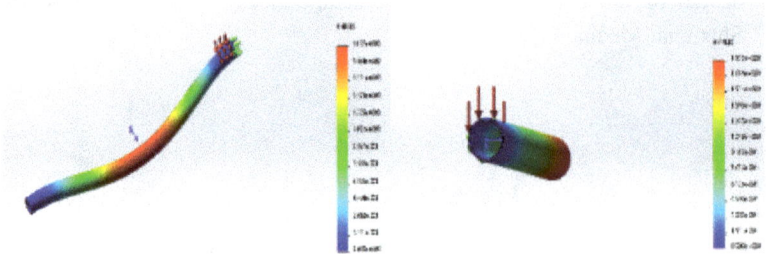

Figure 1. Modal vibration mode of the first prestressing force of the roller

1066

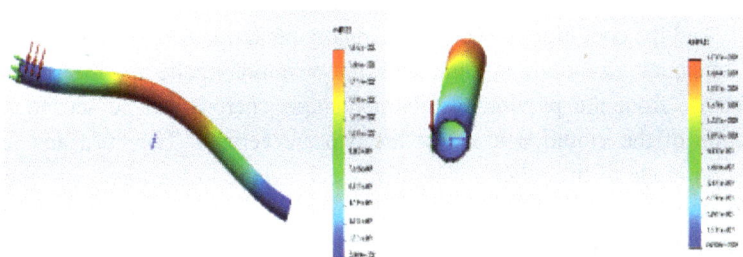

Figure 2. Modal vibration mode of the second prestressing force of the roller

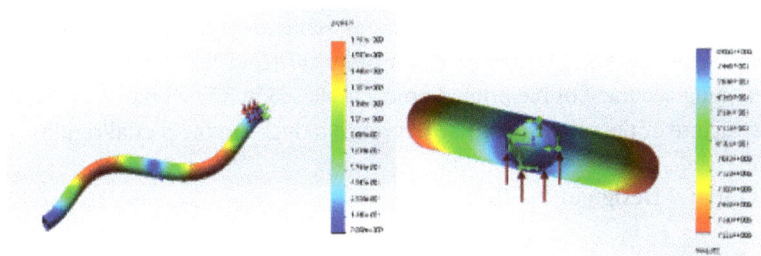

Figure 3. Modal vibration mode of the third prestressing force of the roller

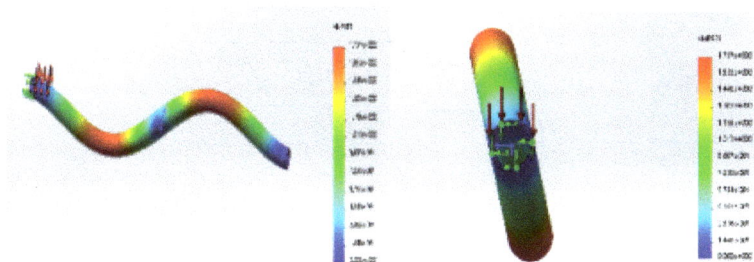

Figure 4. Modal vibration mode of the fourth prestressing force of the roller

From the modal vibration mode of the previous 4 stages of the roller, the first and second vibration mode of the roller is similar which both belong to the torsional vibration mode combined by the vibration of xz plane and the vibration

1067

of yz plane. The difference is that the first torsional vibration dominates the vibration mode on the yz plane; the proportion vibration on the xz plane is very small. And the condition of second torsional vibration is just the opposite of the first torsional vibration; it dominates the vibration mode on the xz plane. Moreover, from the previous 2-vibration mode chart it can be seen that the direction of the vibration mode of the two are vertical. The third and fourth vibration mode of the roller are similar which belong to the torsional vibration mode, compared with the first vibration mode, the vibration mode distribution on the yz and xz plane of the third vibration mode are the same, but the direction of the vibration mode on the xz plane is just the opposite to the direction of the first vibration mode on the xz plane. Compared with the fourth vibration mode, the direction of the third vibration mode is vertical to each other.

From the amplitude of the previous 4 stages of the roller, it can be seen that the vibration mode of the roller appears as stable decay, which belongs to the torsional vibration mode. However, in order to avoid the insufficiency of the positioning accuracy of the printed products caused by the vibration of the roller, the structure of the roller should be improved to increase the partial rigidity.

2. Optimal Design of the Roller

2.1. *Selection for the Material of the Roller*

The improvement on the structure of the roller shall be carried on mainly from the two facets, one is to change the material of the roller; two is to change the structure of the roller.

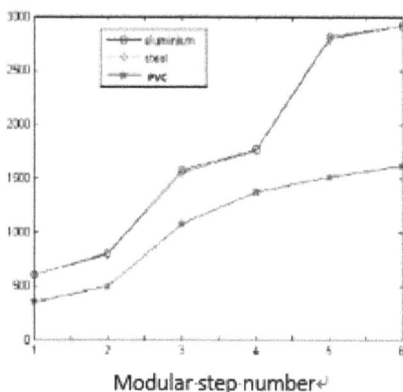

Figure 5. Comparison of natural frequency of different materials

Under the condition of not changing the appearance structure of the roller, use the materials in the classifications of PVC, Q235, aluminum alloy etc. respectively to carry on the finite element modal analysis, the result of the analysis is shown in Figure 5.

From Figure 5, it can be gained that the natural frequency of the aluminum alloy is the highest, the steel (Q235) is the second; the natural frequency of PVC is the lowest. After the change of the material for the roller, the natural frequency changes, but the vibration mode is not changed. The middle part of the roller is still the position with the most apparent amplitude.

The use of aluminum profile for the roller may save the surface processing craft, but due to the cost of aluminum alloy being high, the elasticity modulus being low, under the condition of not changing the original dimension, the rigidity of the roller will be reduced, therefore in order to keep the rigidity of the roller, the dimension of the roller must be changed which will arise the improvement of a series of design on the roller shaft, timing pulley, transmission ratio etc., therefore it is not adequate for the roller to use aluminum alloy.

Due to the rigidity being lower than the Q235 steel, PVC resists wear and corrosion. It is very seldom to be used for the machining materials. The roller must sustain the pressure generated from the pinch roller on the top of it, therefore the requirement for the rigidity of the roller is very high, summing up the above, and Q235 steel is suitable as the material of the roller.

2.2. Dimension Design of the Roller

After the determination of the material for the roller, it is necessary to improve the structural dimension of the roller, the current wall thickness of the roller is 2mm, for the bearing of the pressure from the pinch roller and the precompression on one end of the leather belt, the wall thickness of the roller cannot be too thin. The two ends of the roller are installed with axle bearings, the current type selected for the axle bearing is 61902 deep groove ball bearings, the outside diameter of outer race is 28mm, the inside diameter of outer race is 23.67mm, the inside diameter of inner race is 15mm, which directly cause the wall thickness of the roller being too thin. The roller is fixed on the two sides of the wallboard by the support shaft in the middle. When the wall thickness for the roller is being improved, it must be guaranteed that the support shaft will not have the corresponding change, that is when the bearing type is being changed, it must be assured that the inside diameter of the inner race of the axle bearing remains unchanged. In accordance with the design on the roller, the inside diameter should not exceed the height of 2/3 of the outside race of the axle bearing, therefore the selection of the axle bearing is directly related to the wall thickness of the roller. Under the condition of not changing the outside diameter of the roller, the bearing of type 61802 meets the requirement, the outside diameter of outer race is 24mm, the inside diameter of outer race is 21mm, the inside diameter of inner race is 15mm which is consistent with the original dimension. The wall thickness of the roller is increased from the original 2mm as the maximum to 4mm, with the increase of the wall thickness for the roller; its weight will also be increased. It causes the increase of the bending deflection of the roller that affects the positioning accuracy.

2.3. Force Analysis of the Roller

The roller is driven by the friction between the tensioning belt and its surface. The diagram for the force analysis of the roller is shown as Figure 6. The width of the leather belt is (b-a), the pinch roller is located in the middle of the roller. Suppose that the roller is evenly loaded with the tensioning belt,

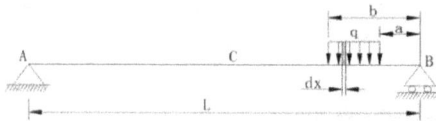

Figure 6. Diagram for the pressure of the roller on the positioning part

Load density is q, the deflection in the middle of the roller w_c is [3-4]:

$$w_C = -\frac{q}{48EI}\int_a^b x(3L_{\text{roller}}^2 - 4x^2)dx =$$
$$-\frac{q}{48EI}[1.5L_{\text{roller}}^2(b^2 - a^2) + a^4 - b^4] \tag{1}$$

The material of the roller is Q235 of which the size for the elasticity modulus (E) of the roller is 2.06×10^{11} Pa. The cross section of the roller is the cylinder with thin wall, suppose the wall thickness of the cylinder is x, the product I of inertia of the cross section can be indicated as:

$$I = \frac{\pi}{32}[d^4 - (d - 2x)^4] = \frac{\pi}{4}x(d^3 - 3d^2x + 4dx^2 - 2x^3) \tag{2}$$

Brought into the equation (1) and sorted to get:

$$w_C = -\frac{q(1.5L_{roller}^2(b^2 - a^2) + a^4 - b^4)}{12\pi Ex(d^3 - 3d^2x + 4dx^2 - 2x^3)}$$

(3)

After the determination of the deflection, it shall also be guaranteed of the lightweight of the roller, the mass of the roller m is:

$$m = 0.54\rho[\frac{\pi d^2}{4} - \frac{\pi(d - 2x)^2}{4}] = 0.54\pi\rho x(d - x)$$

(4)

It can be determined in accordance with the type of the axle bearing for the boundary conditions of the roller wall thickness x:

$$0.003 < x < 0.004$$

(5)

See table 1 for the values of the respective parameters.

Table 1 Value of the parameters for the optimal design of the roller

Parameter	q (N/mm)	a (mm)	b (mm)	E (Pa)	d (mm)	L_{roller} (mm)
Numeric value	1.137	10	40	2.06e11	30	540

Therefore, the optimal design for the wall thickness of the roller must meet the requirements of both deflection and mass which belongs to the multi-objective unconstrained optimization, use the function of fgoalattain in Matlab to conduct the optimization analysis [5], the target function in the optimization analysis is:

$$f(1) = w_C =$$
$$0.7436/(7.766017 \times 10^{-12}x(2.7 \times 10^{-5} - 0.0027x + 0.12x^2 - 2x^3))$$

(6)

$$f(2) = m = 13317.2x(0.03 - x)$$

(7)

The deflection and mass belong to the point target function; the weight of the two target functions in the whole optimization process is different. With the actual production condition, the weighting factor. $Q = [0.6, 0.4]$.

The result of the optimization analysis is:

$$x = 0.00343 \qquad fopt = 0.00000148 \qquad 1.195 \tag{8}$$

In accordance with the actual processing condition, do the round process for the results and the result is as follows:

$$x_y = 0.0035 \qquad fopt_y = 0.00000144 \qquad 1.235 \tag{9}$$

3. Structural Optimization Analyses

The wall thickness of the roller is selected as 3.5mm in final, the deflection in the middle of the roller is 1.44e-6m, and the mass of the roller is 1.235Kg. The result is shown as Figure 7 to Figure 12.

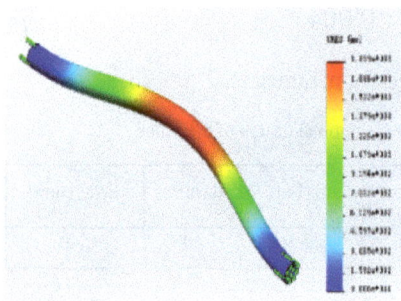

Figure 7. The first modal vibration mode

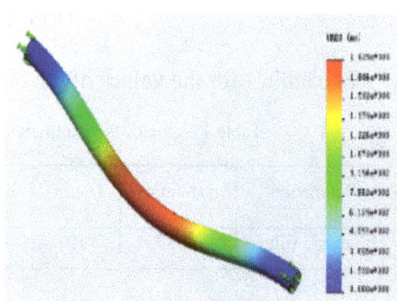

Figure 8. The second modal vibration mode

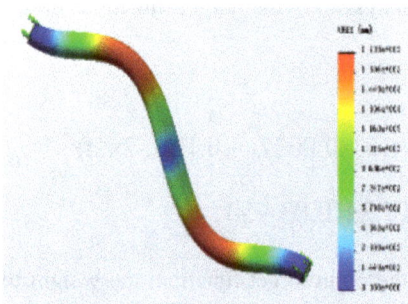

Figure 9. The third modal vibration mode

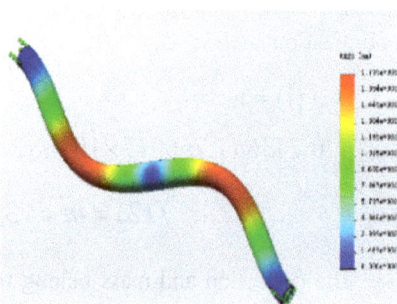

Figure 10. The fourth modal vibration mode

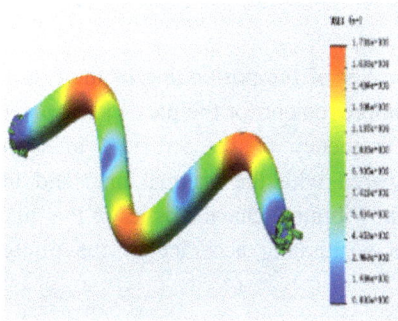

Figure 11. The fifth modal vibration mode   Figure 12. The sixth modal vibration mode

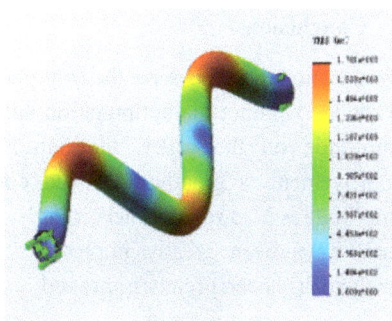

Table 2. Before and after improvement comparison of the natural frequencies of the roller

Modal order	Natural frequency before the improvement (Hz)	Natural frequency after the improvement (Hz)	Amplitude peak before the improvement (mm)	Amplitude peak after the improvement (mm)
1	596.86	854.76	0.428	0.405
2	798.74	912.37	0.419	0.393
3	1570.21	1812.45	0.396	0.381
4	1771.33	2087.36	0.384	0.376
5	2813.72	2994.67	0.362	0.352
6	2915.64	3105.29	0.361	0.349

It can be seen from table 2 that the vibration mode of the roller is the same with the original vibration mode which are both the torsional vibration mode. The natural frequency after the improvement is higher than the original natural frequency of the roller, which indicates that the rigidity of the roller is strengthened to some extent. Compared with the original amplitude, it also has the tendency of weakening. Conduct the positioning test of the printed products for the rollers after the improvement; the testing results indicate that the positioning accuracy of the printed products has been improved to ±0.15 mm.

4. Conclusion

By the modal analysis on the key part roller of the positioning of the printed products; conduct the optimization and improvement for the part with apparent amplitude on the roller, the natural frequency of the structure after the improvement is greatly improved compared with the original one, and the amplitude is also reduced accordingly. The vibration characteristics of the roller itself have been greatly improved; the positioning accuracy of the printed products has been greatly improved.

Acknowledgements

This work was funded by the key project (project no. TJSHG201510015011) of enhancing the innovation ability of Beijing municipal colleges and universities, the project (project no. 04190116008/001) of Municipal scientific research projects and the project (project no. 23190114014) of Beijing Institute of Graphic Communication.

References

1. Lv Duan. Finite Element Modal Analysis of the Crankshaft on the V8 Engine Based on ANSYS Workbench [J]. Machinery Design and Manufacture, 2012, 8(8): 11-13.
2. Mou Weijie. Modal Analysis on the Frame of the Power Based on ANSYS Workbench [J]. Science Technology and Engineering, 2010,22(10): 5592-5594.
3. Zhang Hongcai. ANSYS14.0 Example for the Theoretical Analysis and Engineering Application [M]. Beijing: China Machine Press, 2014:327-328.
4. Liu Hongwen. Mechanics of Material I (Version 5) [M]. Beijing: Higher Education Press, 2010:185-189.
5. Han Min. The Optimization Design of the Transmission of Star Gear Based on the Optimization Tool Box of Matlab [J]. Machinery Design and Manufacture, 2009, 9(9): 31-32.

A New Correction Device for Laminating Machine

Pan-pan Tian[†], Yang Zhang, Li-fei Ma and Shi-bo Wei

School of Mechanical Engineering,
Beijing Institute of Graphic Communication,
Beijing, 102600, China
[†]Email:1205692884@qq.com

In order to solve the previous problems of big paper-feeding errors and low positioning precision, a new correction positioning device for facial tissue has been proposed. Considering application of traditional mechanical correction devices during laminating and comparison on its advantages and disadvantages, it makes an electrification improvement in structure. Being controlled by motion control cards, utilize servo electromotor drives facial paper tissue correction to make sure that when the rate of paper is less than 12000 piece/h, the fitting error can be controlled within±0.5mm and when the rate is between 12000-15000 piece/h, the fitting error can be controlled within±1.0mm. The test results show that this device effectively improves the precision of adjusting and poisoning of facial tissue.

Keywords: Laminating machine; Precision; Correction; Control.

1. Introduction

Slant fitting errors, also called as deflection or slanting paper, refers to the fitting errors of facial tissue and fluting paper in angle deviation from the straight moving direction fitting of paper. The inconsistent space or pressure between rubber wheels of the paper-feeding assembly of facial tissue and paper-feeding belts will cause inconsistent pressure on the paper of feed facial tissue, which is the primary cause of slanting fitting errors of paper [1]. Generally speaking, slanting fitting errors can be reduced by readjusting the space or pressure between rubber wheels and paper-feeding belts. [2,3] However, in engineering practice, this adjustment cannot meet actual needs, which is because different kinds of facial tissue require different pressure and the pressure on each paper-pressing wheel cannot be adjusted to the same value. Therefore, additional correction devices should be added during paper-feeding to reduce slanting errors of facial tissue.

2. Several Typical Correction Devices

At present, there are three methods to control paper slanting errors generally used in laminating machine at home and abroad.

1) Side lay correction device. Side lay, also called side pull lay (Fig. 1), is a widely used gear for adjusting paper slanting in laminating machine, printing machine and film mulching machine with advantages of structural simplicity and better correction effects. However, it is a purely mechanical drive with noise and not suitable to paper of large breadth.

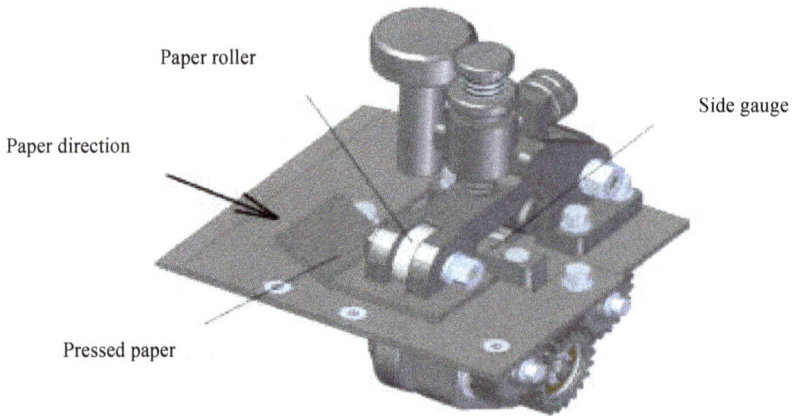

Fig. 1. Side pull lay

2) Front lay correction advice. The structural chart of front lay (working mechanism see Fig. 2) is indicted as Fig. 3. The correction device has advantages of no advancing body paper, high precision, and better process effects. However, paper jogging occurs during paper-feeding with loud noise and easy paper damage. And combined with its intermittent motion, the rate of paper-feeding will be affected.

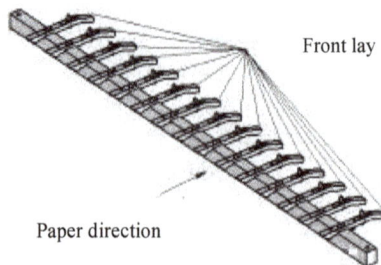

Fig. 2. The working mechanism of front lay

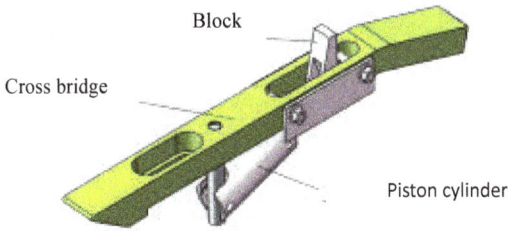

Fig. 3. The structured chart of front lay

3) Push lay correction device. Push lay reduces slanting errors by backend aligning and its structural chart is indicated as Fig. 4. It has advantages of structural simplicity, good effects of aligning, and economy. However, there are phenomena of loud noise, low rate, and paper damage caused by overweight paper.

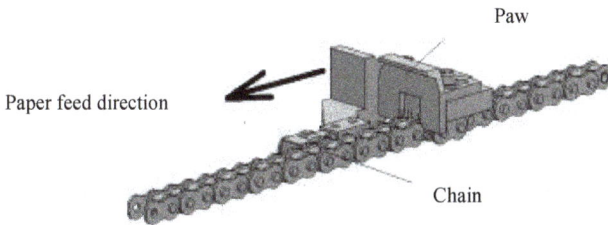

Fig. 4. The structured chart of push lay

3. New Servo Control Correction Devices

3.1. *The Composition of Servo Control Correction Devices*

The servo control correction device is composed of the following parts:

1) Controlled object. Facial tissue, actually, refers to the position of facial tissue. Its status is depended on the friction of belts driven by the roller and on the point power delivered by servo electromotor.

2) Detecting component. Optoelectronic switch, which is photoelectric sensor, usually selects photoelectric sensor of the diffuse type.

3) Controlling components. Besides common switch components, over-heat protector, relay, A.C. contactor, and control transformer, we select motion control cards. Motion control cards is based on PC interface to form the

most functioning motion controller with Levelμs reaction time. Considering laminating machine's rate of thousands of paper per hour, controlling components require a higher reaction rate. Therefore, motion control cards can only be the option.

4) Executing component. Consider factors of controllability, flexibility, preciseness and reaction rate, and select AC servomotor as executing components.

In addition, as a part of modernized control technology, the humanized operationable human-computer interaction interface is an indispensable part. Fig. 5 is the scheme framework of system composition.

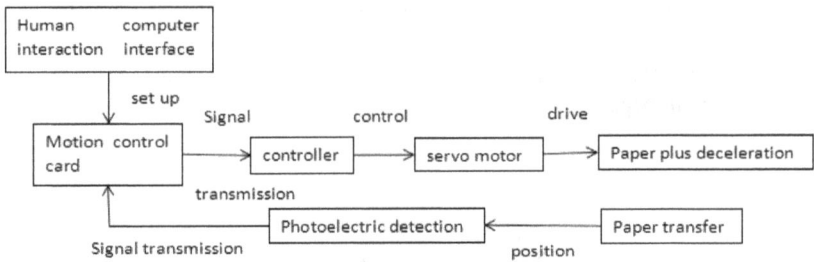

Fig. 5. The scheme framework of system composition

4. Correction Principle of Servo Control Correction Devices

Servo control correction devices have similarities with front lay correction in principles: they both make paper aligned at the front so as to realize correction effects. However, they have the totally difference: front lay correction is depended on the friction of belts to make paper move forward and aligned with the fixed front lay to realize correction and it is intermittent during the process. With servo correction (see Fig. 6), the feeding of facial tissue is depended on the friction of belts. Detected by optoelectronic switch, send the position signal of paper to motion control cards through which read and identify the signal, [6] According to relative programs, make reasonable judgment and manage servo electromotor to accelerate rotating. Impose appropriate point power on paper to facilitate it to move to specific position at the specific rate of paper finally to make paper surface aligned at the front, realizing the purpose of correction. During the whole correction process, servo correction does not interrupt the continuity of paper moving hence to largely increase paper-feeding rate.

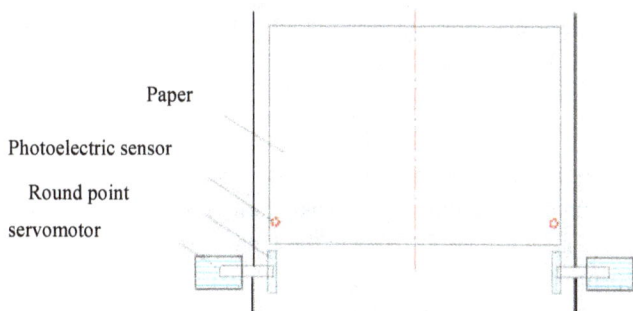

Fig. 6. Principle of servo deviation correction

During servo control correction process, there are mainly three conditions.

1) No paper slanting errors occur (the ideal situation). Under this status, paper can move from paper-feeding board to paper-feeding belts and finally to paper-feeding position with accordance to the predetermined paper-feeding effects without or without final paper twist. At this time, although there are signals sent and transmitted through servo correction system, there is no need to conduct correction on paper.

2) Left-end advancing and right-end lagging of paper occur. In this condition, two photoelectric switches respectively fixed on both sides of paper will send two different pulse signals to motor control cards which make real-time judgment according to input conditions hence to control the servo electromotor on both sides of paper, realizing correction on paper.

3) Left-end lagging and right-end advancing of paper occur. Motor control cards manage servo electromotor to correct paper according to input conditions.

When "lagging" and "advancing" of paper occur, the specific correction principles are as follows:

There are two parallel photoelectric switches and servo electro motors fixed on both sides of the moving paper. Under normal operation of machine, paper's moving rate is v_0, reaction time of photoelectric switch is t_{11}, t_{12}, reaction time of servo electromotor is t_{21}, t_{22} and reaction time of motor control cards is t_3, which are known. When paper slanting errors occur, two photoelectric switches will send two different pulse signals to motor control cards through which the difference Δt between the position and time of paper moving can be calculated.

Hence, the paper-feeding deviation distance Δl of both sides of paper can be calculated as formula (1).

$$\Delta l = v_0 \cdot (\Delta t + t_{12} - t_{11})$$

(1)

Next, we need to maintain the advancing servo electromotor feed paper at the rate v_0 and accelerate the lagging one to make it align with the advancing in operating distance l_0. The constraint l_0 refers to the center distance between servo electromotor and paper-pressing fitting roller.

If the advancing end is regarded as the starting point when detected and as the stop when paper is aligned, operating distance of the advancing point is s_1 and that of the lagging point is s_2 during this process. From the above conditions, formula (2) can be obtained:

$$s_1 = v_0 \cdot (t_3 + t_{22} + t)$$
$$s_2 = v_0 \cdot (t_3 + t_{12}) + 1/2at^2$$
$$\Delta l = s_2 - s_1$$
$$v_0 \cdot t < l_0$$

(2)

After the above calculation, relation formula (3) can be obtained:

$$a = v_0 \cdot (\Delta t + t_{12} + t_{22} - t_{11} - t_{12})/t^2, t < l_0/v_0$$

(3)

On this basis, using MATLAB software, make a functional image of acceleration and time t as indicted in Fig. 7.

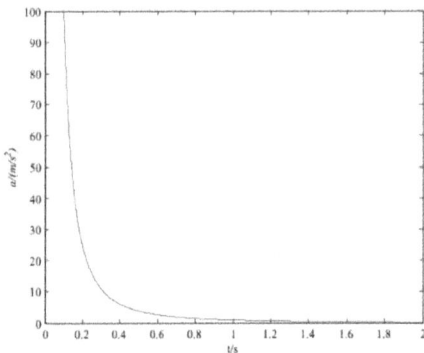

Fig. 7. Functional image of acceleration a and time t

Use this algorithm and C++ make a program and set parameters and realize the correction on servo control system after slight amendment.

5. Correction Effects of Servo Correction Devices

Conduct a field test experiment according to correction principles. On Laminating Machine STM1300, make an experiment respectively on facial tissue of 520 x 368mm and 1092 x 787mm. When the rate of paper is less than 12000 piece/h, the fitting error of both facial tissues can be controlled within ±0.5mm and when the rate is between 12000-15000 piece/h, the fitting error can be controlled within±0.5mm. In contrast, errors of other correction devices can only be maintained within±1.5mm.

6. Conclusion

In order to solve the problems of big paper-feeding errors and low positioning precision, a servo correction device has been proposed. Through comparison with different correction methods, it is found that although with relatively higher costs, servo correction has a wider application scopes, higher correction precision, simpler operation and more practicability, increasing the precision of the adjusting and positioning of facial paper. Servo correction is a new development trend for mechanic paper aligning and positioning after printed.

Acknowledgement

This work was funded by the key project (project no: TJSHG201510015011) of enhancing the innovation ability of Beijing municipal colleges and universities, the project (project no: 04190116008/001) of Municipal scientific research projects and the project (project no: KM201510015005) of Beijing City Board of Education Science and Technology and the strategic project.

References

1. Cai Jifei, Ye Liming. Mulch Applicator Principles and Techniques [M]. Beijing: Beijing Electronic Science and Technology Press, 2013:17-35.
2. Wang Jinjun, Chen Guangzhu, Lin Zhuang. Design and Analysis on the Automatic Correction Device of New Belt Conveyer [J]. Coal Electrical Machinery, 2010 (4) : 13-15.
3. Wang Xiaomei. Application of Full-auto Mechanic Correction Devices [J] Enterprise Science and Development, 2014(20): 12-14.
4. Chen Jialuo, Li Jinyao, Yang Ming. Improvement of Facial Tissue Conveying System of Laminating Machine [J]. Beijing Institute of Graphic Communication, 2013, 21(6): 56-59.

5. Van de Straete 11, J, Degezelle 1', De Schutter, J, et al. Servo Motor Selection Criterion for Mechatronic Applications [J]. IEEE/ASME. Transactions on Mechatronics, 1998, 3(1): 43-50.

6. Kang Jing. Design Research on Photoelectric Correction Controller [D]. Xi'an: Xi'an University of Technology, 2008.

On Lateral Positioning Accuracy of the Die-Cutting Machine

Ying-min Han*, Ning-ning Guo, Ju-ying Qian and Pan-pan Tian

School of Mechanical Engineering,
Beijing Institute of Graphic Communication,
Beijing, 102600, China
*Email:1023745440qq.com

This article analyzes the strengths and weaknesses of the traditional roll pull guide, raises the necessity of design for a new cylindrical cam roll pull guide, conduct the contour modeling of the swing cam, in accordance with the principle of meshing curved surface of the envelope, and carry on the cam solid modeling in solidworks. This cylinder cam roll pull guide meets the requirement of design and is applied to the lateral positioning process of the die-cutting machine which greatly improves the accuracy of die cutting.

Keywords: Lateral; Accuracy; Cylinder cam; Die-cutting Machine.

1. Introduction

Side gauge also known as the roll pull guide, is a mechanism that is widely used in the adjustment of the paper skew of the machineries in the classification of printing like the printer and die-cutting machine die-cutting machine etc. The traditional roll pull guide is affordable and durable with simple structure which is widely used in the laminating machines, but its accuracy is not high neither is it applicable to the correction of papers with large mass, it is not applied in the die-cutting machine. In order to meet the accuracy of the die-cutting machine, under the condition of doing the research on the process of paper running for the die-cutting machine, with the help of the design software SolidWorks, a new type of cylinder cam roll pull guide with side lateral positioning. Compared with the ordinary gauge, the cylinder cam gauge is more accurate on the control of the driven member. The lateral movement of the side block may reduce the horizontal pulling of the paper [1,2], which is more advantageous to achieve the speedup of the die cutting. The work contour of the cylinder cam [3] belongs to the scope of space contour, its design accordance is: The principle of meshing curved surface of the envelope in space [4 – 6].

2. Contour Modeling of the Swing Cam

2.1. *Basic Geometric Dimensioning of the Swing Cam*

The geometric dimensioning of the swing cam mainly includes:
- Center distance C: refers to the vertical distance between the panning axis Z2 and the rotation axis of the cam Z1;
- Basal spur A: the distance from the inner end surface of the wobble plate to the axis of the cam Z1;
- Pitch radius of the wobble plate rp: the distance from the axis of the wobble plate to the axis of the running pulley;
- Distance between points of contact r: the distance from the contact points of the running pulley and the cam to the inner end surface of the wobble plate.

2.2. *Establish the Coordinate System*

- Static system S0-O0X0Y0Z0, the origin of coordinate O0 is the center of rotation of the inner end surface of the wobble plate, take the rotation axis of the wobble plate that passes the point O0 as the Z0 axis, the direction pointed to the cam is positive, take the direction that is parallel to the axis of the cam which passes the O0 as the Y0 axis, the one that is vertical to the Y0O0Z0 plane which passes the point O0 and points to the direction of the cam is positive, the positive direction of axis Y0 shall be determined with the right hand rule.
- Static system S'0-O'0X'0Y'0Z'0, X'0 axis is parallel to the axis of X0 which all point to the same direction, Z'0 axis coincides with the axis of the cam, Y'0 axis shall be determined with the right hand rule, the three axis cross at the point of O'0, point O'0 is on the plane of Y0O0Z0.
- Dynamic system S1-O1X1Y1Z1, its origin of coordinate O1 coincides with O'0, Z1 axis coincides with Z'0 and point to the same direction; The dihedral angle between O_1X_1 and $O'_0X'_0$ is θ, when $\theta=0$, O_1X_1 coincides with $O'_0X'_0$. Similarly, the dihedral angle between O_1Y_1 and $O'_0Y'_0$ is also θ, when $\theta=0$, O_1Y_1 coincides with $O'_0Y'_0$.
- Dynamic system S2-O2X2Y2Z2, its origin of coordinate O2 coincides with O0, Z2 axis coincides with Z0 and point to the same direction; the dihedral angle between O_2X_2 and O_0X_0 is φ, when $\varphi=0$, O_2X_2 coincides with O_0X_0; similarly, the dihedral angle between O_2Y_2 and O_0Y_0 is also φ, when $\varphi=0$, O_2Y_2 coincides with O_0Y_0.

2.3. Derivation of Cylinder Cam Contour Equation

In the dynamic system S2, the point on the axis of the running pulley can be indicated as:

$$\begin{cases} x_{20} = r_p \\ y_{20} = 0 \\ z_{20} = r \end{cases} \tag{1}$$

Use the vector to indicate it as:

$$\left(r_{k20}\right)_2 = \left(r_p, 0, r\right)^T \tag{2}$$

And then conduct the space coordinate transformation, set the center point of the running pulley in the dynamic system S1 as K1, the vector is indicated as (rk1o)1; In the dynamic system S2, it is indicated as k2, the vector is indicated as (rk2o)2'; In the static system S0, it is indicated as k0, the vector is indicated as (rko)0; In the static system S'0, it is indicated as K_0', the vector is indicated as (rko)0'. The rotational angle that circles Z0 of the dynamic system S2 relative to the static system S0, it can be gained the coordinate transformation relationship between the two vectors of (rko)0 and (rk2o)2 as:

$$\left(r_{k0}\right)_0 = R_{02}\left(r_{k20}\right)_2 \tag{3}$$

Of which, $R_{02} = \begin{pmatrix} \cos\cos\phi & -\sin\sin\phi & 0 \\ \sin\sin\phi & \cos\cos\phi & 0 \\ 0 & 0 & 1 \end{pmatrix}$

Substitute (r$_{k2o}$)$_2$, R$_{02}$ into the equation (3):

$$\left(r_{k0}\right)_0 = \left(r_p\cos\cos\phi, r_p\sin r_p\sin\phi, r\right)^T \tag{4}$$

Similarly, it can be gained for:

$$\left(r_{k0}\right)_{0'} = R_{0'1}\left(r_{k10}\right)_1. \tag{5}$$

Of which,

$$R_{0'1} = \begin{pmatrix} \cos\cos\theta & -\sin\sin\theta & 0 \\ \sin\sin\theta & \cos\cos\theta & 0 \\ 0 & 0 & 1 \end{pmatrix} \tag{6}$$

The coordinate transformation relationship between the two vectors of $(r_{ko})_0$ and $(r_{ko})_{0'}$;

$$\left(r_{k0}\right)_0 = R_{00'} \left(r_{k0}\right)_{0'} + \left(r_{o_0'}^{o_0}\right)_0 \tag{7}$$

Of which, $\left(r_{o_0'}^{o_0}\right)_0$ refers to the distance from O_0 to O_0' in the static system S_0:

$$\left(r_{o_0'}^{o_0}\right)_0 = (C,0,A)^T \tag{8}$$

The rotation transformation matrix $R_{00'}$ is:

$$R_{00'} = \begin{pmatrix} 1 & 0 & 0 \\ 0 & \cos\cos\left(\dfrac{\pi}{2}\right) & -\sin-\sin\left(\dfrac{\pi}{2}\right) \\ 0 & \sin\sin(\dfrac{\pi}{2}) & \cos\cos(\dfrac{\pi}{2}) \end{pmatrix} = \begin{pmatrix} 1 & 0 & 0 \\ 0 & 0 & -1 \\ 0 & 1 & 0 \end{pmatrix} \tag{9}$$

Sort the equations of (5), (6), (7) and (8), it can be gained $\left(r_{k10}\right)_1$ as:

$$\left(r_{k10}\right)_1 = \left[R_{00'}R_{0'1}\right]^T \left[\left(r_{k0}\right)_0 - \left(r_{o_0'}^{o_0}\right)_0\right] \tag{10}$$

Of which, $\left[R_{00'}R_{0'1}\right]^T$ is:

$$\left[R_{00'}R_{0'1}\right]^T = \begin{pmatrix} \cos\cos\theta & 0 & \sin\sin\theta \\ -\sin\sin\theta & 0 & \cos\cos\theta \\ 0 & -1 & 0 \end{pmatrix} \tag{11}$$

Substitute (4), (11) into the equation of (10), the contour path equation of the center of the rolling pulley on the cylinder can be gained as:

$$\begin{cases} x_{10} = \left(r_p\cos\phi - C\right)\cos\cos\theta + (r - A)\sin\sin\theta \\ y_{10} = (-r_p\cos\cos\phi + C)\sin\sin\theta + (r - A)\cos\cos\theta \\ z_{10} = -r_p\sin\phi \end{cases} \tag{12}$$

On the base of that, the curvilinear equation of the work contours of the cam of the rolling pulley on the cylinder can be indicated as:

$$
\begin{pmatrix} x_1 \\ y_1 \\ z_1 \end{pmatrix} = \begin{pmatrix} x_{10} \\ y_{10} \\ z_{10} \end{pmatrix} - \rho \bullet n \tag{13}
$$

In the equation, ρ is the radius of the rolling pulley, n is the normal vector of the mesh point of contact, its equation for solution is as follows:

$$
\begin{vmatrix}
\\
\\
\dfrac{\partial s_{10}}{\partial \theta} = \\
\\
\\
\end{vmatrix}
\begin{aligned}
n &= \dfrac{\dfrac{\partial s_{10}}{\partial \theta} \times \dfrac{\partial s_{10}}{\partial r}}{\left| \dfrac{\partial s_{10}}{\partial \theta} \times \dfrac{\partial s_{10}}{\partial r} \right|} \\[2mm]
&\begin{pmatrix} -r_p \dfrac{\omega_2}{\omega_1} \sin\phi\cos\theta - \left(r_p\cos\phi - C \right)\sin\sin\theta + \left(r - A \right)\cos\cos\theta \\[2mm] r_p \dfrac{\omega_2}{\omega_1} \sin\phi\sin\sin\theta - \left(r_p\cos\cos\phi - C \right)\cos\cos\theta - \left(r - A \right)\sin\sin\theta \\[2mm] -r_p \dfrac{\omega_2}{\omega_1} \cos\phi \end{pmatrix} \\[2mm]
\dfrac{\partial s_{10}}{\partial r} &= \begin{pmatrix} \sin\sin\theta \\ \cos\cos\theta \\ 0 \end{pmatrix}
\end{aligned} \tag{14}
$$

Simplify the (14) equation, it can be gained that:

$$
n = \dfrac{1}{\sqrt{\left(r_p \dfrac{\omega_2}{\omega_1} \right)^2 + \left(r - A \right)^2 - 2\left(r - A \right) r_p \dfrac{\omega_2}{\omega_1}\sin\phi}} \begin{pmatrix} r_p \dfrac{\omega_2}{\omega_1}\cos\phi\cos\cos\theta \\[2mm] -r_p \dfrac{\omega_2}{\omega_1}\cos\phi\sin\sin\theta \\[2mm] \left(r - A \right) - r_p \dfrac{\omega_2}{\omega_1}\sin\phi \end{pmatrix}
$$

Substitute the normal vector n into the equation of (13), it can be gained that:

$$
\left\{
\begin{array}{l}
x_1 = \left(r_p cos\phi - C\right)\cos\cos\theta + (r - A)\sin\sin\theta - \dfrac{\rho r_p \dfrac{\omega_2}{\omega_1}\cos\phi\cos\cos\theta}{\sqrt{\left(r_p\dfrac{\omega_2}{\omega_1}\right)^2 + (r-A)^2 - 2(r-A)r_p\dfrac{\omega_2}{\omega_1}\sin\phi}} \\[30pt]
y_1 = (-r_p\cos\cos\phi + C)\sin\sin\theta + (r-A)\cos\cos\theta - \dfrac{-\rho r_p\dfrac{\omega_2}{\omega_1}\cos\phi\sin\sin\theta}{\sqrt{\left(r_p\dfrac{\omega_2}{\omega_1}\right)^2 + (r-A)^2 - 2(r-A)r_p\dfrac{\omega_2}{\omega_1}\sin\phi}} \\[30pt]
z_1 = -r_p\sin\phi - \dfrac{\rho\left[(r-A) - r_p\dfrac{\omega_2}{\omega_1}\sin\phi\right]}{\sqrt{\left(r_p\dfrac{\omega_2}{\omega_1}\right)^2 + (r-A)^2 - 2(r-A)r_p\dfrac{\omega_2}{\omega_1}\sin\phi}}
\end{array}
\right.
$$

(15)

Where: ω_1 is the rotational speed of the cylinder cam, ω_2 is the rotational speed of the wobble plate.

In the equation of (15) and following the law of motion of the driven member, w2 is the first derivative ofto time; The independent variables are the angular velocity of the cam w1 and r, under the condition of the geometric dimension rp, C, A, in the equation being determined, with the determination of the angular velocity w1 of the cam and a series of r value, it can be gained for the work contour of the swing cam.

3. Contour Modeling of the Translation Cam

3.1. *Basic Geometric Dimensioning of the Translation Cam*

The geometric dimensioning of the translation cam mainly includes:
- Center distance D: refers to the vertical distance between the lower end surface of the pulling pulley and the rotation axis of the cam Z_1;
- Distance between points of contact r: the distance from the contact points of the lower end surface of the pulling pulley and the cam to the inner end surface of the wobble plate.

3.2. *Establish the Coordinate System*

- Static system S_0-$O_0X_0Y_0Z_0$, the origin of coordinate O_0 is at the center of gyration on the cam and it is parallel and level to the lowest contour surface of the groove on the cam, make Z_0 axis by passing the point O_0, it shall be coincide with the axis of the cam, the one with back against the rolling pulley is positive; Make X_0 axis by passing the point O_0, it shall be parallel and level to the axis of the rolling pulley, the one with the back against the direction of the rolling pulley is positive; The one that passes O_0 and the vertical direction of the axis of the rolling pulley is Y_0 axis, the positive direction of Y_0 axis shall be determined by the right hand rules.

- Dynamic system S_1-$O_1X_1Y_1Z_1$, its origin of coordinate O_1 coincides with O'_0, Z_1 axis coincides with Z'_0 and point to the same direction; The dihedral angle between O_1X_1 and $O'_0X'_0$ is θ, when $\theta=0$, O_1X_1 coincides with $O'_0X'_0$. Similarly, the dihedral angle between O_1Y_1 and $O'_0Y'_0$ is also θ, when $\theta=0$, O_1Y_1 coincides with $O'_0Y'_0$.
- Dynamic system S_2-$O_2X_2Y_2Z_2$, the origin of coordinate O_2 is located at the intersection point of the axial of the rolling pulley and the lower end surface of the pulling pulley, X_2 axis coincides with the axis of the rolling pulley with the direction pointing to the cam being positive; Y_2 axis passes the point O_2 and is parallel to the cam axis with its direction being opposite to Z_1; Z_1 axis passes point O_2, following the right hand rules.

3.3. *Derivation of Cylinder Cam Contour Equation*

Consistent with the derivation process of the swing cam, in the dynamic system of S_2, the point on the axis of the rolling pulley can be indicated as:

$$\begin{cases} x_2 = r \\ y_2 = 0 \\ z_2 = 0 \end{cases}$$

Use the vector to indicate it as: $\left(r_{k2}\right)_2 = \left(r,0,0\right)^T$;

In the static system S_0, the point on the axis of the running pulley can beindicated as: $\left(r_{k2}\right)_0 = \left(r-D,0,-h\right)^T$;

In the dynamic system S_1, the point on the axis of the running pulley can be indicated as:

$$\left(r_{k2}\right)_1 = \begin{pmatrix} (r-D)\cos\cos\theta \\ (D-r)\sin\sin\theta \\ -h \end{pmatrix}$$

Similarly, the curvilinear equation of the work contour of the translation cam can be indicated as:

1089

$$\begin{pmatrix} x \\ y \\ z \end{pmatrix} = \begin{pmatrix} x_1 \\ y_1 \\ z_1 \end{pmatrix} \pm \rho \bullet n$$

Of which, $n = \left(-\dfrac{v}{\omega_1}\sin\sin\theta, -\dfrac{v}{\omega_1}\cos\cos\theta, r - D \right)^T$;

So the equation of the work contour of the translation cam can be gained as:

$$\left\{ \begin{aligned} x &= (r-D)\cos\cos\theta \pm \frac{-\rho\dfrac{v}{\omega_1}\sin\sin\theta}{\sqrt{\left(\dfrac{v}{\omega_1}\right)^2 + (r-D)^2}} \\ y &= (D-r)\sin\sin\theta \pm \frac{-\rho\dfrac{v}{\omega_1}\cos\cos\theta}{\sqrt{\left(\dfrac{v}{\omega_1}\right)^2 + (r-D)^2}} \\ z &= -h \pm \frac{\rho(r-D)}{\sqrt{\left(\dfrac{v}{\omega_1}\right)^2 + (r-D)^2}} \end{aligned} \right. \tag{16}$$

In the equation of (16), the travel is related to the cam angular velocity w1, following the cycloidal motion; v is the first derivative of h to the time; The independent variable is r, under the condition of the geometric dimensioning D and in the equation are determined, with the determination of the cam angular velocity w1 and a series of r value, the work contour of the translation cam can be gained.

4. Solid Modeling of the Cylinder Cam based on SoldWorks

In the process of 3D modeling for the cylinder cam, it mainly uses the SolidWorks 3D solid modeling software.

4.1. *Generation of Data Documents*

The generation of data is a process of programming with the use of the MATLAB software to the equation, function value etc. involved in the process of mathematical reasoning in the last section. Take the swing cam 1 as an example, under the condition of the rotational speed of the fixed cam $n = 200r\,/\min$, the length of the pitch radius of the wobble plate

$r_p = 25mm$, center distance $C = 30mm$, basal spur $A = 25mm$, radius of the rolling pulley $\rho = 7.5mm$ being determined, with the selection of sine motion law for the lift and return, set the maximum pivot angle $\phi_{max} = 20°$, take a series of values in the scope of 1~10mm for the point margin r to operate the program, the data document for the 11 pieces of space meshing line coordinates on the cylinder cam can be gained.

4.2. 3D Modeling

Take advantage of the above data documents, use SolidWorks software, the 3D modeling [1-4] can be achieved with the contour order, the detailed method is as follows:

First of all, apply the order of "curve that passes through the point XYZ"in the curve of SolidWorks, preserve the 11 pieces of curve data for the cylinder cam as the format of text for the import and generate the cam meshing curve.

Secondary, use the order of "border- contour", build the cam space mesh surface for the 11 pieces of space curve.

Thirdly, build the cam substance, use the order of "stretching, curved surface resection", in accordance with the requirement of height dimension, and complete the 3D manipulation of the swing cam substance.

Use the same method, set the parameters→import the data→build the curve and curved surface→complete the design for another two cams. Lastly, build the assembly; assemble the swing cam 1 in accordance with the sequence of motion of the driven member before the adjustment to the initial phase angle, the assembly of the cam is shown as Figure 1, and processing entity in Figure 2.

Figure 1 Cam assembly drawing

Figure 2 Processing entity drawing

5. Accuracy test comparison results

Lateral accuracy test is carried out in the original MY-1080E automatic die-cutting machine, select the die size 1070 * 760mm, 2mm thick corrugated paper format for the 1280 * 900mm as shown in Figure 3.

Figure 3 Die-cutting plate and die-cutting printing

When the die-cutting speed is 2000r / h, the lateral accuracy is ± 0.14mm; when the speed reached 4000r / h, the lateral accuracy is ± 0.17mm; die-cutting speed of 6000r / h, the accuracy error is ± 0.21mm. The same test was carried out on a MY-1080E die-cutting machine equipped with the cylinder cam roll pull guide. When the cutting speed is 2000r / h, accuracy error down to ± 0.11mm; cutting speed is 4000r / h, the accuracy error is reduced to ± 0.13mm; when the cutting speed is 6000r / h, accuracy error is ± 0.17mm.

This cam meets the requirement of design and is applied to the lateral positioning process of the die-cutting machine, which greatly improves the accuracy of die cutting shown in Table 1.

Table 1. Comparison of Die Cutting Accuracy

Speed (r/h)	Model	Lateral Accuracy (mm)
2000	MY-1080E	±0.14
	Improved Model	±0.11
4000	MY-1080E	±0.17
	Improved Model	±0.13
6000	MY-1080E	±0.21
	Improved Model	±0.17

It can be seen from the lateral accuracy comparation date about the original models and improved models running under2000 r / h, 4000r / h and 6000r / h, after the cylinder cam roll pull guide was applied to the MY-1080E automatic die-cutting machine, the lateral accuracy of the die-cutting machine has been greatly improved, thus confirming the feasibility of the design of the device.

6. Conclusion

In this investigation, we have tested and conclude that this cam meets the requirement of design and is applied to the lateral positioning process of the die-cutting machine, which greatly improves the accuracy of die cutting.

Acknowledgement

This work is funded by the key project (project no. TJSHG201510015011) of enhancing the innovation ability of Beijing municipal colleges and universities, the project (project no. 04190116008/001) of Municipal scientific research projects and the project (project no. 23190114014) of Beijing Institute of Graphic Communication.

References

1. Kang Qilai. How To Improve the Quality of the Collage of the High Corrugated Paperboard [J]. Printing Quality and Standardization, 2010, 8:42-43.
2. Zhang Yizhi, Guo Liankao. The Design of the Conjugate Cam for the Suction Nozzle of Paper-feeding Mechanism of the Paper Feeder [J]. Packaging Engineering, 2007, 28(1): 104-105
3. Liu Jie, Li Qiang. Analysis and Research on the Method of Modeling for the Cylinder Cam [J]. Mechanical Drive, 2012,36(11): 74-76.
4. Jiang Meng, Ge Zhenghao, Qu Yi etc. CAD/CAM Based on the Cylindrical Cam Mechanism of the Reverse Engineering Technology [J]. Mechanical Design, 2011, 28(5): 18-21.

5. Sun Huan, Chen Zuomo, Ge Wenjie. Principle of Machinery (Version VIII) [M]. Xi'an, Northwestern Polytechnical University, 2013.
6. Liu Huran, Zhao Dongfu, Song Deyu. Modern Meshing Theory [M]. Hangzhou, Zhejiang University Press, 2008.

Author Index

www.ingramcontent.com/pod-product-compliance
Lightning Source LLC
Chambersburg PA
CBHW050533190326
41458CB00007B/1769